Algebra

Properties of Real Numbers

Commutative: $\quad a + b = b + a; \quad ab = ba$

Associative: $\quad a + (b + c) = (a + b) + c;$
$a(bc) = (ab)c$

Additive Identity: $\quad a + 0 = 0 + a = a$

Additive Inverse: $\quad -a + a = a + (-a) = 0$

Multiplicative Identity: $\quad a \cdot 1 = 1 \cdot a = a$

Multiplicative Inverse: $\quad a \cdot \dfrac{1}{a} = 1, \ a \neq 0$

Distributive: $\quad a(b + c) = ab + ac$

Exponents and Radicals

$$a^m \cdot a^n = a^{m+n} \qquad \frac{a^m}{a^n} = a^{m-n}$$

$$(a^m)^n = a^{mn} \qquad (ab)^m = a^m b^m$$

$$\left(\frac{a}{b}\right)^m = \frac{a^m}{b^m} \qquad a^{-n} = \frac{1}{a^n}$$

If n is even, $\sqrt[n]{a^n} = |a|$.

If n is odd, $\sqrt[n]{a^n} = a$.

$$\sqrt[n]{a} \cdot \sqrt[n]{b} = \sqrt[n]{ab}, \ a,b \geq 0$$

$$\sqrt[n]{\frac{a}{b}} = \frac{\sqrt[n]{a}}{\sqrt[n]{b}}$$

$$\sqrt[n]{a^m} = (\sqrt[n]{a})^m = a^{m/n}$$

Special-Product Formulas

$$(a + b)(a - b) = a^2 - b^2$$
$$(a + b)^2 = a^2 + 2ab + b^2$$
$$(a - b)^2 = a^2 - 2ab + b^2$$
$$(a + b)^3 = a^3 + 3a^2b + 3ab^2 + b^3$$
$$(a - b)^3 = a^3 - 3a^2b + 3ab^2 - b^3$$

$$(a + b)^n = \sum_{k=0}^{n} \binom{n}{k} a^{n-k} b^k, \quad \text{where}$$

$$\binom{n}{k} = \frac{n!}{k!\,(n - k)!}$$
$$= \frac{n(n - 1)(n - 2) \cdots [n - (k - 1)]}{k!}$$

Factoring Formulas

$$a^2 - b^2 = (a + b)(a - b)$$
$$a^2 + 2ab + b^2 = (a + b)^2$$
$$a^2 - 2ab + b^2 = (a - b)^2$$
$$a^3 + b^3 = (a + b)(a^2 - ab + b^2)$$
$$a^3 - b^3 = (a - b)(a^2 + ab + b^2)$$

Interval Notation

$(a, b) = \{x | a < x < b\} \qquad [a, b] = \{x | a \leq x \leq b\}$

$(a, b] = \{x | a < x \leq b\} \qquad [a, b) = \{x | a \leq x < b\}$

$(-\infty, a) = \{x | x < a\} \qquad (a, \infty) = \{x | x > a\}$

$(-\infty, a] = \{x | x \leq a\} \qquad [a, \infty) = \{x | x \geq a\}$

Absolute Value

$|a| \geq 0 \qquad\qquad |-a| = |a|$

$|ab| = |a| \cdot |b| \qquad \left|\dfrac{a}{b}\right| = \dfrac{|a|}{|b|}, \ b \neq 0$

For $a > 0$,

$$|x| = a \rightarrow x = -a \quad \text{or} \quad x = a,$$
$$|x| < a \rightarrow -a < x < a,$$
$$|x| > a \rightarrow x < -a \quad \text{or} \quad x > a.$$

Equation-Solving Principles

$$a = b \rightarrow a + c = b + c$$
$$a = b \rightarrow ac = bc$$
$$a = b \rightarrow a^n = b^n$$
$$ab = 0 \leftrightarrow a = 0 \quad \text{or} \quad b = 0$$
$$x^2 = k \rightarrow x = \sqrt{k} \quad \text{or} \quad x = -\sqrt{k}$$

Inequality-Solving Principles

$$a < b \rightarrow a + c < b + c$$
$$a < b \text{ and } c > 0 \rightarrow ac < bc$$
$$a < b \text{ and } c < 0 \rightarrow ac > bc$$

(Algebra continued)

College Algebra

GRAPHS AND MODELS

College Algebra

Marvin L. Bittinger
Indiana University–Purdue University at Indianapolis

Judith A. Beecher
Indiana University–Purdue University at Indianapolis

David Ellenbogen
St. Michael's College

Judith A. Penna
Indiana University–Purdue University at Indianapolis

ADDISON-WESLEY

An imprint of Addison Wesley Longman, Inc.

Reading, Massachusetts • Menlo Park, California • New York • Harlow, England
Don Mills, Ontario • Sydney • Mexico City • Madrid • Amsterdam

<table>
<tr><td>Sponsoring Editor</td><td>Jason Jordan</td></tr>
<tr><td>Project Manager</td><td>Kari Heen</td></tr>
<tr><td>Managing Editor</td><td>Karen Guardino</td></tr>
<tr><td>Text Design</td><td>Geri Davis/The Davis Group, Inc.</td></tr>
<tr><td>Art, Design, Editorial, and Production Services</td><td>Martha Morong/Quadrata, Inc.
Geri Davis/The Davis Group, Inc.</td></tr>
<tr><td>Marketing Manager</td><td>Michelle Babinec</td></tr>
<tr><td>Illustrators</td><td>Scientific Illustrators, J. A. K. Graphics, Parrot Graphics, and Gail Hayes</td></tr>
<tr><td>Compositor</td><td>The Beacon Group</td></tr>
<tr><td>Cover Designer</td><td>Karen Rappaport</td></tr>
<tr><td>Cover Photograph</td><td>Takeshi Odawara/PHOTONICA</td></tr>
</table>

Photo Credits

134, PF Sports Images   **140 (left),** SuperStock   **309,** SuperStock
314, The Topps Company, Inc.   **502,** © Mary Dyer

2, Graph on number of births: From THE NEW YORK TIMES, 2/12/95, p. A1. Copyright © 1995 by The New York Times Company. Reprinted with permission.

Library of Congress Cataloging-in-Publication Data

College algebra: graphs & models/Marvin L. Bittinger . . . [et al.].
      p. cm.
Includes index.
ISBN 0-201-84890-2
    1. Algebra. 2. Algebra—Graphic methods. I. Bittinger, Marvin L.
QA154.2.C643 1996
512.9—dc20   96-21890
                                  CIP

2 3 4 5 6 7 8 9 10—DOW—999897

Contents

6 Sequences, Series, and Combinatorics

Preface

College Algebra: Graphs and Models covers college-level algebra and is appropriate for a one- or two-term course in precalculus mathematics. The approach of this text is more interactive than most college algebra texts. Our goal is to enhance the learning process through the use of technology and to provide as much support and help for students as possible in their study of algebra. A course in intermediate algebra is a prerequisite for the text, although Chapter R provides sufficient review to unify the diverse backgrounds of most students.

Content Features

- **Integrated Technology** The technology of the graphing calculator is completely integrated throughout the text to provide a visual means of increasing understanding. In this text, we use the term "grapher" to refer to all graphing calculator technology. The use of the grapher is woven throughout the exposition, the exercise sets, and the testing program without sacrificing algebraic skills. We use the grapher technology to enhance, not to replace, the students' mathematical skills and to alleviate the tedium associated with certain procedures. It is assumed that each student is required to have a grapher (or at least access to one) while enrolled in this course.

- **Learning to Use the Technology** To minimize the need for valuable class time to teach students how to use a grapher, we have provided several features that shorten the learning curve while increasing the students' knowledge of the fundamentals of a grapher. The first of these is the section entitled "Introduction to Graphs and Graphers" found at the beginning of the text. It introduces students to the basic functions of the grapher. The others are the *Graphing Calculator Manual* (see p. xiv) and the video series entitled *Graphing Calculator Instructional Videos* (see p. xiv). In addition, a set of programs has been included in the *Graphing Calculator Manual*. All of these features have been specifically written and produced for this text.

- **Interactive Discoveries** The grapher provides an exciting teaching opportunity in which a student can discover and further investigate mathematical concepts. This unique Interactive Discovery feature is used to introduce new topics and provides a vehicle for students to "see" a concept quickly. This feature reinforces the idea that grapher technology is an integral part of the course as well as an important learning tool. It invites the student to develop analytic and reasoning skills while taking an active role in the learning process. (See pp. 114, 226, and 350.)

- **Function Emphasis** The use of technology with its immediate visualization of a concept encourages the early presentation of functions. Graphing and functions are both introduced in the first section of Chapter 1. The study of the family of functions (linear, quadratic, higher-degree polynomial, rational, exponential, and logarithmic) has been enhanced and streamlined with the inclusion of the grapher. Applications with graphs are incorporated throughout to amplify and add relevance to the study of functions. (See pp. 84, 205, and 261.)

- **Variety in Approaches to Solutions** Skill in solving mathematical problems is expanded when a student is exposed to a variety of approaches to finding a solution. We have carefully incorporated three solution approaches throughout the text: algebraic, graphical, and numerical. Chapter openers illustrate an application with a concurrent grapher presentation of both a table and a graph (see pp. 171 and 321). The TABLE feature on a grapher provides a numerical display or check of the solution (see pp. 198 and 230).

    To highlight both the algebraic- and graphical-solution approaches in solving equations, we have used a two-column solution format in numerous examples (see pp. 184, 297, and 418). In the algebraic/graphical side-by-side features, both methods are presented together; each method provides a complete solution. This feature emphasizes that there is more than one way to obtain a result and illustrates the comparative efficiency and accuracy of the two methods.

- **Real-Data Applications** Throughout the writing process, we conducted an energetic search for real-data applications. The result of that effort is a variety of examples and exercises that connect the mathematical content with the real world. Source lines appear with most real-data applications and charts and graphs are frequently included. Many applications are drawn from the fields of health, business and economics, life and physical sciences, social science, and areas of general interest such as sports and daily life. We encourage students to "see" and interpret the mathematics that appears around them every day. (See pp. 134, 304, 339, and I-1.)

- **Regression** Using regression or curve fitting to model data is introduced in Chapter 1 with linear functions. This visual theme is continued with quadratic, cubic, quartic, higher-degree polynomial, exponential, logarithmic, and logistic functions. Although the theoretic aspects of curve fitting cannot be developed in this course, the power of the grapher is very apparent in this area as the technique is applied to real data. Students can quickly make the "what is this used for?" connection between real data and the extrapolated results of the curve fitting, thus giving them a better conceptual understanding of the material. (See pp. 129 and 312.)

- **Verifying Identities** Identities can be partially verified with a grapher using both the GRAPH and TABLE features. This use of the grapher is first seen in the Introduction to Graphs and Graphers and is continued in later chapters in discovery and verification of possible identities (see pp. 11 and 287). This content feature allows a visual answer to such frequent questions as "Why isn't $(x + 2)^2$ equal to $x^2 + 4$?" This approach also provides a unique lead-in to the development of the properties of exponents and logarithms.

- **Optional Review Chapter** Chapter R provides an optional review of intermediate algebra. We purposely placed the Introduction to Graphs and Graphers before this chapter to allow the grapher to be used in the review. The incorporation of technology gives these topics a fresh perspective and sets the tone for the rest of the course (see pp. 41 and 60). Chapter R can also be used as a convenient source of information for a student who needs a quick review of a particular topic.

Pedagogical Features

- **Use of Color** The text uses full color in an extremely functional way, as seen in the design elements and artwork on nearly every page. The choice of color has been carried out in a methodical and precise manner so that its use carries a consistent meaning, which enhances the readability of the text for both student and instructor. (See pp. 93 and 154.)

- **Art Package** The text contains over 1500 art pieces including a new form of art called photorealism. Photorealism superimposes mathematics on a photograph and encourages students to "see mathematics" in familiar settings (see pp. 115 and 460). The exceptional situational art and statistical graphs throughout the text highlight the abundance of real-world applications while helping students visualize the mathematics (see pp. 74, 205, and 447). The design and use of color with the grapher windows exemplifies the impact that technology has in today's mathematical curriculum (see pp. 13, 159, and 231).

- **Annotated Examples** Over 560 examples fully prepare the student for the exercise sets. Learning is carefully guided with numerous color-coded art pieces and step-by-step annotations, with substitutions and annotations highlighted in red (see pp. 182 and 305). The basis for problem solving is a five-step process established early in the text to aid the student in strategically approaching and solving applications. (See pp. 72 and 327.)

- **Variety of Exercises** There are over 4100 exercises in this text. The exercise sets are enhanced not only by the inclusion of real-data applications with source lines, detailed art pieces, and technology windows that include both tables and graphs, but also by the following features.

  Technology Exercises Since the grapher is totally integrated in this text, exercise sets include both grapher and nongrapher exercises. In some cases, detailed instruction lines indicate the approach the student is expected to use. In others, the student is left to choose the approach that seems best, thereby encouraging critical thinking. (See pp. 194 and 302.)

  Skill Maintenance The exercises in this section have been specifically selected to review concepts previously taught in the text that are foundations for the material presented in the following section. They are chosen to prepare the student for the new concept(s) that will be covered next. (See p. 97.)

  Synthesis Exercises These exercises, which appear at the end of each exercise set, encourage critical thinking by requiring students to synthesize concepts from several sections or to take a concept a step further than in the regular exercises. (See p. 195.)

  Thinking and Writing Exercises for thinking and writing, at the beginning of the synthesis exercises, are denoted with a maze icon ◈. They encourage students to both consider and write about key mathematical ideas in the chapter. Many of these exercises are open-ended, making them particularly suitable for use in class discussions or as collaborative activities. (See p. 260.)

- **Chapter Openers** Each chapter opens with an application illustrated with both technology windows and situational art. The openers also include a table of contents listing section titles. (See pp. 83 and 249.)

- **Section Objectives** Content objectives are listed at the beginning of each section. These, together with subheadings throughout the section, provide a useful outline for both instructors and students. (See pp. 111 and 322.)

- **Highlighted Information** Important definitions, properties, and rules are displayed in screened boxes. Summaries and procedures are listed in color-outlined boxes. Both of these design features present and organize the material for efficient learning and review. (See pp. 116 and 123.)

- **Summary and Review** The Summary and Review at the end of each chapter provides an extensive set of review exercises along with a list of important properties and formulas covered in that chapter. This feature provides an excellent preparation for chapter tests and the final examination. Answers to all review exercises appear in the text along with section references that direct students to material to reexamine if they have difficulty with a particular exercise. (See pp. 167 and 318.)

Supplements for the Instructor

Instructor's Solutions Manual

The Instructor's Solutions Manual by Judith A. Penna contains worked-out solutions to all exercises in the exercise sets, including the thinking and writing exercises. It also includes a sample test with answers for each chapter and answers to the exercises in the appendixes. The sample tests are also included in the Student's Solutions Manual.

Printed Test Bank/Instructor's Manual

Prepared by Donna DeSpain, the Printed Test Bank/Instructor's Manual contains the following:

- 4 free-response test forms for each chapter, following the format and with the same level of difficulty as the tests in the Student's Solutions Manual.
- 2 multiple-choice test forms for each chapter.
- 6 alternate forms of the final examination, 4 with free-response questions and 2 with multiple choice questions.
- Index to the Graphing Calculator Instructional Videos.

Testgen EQ

Testgen EQ is a computerized test generator that allows instructors to select test questions manually or randomly from selected topics or to use a ready-made test for each chapter. The test questions are algorithm-driven so that regenerated number values maintain problem types and provide a large number of test items in both multiple-choice and open-ended formats for one or more test forms. Test items can be viewed on screen, and the built-in question editor lets instructors modify existing questions or add new ones that include pictures, graphs, accurate math symbols, and variable text and numbers.

Additional features in the new Windows and Macintosh Test Generators allow the instructor to customize both the look and content of test-banks and tests. Test questions are easily transferred from the testbank to a test and can be sorted, searched, and displayed in various ways. Testgen EQ is free to adopters.

Course Management and Testing System

InterAct Math Plus for Windows and Macintosh (available from Addison Wesley Longman) combines course management and on-line testing with the features of the basic tutorial software (see "Supplements for the Student") to create an invaluable teaching resource. Consult your local Addison Wesley Longman sales consultant for details.

Supplements for the Student

Graphing Calculator Manual

The Graphing Calculator Manual by Judith A. Penna, with the assistance of John Garlow and Mike Rosenborg, contains keystroke level instruction for the Texas Instruments TI-82, TI-83, TI-85, and Hewlett Packard HP 38G graphing calculators. Modules for the Casio 9850, the Sharp 9200 and 9300, and the HP48G are available on request. Contact your local Addison Wesley Longman sales consultant for details.

Bundled free with every copy of the text, the Graphing Calculator Manual uses actual examples and exercises from *College Algebra: Graphs and Models* to help teach students to use their graphing calculator. The order of topics in the Graphing Calculator Manual mirrors that of the texts, providing a just-in-time mode of instruction.

Student's Solutions Manual

The Student's Solutions Manual by Judith A. Penna contains completely worked-out solutions with step-by-step annotations for all the odd-numbered exercises in the exercise sets in the text, with the exception of the thinking and writing exercises. It also includes a self-test with answers for each chapter and a final examination.

The Student's Solutions Manual can be purchased by your students from Addison Wesley Longman.

Graphing Calculator Instructional Videos

Designed and produced specifically for *College Algebra: Graphs and Models*, the Graphing Calculator Instructional Videos take students through procedures on the graphing calculator using content from the text. These videos include most topics covered in the Graphing Calculator Manual, as well as several additional topics.

Every video section uses actual text examples or odd-numbered exercises from the text to help motivate students while they learn to use the graphing calculator. In the videos, an instructor shows students keystroke level procedures that they will need to succeed with the grapher as they proceed through the course. The videos are correlated to the sections of the text.

A complete set of Graphing Calculator Instructional Videos is free to qualifying adopters.

InterAct Math Tutorial Software

InterAct Math Tutorial Software has been developed and designed by professional software engineers working closely with a team of experienced math educators.

InterAct Math Tutorial Software includes exercises that are linked with every objective in the textbook and require the same computational and problem-solving skills as their companion exercises in the text. Each exercise has an example and an interactive guided solution that are designed to involve students in the solution process and to help them identify precisely where they are having trouble. In addition, the software recognizes common student errors and provides students with appropriate customized feedback.

With its sophisticated answer recognition capabilities, InterAct Math Tutorial Software recognizes appropriate forms of the same answer for any kind of input. It also tracks student activity and scores for each section, which can then be printed out.

Available for both Windows and Macintosh computers, the software is free to qualifying adopters.

Acknowledgments

We wish to express our genuine appreciation to a number of people who contributed in special ways to the development of this textbook. Jason Jordan and Greg Tobin, our editors at Addison Wesley Longman, shared our vision and provided encouragement and motivation. In addition, the production and marketing departments of Addison Wesley Longman brought to the project their unsurpassed commitment to excellence. The unwavering support from the Higher Education Group has been a continuing source of strength for this author team. For this we are most grateful. Mike Rosenborg, Barbara Johnson, Patty Slipher, Irene Doo, and Larry Bittinger provided many constructive comments as well as accuracy checks to the manuscript.

Finally, Professor Bittinger would like to thank his MA 153 students at IUPUI for their productive response to parts of the manuscript that were class-tested in the spring of 1996 using the graphing calculator. This teaching approach resulted in the most satisfying class he has taught at IUPUI in 28 years. Further information regarding this class can be obtained from Professor Bittinger at his e-mail address, exponent@aol.com, or through his home page (see Web Connection that follows).

We would also like to thank the following reviewers for their invaluable contribution to the development of this text:

Sandra Beken, *Horry-Georgetown Technical College*

Diane Benjamin, *University of Wisconsin — Platteville*

Robert Bohac, *Northwest College*

Diane W. Burleson, *Central Piedmont Community College*

John W. Coburn, *St. Louis Community College at Florissant Valley*

Elaine N. Daniels, *Salve Regina University*

Donna DeSpain, *Benedictine University*

Eunice F. Everett, *Seminole Community College*

Rob Farinelli, *Community College of Allegheny County South Campus*

Betty P. Givan, *Eastern Kentucky University*

Allen R. Hesse, *Rochester Community College*

Heidi A. Howard, *Florida Community College at Jacksonville*
Steve Kahn, *Anne Arundel Community College*
Timothy A. Loughlin, *New York Institute of Technology*
Larry Luck, *Anoka-Ramsey Community College*
Joseph D. Mahoney, *Paducah Community College*
Peggy I. Miller, *University of Nebraska at Kearney*
John A. Oppelt, *Bellarmine College*
Tom Schaffter, *Fort Lewis College*
Eric Schulz, *Walla Walla Community College*
Judith D. Smalling, *St. Petersburg Junior College*
Craig M. Steenberg, *Lewis-Clark State College*
Kathryn C. Wetzel, *Amarillo College*
Kemble Yates, *Southern Oregon State College*

M.L.B.
J.A.B.
D.J.E.
J.A.P.

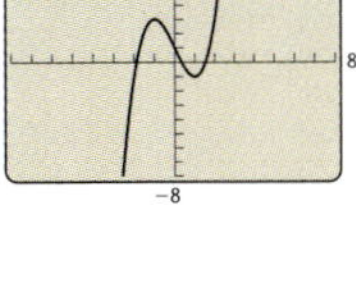

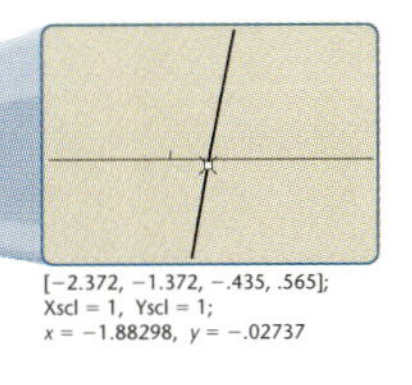

To find the solution to the nearest thousandth, we trace and zoom until the cursor's x-value just to the left of the intercept and the cursor's x-value just to the right of the intercept are the same, when rounded to the nearest thousandth. By using the TRACE and ZOOM features three more times, we find the solution to be about −1.879. In a similar manner, we find that the other solutions of the equation $x^3 - 3x + 1 = 0$ are about 0.347 and 1.532. If available, we might also use the SOLVE or ROOT features to approximate solutions.

There are many ways in which the ZOOM feature can be used. For example, we can zoom in for more precision, as in Example 10, or zoom out on a graph—say, to reveal more of its curvature—adjusting the factors of the zoom in any way we choose. In Example 10, we used zoom factors of 4. We may also be able to use a ZOOM-BOX feature to zoom in on a boxed region of our choosing. All such details can be found by consulting the manual for your particular grapher or the Graphing Calculator Manual that accompanies this book.

- **Interactive Discoveries:**
Throughout the exposition, students are directed to investigate new concepts before they are formally developed. This design invites students to be actively involved with the material in order to identify a mathematical pattern or form an intuitive understanding of a new topic.

Interactive Discovery

With a square viewing window (see the Introduction to Graphs and Graphers), graph the following equations:

$$y_1 = x, \qquad y_2 = 2x, \qquad y_3 = 5x, \quad \text{and} \quad y_4 = 10x.$$

What do you think the graph of $y = 128x$ will look like?

Clear the screen and graph the following equations:

$$y_1 = x, \qquad y_2 = \tfrac{3}{4}x, \qquad y_3 = 0.48x, \quad \text{and} \quad y_4 = \tfrac{3}{25}x.$$

What do you think the graph of $y = 0.000029x$ will look like?

Again clear the screen and graph each set of equations:

$$y_1 = -x, \qquad y_2 = -2x, \qquad y_3 = -4x, \quad \text{and} \quad y_4 = -10x$$

and $\quad y_1 = -x, \qquad y_2 = -\tfrac{2}{3}x, \qquad y_3 = -0.35x, \quad \text{and} \quad y_4 = -\tfrac{1}{10}x.$

From your observations, what do you think the graphs of $y = -200x$ and $y = -0.000017x$ will look like?

If a line slants up from left to right, the change in x and the change in y have the same sign, so the line has a positive slope. The larger the slope is, the steeper the line. If a line slants down from left to right, the change in x and the change in y are of opposite signs, so the line has a negative slope. The larger the absolute value of the slope, the steeper the line.

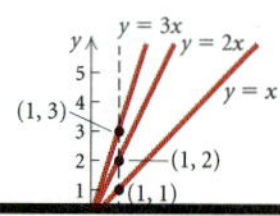
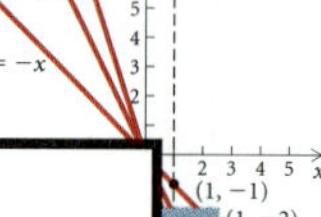
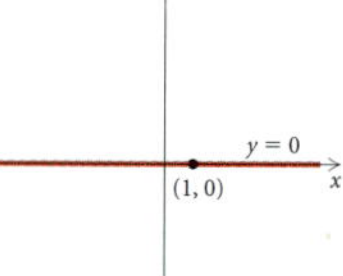

, or $y = 0$, as shown in the graph on the right s both the x-axis and a horizontal line.

ical Lines

al, the change in y for any two points is 0. ne has slope 0.

the change in x for any two points is 0. Thus *ned* because we cannot divide by 0.

be and an undefined slope are two very different

The Quadratic Formula

The solutions of $ax^2 + bx + c = 0$, $a \neq 0$, are given by

$$x = \frac{-b \pm \sqrt{b^2 - 4ac}}{2a}.$$

Example 3 Solve $3x^2 + 2x = 7$. Find exact solutions and approximate solutions rounded to the nearest thousandth.

We show both algebraic and graphical solutions. Note that only the algebraic approach yields the exact solutions.

ALGEBRAIC SOLUTION

After finding standard form, we are unable to factor, so we identify a, b, and c in order to use the quadratic formula:

$$3x^2 + 2x - 7 = 0;$$
$$a = 3, \quad b = 2, \quad c = -7.$$

We then use the quadratic formula:

$$x = \frac{-b \pm \sqrt{b^2 - 4ac}}{2a}$$
$$= \frac{-2 \pm \sqrt{2^2 - 4(3)(-7)}}{2(3)} \qquad \text{Substituting}$$
$$= \frac{-2 \pm \sqrt{4 + 84}}{6} = \frac{-2 \pm \sqrt{88}}{6}$$
$$= \frac{-2 \pm \sqrt{4 \cdot 22}}{6} = \frac{-2 \pm 2\sqrt{22}}{6}$$
$$= \frac{2}{2} \cdot \frac{-1 \pm \sqrt{22}}{3} = \frac{-1 \pm \sqrt{22}}{3}.$$

The exact solutions are

$$\frac{-1 - \sqrt{22}}{3} \quad \text{and} \quad \frac{-1 + \sqrt{22}}{3}.$$

Using the scientific keys on a grapher, we approximate the solutions to be -1.897 and 1.230.

GRAPHICAL SOLUTION

To solve $3x^2 + 2x = 7$, or $3x^2 + 2x - 7 = 0$, we first graph the function $f(x) = 3x^2 + 2x - 7$. Then we look for points where the graph crosses the x-axis. It appears that there are two possible zeros, one near -2 and one near 1. We can use TRACE and ZOOM to approximate these zeros, or we can use a SOLVE or POLY feature.

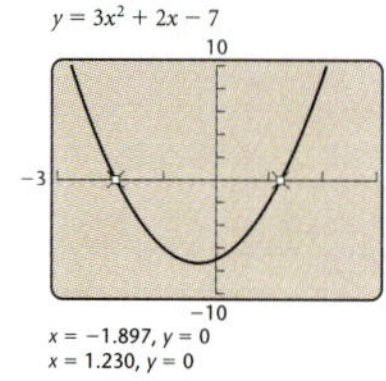

We get the approximate zeros -1.896805 and 1.2301386, or -1.897 and 1.230, rounded to three decimal places. The zeros of the function are the solutions of the equation $3x^2 + 2x = 7$.

- **Side-by-Side Algebraic and Graphical Solutions:**
Many examples in the text are presented in a two-column format that shows simultaneous algebraic and graphical solution methods. This balanced approach allows students to compare the efficiency and appropriateness of each method.

- **Art:** Generous amounts of color-coded technical and situational art appear throughout the text to enhance understanding of an example or exercise, to interest students, and to aid in the visualization of concepts.

Applications of Slope

Slope has many real-world applications. For example, numbers like 2%, 4%, and 7% are often used to represent the **grade** of a road. Such a number is meant to tell how steep a road is on a hill or mountain. For example, a 4% grade means that the road rises 4 ft for every horizontal distance of 100 ft if a vehicle is going up; and -4% means that the road is dropping 4 ft for every 100 ft, if the vehicle is going down.

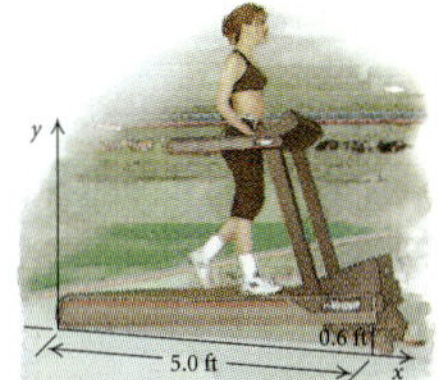

The concept of grade is also used in cardiology when a person runs on a treadmill. A physician may change the slope, or grade, of a treadmill to measure its effect on heart rate. Another example occurs in hydrology. When a river flows, the strength or force of the river depends on how far the river falls vertically compared to how far it flows horizontally.

Example 2 *Ramps for the Handicapped.* Construction laws regarding for the handicapped state that every vertical rise of 1 ft izontal run of 12 ft. What is the grade, or slope, of such

ade, or slope, is given by

$0.083 \approx 8.3\%.$

- To assist students with understanding, **graphs with multiple curves** use different colors for each curve. Movement on the graph may be indicated by a gradual shift in color on an accompanying shift arrow.

SOLUTION

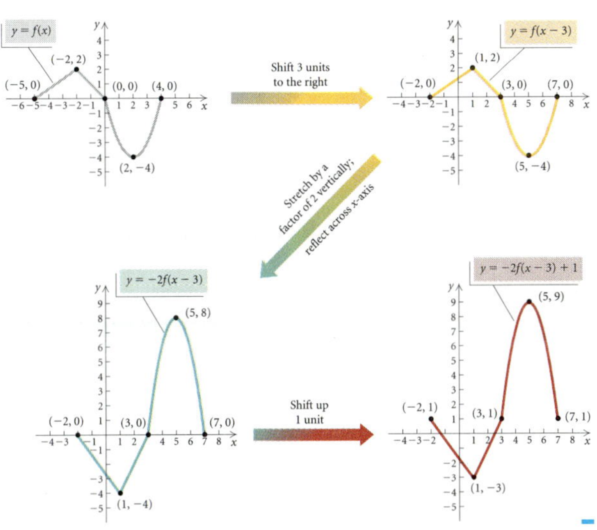

Vertical or Horizontal Translation

For $b > 0$,

the graph of $y_2 = f(x) + b$ is the graph of $y_1 = f(x)$ shifted *up b units*;

the graph of $y_2 = f(x) - b$ is the graph of $y_1 = f(x)$ shifted *down b units*.

For $d > 0$,

the graph of $y_2 = f(x - d)$ is the graph of $y_1 = f(x)$ shifted *right d units*;

the graph of $y_2 = f(x + d)$ is the graph of $y_1 = f(x)$ shifted *left d units*.

- **Data Analysis and Modeling:** The author team highlights and reinforces the theme of data analysis throughout the text.

1.4
Data Analysis, Curve Fitting, and Linear Regression

- *Analyze a set of data to determine whether it can be modeled by a linear function.*
- *Fit a regression line to a set of data; then use the linear model to make predictions.*

Mathematical Models

When a real-world problem can be described in mathematical language, we have a **mathematical model**. For example, the natural numbers constitute a mathematical model for situations in which counting is essential. Situations in which algebra can be brought to bear often require the use of functions.

Mathematical models are abstracted from real-world situations. Procedures within the mathematical model then give results that allow one to predict what will happen in that real-world situation. If the predictions are inaccurate or the results of experimentation do not conform to the model, the model needs to be changed or discarded.

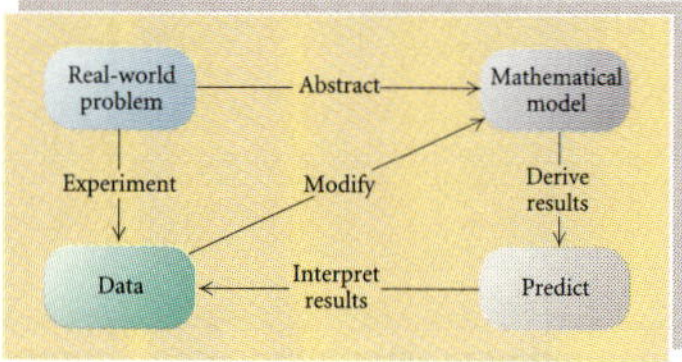

Mathematical modeling can be an ongoing process. For example, finding a mathematical model that will enable an accurate prediction of population growth is not a simple problem. Any population model that one might devise would need to be reshaped as further information is acquired.

Curve Fitting

We will develop and use many kinds of mathematical models in this text. In this chapter, we have considered many functions that can be used as models. Let's look at four of them.

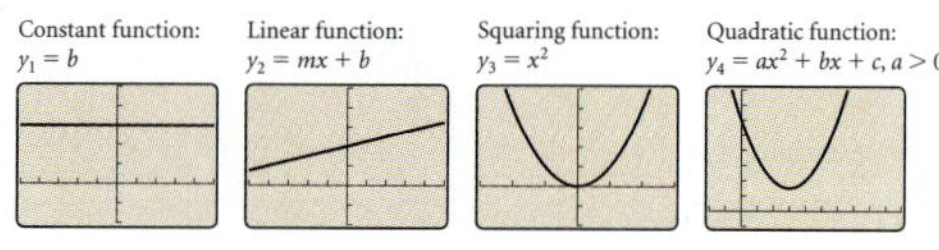

For the scatterplots and graphs in Exercises 17–25, determine which, if any, of the following functions might be used as a model for the data.

a) *Linear,* $f(x) = mx + b$
b) *Quadratic,* $f(x) = ax^2 + bx + c, a > 0$
c) *Quadratic,* $f(x) = ax^2 + bx + c, a < 0$
d) *Polynomial, not quadratic or linear*

17.

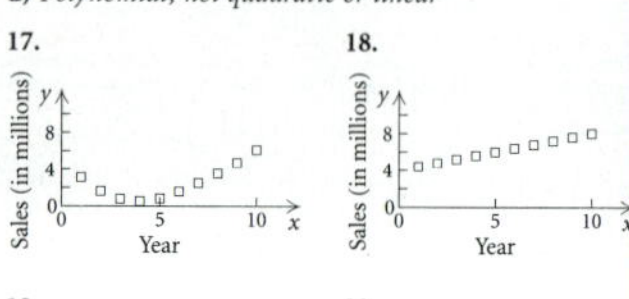

18.

19.

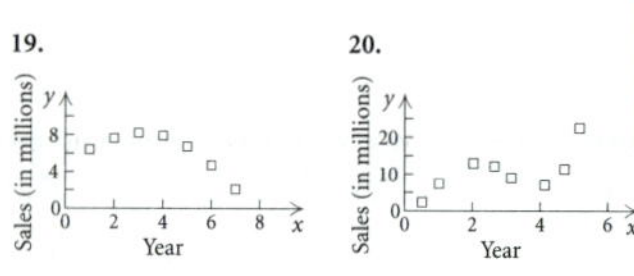

20.

21.

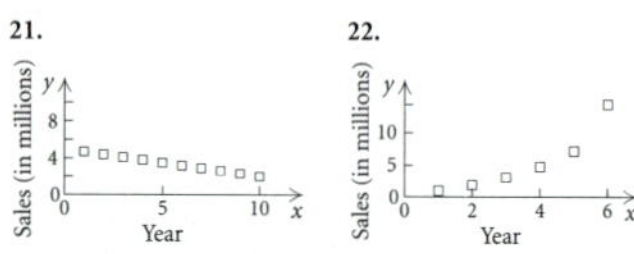

22.

23.

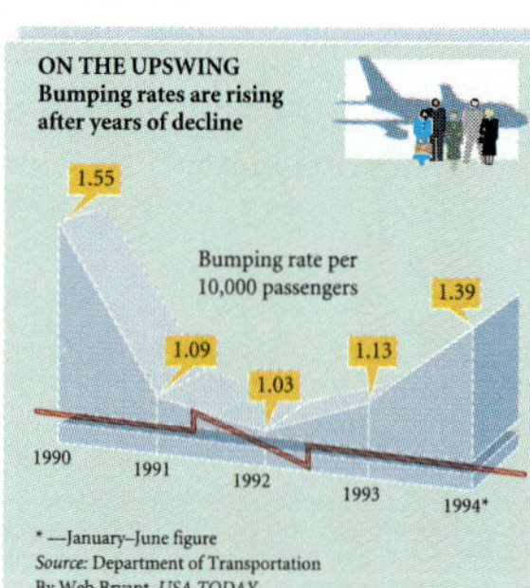

25.

DRIVER FATALITIES BY AGE
Number of licensed drivers per 100,000 who died in motor vehicle accidents in 1990. The fatality rates for both the 70–79 group and 80+ age group were lower than for the 15- to 24-year-olds.

- By fitting curves to data and discussing **mathematical models**, students develop an understanding of mathematical patterns as they are seen in the real world and gain insight into using models to make predictions.

Direction lines

• **Direction lines** in the exercise sets instruct students to use both algebraic and graphical methods to solve problems. Sometimes the student is asked to solve with a grapher and check algebraically. In other problems, the student is directed to solve algebraically and check the work with a grapher. Many exercises do not specify a solution method and allow the student to determine the method, thus encouraging critical thinking and analytical skills.

3.5 Exercise Set

Solve each exponential equation algebraically. Then check on a grapher.

1. $3^x = 81$
2. $2^x = 32$
3. $2^{2x} = 8$
4. $3^{7x} = 27$
5. $2^x = 33$
6. $2^x = 40$
7. $5^{4x-7} = 125$
8. $4^{3x-5} = 16$
9. $27 = 3^{5x} \cdot 9^{x^2}$
10. $3^{x^2+4x} = \frac{1}{27}$
11. $84^x = 70$
12. $28^x = 10^{-3x}$
13. $e^t = 1000$
14. $e^{-t} = 0.04$
15. $e^{-0.03t} = 0.08$
16. $1000e^{0.09t} = 5000$
17. $3^x = 2^{x-1}$
18. $5^{x+2} = 4^{1-x}$
19. $(3.9)^x = 48$
20. $250 - (1.87)^x = 0$
21. $e^x + e^{-x} = 5$
22. $e^x - 6e^{-x} = 1$
23. $\dfrac{e^x + e^{-x}}{e^x - e^{-x}} = 3$
24. $\dfrac{5^x - 5^{-x}}{5^x + 5^{-x}} = 8$

Solve each logarithmic equation algebraically. Then check on a grapher.

25. $\log_5 x = 4$
26. $\log_2 x = -3$
27. $\log x = -4$
28. $\log x = 1$
29. $\ln x = 1$
30. $\ln x = -2$
31. $\log_2 (10 + 3x) = 5$
32. $\log_5 (8 - 7x) = 3$
33. $\log x + \log (x - 9) = 1$
34. $\log_2 (x + 1) + \log_2 (x - 1) = 3$
35. $\log_8 (x + 1) - \log_8 x = 2$
36. $\log x - \log (x + 3) = -1$
37. $\log_4 (x + 3) + \log_4 (x - 3) = 2$
38. $\ln (x + 1) - \ln x = \ln 4$
39. $\log (2x + 1) - \log (x - 2) = 1$
40. $\log_5 (x + 4) + \log_5 (x - 4) = 2$

Use only a grapher. Find approximate solutions of each equation or approximate the point(s) of intersection of a pair of equations.

41. $e^{7.2x} = 14.009$
42. $0.082e^{0.05x} = 0.034$
43. $xe^{3x} - 1 = 3$
44. $5e^{5x} + 10 = 3x + 40$
45. $4 \ln (x + 3.4) = 2.5$
46. $\ln x^2 = -x^2$
47. $\log_8 x + \log_8 (x + 2) = 2$
48. $\log_3 x + 7 = 4 - \log_5 x$
49. $\log_5 (x + 7) - \log_5 (2x - 3) = 1$
50. $y = \ln 3x, \ y = 3x - 8$

51. $2.3x + 3.8y = 12.4, \ y = 1.1 \ln (x - 2.05)$

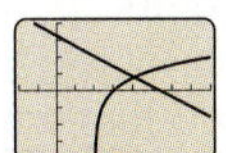

52. $y = 2.3 \ln (x + 10.7), \ y = 10e^{-0.07x^2}$

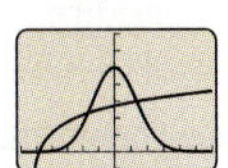

53. $y = 2.3 \ln (x + 10.7), \ y = 10e^{-0.007x^2}$

Skill Maintenance

54. Solve $K = \frac{1}{2}mv^2$ for v.

Solve.

55. $x^4 + 5x^2 = 36$
56. $t^{2/3} - 10 = 3t^{1/3}$

57. *Total Sales of Goodyear.* The following table shows factual data regarding total sales of The Goodyear Tire and Rubber Company.

YEAR, x	TOTAL SALES, y (IN MILLIONS)
1. 1991	$10,906.8
2. 1992	11,784.9
3. 1993	11,643.4
4. 1994	12,288.2

Source: The Goodyear Tire and Rubber Company Annual Report.

a) Use linear regression on a grapher to fit an equation $y = mx + b$, where $x = 1$ corresponds to 1991, to the data points. Predict total sales in 1999.

b) Use quadratic regression on a grapher to fit a quadratic equation $y = ax^2 + bx + c$ to the data points. Predict total sales in 1999.

Synthesis

58. ◆ In Example 3, we took the natural logarithm on both sides. What would have happened had we taken the common logarithm? Explain which seems best to you and why.

59. ◆ Explain how Exercises 29 and 30 could be solved using the graph of $f(x) = \ln x$.

---

3. Carry out.

ALGEBRAIC SOLUTION

We solve the equation:

$$x + 2x = 30$$
$$3x = 30$$
$$x = 10.$$

GRAPHICAL SOLU…

Graph $y_1 = x$ …
point of inters…

4. Check. When …
the Australian …
10 + 20, or 30 days …

5. State. After one ye… …
10 vacation days and Australian employees get an average of 20 vacation days per year.

In some applications we need to use a formula that describes the relationship between variables. When a situation involves distance, speed, and time, for example, we need to recall the **distance formula**:

$d = rt,$ where d = distance, r = rate (or speed), and t = time.

Example 2 *Speed.* A 1996 BMW M3 leaves a town on the Autobahn traveling at its top speed of 237 km/h. Fifteen minutes later, a 1995 Aston Martin DB7 leaves the same town and follows the same route at its top speed of 266 km/h. How long will it take the Aston Martin to overtake the BMW? (*Sources: Car and Driver,* August 1995 and February 1995)

SOLUTION

1. Familiarize. We make a drawing showing both the known and the unknown information. We let t = the time, in hours, that the BMW

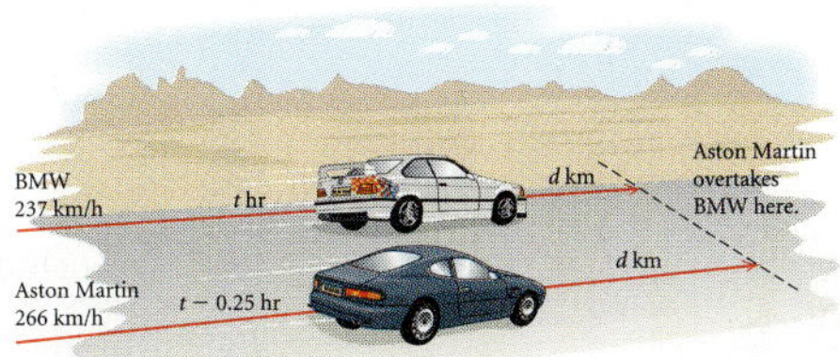

• **Examples/Exercises:** The examples and exercises reflect the text's focus. Many use situational or grapher art and many also incorporate recent, source-based data to illustrate applications and concepts. An Index of Applications is included at the end of the text.

- **Synthesis/Skill Maintenance exercises:** Skill Maintenance exercises prepare students for the next section by reviewing and reinforcing skills that are used in the following section. Synthesis exercises require additional thought and encourage students to take the material one step further by combining concepts. Exercises marked with the maze icon are designed to foster critical thinking and writing skills; they may also be completed by small groups.

---

the formula

$$C(x) = \frac{100 + 5x}{x},$$

where x = the number of people in the group and $C(x)$ is in dollars. Determine $C^{-1}(x)$ and explain what it represents.

69. *Reaction Distance.*   You are driving a car when a deer suddenly darts across the road in front of you. Your brain registers the emergency and sends a signal to your foot to hit the brake. The car travels a distance D, in feet, during this time, where D is a function of the speed r, in miles per hour, that the car is traveling when you see the deer. That reaction distance D is a linear function given by

$$D(r) = \frac{11r + 5}{10}.$$

a) Find $D(0)$, $D(10)$, $D(20)$, $D(50)$, and $D(65)$.
b) Graph $D(r)$.
c) Find $D^{-1}(r)$ and explain what it represents.
d) Graph the inverse.

Graph each of the following.

71. $y = x^3 - x$

72. $x = y^3 - y$

73. $f(x) = \sqrt[3]{x}$

74. $f(x) = \dfrac{8}{x^2 - 4}$

Synthesis

75. ◆ Suppose that you have graphed a function using a grapher and you see that it is one-to-one. How could you then use the TRACE feature to make a hand-drawn graph of the inverse?

76. ◆ The following formulas for the conversion between Fahrenheit and Celsius temperatures have been considered several times in this text:

$$C = \tfrac{5}{9}(F - 32)$$

and

$$F = \tfrac{9}{5}C + 32.$$

Discuss these formulas from the standpoint of inverses.

Using only a grapher, determine whether the functions are inverses of each other.

77. $f(x) = \sqrt[3]{\dfrac{x - 3.2}{1.4}}$, $g(x) = 1.4x^3 + 3.2$

78. $f(x) = \dfrac{2x - 5}{4x + 7}$, $g(x) = \dfrac{7x - 4}{5x + 2}$

79. $f(x) = \dfrac{2}{3}$, $g(x) = \dfrac{3}{2}$

80. $f(x) = x^4$, $x \geq 0$; $g(x) = \sqrt[4]{x}$

. Find three examples of functions that are their own inverses, that is, $f = f^{-1}$.

. Consider the function f given by

$$f(x) = \begin{cases} x^3 + 2, & x \leq -1, \\ x^2, & -1 < x < 1, \\ x + 1, & x \geq 1. \end{cases}$$

Does f have an inverse that is a function? Why or why not?

CHAPTER

3 Summary and Review

Important Properties and Formulas

One-to-One Function:	$f(a) = f(b) \rightarrow a = b$
Exponential Function:	$f(x) = a^x$
The Number e =	$2.7182818284\ldots$
Logarithmic Function:	$f(x) = \log_a x$
A Logarithm is an Exponent:	$\log_a x = y \leftrightarrow x = a^y$
The Change-of-Base Formula:	$\log_b M = \dfrac{\log_a M}{\log_a b}$
The Product Rule:	$\log_a MN = \log_a M + \log_a N$
The Power Rule:	$\log_a M^p = p \log_a M$
The Quotient Rule:	$\log_a \dfrac{M}{N} = \log_a M - \log_a N$
Other Properties:	$\log_a a = 1, \qquad \log_a 1 = 0,$
	$\log_a a^x = x, \qquad a^{\log_a x} = x$
Base–Exponent Property:	$a^x = a^y \leftrightarrow x = y$
Exponential Growth Model:	$P(t) = P_0 e^{kt}$
Exponential Decay Model:	$P(t) = P_0 e^{-kt}$
Interest Compounded Continuously:	$P(t) = P_0 e^{kt}$
Limited Growth:	$P(t) = \dfrac{a}{1 + be^{-kt}}$

REVIEW EXERCISES

1. Find the inverse of the relation
   $\{(1.3, -2.7), (8, -3), (-5, 3), (6, -3), (7, -5)\}$.

2. Find an equation of the inverse relation.
   a) $y = 3x^2 + 2x - 1$
   b) $0.8x^3 - 5.4y^2 = 3x$

Given each function:
a) *Determine whether it is one-to-one, using a grapher if desired.*
b) *If it is one-to-one, find a formula for the inverse.*
   3. $f(x) = \sqrt{x - 6}$ **4.** $f(x) = x^3 - 8$

5. $f(x) = 3x^2 + 2x - 1$ **6.** $f(x) = e^x$

7. Find $f(f^{-1}(657))$: $f(x) = \dfrac{4x^5 - 16x^{37}}{119x}$, $x > 0$.

In Exercises 8–13, match the equation with one of figures (a)–(f), which follow. If needed, use a grapher.

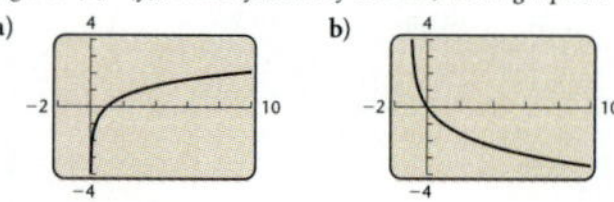

- **End-of-Chapter material:** End-of-Chapter material includes a summary and review of properties and formulas along with a complete set of Review Exercises. Review Exercises also include Synthesis exercises, as well as critical thinking and writing exercises. The answers to the Review Exercises, which appear at the back of the text, have text section references to further aid students.

College Algebra

GRAPHS AND MODELS

Introduction to Graphs and Graphers

- *Review hand-drawn graphs of equations.*
- *Use a grapher to create graphs with various viewing windows, to create x, y tables for equations, and to determine whether an equation is an identity.*
- *Use a grapher to find coordinates of points on a graph, to solve equations, and to find the points of intersection of two graphs.*

Graphing calculators and computers equipped with graphing software can alleviate much of the toil of graph making. We will henceforth refer to all such graphing utilities simply as **graphers**. As the equations we encounter become more complicated, it becomes increasingly difficult to produce accurate hand-drawn graphs. It can take considerable time to calculate just a few ordered pairs that are solutions. In addition, many ordered pairs are often required to produce an accurate graph. The use of a grapher can not only shorten this process but also perform many other mathematical procedures efficiently. Also, many equations arise that are difficult to analyze without machine assistance. Throughout this text, we will use graphers to enhance the learning process. The following is our philosophy regarding the use of graphers.

The Use of the Grapher

The grapher creates a visual presentation that *increases* understanding and *saves* time. It is used to enhance the mathematics, not to replace it!

Most of our discussion of the grapher is presented in a relatively generic form. Expanded discussion specific to certain graphers is included in the Graphing Calculator Manual that accompanies this book. To determine specific procedures, you should consult this manual, your user's manual, or your instructor.

Graphs

Hand-drawn graphs are reviewed first. With this groundwork established, we then consider the use of graphers. This introduction to graphs and graphers provides the foundation for the remainder of the text.

Graphs provide a natural way to link algebra and geometry. It is not uncommon to open a newspaper or magazine and encounter graphs. Shown below are examples of bar, circle, and line graphs.

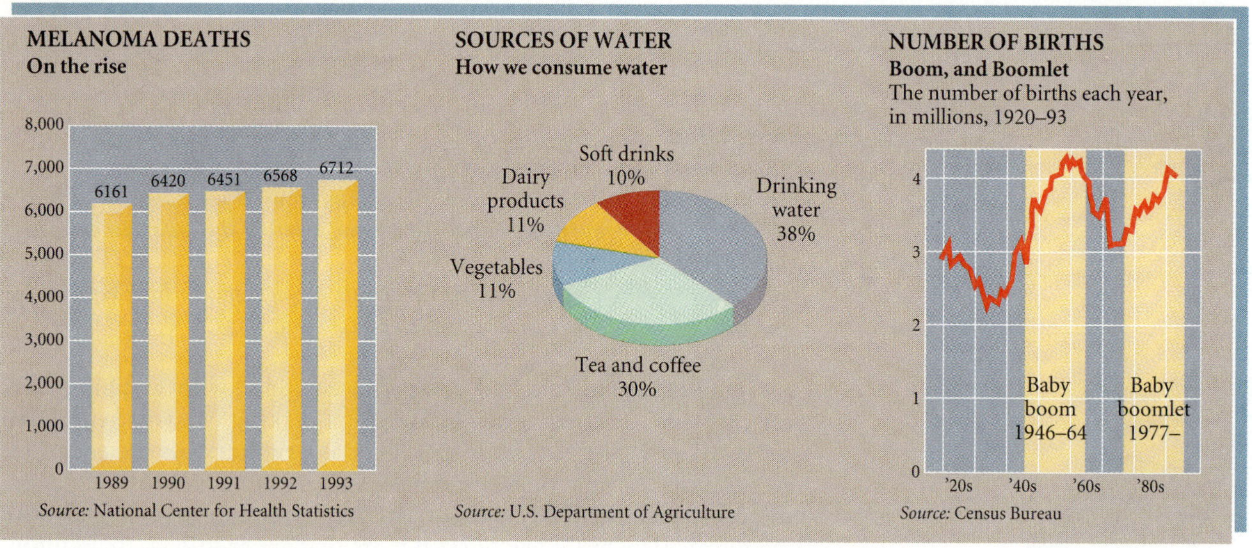

Many real-world situations can be modeled using equations in which two variables appear. We use a plane to graph a pair of numbers. To locate points on a plane, we use two perpendicular number lines, called **axes**, which intersect at 0. We call this point the **origin**. The horizontal axis is called the **x-axis**, and the vertical axis is called the **y-axis**. (Other variables can be used.) The axes divide the plane into four regions, called **quadrants**, denoted by Roman numerals, numbered counterclockwise from the upper right. Arrows show the positive direction of each axis.

Each point (a, b) in the plane is called an **ordered pair**. The first number, a, indicates the point's horizontal location with respect to the y-axis, and the second number, b, indicates the point's vertical location with respect to the x-axis. We call a the **first coordinate**, **x-coordinate**, or **abscissa**. We call b the **second coordinate**, **y-coordinate**, or **ordinate**. Such a representation is called the **Cartesian coordinate system** in honor of the great French mathematician and philosopher René Descartes (1596–1650).

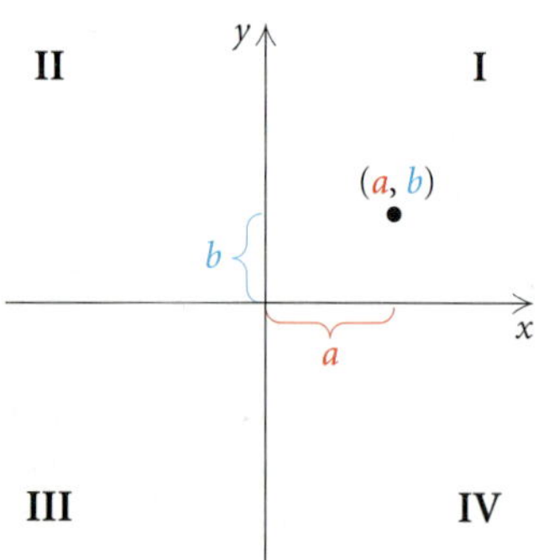

Example 1   Graph and label the points in the set

$\{(-3, 5), (4, 3), (3, 4), (-4, -2), (3, -4), (0, 4), (-3, 0), (0, 0)\}.$

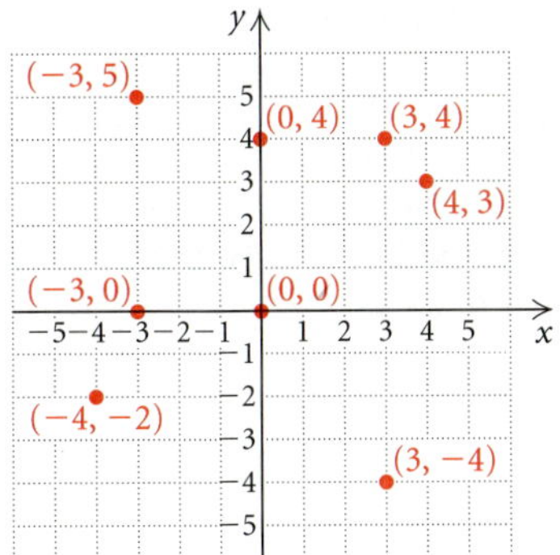

SOLUTION To graph or **plot** $(-3, 5)$, we note that the x-coordinate tells us to move from the origin 3 units to the left of the y-axis. Then we move 5 units up from the x-axis. To graph the other points, we proceed in a similar manner. (See the graph at left.) Note that the origin, $(0, 0)$, is located at the intersection of the axes and that the point $(4, 3)$ is different from the point $(3, 4)$.

A graph of a set of points, as in Example 1, is called a **scatterplot**.

Solutions of Equations

Equations in two variables, like $2x + 3y = 18$, have solutions that are ordered pairs such that when the first coordinate is substituted for x and the second coordinate is substituted for y, the result is a true equation.

Example 2 Determine whether each ordered pair is a solution of $2x + 3y = 18$.

a) $(-5, 7)$ b) $(3, 4)$

SOLUTION We check as follows.

a)
$$2x + 3y = 18$$
$$2(-5) + 3(7) \; ? \; 18 \qquad \text{We substitute } -5 \text{ for } x \text{ and } 7 \text{ for } y \text{ (alphabetical order).}$$
$$-10 + 21$$
$$11 \;\bigm|\; 18 \qquad \text{FALSE}$$

The equation $11 = 18$ is false, so $(-5, 7)$ is not a solution.

b)
$$2x + 3y = 18$$
$$2(3) + 3(4) \; ? \; 18 \qquad \text{We substitute 3 for } x \text{ and 4 for } y.$$
$$6 + 12$$
$$18 \;\bigm|\; 18 \qquad \text{TRUE}$$

The equation $18 = 18$ is true, so $(3, 4)$ is a solution.

Graphs of Equations

The equation considered in Example 2 actually has an infinite number of solutions. Since we cannot list all of the solutions, we will make a drawing, called a **graph**, that represents them.

To Graph an Equation

To graph an equation is to make a drawing that represents the solutions of that equation.

Suggestions for Hand-Drawn Graphs

1. Use graph paper.

2. Draw axes and label them with the variables.

3. Use arrows on the axes to indicate positive directions.

4. Scale the axes, that is, mark numbers on the axes.

5. Calculate solutions and list the ordered pairs in a table.

6. Plot solutions, look for patterns, and complete the graph. Label the graph with the equation being graphed.

At left are some suggestions for making hand-drawn graphs.

Example 3   Graph: $2x + 3y = 18$.

SOLUTION   To find ordered pairs that are solutions of this equation, we can replace either x or y with any number and then solve for the other variable. For instance, if x is replaced with 0, then

$$2 \cdot 0 + 3y = 18$$
$$3y = 18$$
$$y = 6. \qquad \text{Dividing by 3}$$

Thus $(0, 6)$ is a solution. If x is replaced with 5, then

$$2 \cdot 5 + 3y = 18$$
$$10 + 3y = 18$$
$$3y = 8 \qquad \text{Subtracting 10}$$
$$y = \tfrac{8}{3}. \qquad \text{Dividing by 3}$$

Thus $\left(5, \tfrac{8}{3}\right)$ is a solution. If y is replaced with 0, then

$$2x + 3 \cdot 0 = 18$$
$$2x = 18$$
$$x = 9. \qquad \text{Dividing by 2}$$

Thus $(9, 0)$ is a solution.

We continue choosing values for one variable and finding the corresponding values of the other. We list the solutions in a table, and then plot the points. Note that the points appear to lie on a straight line.

x	y	(x, y)
0	6	$(0, 6)$
5	$\tfrac{8}{3}$	$\left(5, \tfrac{8}{3}\right)$
9	0	$(9, 0)$
-1	$\tfrac{20}{3}$	$\left(-1, \tfrac{20}{3}\right)$

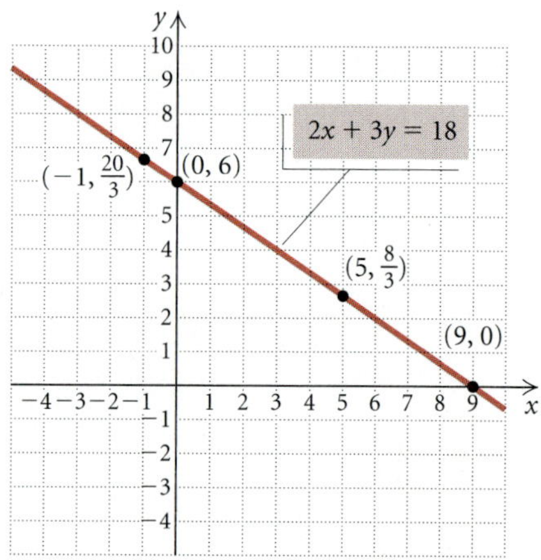

In fact, were we to graph additional solutions of $2x + 3y = 18$, they would be on the same straight line. Thus, to complete the graph, we use a straightedge to draw a line as shown in the figure. That line represents all solutions of the equation.

Graphs of lines are examined in detail in Section 1.3.

Example 4   Graph: $y = x^2 - 5$.

SOLUTION   Since y is expressed in terms of x, the easiest way to find solutions of this equation is to replace x with various values and then calculate the corresponding values for y. In cases like this, we say that x is the **independent variable** (because we *choose* its value) and y is the **dependent variable** (because its value is then calculated).

x	y	(x, y)
0	−5	$(0, -5)$
−1	−4	$(-1, -4)$
1	−4	$(1, -4)$
−2	−1	$(-2, -1)$
2	−1	$(2, -1)$
−3	4	$(-3, 4)$
3	4	$(3, 4)$

① Select values for x.
② Compute values for y.

Next, we plot these points and connect them. We note that as the absolute value of x increases, $x^2 - 5$ also increases. Thus the graph is a curve that rises on either side of the y-axis.

Graphs of curves similar to the one in Example 4 are examined in more detail in Section 2.2.

Example 5   Graph: $x = y^2 + 1$.

SOLUTION   Since x is expressed in terms of y, we select values for y and then find the corresponding values for x. In this case, y is the independent variable and x the dependent variable.

x	y	(x, y)
1	0	$(1, 0)$
2	−1	$(2, -1)$
2	1	$(2, 1)$
5	−2	$(5, -2)$
5	2	$(5, 2)$

① Select values for y.
② Compute values for x.

When plotting, we must remember that x is the first coordinate and y is the second coordinate. Note in the figure that the graph moves farther from the x-axis as it extends farther to the right.

Example 6 Graph: $xy = 12$.

SOLUTION To find numbers that satisfy the equation, it helps to solve for y (that is, $y = 12/x$) or solve for x (that is, $x = 12/y$). Typically, we solve for y if it is convenient. Then we make substitutions.

x	y	(x, y)
2	6	$(2, 6)$
-2	-6	$(-2, -6)$
3	4	$(3, 4)$
-3	-4	$(-3, -4)$
4	3	$(4, 3)$
-4	-3	$(-4, -3)$
6	2	$(6, 2)$
-6	-2	$(-6, -2)$

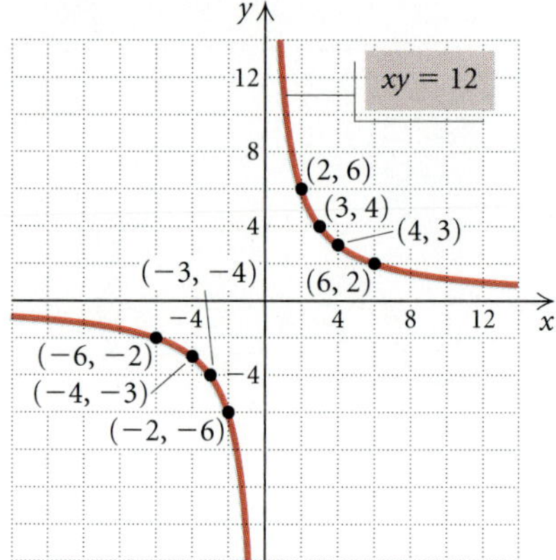

We plot these points and connect them. As the absolute value of x becomes small, the absolute value of y becomes large, and vice versa. Since neither x nor y can be 0, the graph does not cross either axis. Thus the graph consists of two branches.

Example 7 Graph: $y = |x|$.

SOLUTION We find numbers that satisfy the equation. For example, when $x = -3$, $y = |-3| = 3$. When $x = 2$, $y = |2| = 2$, and when $x = 0$, $y = |0| = 0$.

x	y	(x, y)
0	0	$(0, 0)$
-1	1	$(-1, 1)$
1	1	$(1, 1)$
-2	2	$(-2, 2)$
2	2	$(2, 2)$
-3	3	$(-3, 3)$
3	3	$(3, 3)$

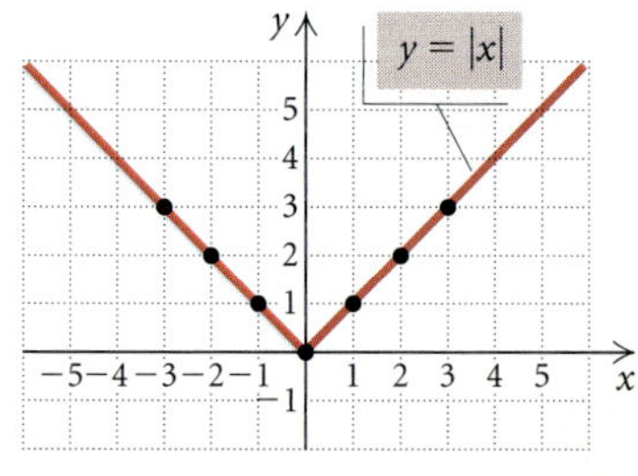

We plot these points and connect them. As the points fall farther from the origin, to the left or right, the absolute value of x increases. Thus the graph is a V-shaped curve that rises to the left and right of the y-axis.

Graphers and Viewing Windows

We now consider the use of graphers. One feature common to all graphers is the **viewing window**. This refers to the rectangular portion of the xy-plane that appears on the screen of a grapher. The notation we

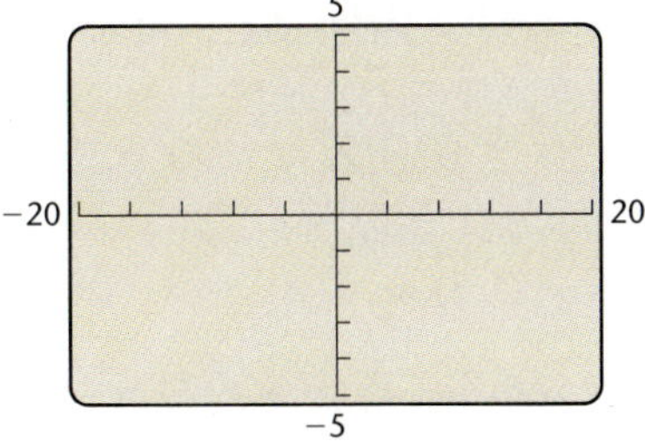

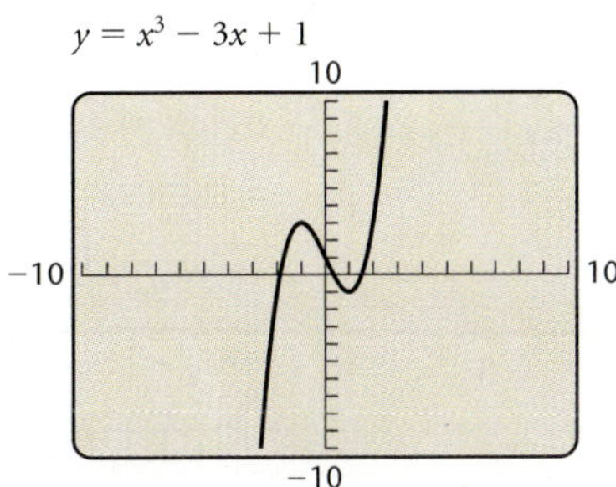

will use to describe viewing windows consists of four numbers, $[L, R, B, T]$ in brackets, which represent the *Left* and *Right* endpoints of the x-axis and the *Bottom* and *Top* endpoints of the y-axis. A WINDOW feature might be used to set these dimensions, but the method varies for each grapher. The screen at top left is a window setting of $[-20, 20, -5, 5]$ with axis scaling denoted as Xscl = 4 and Yscl = 1, which means that there are 4 units between tick marks on the x-axis and 1 unit between tick marks on the y-axis. The screen at bottom left shows the results from this setting.

Axis scaling must be chosen with care, because tick marks become blurred and indistinguishable when too many appear. On some graphers, a setting of $[-10, 10, -10, 10]$, Xscl = 1, Yscl = 1 is considered the **standard window**. There is usually a procedure to obtain a standard setting quickly. Consult your manual.

The primary use of graphers is to graph equations. As an example, let's graph the equation $y = x^3 - 3x + 1$. The equation might be entered using the notation $y = x\text{^}3 - 3x + 1$. Some software use BASIC notation, in which case the equation would be entered as $y = x\text{^}3 - 3*x + 1$. We obtain the following graph.

You may need to change the viewing window in order to clearly reveal the curvature of a graph. For example, each of the following is a graph of $y = 3x^5 - 20x^3$. The graph on the right best represents the curve.

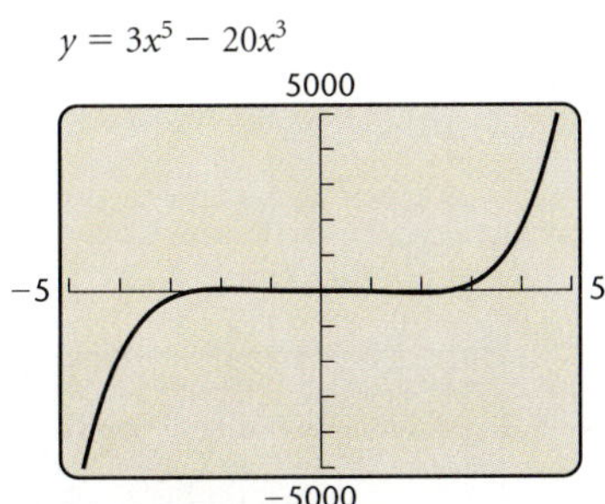

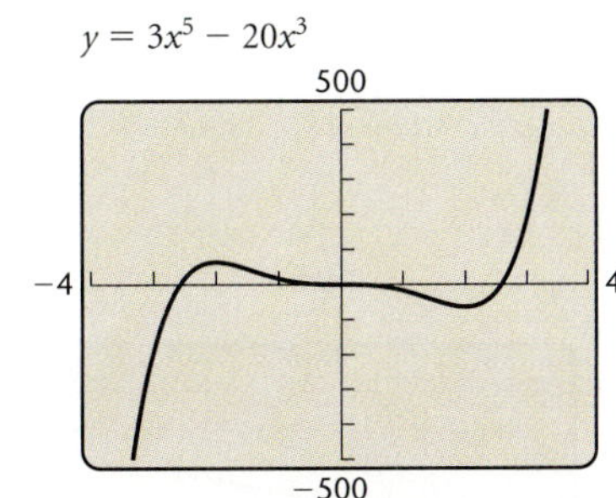

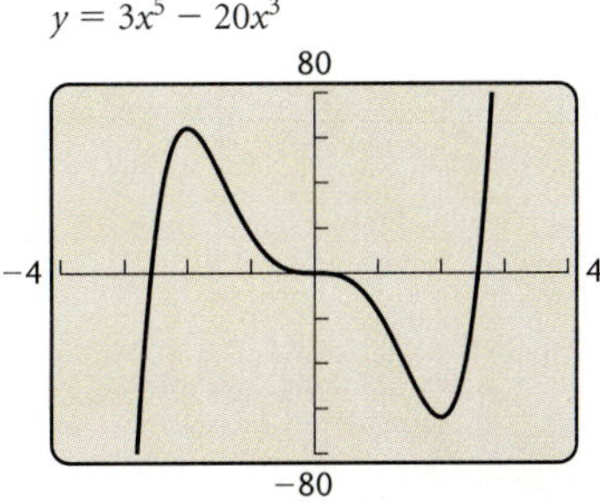

To graph an equation like $2x + 3y = 18$, most graphers require that the equation be solved for y, that is, "$y = \ldots$". When we solve $2x + 3y = 18$ for y, we obtain $y = (18 - 2x)/3$, or $y = 6 - \frac{2}{3}x$. As another example, we can write an equation like $x = y^2 + 1$ as

$y = \pm\sqrt{x - 1}$ and graph the individual equations $y_1 = \sqrt{x - 1}$ and $y_2 = -\sqrt{x - 1}$. The combination of the two graphs yields the graph of $x = y^2 + 1$.

Just as the form in which an equation is entered varies among graphers, so too do other types of notation. Some examples of commonly used grapher notation follow.

EQUATION	POSSIBLE GRAPHER NOTATION
$y = 4x^3 - 7x^2 + 5x - 10$	$y = 4x\wedge3 - 7x^2 + 5x - 10$ or $4*x\wedge3 - 7*x\wedge2 + 5*x - 10$
$y = \sqrt{x - 1}$	$y = (x - 1)\wedge(1/2)$ or $\sqrt{}(x - 1)$
$y = \lvert x - 5 \rvert$	$y = abs(x - 5)$
$y = \sqrt[3]{x}$	$y = x\wedge(1/3)$ or $\sqrt[3]{}x$
$y = \dfrac{7.8x^2 - 1}{x^5 + 3}$	$y = (7.8x^2 - 1)/(x\wedge5 + 3)$ or $(7.8*x\wedge2 - 1)/(x\wedge5 + 3)$

Example 8   Graph each of the following equations choosing a viewing window that best reveals the curvature of the graph. Answers may vary.

a) $2x + 3y = 18$ **b)** $y = x^2 - 5$ **c)** $x = y^2 + 1$

d) $xy = 12$ **e)** $y = \lvert x \rvert$ **f)** $y = x^4 - 2x^2 - 3$

SOLUTION

a) $2x + 3y = 18$

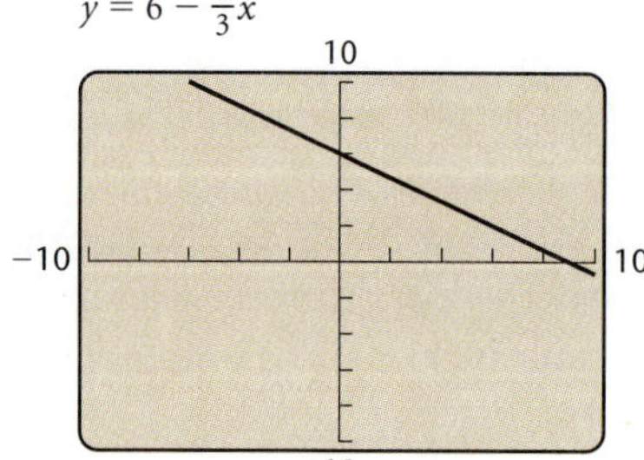

b) $y = x^2 - 5$

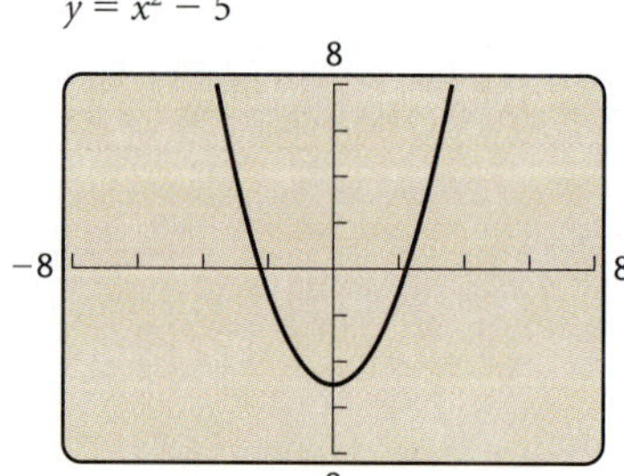

c) $x = y^2 + 1$

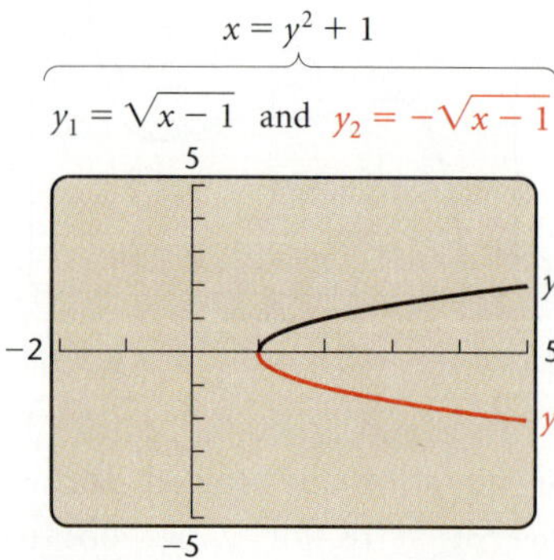

d) $xy = 12$

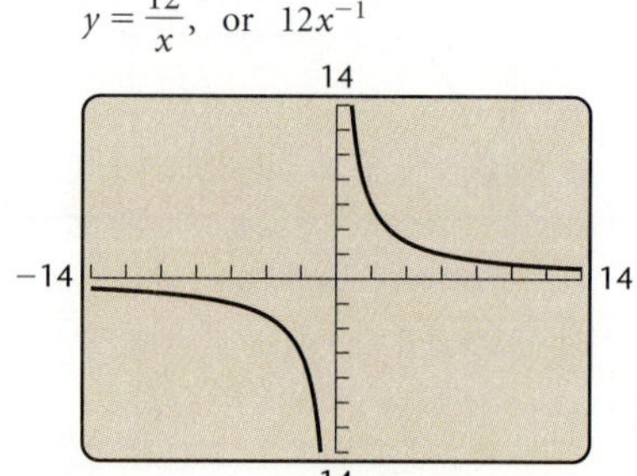

e) $y = |x|$

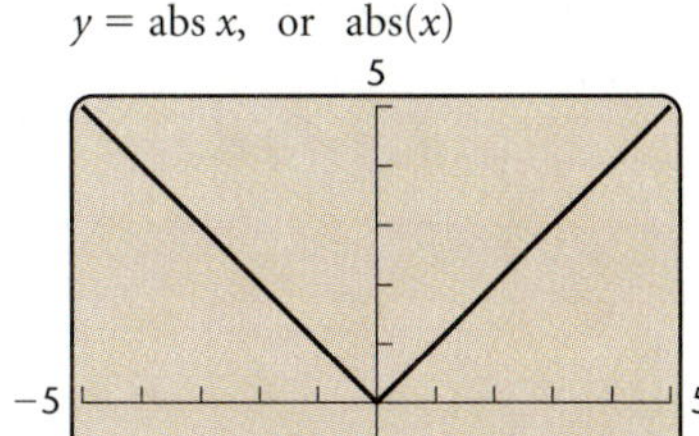

f) $y = x^4 - 2x^2 - 3$

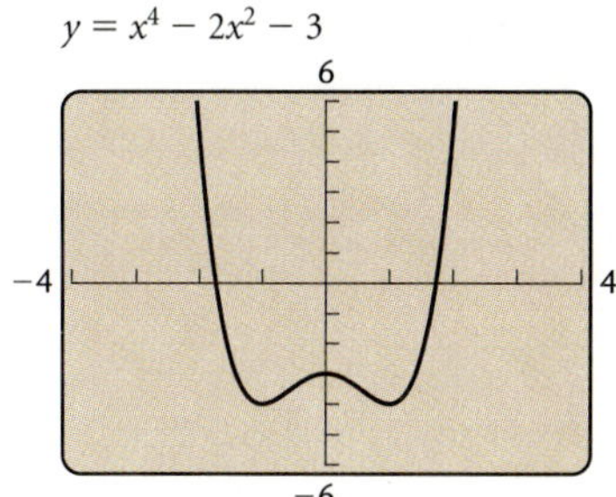

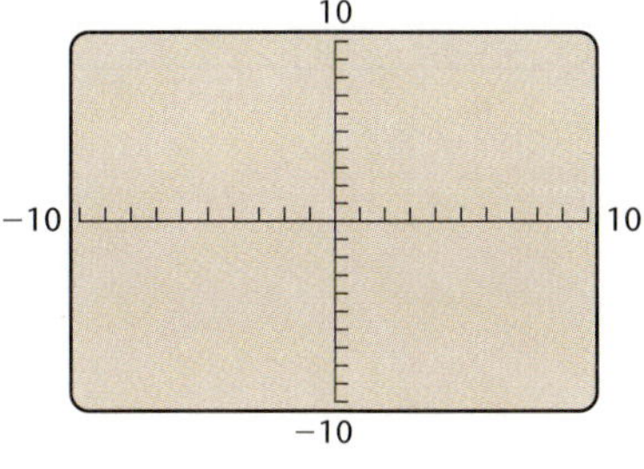

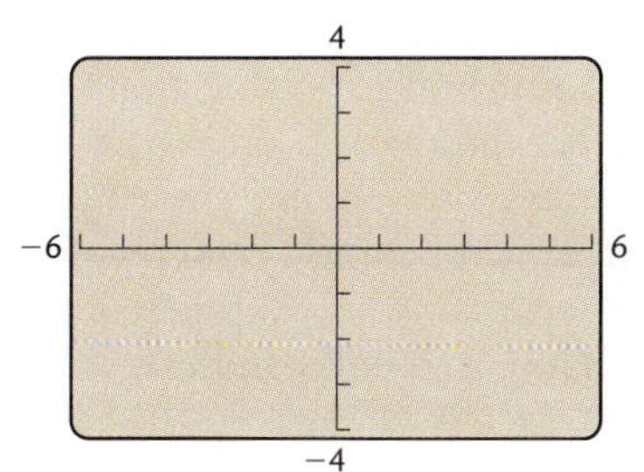

Squaring a Viewing Window

Consider the $[-10, 10, -10, 10]$ viewing window shown first at left. Note that the distance between units is not visually the same on both axes. In this case, the length of the interval shown on the y-axis is about $\frac{2}{3}$ of the length of the interval shown on the x-axis. If we change to the window shown second at left, with dimensions $[-6, 6, -4, 4]$, we get a graph for which the units are visually about the same on both axes. We choose these dimensions so that the length of the y-axis is $\frac{2}{3}$ the length of the x-axis. The window has been *squared*. Any dimensions in this ratio will produce the desired effect with this grapher.

On another grapher, the ratio might be $\frac{8}{15}$, and the window $[-7.5, 7.5, -4, 4]$ would show the units visually the same on both axes.* Creating such a window is called **squaring**. On some graphers, there is a ZSQUARE feature for automatic window squaring. Consult your manual about how to square a window.

Each of the following is a graph of the line $y = 2x - 3$, but the viewing windows are different.

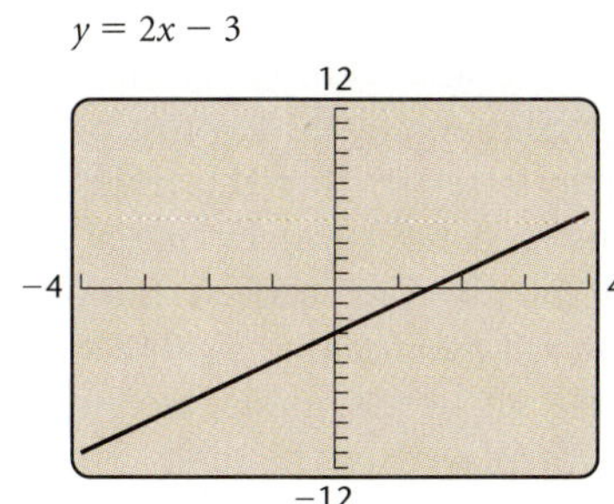

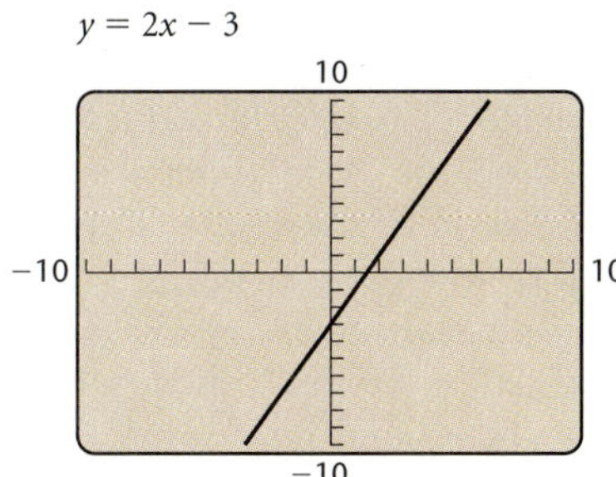

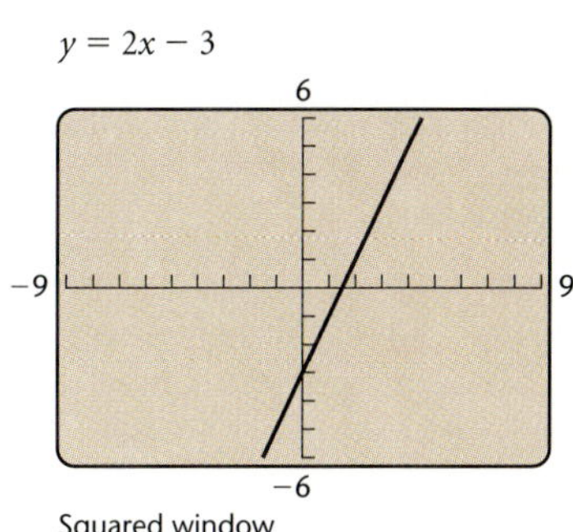

Squared window

When the x-units and the y-units are visually the same, we get an accurate representation of the *slope* of the line. We will study slope in Section 1.3.

---

*The exact ratio is 63/95 on a TI-82 or a TI-83, 63/127 on a TI-85, and 103/239 on a TI-92.

A squared window also eliminates distortion of the graph. Compare the graph of the circle $x^2 + y^2 = 4$ shown here in both a nonsquared and a squared window.

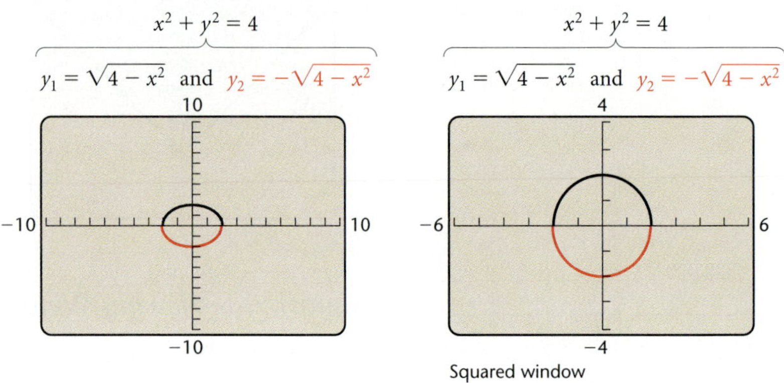

Circles will be studied in Section 1.5.

The Table Feature

Many graphers can display a table of values similar to the tables we used for hand-drawn graphs. For an equation entered on the "$y =$" screen, we can use the table set-up screen to choose a minimum x-value along with a step-value or increment. The grapher then produces a table of x- and y-values for the given equation. For example, if we enter the equation $y = 1/(x - 3)$ along with a minimum x-value of 2 and an increment of 0.25, the table at left is produced. We see that $y = -1$ when $x = 2$, $y = -1.3333$ when $x = 2.25$, and so on. The ERROR entry indicates that 3 is not an acceptable value for x.

We can use the arrow keys to scroll through more x, y-values for this equation. Some graphers can display both a graph and a table on a split screen.

The TABLE feature can also be used to evaluate an expression in one variable for particular values of the variable. Again the expression is entered on the "$y =$" screen and the table is set so that the independent variable is in ASK mode. This allows us to enter x-values one at a time and see the corresponding y-values for the given expression. Suppose we want to evaluate $2x^5 - 6x^3 + 7$ for $x = -8$ and for $x = 5$. First we enter $y = 2x^5 - 6x^3 + 7$; then, with the table set in ASK mode, we enter -8. The corresponding y-value $-62{,}457$ is displayed in the table. Similarly, we enter 5 and read the y-value 5507. We can continue to enter x-values as desired.

TblMin = 2, ΔTbl = .25

X	Y1
2	−1
2.25	−1.333
2.5	−2
2.75	−4
3	ERROR
3.25	4
3.5	2

X = 2.25

X	Y1
−8	−62457
5	5507

X =

Program

The Graphing Calculator Manual that accompanies this text contains programs that can be entered in the TI-82, TI-83, TI-85, or HP38G graphing calculators. For a complete list of these program references, see the Program heading in the index. The first program is designed for the TI-85.

TABLE: This program provides the TI-85 with a TABLE feature.

Identities

Consider the equation

$$(x + 1)^2 = x^2 + 2x + 1.$$

We know from the laws of algebra that this equation is true for every possible substitution of x by a real number. Let's look at this equation graphically. We consider two equations of the form "$y = \ldots$", one using the expression on the left side of the equation and one using the expression on the right:

$$y_1 = (x + 1)^2 \quad \text{and} \quad y_2 = x^2 + 2x + 1.$$

If we graph these two equations using the same viewing window, $[-5, 5, -1, 6]$, we see that the graphs seem to coincide. You might try other viewing windows to see that they do not differ. The TABLE feature will also confirm that y_1 and y_2 appear to have the same value for a given value of x.

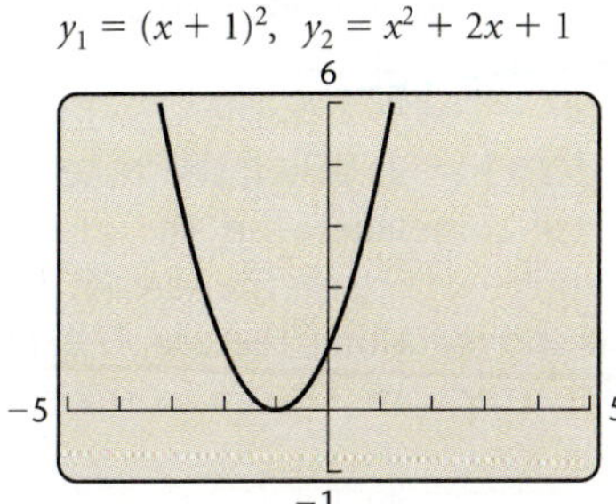

X	Y1	Y2
-2	1	1
-1	0	0
0	1	1
1	4	4
2	9	9
3	16	16
4	25	25

The equation $(x + 1)^2 = x^2 + 2x + 1$ is called an **identity**. As illustrated above, it is true for every possible substitution. The graphs are the same for the values of x for which both expressions are defined. One way to get a partial check of a potential identity, or to make a conjecture that an equation is an identity, is to graph each side and see if the graphs coincide. Although this procedure can be fruitful, one must be aware of its shortcomings. First, we rarely can draw all or even most of a graph. Thus there is always the possibility that somewhere outside the window, the graphs differ. The second limitation is that the size and scale of the window may deceive us about whether graphs really do coincide. An alternate partial check can also be obtained using the TABLE feature or other function evaluator. Many values of a pair of expressions can be compared very quickly.

Example 9 Determine graphically whether each of the following seems to be an identity.

a) $(x^2)^3 = x^6$ **b)** $\sqrt{x + 4} = \sqrt{x} + 2$

a) We leave it to the student to graph $y_1 = (x^2)^3$ and $y_2 = x^6$. The graphs appear to coincide no matter what the viewing window. Thus, $(x^2)^3 = x^6$ seems to be an identity.

b) We graph $y_1 = \sqrt{x + 4}$ and $y_2 = \sqrt{x} + 2$ using the viewing window $[-6, 8, -1, 6]$. We see that the graphs differ, so the equation is *not* an identity. Comparing y-values using the TABLE feature also illustrates that this equation is *not* an identity.

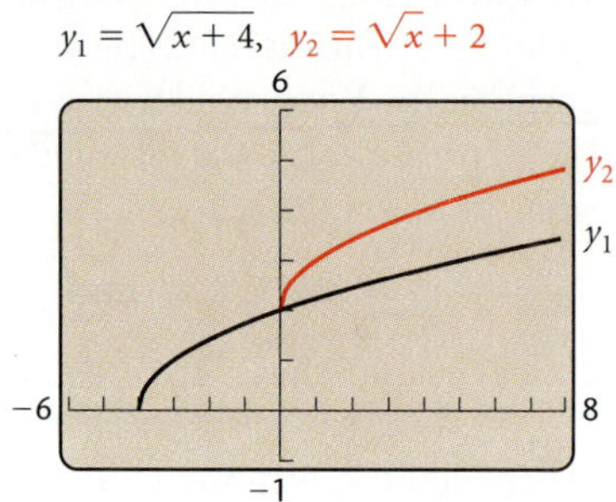

The Trace Feature

Once a graph has been created, we can investigate some of its points by using the TRACE feature that most graphers offer. Consider the graph of $y = x^3 - 3x + 1$ in the viewing window $[-10, 10, -10, 10]$ with Xscl = 2 and Yscl = 2. When the TRACE key is pressed, a cursor (often blinking) appears somewhere on the graph, while the x- and y-coordinates are shown elsewhere on the screen. These coordinates will change as the cursor is moved along the graph using the arrow keys ($\triangleleft, \triangleright$). Some points on the graph are shown in the figures below.

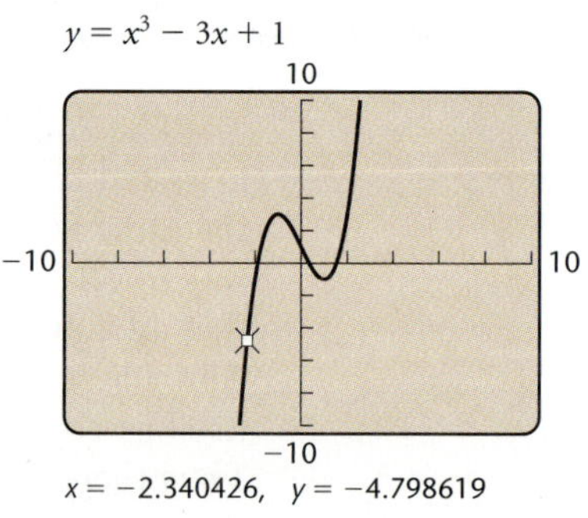

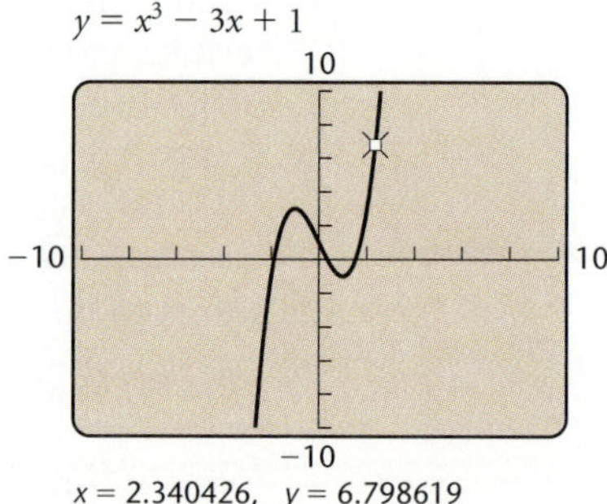

Equation Solving

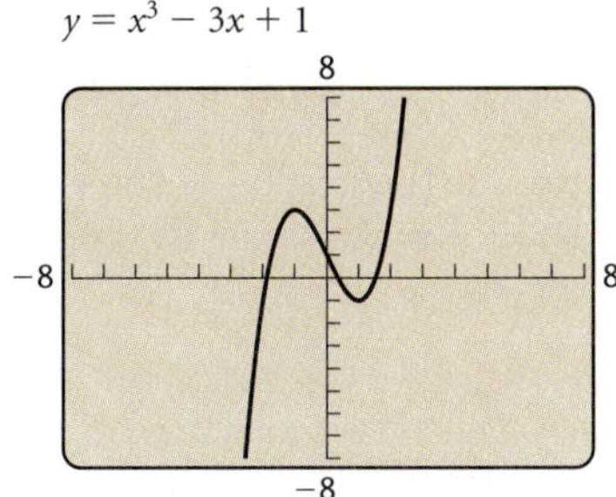

Example 10 Solve $x^3 - 3x + 1 = 0$ graphically. Approximate solutions to three decimal places.

SOLUTION The solutions of the equation $x^3 - 3x + 1 = 0$ are the first coordinates of the points where the graph of $y = x^3 - 3x + 1$ crosses the x-axis. These points are called the **x-intercepts** of the graph. An x-intercept has the general form $(a, 0)$ and occurs at an x-value a for which $y = 0$. It appears from the graph shown at left that there are three such x-values, one near -2, one near 0, and one near 1.5.

To acquire more precision, we can change the viewing window. A fast way to do this on many graphers is to use a feature called ZOOM. The zoom-in feature allows a portion of the graph to be magnified; that is, a small section of the graph is enlarged in the viewing window. Let's examine the graph near $x = -2$. We trace to the y-value closest to 0 and zoom in on the graph at that point. We continue to zoom in until the desired accuracy is achieved.

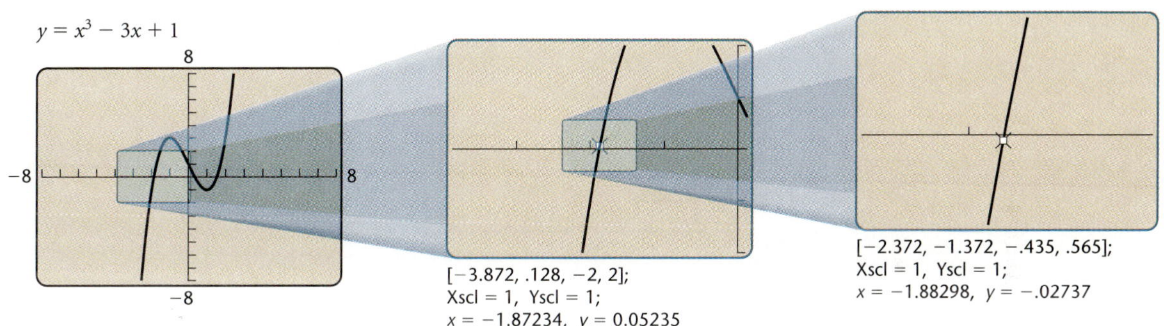

To find the solution to the nearest thousandth, we trace and zoom until the cursor's x-value just to the left of the intercept and the cursor's x-value just to the right of the intercept are the same, when rounded to the nearest thousandth. By using the TRACE and ZOOM features three more times, we find the solution to be about -1.879. In a similar manner, we find that the other solutions of the equation $x^3 - 3x + 1 = 0$ are about 0.347 and 1.532. If available, we might also use the SOLVE or ROOT features to approximate solutions.

There are many ways in which the ZOOM feature can be used. For example, we can zoom in for more precision, as in Example 10, or zoom out on a graph—say, to reveal more of its curvature—adjusting the factors of the zoom in any way we choose. In Example 10, we used zoom factors of 4. We may also be able to use a ZOOM-BOX feature to zoom in on a boxed region of our choosing. All such details can be found by consulting the manual for your particular grapher or the Graphing Calculator Manual that accompanies this book.

Finding Points of Intersection

There are many situations in which we want to determine the point(s) of intersection of two graphs.

Example 11   Find the points of intersection of the graphs of the equations $y_1 = 3x^5 - 20x^3$ and $y_2 = 34.7 - 1.28x^2$. Approximate the coordinates to three decimal places.

SOLUTION

Method 1. We use a grapher, trying to create graphs that clearly show the curvature. Then we look for the coordinates of the points at which the graphs cross each other. It appears that there are three points of intersection. We can use TRACE and ZOOM to approximate their coordinates, or we can use the INTERSECT feature, if available. Experiment to see which you prefer.

$$y_1 = 3x^5 - 20x^3, \quad y_2 = 34.7 - 1.28x^2$$

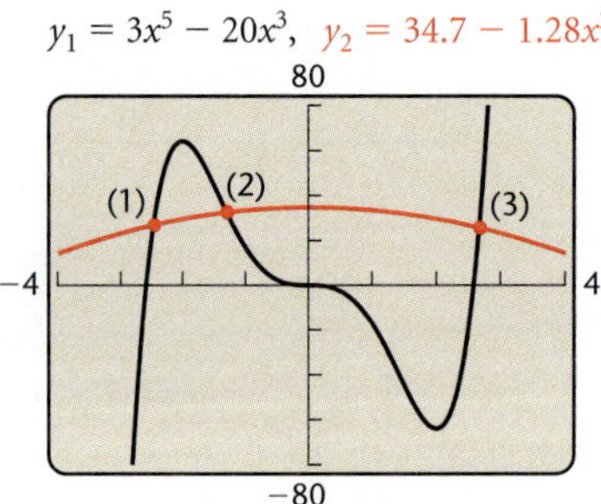

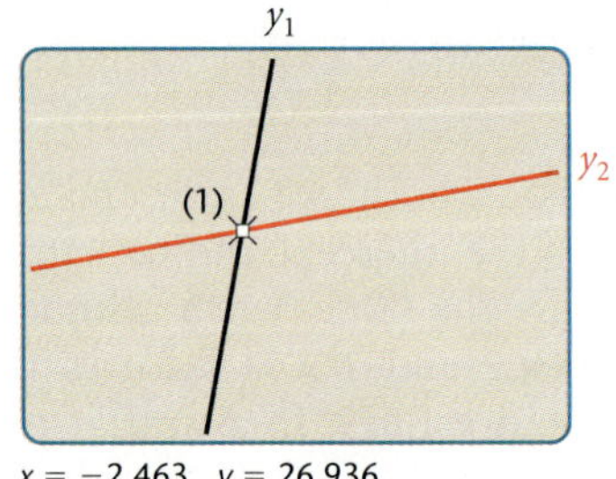

$x = -2.463, \ y = 26.936$

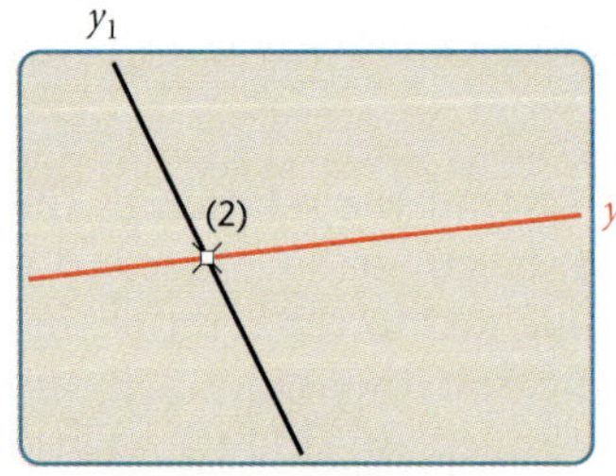

$x = -1.296, \ y = 32.551$

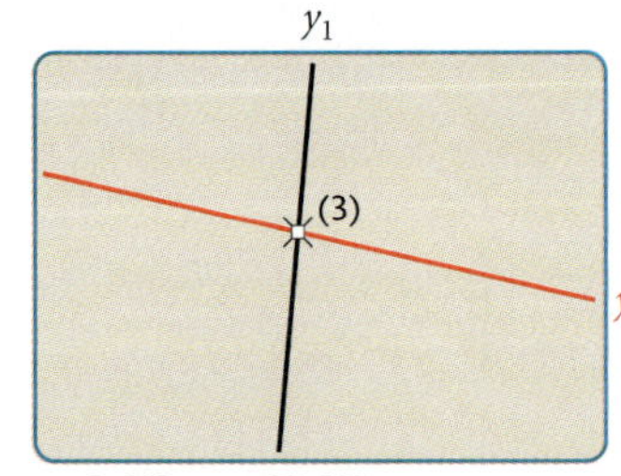

$x = 2.668, \ y = 25.591$

The values found can be checked by substitution. The points of intersection are about $(-2.463, 26.936)$, $(-1.296, 32.551)$, and $(2.668, 25.591)$.

Method 2. Another method is to use the SOLVE feature. We find the x-values that are solutions of the equation $y_1 = y_2$. To do so, we solve

$$y_1 - y_2 = 0, \quad \text{or} \quad 3x^5 - 20x^3 + 1.28x^2 - 34.7 = 0.$$

On some graphers there is actually a POLY or ROOT feature, which allows you to enter this polynomial and gives all the solutions of this equation directly. We round each x-value to three decimal places. Then, to compute the second coordinate, we substitute the x-value into either y_1 or y_2.

Example 12   Solve: $\frac{2}{3}x - 7 = 5$.

SOLUTION

Method 1. One way to solve this equation is to ask ourselves, "For what x-values will the expression $\frac{2}{3}x - 7$ equal 5?" This can be visualized by asking, "Where will the graph of $y = \frac{2}{3}x - 7$ cross the horizontal line $y = 5$?" Thus we graph the equations $y_1 = \frac{2}{3}x - 7$ and $y_2 = 5$ and look for the _first coordinate_ of their point of intersection. The INTERSECT feature provides us with an efficient means of finding this point. TRACE and ZOOM could also be used.

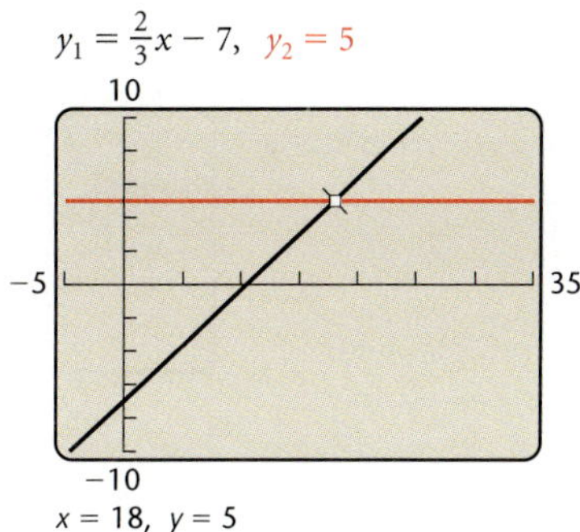

X	Y1	Y2
17.7	4.8	5
17.8	4.8667	5
17.9	4.9333	5
18	5	5
18.1	5.0667	5
18.2	5.1333	5
18.3	5.2	5

X = 18

We see that the point of intersection is (18, 5). Thus the solution is 18.

We can also confirm the solution numerically, using the table shown at left. The step-value in the table is 0.1. We choose the minimum x-value to be 17.7 and observe the y-values as the x-values increase to 18. The solution can also be checked using the table set in ASK mode.

Method 2. Another way to solve this equation is to solve

$$y_1 - y_2 = 0, \quad \text{or} \quad \tfrac{2}{3}x - 7 - 5 = 0, \quad \text{or} \quad \tfrac{2}{3}x - 12 = 0.$$

We can do this in a manner similar to Example 10. We graph the equation $y = \frac{2}{3}x - 12$ and see that it intersects the x-axis at (18, 0). Thus the solution is 18. We could also use the SOLVE or ROOT features.

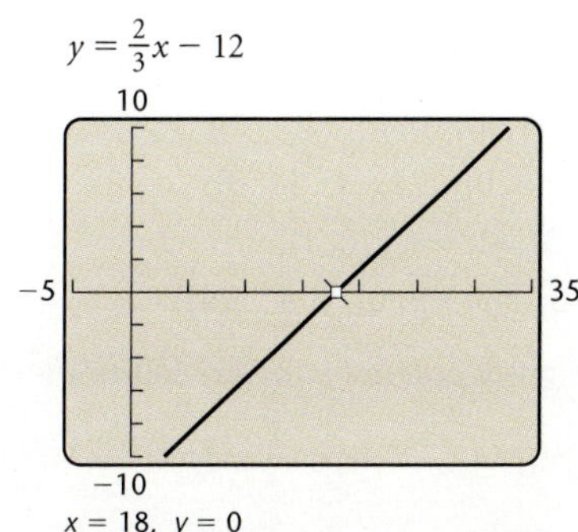

A Special Case

Graphers, for all their advantages, also have their disadvantages, one of which can occur with certain kinds of rational equations. Consider the equation $y = 8/(x^2 - 4)$. The numbers -2 and 2 cannot be substituted into the equation because they result in division by 0. Thus no point with either of these numbers as a first coordinate can be part of the graph. Now look at the two graphs of $y = 8/(x^2 - 4)$ shown below.

CONNECTED MODE

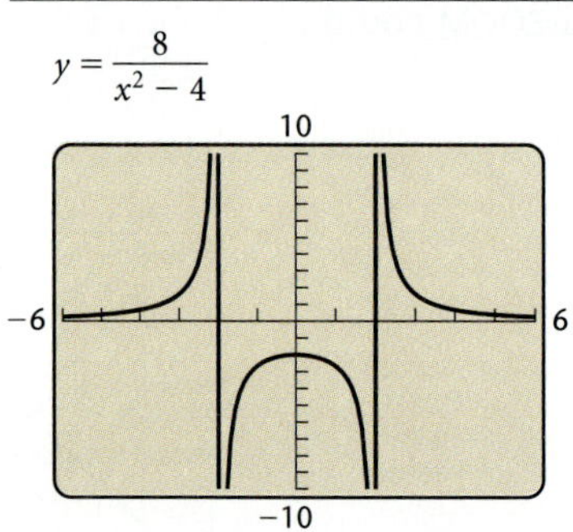

DOT MODE

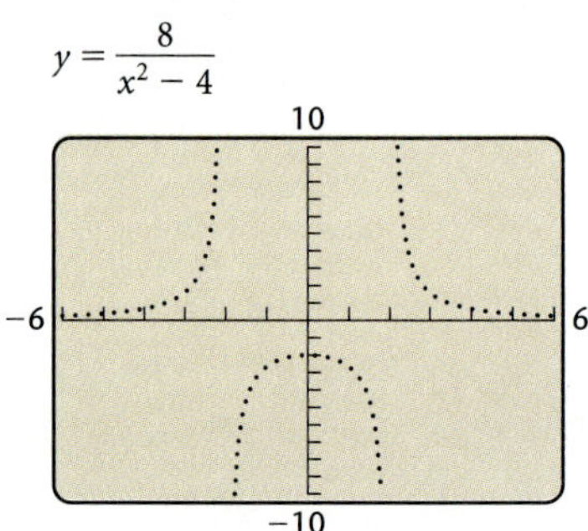

In CONNECTED mode, a grapher connects plotted points with line segments. In DOT mode, it simply plots unconnected points. In the graph on the left, the grapher has connected the points plotted on either side of the x-values -2 and 2 with vertical lines. These lines, however, cannot be part of the graph because substitutions of -2 and 2 are not allowed in this equation. Thus this is not a correct graph of this equation. The graph on the right is more accurate. When graphing equations in which a variable appears in a denominator, use DOT mode if you have a choice.

This kind of difficulty should not deter you from using a grapher. As we will see throughout the text, we will often need to apply our math knowledge to make the best use of a grapher.

Exercise Set

Graph and label each set of points. Make a hand-drawn graph.

1. $\{(4, 0), (-3, -5), (-1, 4), (0, 2), (2, -2)\}$

2. $\{(1, 4), (-4, -2), (-5, 0), (2, -4), (4, 0)\}$

3. $\{(-5, 1), (5, 1), (2, 3), (2, -1), (0, 1)\}$

4. $\{(4, 0), (4, -3), (-5, 2), (-5, 0), (-1, -5)\}$

Determine whether the given ordered pairs are solutions of the given equation.

5. $(1, -1), (0, 3); \ y = 2x - 3$

6. $(2, 5), (-2, -5); \ y = 3x - 1$

7. $\left(-\frac{1}{2}, -\frac{4}{5}\right), \left(0, \frac{3}{5}\right); \ 2a + 5b = 3$

8. $\left(0, \frac{3}{2}\right), \left(\frac{2}{3}, 1\right); \ 3m + 4n = 6$

9. $(-0.75, 2.75), (2, -1); \ x^2 - y^2 = 3$

10. $(2, -4), (4, -5); \ 5x + 2y^2 = 70$

Make a hand-drawn graph of each equation. Use the indicated x-values along with any others you are curious about.

11. $y = 3 - 2x; \ x = -3, -2, -1, 0, 1, 2, 3$

12. $x + y = 4; \ x = -3, -2, -1, 0, 1, 2, 3$

13. $3x - 4y = 12; \ x = -3, -2, -1, 0, 1, 2, 3$

14. $y = x^2$; $x = -3, -2, -1, 0, 1, 2, 3$

15. $y = -x^2$; $x = -3, -2, -1, 0, 1, 2, 3$

16. $y = \sqrt{x}$; $x = 0, 1, 4, 9$

17. $y = |x| - 2$; $x = -3, -2, -1, 0, 1, 2, 3$

18. $y = x^3$; $x = -2, -1, 0, 1, 2$

19. $y = \sqrt{x} - 4$; $x = 0, 1, 4, 9$

20. $y = |x| + 2$; $x = -3, -2, -1, 0, 1, 2, 3$

21. $y = -\dfrac{2}{x}$; $x = -4, -2, -1, -\dfrac{1}{2}, \dfrac{1}{2}, 1, 2, 4$

22. $y = 4 - x^2$; $x = -3, -2, -1, 0, 1, 2, 3$

In Exercises 23–30, use a grapher to match the equations with graphs (a)–(h).

a)
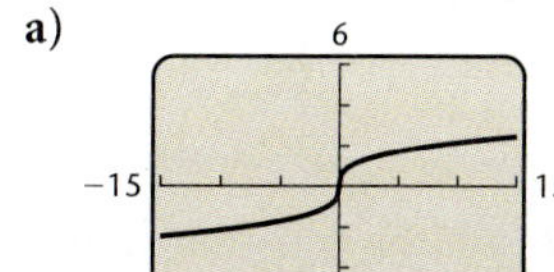

b)
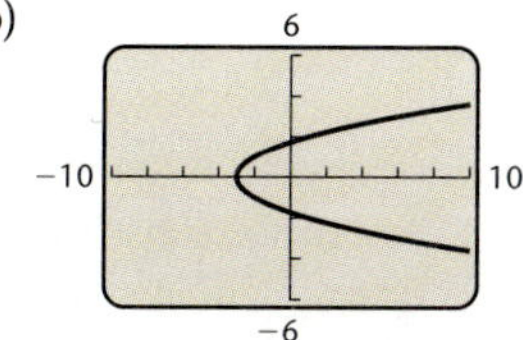

c)
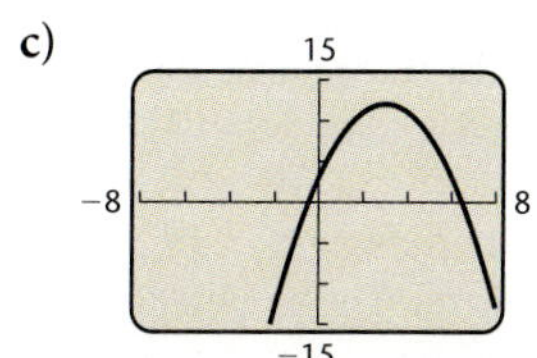

d)
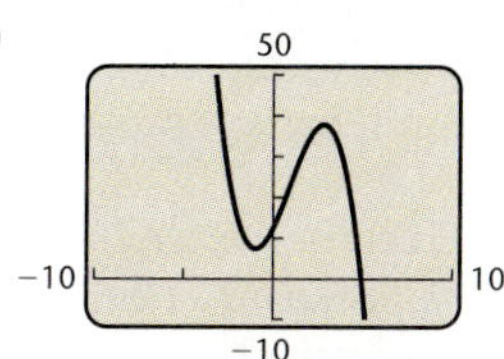

e)
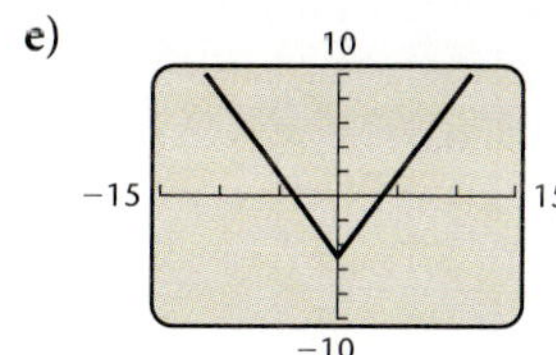

f)
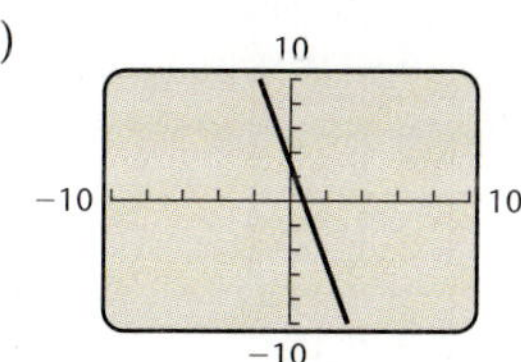

g)
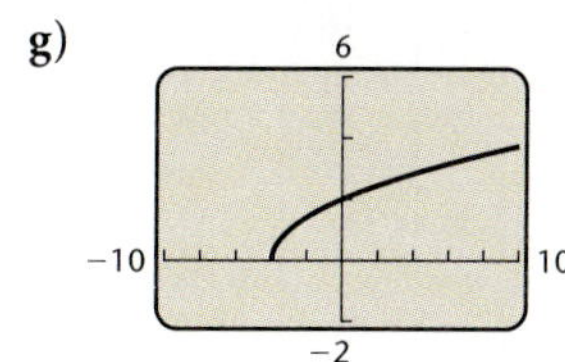

h)
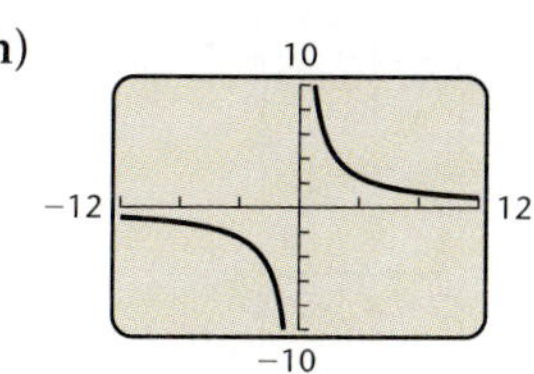

23. $y = 3 - 4x$

24. $y = |x| - 5$

25. $y = 3 + 6x - x^2$

26. $y = \sqrt{x + 4}$

27. $x = y^2 - 3$

28. $y = \sqrt[3]{x}$

29. $y = 12.4 + 9.1x + 3.07x^2 - 1.1x^3$

30. $xy = 10$

Graph each of the following equations, using the standard viewing window, $[-10, 10, -10, 10]$. Then try other windows until you find one that best represents the graph. Answers may vary.

31. $y = \frac{1}{5}x + 2$

32. $4x - 5y = 20$

33. $x = y^2 + 2$

34. $y = |x - 1|$

35. $xy = -18$

36. $y = (x + 2)^2$

37. $y = x^2 - 4x + 3$

38. $x = 4 - y^2$

39. $y = \sqrt{x} - 2$

40. $y = x^3 + 2x^2 - 4x - 13$

41. $y^2 = 4 + x$

42. $y = \sqrt[3]{x + 4}$

In Exercises 43–46, an equation is given together with four choices of viewing windows and scaling conditions. Choose the best option. If you reject a situation, explain why you did so.

43. $y = 2x^3 - x^4$

   a) $[-1, 1, -1, 1]$, Xscl $= 1$, Yscl $= 1$

   b) $[-2, 3, -4, 3]$, Xscl $= 1$, Yscl $= 1$

   c) $[-2, 3, -4, 3]$, Xscl $= 0.01$, Yscl $= 0.1$

   d) $[-20, 30, -40, 30]$, Xscl $= 1$, Yscl $= 1$

44. $y = x^4 - 8x^3 + 18x^2$

   a) $[-10, 10, -10, 10]$, Xscl $= 1$, Yscl $= 1$

   b) $[-1, 1, -10, 10]$, Xscl $= 1$, Yscl $= 1$

   c) $[-3, 6, -10, 60]$, Xscl $= 1$, Yscl $= 10$

   d) $[-10, 60, -3, 6]$, Xscl $= 1$, Yscl $= 10$

45. $y = \dfrac{8x}{x^2 + 1}$

   a) $[-1, 1, -10, 10]$, Xscl $= 1$, Yscl $= 1$

   b) $[-5, 5, -7, 7]$, Xscl $= 1$, Yscl $= 1$

   c) $[-3, 6, -10, 60]$, Xscl $= 1$, Yscl $= 10$

   d) $[-100, 100, -100, 100]$, Xscl $= 10$, Yscl $= 10$

46. $y = 4x - x^3$

   a) $[-0.1, 10, -10, 10]$, Xscl $= 1$, Yscl $= 1$

   b) $[-10, 0.1, -10, 10]$, Xscl $= 1$, Yscl $= 1$

   c) $[-5, 5, -0.1, 1]$, Xscl $= 1$, Yscl $= 1$

   d) $[-3, 3, -10, 10]$, Xscl $= 1$, Yscl $= 1$

47. Using a grapher, complete this table for the equations

$$y_1 = \sqrt{10 - x^2} \quad \text{and} \quad y_2 = \frac{16}{x + 2.8}.$$

Then extend the table from -2.6 to 3.3.

X	Y1	Y2
−3.3		
−3.2		
−3.1		
−3		
−2.9		
−2.8		
−2.7		

X = −3.3

Determine with a grapher whether each of the following seems to be an identity.

48. $7x = x \cdot 7$

49. $(x - 5)^2 = x^2 - 10x + 25$

50. $(x + 4)^2 = x^2 + 16$

51. $x^3 + x^4 = x^7$

52. $(x - 9)^2 = (x + 3)(x - 3)$

53. $x^3 - 1 = (x - 1)(x^2 + x + 1)$

54. $(x + 2)^3 = x^3 + 8$

55. $\sqrt{x^2 - 16} = x - 4$

56. $(x^2)^3 = x^6$

57. $\dfrac{x^5}{x^2} = x^3$

58. $\sqrt{1 + x} = 1 + \dfrac{x}{2} - \dfrac{x^2}{8} + \dfrac{x^3}{16}$

Solve with a grapher.

59. $\frac{3}{4}x + 2 = -4$

60. $-5 = -\frac{3}{2}x - 3$

61. $3x + 4 = -\frac{2}{5}x + 1$

62. $15 - 2x = \dfrac{2x + 5}{3}$

63. $1.4x + 0.7 = 0.9x - 2.2$

64. $-2.6(x - 8.4) + 1.92 = 23x - 0.93(8x + 11.3)$

65. $x - 7.4 = 2.8\sqrt{x + 1.1}$

66. $\sqrt{x - 3.2} + \sqrt{x + 5.03} = 4.91$

67. $x^3 - 6x^2 = -9x - 1$

68. $2.1x^4 - 4.3x^2 = 5$

69. $1.09x^2 - 0.8x^4 = -7.6$

70. $144x + 140 = x^3 - 3x^2$

71. $x^4 + 4x^3 + 300 = 36x^2 + 160x$

72. $x^3 = 2x^2 + 2$

73. $|x + 1| + |x - 2| = 5$

74. $|x + 1| + |x - 2| = 0$

Find the points of intersection of the graphs of each pair of equations using a grapher.

75. $y_1 = x^3 + 3x^2 - 9x - 13, \ y_2 = 0$

76. $y_1 = \sqrt{7 - x^2}, \ y_2 = 2.5$

77. $y = x^3 - 3x^2, \ 4x - 7y = 20$

78. $y = x^4 - 2x^3, \ y = -0.7x + 3$

79. $y = \dfrac{8x}{x^2 + 1}, \ y = 0.9x$

80. $y = x\sqrt{8 - x^2}, \ 93x - 100y = 1$

81. *A Hole in a Graph.*   Consider the graph of
$$y = \frac{x^2 - 4}{x - 2}.$$

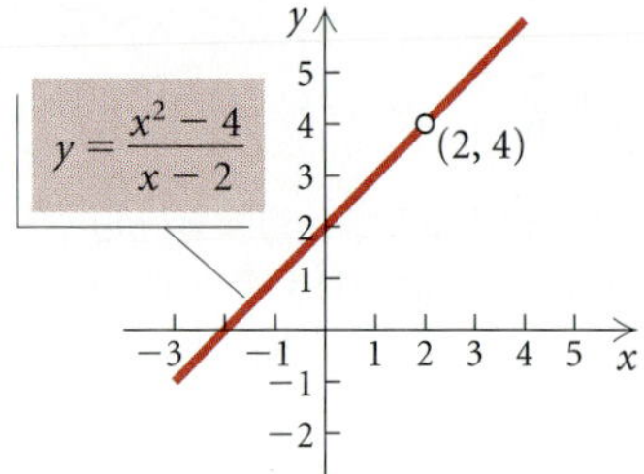

We cannot substitute $x = 2$ into the equation. This creates a "hole" in the graph. The point $(2, 4)$ is not part of the graph. On some graphers it may be very difficult to find the hole. Graph this function and try to find the hole. Experiment with some of the grapher's features.

In Exercises 82 and 83, make a hand-drawn graph. Look for a pattern to find the missing data. Try to determine an equation that fits the data.

82.

x	y
1	4
2	7
3	10
4	13
5	
6	
	31
	34

83.

x	y
0	1
1	2
2	5
3	10
4	17
5	
	82
	122

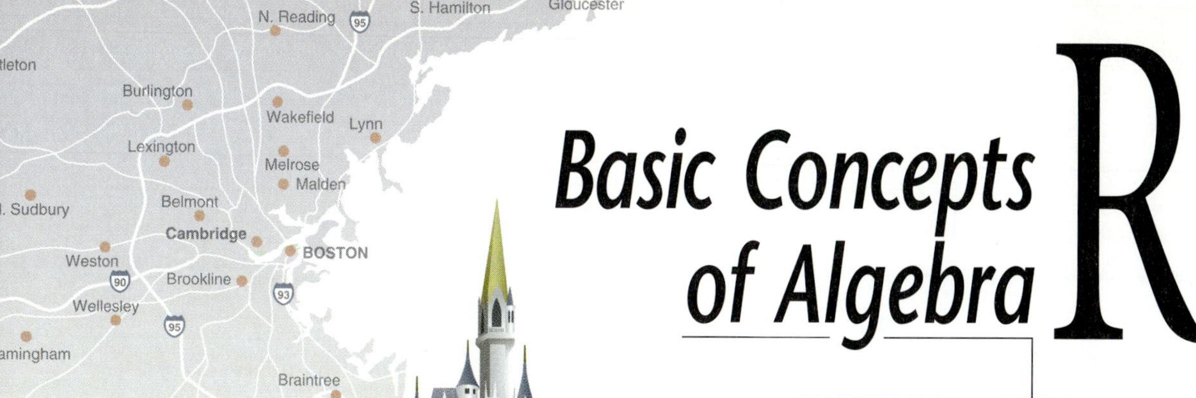

Basic Concepts of Algebra R

The solution of the equation $x + 2x = 30$ gives the average number of vacation days a U.S. worker receives each year.

X	Y1	Y2
7	21	30
8	24	30
9	27	30
10	30	30
11	33	30
12	36	30
13	39	30

X = 10

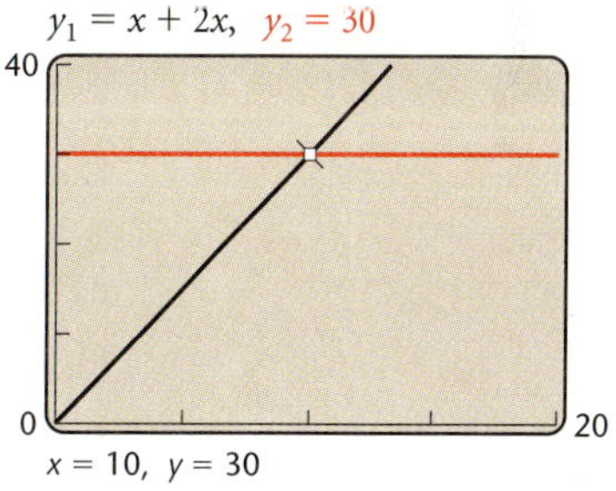

The algebraic concepts reviewed in this chapter provide the foundation for your further study in this text. They include the properties of the real-number system, manipulations of algebraic expressions, the solution of equations and inequalities, and the solution of applied problems, like the one shown concerning vacation time. Grapher technology is used to enhance the understanding and visualization of these topics.

R.1 The Real-Number System

R.2 Integer Exponents, Scientific Notation, and Order of Operations

R.3 Addition, Subtraction, and Multiplication of Polynomials

R.4 Factoring

R.5 Rational Expressions

R.6 Radical Notation and Rational Exponents

R.7 Solving Equations

R.8 Solving Inequalities

R.9 Modeling and Applications

SUMMARY AND REVIEW

R.1

The Real-Number System

- *Identify various kinds of real numbers.*
- *Find the absolute value of an algebraic expression.*
- *Identify the properties of real numbers.*

Real Numbers

In applications of algebraic concepts, we use real numbers to represent quantities such as distance, time, speed, area, profit, loss, and temperature. Some frequently used sets of real numbers are listed below.

Sets of Real Numbers

Natural numbers. The numbers used for counting: $\{1, 2, 3, \ldots\}$.

Whole numbers. The natural numbers and 0: $\{0, 1, 2, 3, \ldots\}$.

Integers. The whole numbers and their opposites:
$\{\ldots, -3, -2, -1, 0, 1, 2, 3, \ldots\}$.

Rational numbers. The numbers that can be expressed in the form p/q, where p and q are integers and $q \neq 0$. Some examples are $\frac{1}{4}$, $-\frac{12}{5}$, $0.125 \left(0.125 = \frac{1}{8}\right)$, $106 \left(106 = \frac{106}{1}\right)$, and $-0.\overline{3} \left(-0.\overline{3} = -0.333\ldots = -\frac{1}{3}\right)$.

Irrational numbers. The real numbers that are not rational. Some examples are $-\sqrt{5}$, π, $\sqrt[3]{19}$, and $-0.2020020002\ldots$.

Decimal notation for rational numbers either *terminates* (ends) or *repeats.* Each of the following is a rational number.

a) 0 $\qquad\qquad 0 = \dfrac{0}{a}$ for any nonzero integer a

b) $-7 \qquad\qquad -7 = \dfrac{-7}{1}$, or $\dfrac{7}{-1}$, or $-\dfrac{7}{1}$

c) $\dfrac{1}{4} = 0.25 \qquad\qquad$ Terminating decimal

d) $-\dfrac{5}{11} = -0.\overline{45} \qquad\qquad$ Repeating decimal

Decimal notation for irrational numbers neither terminates nor repeats. Each of the following is an irrational number.

a) $\pi = 3.1415926535\ldots \qquad$ There is no repeating block of digits.

$\left(\frac{22}{7}$ and 3.14 are rational *approximations* of the irrational number $\pi.\right)$

b) $\sqrt{2} = 1.414213562\ldots \qquad$ There is no repeating block of digits.

c) $-6.12122122212222\ldots \qquad$ Although there is a pattern, there is no repeating block of digits.

The set of all rational numbers combined with the set of all irrational numbers gives us the set of **real numbers.** The relationships among the various sets of real numbers are shown below.

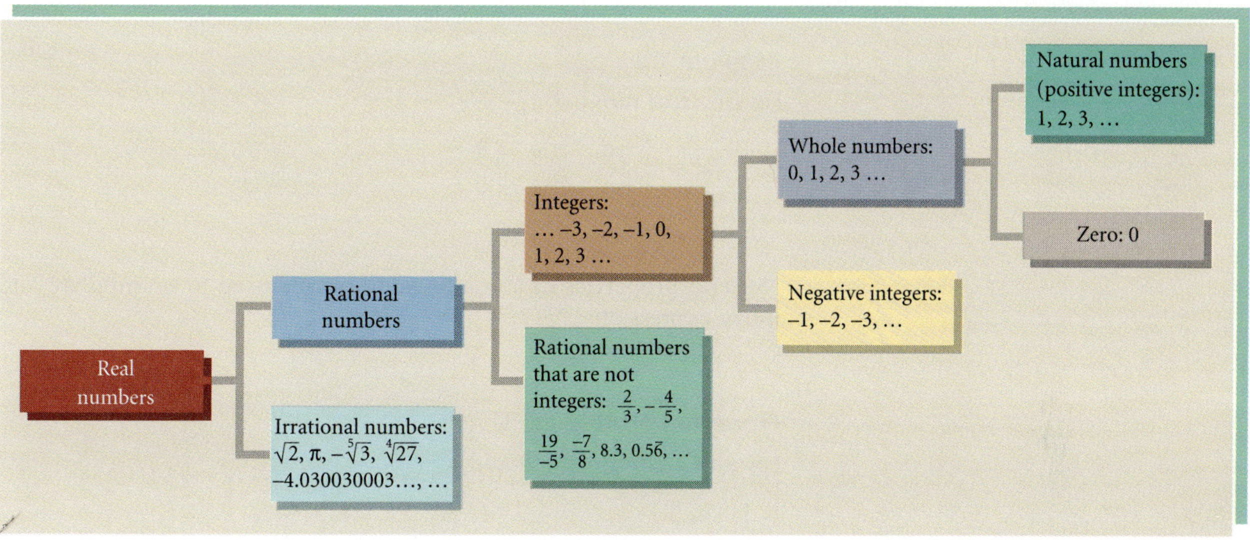

The symbol $\in$ is used to indicate that a member, or *element*, belongs to a set. Thus if we let $\mathbb{Q}$ represent the set of rational numbers, we can see from the diagram that $0.5\overline{6} \in \mathbb{Q}$. We can also write $\sqrt{2} \notin \mathbb{Q}$ to indicate that $\sqrt{2}$ is *not* an element of the set of rational numbers.

When *all* the elements of one set are elements of a second set, we say that the first set is a **subset** of the second set. The symbol $\subseteq$ is used to denote this. For instance, if we let $\mathbb{R}$ represent the set of real numbers, we can see from the diagram that $\mathbb{Q} \subseteq \mathbb{R}$ (read "$\mathbb{Q}$ is a subset of $\mathbb{R}$").

The real numbers are *modeled* using a **number line,** as shown below.

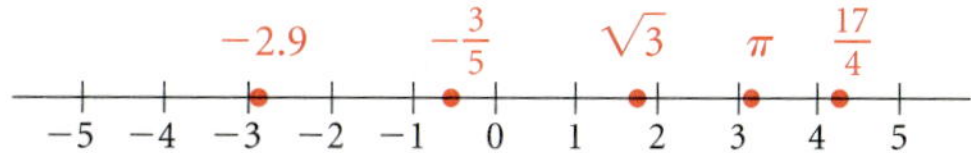

The order of the real numbers can be determined from the number line. If a number a is to the left of a number b, then a **is less than** b ($a < b$). Similarly, a **is greater than** b ($a > b$), if a is to the right of b on the number line. For example, we see from the number line above that $-2.9 < -\frac{3}{5}$, because -2.9 is to the left of $-\frac{3}{5}$. Also, $\frac{17}{4} > \sqrt{3}$, because $\frac{17}{4}$ is to the right of $\sqrt{3}$.

The statement $a \leq b$, read "a is less than or equal to b," is true if either $a < b$ is true or $a = b$ is true.

Absolute Value

The number line can also be used to provide a geometric interpretation of *absolute value*. The absolute value of a number a, denoted $|a|$, is its distance from 0 on the number line. For example, $|-5| = 5$, because the distance of -5 from 0 is 5. Similarly, $\left|\frac{3}{4}\right| = \frac{3}{4}$, because the distance of $\frac{3}{4}$ from 0 is $\frac{3}{4}$.

Absolute Value

For any real number a,

$$|a| = \begin{cases} a, & \text{if } a \geq 0 \\ -a, & \text{if } a < 0. \end{cases}$$

Several properties of absolute value can be used to manipulate and simplify expressions.

Properties of Absolute Value

For any real numbers a and b:

1. $|a| \geq 0$.
2. $|ab| = |a| \cdot |b|$.
3. $\left|\dfrac{a}{b}\right| = \dfrac{|a|}{|b|} \ (b \neq 0)$.
4. $|-a| = |a|$.

We can use a grapher to confirm the first property. We might need to enter $y = |x|$ as $y = \text{abs } x$.

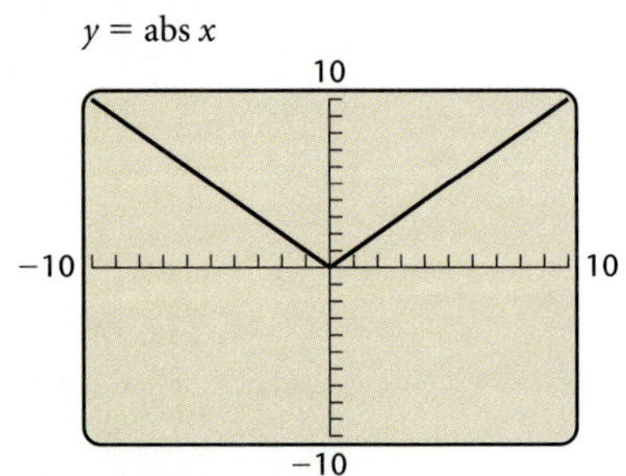

We see from the graph that all y-values are nonnegative.

Absolute value can be used to find the distance between two points on the number line.

Distance Between Two Points on the Number Line

For any real numbers a and b, the distance between a and b is $|a - b|$, or, equivalently, $|b - a|$.

To see that $|a - b|$ and $|b - a|$ are equivalent, note that by the fourth property of absolute value above, $|a - b| = |-(a - b)| = |b - a|$. We can also use a grapher to demonstrate this equivalence by graphing $y_1 = |x - a|$ and $y_2 = |a - x|$ for any nonzero real number a. For example, let's graph $y_1 = |x - 4|$ and $y_2 = |4 - x|$.

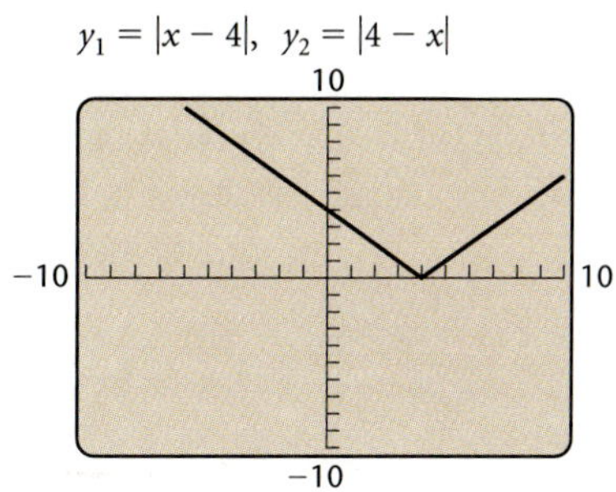

The graphs appear to coincide. Try this for several other values of a.

Example 1   Find the distance between -2 and 3.

SOLUTION

$$|-2 - 3| = |-5| = 5, \quad \text{or, equivalently,}$$
$$|3 - (-2)| = |3 + 2| = |5| = 5.$$

Properties of the Real Numbers

The following properties can be used to manipulate algebraic expressions as well as real numbers.

Properties of the Real Numbers

For any real numbers a, b, and c:

$a + b = b + a$ and $ab = ba$	Commutative properties of addition and multiplication
$a + (b + c) = (a + b) + c$ and $a(bc) = (ab)c$	Associative properties of addition and multiplication
$a + 0 = 0 + a = a$	Additive identity property
$-a + a = a + (-a) = 0$	Additive inverse property
$a \cdot 1 = 1 \cdot a = a$	Multiplicative identity property
$a \cdot \dfrac{1}{a} = 1 \ (a \neq 0)$	Multiplicative inverse property
$a(b + c) = ab + ac$	Distributive property

Note that the distributive property is also true for subtraction since $a(b - c) = a[b + (-c)] = ab + a(-c) = ab - ac$.

Example 2 State the property being illustrated in each sentence.

a) $8 \cdot 5 = 5 \cdot 8$

b) $5 + (m + n) = (5 + m) + n$

c) $x + 0 = x$

d) $14 + (-14) = 0$

e) $2(a - b) = 2a - 2b$

SOLUTION

SENTENCE	PROPERTY
a) $8 \cdot 5 = 5 \cdot 8$	Commutative property of multiplication: $ab = ba$
b) $5 + (m + n) = (5 + m) + n$	Associative property of addition: $a + (b + c) = (a + b) + c$
c) $x + 0 = x$	Additive identity property: $a + 0 = a$
d) $14 + (-14) = 0$	Additive inverse property: $a + (-a) = 0$
e) $2(a - b) = 2a - 2b$	Distributive property: $a(b + c) = ab + ac$

R.1 *Exercise Set*

In Exercises 1–6, consider the numbers -12, $\sqrt{7}$, $5.\overline{3}$, $-\frac{7}{3}$, $\sqrt[3]{8}$, 0, $5.242242224\ldots$, $-\sqrt{14}$, $\sqrt[5]{5}$, -1.96, 9, $4\frac{2}{3}$, $\sqrt{25}$, $\sqrt[3]{4}$, $\frac{5}{7}$.

1. Which are whole numbers?

2. Which are integers?

3. Which are irrational numbers?

4. Which are natural numbers?

5. Which are rational numbers?

6. Which are real numbers?

Find an example of each of the following. Answers may vary.

7. A rational number that is not an integer

8. A real number that is not a rational number

9. An integer that is not a whole number

10. A real number that is not an irrational number

In Exercises 11–28, the following notation is used: $\mathbb{N}$ = the set of natural numbers, $\mathbb{W}$ = the set of whole numbers, $\mathbb{Z}$ = the set of integers, $\mathbb{Q}$ = the set of rational numbers, $\mathbb{I}$ = the set of irrational numbers, and $\mathbb{R}$ = the set of real numbers. Classify each statement as true or false.

11. $6 \in \mathbb{N}$

12. $0 \notin \mathbb{N}$

13. $3.2 \in \mathbb{Z}$

14. $-10.\overline{1} \in \mathbb{R}$

15. $-\dfrac{11}{5} \in \mathbb{Q}$

16. $-\sqrt{6} \in \mathbb{Q}$

17. $\sqrt{11} \notin \mathbb{R}$

18. $-1 \in \mathbb{W}$

19. $24 \notin \mathbb{W}$

20. $1 \in \mathbb{Z}$

21. $1.089 \notin \mathbb{I}$

22. $\mathbb{N} \subseteq \mathbb{W}$

23. $\mathbb{W} \subseteq \mathbb{Z}$

24. $\mathbb{Z} \subseteq \mathbb{N}$

25. $\mathbb{Q} \subseteq \mathbb{R}$

26. $\mathbb{Z} \subseteq \mathbb{Q}$

27. $\mathbb{R} \subseteq \mathbb{Z}$

28. $\mathbb{Q} \subseteq \mathbb{I}$

Simplify.

29. $|-7.1|$

30. $|0|$

31. $\left|\dfrac{5}{4}\right|$

32. $|-\sqrt{3}|$

33. $|-8b|$

34. $|-13.8a|$

35. $|-5xz|$

36. $\left|-\dfrac{3}{4}ab\right|$

37. $\left|\dfrac{0.02x}{y}\right|$

38. $\left|\dfrac{6a}{b}\right|$

Find the distance between each pair of points on the number line.

39. -5, 6

40. 0, -2.5

41. $-2, -8$

42. $\dfrac{15}{8}, \dfrac{23}{12}$

43. $12.1, 6.7$

44. $-3, -14$

Name the property illustrated by each sentence.

45. $6x = x6$

46. $3 + (x + y) = (3 + x) + y$

47. $-3 \cdot 1 = -3$

48. $x + 4 = 4 + x$

49. $5(ab) = (5a)b$

50. $4(y - z) = 4y - 4z$

51. $2(a + b) = (a + b)2$

52. $-7 + 7 = 0$

53. $-6(m + n) = -6(n + m)$

54. $t + 0 = t$

55. $8 \cdot \dfrac{1}{8} = 1$

56. $9x + 9y = 9(x + y)$

Determine graphically whether each of the following appears to be an identity. (See the Introduction to Graphs and Graphers.)

57. $|x| = |-x|$

58. $|x^2| = x^2$

59. $|-5a| = 5a$

60. $\left|\dfrac{x}{3}\right| = \dfrac{|x|}{3}$

61. $x + 7 = 7 + x$

62. $8(x + 1) = 8x + 1$

Synthesis

To the student and the instructor: The synthesis exercises found at the end of every exercise set challenge students to combine concepts or skills studied in that section or in preceding parts of the text. Writing exercises, denoted by the ◆ icon, are meant to be answered with one or more sentences.

63. ◆ How would you convince a classmate that division is not associative?

64. ◆ Under what circumstances is $\sqrt{a}$ a rational number?

Between any two (different) real numbers there are many other real numbers. Find each of the following. Answers may vary.

65. An irrational number between 0.124 and 0.125

66. A rational number between $-\sqrt{2.01}$ and $-\sqrt{2}$

67. A rational number between $-\dfrac{1}{101}$ and $-\dfrac{1}{100}$

68. An irrational number between $\sqrt{5.99}$ and $\sqrt{6}$

69. The hypotenuse of an isosceles right triangle with legs of length 1 unit can be used to "measure" a value for $\sqrt{2}$ by using the Pythagorean theorem, as shown.

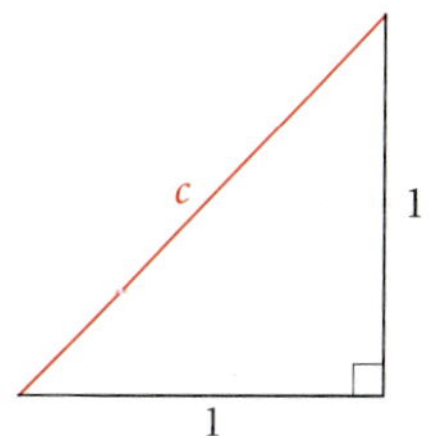

Draw a right triangle that could be used to "measure" $\sqrt{10}$ units.

R.2
Integer Exponents, Scientific Notation, and Order of Operations

- *Simplify expressions with integer exponents.*
- *Solve problems using scientific notation.*
- *Use the rules for order of operations.*

Integers as Exponents

When a positive integer is used as an *exponent*, it indicates the number of times a factor appears in a product. For example, $7 \cdot 7 \cdot 7$ can be written as 7^3.

For any positive integer n,

$$a^n = \underbrace{a \cdot a \cdot a \cdots a}_{n \text{ factors}},$$

where a is the *base* and n is the *exponent*.

Zero and negative-integer exponents are defined as follows.

For any nonzero real number a and any integer n,

$$a^0 = 1 \quad \text{and} \quad a^{-n} = \frac{1}{a^n}.$$

Example 1   Simplify each of the following.

a) 6^0 **b)** $(-3.4)^0$ **c)** 4^{-5} **d)** $\dfrac{1}{(0.82)^{-7}}$

SOLUTION

a) $6^0 = 1$ **b)** $(-3.4)^0 = 1$

c) $4^{-5} = \dfrac{1}{4^5}$ **d)** $\dfrac{1}{(0.82)^{-7}} = (0.82)^{-(-7)} = (0.82)^7$

The following properties of exponents can be used to simplify expressions.

Properties of Exponents

For any real numbers a and b and any integers m and n, assuming 0 is not raised to a nonpositive power:

$a^m \cdot a^n = a^{m+n}$	Product rule
$\dfrac{a^m}{a^n} = a^{m-n} \quad (a \neq 0)$	Quotient rule
$(a^m)^n = a^{mn}$	Power rule
$(ab)^m = a^m b^m$	Raising a product to a power
$\left(\dfrac{a}{b}\right)^m = \dfrac{a^m}{b^m} \quad (b \neq 0)$	Raising a quotient to a power

We can use a grapher for partial confirmation of the first three properties. In the case of the product rule, for instance, we graph $y_1 = x^3 \cdot x^2$ and $y_2 = x^5$.

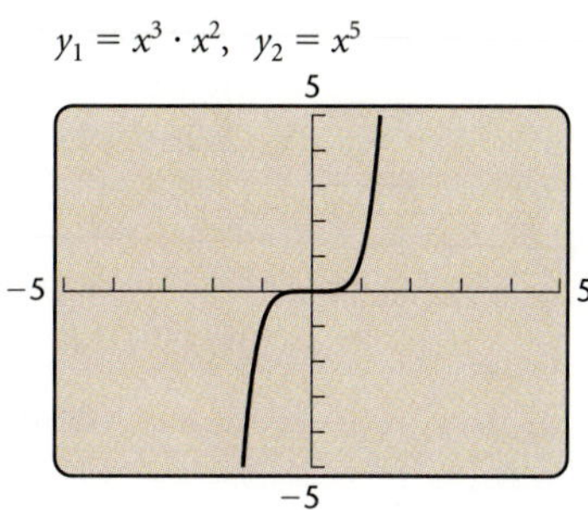

The graphs appear to coincide. We can perform similar confirmations of the quotient rule and the power rule.

Example 2   Simplify each of the following.

a) $y^{-5} \cdot y^3$

b) $\dfrac{48x^{12}}{16x^4}$

c) $(t^{-3})^5$

d) $(2s^{-2})^5$

e) $\left(\dfrac{45x^{-4}y^2}{9z^{-8}}\right)^{-3}$

SOLUTION

a) $y^{-5} \cdot y^3 = y^{-5+3} = y^{-2}$, or $\dfrac{1}{y^2}$

b) $\dfrac{48x^{12}}{16x^4} = \dfrac{48}{16}x^{12-4} = 3x^8$

c) $(t^{-3})^5 = t^{-3 \cdot 5} = t^{-15}$, or $\dfrac{1}{t^{15}}$

d) $(2s^{-2})^5 = 2^5(s^{-2})^5 = 32s^{-10}$, or $\dfrac{32}{s^{10}}$

e) $\left(\dfrac{45x^{-4}y^2}{9z^{-8}}\right)^{-3} = \left(\dfrac{5x^{-4}y^2}{z^{-8}}\right)^{-3} = \dfrac{5^{-3}x^{12}y^{-6}}{z^{24}} = \dfrac{x^{12}y^{-6}}{125z^{24}}$, or $\dfrac{x^{12}}{125y^6z^{24}}$

Scientific Notation

We can use exponential, or scientific, notation to name very large and very small positive numbers and to perform computations.

> **Scientific Notation**
>
> *Scientific notation* for a number is an expression of the type
>
> $$N \times 10^n,$$
>
> where $1 \le N < 10$, N is in decimal notation, and n is an integer.

Keep in mind that in scientific notation, positive exponents are used for numbers greater than 10 and negative exponents for numbers between 0 and 1.

Example 3 *National Debt.* In a recent month, the national debt increased by \$38,500,000,000. Convert this number to scientific notation. (*Source*: U.S. Treasury Department)

SOLUTION   We want the decimal point to be positioned between the 3 and the 8, so we move it 10 places to the left. Since the number to be converted is greater than 10, the exponent must be positive.

$$38{,}500{,}000{,}000 = 3.85 \times 10^{10}$$

Example 4 *Mass of a Neutron.* The mass of a neutron is about 0.00000000000000000000000000167 kg. Convert this number to scientific notation.

SOLUTION   We want the decimal point to be positioned between the 1 and the 6, so we move it 27 places to the right. Since the number to be converted is between 0 and 1, the exponent must be negative.

$$0.00000000000000000000000000167 = 1.67 \times 10^{-27}$$

Example 5 Convert each of the following to decimal notation.

a) 7.632×10^{-4} **b)** 9.4×10^{5}

SOLUTION

a) The exponent is negative, so the number is between 0 and 1. We move the decimal point 4 places to the left.

$$7.632 \times 10^{-4} = 0.0007632$$

b) The exponent is positive, so the number is greater than 10. We move the decimal point 5 places to the right.

$$9.4 \times 10^{5} = 940{,}000$$

Most calculators make use of scientific notation. For example, a number like 48,000,000,000,000 might be expressed as

| 4.8 E 13 | or | 4.8 13 |

Example 6 *Distance to a Star.* Alpha Centauri is about 4.3 light-years from the earth. One **light-year** is the distance that light travels in one year and is about 5.88×10^{12} miles. How many miles is it from earth to Alpha Centauri? Express your answer in scientific notation.

SOLUTION

$$
\begin{aligned}
4.3 \times (5.88 \times 10^{12}) &= (4.3 \times 5.88) \times 10^{12} \\
&= 25.284 \times 10^{12} \qquad \text{This is not scientific notation.} \\
&= (2.5284 \times 10^{1}) \times 10^{12} \\
&= 2.5284 \times 10^{13} \text{ miles} \qquad \text{Writing scientific notation}
\end{aligned}
$$

Order of Operations

Recall that to simplify the expression $3 + 4 \cdot 5$, we multiply 4 and 5 first to get 20 and then add 3 to get 23. Mathematicians have agreed on the following procedure, or rules for order of operations.

Rules for Order of Operations

1. Do all calculations within grouping symbols before operations outside. When nested grouping symbols are present, work from the inside out.
2. Evaluate all exponential expressions.
3. Do all multiplications and divisions in order from left to right.
4. Do all additions and subtractions in order from left to right.

Most graphers use the order of operations above. Thus it is essential to remember these rules when entering a computation.

Example 7   Simplify each of the following.

a) $8(5 - 3)^3 - 20$

b) $\dfrac{10 \div (8 - 6) + 9 \cdot 4}{2^5 + 3^2}$

SOLUTION

a)
$$\begin{aligned}
8(5 - 3)^3 - 20 &= 8 \cdot 2^3 - 20 &&\text{Doing the calculation within parentheses}\\
&= 8 \cdot 8 - 20 &&\text{Evaluating the exponential expression}\\
&= 64 - 20 &&\text{Multiplying}\\
&= 44 &&\text{Subtracting}
\end{aligned}$$

On a grapher this computation would be entered using the scientific keys, as follows.

```
8(5 − 3)^3 − 20               44
```

b)
$$\frac{10 \div (8 - 6) + 9 \cdot 4}{2^5 + 3^2} = \frac{10 \div 2 + 9 \cdot 4}{32 + 9} = \frac{5 + 36}{41} = \frac{41}{41} = 1$$

Note that fraction bars act as grouping symbols. That is, the given expression is equivalent to $[10 \div (8 - 6) + 9 \cdot 4] \div (2^5 + 3^2)$. On a grapher we would enter the following.

```
(10/(8 − 6) + 9*4)/(2^5 + 3^2)
```

To confirm that parentheses are essential in Example 6(b), enter the computation on a grapher without using parentheses. What is the result?

Example 8   *Compound Interest.* If a principal P is invested at an interest rate i, compounded n times per year, in t years it will grow to an amount A given by

$$A = P\left(1 + \frac{i}{n}\right)^{nt}.$$

Thus when \$1250 is invested at 4.6% interest, compounded quarterly, the amount in the account at the end of 8 years is given by

$$A = 1250\left(1 + \frac{0.046}{4}\right)^{32}.$$

Evaluate this expression.

SOLUTION

$$A = 1250\left(1 + \frac{0.046}{4}\right)^{32}$$

$$= 1250(1 + 0.0115)^{32} \qquad \text{Dividing}$$

$$= 1250(1.0115)^{32} \qquad \text{Adding}$$

$$\approx 1250(1.441811175) \qquad \text{Evaluating the exponential expression}$$

$$\approx 1802.263969 \qquad \text{Multiplying}$$

$$\approx 1802.26 \qquad \text{Rounding to the nearest cent}$$

We can also use the scientific keys on a grapher.

```
1250(1 + 0.046/4)^32
                1802.263969
```

The amount in the account at the end of 8 years is \$1802.26.

R.2 Exercise Set

Simplify.

1. $5^8 \cdot 5^{-6}$

2. $6^2 \cdot 6^{-7}$

3. $m^{-5} \cdot m^5$

4. $n^9 \cdot n^{-9}$

5. $7^3 \cdot 7^{-5} \cdot 7$

6. $3^6 \cdot 3^{-5} \cdot 3^4$

7. $2x^3 \cdot 3x^2$

8. $3y^4 \cdot 4y^3$

9. $(5a^2b)(3a^{-3}b^4)$

10. $(4xy^2)(3x^{-4}y^5)$

11. $(2x)^3(3x)^2$

12. $(4y)^2(3y)^3$

13. $\dfrac{b^{40}}{b^{37}}$

14. $\dfrac{a^{39}}{a^{32}}$

15. $\dfrac{x^2y^{-2}}{x^{-1}y}$

16. $\dfrac{x^3y^{-3}}{x^{-1}y^2}$

17. $\dfrac{32x^{-4}y^3}{4x^{-5}y^8}$

18. $\dfrac{20a^5b^{-2}}{5a^7b^{-3}}$

19. $(2ab^2)^3$

20. $(4xy^3)^2$

21. $(-2x^3)^4$

22. $(-3x^2)^4$

23. $(-5c^{-1}d^{-2})^{-2}$

24. $(-4x^{-5}z^{-2})^{-3}$

25. $\left(\dfrac{24a^{10}b^{-8}c^7}{3a^6b^{-3}c^5}\right)^5$

26. $\left(\dfrac{125p^{12}q^{-14}r^{22}}{25p^8q^6r^{-15}}\right)^{-4}$

Convert to scientific notation.

27. 405,000

28. 1,670,000

29. 0.00000039

30. 0.00092

31. One cubic inch is approximately equal to 0.000016 m^3.

32. The average *Fortune 500* company has an annual telephone bill of $34,000,000. (*Source:* Gallup survey of *Fortune 500* companies for Pitney Bowes)

Convert to decimal notation.

33. 8.3×10^{-5}

34. 4.1×10^6

35. 2.07×10^7

36. 3.15×10^{-6}

37. In 1994, the national debt was $\$4.69 \times 10^{12}$. (*Source:* U.S. Treasury Department)

38. The mass of a proton is about 1.67×10^{-24} g.

Compute. Write scientific notation for the answer.

39. $(3.1 \times 10^5)(4.5 \times 10^{-3})$

40. $(9.1 \times 10^{-17})(8.2 \times 10^3)$

41. $\dfrac{6.4 \times 10^{-7}}{8.0 \times 10^6}$

42. $\dfrac{1.1 \times 10^{-40}}{2.0 \times 10^{-71}}$

Solve. Write the answers using scientific notation.

43. *Nuclear Disintegration.*   One gram of radium produces 37 billion disintegrations per second. How many disintegrations are produced in 1 hr?

44. *Length of Earth's Orbit.*   The average distance from the earth to the sun is 93 million mi. About how far does the earth travel in a yearly orbit? (Assume a circular orbit. Use 3.14 for π.)

45. *Chesapeake Bay Bridge-Tunnel.*   The 17.6-mile-long Chesapeake Bay Bridge-Tunnel was completed in 1964. Construction costs were $210 million. Find the average cost per mile.

46. *Retail Hardware Sales.*   In a recent year, America's largest hardware retailer had about $7.2 billion in sales in its 240 stores. Find the average sales per store.

Calculate. Use a grapher to check each answer.

47. $3 \cdot 2 - 4 \cdot 2^2 + 6(3 - 1)$

48. $3[(2 + 4 \cdot 2^2) - 6(3 - 1)]$

49. $16 \div 4 \cdot 4 \div 2 \cdot 256$

50. $2^6 \cdot 2^{-3} \div 2^{10} \div 2^{-8}$

51. $\dfrac{4(8 - 6)^2 - 4 \cdot 3 + 2 \cdot 8}{3^1 + 19^0}$

52. $\dfrac{[4(8 - 6)^2 + 4](3 - 2 \cdot 8)}{2^2(2^3 + 5)}$

Compound Interest.   *Use the compound interest formula in Exercises 53–56. Round to the nearest cent.*

53. Suppose $2125 is invested at 6.2%, compounded semiannually. How much is in the account at the end of 5 yr?

54. Suppose $9550 is invested at 5.4%, compounded semiannually. How much is in the account at the end of 7 yr?

55. Suppose $6700 is invested at 4.5%, compounded quarterly. How much is in the account at the end of 6 yr?

56. Suppose $4875 is invested at 5.8%, compounded quarterly. How much is in the account at the end of 9 yr?

Synthesis

57. ◆ Are parentheses necessary in the expression $4 \cdot 25 \div (10 - 5)$? Why or why not?

58. ◆ Is $x^{-2} < x^{-1}$ for any negative value(s) of x? Why or why not?

Simplify. Assume that all exponents are integers, all denominators are nonzero, and zero is not raised to a nonpositive power.

59. $(x^t \cdot x^{3t})^2$

60. $(x^y \cdot x^{-y})^3$

61. $(t^{a+x} \cdot t^{x-a})^4$

62. $(m^{x-b} \cdot n^{x+b})^x (m^b n^{-b})^x$

63. $\left[\dfrac{(3x^a y^b)^3}{(-3x^u y^b)^2}\right]^2$

64. $\left[\left(\dfrac{x^r}{y^t}\right)^2 \left(\dfrac{x^{2r}}{y^{4t}}\right)^{-2}\right]^{-3}$

65. *Mortgage Payments.*   The formula

$$M = P\left[\dfrac{\dfrac{i}{12}\left(1 + \dfrac{i}{12}\right)^n}{\left(1 + \dfrac{i}{12}\right)^n - 1}\right]$$

gives the monthly mortgage payment M on a home loan of P dollars at interest rate i, where n is the total number of payments (12 times the number of years). The cost of a house is $98,000. The down payment is $16,000, the interest rate is $8\frac{1}{2}$%, and the loan period is 25 yr. What is the monthly payment?

Use a grapher to graph $y_1 = x^2$, $y_2 = x^3$, and $y_3 = x^4$. Then do Exercises 66 and 67.

66. For what values of x is $x^2 > x^3$?

67. For what values of x is $x^4 > x^2$?

R.3

Addition, Subtraction, and Multiplication of Polynomials

- *Identify the terms, coefficients, and degree of a polynomial.*
- *Add, subtract, and multiply polynomials.*

Polynomials

Polynomials are a type of algebraic expression that you will often encounter in your study of algebra. Some examples of polynomials are

$$3x - 4y, \quad 5y^3 - \tfrac{7}{3}y^2 + 3y - 2, \quad 0, \quad -2.3a^4, \quad \text{and} \quad z^6 - \sqrt{5}.$$

All but the first are polynomials in one variable.

> **Polynomials in One Variable**
>
> A *polynomial in one variable* is any expression of the type
>
> $$a_n x^n + a_{n-1} x^{n-1} + \cdots + a_2 x^2 + a_1 x + a_0,$$
>
> where n is a nonnegative integer, $a_n, \ldots, a_0$ are real numbers, called *coefficients*, and $a_n \neq 0$. The parts of a polynomial separated by plus signs are called *terms*. The *degree* of the polynomial is n, the *leading coefficient* is a_n, and the *constant term* is a_0. The polynomial is said to be written in *descending order*, because the exponents decrease from left to right.

Example 1   Identify the terms of the polynomial

$$2x^4 - 7.5x^3 + x - 12.$$

SOLUTION

$$2x^4 - 7.5x^3 + x - 12 = 2x^4 + (-7.5x^3) + x + (-12),$$

so the terms are

$$2x^4, \quad -7.5x^3, \quad x, \quad \text{and} \quad -12.$$

A polynomial consisting of only a nonzero constant term, like 23, has degree 0. It is agreed that the polynomial consisting only of 0 has *no* degree.

Example 2   Find the degree of each polynomial.

a) $2x^3 - 9$ **b)** $y^2 - \tfrac{3}{2} + 5y^4$ **c)** 7

SOLUTION

POLYNOMIAL	DEGREE
a) $2x^3 - 9$	3
b) $y^2 - \tfrac{3}{2} + 5y^4 = 5y^4 + y^2 - \tfrac{3}{2}$	4
c) $7 = 7x^0$	0

Algebraic expressions like $3ab^3 - 8$ and $5x^4y^2 - 3x^3y^8 + 7xy^2 + 6$ are **polynomials in several variables.** The **degree of a term** is the sum of the exponents of the variables in that term. The **degree of a polynomial** is the degree of the term of highest degree.

Example 3   Find the degree of the polynomial $7ab^3 - 11a^2b^4 + 8$.

SOLUTION   The degrees of the terms of $7ab^3 - 11a^2b^4 + 8$ are 4, 6, and 0, respectively, so the degree of the polynomial is 6.   ▬

A polynomial with just one term, like $-9y^6$, is a *monomial*. If a polynomial has two terms, like $x^2 + 4$, it is a *binomial*. A polynomial with three terms, like $4x^2 - 4xy + 1$, is a *trinomial*.

Expressions like

$$2x^2 - 5x + \frac{3}{x}, \qquad 9 - \sqrt{x}, \quad \text{and} \quad \frac{x + 1}{x^4 + 5}$$

are not polynomials, because they cannot be written in the form $a_nx^n + a_{n-1}x^{n-1} + \cdots + a_1x + a_0$, where the exponents are all nonnegative integers and the coefficients are all real numbers.

Addition and Subtraction

If two terms of an expression have the same variables raised to the same powers, they are called **like terms,** or **similar terms.** We can **combine,** or **collect like terms** using the distributive property. For example, $3y^2$ and $5y^2$ are like terms and

$$3y^2 + 5y^2 = (3 + 5)y^2 = 8y^2.$$

We add or subtract polynomials by combining like terms.

Example 4   Add or subtract each of the following.

a) $(-5x^3 + 3x^2 - x) + (12x^3 - 7x^2 + 3)$
b) $(6x^2y^3 - 9xy) - (5x^2y^3 - 4xy)$

SOLUTION

a) $(-5x^3 + 3x^2 - x) + (12x^3 - 7x^2 + 3)$

$= (-5x^3 + 12x^3) + (3x^2 - 7x^2) - x + 3$ Rearranging using the commutative and associative properties

$= (-5 + 12)x^3 + (3 - 7)x^2 - x + 3$ Using the distributive property

$= 7x^3 - 4x^2 - x + 3$

b) We can subtract by adding an opposite:

$(6x^2y^3 - 9xy) - (5x^2y^3 - 4xy)$

$= (6x^2y^3 - 9xy) + (-5x^2y^3 + 4xy)$ Adding the opposite of $5x^2y^3 - 4xy$

$= 6x^2y^3 - 9xy - 5x^2y^3 + 4xy$

$= x^2y^3 - 5xy.$ Combining like terms   ▬

Multiplication

Multiplication of polynomials is based on the distributive property. For example,

$$(x + 4)(x + 3) = (x + 4)x + (x + 4)3 \quad \text{Using the distributive property}$$
$$= x^2 + 4x + 3x + 12 \quad \text{Using the distributive property two more times}$$
$$= x^2 + 7x + 12. \quad \text{Combining like terms}$$

We can find the product of two binomials, as we did above, by multiplying the **F**irst terms, then the **O**uter terms, then the **I**nner terms, then the **L**ast terms. Then we collect like terms, if possible. This procedure is sometimes called **FOIL**.

Example 5   Multiply: $(2x - 7)(3x + 4)$.

SOLUTION

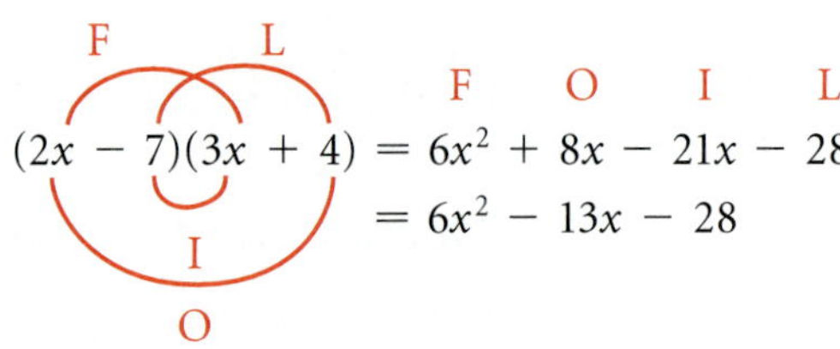

$$(2x - 7)(3x + 4) = 6x^2 + 8x - 21x - 28$$
$$= 6x^2 - 13x - 28$$

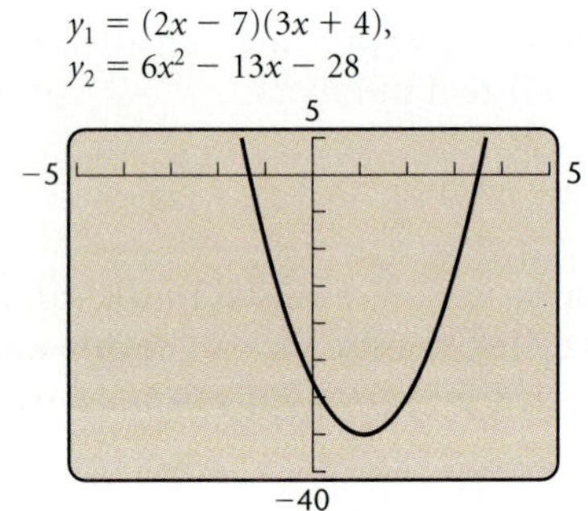

We can use a grapher to do a partial check of this result. Graph $y_1 = (2x - 7)(3x + 4)$ and $y_2 = 6x^2 - 13x - 28$. (See the graph at left.) The graphs appear to coincide.

In general, to multiply two polynomials, we multiply each term of one by each term of the other and add the products. One way to do this is to use columns.

Example 6   Multiply: $(4x^4y - 7x^2y + 3y)(2y - 3x^2y)$.

SOLUTION   We can use columns to organize our work.

$$
\begin{array}{r}
4x^4y - 7x^2y + 3y \\
2y - 3x^2y \\
\hline
-12x^6y^2 + 21x^4y^2 - 9x^2y^2 \\
8x^4y^2 - 14x^2y^2 + 6y^2 \\
\hline
-12x^6y^2 + 29x^4y^2 - 23x^2y^2 + 6y^2
\end{array}
$$

Multiplying by $-3x^2y$
Multiplying by $2y$
Adding

Interactive Discovery

Graph $y_1 = (x + 3)^2$, $y_2 = x^2 + 9$, $y_3 = x^2 - 9$, $y_4 = x^2 + 6x + 9$, and $y_5 = (x - 3)(x + 3)$. Can you discover any identities?

We can use FOIL to find some special products.

Special Products of Binomials

$$(A + B)^2 = A^2 + 2AB + B^2 \qquad \text{Square of a sum}$$
$$(A - B)^2 = A^2 - 2AB + B^2 \qquad \text{Square of a difference}$$
$$(A + B)(A - B) = A^2 - B^2 \qquad \text{Product of a sum and a difference}$$

Example 7   Multiply each of the following.

a) $(4x + 1)^2$ **b)** $(3y^2 - 2)^2$ **c)** $(x^2 + 3y)(x^2 - 3y)$

SOLUTION

a) $(4x + 1)^2 = (4x)^2 + 2 \cdot 4x \cdot 1 + 1^2 = 16x^2 + 8x + 1$
b) $(3y^2 - 2)^2 = (3y^2)^2 - 2 \cdot 3y^2 \cdot 2 + 2^2 = 9y^4 - 12y^2 + 4$
c) $(x^2 + 3y)(x^2 - 3y) = (x^2)^2 - (3y)^2 = x^4 - 9y^2$

R.3 Exercise Set

Determine the terms and the degree of each polynomial.

1. $-5y^4 + 3y^3 + 7y^2 - y - 4$

2. $2m^3 - m^2 - 4m + 11$

3. $3a^4b - 7a^3b^3 + 5ab - 2$

4. $6p^3q^2 - p^2q^4 - 3pq^2 + 5$

Perform the operations indicated.

5. $(5x^2y - 2xy^2 + 3xy - 5) + (-2x^2y - 3xy^2 + 4xy + 7)$

6. $(6x^2y - 3xy^2 + 5xy - 3) + (-4x^2y - 4xy^2 + 3xy + 8)$

7. $(2x + 3y + z - 7) + (4x - 2y - z + 8) + (-3x + y - 2z - 4)$

8. $(2x^2 + 12xy - 11) + (6x^2 - 2x + 4) + (-x^2 - y - 2)$

9. $(3x^2 - 2x - x^3 + 2) - (5x^2 - 8x - x^3 + 4)$

10. $(5x^2 + 4xy - 3y^2 + 2) - (9x^2 - 4xy + 2y^2 - 1)$

11. $(x^4 - 3x^2 + 4x) - (3x^3 + x^2 - 5x + 3)$

12. $(2x^4 - 3x^2 + 7x) - (5x^3 + 2x^2 - 3x + 5)$

13. $(a - b)(2a^3 - ab + 3b^2)$

14. $(n + 1)(n^2 - 6n - 4)$

15. $(x + 5)(x - 3)$

16. $(y - 4)(y + 1)$

17. $(2a + 3)(a + 5)$

18. $(3b + 1)(b - 2)$

19. Use a grapher to check your answer to Exercise 17.

20. Use a grapher to check your answer to Exercise 18.

21. $(2x + 3y)(2x + y)$ **22.** $(2a - 3b)(2a - b)$

23. $(x + 5)^2$ **24.** $(x + 7)^2$

25. $(5y - 3)^2$ **26.** $(3y - 2)^2$

27. Use a grapher to check your answer to Exercise 25.

28. Use a grapher to check your answer to Exercise 26.

29. $(2x + 3y)^2$ **30.** $(5x + 2y)^2$

31. $(2x^2 - 3y)^2$ **32.** $(4x^2 - 5y)^2$

33. $(a + 3)(a - 3)$ **34.** $(b + 4)(b - 4)$

35. Use a grapher to check your answer to Exercise 33.

36. Use a grapher to check your answer to Exercise 34.

37. $(3x - 2y)(3x + 2y)$

38. $(3x + 5y)(3x - 5y)$

39. $(2x + 3y + 4)(2x + 3y - 4)$

40. $(5x + 2y + 3)(5x + 2y - 3)$

41. $(x + 1)(x - 1)(x^2 + 1)$

42. $(y - 2)(y + 2)(y^2 + 4)$

Synthesis

43. ◆ Is the sum of two polynomials of degree n always a polynomial of degree n? Why or why not?

44. ◆ Explain how you would convince a classmate that $(A + B)^2 \neq A^2 + B^2$.

Multiply. Assume that all exponents are natural numbers.

45. $(a^n + b^n)(a^n - b^n)$

46. $(t^a + 4)(t^a - 7)$

47. $(a^n + b^n)^2$

48. $(x^{3m} - t^{5n})^2$

49. $(x - 1)(x^2 + x + 1)(x^3 + 1)$

50. $[(2x - 1)^2 - 1]^2$

51. $(x^{a-b})^{a+b}$

52. $(t^{m+n})^{m+n} \cdot (t^{m-n})^{m-n}$

53. $(a + b + c)^2$

54. $(a + b + c)^3$

55. $(a + b)^4$

R.4

Factoring

• *Factor polynomials by removing a common factor.*
• *Factor special products of polynomials.*

To factor a polynomial, we do the reverse of multiplying; that is, we find an equivalent expression that is a product.

Terms with Common Factors

When a polynomial is to be factored, we should always look first to factor out a factor that is common to all the terms using the distributive law. We usually look for the constant common factor with the largest absolute value and for variables with the largest exponent common to all the terms. In this sense, we factor out the "largest" common factor.

Example 1   Factor each of the following.

a) $15 + 10x - 5x^2$

b) $12x^2y^2 - 20x^3y$

SOLUTION

a) $15 + 10x - 5x^2 = 5 \cdot 3 + 5 \cdot 2x - 5 \cdot x^2 = 5(3 + 2x - x^2)$

We can check by multiplying: $5(3 + 2x - x^2) = 15 + 10x - 5x^2$.

b) $12x^2y^2 - 20x^3y = 4x^2y(3y - 5x)$

Note that there are several factors common to the terms of $12x^2y^2 - 20x^3y$, but $4x^2y$ is the "largest" of these.   ▬

In some polynomials, pairs of terms have a common factor that can be removed in a process called **factoring by grouping**.

Example 2   Factor: $x^3 + 3x^2 - 5x - 15$.

SOLUTION   We have

$$x^3 + 3x^2 - 5x - 15 = x^2(x + 3) - 5(x + 3)$$
$$= (x + 3)(x^2 - 5).$$

To check graphically, graph $y_1 = x^3 + 3x^2 - 5x - 15$ and $y_2 = (x + 3)(x^2 - 5)$.

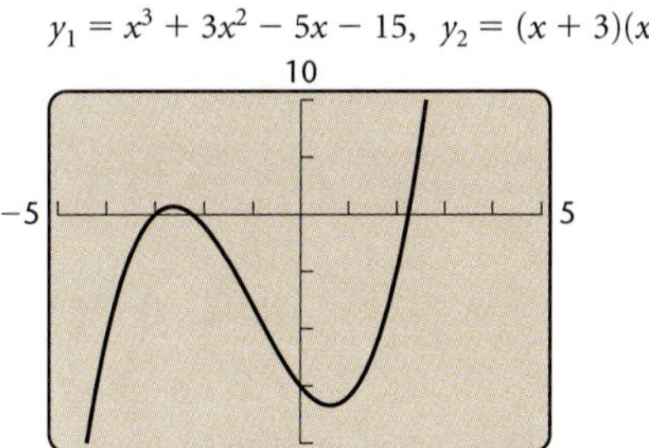

The graphs appear to coincide.

Trinomials

Some trinomials can be factored into the product of two binomials. To factor a trinomial of the form $x^2 + bx + c$, we look for two numbers with a product of c and a sum of b.

Example 3 Factor each of the following.

a) $x^2 + 5x + 6$
b) $y^2 - 2y - 24$

SOLUTION

a) We look for two numbers with a product of 6 and a sum of 5. They are 2 and 3. Then

$$x^2 + 5x + 6 = (x + 2)(x + 3). \qquad \text{To check, multiply or graph.}$$

b) We look for two numbers with a product of -24 and a sum of -2:

$$y^2 - 2y - 24 = (y - 6)(y + 4). \qquad \text{Note that } (-6)(4) = -24 \text{ and } -6 + 4 = -2.$$

The next example illustrates a method for factoring a trinomial of the form $ax^2 + bx + c, a \neq 1$.

Example 4 Factor: $3x^2 - 10x - 8$.

SOLUTION We look for factors $px + q$ and $rx + s$ for which $px \cdot rx = 3x^2$ and $q \cdot s = -8$. When we multiply the inside terms, then the outside terms, and add, we must have $-10x$. By trial, we determine the factorization:

$$3x^2 - 10x - 8 = (3x + 2)(x - 4). \qquad \text{To check, multiply or graph.}$$

Special Factorizations

We reverse the equation $(A + B)(A - B) = A^2 - B^2$ to factor a **difference of squares**.

$$A^2 - B^2 = (A + B)(A - B)$$

Example 5 Factor each of the following.

a) $x^2 - 16$ **b)** $9a^2 - 25$ **c)** $6x^4 - 6y^4$

SOLUTION

a) $x^2 - 16 = (x + 4)(x - 4)$

To check graphically, graph $y_1 = x^2 - 16$ and $y_2 = (x + 4)(x - 4)$.

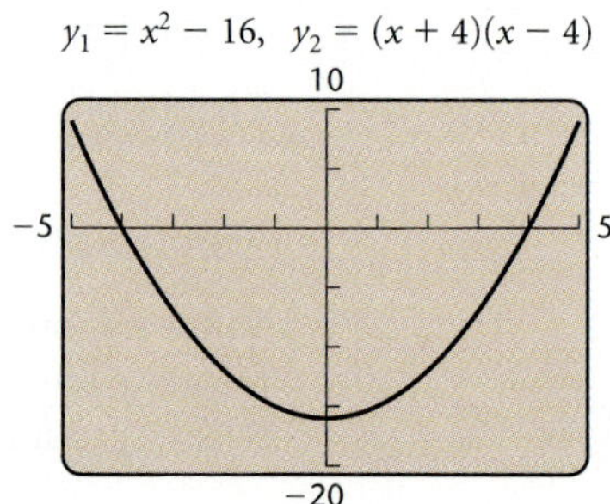

The graphs appear to coincide.

b) $9a^2 - 25 = (3a + 5)(3a - 5)$ To check, multiply or graph.

c) $6x^4 - 6y^4 = 6(x^4 - y^4)$
$$= 6(x^2 + y^2)(x^2 - y^2)$$
$$= 6(x^2 + y^2)(x + y)(x - y)$$

Because none of these factors can be factored further, we have *factored completely.*

The rules for squaring binomials can be reversed to factor trinomial squares.

$$A^2 + 2AB + B^2 = (A + B)^2$$
$$A^2 - 2AB + B^2 = (A - B)^2$$

Example 6 Factor each of the following.

a) $x^2 + 8x + 16$

b) $25y^2 - 30y + 9$

SOLUTION

a) $x^2 + 8x + 16 = (x + 4)^2$ See the graphical check in Fig. 1.

b) $25y^2 - 30y + 9 = (5y - 3)^2$ To check, multiply or graph.

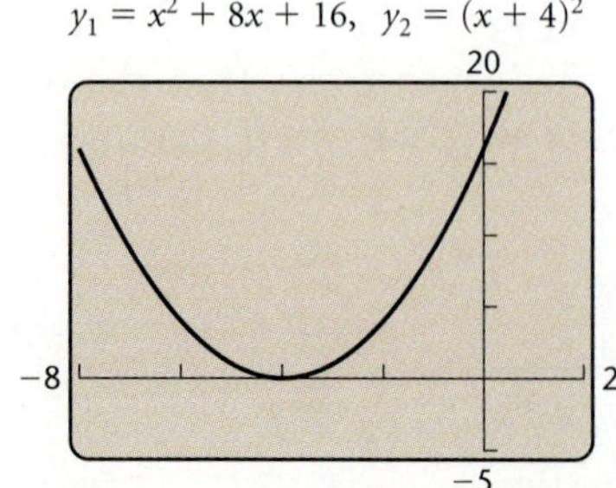

FIGURE 1

We can use the following rules to factor a **sum** or a **difference of cubes**:

$$A^3 + B^3 = (A + B)(A^2 - AB + B^2);$$
$$A^3 - B^3 = (A - B)(A^2 + AB + B^2).$$

These rules can be verified by multiplying.

Example 7   Factor each of the following.

a) $x^3 + 27$ **b)** $16y^3 - 250$

SOLUTION

a) $x^3 + 27 = x^3 + 3^3 = (x + 3)(x^2 - 3x + 9)$ See the graphical check in Fig. 2.

b) $16y^3 - 250 = 2(8y^3 - 125)$
$$= 2[(2y)^3 - 5^3]$$
$$= 2(2y - 5)(4y^2 + 10y + 25)$$ To check, multiply or graph.

Not all polynomials can be factored into polynomials with integer coefficients. An example is $x^2 - x + 7$. There are no real factors of 7 whose sum is -1. In such a case we say that the polynomial is "not factorable," or **prime**.

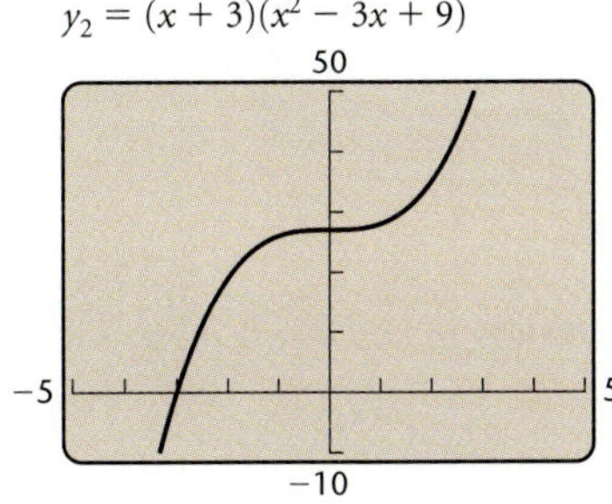

FIGURE 2

R.4   *Exercise Set*

Factor out a common factor.

1. $2x - 10$

2. $7y + 42$

3. $3x^4 - 9x^2$

4. $20y^2 - 5y^5$

5. $4a^2 - 12a + 16$

6. $6n^2 + 24n - 18$

7. $a(b - 2) + c(b - 2)$

8. $a(x^2 - 3) - 2(x^2 - 3)$

Factor by grouping.

9. $x^3 + 3x^2 + 6x + 18$

10. $3x^3 + x^2 - 18x - 6$

11. $y^3 - 3y^2 - 4y + 12$

12. $p^3 - 2p^2 - 9p + 18$

13. Use a grapher to check your answer to Exercise 9.

14. Use a grapher to check your answer to Exercise 10.

Factor the trinomial.

15. $p^2 + 6p + 8$

16. $w^2 - 7w + 10$

17. $2n^2 + 9n - 56$

18. $3y^2 + 7y - 20$

19. $y^4 - 4y^2 - 21$

20. $m^4 - m^2 - 90$

21. Use a grapher to check your answer to Exercise 15.

22. Use a grapher to check your answer to Exercise 16.

Factor the difference of squares.

23. $9x^2 - 25$

24. $16x^2 - 9$

25. $4xy^4 - 4xz^2$

26. $5x^4y - 5yz^4$

Factor the trinomial square.

27. $y^2 - 6y + 9$

28. $x^2 + 8x + 16$

29. $1 - 8x + 16x^2$ **30.** $1 + 10x + 25x^2$

31. Use a grapher to check your answer to Exercise 29.

32. Use a grapher to check your answer to Exercise 30.

Factor the sum or difference of cubes.

33. $x^3 + 8$ **34.** $y^3 - 64$

35. $m^3 - 1$ **36.** $n^3 + 216$

37. Use a grapher to check your answer to Exercise 35.

38. Use a grapher to check your answer to Exercise 36.

Factor completely.

39. $18a^2b - 15ab^2$ **40.** $4x^2y + 12xy^2$

41. $x^3 - 4x^2 + 5x - 20$ **42.** $y^3 + 3y^2 - 3y - 9$

43. $8x^2 - 32$ **44.** $6y^2 - 6$

45. $4x^2 - 5$ **46.** $16x^2 - 7$

47. Use a grapher to check your answer to Exercise 41.

48. Use a grapher to check your answer to Exercise 42.

49. $x^2 + 9x + 20$ **50.** $y^2 + y - 6$

51. $y^2 - 6y + 5$ **52.** $x^2 - 4x - 21$

53. $2a^2 + 9a + 4$ **54.** $3b^2 - b - 2$

55. $6x^2 + 7x - 3$ **56.** $8x^2 + 2x - 15$

57. $y^2 - 18y + 81$ **58.** $n^2 + 2n + 1$

59. $x^2y^2 - 14xy + 49$ **60.** $x^2y^2 - 16xy + 64$

61. $4ax^2 + 20ax - 56a$ **62.** $21x^2y + 2xy - 8y$

63. $3z^3 - 24$ **64.** $4t^3 + 108$

65. $16a^7b + 54ab^7$ **66.** $24a^2x^4 - 375a^8x$

67. Use a grapher to check your answer to Exercise 55.

68. Use a grapher to check your answer to Exercise 56.

Synthesis

69. ◆ Under what circumstances can $A^2 + B^2$ be factored?

70. ◆ Explain how the rule for factoring a sum of cubes can be used to factor a difference of cubes.

Factor.

71. $y^4 - 84 + 5y^2$ **72.** $11x^2 + x^4 - 80$

73. $y^2 - \frac{8}{49} + \frac{2}{7}y$ **74.** $x^2 + \frac{3}{5}x - \frac{4}{25}$

75. $t^2 - 0.27 + 0.6t$ **76.** $0.4m - 0.05 + m^2$

77. $(x + h)^3 - x^3$ **78.** $(x + 0.01)^2 - x^2$

79. $(x + 3)^2 - 2(x + 3) - 35$

80. $(y - 4)^2 + 5(y - 4) - 24$

81. $3(a - b)^2 + 10(a - b) - 8$

82. $6(2p + q)^2 - 5(2p + q) - 25$

Factor. Assume that variables in exponents represent natural numbers.

83. $x^{2n} + 5x^n - 24$ **84.** $4x^{2n} - 4x^n - 3$

85. $x^2 + ax + bx + ab$

86. $bdy^2 + ady + bcy + ac$

87. $\frac{1}{4}t^2 - \frac{2}{5}t + \frac{4}{25}$

88. $\frac{4}{27}r^2 + \frac{5}{9}rs + \frac{1}{12}s^2 - \frac{1}{3}rs$

89. $25y^{2m} - (x^{2n} - 2x^n + 1)$

90. $4x^{4a} + 12x^{2a} + 10x^{2a} + 30$

91. $3x^{3n} - 24y^{3m}$

92. $x^{6a} - t^{3b}$

93. $(y - 1)^4 - (y - 1)^2$

94. $x^6 - 2x^5 + x^4 - x^2 + 2x - 1$

R.5

Rational Expressions

- *Determine the domain of a rational expression.*
- *Simplify rational expressions.*
- *Multiply, divide, add, or subtract rational expressions.*
- *Simplify complex rational expressions.*

A **rational expression** is the quotient of two polynomials. For example,

$$\frac{3}{5}, \quad \frac{2}{x - 3}, \quad \text{and} \quad \frac{x^2 - 4}{x^2 - 4x - 5}$$

are rational expressions.

The Domain of a Rational Expression

The **domain** of an algebraic expression is the set of all real numbers for which the expression is defined. Since division by zero is not defined, any number that makes the denominator zero is not in the domain of a rational expression.

Example 1 Find the domain of each of the following.

a) $\dfrac{2}{x-3}$

b) $\dfrac{x^2-4}{x^2-4x-5}$

SOLUTION

a) Since $x-3=0$ when $x=3$, the domain of $2/(x-3)$ is the set of all real numbers except 3. As a partial check, graph $y=2/(x-3)$ using DOT mode.

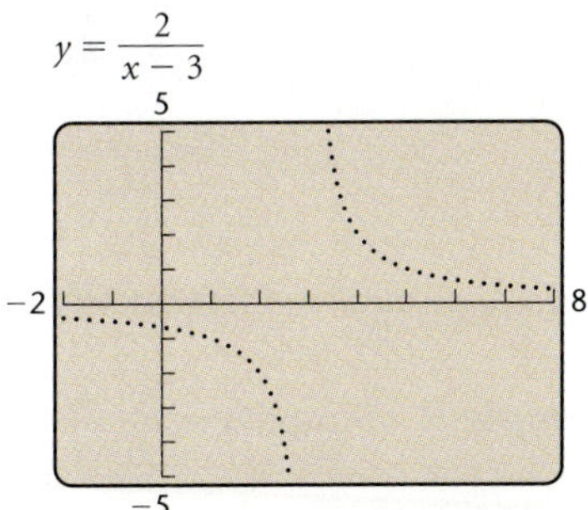

Note that there does not appear to be a point on the graph that has 3 as a first coordinate. Also note that the grapher's TABLE feature produces an error message for $x=3$.

b) To determine the domain of $(x^2-4)/(x^2-4x-5)$, we first factor the denominator:

$$\frac{x^2-4}{x^2-4x-5}=\frac{x^2-4}{(x+1)(x-5)}.$$

The factor $x+1$ is 0 when $x=-1$ and the factor $x-5$ is 0 when $x=5$. Since $(x+1)(x-5)=0$ when $x=-1$ or $x=5$, the domain is the set of all real numbers except -1 and 5. ▬

We can describe the domains found in Example 1 using **set-builder notation**. For example, we write "The set of all real numbers x such that x is not equal to 3" as

$$\{x \mid x \text{ is a real number and } x \neq 3\}.$$

Similarly, we write "The set of all real numbers x such that x is not equal to -1 and x is not equal to 5" as

$$\{x \mid x \text{ is a real number and } x \neq -1 \text{ and } x \neq 5\}.$$

Simplifying, Multiplying, and Dividing Rational Expressions

To simplify rational expressions, we use the fact that

$$\frac{a \cdot c}{b \cdot c} = \frac{a}{b} \cdot \frac{c}{c} = \frac{a}{b} \cdot 1 = \frac{a}{b}.$$

Example 2 Simplify: $\dfrac{9x^2 + 6x - 3}{12x^2 - 12}$.

SOLUTION

$$\frac{9x^2 + 6x - 3}{12x^2 - 12} = \frac{3(x + 1)(3x - 1)}{3 \cdot 4(x + 1)(x - 1)}$$ Factoring the numerator and the denominator

$$= \frac{3(x + 1)}{3(x + 1)} \cdot \frac{3x - 1}{4(x - 1)}$$ Factoring the rational expression

$$= 1 \cdot \frac{3x - 1}{4(x - 1)}$$ $\dfrac{3(x + 1)}{3(x + 1)} = 1$

$$= \frac{3x - 1}{4(x - 1)}$$ Removing a factor of 1

Canceling is a shortcut that is often used to remove a factor of 1.

Example 3 Simplify each of the following.

a) $\dfrac{4x^3 + 16x^2}{2x^3 + 6x^2 - 8x}$

b) $\dfrac{2 - x}{x^2 + x - 6}$

SOLUTION

a) $\dfrac{4x^3 + 16x^2}{2x^3 + 6x^2 - 8x} = \dfrac{2 \cdot 2 \cdot x \cdot x(x + 4)}{2 \cdot x(x + 4)(x - 1)}$ Factoring the numerator and the denominator

$$= \frac{\cancel{2} \cdot 2 \cdot \cancel{x} \cdot x\cancel{(x + 4)}}{\cancel{2} \cdot \cancel{x}\cancel{(x + 4)}(x - 1)}$$ Removing a factor of 1: $\dfrac{2x(x + 4)}{2x(x + 4)} = 1$

$$= \frac{2x}{x - 1}$$

b) $\dfrac{2 - x}{x^2 + x - 6} = \dfrac{2 - x}{(x + 3)(x - 2)}$ Factoring the denominator

$$= \frac{-1(x - 2)}{(x + 3)(x - 2)}$$ $2 - x = -1(x - 2)$

$$= \frac{-1\cancel{(x - 2)}}{(x + 3)\cancel{(x - 2)}}$$ Removing a factor of 1: $\dfrac{x - 2}{x - 2} = 1$

$$= \frac{-1}{x + 3}, \text{ or } -\frac{1}{x + 3}$$

Be careful to cancel only common *factors*. Like terms that are *not* factors cannot be canceled. For example,

$$\frac{x + 4}{x} \neq 4,$$

because x is not a factor of the numerator.

To confirm this graphically, graph $y_1 = \dfrac{x + 4}{x}$ and $y_2 = 4$.

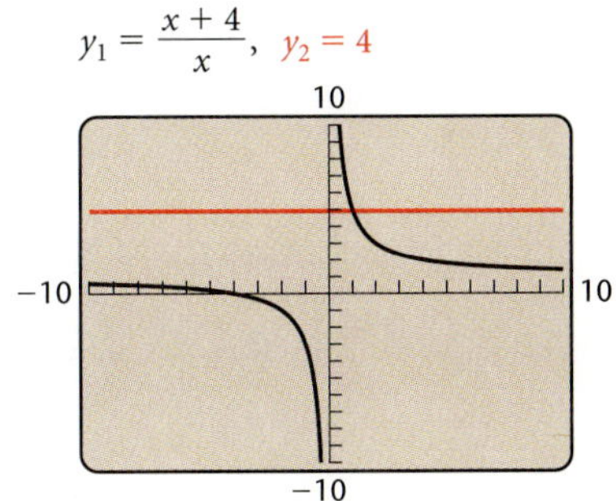

The graphs are different.

In Example 3(b), we saw that

$$\frac{2 - x}{x^2 + x - 6} \quad \text{and} \quad \frac{-1}{x + 3}$$

are **equivalent expressions**. This means that they have the same value for all numbers that are in both domains.

For

$$y_1 = \frac{2 - x}{x^2 + x - 6} \quad \text{and} \quad y_2 = \frac{-1}{x + 3},$$

a table shows that y_1 and y_2 have the same values except for $x = -3$ and $x = 2$. Note that -3 is not in the domain of *either* expression, whereas 2 is not in the domain of *both* expressions.

X	Y1	Y2
-3	ERROR	ERROR
-2	-1	-1
-1	-.5	-.5
0	-.3333	-.3333
1	-.25	-.25
2	ERROR	-.2
3	-.1667	-.1667

X = 2

To multiply rational expressions, we multiply numerators and multiply denominators and, if possible, simplify the result. To divide rational expressions, we multiply the dividend by the reciprocal of the divisor and, if possible, simplify the result.

Example 4   Multiply or divide and simplify each of the following.

a) $\dfrac{x + 4}{x - 3} \cdot \dfrac{x^2 - 9}{x^2 - x - 2}$ b) $\dfrac{y^3 - 1}{y^2 - 1} \div \dfrac{y^2 + y + 1}{y^2 + 2y + 1}$

SOLUTION

a) $\dfrac{x + 4}{x - 3} \cdot \dfrac{x^2 - 9}{x^2 - x - 2} = \dfrac{(x + 4)(x^2 - 9)}{(x - 3)(x^2 - x - 2)}$

            Multiplying the numerators and the denominators

$\qquad\qquad = \dfrac{(x + 4)(x + 3)(x - 3)}{(x - 3)(x - 2)(x + 1)}$ Factoring and removing a factor of 1: $\dfrac{x - 3}{x - 3} = 1$

$\qquad\qquad = \dfrac{(x + 4)(x + 3)}{(x - 2)(x + 1)}$

b) $\dfrac{y^3 - 1}{y^2 - 1} \div \dfrac{y^2 + y + 1}{y^2 + 2y + 1} = \dfrac{y^3 - 1}{y^2 - 1} \cdot \dfrac{y^2 + 2y + 1}{y^2 + y + 1}$ Multiplying by the reciprocal of the divisor

$\qquad\qquad = \dfrac{(y^3 - 1)(y^2 + 2y + 1)}{(y^2 - 1)(y^2 + y + 1)}$

$\qquad\qquad = \dfrac{(y - 1)(y^2 + y + 1)(y + 1)(y + 1)}{(y + 1)(y - 1)(y^2 + y + 1)}$

            Factoring and removing a factor of 1

$\qquad\qquad = y + 1$

Adding and Subtracting Rational Expressions

When rational expressions have the same denominator, we can add or subtract by adding or subtracting the numerators and retaining the common denominator. If the denominators differ, we must find equivalent rational expressions that have a common denominator.

It is usually most efficient to find the **least common denominator** (**LCD**) of the expressions. To do so, we factor each denominator and form the product that uses each factor the greatest number of times that it occurs in any factorization.

Example 5   Add or subtract and simplify each of the following.

a) $\dfrac{x^2 - 4x + 4}{2x^2 - 3x + 1} + \dfrac{x + 4}{2x - 2}$ b) $\dfrac{x}{x^2 + 11x + 30} - \dfrac{5}{x^2 + 9x + 20}$

SOLUTION

a) $\dfrac{x^2 - 4x + 4}{2x^2 - 3x + 1} + \dfrac{x + 4}{2x - 2}$

$\qquad = \dfrac{x^2 - 4x + 4}{(2x - 1)(x - 1)} + \dfrac{x + 4}{2(x - 1)}$ Factoring the denominators

       The LCD is $(2x - 1)(x - 1)(2)$, or $2(2x - 1)(x - 1)$.

$$= \frac{x^2 - 4x + 4}{(2x - 1)(x - 1)} \cdot \frac{2}{2} + \frac{x + 4}{2(x - 1)} \cdot \frac{2x - 1}{2x - 1}$$

Multiplying each term by 1 to get the LCD

$$= \frac{2x^2 - 8x + 8}{(2x - 1)(x - 1)(2)} + \frac{2x^2 + 7x - 4}{2(x - 1)(2x - 1)}$$

$$= \frac{4x^2 - x + 4}{2(x - 1)(2x - 1)}$$

Adding the numerators

b) $\dfrac{x}{x^2 + 11x + 30} - \dfrac{5}{x^2 + 9x + 20}$

$$= \frac{x}{(x + 5)(x + 6)} - \frac{5}{(x + 5)(x + 4)}$$

The LCD is $(x + 5)(x + 6)(x + 4)$.

$$= \frac{x}{(x + 5)(x + 6)} \cdot \frac{x + 4}{x + 4} - \frac{5}{(x + 5)(x + 4)} \cdot \frac{x + 6}{x + 6}$$

Multiplying each term by 1 to get the LCD

$$= \frac{x^2 + 4x}{(x + 5)(x + 6)(x + 4)} - \frac{5x + 30}{(x + 5)(x + 4)(x - 6)}$$

$$= \frac{x^2 + 4x - 5x - 30}{(x + 5)(x + 6)(x + 4)}$$

Be sure to change the sign of *every* term in the numerator of the expression being subtracted

$$= \frac{x^2 - x - 30}{(x + 5)(x + 6)(x + 4)}$$

$$= \frac{(x + 5)(x - 6)}{(x + 5)(x + 6)(x + 4)}$$

Removing a factor of 1: $\dfrac{x + 5}{x + 5} = 1$

$$= \frac{x - 6}{(x + 6)(x + 4)}$$

Complex Rational Expressions

A **complex rational expression** has rational expressions in its numerator or its denominator or both.

To simplify a complex rational expression:

Method 1. Find the LCD of all the denominators within the complex rational expression. Then multiply by 1 using that LCD as the numerator and the denominator of the expression for 1.

Method 2. First add or subtract, if necessary, to get a single rational expression in both the numerator and the denominator. Then divide by multiplying by the reciprocal of the denominator.

Example 6   Simplify: $\dfrac{\dfrac{1}{a} + \dfrac{1}{b}}{\dfrac{1}{a^3} + \dfrac{1}{b^3}}$.

SOLUTION

Method 1. The LCD of the four rational expressions in the numerator and the denominator is a^3b^3.

$$\frac{\dfrac{1}{a} + \dfrac{1}{b}}{\dfrac{1}{a^3} + \dfrac{1}{b^3}} = \frac{\dfrac{1}{a} + \dfrac{1}{b}}{\dfrac{1}{a^3} + \dfrac{1}{b^3}} \cdot \frac{a^3b^3}{a^3b^3} \qquad \text{Multiplying by 1}$$

using $\dfrac{a^3b^3}{a^3b^3}$

$$= \frac{\left(\dfrac{1}{a} + \dfrac{1}{b}\right)(a^3b^3)}{\left(\dfrac{1}{a^3} + \dfrac{1}{b^3}\right)(a^3b^3)}$$

$$= \frac{a^2b^3 + a^3b^2}{b^3 + a^3}$$

$$= \frac{a^2b^2(b + a)}{(b + a)(b^2 - ba + a^2)} \qquad \text{Factoring and removing a}$$

factor of 1: $\dfrac{b + a}{b + a} = 1$

$$= \frac{a^2b^2}{b^2 - ab + a^2}$$

Method 2. We add in the numerator and in the denominator.

$$\frac{\dfrac{1}{a} + \dfrac{1}{b}}{\dfrac{1}{a^3} + \dfrac{1}{b^3}} = \frac{\dfrac{1}{a} \cdot \dfrac{b}{b} + \dfrac{1}{b} \cdot \dfrac{a}{a}}{\dfrac{1}{a^3} \cdot \dfrac{b^3}{b^3} + \dfrac{1}{b^3} \cdot \dfrac{a^3}{a^3}}$$

⟵ The LCD is ab.

⟵ The LCD is a^3b^3.

$$= \frac{\dfrac{b}{ab} + \dfrac{a}{ab}}{\dfrac{b^3}{a^3b^3} + \dfrac{a^3}{a^3b^3}}$$

$$= \frac{\dfrac{b + a}{ab}}{\dfrac{b^3 + a^3}{a^3b^3}} \qquad \text{We have a single rational}$$

expression in both the numerator and the denominator.

$$= \frac{b + a}{ab} \cdot \frac{a^3b^3}{b^3 + a^3} \qquad \text{Multiplying by the reciprocal}$$

of the denominator

$$= \frac{(b + a)(a)(b)(a^2b^2)}{(a)(b)(b + a)(b^2 - ba + a^2)}$$

$$= \frac{a^2b^2}{b^2 - ab + a^2}$$

R.5 | *Exercise Set*

Find the domain of each rational expression.

1. $-\dfrac{3}{4}$

2. $\dfrac{5}{8 - x}$

3. $\dfrac{3x - 3}{x(x - 1)}$

4. $\dfrac{(x^2 - 4)(x + 1)}{(x + 2)(x^2 - 1)}$

5. $\dfrac{7x^2 - 28x + 28}{(x^2 - 4)(x^2 + 3x - 10)}$

6. $\dfrac{7x^2 + 11x - 6}{x(x^2 - x - 6)}$

7. Use the TABLE feature of a grapher to check your answer to Exercise 3.

8. Use the TABLE feature of a grapher to check your answer to Exercise 4.

Multiply or divide and, if possible, simplify.

9. $\dfrac{x^2 - y^2}{(x - y)^2} \cdot \dfrac{1}{x + y}$

10. $\dfrac{r - s}{r + s} \cdot \dfrac{r^2 - s^2}{(r - s)^2}$

11. $\dfrac{x^2 - 2x - 35}{2x^3 - 3x^2} \cdot \dfrac{4x^3 - 9x}{7x - 49}$

12. $\dfrac{x^2 + 2x - 35}{3x^3 - 2x^2} \cdot \dfrac{9x^3 - 4x}{7x + 49}$

13. $\dfrac{a^2 - a - 6}{a^2 - 7a + 12} \cdot \dfrac{a^2 - 2a - 8}{a^2 - 3a - 10}$

14. $\dfrac{a^2 - a - 12}{a^2 - 6a + 8} \cdot \dfrac{a^2 + a - 6}{a^2 - 2a - 24}$

15. $\dfrac{m^2 - n^2}{r + s} \div \dfrac{m - n}{r + s}$

16. $\dfrac{a^2 - b^2}{x - y} \div \dfrac{a + b}{x - y}$

17. $\dfrac{3x + 12}{2x - 8} \div \dfrac{(x + 4)^2}{(x - 4)^2}$

18. $\dfrac{a^2 - a - 2}{a^2 - a - 6} \div \dfrac{a^2 - 2a}{2a + a^2}$

19. $\dfrac{x^2 - y^2}{x^3 - y^3} \cdot \dfrac{x^2 + xy + y^2}{x^2 + 2xy + y^2}$

20. $\dfrac{c^3 + 8}{c^2 - 4} \div \dfrac{c^2 - 2c + 4}{c^2 - 4c + 4}$

21. $\dfrac{(x - y)^2 - z^2}{(x + y)^2 - z^2} \div \dfrac{x - y + z}{x + y - z}$

22. $\dfrac{(a + b)^2 - 9}{(a - b)^2 - 9} \cdot \dfrac{a - b - 3}{a + b + 3}$

23. Use a grapher to check your answer to Exercise 11.

24. Use a grapher to check your answer to Exercise 12.

Add or subtract and, if possible, simplify.

25. $\dfrac{3}{2a + 3} + \dfrac{2a}{2a + 3}$

26. $\dfrac{a - 3b}{a + b} + \dfrac{a + 5b}{a + b}$

27. $\dfrac{y}{y - 1} + \dfrac{2}{1 - y}$

28. $\dfrac{a}{a - b} + \dfrac{b}{b - a}$

29. $\dfrac{x}{2x - 3y} - \dfrac{y}{3y - 2x}$

30. $\dfrac{3a}{3a - 2b} - \dfrac{2a}{2b - 3a}$

31. $\dfrac{3}{x + 2} + \dfrac{2}{x^2 - 4}$

32. $\dfrac{5}{a - 3} - \dfrac{2}{a^2 - 9}$

33. $\dfrac{y}{y^2 - y - 20} - \dfrac{2}{y + 4}$

34. $\dfrac{6}{y^2 + 6y + 9} - \dfrac{5}{y + 3}$

35. $\dfrac{3}{x + y} + \dfrac{x - 5y}{x^2 - y^2}$

36. $\dfrac{a^2 + 1}{a^2 - 1} - \dfrac{a - 1}{a + 1}$

37. $\dfrac{9x + 2}{3x^2 - 2x - 8} + \dfrac{7}{3x^2 + x - 4}$

38. $\dfrac{3y}{y^2 - 7y + 10} - \dfrac{2y}{y^2 - 8y + 15}$

39. $\dfrac{5a}{a - b} + \dfrac{ab}{a^2 - b^2} + \dfrac{4b}{a + b}$

40. $\dfrac{6a}{a-b} - \dfrac{3b}{b-a} + \dfrac{5}{a^2-b^2}$

41. $\dfrac{7}{x+2} - \dfrac{x+8}{4-x^2} + \dfrac{3x-2}{4-4x+x^2}$

42. $\dfrac{6}{x+3} - \dfrac{x+4}{9-x^2} + \dfrac{2x-3}{9-6x+x^2}$

43. $\dfrac{1}{x+1} + \dfrac{x}{2-x} + \dfrac{x^2+2}{x^2-x-2}$

44. $\dfrac{x-1}{x-2} - \dfrac{x+1}{x+2} - \dfrac{x-6}{4-x^2}$

45. Use a grapher to check your answer to Exercise 31.

46. Use a grapher to check your answer to Exercise 32.

Simplify.

47. $\dfrac{\dfrac{x^2-y^2}{xy}}{\dfrac{x-y}{y}}$

48. $\dfrac{\dfrac{a-b}{b}}{\dfrac{a^2-b^2}{ab}}$

49. $\dfrac{a-a^{-1}}{a+a^{-1}}$

50. $\dfrac{a-\dfrac{a}{b}}{b-\dfrac{b}{a}}$

51. $\dfrac{c+\dfrac{8}{c^2}}{1+\dfrac{2}{c}}$

52. $\dfrac{x^{-1}+y^{-1}}{x^{-3}+y^{-3}}$

53. $\dfrac{x^2+xy+y^2}{\dfrac{x^2}{y}-\dfrac{y^2}{x}}$

54. $\dfrac{\dfrac{a^2}{b}+\dfrac{b^2}{a}}{a^2-ab+b^2}$

55. $\dfrac{\dfrac{x}{y}-\dfrac{y}{x}}{\dfrac{1}{y}+\dfrac{1}{x}}$

56. $\dfrac{\dfrac{a}{b}-\dfrac{b}{a}}{\dfrac{1}{a}-\dfrac{1}{b}}$

57. $\dfrac{\dfrac{1}{x-3}+\dfrac{2}{x+3}}{\dfrac{3}{x-1}-\dfrac{4}{x+2}}$

58. $\dfrac{\dfrac{5}{x+1}-\dfrac{3}{x-2}}{\dfrac{1}{x-5}+\dfrac{2}{x+2}}$

59. $\dfrac{\dfrac{a}{1-a}+\dfrac{1+a}{a}}{\dfrac{1-a}{a}+\dfrac{a}{1+a}}$

60. $\dfrac{\dfrac{1-x}{x}+\dfrac{x}{1+x}}{\dfrac{1+x}{x}+\dfrac{x}{1-x}}$

61. $\dfrac{\dfrac{1}{a^2}+\dfrac{2}{ab}+\dfrac{1}{b^2}}{\dfrac{1}{a^2}-\dfrac{1}{b^2}}$

62. $\dfrac{\dfrac{1}{x^2}-\dfrac{1}{y^2}}{\dfrac{1}{x^2}-\dfrac{2}{xy}+\dfrac{1}{y^2}}$

Synthesis

63. ◈ When adding or subtracting rational expressions, we can always find a common denominator by forming the product of all the denominators. Explain why it is usually preferable to find the least common denominator.

64. ◈ How would you determine which method to use for simplifying a particular complex rational expression?

Simplify.

65. $\dfrac{(x+h)^2-x^2}{h}$

66. $\dfrac{\dfrac{1}{x+h}-\dfrac{1}{x}}{h}$

67. $\dfrac{(x+h)^3-x^3}{h}$

68. $\dfrac{\dfrac{1}{(x+h)^2}-\dfrac{1}{x^2}}{h}$

69. $\left[\dfrac{\dfrac{x+1}{x-1}+1}{\dfrac{x+1}{x-1}-1}\right]^5$

70. $1+\dfrac{1}{1+\dfrac{1}{1+\dfrac{1}{1+\dfrac{1}{x}}}}$

Perform the indicated operations and, if possible, simplify.

71. $\dfrac{n(n+1)(n+2)}{2\cdot 3} + \dfrac{(n+1)(n+2)}{2}$

72. $\dfrac{n(n+1)(n+2)(n+3)}{2\cdot 3\cdot 4} + \dfrac{(n+1)(n+2)(n+3)}{2\cdot 3}$

73. $\dfrac{x^2-9}{x^3+27} \cdot \dfrac{5x^2-15x+45}{x^2-2x-3} + \dfrac{x^2+x}{4+2x}$

74. $\dfrac{x^2+2x-3}{x^2-x-12} \div \dfrac{x^2-1}{x^2-16} - \dfrac{2x+1}{x^2+2x+1}$

R.6
Radical Notation and Rational Exponents

- *Simplify radical expressions.*
- *Rationalize denominators or numerators in rational expressions.*
- *Convert between exponential and radical notation.*
- *Simplify expressions with rational exponents.*
- *Factor expressions with rational exponents.*

A number c is said to be a **square root** of a if $c^2 = a$. Thus, 3 is a square root of 9, because $3^2 = 9$, and -3 is also a square root of 9, because $(-3)^2 = 9$. Similarly, 5 is a third root (called a **cube root**) of 125, because $5^3 = 125$. The number 125 has no other real-number cube root.

---

nth Root

A number c is said to be an *nth root* of a if $c^n = a$.

The symbol $\sqrt{a}$ denotes the nonnegative square root of a, and the symbol $\sqrt[3]{a}$ denotes the real-number cube root of a. The symbol $\sqrt[n]{a}$ denotes the nth root of a, that is, a number whose nth power is a. The symbol $\sqrt[n]{}$ is called a **radical**, and the expression under the radical is called the **radicand**. The number n (which is omitted when it is 2) is called the **index**. Examples of roots for $n = 3$, 4, and 2, respectively, are

$$\sqrt[3]{125}, \quad \sqrt[4]{16}, \quad \text{and} \quad \sqrt{3600}.$$

Any real number has only one real-number odd root. Any positive number has two square roots, one positive and one negative. The same is true for fourth roots or roots with any even index. The positive root is called the **principal root**. When an expression such as $\sqrt{4}$ or $\sqrt[6]{23}$ is used, it is understood to represent the principal (nonnegative) root. To denote a negative root, we use $-\sqrt{4}$, $-\sqrt[6]{23}$, and so on.

Example 1   Simplify each of the following.

a) $\sqrt{36}$ b) $-\sqrt{36}$ c) $\sqrt[5]{\dfrac{32}{243}}$

d) $\sqrt[3]{-8}$ e) $\sqrt[4]{-16}$

SOLUTION

a) $\sqrt{36} = 6$, because $6^2 = 36$.

b) $-\sqrt{36} = -6$, because $6^2 = 36$ and $-(\sqrt{36}) = -(6) = -6$.

c) $\sqrt[5]{\dfrac{32}{243}} = \dfrac{2}{3}$, because $\left(\dfrac{2}{3}\right)^5 = \dfrac{2^5}{3^5} = \dfrac{32}{243}$.

d) $\sqrt[3]{-8} = -2$, because $(-2)^3 = -8$.

e) $\sqrt[4]{-16}$ is not a real number, because we cannot find a real number that can be raised to the fourth power to get -16.

We can generalize Example 1(e) and say that when a is negative and n is even, $\sqrt[n]{a}$ is not a real number. For example, $\sqrt{-4}$ and $\sqrt[4]{-81}$ are not real numbers.

Simplifying Radical Expressions

Consider the expression $\sqrt{(-3)^2}$. This is equivalent to $\sqrt{9}$, or 3. Similarly, $\sqrt{3^2} = \sqrt{9} = 3$. This illustrates the first of several properties of radicals, listed below.

Properties of Radicals

Let a and b be any real numbers or expressions for which the given roots exist. For any natural numbers m and n ($n \neq 1$):

1. If n is even, $\sqrt[n]{a^n} = |a|$.

2. If n is odd, $\sqrt[n]{a^n} = a$.

3. $\sqrt[n]{a} \cdot \sqrt[n]{b} = \sqrt[n]{ab}$.

4. $\sqrt[n]{\dfrac{a}{b}} = \dfrac{\sqrt[n]{a}}{\sqrt[n]{b}}$ $(b \neq 0)$.

5. $\sqrt[n]{a^m} = (\sqrt[n]{a})^m$.

To illustrate the first property, graph $y_1 = \sqrt{x^2}$ and $y_2 = |x|$. We can illustrate the second property in a similar manner, graphing $y_1 = \sqrt[3]{x^3}$ and $y_2 = x$.

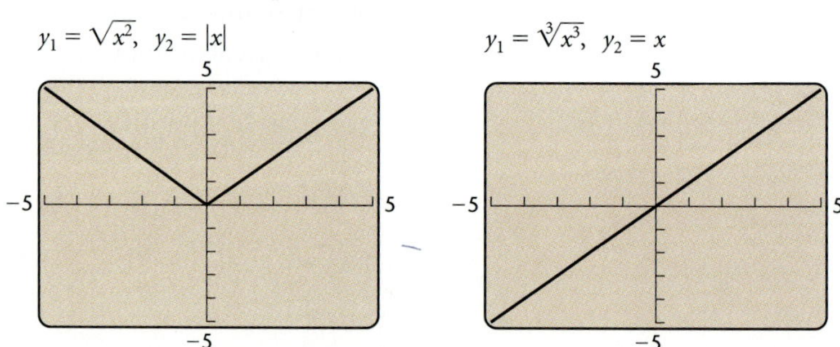

The graphs appear to coincide in each case.

Example 2   Simplify each of the following.

a) $\sqrt{(-5)^2}$ b) $\sqrt[3]{(-5)^3}$ c) $\sqrt[4]{4} \cdot \sqrt[4]{5}$ d) $\sqrt{50}$

e) $\dfrac{\sqrt{72}}{\sqrt{6}}$ f) $\sqrt[3]{8^5}$ g) $\sqrt{216x^5y^3}$ h) $\sqrt{\dfrac{x^2}{16}}$

SOLUTION

a) $\sqrt{(-5)^2} = |-5| = 5$ Using Property 1

b) $\sqrt[3]{(-5)^3} = -5$ Using Property 2

c) $\sqrt[4]{4} \cdot \sqrt[4]{5} = \sqrt[4]{4 \cdot 5} = \sqrt[4]{20}$ Using Property 3

d) $\sqrt{50} = \sqrt{25 \cdot 2} = \sqrt{25} \cdot \sqrt{2} = 5\sqrt{2}$ Using Property 3

e) $\dfrac{\sqrt{72}}{\sqrt{6}} = \sqrt{\dfrac{72}{6}}$ Using Property 4

$\qquad = \sqrt{12} = \sqrt{4 \cdot 3} = \sqrt{4}\sqrt{3}$ Using Property 3

$\qquad = 2\sqrt{3}$

f) $\sqrt[3]{8^5} = (\sqrt[3]{8})^5$ Using Property 5

$\qquad = 2^5 = 32$

g) $\sqrt{216x^5y^3} = \sqrt{36 \cdot 6 \cdot x^4 \cdot x \cdot y^2 \cdot y}$

$\qquad = \sqrt{36x^4y^2}\sqrt{6xy}$ Using Property 3

$\qquad = |6x^2y|\sqrt{6xy}$ Using Property 1

$\qquad = 6x^2|y|\sqrt{6xy}$ $6x^2$ cannot be negative, so absolute value signs are not needed for it.

h) $\sqrt{\dfrac{x^2}{16}} = \dfrac{\sqrt{x^2}}{\sqrt{16}}$ Using Property 4

$\qquad = \dfrac{|x|}{4}$ Using Property 1

We can check Example 2(h) graphically. Graph $y_1 = \sqrt{x^2/16}$ and $y_2 = |x|/4$.

$$y_1 = \sqrt{\dfrac{x^2}{16}}, \quad y_2 = \dfrac{|x|}{4}$$

X	Y₁	Y₂
−3	.75	.75
−2	.5	.5
−1	.25	.25
0	0	0
1	.25	.25
2	.5	.5
3	.75	.75

X = 0

The graphs appear to coincide. We can also use the TABLE feature to confirm that for each value of x, the y-values are the same.

In many situations, radicands are never formed by raising negative quantities to even powers. In such cases, absolute-value notation is not required. For this reason, we will henceforth assume that no radicands are formed by raising negative quantities to even powers. For example, we will write $\sqrt{x^2} = x$ and $\sqrt[4]{a^5b} = a\sqrt[4]{ab}$.

Radical expressions with the same index and the same radicand can be added or subtracted.

Example 3 Perform the operations indicated.

a) $3\sqrt{8x^2} - 5\sqrt{2x^2}$

b) $(4\sqrt{3} + \sqrt{2})(\sqrt{3} - 5\sqrt{2})$

SOLUTION

a) $3\sqrt{8x^2} - 5\sqrt{2x^2} = 3\sqrt{4x^2 \cdot 2} - 5\sqrt{x^2 \cdot 2}$

$$= 3 \cdot 2x\sqrt{2} - 5x\sqrt{2}$$
$$= 6x\sqrt{2} - 5x\sqrt{2}$$
$$= (6x - 5x)\sqrt{2} \qquad \text{Using the distributive property}$$
$$= x\sqrt{2}$$

b) $(4\sqrt{3} + \sqrt{2})(\sqrt{3} - 5\sqrt{2}) = 4(\sqrt{3})^2 - 20\sqrt{6} + \sqrt{6} - 5(\sqrt{2})^2$

$$\text{Multiplying}$$
$$= 4 \cdot 3 + (-20 + 1)\sqrt{6} - 5 \cdot 2$$
$$= 12 - 19\sqrt{6} - 10$$
$$= 2 - 19\sqrt{6}$$

An Application

The Pythagoren theorem relates the lengths of the sides of a right triangle. The side opposite the triangle's right angle is called the **hypotenuse**. The other sides are the **legs**.

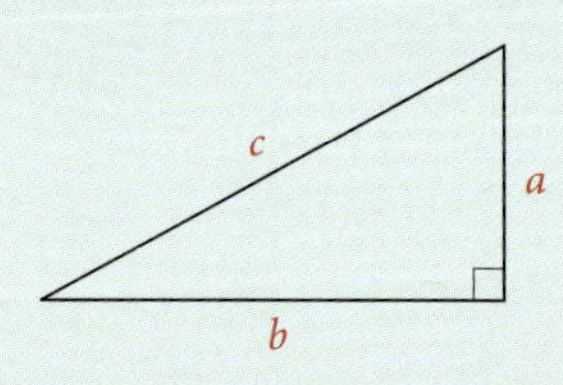

The Pythagorean Theorem

The sum of the squares of the lengths of the legs of a right triangle is equal to the square of the length of the hypotenuse:

$$a^2 + b^2 = c^2.$$

Example 4 A surveyor places poles at points A, B, and C in order to measure the distance across a pond. The distances AC and BC are measured as shown. Find the distance AB across the pond.

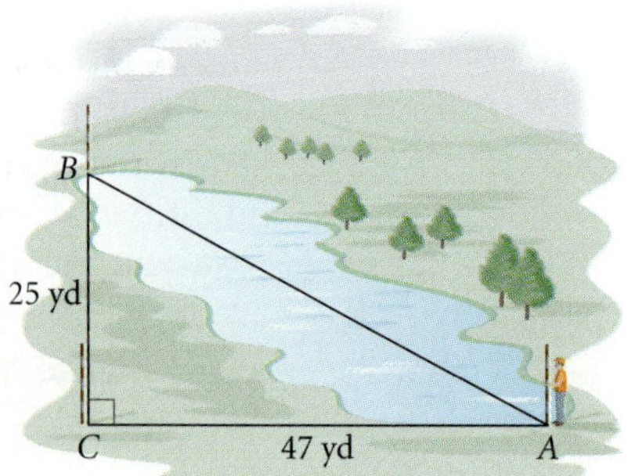

SOLUTION   We see that the lengths of the legs of a right triangle are given. Thus we use the Pythagorean theorem to find the length of the hypotenuse AB:

$$AB^2 = AC^2 + BC^2$$

so

$$\begin{aligned} AB &= \sqrt{AC^2 + BC^2} \\ &= \sqrt{47^2 + 25^2} \\ &= \sqrt{2209 + 625} \\ &= \sqrt{2834} \\ &\approx 53.2. \end{aligned}$$

The distance across the pond is about 53.2 yd.

Rationalizing Denominators or Numerators

There are times when we need to remove the radicals in a denominator or a numerator. This is called **rationalizing the denominator** or **rationalizing the numerator**. It is done by multiplying by 1 in such a way as to obtain a perfect nth power.

Example 5   Rationalize the denominator of each of the following.

a) $\sqrt{\dfrac{3}{2}}$
    b) $\dfrac{\sqrt[3]{7}}{\sqrt[3]{9}}$

SOLUTION

a) $\sqrt{\dfrac{3}{2}} = \sqrt{\dfrac{3}{2} \cdot \dfrac{2}{2}} = \sqrt{\dfrac{6}{4}} = \dfrac{\sqrt{6}}{\sqrt{4}} = \dfrac{\sqrt{6}}{2}$

b) $\dfrac{\sqrt[3]{7}}{\sqrt[3]{9}} = \dfrac{\sqrt[3]{7}}{\sqrt[3]{9}} \cdot \dfrac{\sqrt[3]{3}}{\sqrt[3]{3}} = \dfrac{\sqrt[3]{21}}{\sqrt[3]{27}} = \dfrac{\sqrt[3]{21}}{3}$

Pairs of expressions of the form $a\sqrt{b} + c\sqrt{d}$ and $a\sqrt{b} - c\sqrt{d}$ are called **conjugates**. The product of such a pair contains no radicals and can be used to rationalize a denominator or a numerator.

Example 6   Rationalize the numerator: $\dfrac{\sqrt{x} - \sqrt{y}}{5}$.

SOLUTION

$$\dfrac{\sqrt{x} - \sqrt{y}}{5} = \dfrac{\sqrt{x} - \sqrt{y}}{5} \cdot \dfrac{\sqrt{x} + \sqrt{y}}{\sqrt{x} + \sqrt{y}} \qquad \begin{array}{l}\text{The conjugate of} \\ \sqrt{x} - \sqrt{y} \text{ is} \\ \sqrt{x} + \sqrt{y}.\end{array}$$

$$= \dfrac{(\sqrt{x})^2 - (\sqrt{y})^2}{5\sqrt{x} + 5\sqrt{y}}$$

$$= \dfrac{x - y}{5\sqrt{x} + 5\sqrt{y}}$$

Rational Exponents

We are motivated to define *rational exponents* so that the rules, or laws, for integer exponents hold for them. For example, we must have

$$a^{1/2} \cdot a^{1/2} = a^{1/2+1/2} = a^1 = a.$$

Thus we are led to define $a^{1/2}$ to mean $\sqrt{a}$. Similarly, $a^{1/n}$ would mean $\sqrt[n]{a}$. Again, if the laws of exponents are to hold, we must have

$$(a^{1/n})^m = (a^m)^{1/n} = a^{m/n}.$$

Thus we are led to define $a^{m/n}$ to mean $(\sqrt[n]{a})^m$, or, equivalently, $\sqrt[n]{a^m}$.

Rational Exponents

For any real number a and any natural numbers m and n for which $\sqrt[n]{a}$ exists,

$$a^{1/n} = \sqrt[n]{a},$$
$$a^{m/n} = \sqrt[n]{a^m} = (\sqrt[n]{a})^m, \quad \text{and}$$
$$a^{-m/n} = \frac{1}{a^{m/n}}.$$

We can use the definition of rational exponents to convert between radical and exponential notation.

Example 7 Convert to radical notation and, if possible, simplify each of the following.

a) $7^{3/4}$

b) $8^{-5/3}$

c) $m^{1/6}$

d) $(-32)^{2/5}$

SOLUTION

a) $7^{3/4} = \sqrt[4]{7^3}$, or $(\sqrt[4]{7})^3$

b) $8^{-5/3} = \dfrac{1}{8^{5/3}} = \dfrac{1}{(8^{1/3})^5} = \dfrac{1}{2^5} = \dfrac{1}{32}$

c) $m^{1/6} = \sqrt[6]{m}$

d) $(-32)^{2/5} = \sqrt[5]{(-32)^2} = \sqrt[5]{1024} = 4,$ or
   $(-32)^{2/5} = (\sqrt[5]{-32})^2 = (-2)^2 = 4$

Interactive Discovery

Enter each of the following expressions on a grapher and compare the results with the result of Example 7(d):

$$(-32)^{2/5}, \qquad ((-32)^2)^{1/5}, \qquad ((-32)^{1/5})^2.$$

Next, use the definition of rational exponents to find $(-8)^{5/3}$. Then enter $(-8)^{5/3}$, $((-8)^5)^{1/3}$, and $((-8)^{1/3})^5$ on a grapher and compare your results.

What do you observe about how to use a grapher to evaluate $a^{m/n}$ when a is negative and m is greater than 1?

X	Y1	Y2
-4	ERROR	2.5198
-3	ERROR	2.0801
-2	ERROR	1.5874
-1	ERROR	1
0	0	0
1	1	1
2	1.5874	1.5874

X = -1

When using a grapher to find rational roots of the form $a^{m/n}$, with $a < 0$ and $m > 1$, we must enter the expression in the form $((a)^m)^{1/n}$ or $((a)^{1/n})^m$. Similarly, to graph $y = x^{m/n}$, where $m > 1$, we must enter $y = (x^m)^{1/n}$ or $y = (x^{1/n})^m$ in order to see the graph for negative values of x. The TABLE feature will confirm this. For example, enter $y_1 = x^{2/3}$ and $y_2 = (x^2)^{1/3}$. The table (shown at left) gives an ERROR message for values of y_1 when x is negative. Some graphers will give an ERROR message for negative values of x regardless of the form in which $x^{m/n}$, $m > 1$, is entered.

Example 8   Convert to exponential notation and, if possible, simplify each of the following.

a) $(\sqrt[4]{7xy})^5$ **b)** $\sqrt[6]{x^3}$ **c)** $\sqrt[3]{\sqrt{7}}$

SOLUTION

a) $(\sqrt[4]{7xy})^5 = (7xy)^{5/4}$

b) $\sqrt[6]{x^3} = x^{3/6} = x^{1/2}$

c) $\sqrt[3]{\sqrt{7}} = \sqrt[3]{7^{1/2}} = (7^{1/2})^{1/3} = 7^{1/6}$

We can use the laws of exponents to simplify exponential and radical expressions.

Example 9   Simplify and then write radical notation, if appropriate, for each of the following.

a) $x^{5/6} \cdot x^{2/3}$ **b)** $(x + 3)^{5/2}(x + 3)^{-1/2}$

SOLUTION

a) $x^{5/6} \cdot x^{2/3} = x^{5/6+2/3} = x^{9/6} = x^{3/2} = \sqrt{x^3} = \sqrt{x^2}\sqrt{x} = x\sqrt{x}$

b) $(x + 3)^{5/2}(x + 3)^{-1/2} = (x + 3)^{5/2-1/2} = (x + 3)^2$

Example 10   Write an expression containing a single radical: $a^{1/2}b^{5/6}$.

SOLUTION

$$a^{1/2}b^{5/6} = a^{3/6}b^{5/6} = (a^3b^5)^{1/6} = \sqrt[6]{a^3b^5}$$

Factoring Expressions with Rational Exponents

To factor expressions containing negative exponents, factor out the greatest common numerical factor and common variable factors with the smallest exponents.

Example 11   Factor: $3a^{1/2}b^{-3/4} - a^{-1/2}b^{1/4}$.

SOLUTION   Note that $a^{-1/2}$ has a smaller exponent than $a^{1/2}$ and $b^{-3/4}$ has a smaller exponent than $b^{1/4}$.

$3a^{1/2}b^{-3/4} - a^{-1/2}b^{1/4}$

$= 3 \cdot a \cdot a^{-1/2}b^{-3/4} - a^{-1/2} \cdot b \cdot b^{-3/4}$ $a^{1/2} = a \cdot a^{-1/2},\ b^{1/4} = b \cdot b^{-3/4}$

$= a^{-1/2}b^{-3/4}(3a - b)$ **Factoring out $a^{-1/2}b^{-3/4}$**

R.6 | Exercise Set

Simplify. Assume that variables can represent any real number.

1. $\sqrt{(-11)^2}$

2. $\sqrt{(-1)^2}$

3. $\sqrt{16x^2}$

4. $\sqrt{36t^2}$

5. $\sqrt{(b+1)^2}$

6. $\sqrt{(2c-3)^2}$

7. $\sqrt[3]{-27x^3}$

8. $\sqrt[3]{-8y^3}$

9. $\sqrt{x^2 - 4x + 4}$

10. $\sqrt{y^2 + 16y + 64}$

11. $\sqrt[5]{32}$

12. $\sqrt[5]{-32}$

13. $\sqrt{180}$

14. $\sqrt{48}$

15. $\sqrt[3]{54}$

16. $\sqrt[3]{135}$

17. $\sqrt{128c^2 d^4}$

18. $\sqrt{162c^4 d^6}$

19. Use the TABLE feature of a grapher to check your answer to Exercise 9.

20. Use the TABLE feature of a grapher to check your answer to Exercise 10.

Simplify. Assume that no radicands were formed by raising negative quantities to even powers.

21. $\sqrt{2x^3y}\sqrt{12xy}$

22. $\sqrt{3y^4z}\sqrt{20z}$

23. $\sqrt[3]{3x^2y}\sqrt[3]{36x}$

24. $\sqrt[5]{8x^3y^4}\sqrt[5]{4x^4y}$

25. $\sqrt[3]{2(x+4)}\,\sqrt[3]{4(x+4)^4}$

26. $\sqrt[3]{4(x+1)^2}\,\sqrt[3]{18(x+1)^2}$

27. $\sqrt[6]{\dfrac{m^{12}n^{24}}{64}}$

28. $\sqrt[8]{\dfrac{m^{16}n^{24}}{2^8}}$

29. $\dfrac{\sqrt[3]{40m}}{\sqrt[3]{5m}}$

30. $\dfrac{\sqrt{40xy}}{\sqrt{8x}}$

31. $\dfrac{\sqrt[3]{3x^2}}{\sqrt[3]{24x^5}}$

32. $\dfrac{\sqrt{128a^2b^4}}{\sqrt{16ab}}$

33. $\sqrt[3]{\dfrac{64a^4}{27b^3}}$

34. $\sqrt{\dfrac{9x^7}{16y^8}}$

35. $\sqrt{\dfrac{7x^3}{36y^6}}$

36. $\sqrt[3]{\dfrac{2yz}{250z^4}}$

37. $9\sqrt{50} + 6\sqrt{2}$

38. $11\sqrt{27} - 4\sqrt{3}$

39. $8\sqrt{2x^2} - 6\sqrt{20x} - 5\sqrt{8x^2}$

40. $2\sqrt[3]{8x^2} + 5\sqrt[3]{27x^2} - 3\sqrt{x^3}$

41. $(\sqrt{3} - \sqrt{2})(\sqrt{3} + \sqrt{2})$

42. $(\sqrt{8} + 2\sqrt{5})(\sqrt{8} - 2\sqrt{5})$

43. $(1 + \sqrt{3})^2$

44. $(\sqrt{2} - 5)^2$

45. Use a grapher to check your answer to Exercise 39.

46. Use a grapher to check your answer to Exercise 40.

47. An airplane is flying at an altitude of 3700 ft. The slanted distance directly to the airport is 14,200 ft. How far horizontally is the airplane from the airport?

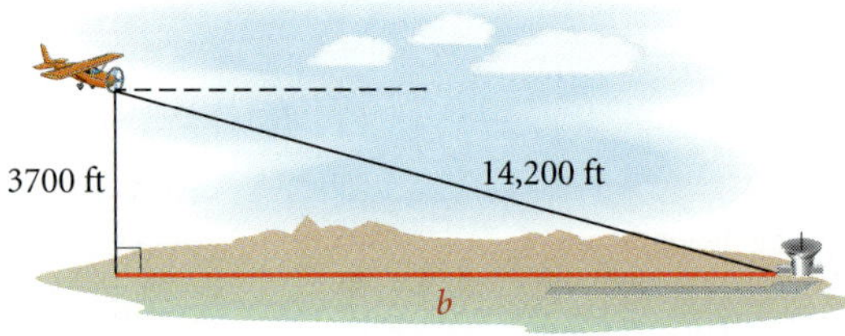

48. During a summer heat wave, a 2-mi bridge expands 2 ft in length. Assuming that the bulge occurs straight up the middle, estimate the height of the bulge. (In reality, bridges are built with expansion joints to control such buckling.)

49. An *equilateral triangle* is shown below.

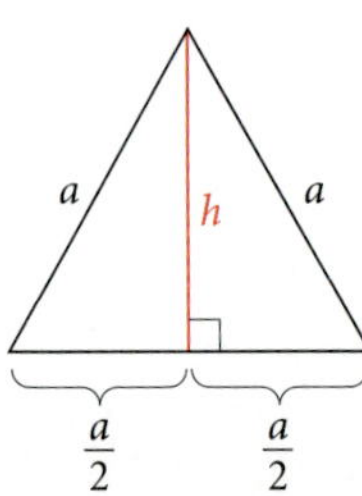

a) Find an expression for its height h in terms of a.

b) Find an expression for its area A in terms of a.

50. An isosceles right triangle has legs of length s. Find an expression for the length of the hypotenuse in terms of s.

51. The diagonal of a square has length $8\sqrt{2}$. Find the length of a side of the square.

52. The area of square $PQRS$ is 100 ft^2, and A, B, C, and D are the midpoints of the sides. Find the area of square $ABCD$.

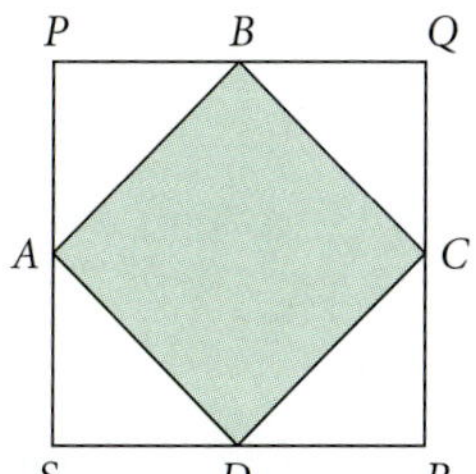

Rationalize the denominator.

53. $\sqrt{\dfrac{2}{3}}$

54. $\sqrt{\dfrac{3}{7}}$

55. $\dfrac{\sqrt[3]{5}}{\sqrt[3]{4}}$

56. $\dfrac{\sqrt[3]{7}}{\sqrt[3]{25}}$

57. $\sqrt[3]{\dfrac{16}{3}}$

58. $\sqrt[3]{\dfrac{3}{5}}$

59. $\dfrac{6}{3 + \sqrt{5}}$

60. $\dfrac{2}{\sqrt{3} - 1}$

61. $\dfrac{6}{\sqrt{m} - \sqrt{n}}$

62. $\dfrac{3}{\sqrt{v} - \sqrt{w}}$

Rationalize the numerator.

63. $\dfrac{\sqrt{12}}{5}$

64. $\dfrac{\sqrt{50}}{3}$

65. $\sqrt[3]{\dfrac{7}{2}}$

66. $\sqrt[3]{\dfrac{2}{5}}$

67. $\dfrac{\sqrt{11}}{\sqrt{3}}$

68. $\dfrac{\sqrt{5}}{\sqrt{2}}$

69. $\dfrac{9 - \sqrt{5}}{3 - \sqrt{3}}$

70. $\dfrac{8 - \sqrt{6}}{5 - \sqrt{2}}$

71. $\dfrac{\sqrt{a} + \sqrt{b}}{3a}$

72. $\dfrac{\sqrt{p} - \sqrt{q}}{1 + \sqrt{q}}$

Convert to radical notation and simplify.

73. $x^{3/4}$

74. $y^{2/5}$

75. $16^{3/4}$

76. $4^{7/2}$

77. $125^{-1/3}$

78. $32^{-4/5}$

79. $a^{5/4}b^{-3/4}$

80. $x^{2/5}y^{-1/5}$

Convert to exponential notation and simplify.

81. $(\sqrt[4]{13})^5$

82. $\sqrt[5]{17^3}$

83. $\sqrt[3]{20^2}$

84. $(\sqrt[5]{12})^4$

85. $\sqrt[3]{\sqrt{11}}$

86. $\sqrt[3]{\sqrt[4]{7}}$

87. $\sqrt{5}\sqrt[3]{5}$

88. $\sqrt[3]{2}\sqrt{2}$

89. $\sqrt[5]{32^2}$

90. $\sqrt[3]{64^2}$

Simplify and then write radical notation, if appropriate.

91. $(2a^{3/2})(4a^{1/2})$

92. $(3a^{5/6})(8a^{2/3})$

93. $\left(\dfrac{x^6}{9b^{-4}}\right)^{1/2}$

94. $\left(\dfrac{x^{2/3}}{4y^{-2}}\right)^{1/2}$

95. $\dfrac{x^{2/3}y^{5/6}}{x^{-1/3}y^{1/2}}$

96. $\dfrac{a^{1/2}b^{5/8}}{a^{1/4}b^{3/8}}$

97. Use a grapher to check your answer to Exercise 91.

98. Use a grapher to check your answer to Exercise 92.

Write an expression containing a single radical and simplify.

99. $\sqrt[3]{6}\sqrt{2}$

100. $\sqrt{2}\sqrt[4]{8}$

101. $\sqrt[4]{xy}\sqrt[3]{x^2y}$

102. $\sqrt[3]{ab^2}\sqrt{ab}$

103. $\sqrt[3]{a^4}\sqrt{a^3}$

104. $\sqrt{a^3}\sqrt[3]{a^2}$

105. $\dfrac{\sqrt{(a + x)^3}\sqrt[3]{(a + x)^2}}{\sqrt[4]{a + x}}$

106. $\dfrac{\sqrt[4]{(x + y)^2}\sqrt[3]{x + y}}{\sqrt{(x + y)^3}}$

Factor and simplify.

107. $a^{-2}b^5 - a^3b^{-5}$

108. $p^8q^{-2} + p^{-5}q^4$

109. $x^{-1/3}y^{3/4} - x^{2/3}y^{-1/4}$

110. $4a^{1/2}b^{-3/4} - 6a^{-1/2}b^{1/4}$

111. $(2x - 3)^{-3}(x + 1)^{5/4} + (2x - 3)^{-2}(x + 1)^{1/4}$

112. $2x(5x + 3)^{2/3} + 3x^2(5x + 3)^{-1/3}$

Synthesis

113. ◈ Explain how you would convince a classmate that $\sqrt{a + b}$ is not equivalent to $\sqrt{a} + \sqrt{b}$, for positive real numbers a and b. Give both an algebraic explanation and a graphical explanation.

114. ◈ Explain how you would determine whether $10\sqrt{26} - 50$ is positive or negative without carrying out the actual computation.

Simplify.

115. $\sqrt{1 + x^2} + \dfrac{1}{\sqrt{1 + x^2}}$

116. $\sqrt{1 - x^2} - \dfrac{x^2}{2\sqrt{1 - x^2}}$

117. $(\sqrt{a\sqrt{a}})\sqrt{a}$

118. $(2a^3 b^{5/4} c^{1/7})^4 \div (54 a^{-2} b^{2/3} c^{6/5})^{-1/3}$

119. Use a grapher to determine all values of x for which:

    a) $x^{1/2} = x^{1/3}$
    b) $x^{1/2} > x^{1/3}$
    c) $x^{1/2} < x^{1/3}$

R.7

Solving Equations

- *Solve linear, quadratic, rational, and radical equations and equations with absolute value.*
- *Solve a formula for a given variable.*

An **equation** is a statement that two expressions are equal. To **solve** an equation in one variable is to find all the values of the variable that make the equation true. Each of these numbers is a **solution** of the equation. The set of all solutions of an equation is its **solution set**. Equations that have the same solution set are called **equivalent equations**.

Linear and Quadratic Equations

> A *linear equation in one variable* is an equation that is equivalent to one of the form $ax + b = 0$, where a and b are real numbers and $a \neq 0$.
>
> A *quadratic equation* is an equation that is equivalent to one of the form $ax^2 + bx + c = 0$, where a, b, and c are real numbers and $a \neq 0$.

The following principles allow us to solve many linear and quadratic equations.

> **Equation-Solving Principles**
>
> For any real numbers a, b, and c:
>
> **The Addition Principle:**   If $a = b$ is true, then $a + c = b + c$ is true.
>
> **The Multiplication Principle:**   If $a = b$ is true, then $ac = bc$ is true.
>
> **The Principle of Zero Products:**   If $ab = 0$ is true, then $a = 0$ or $b = 0$, and if $a = 0$ or $b = 0$, then $ab = 0$.
>
> **The Principle of Square Roots:**   If $x^2 = k$, then $x = \sqrt{k}$ or $x = -\sqrt{k}$.

Example 1 Solve: $3(7 - 2x) = 14 - 8(x - 1)$.

ALGEBRAIC SOLUTION

We have

$3(7 - 2x) = 14 - 8(x - 1)$

$21 - 6x = 14 - 8x + 8$ **Using the distributive property**

$21 - 6x = 22 - 8x$ **Combining like terms**

$21 + 2x = 22$ **Using the addition principle to add $8x$ on both sides**

$2x = 1$ **Using the addition principle to add -21 or subtract 21 on both sides**

$x = \frac{1}{2}.$ **Using the multiplication principle to multiply by $\frac{1}{2}$ or divide by 2 on both sides**

CHECK: $\quad 3(7 - 2x) = 14 - 8(x - 1)$

$$3\left(7 - 2 \cdot \tfrac{1}{2}\right) \ ? \ 14 - 8\left(\tfrac{1}{2} - 1\right) \quad \textbf{Substituting } \tfrac{1}{2} \textbf{ for } x$$

$$3(7 - 1) \ \bigm| \ 14 - 8\left(-\tfrac{1}{2}\right)$$

$$3 \cdot 6 \ \bigm| \ 14 + 4$$

$$18 \ \bigm| \ 18 \qquad\qquad \text{TRUE}$$

The solution is $\frac{1}{2}$. The set of all solutions, the solution set, is $\left\{\frac{1}{2}\right\}$.

We can also use the TABLE feature of a grapher—set in ASK mode, if available— to check the answer. We let $y_1 = 3(7 - 2x)$ and $y_2 = 14 - 8(x - 1)$. When we enter .5 for x, we see that Y_1 and Y_2 are both 18.

X	Y₁	Y₂
.5	18	18

X = .5

GRAPHICAL SOLUTION

We can use a grapher to solve this equation. Graph $y_1 = 3(7 - 2x)$ and $y_2 = 14 - 8(x - 1)$.

$y_1 = 3(7 - 2x), \ y_2 = 14 - 8(x - 1)$

x = .5, y = 18

Using the TRACE and ZOOM features or the INTERSECT feature, we find that the first coordinate of the point of intersection is .5, or $\frac{1}{2}$. The solution set is $\left\{\frac{1}{2}\right\}$.

We can also use the SOLVE feature. It might be necessary to first write the equation as

$$3(7 - 2x) - (14 - 8(x - 1)) = 0.$$

The grapher returns the value .5, or $\frac{1}{2}$.

Note that the equation in Example 1 is not an identity, because it is not true for *all* values of x in its domain. In fact, we see from the graph above that $y_1 = 3(7 - 2x)$ and $y_2 = 14 - 8(x - 1)$ coincide, or intersect, only for $x = \frac{1}{2}$.

Example 2 Solve each of the following.

a) $2x^2 - x = 3$

b) $2x^2 - 10 = 0$

SOLUTION

a) We solve $2x^2 - x = 3$ both algebraically and graphically.

ALGEBRAIC SOLUTION

We have

$$2x^2 - x = 3$$
$$2x^2 - x - 3 = 0 \qquad \text{Subtracting 3 on both sides}$$
$$(2x - 3)(x + 1) = 0 \qquad \text{Factoring}$$
$$2x - 3 = 0 \quad \text{or} \quad x + 1 = 0$$
$$\text{Using the principle of zero products}$$
$$x = \tfrac{3}{2} \quad \text{or} \qquad x = -1.$$

Both numbers check. We can use the TABLE feature of a grapher, set in ASK mode, to confirm this. Let $y = 2x^2 - x$. The y-values in the table should be 3 for both $x = 1.5$ and $x = -1$.

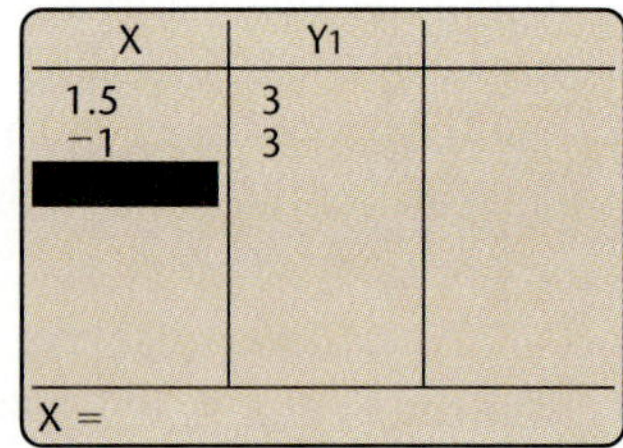

The solution set is $\left\{\tfrac{3}{2}, -1\right\}$.

GRAPHICAL SOLUTION

Graph $y_1 = 2x^2 - x$ and $y_2 = 3$.

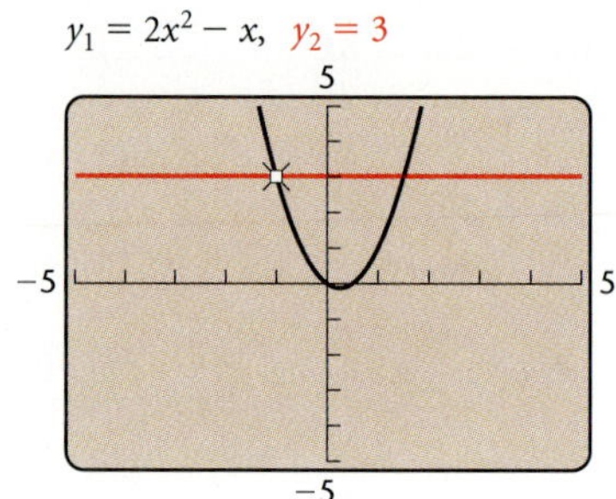

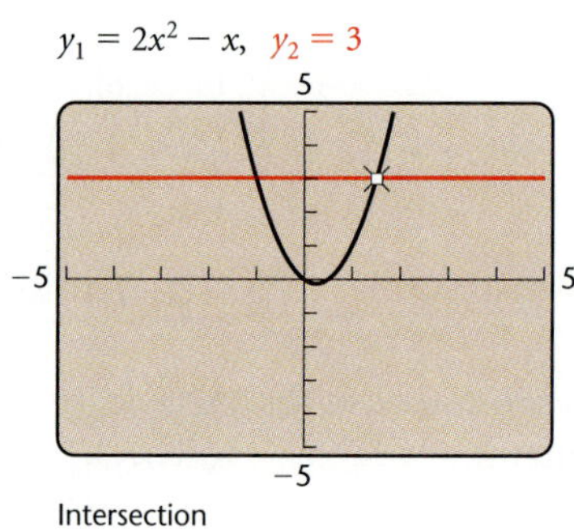

We use the TRACE and ZOOM features or the INTERSECT feature and find that the solutions are -1 and 1.5. We can also use the SOLVE feature to get this result. The solution set is $\{-1, 1.5\}$.

b) We solve $2x^2 - 10 = 0$ algebraically:

$$2x^2 - 10 = 0$$
$$2x^2 = 10 \qquad \text{Adding 10 on both sides}$$
$$x^2 = 5 \qquad \text{Dividing by 2 on both sides}$$
$$x = \sqrt{5} \quad \text{or} \quad x = -\sqrt{5}. \qquad \text{Using the principle of square roots}$$

Both numbers check. The solution set is $\{\sqrt{5}, -\sqrt{5}\}$, or $\{\pm\sqrt{5}\}$. Use the TABLE feature of a grapher, set in ASK mode, to confirm this. Using a grapher, we find that $x \approx 2.236$ or $x \approx -2.236$.

Rational and Radical Equations

Equations containing rational expressions are called **rational equations**. Solving such equations involves multiplying on both sides by the LCD.

Example 3 Solve: $\dfrac{x^2}{x-3} = \dfrac{9}{x-3}$.

SOLUTION The LCD is $x - 3$.

$$(x - 3) \cdot \frac{x^2}{x - 3} = (x - 3) \cdot \frac{9}{x - 3}$$

$$x^2 = 9$$

$$x = -3 \quad or \quad x = 3 \qquad \text{Using the principle of square roots}$$

The possible solutions are 3 and -3. We check.

CHECK: For 3:

$$\frac{x^2}{x - 3} = \frac{9}{x - 3}$$

$$\frac{3^2}{3 - 3} \stackrel{?}{=} \frac{9}{3 - 3}$$

$$\frac{9}{0} \;\Big|\; \frac{9}{0} \qquad \text{UNDEFINED}$$

For -3:

$$\frac{x^2}{x - 3} = \frac{9}{x - 3}$$

$$\frac{(-3)^2}{-3 - 3} \stackrel{?}{=} \frac{9}{-3 - 3}$$

$$\frac{9}{-6} \;\Big|\; \frac{9}{-6} \qquad \text{TRUE}$$

Since division by 0 is undefined, 3 is not a solution. Note that 3 is not in the domain of $x^2/(x - 3)$ or $9/(x - 3)$. (See the table below.) The number -3 checks, so it is a solution. The solution set is $\{-3\}$.

$$y_1 = \frac{x^2}{x - 3}, \; y_2 = \frac{9}{x - 3}$$

X	Y1	Y2
0	0	-3
1	$-.5$	-4.5
2	-4	-9
3	ERROR	ERROR
4	16	9
5	12.5	4.5
6	12	3

X = 3

When we use the multiplication principle to multiply (or divide) both sides of an equation by an expression with a variable, we might not obtain an equivalent equation. We must check possible solutions by substituting in the original equation.

A **radical equation** is an equation in which variables appear in one or more radicands. For example,

$$\sqrt{2x - 5} - \sqrt{x - 3} = 1$$

is a radical equation. The following principle is needed to solve such equations.

The Principle of Powers

For any positive integer n:

If $a = b$ is true, then $a^n = b^n$ is true.

Example 4   Solve: $5 + \sqrt{x + 7} = x$.

We first isolate the radical and then use the principle of powers.

$$5 + \sqrt{x + 7} = x$$
$$\sqrt{x + 7} = x - 5 \qquad \text{Subtracting 5 on both sides}$$
$$(\sqrt{x + 7})^2 = (x - 5)^2 \qquad \text{Using the principle of powers;}$$
$$\text{squaring both sides}$$
$$x + 7 = x^2 - 10x + 25$$
$$0 = x^2 - 11x + 18$$
$$0 = (x - 9)(x - 2)$$
$$x - 9 = 0 \quad \text{or} \quad x - 2 = 0$$
$$x = 9 \quad \text{or} \qquad x = 2$$

The possible solutions are 9 and 2.

CHECK: For 9:

$$5 + \sqrt{x + 7} = x$$
$$\overline{5 + \sqrt{9 + 7} \;?\; 9}$$
$$5 + \sqrt{16}$$
$$5 + 4$$
$$9 \mid 9 \quad \text{TRUE}$$

For 2:

$$5 + \sqrt{x + 7} = x$$
$$\overline{5 + \sqrt{2 + 7} \;?\; 2}$$
$$5 + \sqrt{9}$$
$$5 + 3$$
$$8 \mid 2 \quad \text{FALSE}$$

Since 9 checks but 2 does not, the only solution is 9. The solution set is $\{9\}$.

Graph $y_1 = 5 + \sqrt{x + 7}$ and $y_2 = x$.

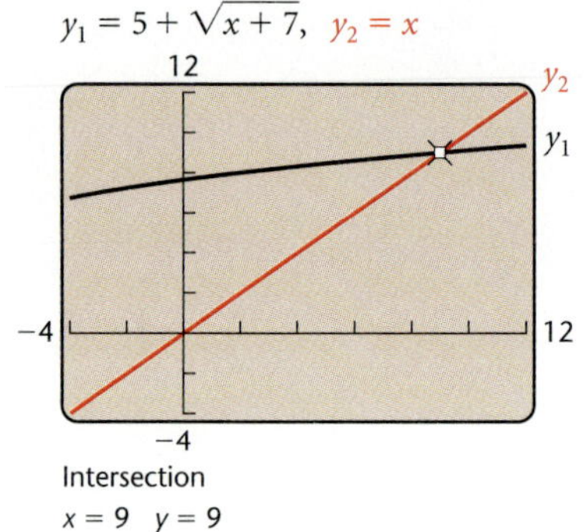

Using the TRACE and ZOOM features or the INTERSECT feature, we see that the only solution is 9. The solution set is $\{9\}$.

We can also use the SOLVE feature to get this result. It might be necessary, however, to first write the equation as

$$5 + \sqrt{x + 7} - x = 0.$$

When we raise both sides of an equation to an even power, the resulting equation can have solutions that the original equation does not. This is because the converse of the principle of powers is not necessarily true. That is, if $a^n = b^n$ is true, we do not know that $a = b$ is true. For example, $(-2)^2 = 2^2$, but $-2 \neq 2$. Thus, as we see in Example 4, it is necessary to check the possible solutions in the original equation when using the principle of powers to raise both sides of an equation to an even power.

When a radical equation has two radical terms on one side, we isolate one of them and then use the principle of powers. If, after doing so, a radical term remains, we repeat these steps.

Example 5   Solve: $\sqrt{x - 3} + \sqrt{x + 5} = 4$.

SOLUTION

$$\sqrt{x - 3} = 4 - \sqrt{x + 5} \qquad \text{Isolating one radical}$$
$$(\sqrt{x - 3})^2 = (4 - \sqrt{x + 5})^2 \qquad \text{Using the principle of powers;}$$
$$\text{squaring both sides}$$

$$x - 3 = 16 - 8\sqrt{x + 5} + (x + 5)$$

$$x - 3 = 21 - 8\sqrt{x + 5} + x \qquad \text{Combining like terms}$$

$$-24 = -8\sqrt{x + 5} \qquad \text{Isolating the remaining radical}$$

$$3 = \sqrt{x + 5} \qquad \text{Dividing by } -8 \text{ on both sides}$$

$$3^2 = (\sqrt{x + 5})^2 \qquad \text{Using the principle of powers; squaring both sides}$$

$$9 = x + 5$$

$$4 = x$$

The number 4 checks and is the solution. Use graphs or the TABLE feature of a grapher to confirm this. The solution set is {4}.

Equations with Absolute Value

Recall that the absolute value of a number is its distance from 0 on the number line. We use this concept to solve equations with absolute value.

For $a > 0$:

$$|x| = a \text{ is equivalent to } x = -a \text{ or } x = a.$$

Example 6 Solve each of the following.

a) $|x| = 5$

b) $|x - 3| - 2$

SOLUTION

a) We solve $|x| = 5$ both algebraically and graphically.

ALGEBRAIC SOLUTION

We have

$$|x| = 5$$

$$x = -5 \quad or \quad x = 5.$$

Writing an equivalent statement

The solution set is {−5, 5}. Use the TABLE feature of a grapher to confirm this.

GRAPHICAL SOLUTION

Graph $y_1 = |x|$ and $y_2 = 5$ and find the first coordinates of the points of intersection using TRACE and ZOOM or INTERSECT. We can also use the SOLVE feature to get this result.

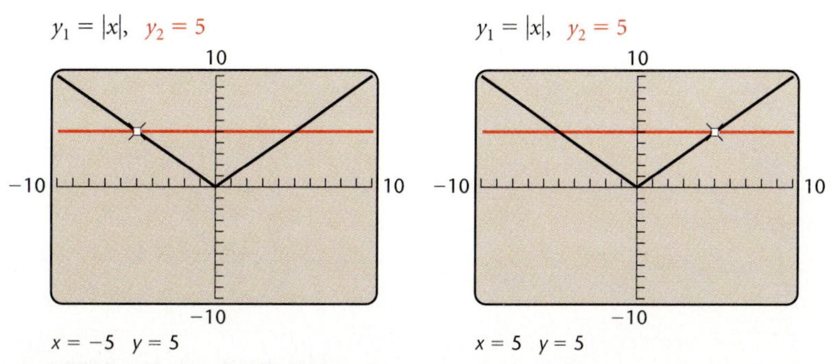

The solution set is {−5, 5}.

b) We solve $|x - 3| = 2$ algebraically:

$$|x - 3| = 2$$
$$x - 3 = -2 \quad or \quad x - 3 = 2 \qquad \text{Writing an equivalent statement}$$
$$x = 1 \quad or \quad x = 5. \qquad \text{Adding 3}$$

The solution set is $\{1, 5\}$.

When $a = 0$, $|x| = a$ is equivalent to $x = 0$. Note that for $a < 0$, $|x| = a$ has no solution, because the absolute value of an expression is never negative. We can use a graph to illustrate the last statement for a specific value of a. For example, let $a = -3$. Graph $y_1 = |x|$ and $y_2 = -3$.

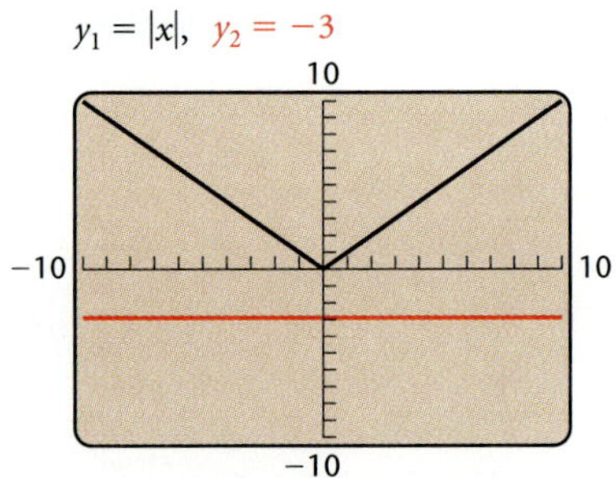

The graphs do not intersect. Thus the equation $|x| = -3$ has no solution. The solution set is the **empty set**, denoted $\varnothing$.

Formulas

A **formula** is an equation that can be used to *model* a situation. For example, the formula $P = 2l + 2w$ gives the perimeter of a rectangle with length l and width w. The equation-solving principles presented earlier can be used to solve a formula for a given variable.

Example 7   Solve $P = 2l + 2w$ for l.

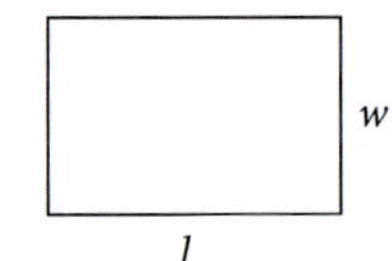

SOLUTION

$$P = 2l + 2w$$
$$P - 2w = 2l \qquad \text{Subtracting } 2w \text{ on both sides}$$
$$\frac{P - 2w}{2} = l \qquad \text{Dividing by 2 on both sides}$$

The formula $l = \dfrac{P - 2w}{2}$ can be used to determine a rectangle's length if we are given its perimeter and its width.

R.7 Exercise Set

Solve using a grapher.

1. $4x + 5 = 21$

2. $2y - 1 = 3$

3. $y + 1 = 2y - 7$

4. $5 - 4x = x - 13$

Solve algebraically. Check your answers on a grapher.

5. $5x - 2 + 3x = 2x + 6 - 4x$

6. $5x - 17 - 2x = 6x - 1 - x$

7. $7(3x + 6) = 11 - (x + 2)$

8. $4(5y + 3) = 3(2y - 5)$

9. $(2x - 3)(3x - 2) = 0$

10. $(5x - 2)(2x + 3) = 0$

11. $3x^2 + x - 2 = 0$

12. $10x^2 - 16x + 6 = 0$

13. $2x^2 = 6x$

14. $18x + 9x^2 = 0$

15. $3y^3 - 5y^2 - 2y = 0$

16. $3t^3 + 2t = 5t^2$

17. $7x^3 + x^2 - 7x - 1 = 0$
(*Hint*: Factor by grouping.)

18. $3x^3 + x^2 - 12x - 4 = 0$
(*Hint*: Factor by grouping.)

19. $\dfrac{1}{4} + \dfrac{1}{5} = \dfrac{1}{t}$

20. $\dfrac{1}{3} - \dfrac{5}{6} = \dfrac{1}{x}$

21. $\dfrac{x + 2}{4} - \dfrac{x - 1}{5} = 15$

22. $\dfrac{t + 1}{3} - \dfrac{t - 1}{2} = 1$

23. $\dfrac{1}{2} + \dfrac{2}{x} = \dfrac{1}{3} + \dfrac{3}{x}$

24. $\dfrac{1}{t} + \dfrac{1}{2t} + \dfrac{1}{3t} = 5$

25. $\dfrac{3x}{x + 2} + \dfrac{6}{x} = \dfrac{12}{x^2 + 2x}$

26. $\dfrac{5x}{x - 4} - \dfrac{20}{x} = \dfrac{80}{x^2 - 4x}$

27. $\dfrac{4}{x^2 - 1} - \dfrac{2}{x - 1} = \dfrac{3}{x + 1}$

28. $\dfrac{3y + 5}{y^2 + 5y} + \dfrac{y + 4}{y + 5} = \dfrac{y + 1}{y}$

29. $\dfrac{490}{x^2 - 49} = \dfrac{5x}{x - 7} - \dfrac{35}{x + 7}$

30. $\dfrac{3}{m + 2} + \dfrac{2}{m} = \dfrac{4m - 4}{m^2 - 4}$

31. $\dfrac{1}{x - 6} - \dfrac{1}{x} = \dfrac{6}{x^2 - 6x}$

32. $\dfrac{8}{x^2 - 4} = \dfrac{x}{x - 2} - \dfrac{2}{x + 2}$

33. $\dfrac{8}{x^2 - 2x + 4} = \dfrac{x}{x + 2} + \dfrac{24}{x^3 + 8}$

34. $\dfrac{18}{x^2 - 3x + 9} - \dfrac{x}{x + 3} = \dfrac{81}{x^3 + 27}$

35. $\sqrt{3x - 4} = 1$

36. $\sqrt[3]{2x + 1} = -5$

37. $\sqrt[4]{x^2 - 1} = 1$

38. $\sqrt{m + 1} - 5 = 8$

39. $\sqrt{y - 1} + 4 = 0$

40. $\sqrt[5]{3x + 4} = 2$

41. $\sqrt[3]{6x + 9} + 8 = 5$

42. $\sqrt{6x + 7} = x + 2$

43. $\sqrt{x - 3} + \sqrt{x + 2} = 5$

44. $\sqrt{x} - \sqrt{x - 5} = 1$

45. $\sqrt{3x - 5} + \sqrt{2x + 3} + 1 = 0$

46. $\sqrt{2m - 3} = \sqrt{m + 7} - 2$

47. $\sqrt{x} - \sqrt{3x - 3} = 1$

48. $\sqrt{2x + 1} - \sqrt{x} = 1$

49. $\sqrt{2y - 5} - \sqrt{y - 3} = 1$

50. $\sqrt{4p + 5} + \sqrt{p + 5} = 3$

51. $x^{1/3} = -2$

52. $t^{1/5} = 2$

53. $t^{1/4} = 3$

54. $m^{1/2} = -7$

Solve.

55. $|x| = 7$

56. $|x| = 4.5$

57. $|x| = -10.7$

58. $|x| = -\dfrac{3}{5}$

59. $|x - 1| = 4$

60. $|x - 7| = 5$

61. $|3x| = 1$

62. $|5x| = 4$

63. $|x| = 0$

64. $|6x| = 0$

65. $|3x + 2| = 1$

66. $|7x - 4| = 8$

67. $\left|\frac{1}{2}x - 5\right| = 17$

68. $\left|\frac{1}{3}x - 4\right| = 13$

69. $|x - 1| + 3 = 6$

70. $|x + 2| - 5 = 9$

Solve.

71. $A = \frac{1}{2}bh$, for b

(Area of a triangle)

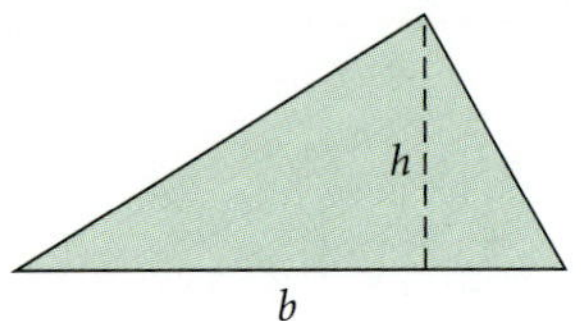

72. $A = \pi r^2$, for π
(Area of a circle)

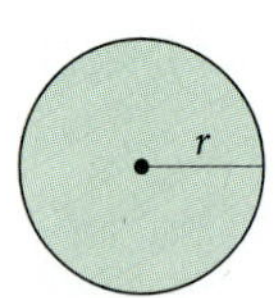

73. $P = 2l + 2w$, for w
(Perimeter of a rectangle)

74. $A = P + Prt$, for P
(Simple interest)

75. $\frac{P_1 V_1}{T_1} = \frac{P_2 V_2}{T_2}$, for T_1
(A chemistry formula for gases)

76. $A = \frac{1}{2}h(b_1 + b_2)$, for h

(Area of a trapezoid)

77. $\frac{1}{R} = \frac{1}{R_1} + \frac{1}{R_2}$, for R_2
(Resistance)

78. $A = P(1 + i)^2$, for i
(Compound interest)

79. $\frac{1}{F} = \frac{1}{m} + \frac{1}{p}$, for p
(A formula from optics)

80. $\frac{1}{F} = \frac{1}{m} + \frac{1}{p}$, for F
(A formula from optics)

Synthesis

81. ◈ Explain why it is necessary to check the possible solutions of a rational equation.

82. ◈ Explain in your own words why it is necessary to check the possible solutions when the principle of powers is used to solve an equation.

83. ◈ Use a graphical argument to explain why the equation $x^2 - 6x + 9 = 1 - x^4$ has no solution.

Solve.

84. $(x + 1)^3 = (x - 1)^3 + 26$

85. $(x - 2)^3 = x^3 - 2$

86. $\dfrac{x + 3}{x + 2} - \dfrac{x + 4}{x + 3} = \dfrac{x + 5}{x + 4} - \dfrac{x + 6}{x + 5}$

87. $(x - 3)^{2/3} = 2$

88. $\sqrt{15 + \sqrt{2x + 80}} = 5$

89. $\sqrt{x + 5} + 1 = \dfrac{6}{\sqrt{x + 5}}$

90. $x^{2/3} = x + 1$

R.8

Solving Inequalities

- *Solve linear inequalities, using interval notation to express solution sets.*
- *Solve compound inequalities.*
- *Solve inequalities with absolute value.*

An **inequality** is a sentence with $<$, $>$, $\leq$, or $\geq$ as its verb. An example is $3x - 5 < 6 - 2x$. To **solve** an inequality is to find all values of the variable that make the inequality true. Each of these numbers is a **solu-**

tion of the inequality, and the set of all such solutions is its **solution set**. Inequalities that have the same solution set are called **equivalent inequalities**.

Linear Inequalities

The principles for solving inequalities are similar to those for solving equations.

Principles for Solving Inequalities

For any real numbers a, b, and c:

The Addition Principle for Inequalities: If $a < b$ is true, then $a + c < b + c$ is true.

The Multiplication Principle for Inequalities: If $a < b$ and $c > 0$ are true, then $ac < bc$ is true. If $a < b$ and $c < 0$ are true, then $ac > bc$ is true. Similar statements hold for $a \leq b$.

Note that when both sides of an inequality are multiplied by a negative number, we must reverse the inequality sign.

First-degree inequalities with one variable, like those in Example 1 below, are **linear inequalities**.

Example 1 Solve each of the following.

a) $3x - 5 < 6 - 2x$ $\qquad\qquad$ **b)** $13 - 7y \geq 10y - 4$

SOLUTION

a) $3x - 5 < 6 - 2x$

$\quad 5x - 5 < 6$ $\qquad$ Using the addition principle for inequalities; adding $2x$ on both sides

$\qquad 5x < 11$ $\qquad$ Using the addition principle for inequalities; adding 5 on both sides

$\qquad x < \frac{11}{5}$ $\qquad$ Using the multiplication principle for inequalities; multiplying by $\frac{1}{5}$ or dividing by 5 on both sides

Any number less than $\frac{11}{5}$ is a solution. To check graphically, graph $y_1 = 3x - 5$ and $y_2 = 6 - 2x$. The graph (shown at left) confirms that for $x < \frac{11}{5}$, $y_1 < y_2$. The solution set is $\left\{x \mid x < \frac{11}{5}\right\}$.

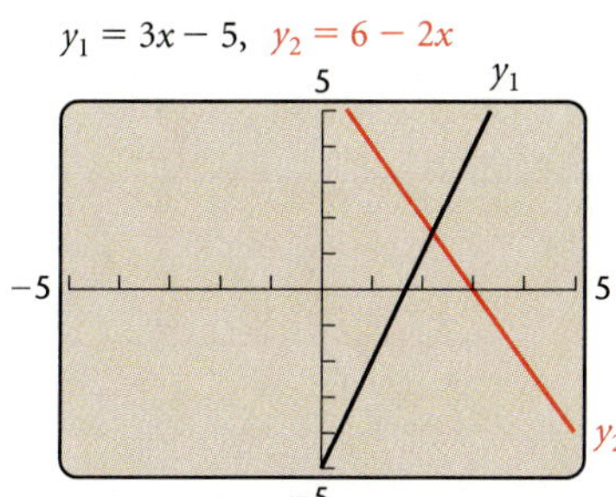

b) $13 - 7y \geq 10y - 4$

$\quad 13 - 17y \geq -4$ $\qquad$ Subtracting $10y$ on both sides

$\qquad -17y \geq -17$ $\qquad$ Subtracting 13 on both sides

$\qquad\quad y \leq 1$ $\qquad$ Dividing by -17 on both sides and reversing the inequality sign

The solution set is $\{y \mid y \leq 1\}$.

Interval Notation

Solutions of inequalities can also be expressed using **interval notation**. For example, for real numbers a and b such that $a < b$, the **open interval** (a, b) is the set of real numbers between, but not including, a and b. That is,

$$(a, b) = \{x \mid a < x < b\}.$$

The points a and b are the **endpoints** of the interval. The parentheses indicate that the endpoints are not included in the interval.

Some intervals extend without bound in one or both directions. The interval $[a, \infty)$, for example, begins at a and extends to the right without bound. That is,

$$[a, \infty) = \{x \mid x \geq a\}.$$

The bracket indicates that a is included in the interval.

The various types of intervals are listed below.

Intervals: Types, Notation, and Graphs

TYPE	INTERVAL NOTATION	SET NOTATION	GRAPH
Open	(a, b)	$\{x \mid a < x < b\}$	
Closed	$[a, b]$	$\{x \mid a \leq x \leq b\}$	
Half-open	$[a, b)$	$\{x \mid a \leq x < b\}$	
Half-open	$(a, b]$	$\{x \mid a < x \leq b\}$	
Open	(a, ∞)	$\{x \mid x > a\}$	
Half-open	$[a, \infty)$	$\{x \mid x \geq a\}$	
Open	$(-\infty, b)$	$\{x \mid x < b\}$	
Half-open	$(-\infty, b]$	$\{x \mid x \leq b\}$	

The interval $(-\infty, \infty)$ names the set of all real numbers, $\mathbb{R}$.

We see that we can express the solutions in Example 1 using interval notation:

$$\left\{x \mid x < \tfrac{11}{5}\right\} = \left(-\infty, \tfrac{11}{5}\right) \quad \text{and} \quad \{y \mid y \le 1\} = (-\infty, 1].$$

Example 2 Write interval notation for each set.

a) $\{x \mid -4 < x < 5\}$ **b)** $\{x \mid x \ge 1.7\}$

c) $\{x \mid -10 < x \le -5\}$ **d)** $\{x \mid x < \sqrt{5}\}$

SOLUTION

a) $\{x \mid -4 < x < 5\} = (-4, 5)$

b) $\{x \mid x \ge 1.7\} = [1.7, \infty)$

c) $\{x \mid -10 < x \le -5\} = (-10, -5]$

d) $\{x \mid x < \sqrt{5}\} = (-\infty, \sqrt{5})$

Compound Inequalities

When two inequalities are joined by the word *and* or the word *or*, a **compound inequality** is formed. A compound inequality like $-3 < 2x + 5$ *and* $2x + 5 \le 7$ is called a **conjunction**, because it uses the word *and*. The sentence $-3 < 2x + 5 \le 7$ is an abbreviation for the preceding conjunction.

Compound inequalities can be solved using the addition and multiplication principles for inequalities.

Example 3 Solve $-3 < 2x + 5 \le 7$. Then graph the solution set.

SOLUTION We have

$$-3 < 2x + 5 \le 7$$
$$-8 < 2x \le 2 \qquad \text{Subtracting 5}$$
$$-4 < x \le 1. \qquad \text{Dividing by 2}$$

The solution set is $\{x \mid -4 < x \le 1\}$, or $(-4, 1]$. The graph of the solution set is as shown at left.

To check, graph $y_1 = -3$, $y_2 = 2x + 5$, and $y_3 = 7$.

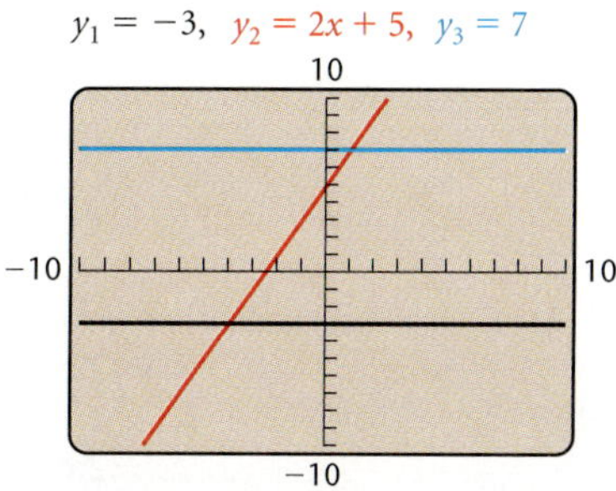

Note that $y_1 < y_2 \le y_3$ for $-4 < x \le 1$.

A compound inequality like $2x - 5 \leq -7$ or $2x - 5 > 1$ is called a **disjunction**, because it contains the word *or*. Unlike some conjunctions, it cannot be abbreviated; that is, it cannot be written without the word *or*.

Example 4 Solve $2x - 5 \leq -7$ or $2x - 5 > 1$. Then graph the solution set.

SOLUTION We have

$$2x - 5 \leq -7 \quad or \quad 2x - 5 > 1$$
$$2x \leq -2 \quad or \qquad 2x > 6 \qquad \text{Adding 5}$$
$$x \leq -1 \quad or \qquad x > 3. \qquad \text{Dividing by 2}$$

The solution set is $\{x \mid x \leq -1 \text{ or } x > 3\}$. We can also write the solution using interval notation and the symbol $\cup$ for the **union** or inclusion of both sets: $(-\infty, -1] \cup (3, \infty)$. The graph of the solution set is as shown at left.

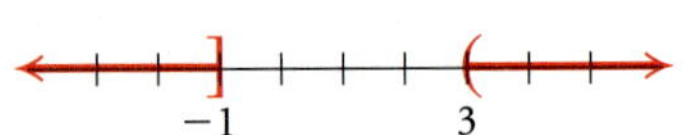

Inequalities with Absolute Value

Inequalities sometimes contain absolute value notation. The following properties are used to solve them.

For $a > 0$:

$|x| < a$ is equivalent to $-a < x < a$.

$|x| > a$ is equivalent to $x < -a$ or $x > a$.

Similar statements hold for $|x| \leq a$ and $|x| \geq a$.

Example 5 Solve each of the following.

a) $|3x + 2| < 5$
b) $|5 - 2x| \geq 1$

SOLUTION

a) $|3x + 2| < 5$
$$-5 < 3x + 2 < 5 \qquad \text{Writing an equivalent inequality}$$
$$-7 < 3x < 3 \qquad \text{Subtracting 2}$$
$$-\tfrac{7}{3} < x < 1 \qquad \text{Dividing by 3}$$
The solution is $\left\{x \mid -\tfrac{7}{3} < x < 1\right\}$, or $\left(-\tfrac{7}{3}, 1\right)$.

b) $|5 - 2x| \geq 1$
$$5 - 2x \leq -1 \quad or \quad 5 - 2x \geq 1 \qquad \text{Writing an equivalent inequality}$$
$$-2x \leq -6 \quad or \qquad -2x \geq -4 \qquad \text{Subtracting 5}$$
$$x \geq 3 \quad or \qquad x \leq 2 \qquad \text{Dividing by } -2 \text{ and reversing the inequality signs}$$

The solution is $\{x \mid x \leq 2 \text{ or } x \geq 3\}$, or $(-\infty, 2] \cup [3, \infty)$.

R.8 | *Exercise Set*

Solve.

1. $x + 6 < 5x - 6$

2. $3 - x < 4x + 7$

3. $3x - 3 + 2x \geq 1 - 7x - 9$

4. $5y - 5 + y \leq 2 - 6y - 8$

5. $14 - 5y \leq 8y - 8$

6. $8x - 7 < 6x + 3$

7. $-\frac{3}{4}x \geq -\frac{5}{8} + \frac{2}{3}x$

8. $-\frac{5}{6}x \leq \frac{3}{4} + \frac{8}{3}x$

9. $4x(x - 2) < 2(2x - 1)(x - 3)$

10. $(x + 1)(x + 2) > x(x + 1)$

11. Use a grapher to check your answer to Exercise 9.

12. Use a grapher to check your answer to Exercise 10.

Write interval notation.

13. $\{x \mid -3 \leq x \leq 3\}$

14. $\{x \mid -4 < x < 4\}$

15. $\{x \mid -14 \leq x < -11\}$

16. $\{x \mid 6 < x \leq 20\}$

17. $\{x \mid x \leq -4\}$

18. $\{x \mid x > -5\}$

19. $\{x \mid x < 3.8\}$

20. $\{x \mid x \geq \sqrt{3}\}$

21. $\{x \mid x \neq 7\}$

22. $\{x \mid x \neq -3\}$

Write interval notation for each graph.

23.

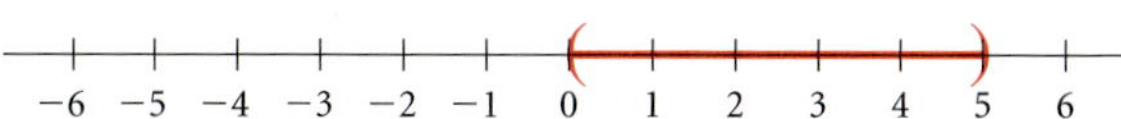

24.

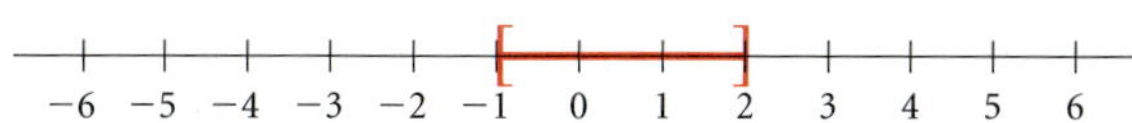

25.

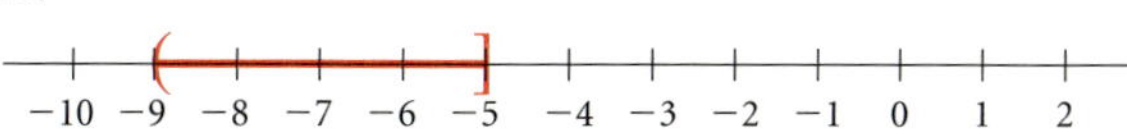

26.

27.

28.

29.

30.

Solve. Write interval notation.

31. $-2 \leq x + 1 < 4$

32. $-3 < x + 2 \leq 5$

33. $5 \leq x - 3 \leq 7$

34. $-1 < x - 4 < 7$

35. $-3 \leq x + 4 \leq 3$

36. $-5 < x + 2 < 15$

37. $-2 < 2x + 1 < 5$

38. $-3 \leq 5x + 1 \leq 3$

39. $-4 \leq 6 - 2x < 4$

40. $-3 < 1 - 2x \leq 3$

41. $-5 < \frac{1}{2}(3x + 1) \leq 7$

42. $\frac{2}{3} \leq -\frac{4}{5}(x - 3) < 1$

43. $3x \leq -6 \text{ or } x - 1 > 0$

44. $2x < 8 \text{ or } x + 3 \geq 10$

45. $2x + 3 \leq -4 \text{ or } 2x + 3 \geq 4$

46. $3x - 1 < -5 \text{ or } 3x - 1 > 5$

47. $2x - 20 < -0.8 \text{ or } 2x - 20 > 0.8$

48. $5x + 11 \leq -4 \text{ or } 5x + 11 \geq 4$

49. $x + 14 \leq -\frac{1}{4} \text{ or } x + 14 \geq \frac{1}{4}$

50. $x - 9 < -\frac{1}{2} \text{ or } x - 9 > \frac{1}{2}$

51. Use a grapher to check your answer to Exercise 39.

52. Use a grapher to check your answer to Exercise 40.

Solve and graph the solution set.

53. $|x| < 7$

54. $|x| \leq 4.5$

55. $|x| \geq 4.5$

56. $|x| > 7$

Solve.

57. $|x + 8| < 9$

58. $|x + 6| \leq 10$

59. $|x + 8| \geq 9$

60. $|x + 6| > 10$

61. $\left| x - \frac{1}{4} \right| < \frac{1}{2}$

62. $|x - 0.5| \leq 0.2$

63. $|3x| < 1$

64. $|5x| \leq 4$

65. $|2x + 3| \leq 9$

66. $|3x + 4| < 13$

67. $|x - 5| > 0.1$

68. $|x - 7| \geq 0.4$

69. $|6 - 4x| \leq 8$

70. $|5 - 2x| > 10$

71. $\left| x + \frac{2}{3} \right| \leq \frac{5}{3}$

72. $\left| x + \frac{3}{4} \right| < \frac{1}{4}$

73. $\left| \dfrac{2x + 1}{3} \right| > 5$

74. $\left| \dfrac{2x - 1}{3} \right| \geq \dfrac{5}{6}$

75. $|2x - 4| < -5$ **76.** $|3x + 5| < 0$

77. Use a grapher to check your answer to Exercise 59.

78. Use a grapher to check your answer to Exercise 60.

Synthesis

79. ◈ Explain why $|x| < p$ has no solution for $p \leq 0$.

80. ◈ Explain why all real numbers are solutions of $|x| > p$, for $p < 0$.

81. ◈ Give a graphical explanation why

$$-\tfrac{1}{2}x + 1 > |x - 5|$$

has no solution.

Solve.

82. $x \leq 3x - 2 \leq 2 - x$

83. $2x \leq 5 - 7x < 7 + x$

84. $|x + 2| \leq |x - 5|$

85. $|3x - 1| > 5x - 2$

86. $|x| + |x + 1| < 10$

87. $(x - 3)^{2/3} = 4^{1/3}$

88. $|x - 3| + |2x + 5| > 6$

89. $|p - 4| + |p + 4| < 8$

R.9

Modeling and Applications

• *Solve applied problems.*

Mathematical techniques are used to answer questions arising from real-world situations. Equations and inequalities *model* many of these applications.

Solving Applied Problems

Although we will not use it in all of our problem solving, the following strategy is of great assistance.

> ### Five Steps for Problem Solving
>
> **1. Familiarize** yourself with the problem situation. If the problem is presented in words, then, of course, this means to read carefully. Some or all of the following can also be helpful.
>
> **a)** Make a drawing, if it makes sense to do so.
>
> **b)** Make a written list of the known facts and a list of what you wish to find out.
>
> **c)** Assign variables to represent unknown quantities.
>
> **d)** Organize the information in a chart or a table.
>
> **e)** Find further information. Look up a formula or consult a reference book or an expert in the field.
>
> **f)** Guess or estimate the answer and check your guess or estimate.

2. **Translate** the problem situation to mathematical language or symbolism. For most of the problems you will encounter in algebra, this means to write one or more equations, but sometimes an inequality or some other mathematical symbolism may be appropriate.

3. **Carry out** some type of mathematical manipulation. Use your mathematical skills to find a possible solution. In algebra, this usually means to solve an equation, an inequality, or a system of equations. The solution might also be done using a grapher.

4. **Check** to see whether your possible solution actually fits the problem situation and is thus really a solution of the problem. You might be able to solve an equation, but the solution(s) of the equation might or might not be solution(s) of the original problem.

5. **State** the answer clearly using a complete sentence.

Example 1 *U.S. versus Australian Vacation Time.* Employees of U.S. businesses receive shorter vacation times than their counterparts in other countries. For example, after one year of service, workers in Australia receive on average twice as many vacation days as U.S. workers. Together, an Australian worker and a U.S. worker have a total of 30 vacation days per year. Find the number of days each employee has. (*Source: Business Week*, September 4, 1995)

SOLUTION

1. **Familiarize.** Suppose the U.S. worker has 8 vacation days per year. Then the Australian worker would have $2 \cdot 8$, or 16 days, and together they have $8 + 16$, or 24 days. This estimate is too low.

   If the U.S. worker has 12 vacation days per year, then the Australian worker has $2 \cdot 12$, or 24 days, and together they have $12 + 24$, or 36 vacation days. This estimate is too high. However, we can determine from these estimates that the correct number of days is between 8 and 12 for the U.S. worker and between 16 and 24 for the Australian worker.

   We let $x =$ the number of vacation days that the U.S. employee has. Then $2x =$ the number of vacation days that the Australian employee has.

2. **Translate.** We translate as follows:

$$
\underbrace{\begin{array}{c}\text{The number of the}\\ \text{U.S. employee's}\\ \text{vacation days}\end{array}}_{x} + \underbrace{\begin{array}{c}\text{The number of the}\\ \text{Australian employee's}\\ \text{vacation days}\end{array}}_{2x} = \underbrace{30 \text{ days.}}_{30}
$$

3. Carry out.

We solve the equation:

$$x + 2x = 30$$
$$3x = 30$$
$$x = 10.$$

Graph $y_1 = x + 2x$ and $y_2 = 30$ and find the first coordinate of the point of intersection of the graphs.

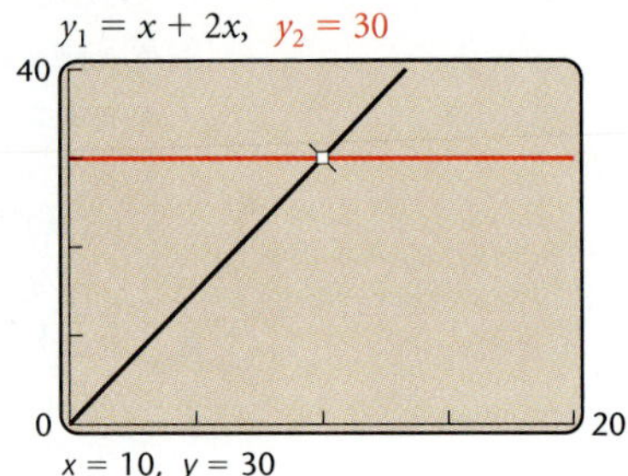

4. Check. When the U.S. employee has 10 vacation days per year, the Australian employee has $2 \cdot 10$, or 20 days. Together they have $10 + 20$, or 30 days. The answer checks.

5. State. After one year of service, U.S. employees get an average of 10 vacation days and Australian employees get an average of 20 vacation days per year.

In some applications we need to use a formula that describes the relationship between variables. When a situation involves distance, speed, and time, for example, we need to recall the **distance formula**:

$d = rt$,   where $d =$ distance, $r =$ rate (or speed), and $t =$ time.

Example 2   *Speed.* A 1996 BMW M3 leaves a town on the Autobahn traveling at its top speed of 237 km/h. Fifteen minutes later, a 1995 Aston Martin DB7 leaves the same town and follows the same route at its top speed of 266 km/h. How long will it take the Aston Martin to overtake the BMW? (*Sources: Car and Driver,* August 1995 and February 1995)

SOLUTION

1. Familiarize. We make a drawing showing both the known and the unknown information. We let $t =$ the time, in hours, that the BMW

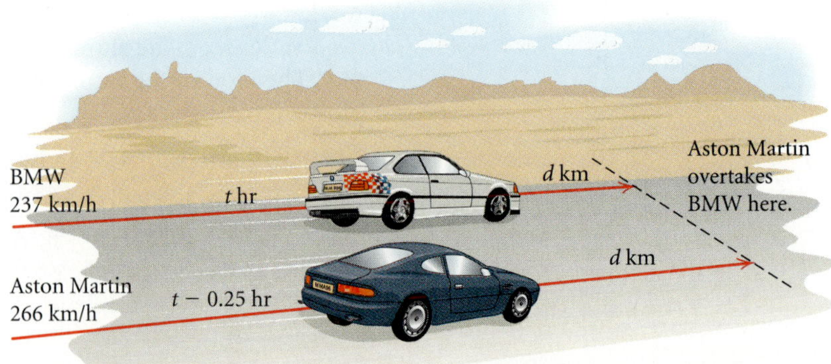

travels before being overtaken by the Aston Martin. Since the Aston Martin leaves 15 min, or 0.25 hr, after the BMW, it will travel for $t - 0.25$ hours before overtaking the BMW. The two cars have traveled the same distance, d, when one overtakes the other.

We can also organize the information in a table.

$$d \;=\; r \;\cdot\; t$$

	DISTANCE	RATE	TIME	
BMW	d	237	t	$\longrightarrow$ $d = 237t$
ASTON MARTIN	d	266	$t - 0.25$	$\longrightarrow$ $d = 266(t - 0.25)$

2. Translate. Using the formula $d = rt$ in each row of the table, we get two expressions for d:

$$d = 237t \quad \text{and} \quad d = 266(t - 0.25).$$

Thus we have the equation

$$237t = 266(t - 0.25).$$

3. Carry out.

We solve the equation:

$$237t = 266(t - 0.25)$$

$$237t = 266t - 66.5 \qquad \text{Using the distributive property}$$

$$-29t = -66.5 \qquad \text{Subtracting } 266t \text{ on both sides}$$

$$t \approx 2.3. \qquad \text{Dividing by } -29$$

We replace t with x and d with y, graph $y_1 = 237x$ and $y_2 = 266(x - 0.25)$, and find the point of intersection of the graphs.

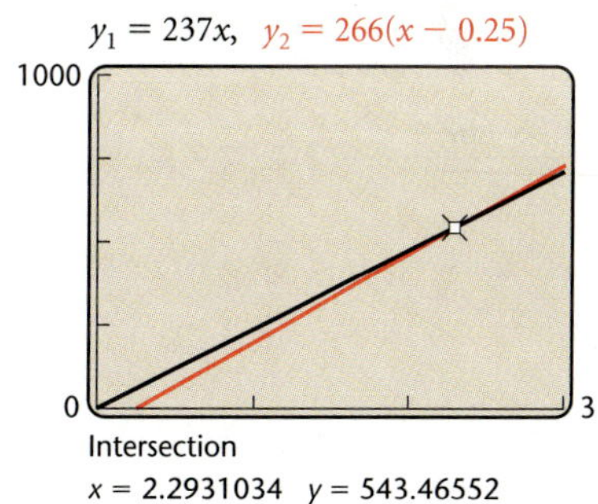

4. Check. In 2.3 hr, the BMW travels $237(2.3)$, or 545.1 km. In $2.3 - 0.25$, or 2.05 hr, the Aston Martin travels $266(2.05)$, or 545.3 km. Since 545.1 km $\approx$ 545.3 km, our answer checks. (The slight difference occurs because we rounded the value of t.)

5. State. It will take the Aston Martin about 2.3 hr to overtake the BMW.

Example 3 *Test Scores.* Kecia is taking an economics course in which there are to be four tests. To get a B or better, she must average at least 80 on the four tests. She has scores of 85, 76, and 74 on the first three tests. What scores on the last test will give her a B or better?

SOLUTION

1. Familiarize. Suppose Kecia gets an 82 on the fourth test. The average of the four scores is their sum divided by the number of tests, 4:

$$\frac{85 + 76 + 74 + 82}{4} = \frac{317}{4} = 79.25.$$

Since the average must be *at least* 80 (that is, greater than or equal to 80), a score of 82 on the fourth test will not give Kecia a B.

Let's make another estimate. If Kecia gets an 88 on the fourth test, her average will be

$$\frac{85 + 76 + 74 + 88}{4} = \frac{323}{4} = 80.75.$$

This score will give her a B.

These two estimates tell us that the lowest score Kecia can get on the fourth test to ensure a grade of B or better is some number between 82 and 88. In order to find *all* scores that will give her a B or better, we translate to an inequality. We let $s =$ Kecia's score on the fourth test.

2. **Translate.** Since the average must be *at least* 80, we translate to the inequality

$$\frac{85 + 76 + 74 + s}{4} \geq 80.$$

3. **Carry out.**

We solve the inequality:

$$4\left(\frac{85 + 76 + 74 + s}{4}\right) \geq 4 \cdot 80$$

   Multiplying by 4 to clear the fraction

$$85 + 76 + 74 + s \geq 320$$

$$235 + s \geq 320$$

   Combining like terms

$$s \geq 85.$$

   Subtracting 235 on both sides

Graph

$$y_1 = \frac{85 + 76 + 74 + x}{4} \quad \text{and} \quad y_2 = 80$$

and determine the values of x for which $y_1 \geq y_2$.

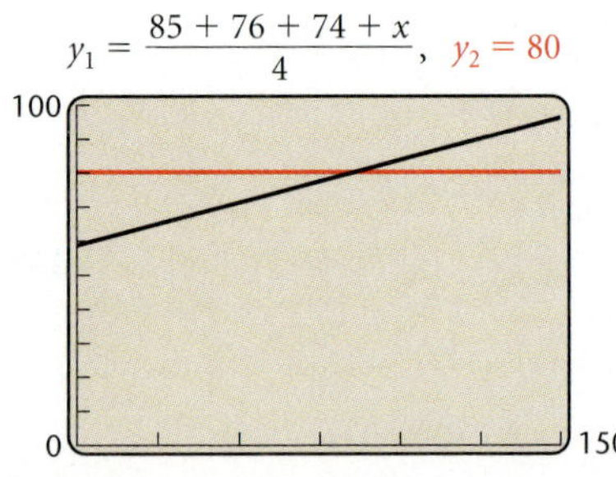

Note that $y_1 \geq y_2$ for $x \geq 85$.

4. **Check.** We can obtain a partial check by substituting a number less than 85 and then a number greater than 85 in the original inequality. We did this in the *Familiarize* step. Our answer appears to be correct.

5. **State.** A score of 85 or higher will give Kecia a B or better.

R.9 Exercise Set

Solve.

1. *Dow Chemical.* In 1994, the Dow Chemical Company spent a total of $4.071 billion on wages and salaries and on employee benefits. The amount spent on wages and salaries was $2.407 billion more than the amount spent on employee benefits. Find the amount spent in each category. (*Source*: Dow Chemical Company, 1994 Annual Report)

2. *Corporate Fax Expense.* The average *Fortune 500* company spends $13 million annually for local and domestic long-distance fax service. The amount spent for local faxes is about one third the amount spent for domestic long-distance faxes. Find the amount spent for each type of fax service. (*Source*: Gallup survey of *Fortune 500* companies for Pitney Bowes)

3. *Test Plot Dimensions.* A flower seed company has a rectangular test plot with a perimeter of 322 m. The length is 25 m more than the width. Find the dimensions of the plot.

4. *Garden Dimensions.* The children at Tiny Tots Day Care plant a rectangular vegetable garden with a perimeter of 39 m. The length is twice the width. Find the dimensions of the garden.

Time of a Free Fall. *The formula $s = 16t^2$ is used to approximate the distance s, in feet, that an object falls freely from rest in t seconds.*

5. The Warszawa Radio Mast in Poland, at 2120 ft, is the world's tallest structure. How long would it take an object falling freely from the top to reach the ground? (*Source*: David Crystal, ed., *The Cambridge Fact Finder.* Cambridge: Cambridge University Press, p. 526)

6. The tallest structure in the United States, at 2063 ft, is the KTHI-TV tower in North Dakota. How long would it take an object falling freely from the top to reach the ground? (*Source*: David Crystal, ed., *The Cambridge Fact Finder.* Cambridge: Cambridge University Press, p. 526)

7. *Box Construction.* An open box is made from a 10-cm by 20-cm piece of tin by cutting a square from each corner and folding up the edges. The area of the resulting base is 96 cm². What is the length of the sides of the squares?

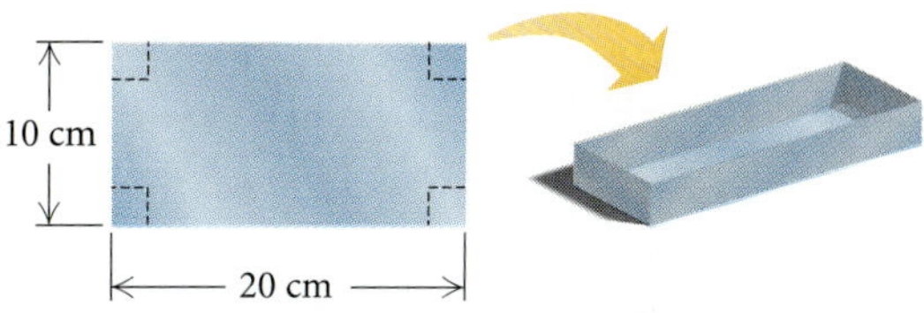

8. *Picture Frame Dimensions.* The frame of a picture is 28 cm by 32 cm outside and is of uniform width. What is the width of the frame if 192 cm² of the picture shows?

9. *Angle Measure.* In triangle ABC, angle B is five times as large as angle A. The measure of angle C is 2° less than that of angle A. Find the measures of the angles. (*Hint*: The sum of the angle measures is 180°.)

10. *Angle Measure.* In triangle ABC, angle B is twice as large as angle A. Angle C measures 20° more than angle A. Find the measures of the angles.

11. *HMO Enrollment.* The number of enrollees in health maintenance organizations (HMOs) in Kentucky in a recent year was 416,300. This was a 48.4% increase over the previous year. What was the former number of enrollees? (*Source*: Marion Merrell Dow Managed Care Department)

12. *Moving Costs.* The average cost of moving a distance of 1255 mi is \$1035 if you handle the move yourself. This is about 52% of the cost of hiring a professional moving firm. What is the cost of the professionally handled move? (*Source:* American Movers Conference)

13. *Bicycling Speed.* Castulo and Linette leave a campsite, Castulo biking due north and Linette biking due east. Castulo bikes 7 km/h slower than Linette. After 4 hr, they are 68 km apart. Find the speed of each bicyclist.

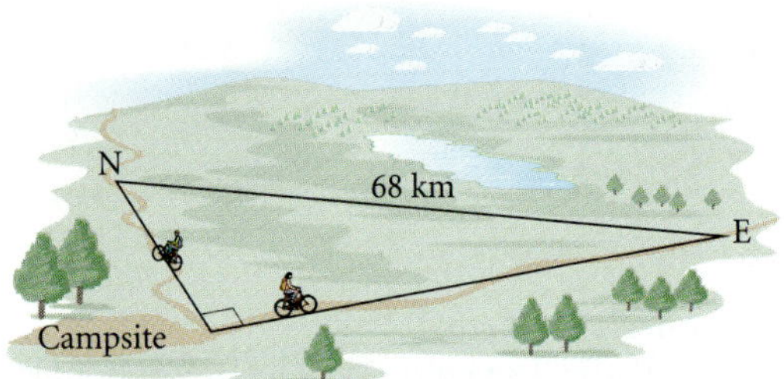

14. *Boat Speed.* A Bayline Cruises sightseeing boat travels 36 mi upriver and 36 mi back in 5 hr. The speed of the river's current is 3 mph. Find the speed of the boat in still water.

15. *Train Speeds.* The speed of an Amtrak passenger train is 14 mph faster than the speed of a Central Railway freight train. The passenger train travels 400 mi in the same time it takes the freight train to travel 330 mi. Find the speed of each train.

16. *Truck Speed.* For the first 80 mi of a trip, Micah's pickup truck traveled 10 mph slower than it did for the remaining 50 mi. The total time for the trip was 3 hr. Find the speed of the truck on each part of the trip.

17. *Distance Traveled.* An airplane leaves Chicago for Cleveland at a speed of 475 mph. Twenty minutes later, a plane going to Chicago leaves Cleveland, which is 350 mi from Chicago, at a speed of 500 mph. When they meet, how far are they from Cleveland?

18. *Distance Traveled.* A private airplane leaves Midway airport and flies due east at a speed of 180 km/h. Two hours later, a jet leaves Midway and flies due east at a speed of 900 km/h. How far from the airport will the jet overtake the private plane?

19. *Compound Interest.* An investment is made at 8%, compounded annually. It grows to \$702 at the end of 1 yr. How much was originally invested?

20. *Compound Interest.* An investment made at 9%, compounded annually, grows to \$926.50 at the end of 1 yr. How much was originally invested?

21. *Test Scores.* Jason is taking a literature course in which there are to be four tests. He has scores of 82, 73, and 79 on the first three tests. To get a B or better, he must average at least 80 on the four tests. What scores on the last test will give Jason a B or better?

22. *Test Scores.* Monica is taking an accounting course in which there are to be five tests, each worth 100 points. She has scores of 93, 89, and 95 on the first three tests. To get an A, she must score a total of at least 450 points. What scores on the fourth test will keep her eligible for an A?

23. *Car Rental.* A car can be rented for \$35 per day with unlimited mileage or for \$30 per day plus 15¢ per mile. For what daily mileages would the unlimited mileage plan cost less?

24. *Car Rental.* A car can be rented for \$28 per day plus 18¢ per mile. A salesperson is on a daily car-rental budget of \$91. What mileages will allow her to stay within the budget?

25. *Cost of a Service Call.* Dennis Appliance Service charges \$50 plus \$20 per hour for a service call. Economy Appliance charges \$45 per hour for a service call. For what lengths of time is Economy Appliance more expensive?

26. *Cost of a Taxi Ride.* A taxi ride with Mason's Cab costs \$1.50 for the first mile and 45¢ for each additional mile. Bell's Taxi charges 60¢ per mile. For what distances is Mason's Cab less expensive?

27. *Salary Plans.* On your new job, you can choose to be paid in one of two ways:

> *Plan A:* A salary of \$1600 per month plus a commission of 5% of gross sales;

> *Plan B:* A salary of \$1800 per month plus a commission of 7% of gross sales over \$8000.

For what gross sales is plan B better than plan A, assuming that gross sales are always more than \$8000?

28. *Insurance Plans.* An insurance company offers two plans. With the first plan, the employee pays the first \$500 of medical bills and the insurance company pays 80% of the rest. With the second plan, the employee pays the first \$900 of medical bills and the insurance company pays 90% of the rest. For what amount of medical bills will the second plan cost the employee less?

Synthesis

29. ◈ Write a problem for a classmate to solve. Devise the problem so that the solution is "The dimensions of the rectangle are 36 in. by 18 in."

30. ◆ Explain why, when solving a problem, it is necessary to check your solution in the original problem situation.

31. *Speed.*   A compact car is driven 144 mi. If it had traveled 4 mph faster, it could have made the trip in $\frac{1}{2}$ hr less time. What was the speed of the car?

32. *Speed.*   A pickup truck is driven 280 mi. If it had traveled 5 mph faster, it could have made the trip in 1 hr less time. What was the speed of the truck?

33. *Average Speed.*   Krista drove 3 hr on a freeway at a speed of 55 mph and then drove 10 mi in the city at 35 mph. What was the average speed? (*Average speed* is defined as total distance divided by total time.)

34. *Average Speed.*   For the first 100 km of a 200-km trip, Bert drove at a speed of 40 km/h. For the second half of the trip, he drove at 60 km/h. What was the average speed for the entire trip? (See Exercise 33.)

35. *Average Speed.*   Luke drove half the distance of a trip at a speed of 40 mph. At what speed would he have to drive for the rest of the distance so that the average speed for the entire trip would be 45 mph and the trip would be completed in 1 hr? (See Exercise 33.)

36. *Distance Traveled.*   Madison drives to work at a speed of 45 mph and arrives 1 min early. At 40 mph, she would arrive 1 min late. How long is her trip to work?

37. *Distribution of Money.*   Suppose your father gives half of all the money in his pockets to your mother. Then he gives one fourth of what is left to your sister, and one third of what is left after that to your brother. He then gives you half of what is left, which happens to be $2. How much was in his pocket at the outset?

38. *Pie Sales.*   Jared walks into a bakery and says to the owner, "I will buy half of all the pies in the store, plus half a pie." The sale is made. Phyllis then comes into the store and makes the same statement and purchase. Then Dexter does likewise. The owner then has exactly one pie left. How many pies did the owner have in the store at the outset?

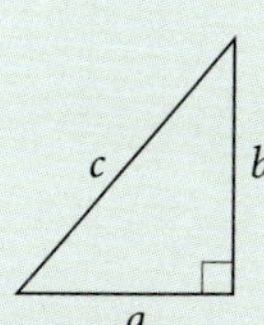

CHAPTER

R *Summary and Review*

Important Properties and Formulas

Properties of Absolute Value

$$|a| \geq 0$$

$$|ab| = |a| \cdot |b|$$

$$\left|\frac{a}{b}\right| = \frac{|a|}{|b|} \quad (b \neq 0)$$

$$|-a| = |a|$$

Pythagorean Theorem

$$a^2 + b^2 = c^2$$

Properties of the Real Numbers

Commutative: $a + b = b + a; \; ab = ba$

Associative: $a + (b + c) = (a + b) + c;$
$a(bc) = (ab)c$

Additive Identity: $a + 0 = 0 + a = a$

Additive Inverse: $-a + a = a + (-a) = 0$

Multiplicative Identity: $a \cdot 1 = 1 \cdot a = a$

Multiplicative Inverse: $a \cdot \dfrac{1}{a} = 1 \quad (a \neq 0)$

Distributive: $a(b + c) = ab + ac$

(continued)

Properties of Exponents

For any real numbers a and b and any integers m and n, assuming 0 is not raised to a nonpositive power:

The Product Rule: $a^m \cdot a^n = a^{m+n}$

The Quotient Rule: $\dfrac{a^m}{a^n} = a^{m-n}$ $(a \neq 0)$

The Power Rule: $(a^m)^n = a^{mn}$

Raising a Product to a Power:
$(ab)^m = a^m b^m$

Raising a Quotient to a Power:
$$\left(\frac{a}{b}\right)^m = \frac{a^m}{b^m} \quad (b \neq 0)$$

Special Products of Binomials

$(A + B)^2 = A^2 + 2AB + B^2$

$(A - B)^2 = A^2 - 2AB + B^2$

$(A + B)(A - B) = A^2 - B^2$

Sum or Difference of Cubes

$A^3 + B^3 = (A + B)(A^2 - AB + B^2)$

$A^3 - B^3 = (A - B)(A^2 + AB + B^2)$

Equation-Solving Principles

The Addition Principle: If $a = b$ is true, then $a + c = b + c$ is true.

The Multiplication Principle: If $a = b$ is true, then $ac = bc$ is true.

The Principle of Zero Products: If $ab = 0$ is true, then $a = 0$ or $b = 0$, and if $a = 0$ or $b = 0$, then $ab = 0$.

The Principle of Square Roots: If $x^2 = k$, then $x = \sqrt{k}$ or $x = -\sqrt{k}$.

The Principle of Powers: For any positive integer n, if $a = b$ is true, then $a^n = b^n$ is true.

Principles for Solving Inequalities

The Addition Principle for Inequalities: If $a < b$ is true, then $a + c < b + c$ is true.

The Multiplication Principle for Inequalities: If $a < b$ and $c > 0$ are true, then $ac < bc$ is true.
If $a < b$ and $c < 0$ are true, then $ac > bc$ is true.

Similar statements hold for $\leq$.

Equations and Inequalities with Absolute Value

For $a > 0$,

$|x| = a \longrightarrow x = -a \quad or \quad x = a$

$|x| < a \longrightarrow -a < x < a$

$|x| > a \longrightarrow x < -a \quad or \quad x > a$

Compound Interest Formula

$$A = P\left(1 + \frac{i}{n}\right)^{nt}$$

Properties of Radicals

Let a and b be any real numbers or expressions for which the given roots exist. For any natural numbers m and n $(n \neq 1)$:

If n is even, $\sqrt[n]{a^n} = |a|$.

If n is odd, $\sqrt[n]{a^n} = a$.

$\sqrt[n]{a} \cdot \sqrt[n]{b} = \sqrt[n]{ab}$.

$\sqrt[n]{\dfrac{a}{b}} = \dfrac{\sqrt[n]{a}}{\sqrt[n]{b}} \quad (b \neq 0)$.

$\sqrt[n]{a^m} = (\sqrt[n]{a})^m$.

Five Steps for Problem Solving

1. Familiarize.
2. Translate.
3. Carry out.
4. Check.
5. State.

REVIEW EXERCISES

Consider the numbers -43.89, 12, -3, $-\frac{1}{5}$, $\sqrt{7}$, $\sqrt[3]{10}$, -1, $-\frac{4}{3}$, $7\frac{2}{3}$, -19, 31, 0.

1. Which are integers?

2. Which are natural numbers?

3. Which are rational numbers?

4. Which are real numbers?

5. Which are irrational numbers?

6. Which are whole numbers?

Simplify.

7. $|-3.5|$ **8.** $\left|-\frac{5}{6}a^2b\right|$

9. Find the distance between -7 and 3 on the number line.

10. Determine graphically whether $|x^3| = x^3$ appears to be an identity.

Compute.

11. $5^3 - [2(4^2 - 3^2 - 6)]^3$

12. $\dfrac{3^4 - (6 - 7)^4}{2^3 - 2^4}$

Convert to decimal notation.

13. 3.261×10^6 **14.** 4.1×10^{-4}

Convert to scientific notation.

15. 0.01432 **16.** $43{,}210$

Compute. Write scientific notation for the answer.

17. $\dfrac{2.5 \times 10^{-8}}{3.2 \times 10^{13}}$

18. $(8.4 \times 10^{-17})(6.5 \times 10^{-16})$

Simplify.

19. $(7a^2b^4)(-2a^{-4}b^3)$

20. $\dfrac{54x^6y^{-4}z^2}{9x^{-3}y^2z^{-4}}$

21. $\sqrt[4]{81}$

22. $\sqrt[5]{-32}$

23. $\dfrac{b - a^{-1}}{a - b^{-1}}$

24. $\dfrac{\dfrac{x^2}{y} + \dfrac{y^2}{x}}{y^2 - xy + x^2}$

25. $(\sqrt{3} - \sqrt{7})(\sqrt{3} + \sqrt{7})$

26. $(5x^2 - \sqrt{2})^2$

27. $8\sqrt{5} + \dfrac{25}{\sqrt{5}}$

28. $(x + t)(x^2 - xt + t^2)$

29. $(5a + 4b)^3$

30. $(5xy^4 - 7xy^2 + 4x^2 - 3) - (-3xy^4 + 2xy^2 - 2y + 4)$

Factor.

31. $x^3 + 2x^2 - 3x - 6$

32. $12a^3 - 27ab^4$

33. $24x + 144 + x^2$

34. $9x^3 + 35x^2 - 4x$

35. $8x^3 - 1$

36. $27x^6 + 125y^6$

37. $6x^3 + 48$

38. $4x^3 - 4x^2 - 9x + 9$

39. $9x^2 - 30x + 25$

40. $18x^2 - 3x + 6$

41. $9x^2 + 6xy + y^2 + 9x + 3y - 4$

42. Divide and simplify:
$$\frac{3x^2 - 12}{x^2 + 4x + 4} \div \frac{x - 2}{x + 2}.$$

43. Subtract and simplify:
$$\frac{x}{x^2 + 9x + 20} - \frac{4}{x^2 + 7x + 12}.$$

Write an expression containing a single radical.

44. $\sqrt{y^5}\,\sqrt[3]{y^2}$

45. $\dfrac{\sqrt{(a + b)^3}\,\sqrt[3]{a + b}}{\sqrt[6]{(a + b)^7}}$

46. Convert to radical notation: $b^{7/5}$.

47. Convert to exponential notation and simplify:
$$\sqrt[8]{\frac{m^{32}n^{16}}{3^8}}.$$

48. Rationalize the numerator:
$$\frac{\sqrt{x} - \sqrt{y}}{\sqrt{x} + \sqrt{y}}.$$

Factor and simplify.

49. $2x^{1/2}y^{-3/4} - 3x^{-1/2}y^{1/4}$

50. $(x - 2)^{-3/4}(3x + 5)^{5/2} + (x - 2)^{1/4}(3x + 5)^{3/2}$

51. How long is a guy wire reaching from the top of a 17-ft pole to a point on the ground 8 ft from the pole?

Solve.

52. $4y - 5 = 11$

53. $5(3x + 1) = 2(x - 4)$

54. $(2y + 5)(3y - 1) = 0$

55. $3x^2 + 2x = 8$

56. $\dfrac{5}{2x + 3} + \dfrac{1}{x - 6} = 0$

57. $\dfrac{3}{8x + 1} + \dfrac{8}{2x + 5} = 1$

58. $\sqrt{5x + 1} - 1 = \sqrt{3x}$

59. $\sqrt{x - 1} - \sqrt{x - 4} = 1$

60. $|x - 4| = 3$

61. $|2y + 7| = 9$

62. $-3 \le 3x + 1 < 5$

63. $-2 < 5x - 4 \le 6$

64. $2x < -1 \ or \ x + 3 > 0$

65. $3x + 7 \le 2 \ or \ 2x + 3 \ge 5$

66. $|6x - 1| < 5$

67. $|x + 4| \ge 2$

68. Solve $v = \sqrt{2gh}$ for h.

69. Solve $\dfrac{1}{a} + \dfrac{1}{b} = \dfrac{1}{t}$ for t.

70. *Legs of a Right Triangle.* The hypotenuse of a triangle is 50 ft. One leg is 10 ft longer than the other. What are the lengths of the legs?

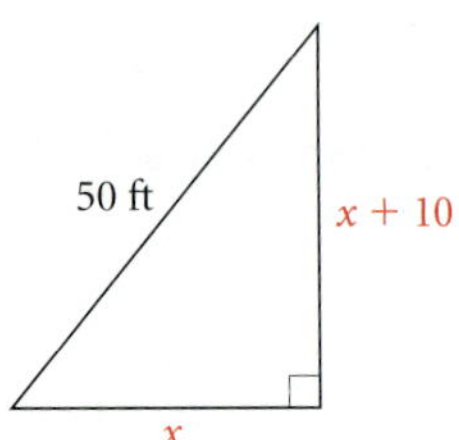

71. *Test Scores.* Jeremiah scores 73 and 79 on two tests. If the third test counts as though it were two tests, what score must he make on the third test so the average will be 85?

72. *Motion.* A Riverboat Cruise Line boat travels 8 mi upstream and 8 mi downstream. The total time for both parts of the trip is 3 hr. The speed of the stream is 2 mph. What is the speed of the boat in still water?

73. *Motion.* Two freight trains leave the same city at right angles. The first train travels at a speed of 60 km/h. In 1 hr, the trains are 100 km apart. How fast is the second train traveling?

74. *Car Rental.* A car can be rented for $32 per day with unlimited mileage or for $28 per day plus 16¢ per mile. For what daily mileages would the unlimited mileage plan cost less?

75. *Cost of a Service Call.* Harris Heating and Cooling charges $40 plus $25 per hour for a service call. Quality Cool charges $45 per hour for a service call. For what lengths of time is Quality Cool less expensive?

Synthesis

76. ◈ Write a problem for a classmate to solve. Devise the problem so that the solution is "Nisha must get a score of at least 84 on the fourth test in order to get a B or better in the course."

77. ◈ Explain the difference between equivalent expressions and equivalent equations.

Multiply. Assume that all exponents are integers.

78. $(x^n + 10)(x^n - 4)$

79. $(t^a + t^{-a})^2$

80. $(y^b - z^c)(y^b + z^c)$

81. $(a^n - b^n)^3$

Factor.

82. $y^{2n} + 16y^n + 64$

83. $x^{2t} - 3x^t - 28$

84. $m^{6n} - m^{3n}$

85. Solve: $\sqrt{\sqrt{\sqrt{x}}} = 2$.

Graphs, Functions, and Models

1

During a thunderstorm, it is possible to calculate how far away, y (in miles), lightning is when the sound of thunder arrives x seconds after the lightning has been sighted. The relationship between x and y can be described by a function, $y = f(x) = \frac{1}{5}x$.

This weather application is just one example of the widespread use of graphs in today's society. In this chapter, we will focus attention on functions and equations with graphs that can be used to model real-life situations mathematically. Such models are invaluable when doing analyses and making predictions in fields ranging from astronomy to zoology.

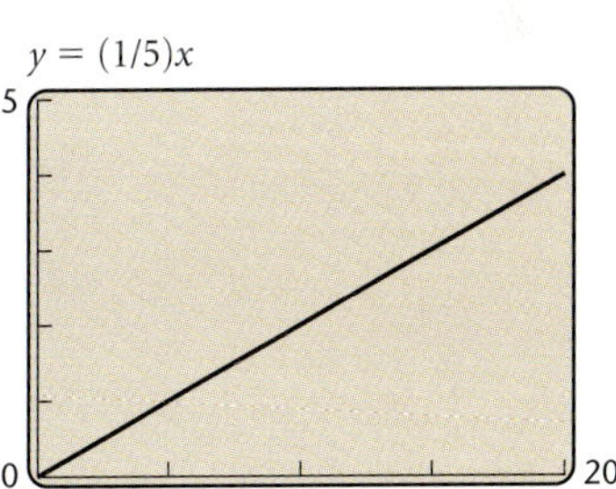

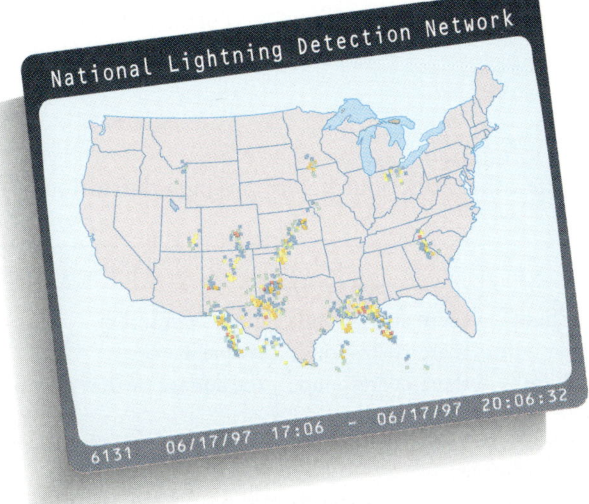

1.1
Functions, Graphs, and Graphers

- *Determine whether a correspondence or a relation is a function.*
- *Find function values, or outputs, using a formula.*
- *Find the domain and the range of a function.*
- *Determine whether a graph is that of a function.*
- *Solve application problems using functions.*

We now focus attention on a concept that is fundamental to many areas of mathematics—the idea of a *function*.

Functions

We first consider an application.

Lightning–Time–Thunder Distance. During a thunderstorm, it is possible to calculate how far away, y (in miles), lightning is when the sound of thunder arrives x seconds after the lightning has been sighted. We can examine the relationship between x and y in several ways:

x	y	ORDERED PAIRS: (x, y)	CORRESPONDENCE
0	0	$(0, 0)$	$0 \longrightarrow 0$
1	$\frac{1}{5}$	$\left(1, \frac{1}{5}\right)$	$1 \longrightarrow \frac{1}{5}$
2	$\frac{2}{5}$	$\left(2, \frac{2}{5}\right)$	$2 \longrightarrow \frac{2}{5}$
5	1	$(5, 1)$	$5 \longrightarrow 1$
10	2	$(10, 2)$	$10 \longrightarrow 2$

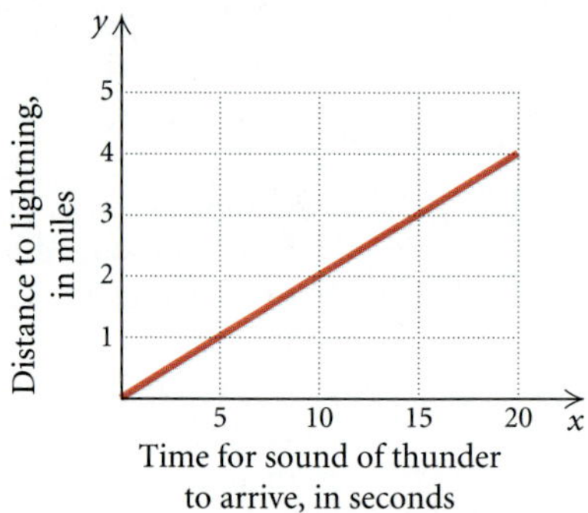

The ordered pairs imply a relationship, or correspondence, between the first and second coordinates. We can see this relationship in the graph as well. There is also an equation that describes the correspondence. It is

$$y = \frac{1}{5}x.$$

This is an example of a *function*.

Let's consider some other functions before giving a definition.

DOMAIN		RANGE
To each registered student	there corresponds	an I. D. number.
To each mountain bike sold	there corresponds	its price.
To each number between -3 and 3	there corresponds	the square of that number.

In each example, the first set is called the **domain** and the second set is called the **range**. For each member, or **element**, in the domain, there is *exactly one* member of the range to which it corresponds. Thus each registered student has exactly *one* I. D. number, each mountain bike has exactly *one* price, and each number between −3 and 3 has exactly *one* square. Each correspondence is a *function*.

> ### Function
>
> A *function* is a correspondence between a first set, called the *domain*, and a second set, called the *range*, such that each member of the domain corresponds to *exactly one* member of the range.

It is important to note that not every correspondence between two sets is a function.

Example 1   Determine whether each of the following correspondences is a function.

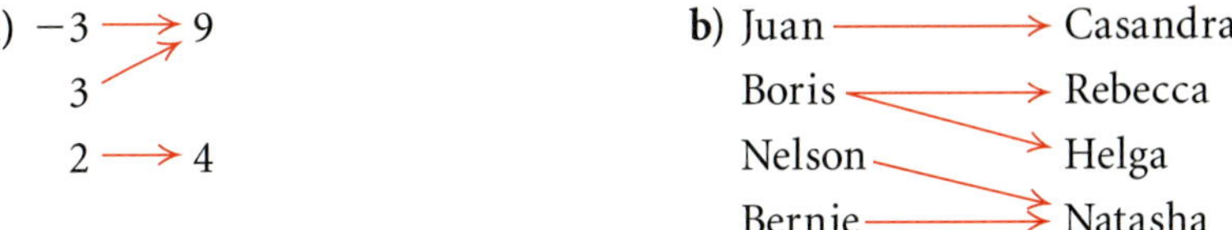

a) −3 → 9
    3
    2 → 4

b) Juan ⟶ Casandra
    Boris ⟷ Rebecca
    Nelson ⟶ Helga
    Bernie ⟶ Natasha

SOLUTION

a) This correspondence *is* a function because each member of the domain corresponds to exactly one member of the range.

b) This correspondence *is not* a function because there is a member of the domain (Boris) that is paired with two different members of the range.

Example 2   Determine whether each of the following correspondences is a function.

DOMAIN	CORRESPONDENCE	RANGE
a) Years in which a president is elected	The person elected	A set of presidents
b) The integers	Each number's cube root	A set of real numbers
c) All states in the United States	A senator from that state	The set of all U.S. senators

SOLUTION

a) This correspondence *is* a function, because in each presidential election *exactly one* president is elected.

b) This correspondence *is* a function, because each integer has *exactly one* cube root.

c) This correspondence *is not* a function, because each state can be paired with *two* different senators.

When a correspondence between two sets is not a function, it is still an example of a **relation**.

> ### Relation
>
> A *relation* is a correspondence between a first set, called the *domain,* and a second set, called the *range,* such that each member of the domain corresponds to *at least one* member of the range.

All the correspondences in Examples 1 and 2 are relations, but, as we have seen, not all are functions. Relations are sometimes written as sets of ordered pairs in which elements of the domain are the first part (coordinate) of each ordered pair and elements of the range are the second part (coordinate). For example, instead of writing $-3 \longrightarrow 9$, we write the ordered pair $(-3, 9)$.

Example 3   Determine whether each of the following relations is a function. Identify the domain and the range.

a) $\{(-2, 5), (5, 7), (0, 1), (4, -2)\}$

b) $\{(9, -5), (9, 5), (2, 4)\}$

SOLUTION

a) The relation *is* a function because no two ordered pairs have the same first coordinate and different second coordinates.

The domain is the set of all first coordinates: $\{-2, 5, 0, 4\}$.

The range is the set of all second coordinates: $\{5, 7, 1, -2\}$.

b) The relation *is not* a function because the ordered pairs $(9, -5)$ and $(9, 5)$ have the same first coordinate and different second coordinates.

The domain is the set of all first coordinates: $\{9, 2\}$.

The range is the set of all second coordinates: $\{-5, 5, 4\}$.

Notation for Functions

Functions used in mathematics are often given by equations. They generally require that certain calculations be performed in order to determine which member of the range is paired with each member of the domain. For example, in the Introduction to Graphs and Graphers, we graphed the function $y = x^2 - 5$ by doing calculations like the following:

for $x = 3$, $y = 3^2 - 5 = 4$,

for $x = 2$, $y = 2^2 - 5 = -1$,

for $x = 1$, $y = 1^2 - 5 = -4$,   and so on.

A more concise notation is often used. For $y = x^2 - 5$, the **inputs** (members of the domain) are values of x substituted into the equation. The **outputs** (members of the range) are the resulting values of y. If we call the function f, we can use x to represent an arbitrary *input* and $f(x)$—read "f of x", or "f at x", or "the value of f at x"—to represent the corresponding *output*. In this notation, the function given by $y = x^2 - 5$ is written as $f(x) = x^2 - 5$ and the above calculations would be

$$f(3) = 3^2 - 5 = 4,$$
$$f(2) = 2^2 - 5 = -1,$$
$$f(1) = 1^2 - 5 = -4.$$

Keep in mind that $f(x)$ *does not* mean $f \cdot x$.

Thus, instead of writing "when $x = 3$, the value of y is 4," we can simply write "$f(3) = 4$," which can also be read as "f of 3 is 4" or "for the input 3, the output of f is 4."

Example 4   A function f is given by $f(x) = 2x^2 - x + 3$. Find each of the following.

a) $f(0)$ **b)** $f(-7)$

c) $f(5a)$ **d)** $f(a - 4)$

SOLUTION   One way to find function values when a formula is given is to think of the formula as follows:

$$f(\blacksquare) = 2(\blacksquare)^2 - (\blacksquare) + 3.$$

Then to find an output for a given input we think: "Whatever goes in the blank on the left goes in the blank(s) on the right." This gives us a "recipe" for finding outputs. We can now solve the example.

a) $f(0) = 2 \cdot (0)^2 - 0 + 3 = 0 - 0 + 3 = 3$

b) $f(-7) = 2(-7)^2 - (-7) + 3 = 2 \cdot 49 + 7 + 3 = 108$

c) $f(5a) = 2(5a)^2 - 5a + 3 = 2 \cdot 25a^2 - 5a + 3 = 50a^2 - 5a + 3$

d) $f(a - 4) = 2(a - 4)^2 - (a - 4) + 3 = 2(a^2 - 8a + 16) - a + 4 + 3$
$$= 2a^2 - 16a + 32 - a + 4 + 3$$
$$= 2a^2 - 17a + 39$$

It is important for students preparing for calculus to be able to simplify rational expressions like

$$\frac{f(a + h) - f(a)}{h}.$$

Example 5   For the function f given by $f(x) = 2x^2 - x - 3$, construct and simplify the expression

$$\frac{f(a + h) - f(a)}{h}.$$

SOLUTION   We find $f(a + h)$ and $f(a)$:

$$f(a + h) = 2(a + h)^2 - (a + h) - 3$$
$$= 2[a^2 + 2ah + h^2] - a - h - 3$$
$$= 2a^2 + 4ah + 2h^2 - a - h - 3,$$
$$f(a) = 2a^2 - a - 3.$$

Then

$$\frac{f(a + h) - f(a)}{h}$$

$$= \frac{[2a^2 + 4ah + 2h^2 - a - h - 3] - [2a^2 - a - 3]}{h}$$

$$= \frac{2a^2 + 4ah + 2h^2 - a - h - 3 - 2a^2 + a + 3}{h} = \frac{4ah + 2h^2 - h}{h}$$

$$= \frac{h(4a + 2h - 1)}{h \cdot 1} = \frac{h}{h} \cdot \frac{4a + 2h - 1}{1} = 4a + 2h - 1.$$

Finding Domains of Functions

When a function is given by a formula, we find function values by making substitutions of numbers or inputs into the formula. When a function f, whose inputs and outputs are real numbers, is given by a formula, the *domain* is understood to be the set of all inputs for which the expression is defined as a real number. When a substitution results in an expression that is not defined as a real number, we say that the function value *does not exist* and that the number being substituted *is not* in the domain of the function.

Example 6   Find the indicated function values. Simplify, if possible.

a) $f(1)$ and $f(3)$, for $f(x) = \dfrac{1}{x - 3}$

b) $g(16)$ and $g(-7)$, for $g(x) = \sqrt{x} + 5$

SOLUTION

a) $f(1) = \dfrac{1}{1 - 3} = \dfrac{1}{-2} = -\dfrac{1}{2};$

$f(3) = \dfrac{1}{3 - 3} = \dfrac{1}{0}$

Since division by 0 is not defined, the number 3 is not in the domain of f. Thus, $f(3)$ does not exist.

b) $g(16) = \sqrt{16} + 5 = 4 + 5 = 9$;

$g(-7) = \sqrt{-7} + 5$

Since $\sqrt{-7}$ is not defined as a real number, the number -7 is not in the domain of g. Thus, $g(-7)$ does not exist.

We can see from Example 7 that inputs that make a denominator 0 or a square-root radicand negative are not in the domain of a function.

Example 7   Find the domain of each of the following functions.

a) $f(x) = \dfrac{1}{x - 3}$

b) $g(x) = \sqrt{x} + 5$

c) $h(x) = \dfrac{3x^2 - x + 7}{x^2 + 2x - 3}$

d) $p(x) = \sqrt{4 - x^2}$

e) $F(x) = x^3 + |x|$

SOLUTION

a) The input 3 results in a denominator of 0. The domain is $\{x \,|\, x \neq 3\}$, or $(-\infty, 3) \cup (3, \infty)$.

b) We can substitute any number for which the radicand is nonnegative, that is, for which $x \geq 0$. Thus the domain is $\{x \,|\, x \geq 0\}$, or the interval $[0, \infty)$.

c) Although we can substitute any real number in the numerator, we must avoid inputs that make the denominator 0. To find those inputs, we solve $x^2 + 2x - 3 = 0$, or $(x + 3)(x - 1) = 0$. Thus the domain consists of the set of all real numbers except -3 and 1, or $\{x \,|\, x \neq -3$ *and* $x \neq 1\}$, or $(-\infty, -3) \cup (-3, 1) \cup (1, \infty)$.

d) We must avoid inputs for which the radicand is negative. Thus the domain is all real numbers for which $4 - x^2 \geq 0$, or $x^2 \leq 4$. These inputs are all numbers in the interval $[-2, 2]$. You can check this by making some substitutions or, on a grapher, by looking at the graphs of $y_1 = x^2$ and $y_2 = 4$.

e) All substitutions are suitable. The domain is the set of all real numbers, $\mathbb{R}$.

Function Values and Tables on a Grapher

A grapher can be used in many different ways to find function values. We can simply do a calculation. The process is the same as that described in the Introduction to Graphs and Graphers.

With a TABLE feature, function values can be displayed for a string of numbers like $x = -3, -2.75, -2.5, -2.25, -2, -1.75$, and so on. Consider the function $f(x) = \sqrt{4 - x^2}$. Shown at left is part of a table for this function. Note that we entered the function as $y = \sqrt{(4 - x^2)}$, or $(4 - x^2)^{\wedge}0.5$. An ERROR message tells us when a number is not in the domain.

X	Y₁	
-3	ERROR	
-2.75	ERROR	
-2.5	ERROR	
-2.25	ERROR	
-2	0	
-1.75	.96825	
-1.50	1.3229	

$X = -3$

It is not easy, maybe not even possible, to determine a domain using a table, but it may point out certain numbers that are not in the domain.

Graphs of Functions

Most of the functions we study in this course and in calculus are given by formulas rather than sets of ordered pairs. We graph functions in much the same way as we do equations. We find ordered pairs (x, y) or $(x, f(x))$, look for patterns, and complete the graph.

Example 8   Graph each of the following functions.

a) $f(x) = x^2 - 5$ $\qquad$ **b)** $f(x) = x^3 - x$ $\qquad$ **c)** $f(x) = \sqrt{x + 4}$

SOLUTION   Most graphers do not use function notation "$f(x) = \ldots$" to enter a function formula. Instead, we must enter the function using "$y = \ldots$". The graphs follow.

a) $f(x) = x^2 - 5$

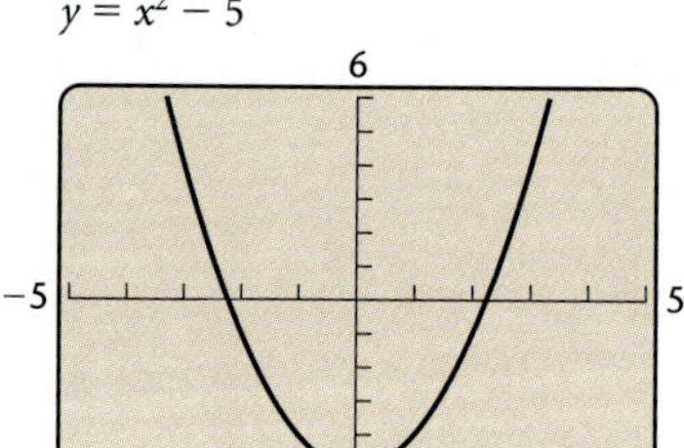

b) $f(x) = x^3 - x$

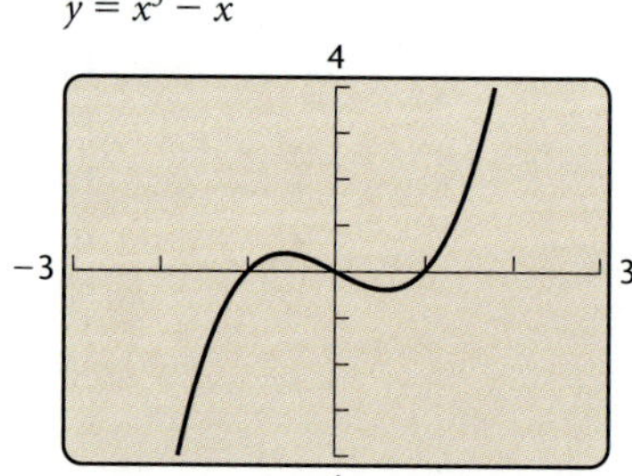

c) $f(x) = \sqrt{x + 4}$

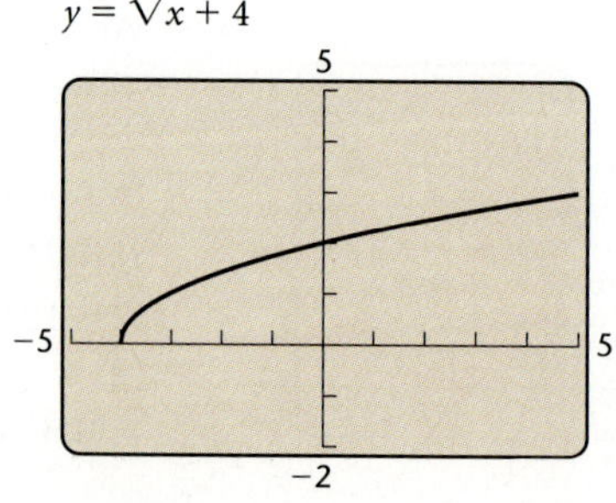

To find a function value, like $f(3)$, from a graph, we locate the input 3 on the horizontal axis, move vertically to the graph of the function, and then horizontally to find the output on the vertical axis. See the graph on the left below.

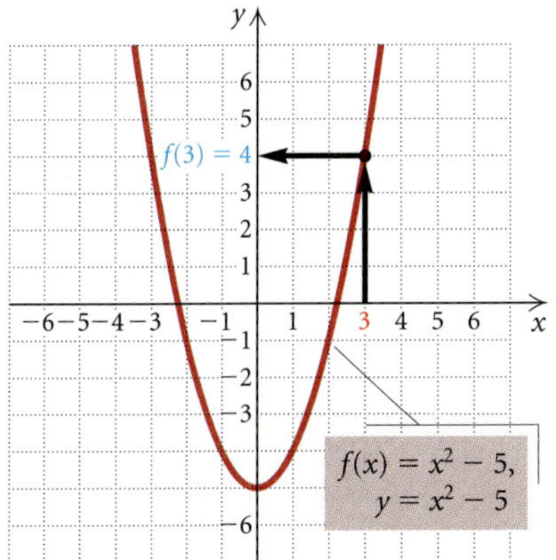

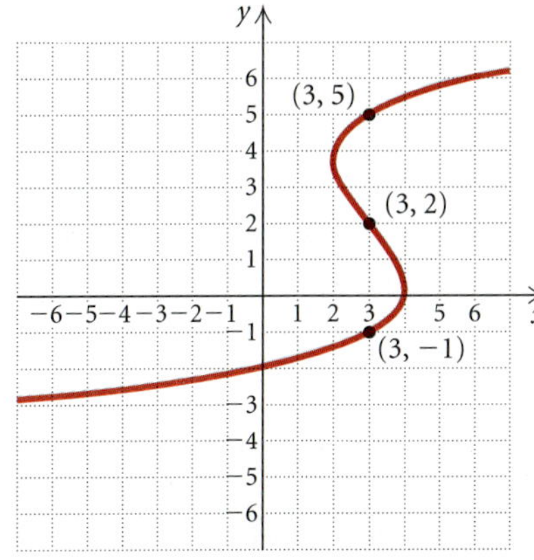

Since 3 is paired with more than one member of the range, the graph does not represent a function.

When one member of the domain is paired with two or more different members of the range, the correspondence *is not* a function. Thus, when a graph contains two or more different points with the same first coordinate, the graph cannot represent a function (see the graph on the right above). Points sharing a common first coordinate are vertically above or below each other. This leads us to the *vertical-line test*.

The Vertical-Line Test

If it is possible for a vertical line to cross a graph more than once, then the graph is not the graph of a function.

To apply the vertical-line test, we try to find a vertical line that crosses the graph more than once. If we do, then the graph is not that of a function. If we do not, then the graph is that of a function.

Example 9   Which of graphs (a) through (f) (in red) are graphs of functions?

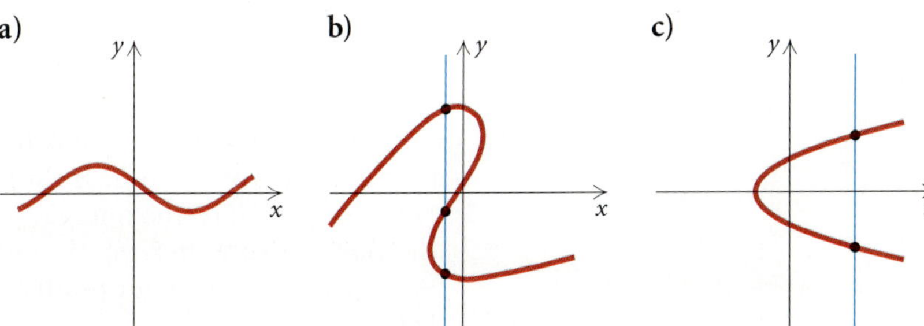

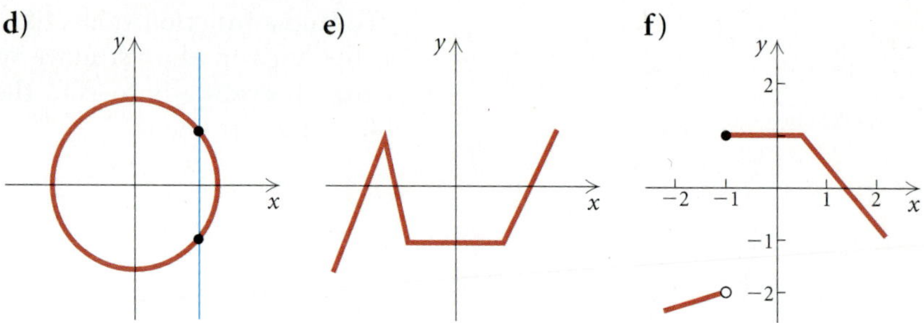

In graph (f), the solid dot shows that $(-1, 1)$ belongs to the graph. The open circle shows that $(-1, -2)$ does *not* belong to the graph.

SOLUTION   Graphs (a), (e), and (f) are graphs of functions because we cannot find a vertical line that crosses any of them more than once. In (b) the vertical line crosses the graph in three points and so it is not a function. Also, in (c) and (d), we can find a vertical line that crosses the graph more than once.

The Domain and the Range Using a Grapher

Keep the following in mind regarding the *graph* of a function:

> **Domain** $=$ the set of a function's inputs, found on the horizontal axis;
>
> **Range** $=$ the set of a function's outputs, found on the vertical axis.

By carefully examining the graph of a function, we may be able to determine the function's domain as well as its range. Consider the graph of $f(x) = \sqrt{4 - (x - 1)^2}$, shown below. We look for the inputs on the x-axis that correspond to a point on the graph. We see that they extend from -1 to 3, inclusive. Thus the domain is $\{x \,|\, -1 \leq x \leq 3\}$, or the interval $[-1, 3]$.

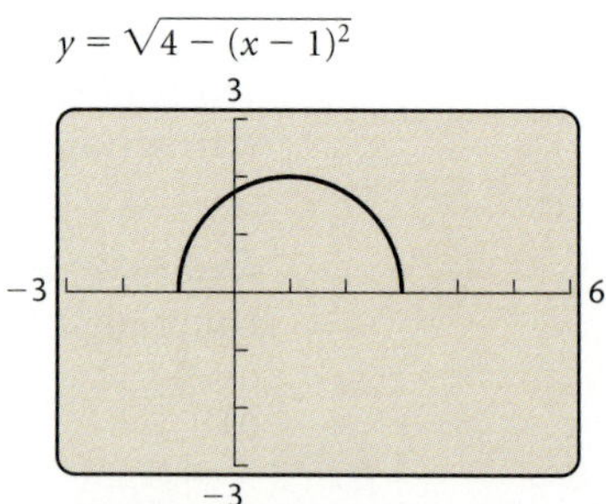

To find the range, we look for the outputs on the y-axis. We see that they extend from 0 to 2, inclusive. Thus the range of this function is $\{y \,|\, 0 \leq y \leq 2\}$, or the interval $[0, 2]$. We can confirm our results using the TRACE feature, moving the cursor from left to right along the curve. We can also confirm our results using a TABLE feature. (See Exercise 63 in Exercise Set 1.1.)

Example 10 Use a grapher to graph each of the following functions. Then estimate the domain and the range of each.

a) $f(x) = \sqrt{x + 4}$

b) $f(x) = x^3 - x$

c) $f(x) = \dfrac{12}{x}$

d) $f(x) = x^4 - 2x^2 - 3$

SOLUTION

a)

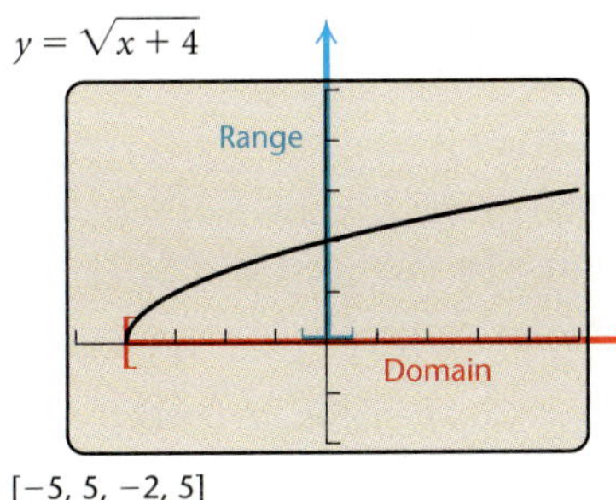

Domain $= [-4, \infty)$;
range $= [0, \infty)$

b)

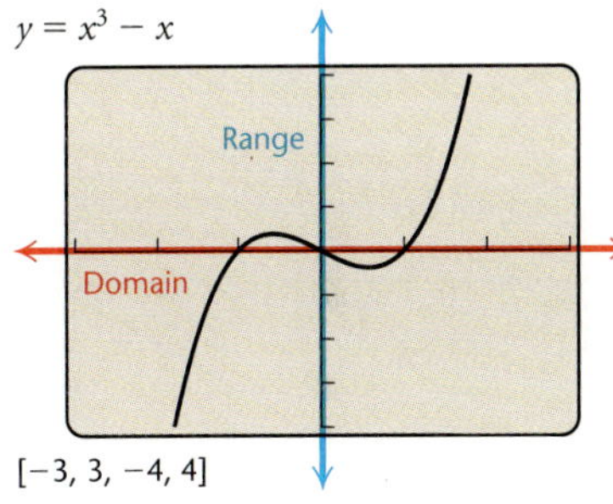

Domain $=$ all real numbers, $\mathbb{R}$;
range $=$ all real numbers, $\mathbb{R}$

c)

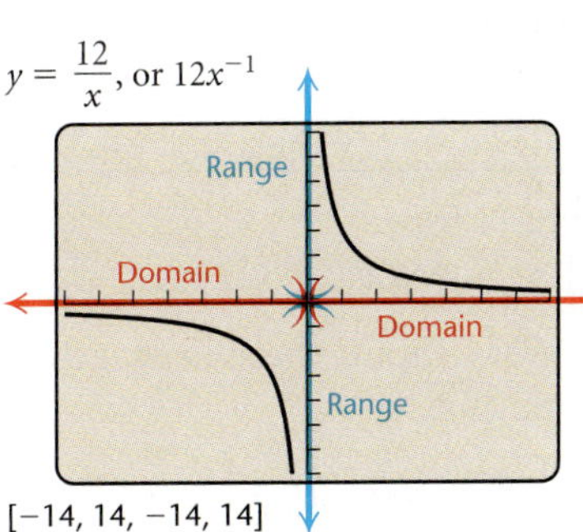

Since the graph does not cross the y-axis, 0 is excluded as an input.
Domain $= (-\infty, 0) \cup (0, \infty)$;
range $= (-\infty, 0) \cup (0, \infty)$

d)

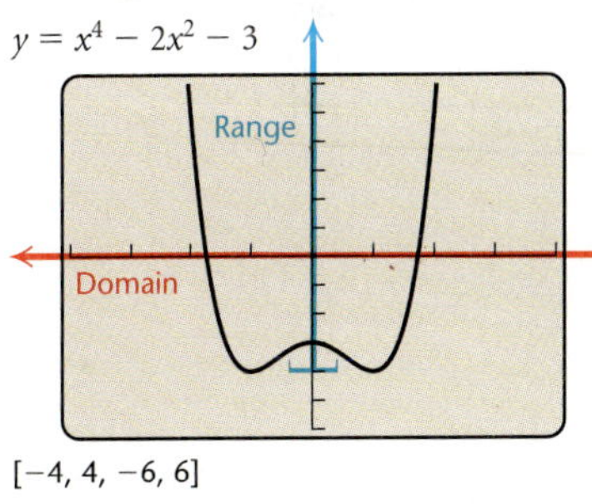

Domain $=$ all real numbers, $\mathbb{R}$;
range $= [-4, \infty)$

Always consider adding the reasoning of Example 7 to a graphical analysis. Think, "What can I substitute?" to find the domain. Think, "What do I get out?" to find the range. Thus, in Example 10(d), it might not look like the domain is all real numbers because the graph rises steeply, but by reexamining the formula we see that we can indeed substitute any real number.

Applications of Functions

Example 11   *Speed of Sound in Air.* The speed S of sound in air is a function of the temperature t, in degrees Fahrenheit, and is given by

$$S(t) = 1087.7 \sqrt{\frac{5t + 2457}{2457}},$$

where S is in feet per second.

a) Graph the function using the viewing window $[-50, 150, 1000, 1250]$ with Xscl = 50 and Yscl = 50.

b) Find the speed of sound in air when the temperature is 0°, 32°, and 70° Fahrenheit.

SOLUTION

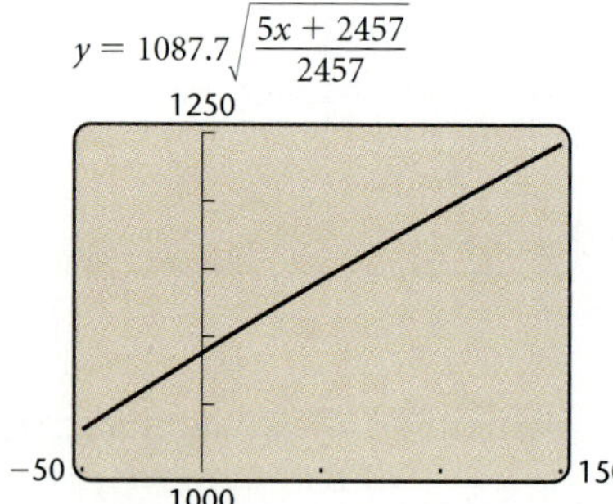

a) The graph is shown at left. Note that $S(t)$ must be changed to y and t must be changed to x.

b) We use a grapher to compute the function values. We find that

$$S(0) = 1087.7 \text{ ft/sec},$$
$$S(32) \approx 1122.6 \text{ ft/sec}, \quad \text{and}$$
$$S(70) \approx 1162.6 \text{ ft/sec}.$$

We can also use the TRACE feature to approximate these function values, or we can use the TABLE feature, set in ASK mode, to find them.

The following is a review of several of the function concepts considered in this section.

FUNCTION CONCEPTS	GRAPH

Formula for f:
$f(x) = 5 + 2x^2 - x^4$.

For every input, there is exactly one output.

$(1, 6)$ is on the graph.

1 is an input. 6 is an output.

$f(1) = 6$

Domain: set of all inputs = $\mathbb{R}$.

Range: set of all outputs = $(-\infty, 6]$.

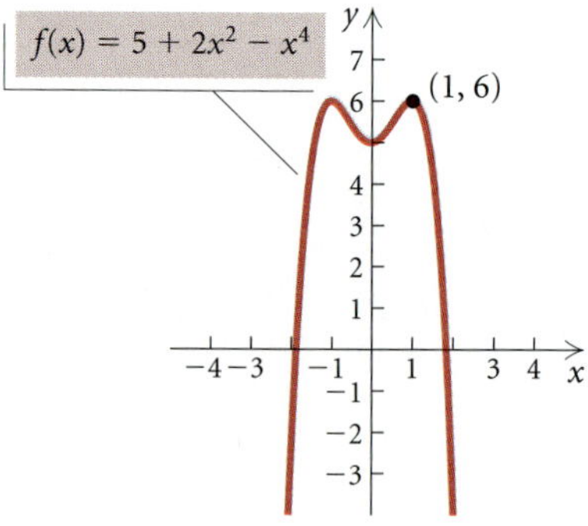

1.1 Exercise Set

Determine whether each correspondence is a function.

1. 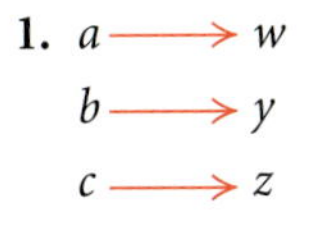

2. 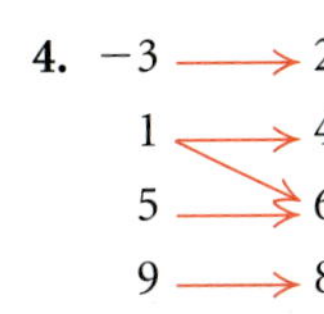

3. 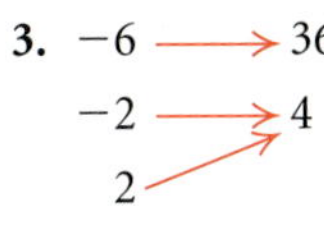

4. 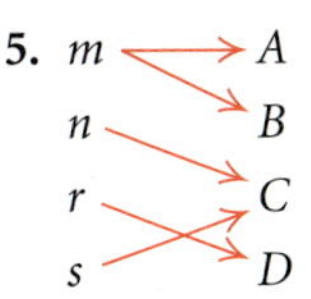

5. 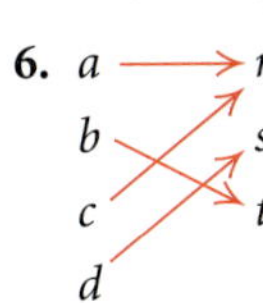

6.

7. NAME OF SPACE MISSION   YEAR OF LAUNCH

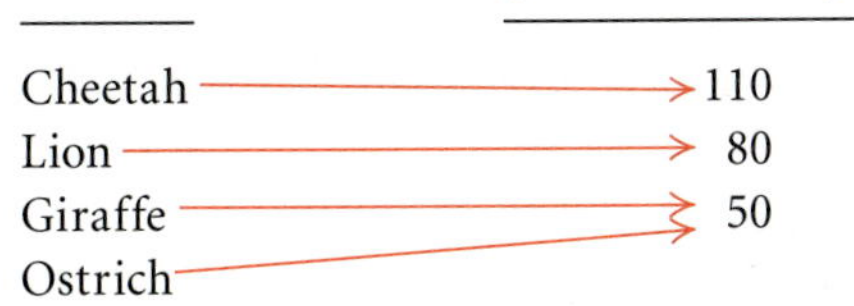

Sputnik 1 ⟶ 1957
Mercury ⟶ 1962
Mariner 2 ⟶ 1969
Apollo 11

(*Source: Cambridge Factfinder, 1993*)

8.

ANIMAL	SPEED ON THE GROUND (IN KILOMETERS/HOUR)
Cheetah	110
Lion	80
Giraffe	50
Ostrich	

(*Source: Cambridge Factfinder, 1993*)

DOMAIN	CORRESPONDENCE	RANGE
9. A set of cars in a parking lot	Each car's license number	A set of numbers
10. A set of people in a town	A doctor a person uses	A set of doctors
11. A set of members of a family	Each person's eye color	A set of colors
12. A set of members of a rock band	An instrument each person plays	A set of instruments
13. A set of students in a class	A student sitting in a neighboring seat	A set of students
14. A set of bags of chips on a shelf	Each bag's weight	A set of weights

Determine whether each relation is a function. Identify the domain and the range.

15. $\{(2, 10), (3, 15), (4, 20)\}$

16. $\{(3, 1), (5, 1), (7, 1)\}$

17. $\{(-7, 3), (-2, 1), (-2, 4), (0, 7)\}$

18. $\{(1, 3), (1, 5), (1, 7), (1, 9)\}$

19. $\{(-2, 1), (0, 1), (2, 1), (4, 1), (-3, 1)\}$

20. $\{(5, 0), (3, -1), (0, 0), (5, -1), (3, -2)\}$

21. A graph of a function f is shown below. Find $f(-1)$, $f(0)$, and $f(1)$.

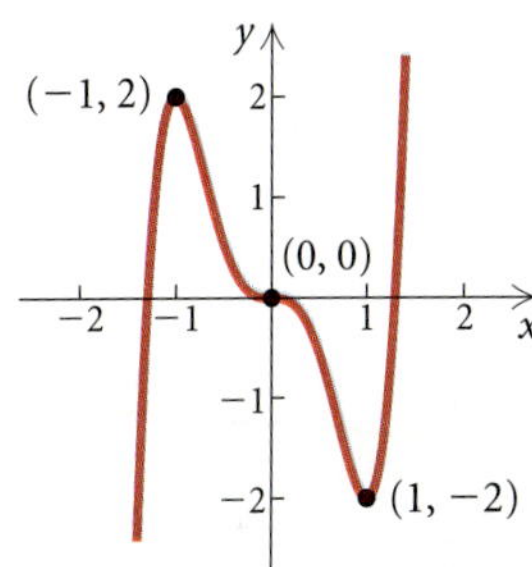

22. Given that $f(x) = 5x^2 + 4x$, find:

a) $f(0)$;   b) $f(-1)$;
c) $f(3)$;   d) $f(t)$;
e) $f(t - 1)$;   f) $\dfrac{f(a + h) - f(a)}{h}$.

23. Given that $g(x) = 3x^2 - 2x + 1$, find:

a) $g(0)$;   b) $g(-1)$;
c) $g(3)$;   d) $g(-x)$;
e) $g(1 - t)$;   f) $\dfrac{g(a + h) - g(a)}{h}$.

24. Given that $f(x) = 2|x| + 3x$, find:

a) $f(1)$;   b) $f(-2)$;
c) $f(-x)$;   d) $f(2y)$;
e) $f(2 - h)$;   f) $\dfrac{f(a + h) - f(a)}{h}$.

25. Given that $g(x) = x^3$, find:

a) $g(2)$;   b) $g(-2)$;
c) $g(-x)$;   d) $g(3y)$;
e) $g(2 + h)$;   f) $\dfrac{g(a + h) - g(a)}{h}$.

26. Given that $f(x) = \dfrac{x}{2 - x}$, find:

    a) $f(2)$;   **b)** $f(1)$;
    c) $f(-16)$;   **d)** $f(-x)$;
    e) $f\left(-\dfrac{2}{3}\right)$ **f)** $\dfrac{f(x + h) - f(x)}{h}$.

27. Given that $g(x) = \dfrac{x - 4}{x + 3}$, find:

    a) $g(5)$;   **b)** $g(4)$;
    c) $g(-3)$;   **d)** $g(-16.25)$;
    e) $g(x + h)$;   **f)** $\dfrac{g(x + h) - g(x)}{h}$.

28. Given that $f(x) = x^2$, find

$$\frac{f(a + h) - f(a)}{h}.$$

29. Find $g(0)$, $g(-1)$, $g(5)$, and $g\left(\tfrac{1}{2}\right)$ for

$$g(x) = \frac{x}{\sqrt{1 - x^2}}.$$

30. Find $h(0)$, $h(2)$, and $h(-x)$ for
$$h(x) = x + \sqrt{x^2 - 1}.$$

Find the domain of each function. Do not use a grapher.

31. $f(x) = 7x + 4$ **32.** $f(x) = |3x - 2|$

33. $f(x) = 4 - \dfrac{2}{x}$ **34.** $f(x) = \dfrac{1}{x^4}$

35. $f(x) = \sqrt{7x + 4}$ **36.** $f(x) = \sqrt{x - 3}$

37. $f(x) = \dfrac{1}{x^2 - 4x - 5}$ **38.** $f(x) = \dfrac{x^4 - 2x^3 + 7}{3x^2 - 10x - 8}$

39. $f(x) = \sqrt{9 - x^2}$ **40.** $f(x) = \dfrac{6x}{\sqrt{9 - x^2}}$

Use a grapher to graph each function. Then visually estimate the domain and the range.

41. $f(x) = |x|$ **42.** $f(x) = |x| - 10.3$

43. $f(x) = \sqrt{9 - x^2}$ **44.** $f(x) = -\sqrt{25 - x^2}$

45. $f(x) = (x - 1)^3 + 2$ **46.** $f(x) = (x - 2)^4 + 1$

47. $f(x) = \sqrt{7 - x}$ **48.** $f(x) = \sqrt{x + 8}$

49. $f(x) = \dfrac{18}{x}$ **50.** $f(x) = 2x^2 - x^4 + 5$

Determine whether each graph is that of a function. An open dot indicates that the point does not belong to the graph.

51.
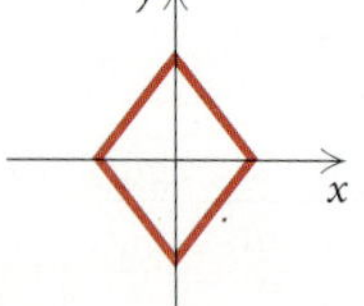

52.
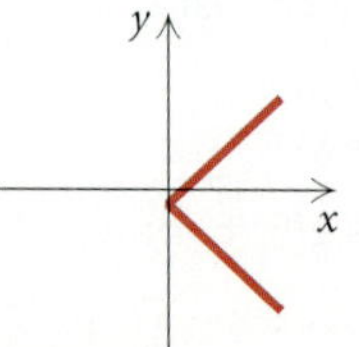

53.
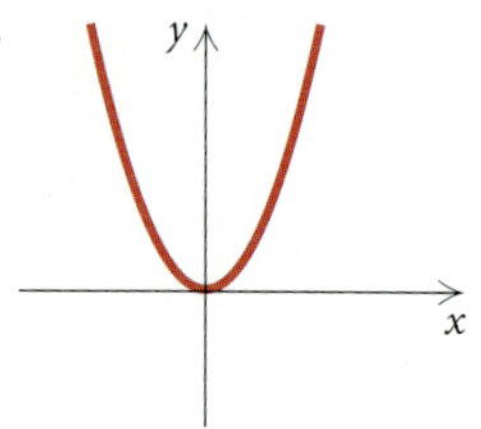

54.
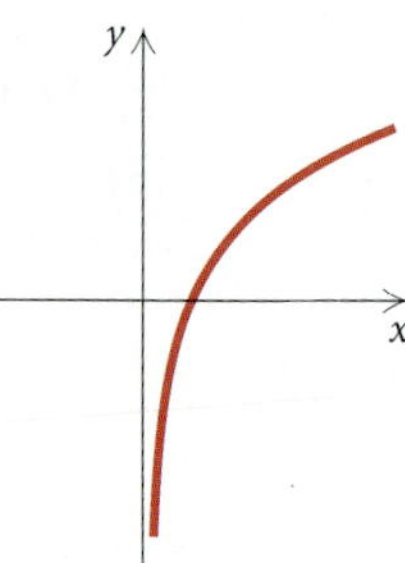

55.
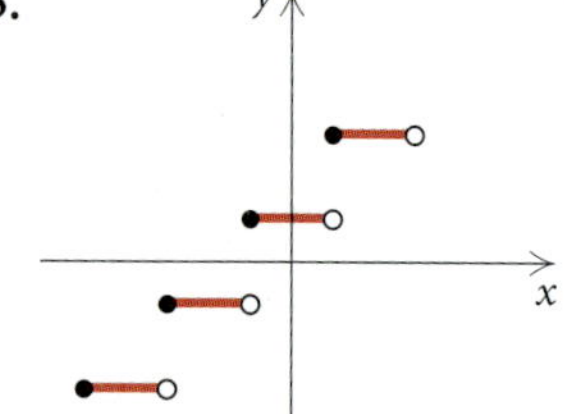

56.
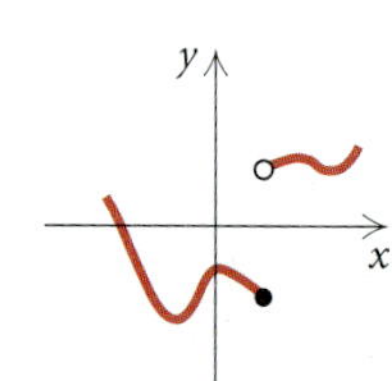

57.
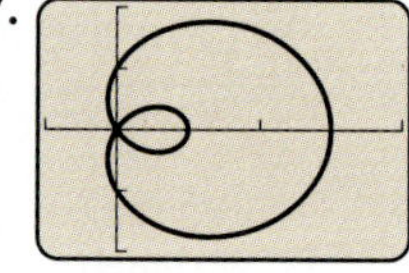

58.
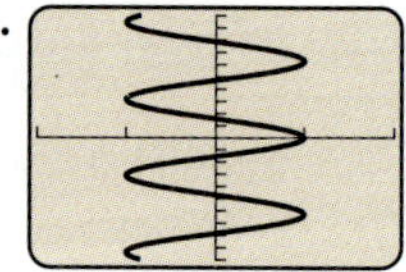

In Exercises 59–62, determine whether the graph is that of a function. Find the domain and the range.

59.
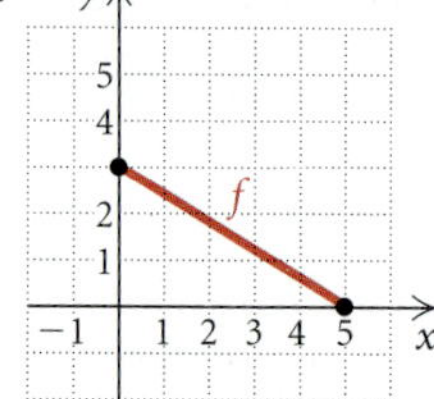

60.
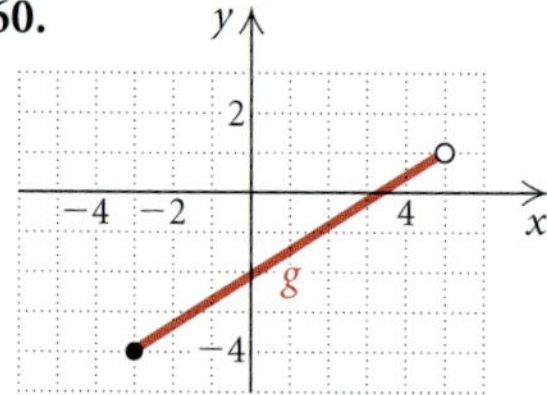

61.
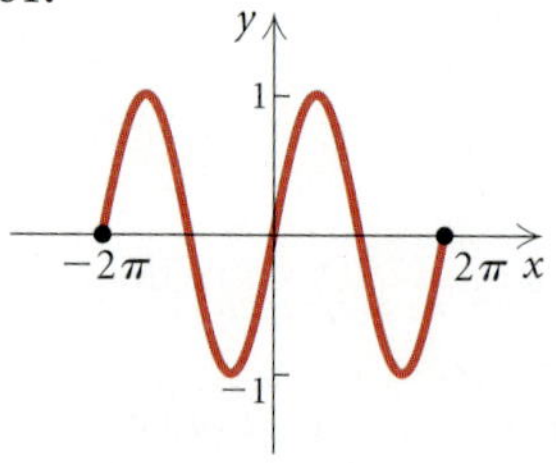

62.
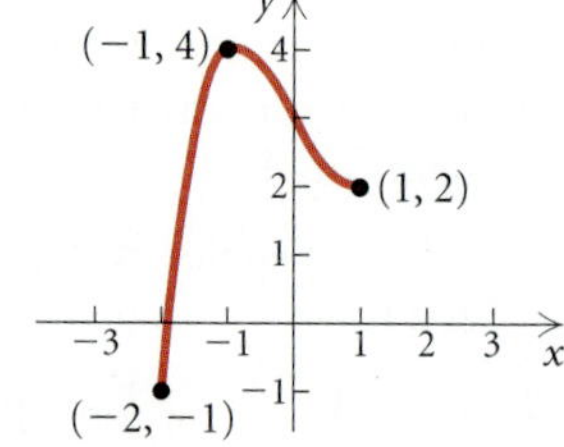

63. Use the TABLE feature to complete the input–output table for the equations

$$y_1 = \sqrt{4 - (x - 1)^2} \quad \text{and} \quad y_2 = \frac{23x}{x - 3.2}.$$

Then use the TABLE feature to consider inputs from -2.0 to 4.8.

X	Y1	Y2
2.2		
2.4		
2.6		
2.8		
3.0		
3.2		
3.4		

$X = 2.2$

a) What does the table tell you about the domain of each function? the range?
b) What seems to be the largest output of each function? the smallest?

64. *Average Price of a Movie Ticket.* The average price of a movie ticket, in dollars, can be estimated by the function P given by

where $x =$ the year. Thus, $P(1998)$ is the average price of a movie ticket in 1998. The price is lower than what might be expected due to lower prices for matinees, senior citizens' discounts, and so on.

a) Use the function to predict the average price in 1998, 2000, and 2010.
b) When will the average price be $8.00?

65. *Boiling Point and Elevation.* The elevation E, in meters, above sea level at which the boiling point of water is t degrees Celsius is given by the function

$$E(t) = 1000(100 - t) + 580(100 - t)^2.$$

At what elevation is the boiling point $99.5°$? $100°$?

66. *Territorial Area of an Animal.* The territorial area of an animal is defined to be its defended, or exclusive region. For example, a lion has a certain region over which it is considered ruler. It has been shown that the territorial area I, in acres, of predatory animals is a function of body weight w, in pounds, and is given by the function

$$T(w) = w^{1.31}.$$

Find the territorial area of animals whose body weights are 0.5 lb, 10 lb, 20 lb, 100 lb, and 200 lb.

Skill Maintenance

To the student and the instructor: The skill maintenance exercises review skills covered previously in the text and anticipate the learning in the next section. You can expect these kinds of exercises in almost every exercise set.

Solve.

67. $\frac{2}{3}x + 7 = 12$

68. $\frac{4}{5}x + 1 = -\frac{2}{3}x + 7$

69. $\sqrt{2x - 4} = 7$

70. $3x^2 - 13x = 10$

Synthesis

To the student and the instructor: The synthesis exercises, found at the end of every exercise set, challenge students to combine concepts or skills developed in that section or in preceding parts of the text. Writing Exercises, denoted by the ◆ icon, are meant to be answered with one or more sentences.

71. ◆ Explain in your own words what a function is.

72. ◆ Explain in your own words the difference between the domain of a function and the range.

73. Construct

$$\frac{f(x + h) - f(x)}{h}$$

for $f(x) = \sqrt{x}$ and rationalize the numerator.

74. Make a hand-drawn graph of a function for which the domain is $[-4, 4]$ and the range is $[1, 2] \cup [3, 5]$. Answers may vary.

75. Give an example of two different functions that have the same domain and the same range, but have no pairs in common. Answers may vary.

76. Make a hand-drawn graph of a function for which the domain is $[-3, -1] \cup [1, 5]$ and the range is $\{1, 2, 3, 4\}$. Answers may vary.

77. Suppose that for some function f, $f(x - 1) = 5x$. Find $f(6)$.

1.2
Functions and Applications

- *Find zeros of functions.*
- *Graph functions, looking for intervals on which the function is increasing, decreasing, or constant, and determine relative maxima and minima.*
- *Graph functions defined piecewise.*
- *Given an application, find a function formula that models the application; find the domain of the function and function values, and then graph the function.*

Because functions occur in so many real-world situations, it is important to be able to analyze them carefully. In this section, we examine a variety of functions and formulate some mathematical models.

Finding Zeros on a Grapher

An input for which a function's output is 0 is called a **zero** of the function.

Zeros of Functions

An input c of a function f is called a *zero* of the function, if the output for c is 0. That is, $f(c) = 0$.

For the function given by $f(x) = x^4 - 4x^2 + 3$, the zeros are $-\sqrt{3}$, -1, 1, and $\sqrt{3}$.

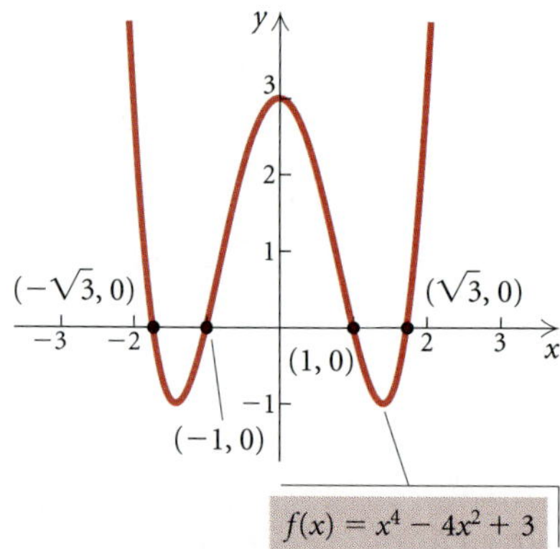

Note that these numbers are the first coordinates of the x-intercepts, $(-\sqrt{3}, 0)$, $(-1, 0)$, $(1, 0)$, and $(\sqrt{3}, 0)$. The **x-intercepts** are the points where the graph crosses the x-axis.

Finding exact values of the zeros of a function can be difficult. We can find approximations using a grapher in much the same way that we solve equations.

Example 1   Find the zeros of the function f given by
$$f(x) = 0.1x^3 - 0.6x^2 - 0.1x + 2.$$
Approximate the zeros to three decimal places.

We have not developed an algebraic procedure that will yield the zeros. We will learn some procedures in Chapter 2.

We use a grapher, trying to create a graph that clearly shows the curvature. Then we look for points where the graph crosses the x-axis. It appears that there are three zeros, one near -2, one near 2, and one near 6. We use TRACE and ZOOM to approximate these zeros, or we can use a SOLVE or ROOT feature.

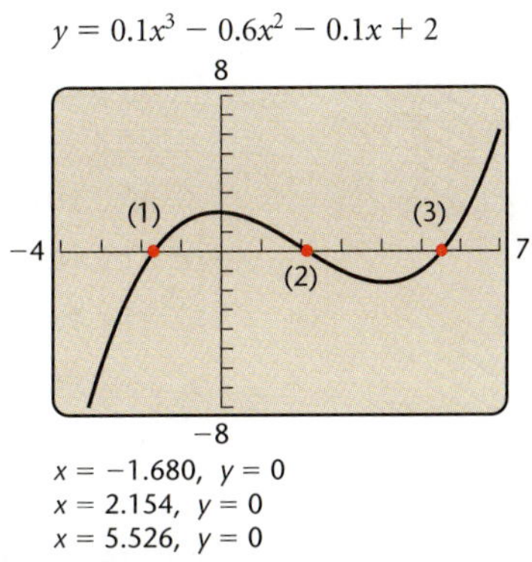

The zeros are about -1.680, 2.154, and 5.526.

Increasing, Decreasing, and Constant Functions

On a given interval, if the graph of a function rises from left to right, it is said to be **increasing** on that interval. If the graph drops from left to right, it is said to be **decreasing**. If the function stays the same from left to right, it is said to be **constant**.

Interactive Discovery

Graph each of the following functions on a grapher using the given viewing window:

$$y = |x + 1| + |x - 2| - 5, \quad [-6, 6, -6, 6];$$
$$y = 5 - (x + 2)^2, \quad [-2, 4, -10, 6];$$
$$y = 4, \quad [-20, 20, -3, 6].$$

Then using the TRACE feature, move the cursor along the graph from left to right, observing what happens to the second coordinate. Confirm the results with a TABLE feature.

We are led to the following definitions.

Increasing, Decreasing, and Constant Functions

A function f is said to be *increasing* on an interval if for all a and b in that interval, $a < b$ implies $f(a) < f(b)$.

A function f is said to be *decreasing* on an interval if for all a and b in that interval, $a < b$ implies $f(a) > f(b)$.

A function f is said to be *constant* on an interval if for all a and b in that interval, $f(a) = f(b)$.

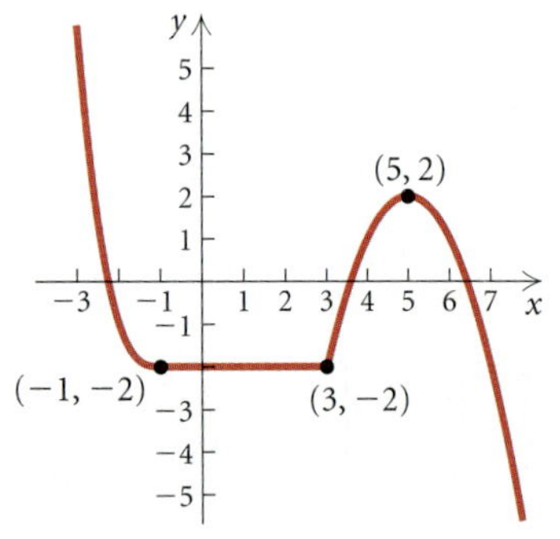

Example 2 Determine the intervals on which the function is each of the following.

a) Increasing **b)** Decreasing **c)** Constant

SOLUTION

a) The function is increasing on the interval $[3, 5]$.

b) The function is decreasing on the intervals $(-\infty, -1]$ and $[5, \infty)$.

c) The function is constant on the interval $[-1, 3]$.

Relative Maximum and Minimum Values

Consider the graph shown below. Note the "peaks" and "valleys" at the points c_1, c_2, and c_3. The function value $f(c_2)$ is called a **relative maximum** (plural, **maxima**). Each of the function values $f(c_1)$ and $f(c_3)$ is called a **relative minimum** (plural, **minima**).

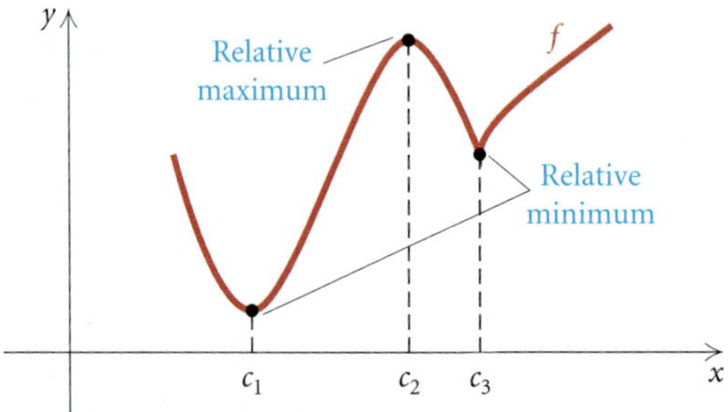

Relative Maxima and Minima

Suppose that f is a function for which $f(c)$ exists for some c in the domain of f. Then:

$f(c)$ is a *relative maximum* if there exists an open interval I containing c in the domain such that $f(c) \geq f(x)$, for all x in I; and

$f(c)$ is a *relative minimum* if there exists an open interval I containing c in the domain such that $f(c) \leq f(x)$, for all x in I.

Simply stated, $f(c)$ is a relative maximum if $f(c)$ is a "high point" in some interval, and $f(c)$ is a relative minimum if $f(c)$ is a "low point" in some interval.

We can use a grapher to approximate relative maxima or minima. On some graphers, this is done using the TRACE and ZOOM features. On other graphers, there may be a MAX or MIN feature that can be used. If you take a calculus course, you will learn a method for determining exact values of relative maxima and minima.

Example 3 Use a grapher to determine any relative maxima or minima of the function $f(x) = 0.1x^3 - 0.6x^2 - 0.1x + 2$, and to determine intervals on which the function is increasing or decreasing.

SOLUTION   We first graph the function, experimenting with the window as needed. The curvature is seen fairly well with $[-4, 6, -3, 3]$.

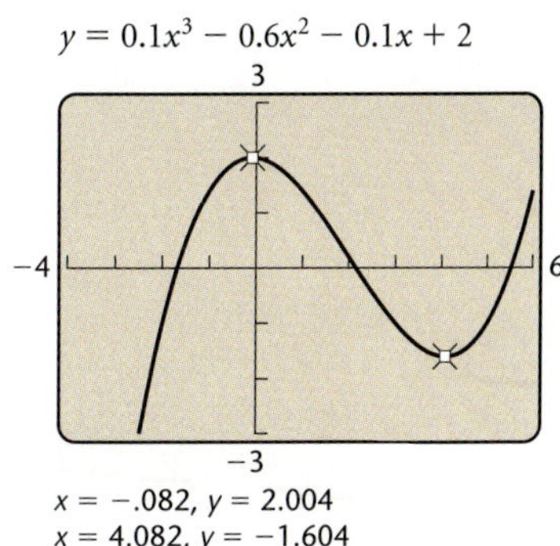

The graph starts rising, or increasing, from the left. We use the TRACE feature and move the cursor from left to right, noting whether the second coordinate increases. We note that the graph tends to stop increasing when x is somewhere around 0. Using the ZOOM feature, we get a good approximation of the relative maximum and where it occurs: 2.004 at $x = -0.082$.

From this relative maximum, the graph decreases to a point near $x = 4$ before it starts increasing again. We zoom in and get a good approximation of the relative minimum: -1.604 at $x = 4.082$.

The function is increasing on the intervals $(-\infty, -0.082]$ and $[4.082, \infty)$. It is decreasing on the interval $[-0.082, 4.082]$.   —

Interactive Discovery

If your grapher has a TABLE feature, try to discover a way to use it to check the results of Example 3. Consider a small interval around each input and a small step value. Are there any difficulties?

Functions Defined Piecewise

Sometimes functions are defined **piecewise** using different output formulas for different parts of the domain.

Example 4 Make a hand-drawn graph of the function defined as

$$f(x) = \begin{cases} 4, & \text{for } x \le 0, \\ 4 - x^2, & \text{for } 0 < x \le 2, \\ 2x - 6, & \text{for } x > 2. \end{cases}$$

SOLUTION We create the graph in three parts, as shown and described below.

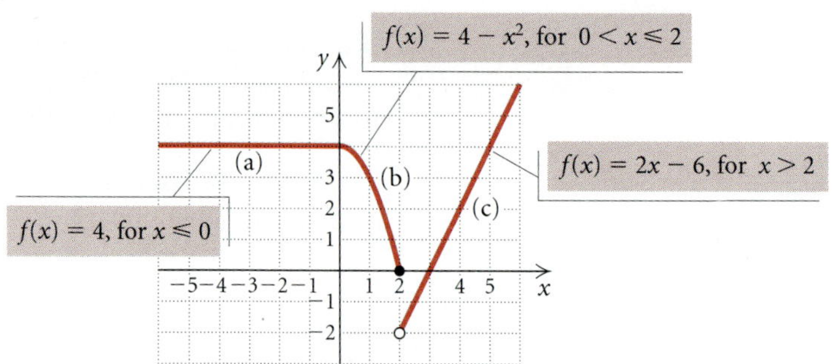

a) We graph $f(x) = 4$ *only* for inputs x less than or equal to 0 (that is, on the interval $(-\infty, 0]$).

b) We graph $f(x) = 4 - x^2$ *only* for inputs x greater than 0 and less than or equal to 2 (that is, on the interval $(0, 2]$).

c) We graph $f(x) = 2x - 6$ *only* for inputs x greater than 2 (that is, on the interval $(2, \infty)$).

   To graph a function defined piecewise on a grapher, consult your manual. For piecewise-defined functions, you should select the DOT feature. Although some do not have the capability, you might try the following way to enter the function formula for Example 4. It incorporates parenthetic descriptions of the intervals:

$$f(x) = 4 \;(x \le 0) + (4 - x^2) \;(x > 0)(x \le 2) + (2x - 6) \;(x > 2).$$

If you wish, you might enter this equation as three separate functions:

$$y_1 = 4 \;(x \le 0), \qquad y_2 = (4 - x^2) \;(x > 0)(x \le 2),$$

$$y_3 = (2x - 6) \;(x > 2).$$

Example 5 Make a hand-drawn graph of the function defined as

$$f(x) = \begin{cases} \dfrac{x^2 - 4}{x + 2}, & \text{for } x \ne -2, \\ 3, & \text{for } x = -2. \end{cases}$$

SOLUTION When $x \ne -2$, the denominator of $(x^2 - 4)/(x + 2)$ is nonzero, so we can simplify:

$$\frac{x^2 - 4}{x + 2} = \frac{(x + 2)(x - 2)}{x + 2} = x - 2.$$

Thus,

$$f(x) = x - 2, \quad \text{for } x \neq -2.$$

The graph of this part of the function consists of a line with a "hole" at the point $(-2, -4)$, indicated by the open dot. At $x = -2$, we have $f(-2) = 3$, so the point $(-2, 3)$ is plotted above the open dot.

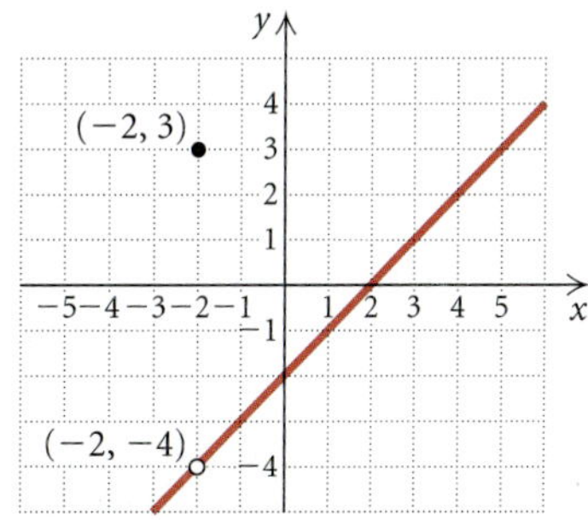

When $y = (x^2 - 4)/(x + 2)$ is graphed using a grapher, the hole may not be visible. If a TABLE feature is available, the following table can be created. The ERROR message indicates that -2 is not in the domain of the function g given by $g(x) = (x^2 - 4)/(x + 2)$. However, -2 *is* in the domain of f because $f(-2)$ is defined to be 3.

X	Y₁
-2.3	-4.3
-2.2	-4.2
-2.1	-4.1
-2	ERROR
-1.9	-3.9
-1.8	-3.8
-1.7	-3.7

X = -2

A function with importance in calculus and computer programming is the *greatest integer function, f*, denoted as $f(x) = \text{INT}(x)$, or $[\![x]\!]$.

Greatest Integer Function

$f(x) = \text{INT}(x) = $ the greatest integer less than or equal to x.

The greatest integer function pairs each input with the greatest integer less than or equal to that input. To check this, note that 1, 1.2, and 1.9 are all paired with the y-value 1. Similarly, 2, 2.3, and 2.8 are all paired with 2, and -3.4, -3.06, and -3.99027 are all paired with -4.

Example 6   Graph: $f(x) = \text{INT}(x)$.

SOLUTION   The greatest integer function can also be defined by a piecewise function with an infinite number of statements:

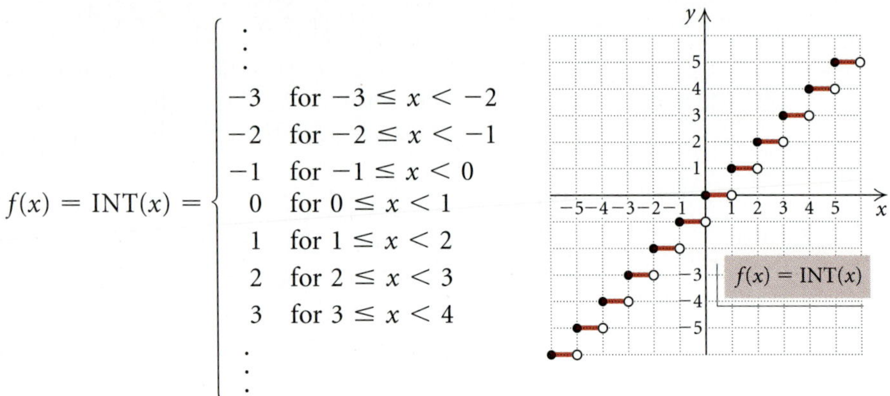

$$f(x) = \text{INT}(x) = \begin{cases} \vdots & \\ -3 & \text{for } -3 \le x < -2 \\ -2 & \text{for } -2 \le x < -1 \\ -1 & \text{for } -1 \le x < 0 \\ 0 & \text{for } 0 \le x < 1 \\ 1 & \text{for } 1 \le x < 2 \\ 2 & \text{for } 2 \le x < 3 \\ 3 & \text{for } 3 \le x < 4 \\ \vdots & \end{cases}$$

On a grapher, we would see the graph on the left below if we used CON-NECTED mode, which connects points (or rectangles called **pixels** on a grapher) with line segments. We would see the graph on the right if we used DOT mode. The DOT mode is preferable, though even it may not show the open dots at the endpoints of the segments.

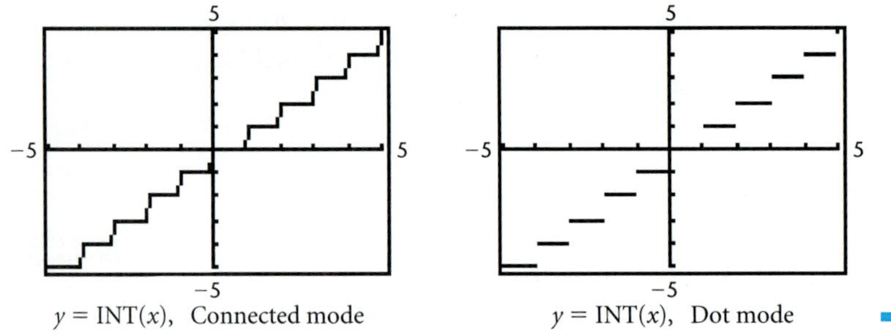

$y = \text{INT}(x)$, Connected mode $y = \text{INT}(x)$, Dot mode

Applications of Functions

Many real-world situations can be modeled by functions.

Example 7   *Car Distance.* Jamie and Ruben drive away from a restaurant at right angles to each other. Jamie's speed is 65 mph and Ruben's is 55 mph.

a) Express the distance between the cars as a function of time.

b) Find the domain of the function.

c) Graph the function.

SOLUTION

a) Suppose 1 hr goes by. At that time, Jamie has traveled 65 mi and Ruben has traveled 55 mi. We can use the Pythagorean theorem then to find the distance between them. This distance would be the length of the hypotenuse of a triangle with legs measuring 65 mi and 55 mi. After 2 hr, the triangle's legs would measure 130 mi and 110 mi.

Observing that the distances will always be changing, we make a sketch and let $t =$ the time, in hours, that Jamie and Ruben have been driving since leaving the restaurant.

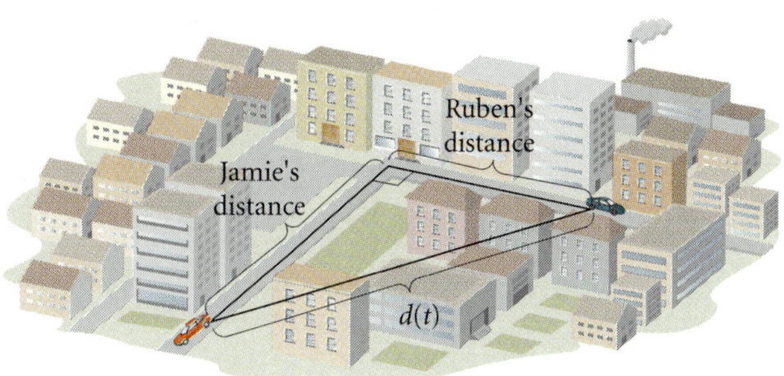

After t hours, Jamie has traveled $65t$ miles and Ruben $55t$ miles. We can use the Pythagorean theorem again:

$$[d(t)]^2 = (65t)^2 + (55t)^2.$$

Because distance must be nonnegative, we need only consider the positive square root when solving for $d(t)$:

$$\begin{aligned}
d(t) &= \sqrt{(65t)^2 + (55t)^2} \\
&= \sqrt{4225t^2 + 3025t^2} \\
&= \sqrt{7250t^2} \\
&= \sqrt{7250}\,\sqrt{t^2} \\
&\approx 85.15|t| \qquad \text{Approximating the root to two decimal places} \\
&\approx 85.15t. \qquad \text{Since } t \geq 0,\ |t| = t.
\end{aligned}$$

Thus, $d(t) = 85.15t$, $t \geq 0$.

b) Since the time traveled must be nonnegative, the domain is the set of nonnegative real numbers $[0, \infty)$.

c) Because of the ease of using a grapher, we can almost always visualize a problem by making a graph. Making such graphs should become a habit as you do applications and problem solving.

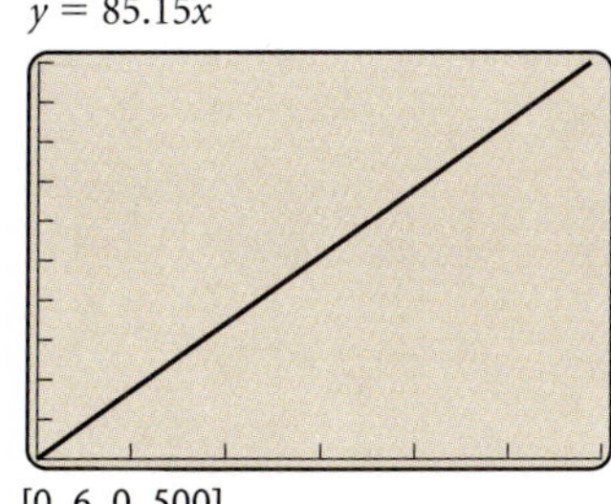

Example 8 *Storage Area.* The Sound Shop has 20 ft of dividers with which to set off a rectangular area for the storage of overstock. If a corner of the store is used, the partition need only form two sides of a rectangle.

a) Express the floor area as a function of the length of the partition.

b) Find the domain of the function.

c) Graph the function.

d) Find the dimensions that maximize the area.

SOLUTION

a) Note that the dividers will form two sides of a rectangle. If, for example, 14 ft of dividers are used for the length of the rectangle, that would leave $20 - 14$, or 6 ft of dividers for the width. Thus if $x =$ the length, in feet, of the rectangle, then $20 - x =$ the width. We represent this information in a sketch.

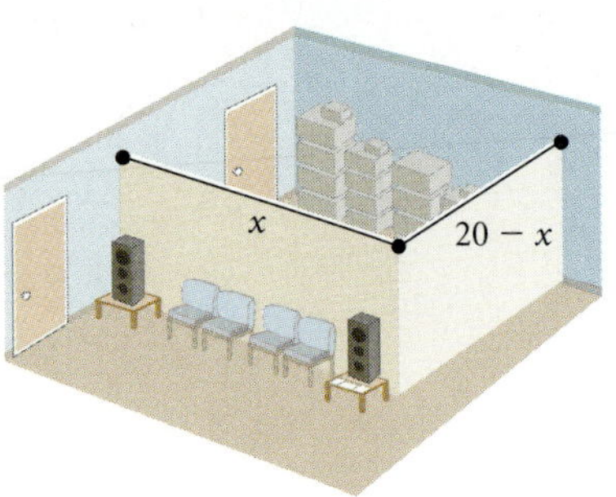

The area, $A(x)$, is given by

$$A(x) = x(20 - x) \qquad \textbf{Area = length · width}$$
$$= 20x - x^2.$$

The function $A(x) = 20x - x^2$ can be used to express the rectangle's area as a function of the length.

b) Because the rectangle's length must be positive and less than 20 ft (why?), we restrict the domain of A to $\{x \mid 0 < x < 20\}$, that is, the interval $(0, 20)$.

c) The graph is shown at left with the viewing window $[-5, 25, 0, 120]$.

d) We use the TRACE and ZOOM features or the MAX–MIN feature as in Example 3. The maximum value (overall largest value) of the area function is 100 when $x = 10$. Thus the dimensions that maximize the area are

$$\text{Length} = x = 10 \text{ ft} \quad \text{and}$$
$$\text{Width} = 20 - x = 20 - 10 = 10 \text{ ft.}$$

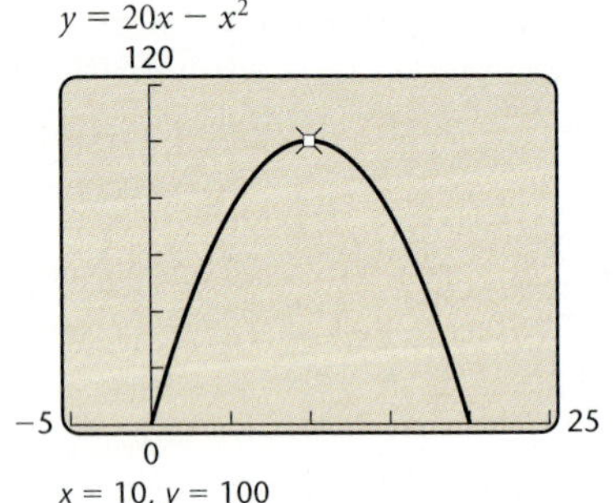

$y = 20x - x^2$
$x = 10, y = 100$

Summary

Let's summarize our analysis of functions by considering the function f given by $f(x) = x^4 - 6x^3 - 4x^2 + 53.2x - 42.6$.

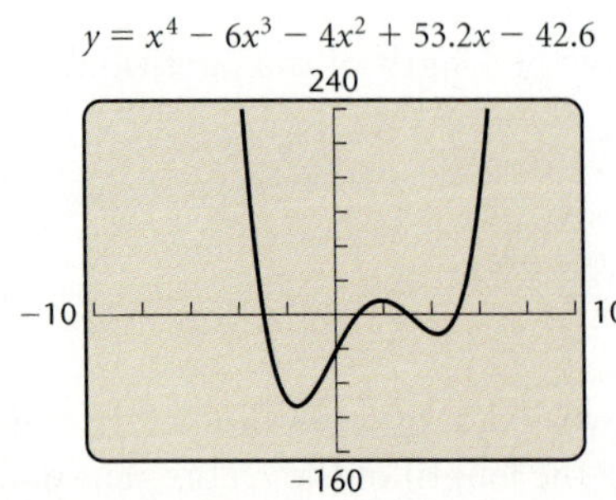

$y = x^4 - 6x^3 - 4x^2 + 53.2x - 42.6$

Function:	$f(x) = x^4 - 6x^3 - 4x^2 + 53.2x - 42.6$
Zeros:	$-2.975, 0.950, 3, 5.025$
Relative maximum:	15.921 at $x = 1.914$
Relative minima:	-106.907 at $x = -1.643$,
	-23.101 at $x = 4.229$
Increasing on:	$[-1.643, 1.914], [4.229, \infty)$
Decreasing on:	$(-\infty, -1.643], [1.914, 4.229]$
Domain:	All real numbers, $\mathbb{R}$
Range:	$[-106.907, \infty)$

1.2 | Exercise Set

For each function, find the zeros.

1. $f(x) = x^3 - 3x - 1$

2. $f(x) = x^3 + 3x^2 - 9x - 13$

3. $f(x) = x^4 - 2x^2$

4. $f(x) = x^4 - 2x^3 - 5.6$

5. $f(x) = x^3 - x$

6. $f(x) = 2x^3 - x^2 - 14x - 10$

7. $f(x) = \frac{1}{2}(|x - 4| + |x - 7|) - 4$

8. $f(x) = \sqrt{7 - x^2}$

9. $f(x) = |x + 1| + |x - 2| - 5$

10. $f(x) = |x + 1| + |x - 2|$

11. $f(x) = |x + 1| + |x - 2| - 3$

12. $f(x) = x^8 + 8x^7 - 28x^6 - 56x^5 + 70x^4$
$+ 56x^3 - 28x^2 - 8x + 1$

Determine intervals on which each function is (a) increasing, (b) decreasing, and (c) constant.

13.

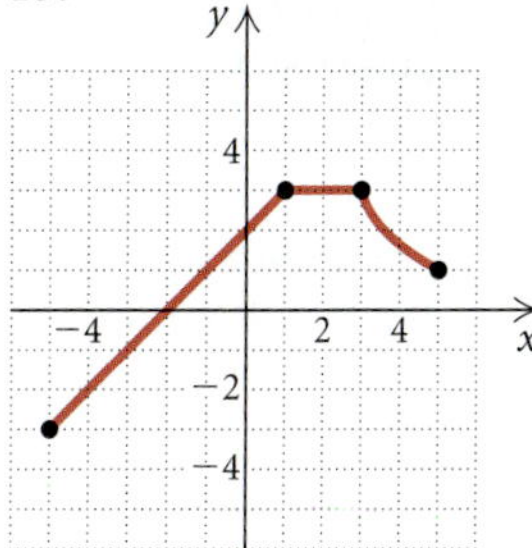

14.

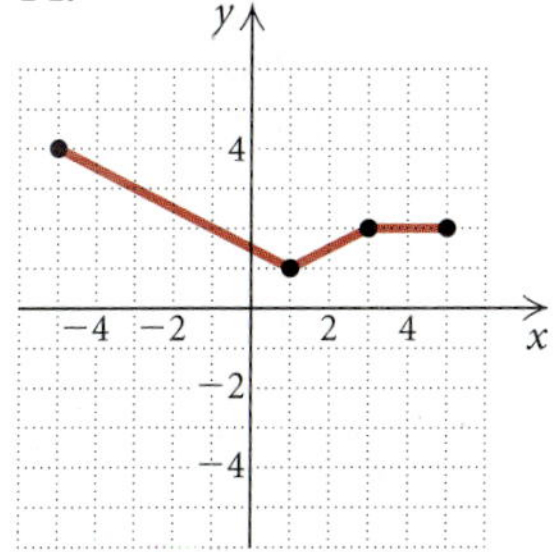

15.

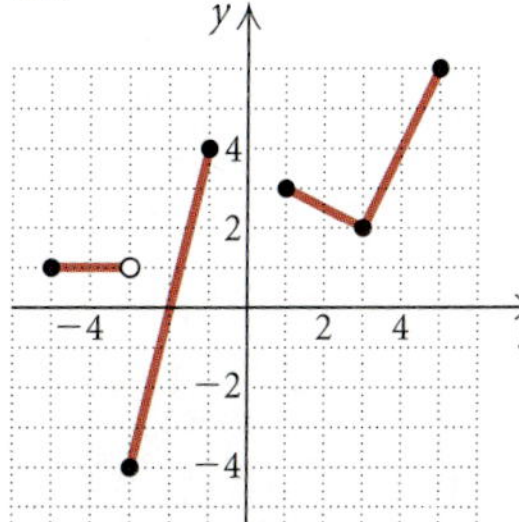

16.

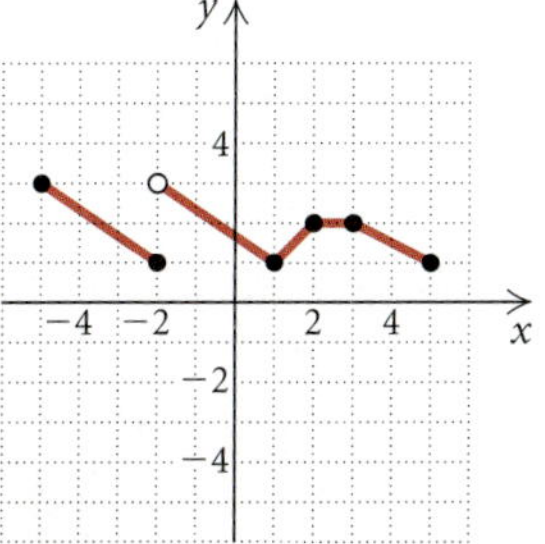

Graph each function using the given viewing window. Find where the function is increasing or decreasing and any relative maxima or minima. Change viewing windows if it seems appropriate for further analysis.

17. $f(x) = x^2$,
$[-4, 4, -1, 10]$

18. $f(x) = 4 - x^2$,
$[-5, 5, -5, 5]$

19. $f(x) = 5 - |x|$,
$[-10, 10, -10, 10]$

20. $f(x) = |x + 3| - 5$,
$[-10, 10, -10, 10]$

21. $f(x) = x^2 - 6x + 10$,
$[-4, 10, -1, 20]$

22. $f(x) = -x^2 - 8x - 9$,
$[-10, 2, -10, 10]$

23. $f(x) = -x^3 + 6x^2 - 9x - 4$,
$[-3, 7, -20, 15]$

24. $f(x) = 0.2x^3 - 0.2x^2 - 5x - 4$,
$[-10, 10, -30, 20]$

25. $f(x) = 1.1x^4 - 5.3x^2 + 4.07$,
$[-4, 4, -4, 8]$

26. $f(x) = 1.2(x + 3)^4 + 10.3(x + 3)^2 + 9.78$,
$[-9, 3, -40, 100]$

27. *Temperature During an Illness.* The temperature of a patient during an illness is given by the function

$$T(t) = -0.1t^2 + 1.2t + 98.6, \quad 0 \le t \le 12,$$

where T = the temperature, in degrees Fahrenheit, at time t, in days, after the onset of the illness.

a) Graph the function using a grapher.

b) Find the relative maximum.

c) At what time was the patient's temperature the highest? What was the highest temperature?

28. *Advertising Effect.* A software firm estimates that it will sell N units of a new CD-ROM video game after spending a dollars on advertising, where

$$N(a) = -a^2 + 300a + 6, \quad 0 \le a \le 300,$$

and a is measured in thousands of dollars.

a) Graph the function using a grapher.

b) Find the relative maximum.

c) For what advertising expenditure will the greatest number of games be sold? How many games will be sold for that amount?

Use a grapher. Find where each function is increasing or decreasing. Consider the entire set of real numbers if no domain is given.

29. $f(x) = \dfrac{8x}{x^2 + 1}$

30. $f(x) = \dfrac{-4}{x^2 + 1}$

31. $f(x) = x\sqrt{4 - x^2}$, for $-2 \le x \le 2$

32. $f(x) = -0.8x\sqrt{9 - x^2}$, for $-3 \le x \le 3$

Make a hand-drawn graph of each of the following. Check your results on a grapher.

33. $f(x) = \begin{cases} \frac{1}{2}x, & \text{for } x < 0, \\ x + 3, & \text{for } x \ge 0 \end{cases}$

34. $f(x) = \begin{cases} -\frac{1}{3}x + 2, & \text{for } x \le 0, \\ x - 5, & \text{for } x > 0 \end{cases}$

35. $f(x) = \begin{cases} -\frac{3}{4}x + 2, & \text{for } x < 4, \\ -1, & \text{for } x \ge 4 \end{cases}$

36. $f(x) = \begin{cases} 4, & \text{for } x \le -2, \\ x + 1, & \text{for } -2 < x < 3, \\ -x, & \text{for } x \ge -3 \end{cases}$

37. $f(x) = \begin{cases} x + 1, & \text{for } x \le -3, \\ -1, & \text{for } -3 < x < 4, \\ \frac{1}{2}x, & \text{for } x \ge 4 \end{cases}$

38. $f(x) = \begin{cases} \dfrac{x^2 - 9}{x + 3}, & \text{for } x \ne -3, \\ 5, & \text{for } x = -3 \end{cases}$

39. $f(x) = \begin{cases} 2, & \text{for } x = 5, \\ \dfrac{x^2 - 25}{x - 5}, & \text{for } x \ne 5 \end{cases}$

40. $f(x) = \begin{cases} \dfrac{x^2 + 3x + 2}{x + 1}, & \text{for } x \ne -1, \\ 7, & \text{for } x = -1 \end{cases}$

41. $f(x) = \text{INT}(x)$ **42.** $f(x) = 1 + \text{INT}(x)$

43. $f(x) = \text{INT}(x) - 1$ **44.** $f(x) = [\![x]\!] - 2$

Graph each of the following with a grapher, if possible.

45. $f(x) = \begin{cases} \sqrt[3]{x}, & \text{for } x \le -1, \\ x^2 - 3x, & \text{for } -1 < x < 4, \\ \sqrt{x - 4}, & \text{for } x \ge 4 \end{cases}$

46. $f(x) = \begin{cases} \left| \dfrac{x}{3} + 2 \right|, & \text{for } x < 5, \\ \sqrt[3]{x - 8}, & \text{for } x \ge 5 \end{cases}$

47. $f(x) = \begin{cases} \sqrt[3]{x + 27}, & \text{for } x < 1, \\ \left| 2 - \dfrac{x}{5} \right|, & \text{for } x \ge 1 \end{cases}$

48. *Airplane Distance.*   An airplane is flying at an altitude of 3700 ft. The slanted distance directly to the airport is d feet. Express the horizontal distance h as a function of d.

49. *Rising Balloon.*   A hot-air balloon rises straight up from the ground at a rate of 120 ft/min. The balloon is tracked from a rangefinder at point P, which is 400 ft from the release point Q of the balloon. Let $d =$ the distance from the balloon to the rangefinder at point P and $t =$ the time, in minutes, since the balloon was released. Express d as a function of t.

50. *Triangular Flag.*   A scout troop is designing a triangular flag so that the length of its base, in inches, is 7 less than twice the height, h. Express the area of the flag as a function of the height.

51. *Garden Area.*   Yardbird Landscaping has 48 m of fencing with which to enclose a rectangular garden. If the garden is x meters long, express the garden's area as a function of the length.

52. *Tablecloth Area.*   A tailor uses 16 ft of lace to trim the edges of a rectangular tablecloth. If the tablecloth is w feet wide, express its area as a function of the width.

53. *Inscribed Rhombus.*   A rhombus is inscribed in a rectangle that is w meters wide with a perimeter of 40 m. Each vertex of the rhombus is a midpoint of a side of the rectangle. Express the area of the rhombus as a function of the rectangle's width. (*Hint:* Consider the area of the rectangle.)

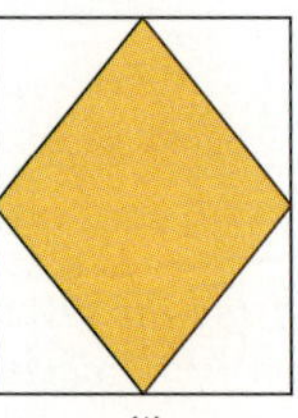

54. *Gas Tank Volume.* A gas tank has ends that are hemispheres of radius r feet. The cylindrical midsection is 6 ft long. Express the volume of the tank as a function of r.

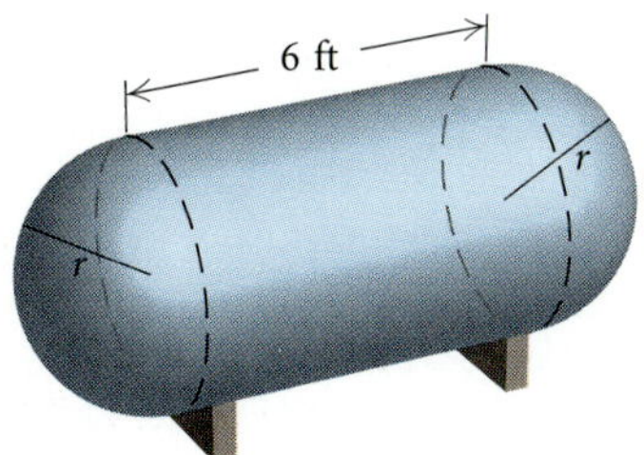

55. *Golf Distance Finder.* A device used in golf to estimate the distance d, in yards, to a hole measures the size s, in inches, that the 7-ft pin appears to be in a viewfinder. Express the distance d as a function of s.

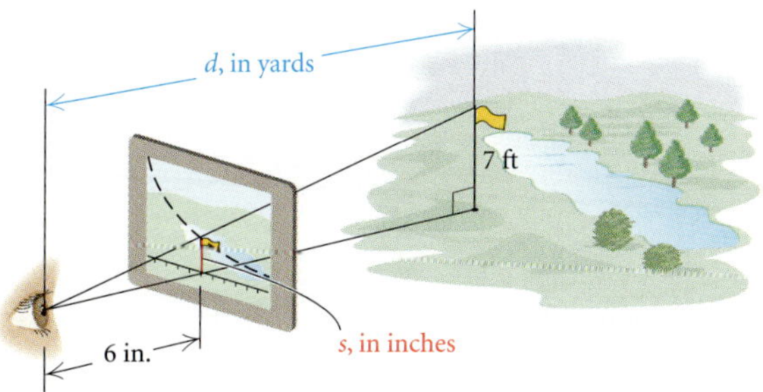

56. *Play Space.* A daycare center has 30 ft of dividers with which to enclose a rectangular play space in a corner of a large room. The sides against the wall require no partition. Suppose the play space is x feet long.

a) Express the area of the play space as a function of x.
b) Find the domain of the function.
c) Graph the function.
d) What dimensions yield the maximum area?

57. *Volume of a Box.* From a 12-cm by 12-cm piece of cardboard, square corners are cut out so that the sides can be folded up to make a box.

a) Express the volume of the box as a function of the length, x, in centimeters, of a cut-out square.
b) Find the domain of the function.
c) Graph the function.
d) What dimensions yield the maximum volume?

58. *Cost of Material.* A rectangular box with volume 320 ft^3 is built with a square base and top. The cost is \$1.50/ft^2 for the bottom, \$2.50/ft^2 for the sides, and \$1/ft^2 for the top. Let $x =$ the length of the base, in feet.

a) Express the cost of the box as a function of x.
b) Find the domain of the function.
c) Graph the function.
d) What dimensions minimize the cost of the box?

59. *Area of an Inscribed Rectangle.* A rectangle that is x feet wide is inscribed in a circle of radius 8 ft.

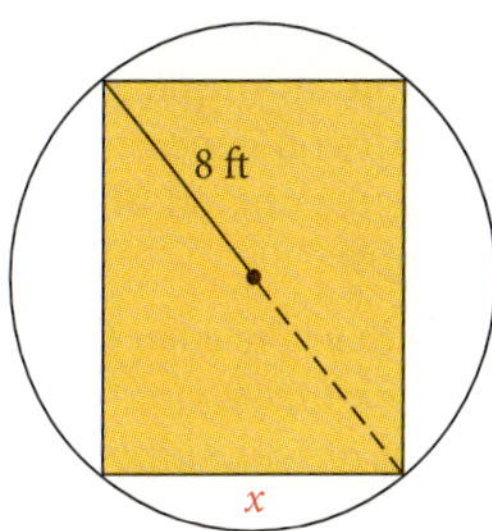

a) Express the area of the rectangle as a function of x.
b) Find the domain of the function.
c) Graph the function.
d) What dimensions maximize the area of the rectangle?

60. *Area of an Inscribed Rectangle.*   A rectangle that is x meters wide is inscribed in a circle of diameter 20 m.

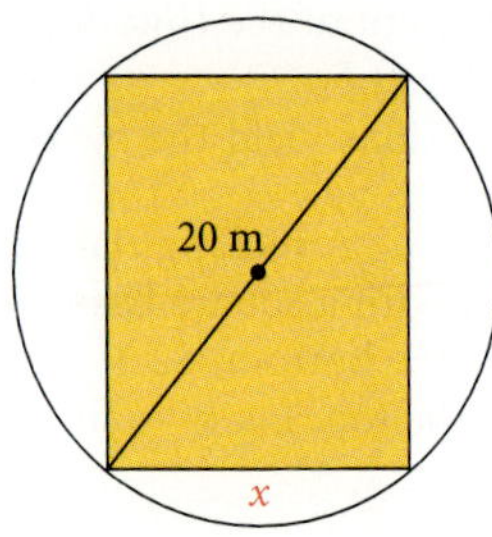

a) Express the area of the rectangle as a function of x.
b) Find the domain of the function.
c) Graph the function.
d) What dimensions maximize the area of the rectangle?

Skill Maintenance

If $f(x) = 7x + 2$, find each of the following.

61. $\dfrac{f(b) - f(a)}{b - a}$

62. $\dfrac{f(a + h) - f(a)}{h}$

Synthesis

63. ◆ Describe a real-world situation that could be modeled by a function that is, in turn, increasing, then constant, and finally decreasing.

64. ◆ Simply stated, a *continuous function* is a function whose graph can be drawn without lifting the pencil from the paper. Examine several functions in this exercise set to see if they are continuous. Then explore the continuous functions to find the relative maxima and minima. For continuous functions, how can you connect the ideas of increasing and decreasing on an interval to relative maxima and minima?

Use a grapher. Estimate where each function is increasing or decreasing and any relative maxima or minima. Consider the entire set of real numbers if no domain is given.

65. $f(x) = 20.17x^3 - 3.24x^5$

66. $f(x) = 3.22x^5 - 5.208x^3 - 11$

67. $f(x) = x^4 + 4x^3 - 36x^2 - 160x + 400$

68. $f(x) = -x^6 - 4x^5 + 54x^4 + 160x^3 - 641x^2 - 828x + 1200$

69. *Parking Costs.*   A parking garage charges \$2 for up to (but not including) 1 hr of parking, \$4 for up to 2 hr of parking, \$6 for up to 3 hr of parking, and so on. Let $C(t) = $ the cost of parking for t hours.

a) Graph the function.
b) Write an equation for $C(t)$ using the greatest integer notation INT.

70. If $\text{INT}(x + 2) = -3$, what are the possible inputs for x?

71. If $[\text{INT}(x)]^2 = 25$, what are the possible inputs for x?

72. *Minimizing Power Line Costs.*   A power line is constructed from a power station at point A to an island at point C, which is 1 mi directly out in the water from a point B on the shore. Point B is 4 mi downshore from the power station at A. It costs \$5000 per mile to lay the power line under water and \$3000 per mile to lay the power line under ground. The line comes to the shore at point S downshore from A. Note that S could very well be A or B. Let $x = $ the distance from B to S.

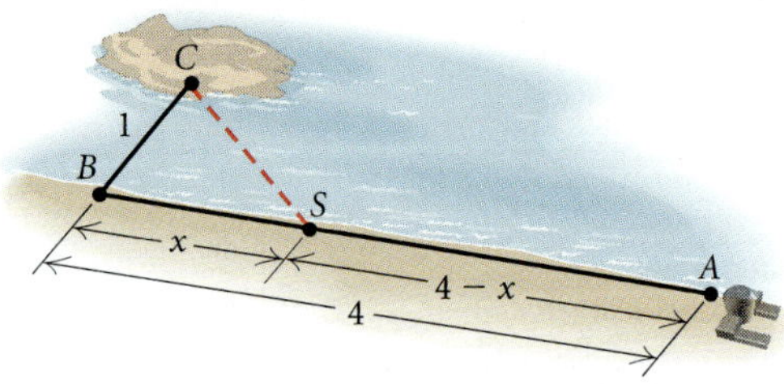

a) Express the cost C of laying the line as a function of x.
b) At what distance x should the line come to shore in order to minimize cost?

73. *Volume of an Inscribed Cylinder.*   A right circular cylinder of height h and radius r is inscribed in a right circular cone with a height of 10 ft and a base with a radius of 6 ft.

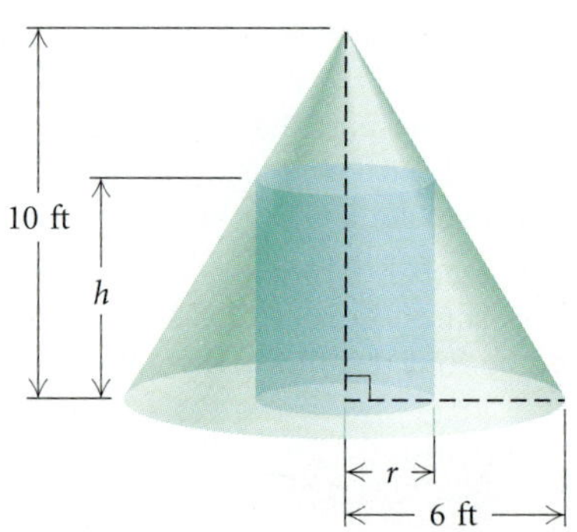

a) Express the height h of the cylinder as a function of r.
b) Express the volume V as a function of r.
c) Express the volume V as a function of h.

1.3

Linear Functions and Applications

- *Graph linear functions and equations, finding the slope and the y-intercept.*
- *Determine equations of lines.*
- *Solve applied problems involving linear functions.*

In real-life situations, we often need to make decisions on the basis of limited information. When the given information is used to formulate an equation or inequality that at least approximates the situation mathematically, we have created a **model**. The most frequently used mathematical models are *linear*—the graphs of these models are straight lines.

Linear Functions

Let's begin to examine the connections among equations, functions, and graphs that are straight lines.

Interactive Discovery

Graph each of the following equations on a grapher. Which have graphs that are lines? Look for patterns.

$$4x + 5y = 20, \qquad y = 5 - x^2,$$

$$y = \frac{x}{2} + 12, \qquad y = 1.5x - 0.05,$$

$$y = -2x - 5, \qquad y = \frac{1}{x},$$

$$x = -3, \qquad y = 6.2,$$

$$y = \sqrt{x}, \qquad x = 4.9.$$

(Some graphers are not able to graph equations of the type $x = a$. Consult your manual.) Which of these equations have graphs that are lines and are also functions?

We have the following results and related terminology.

Linear Functions

A function f is a *linear function* if it is given by $f(x) = mx + b$, where m and b are constants.

If $m = 0$, the function simplifies to the *constant function* $f(x) = b$. If $m = 1$ and $b = 0$, the function simplifies to the *identity function* $f(x) = x$.

Horizontal and Vertical Lines

Horizontal lines are given by equations of the type $y = b$ or $f(x) = b$. (They are functions.)

Vertical lines are given by equations of the type $x = a$. (They are not functions.)

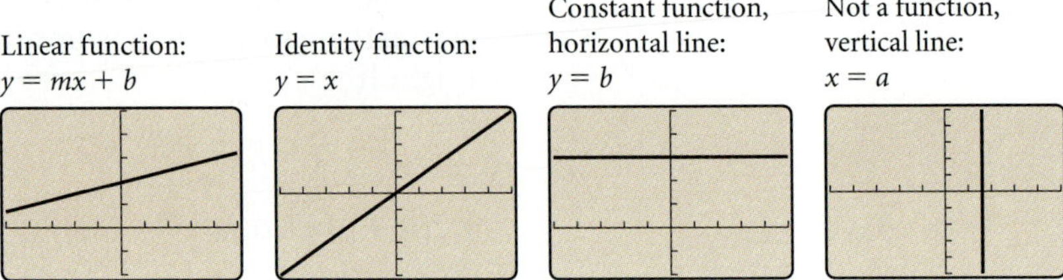

The Linear Function $f(x) = mx + b$ and Slope

To attach meaning to the constant m in the equation $f(x) = mx + b$, we first consider an application. FaxMax is an office machine business. Its total costs for two different time periods are given by two functions shown in the tables and graphs that follow. The variable x represents time, in months. The variable y represents total costs, in thousands of dollars, over that amount of time. Look for a pattern.

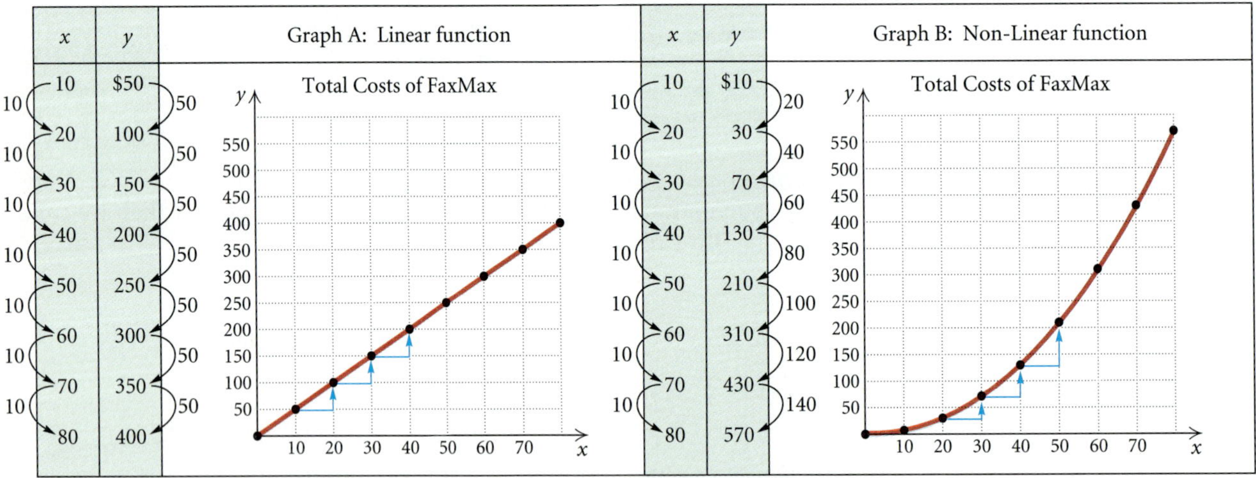

We see in graph A that every change of 10 months results in a $50 thousand change in total costs. But in graph B, changes of 10 months do not result in constant changes in total costs. This is a way to distinguish linear from nonlinear functions. The rate at which a linear function changes, or the steepness of its graph, is constant.

Mathematically, we define a line's steepness, or **slope**, as the ratio of its vertical change (rise) to the corresponding horizontal change (run).

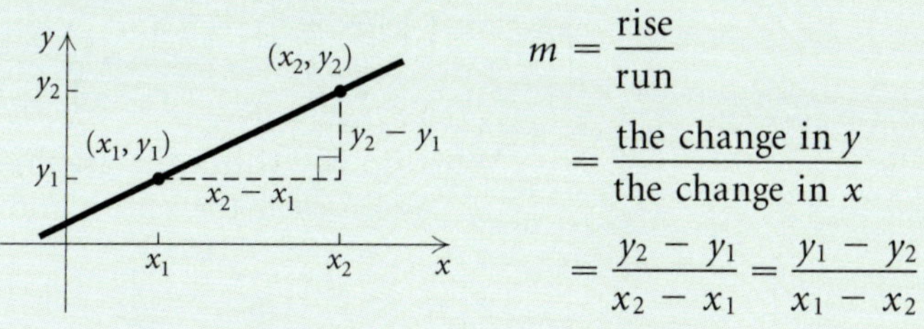

Slope

The *slope* m of a line containing points (x_1, y_1) and (x_2, y_2) is given by

$$m = \frac{\text{rise}}{\text{run}}$$

$$= \frac{\text{the change in } y}{\text{the change in } x}$$

$$= \frac{y_2 - y_1}{x_2 - x_1} = \frac{y_1 - y_2}{x_1 - x_2}.$$

Example 1 Make a hand-drawn graph of the function $f(x) = -\frac{2}{3}x + 1$ and determine its slope.

SOLUTION Since the equation for f is in the form $f(x) = mx + b$, we know it is a linear function. We can graph it by connecting two points on the graph with a straight line. We calculate two ordered pairs, plot the points, graph the function, and determine the slope:

$$f(3) = -\frac{2}{3} \cdot 3 + 1 = -1;$$

$$f(9) = -\frac{2}{3} \cdot 9 + 1 = -5;$$

Pairs: $(3, -1)$, $(9, -5)$;

$$\text{Slope} = m = \frac{y_2 - y_1}{x_2 - x_1}$$

$$= \frac{-5 - (-1)}{9 - 3}$$

$$= \frac{-4}{6} = -\frac{2}{3}.$$

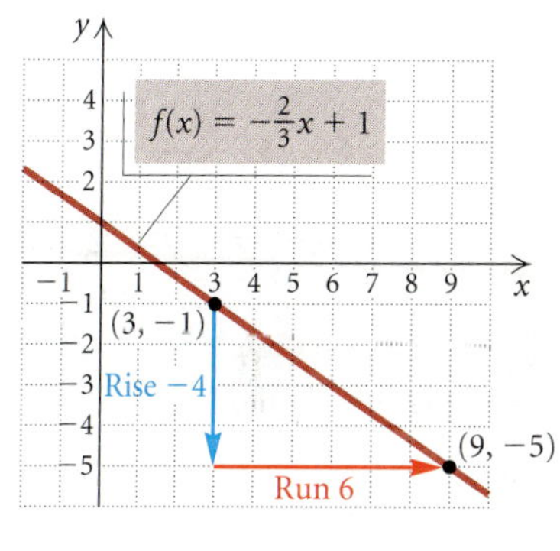

The slope is the same for any two points on a line. Thus, to check our work, note that $f(6) = -\frac{2}{3} \cdot 6 + 1 = -3$. Using the points $(6, -3)$ and $(3, -1)$, we have

$$m = \frac{-3 - (-1)}{6 - 3} = \frac{-2}{3} = -\frac{2}{3}.$$

We can also use the points in the opposite order when computing slope, so long as we are consistent. Note also that the slope of the line is indeed the number m in the equation for the function $f(x) = -\frac{2}{3}x + 1$.

Let's explore the effect of the slope m in linear equations of the type $f(x) = mx$.

Interactive Discovery

With a square viewing window (see the Introduction to Graphs and Graphers), graph the following equations:

$$y_1 = x, \qquad y_2 = 2x, \qquad y_3 = 5x, \quad \text{and} \quad y_4 = 10x.$$

What do you think the graph of $y = 128x$ will look like?

Clear the screen and graph the following equations:

$$y_1 = x, \qquad y_2 = \tfrac{3}{4}x, \qquad y_3 = 0.48x, \quad \text{and} \quad y_4 = \tfrac{3}{25}x.$$

What do you think the graph of $y = 0.000029x$ will look like?

Again clear the screen and graph each set of equations:

$$y_1 = -x, \qquad y_2 = -2x, \qquad y_3 = -4x, \quad \text{and} \quad y_4 = -10x$$

and

$$y_1 = -x, \qquad y_2 = -\tfrac{2}{3}x, \qquad y_3 = -0.35x, \quad \text{and} \quad y_4 = -\tfrac{1}{10}x.$$

From your observations, what do you think the graphs of $y = -200x$ and $y = -0.000017x$ will look like?

If a line slants up from left to right, the change in x and the change in y have the same sign, so the line has a positive slope. The larger the slope is, the steeper the line. If a line slants down from left to right, the change in x and the change in y are of opposite signs, so the line has a negative slope. The larger the absolute value of the slope, the steeper the line.

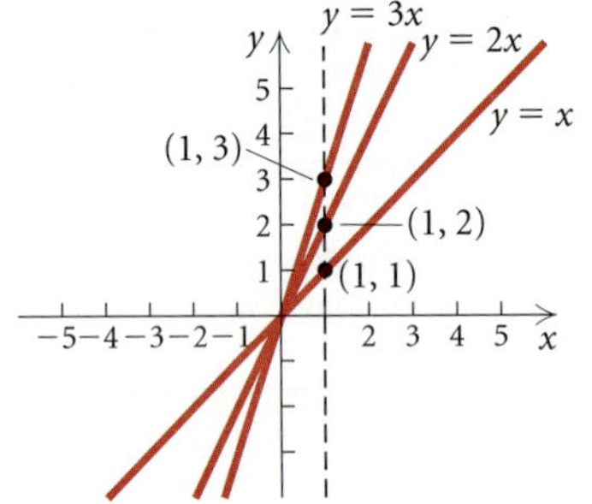

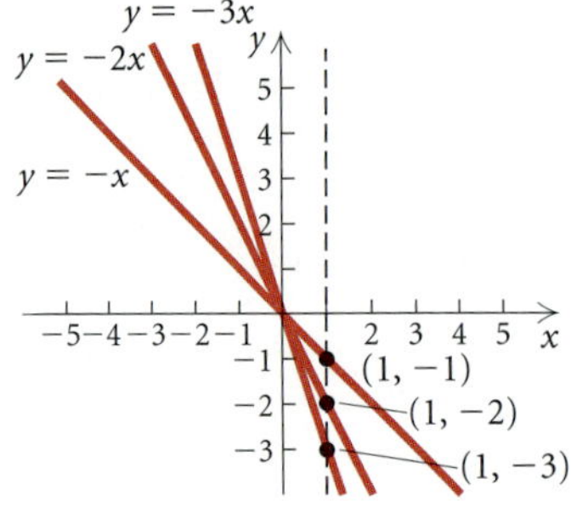

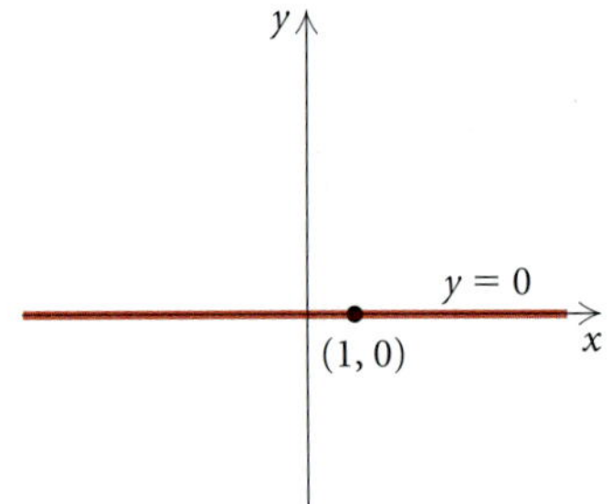

When $m = 0$, $y = 0x$, or $y = 0$, as shown in the graph on the right above. Note that this is both the x-axis and a horizontal line.

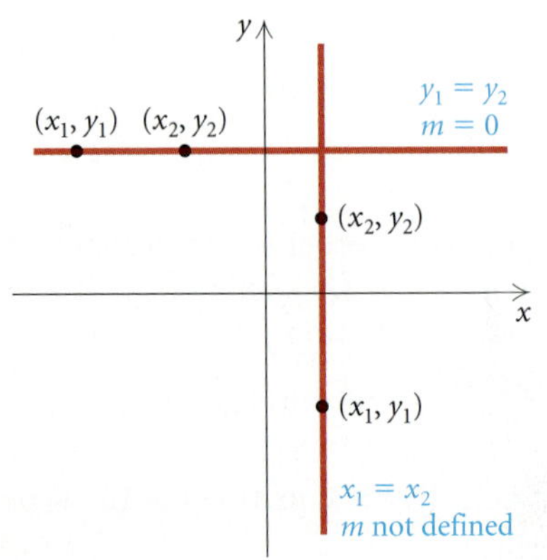

Horizontal and Vertical Lines

If a line is horizontal, the change in y for any two points is 0. Thus a horizontal line has slope 0.

If a line is vertical, the change in x for any two points is 0. Thus the slope is *not defined* because we cannot divide by 0.

Note that zero slope and an undefined slope are two very different concepts.

Applications of Slope

Slope has many real-world applications. For example, numbers like 2%, 4%, and 7% are often used to represent the **grade** of a road. Such a number is meant to tell how steep a road is on a hill or mountain. For example, a 4% grade means that the road rises 4 ft for every horizontal distance of 100 ft if a vehicle is going up; and -4% means that the road is dropping 4 ft for every 100 ft, if the vehicle is going down.

The concept of grade is also used in cardiology when a person runs on a treadmill. A physician may change the slope, or grade, of a treadmill to measure its effect on heart rate. Another example occurs in hydrology. When a river flows, the strength or force of the river depends on how far the river falls vertically compared to how far it flows horizontally.

Example 2   *Ramps for the Handicapped.*   Construction laws regarding access ramps for the handicapped state that every vertical rise of 1 ft requires a horizontal run of 12 ft. What is the grade, or slope, of such a ramp?

SOLUTION   The grade, or slope, is given by

$$m = \frac{1}{12} \approx 0.083 \approx 8.3\%.$$

Slope–Intercept Equations of Lines

Let's explore the effect of the constant b in linear equations of the type $f(x) = mx + b$.

Interactive Discovery

Begin with the graph of $y = x$ and a square viewing window. Now graph the lines $y = x + 3$ and $y = x - 4$ in the same viewing window. How do the lines differ from $y = x$? What do you think the line $y = x - 6$ will look like?

   Try graphing $y = -0.5x$, $y = -0.5x - 4$, and $y = -0.5x + 3$ in the same viewing window. Describe what happens to the graph of $y = -0.5x$ when a number b is added.

Compare the graphs of the equations

$$y = 3x \quad \text{and} \quad y = 3x - 2.$$

Note that the graph of $y = 3x - 2$ is a shift down of the graph of $y = 3x$, and that $y = 3x - 2$ has y-intercept $(0, -2)$. That is, the graph crosses the y-axis at $(0, -2)$.

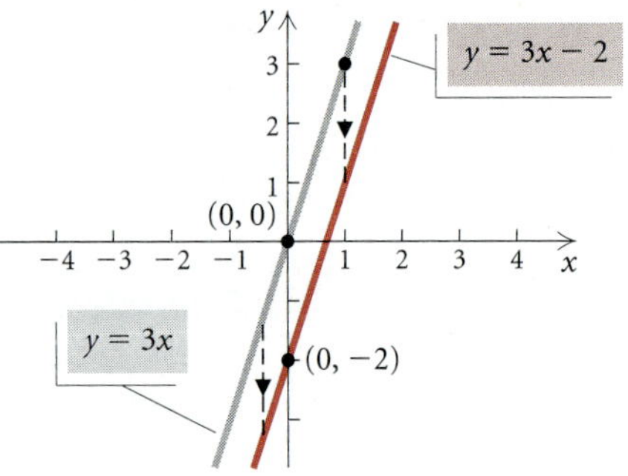

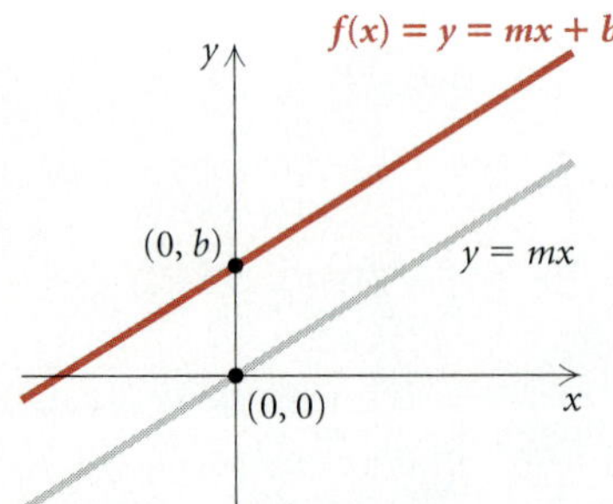

The Slope–Intercept Equation

The linear function f given by

$$f(x) = mx + b$$

has a graph that is a straight line parallel to $y = mx$. The constant m is called the slope, and the y-intercept is $(0, b)$, or b.

   We know that a nonvertical line has slope m and y-intercept $(0, b)$ and that its equation is given by $y = mx + b$. The advantage of $y = mx + b$ is that we can read the slope m and the y-intercept b directly from the equation.

Example 3 Find the slope and the y-intercept of the line with equation $y = -0.25x - 3.8$.

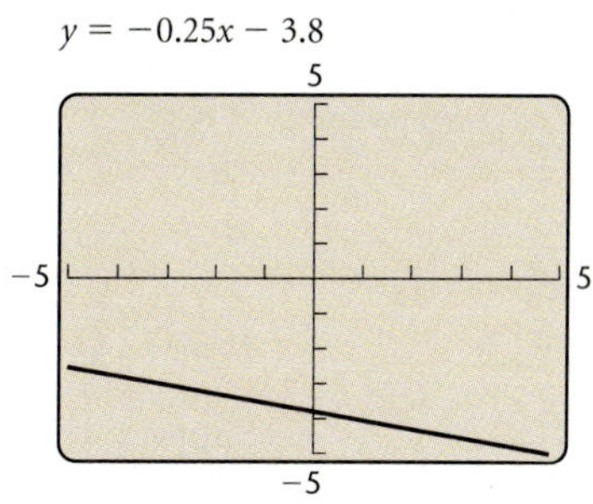

SOLUTION

$$y = \underbrace{-0.25}x \; \underbrace{- \; 3.8}$$

Slope $= -0.25$; y-intercept $= (0, -3.8)$, or -3.8

Example 4 Find the slope and the y-intercept of the line with equation $y = 8$.

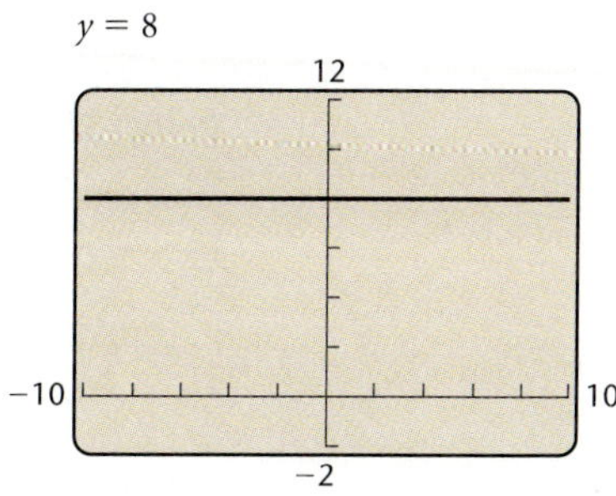

SOLUTION We can rewrite this equation as $y = 0x + 8$. We then see that the slope is 0 and the y-intercept is $(0, 8)$, or 8. The graph is a horizontal line.

Any equation whose graph is a straight line is a **linear equation.** To find the slope and the y-intercept of the graph of a linear equation, we can solve for y, and then read the information from the equation.

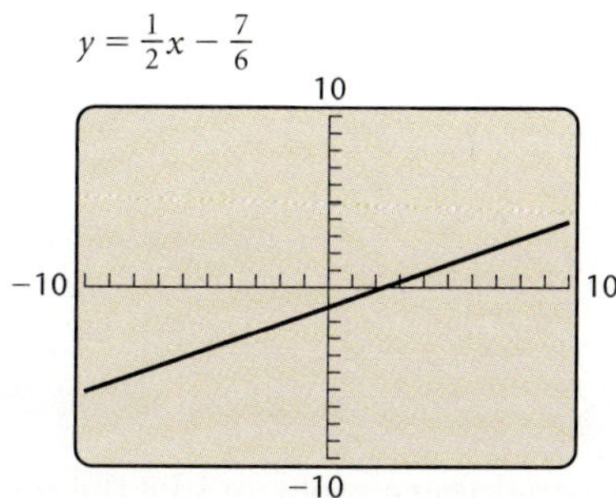

Example 5 Find the slope and the y-intercept of the line with equation $3x - 6y - 7 = 0$.

SOLUTION We solve for y, obtaining $y = \frac{1}{2}x - \frac{7}{6}$. Thus the slope is $\frac{1}{2}$ and the y-intercept is $\left(0, -\frac{7}{6}\right)$, or $-\frac{7}{6}$.

Example 6   A line has slope $-\frac{7}{9}$ and y-intercept $(0, 16)$. Find an equation of the line.

SOLUTION   We use the slope–intercept equation and substitute $-\frac{7}{9}$ for m and 16 for b:

$$y = mx + b$$
$$y = -\tfrac{7}{9}x + 16.$$

Point–Slope Equations of Lines

Suppose that we have a nonvertical line and that the coordinates of point P_1 are (x_1, y_1). We can think of P_1 as fixed and imagine a movable point P on the line with coordinates (x, y). Thus the slope is given by

$$\frac{y - y_1}{x - x_1} = m.$$

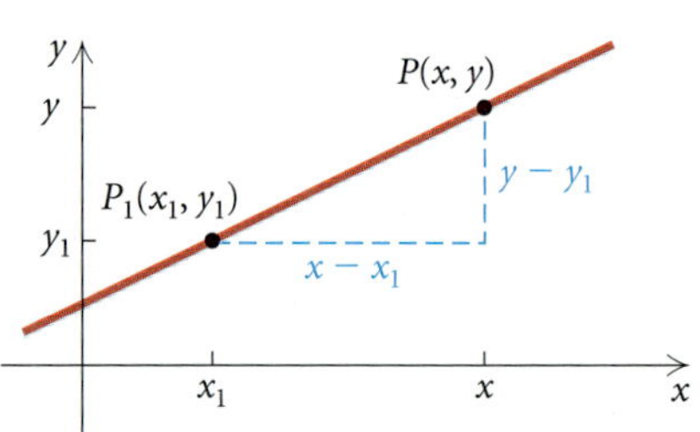

Multiplying on both sides by $x - x_1$, we get the *point–slope equation.*

> **Point–Slope Equation**
>
> The *point–slope equation* of the line with slope m passing through (x_1, y_1) is
>
> $$y - y_1 = m(x - x_1).$$

Thus if we know the slope of a line and the coordinates of one point on the line, we can find an equation of the line.

Example 7   Find an equation of the line containing the point $\left(\frac{1}{2}, -1\right)$ and with slope 5.

SOLUTION   If we substitute in $y - y_1 = m(x - x_1)$, we get

$$y - (-1) = 5\left(x - \tfrac{1}{2}\right),$$

which simplifies as follows:

$$y + 1 = 5x - \tfrac{5}{2}$$
$$y = 5x - \tfrac{5}{2} - 1$$
$$y = 5x - \tfrac{7}{2}. \qquad \textbf{Slope-intercept equation}$$

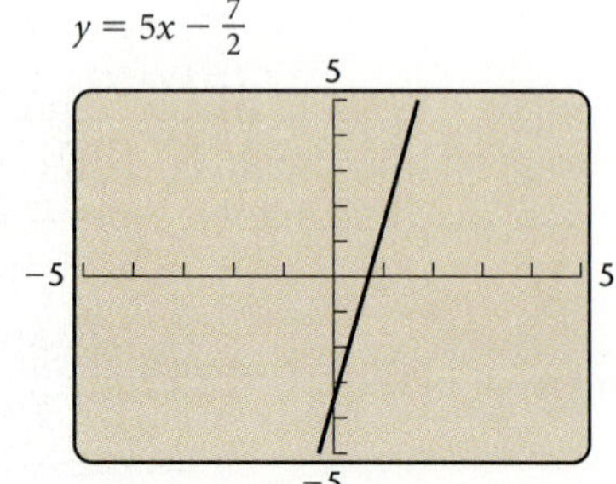

The advantage of having a grapher when doing a problem like the one in Example 7 is that you have a quick, visual way of seeing your result.

Two-Point Equations of Lines

Suppose that a nonvertical line contains the points $P_1(x_1, y_1)$ and $P_2(x_2, y_2)$. The slope of the line is

$$m = \frac{y_2 - y_1}{x_2 - x_1}.$$

If we substitute $\dfrac{y_2 - y_1}{x_2 - x_1}$ for m in the point–slope equation,

$$y - y_1 = m(x - x_1),$$

we get the *two-point equation*.

Two-Point Equation

The *two-point equation* of the line passing through the points (x_1, y_1) and (x_2, y_2) is

$$y - y_1 = \frac{y_2 - y_1}{x_2 - x_1}(x - x_1).$$

Example 8   Find an equation of the line containing the points $(2, 3)$ and $(1, -4)$.

SOLUTION   If we take $(2, 3)$ as P_1 and $(1, -4)$ as P_2 and use the two-point equation, we get

$$y - 3 = \frac{-4 - 3}{1 - 2}(x - 2),$$

which simplifies to $y = 7x - 11$.

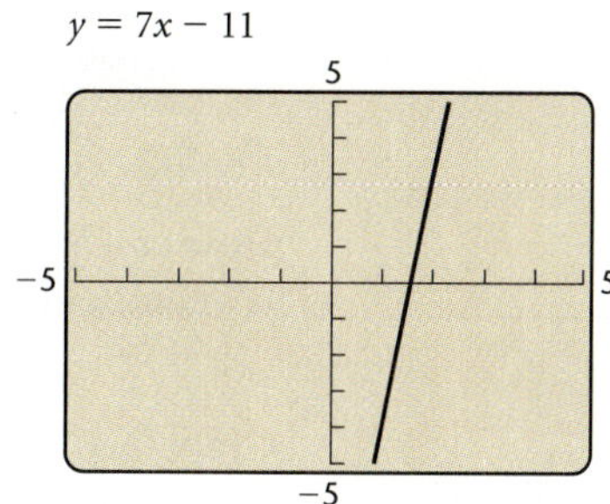

Parallel Lines

How can we examine a pair of equations to determine whether their graphs are parallel—that is, they do not intersect? We can explore this with a grapher.

Interactive Discovery

Graph each of the following pairs of lines. Try to determine whether they are parallel. You will probably need to change viewing windows and either zoom in or zoom out to decide for sure. Complete the table and look for a pattern.

	SLOPES		
PAIRS OF EQUATIONS	m_1	m_2	PARALLEL?
$y = -0.38x + 4.2,\ y = -0.37x - 5.1$			
$y = -0.38x + 4.2,\ 50y + 19x = 21$			
$2x + 1 = y,\ y + 3x = 4$			
$y = 8.02x - 11.3,\ y = 7.98x - 11.3$			
$y = -3,\ y = 4.2$			
$x = -5,\ x = 2$			

Can you find a way to know whether two lines are parallel without graphing?

If two different lines are vertical, then they are parallel. Thus two equations such as $x = a_1$, $x = a_2$, where $a_1 \neq a_2$, have graphs that are *parallel lines*. Two nonvertical lines such as $y = mx + b_1$, $y = mx + b_2$, where $b_1 \neq b_2$, also have graphs that are *parallel lines*.

Parallel Lines

Vertical lines are *parallel*. Nonvertical lines are *parallel* if and only if they have the same slope and different y-intercepts.

Perpendicular Lines

How can we examine a pair of equations to determine whether they are perpendicular? Let's explore this with a grapher.

Interactive Discovery

Graph each of the following pairs of lines (see the table at the top of the following page). Try to determine whether they are perpendicular. You will probably need to change viewing windows and either zoom in or zoom out to decide for sure. However, you must use square viewing windows so that you can see or measure whether angles between lines seem to be 90°. Complete the table and look for a pattern.

(continued)

| | | SLOPES | | PRODUCT OF SLOPES | |
PAIRS OF EQUATIONS		m_1	m_2	$m_1 \cdot m_2$	PERPENDICULAR?
$y = -\frac{2}{5}x + 3, \ y = \frac{5}{2}x + 3$					
$y = 0.1875x - 2, \ y = -\frac{16}{3}x - 5.1$					
$3y + 4x = -21, \ 4y - 3x = 8$					
$y = -3x + 4, \ y = 2x - 7$					
$x = -5, \ y = 2$					

Can you find a way to determine whether two lines are perpendicular without graphing?

Perpendicular Lines

Two lines with slopes m_1 and m_2 are *perpendicular* if and only if the product of their slopes is -1:

$$m_1 m_2 = -1.$$

Lines are also *perpendicular* if one is vertical ($x = a$) and the other is horizontal ($y = b$).

If a line has slope m_1, the slope m_2 of a line perpendicular to it is $-1/m_1$ (the slope of one line is the opposite of the reciprocal of the other).

Example 9 Determine whether each of the following pairs of lines is parallel, perpendicular, or neither.

a) $y + 2 = 5x, \ 5y + x = -15$

b) $2y + 4x = 8, \ 5 + 2x = -y$

c) $2x + 1 = y, \ y + 3x = 4$

SOLUTION We use an algebraic procedure to know for sure. We can create graphs on a grapher as a check.

a) We solve each equation for y:

$$y = 5x - 2, \qquad y = -\frac{1}{5}x - 3.$$

The slopes are 5 and $-\frac{1}{5}$. Their product is -1, so the lines are perpendicular (see the figure at left).

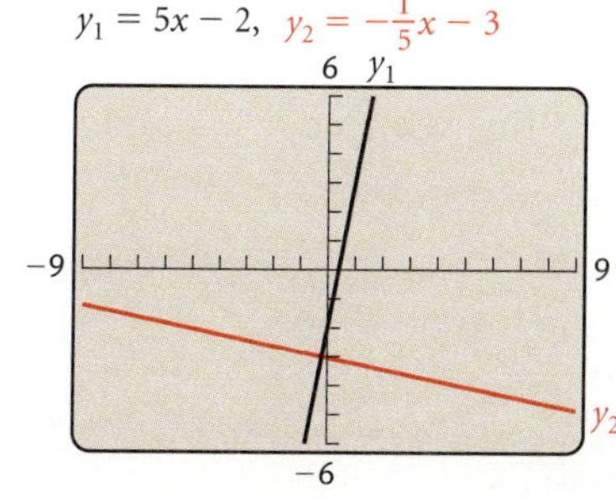

b) Solving each equation for y, we get

$$y = -2x + 4, \qquad y = -2x - 5.$$

We see that $m_1 = -2$ and $m_2 = -2$. Since the slopes are the same and the y-intercepts, 4 and -5, are different, the lines are parallel (see the figure on the left below).

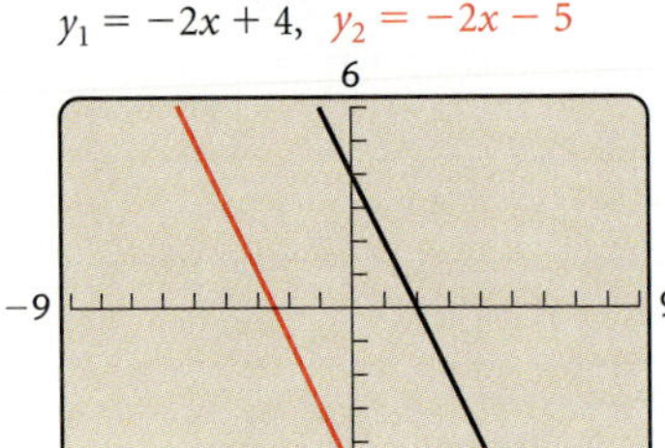

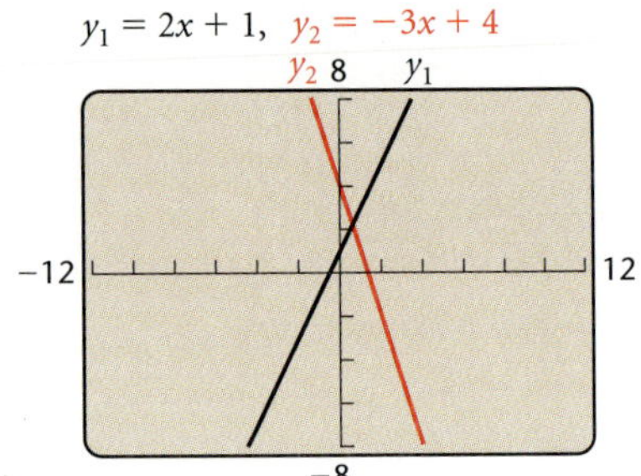

c) Solving each equation for y, we get

$$y = 2x + 1, \qquad y = -3x + 4.$$

Since $m_1 = 2$ and $m_2 = -3$, we know that $m_1 \neq m_2$ and that $m_1 m_2 \neq -1$. It follows that the lines are neither perpendicular nor parallel (see the figure on the right above).

Example 10   Write equations of the lines (a) parallel and (b) perpendicular to the graph of the line $4y - x = 20$ and both containing the point $(2, -3)$.

SOLUTION   We first solve $4y - x = 20$ for y to get $y = \frac{1}{4}x + 5$. Thus the slope is $\frac{1}{4}$.

a) The line parallel to the given line will have slope $\frac{1}{4}$. We use the point–slope equation for a line with slope $\frac{1}{4}$ and containing the point $(2, -3)$:

$$y - y_1 = m(x - x_1)$$
$$y - (-3) = \tfrac{1}{4}(x - 2)$$
$$y = \tfrac{1}{4}x - \tfrac{7}{2}.$$

b) The slope of the perpendicular line is -4. Now we use the point–slope equation to write an equation for a line with slope -4 and containing the point $(2, -3)$:

$$y - y_1 = m(x - x_1)$$
$$y - (-3) = -4(x - 2)$$
$$y = -4x + 5.$$

We can visualize our work on a grapher.

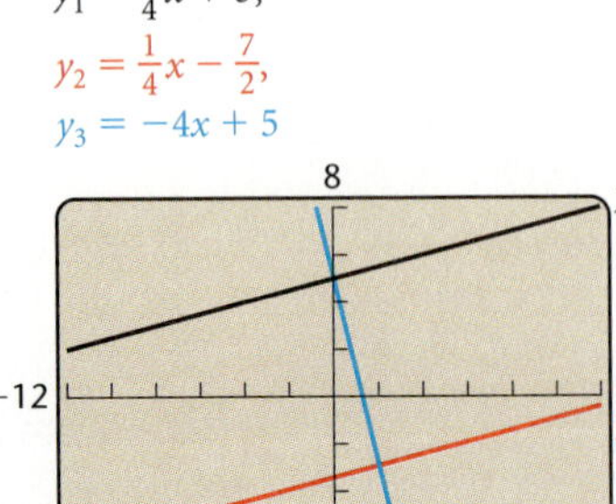

Summary of Terminology About Lines

TERMINOLOGY	MATHEMATICAL INTERPRETATION
Slope	$m = \dfrac{y_2 - y_1}{x_2 - x_1}$
Slope–intercept equation	$y = mx + b$
Point–slope equation	$y - y_1 = m(x - x_1)$
Two-point equation	$y - y_1 = \dfrac{y_2 - y_1}{x_2 - x_1}(x - x_1)$
Horizontal lines	$y = b$
Vertical lines	$x = a$
Parallel lines	$m_1 = m_2,\ b_1 \neq b_2$
Perpendicular lines	$m_1 m_2 = -1$

Applications of Linear Functions

We now consider an application of linear functions.

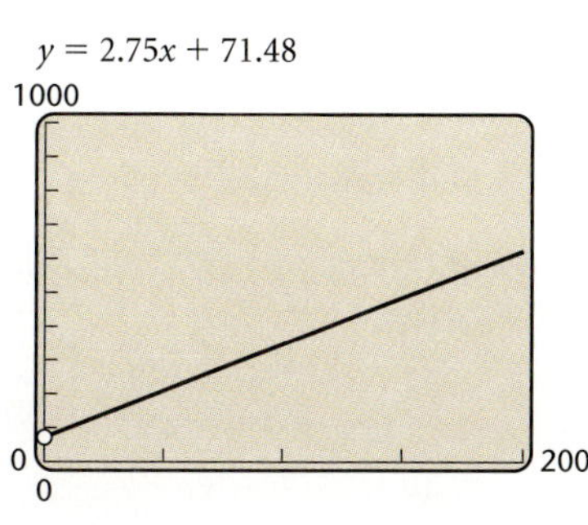

Example 11 *Anthropology Estimates.* An anthropologist can use certain linear functions to estimate the height of a male or a female, given the length of certain bones. The humerus is the bone from the elbow to the shoulder. Let $x = $ the length of the humerus, in centimeters. Then the height, in centimeters, of an adult male with a humerus of length x is given by the function

$$M(x) = 2.89x + 70.64.$$

The height, in centimeters, of an adult female with a humerus of length x is given by the function

$$F(x) = 2.75x + 71.48.$$

a) A 26-cm humerus was uncovered in a ruins. Assuming it was from a female, how tall was she?

b) Graph F.

c) Find the domain of F.

SOLUTION

a) We substitute into the function:

$$F(26) = 2.75(26) + 71.48 = 142.98.$$

Thus the female was 142.98 cm tall.

b) The graph is shown at left.

c) Theoretically, the domain of the function is the set of all real numbers. However, the context of the problem dictates a different domain. One could not find a bone with a length of 0 or less. Thus the domain consists of positive real numbers, that is, the interval $(0, \infty)$.

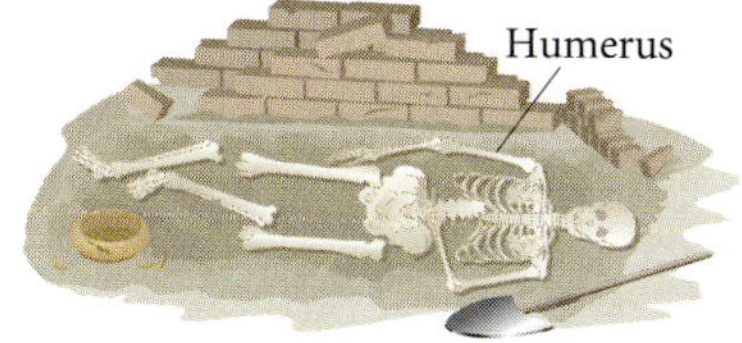

1.3 Exercise Set

Each table of data contains input–output values for a function. Answer the following questions for each table.

a) *Is the change in the inputs x the same?*
b) *Is the change in the outputs y the same?*
c) *Is the function linear?*

1.

x	y
-3	7
-2	10
-1	13
0	16
1	19
2	22
3	25

2.

x	y
20	12.4
30	24.8
40	49.6
50	99.2
60	198.4
70	396.8
80	793.6

3.

x	y
11	3.2
26	5.7
41	8.2
56	9.3
71	11.3
86	13.7
101	19.1

4.

x	y
2	-8
4	-12
6	-16
8	-20
10	-24
12	-28
14	-36

Find the slope of the line containing the given points.

5.
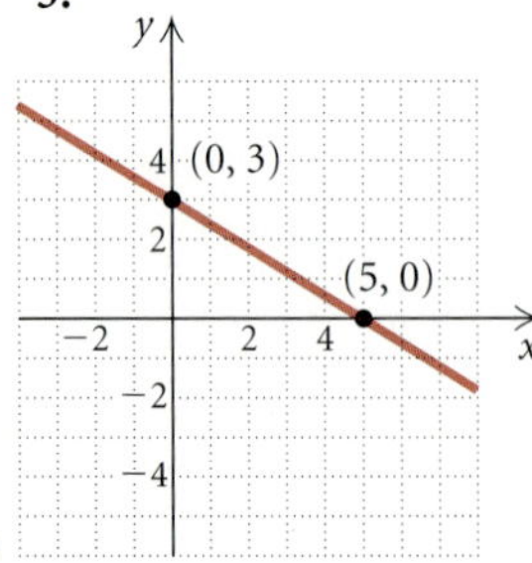

6.
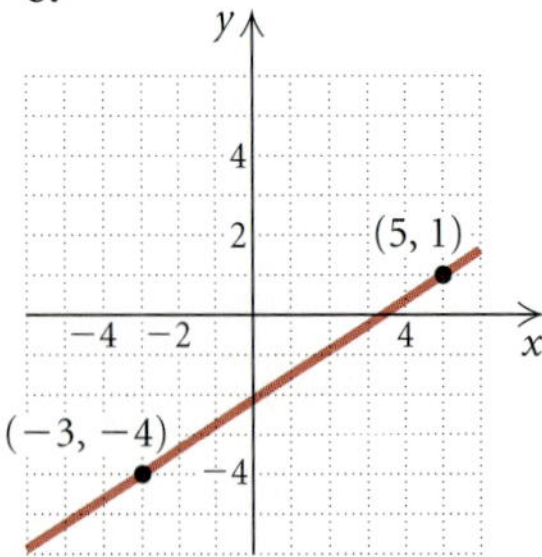

7. $(9, 4)$ and $(-1, 2)$

8. $(-3, 7)$ and $(5, -1)$

9. $(4, -9)$ and $(-5, 6)$

10. $(-6, -1)$ and $(2, -13)$

11. $(\pi, -3)$ and $(\pi, 2)$

12. $(\sqrt{2}, -4)$ and $(0.56, -4)$

13. (a, a^2) and $(a + h, (a + h)^2)$

14. $(a, 3a + 1)$ and $(a + h, 3(a + h) + 1)$

15. *Road Grade.*   Using the following figure, find the road grade and an equation giving the height y as a function of the horizontal distance x.

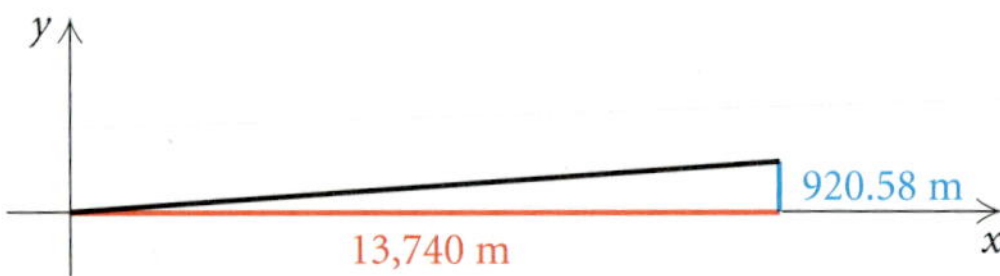

16. *Heart Arrhythmia.*   A treadmill is 5 ft long and is set at an 8% grade when a heart arrhythmia occurs. How high is the end of the treadmill?

Find the slope and the y-intercept of the equation.

17. $y = \frac{3}{5}x - 7$

18. $f(x) = -2x + 3$

19. $f(x) = 5 - \frac{1}{2}x$

20. $y = 2 + \frac{3}{7}x$

21. $3x + 2y = 10$

22. $2x - 3y = 12$

23. $4x - 3f(x) - 15 = 6$

24. $9 = 3 + 5x - 2f(x)$

Write a slope–intercept equation for a line with the given characteristics.

25. $m = \frac{2}{9}$, y-intercept $(0, 4)$

26. $m = -\frac{3}{8}$, y-intercept $(0, 5)$

27. $m = -4$, y-intercept $(0, -7)$

28. $m = \frac{2}{7}$, y-intercept $(0, -6)$

29. $m = -4.2$, y-intercept $(0, \frac{3}{4})$

30. $m = -4$, y-intercept $(0, -\frac{3}{2})$

31. $m = \frac{2}{9}$, passes through $(3, 7)$

32. $m = -\frac{3}{8}$, passes through $(5, 6)$

33. $m = 3$, passes through $(1, -2)$

34. $m = -2$, passes through $(-5, 1)$

35. $m = -\frac{3}{5}$, passes through $(-4, -1)$

36. $m = \frac{2}{3}$, passes through $(-4, -5)$

37. Passes through $(-1, 5)$ and $(2, -4)$

38. Passes through $(2, -1)$ and $(7, -11)$

39. Passes through $(7, 0)$ and $(-1, 4)$

40. Passes through $(-3, 7)$ and $(-1, -5)$

Determine whether each of the following pairs of lines is parallel, perpendicular, or neither.

41. $x + 2y = 5$,
$\quad 2x + 4y = 8$

42. $2x - 5y = -3$,
$\quad 2x + 5y = 4$

43. $y = 4x - 5$,
$\quad 4y = 8 - x$

44. $y = 7 - x$,
$\quad y = x + 3$

Write a slope–intercept equation for a line passing through the given point that is parallel to the given line. Then write a second equation for a line passing through the given point that is perpendicular to the given line.

45. $(3, 5)$, $y = \frac{2}{7}x + 1$

46. $(-1, 6)$, $f(x) = 2x + 9$

47. $(-7, 0)$, $y = -0.3x + 4.3$

48. $(-4, -5)$, $2x + y = -4$

49. $(3, -2)$, $3x + 4y = 5$

50. $(8, -2)$, $y = 4.2(x - 3) + 1$

51. $(3, -3)$, $x = -1$

52. $(4, -5)$, $y = -1$

53. *Ideal Weight.* One formula to estimate the ideal weight of a woman is to multiply her height by 3.5 and subtract 110. Let $W =$ the ideal weight, in pounds, and $h =$ height, in inches.

a) Express W as a linear function of h.
b) Find the ideal weight of a woman whose height is 62 in.
c) Graph W.
d) Find the domain of the function.

54. *Pressure at Sea Depth.* The function P, given by
$$P(d) = \tfrac{1}{33}d + 1,$$
gives the pressure, in atmospheres (atm), at a depth d, in feet, under the sea.

a) Find $P(0)$, $P(5)$, $P(10)$, $P(33)$, and $P(200)$.
b) Graph P.
c) Find the domain of the function.

55. *Stopping Distance on Glare Ice.* The stopping distance (at some fixed speed) of regular tires on glare ice is a function of the air temperature F, in degrees Fahrenheit. This function is estimated by
$$D(F) = 2F + 115,$$
where $D(F) =$ the stopping distance, in feet, when the air temperature is F, in degrees Fahrenheit.

a) Find $D(0°)$, $D(-20°)$, $D(10°)$, and $D(32°)$.
b) Graph D.
c) Explain why the domain should be restricted to $[-57.5°, 32°]$.

56. *Anthropology Estimates.* Consider Example 10 and the function
$$M(x) = 2.89x + 70.64$$
for estimating the height of a male.

a) Find the height of a male if a 26-cm humerus is found in an archeological dig.
b) Graph M.
c) Find the domain of M.

57. *Reaction Time.* While driving a car, you suddenly see a school crossing guard. Your brain registers the information and sends a signal to your foot to hit the brake. The car travels a distance D, in feet, during this time, where D is a function of the speed r, in miles per hour, that the car is traveling when you see the crossing guard. That reaction distance is a linear function given by
$$D(r) = \frac{11r + 5}{10}.$$

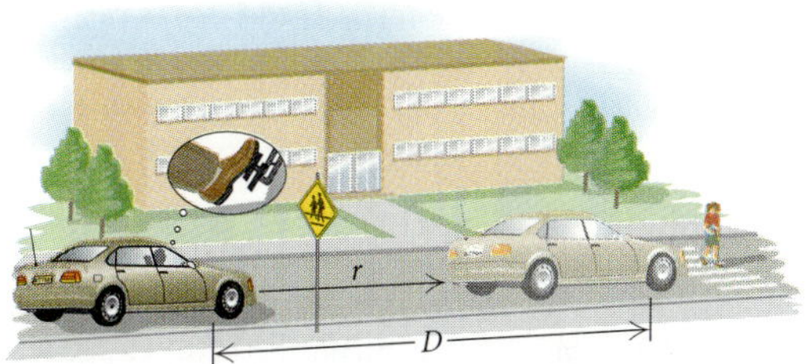

a) Find the slope of this line and interpret its meaning in this application.
b) Find $D(5)$, $D(10)$, $D(20)$, $D(50)$, and $D(65)$.
c) Graph D.
d) What is the domain of this function? Explain.

58. *Straight-line Depreciation.* A company buys a new color printer for $5200 to print banners for a sales campaign. The printer is purchased on January 1 and is expected to last 8 yr, at the end of which time its *trade-in*, or *salvage value*, will be $1100. If the company figures the decline or depreciation in value to be the same each year, then the salvage value V, after t years, is given by the linear function
$$V(t) = \$5200 - \$512.50t, \quad \text{for } 0 \le t \le 8.$$

a) Find $V(0)$, $V(1)$, $V(2)$, $V(3)$, and $V(8)$.
b) Graph V.
c) Find the domain and the range of this function.

For Exercises 59 and 60, express the slope as a ratio of two quantities.

59.

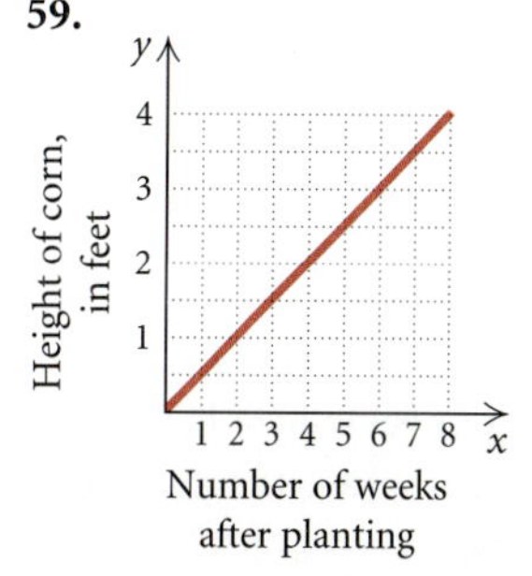

60.

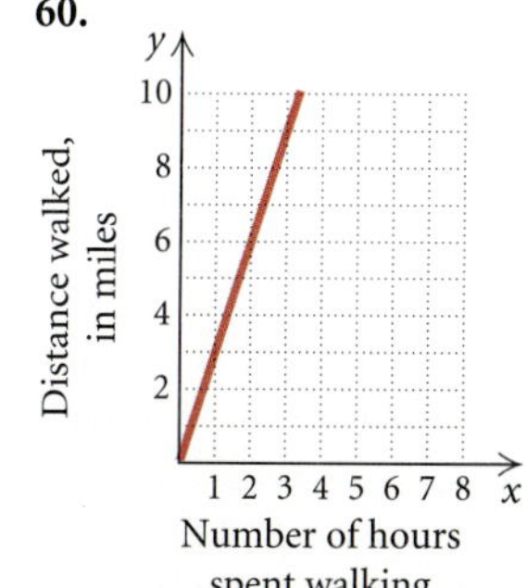

61. *Total Cost.*   The Cellular Connection charges $60 for a phone and $40 per month under its economy plan. Write and graph an equation that can be used to determine the total cost, $C(t)$, of operating a Cellular Connection phone for t months.

62. *Total Cost.*   Twin Cities Cable Television charges a $35 installation fee and $30 per month for "deluxe" service. Write and graph an equation that can be used to determine the total cost, $C(t)$, for t months of deluxe cable television service. Find the total cost for 8 months of service.

*In Exercises 63 and 64, the term **fixed costs** refers to the start-up costs of operating a business. This includes machinery and building costs. The term **variable costs** refers to what it costs a business to produce or service one item.*

63. Kara's Custom Tees experienced fixed costs of $800 and variable costs of $3 per shirt. Write an equation that can be used to determine the total expenses encountered by Kara's Custom Tees. Then graph the equation.

64. It's My Racquet experienced fixed costs of $500 and variable costs of $2 for each tennis racquet that is restrung. Write an equation that can be used to determine the total expenses encountered by It's My Racquet. Then graph the equation.

Direct Variation.   Many applications can be modeled by linear functions like

$$y = kx, \quad k > 0,$$

*where the variables are nonnegative: These are functions of **direct variation**. The constant k is called a **variation constant**. Note that the function is increasing over the interval $[0, \infty)$.*

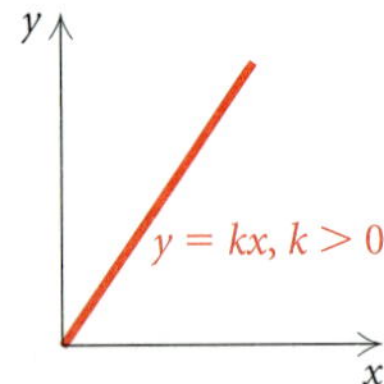

65. *House of Representatives.*   The number of representatives N that each state has varies directly as the number of people P living in the state. If New York, with 18,119,416 residents, has 31 representatives, how many representatives does Michigan, with a population of 9,436,628, have?

66. *Hooke's Law.*   Hooke's law states that the distance d that a spring will stretch varies directly as the mass m of an object hanging from the spring.

Suppose that a 3-kg mass stretches a spring 40 cm. Find a function of direct variation. Then use the equation to predict how far the spring will stretch with a 5-kg mass.

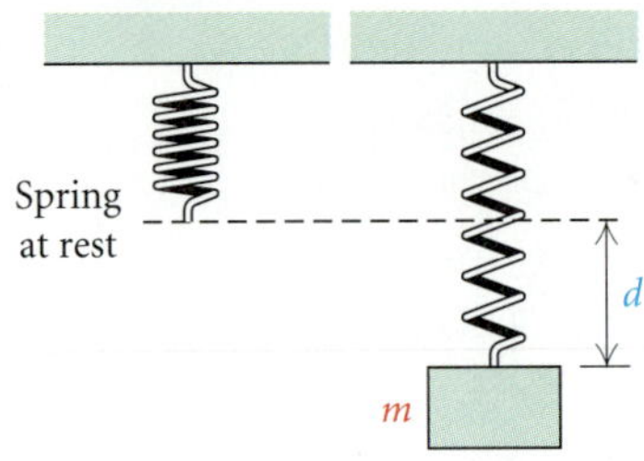

Skill Maintenance

If $f(x) = x^2 - 3x$, find each of the following.

67. $f(-5)$ **68.** $f(5)$

69. $f(-a)$ **70.** $f(a + h)$

Synthesis

71. ◆ Discuss why the graph of a vertical line $x = a$ cannot be that of a function.

72. ◆ Explain as you would to a fellow student how the idea of slope can be used to describe the slant and the steepness of a line using a number.

73. Find k so that the line containing the points $(-3, k)$ and $(4, 8)$ is parallel to the line containing the points $(5, 3)$ and $(1, -6)$.

74. Find an equation of the line passing through the point $(4, 5)$ perpendicular to the line passing through the points $(-1, 3)$ and $(2, 9)$.

75. *Fahrenheit and Celsius Temperatures.*   Fahrenheit temperature F is a linear function of Celsius temperature C. When C is 0, F is 32; and when C is 100, F is 212. Use these data to express C as a function of F and to express F as a function of C.

Suppose that f is a linear function. Then $f(x) = mx + b$. Determine whether each of the following is true or false.

76. $f(c + d) = f(c) + f(d)$

77. $f(cd) = f(c)f(d)$

78. $f(kx) = kf(x)$

79. $f(c - d) = f(c) - f(d)$

Let $f(x) = mx + b$. Find a formula for $f(x)$ given each of the following.

80. $f(3x) = 3f(x)$ **81.** $f(x + 2) = f(x) + 2$

1.4
Data Analysis, Curve Fitting, and Linear Regression

- *Analyze a set of data to determine whether it can be modeled by a linear function.*
- *Fit a regression line to a set of data; then use the linear model to make predictions.*

Mathematical Models

When a real-world problem can be described in mathematical language, we have a **mathematical model**. For example, the natural numbers constitute a mathematical model for situations in which counting is essential. Situations in which algebra can be brought to bear often require the use of functions.

Mathematical models are abstracted from real-world situations. Procedures within the mathematical model then give results that allow one to predict what will happen in that real-world situation. If the predictions are inaccurate or the results of experimentation do not conform to the model, the model needs to be changed or discarded.

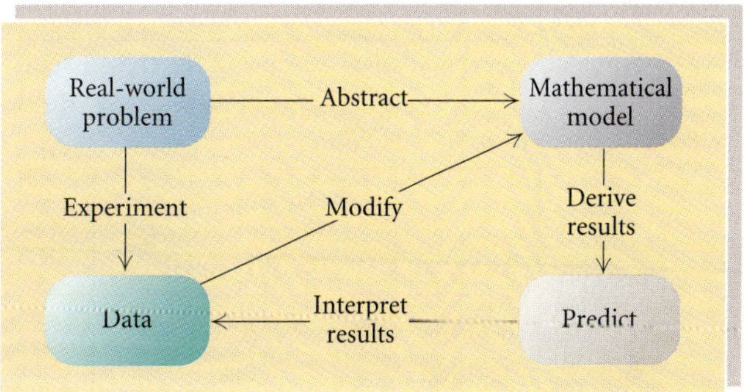

Mathematical modeling can be an ongoing process. For example, finding a mathematical model that will enable an accurate prediction of population growth is not a simple problem. Any population model that one might devise would need to be reshaped as further information is acquired.

Curve Fitting

We will develop and use many kinds of mathematical models in this text. In this chapter, we have considered many functions that can be used as models. Let's look at four of them.

Constant function:
$$y_1 = b$$

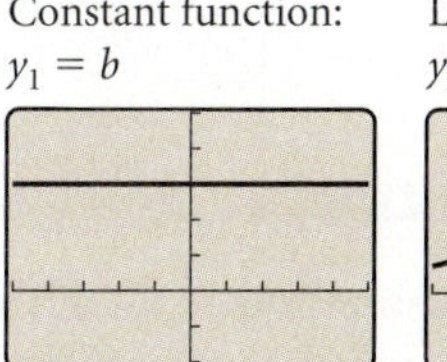

Linear function:
$$y_2 = mx + b$$

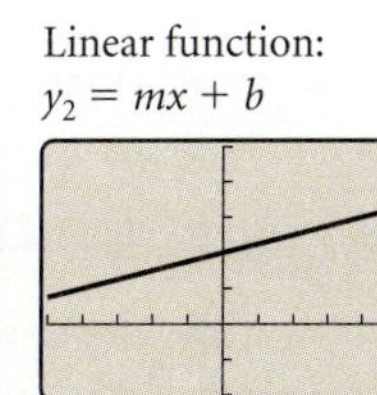

Squaring function:
$$y_3 = x^2$$

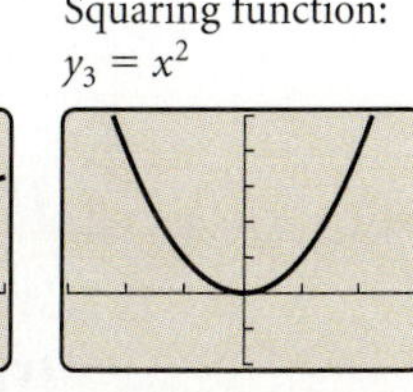

Quadratic function:
$$y_4 = ax^2 + bx + c, a > 0$$

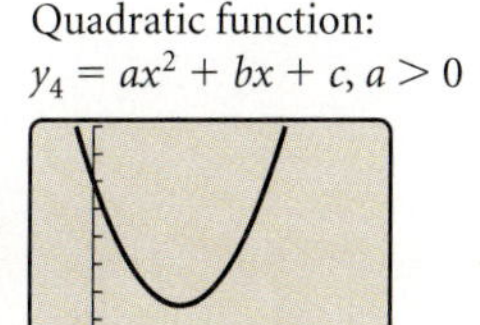

In general, we try to find a function that fits, as well as possible, observations (data), theoretical reasoning, and common sense. We call this **curve fitting**; it is one aspect of mathematical modeling.

Let's first look at some data and related graphs or scatterplots. Do any of the above functions seem to fit either set of data?

Life Expectancy of Women

YEAR, x	LIFE EXPECTANCY, y (IN YEARS)	SCATTERPLOT
0. 1900	49.1	
1. 1910	53.7	
2. 1920	56.3	
3. 1930	61.4	
4. 1940	65.3	
5. 1950	70.9	
6. 1960	73.2	
7. 1970	74.8	
8. 1980	77.5	
9. 1990	78.6	

Source: Statistical Abstract of the United States.

Number of Cellular Phones

YEAR, x	NUMBER OF CELLULAR PHONES, y (IN MILLIONS)	SCATTERPLOT
0. 1985	0.2	
1. 1986	0.4	
2. 1987	1.0	
3. 1988	1.8	
4. 1989	2.8	
5. 1990	4.1	
6. 1991	6.2	
7. 1992	8.6	
8. 1993	13.0	
9. 1994	19.3	

Source: Cellular Telecommunications Industry Association.

Looking at the scatterplots, we see that the life expectancy data seem to be rising in a manner to suggest that a linear function might fit, although a "perfect" straight line cannot be drawn through the data points. A linear function does not seem to fit the cellular phone data. Part of a quadratic function might be close. In fact, it can be modeled by an exponential function that we will consider in Chapter 3.

The Regression Line

We now consider a method to fit a linear function to a set of data: **linear regression**. Although discussion leading to a complete understanding of this method belongs in a statistics course, we present the procedure here because we can carry it out easily using technology. The grapher gives us the powerful capability to find linear models and to make predictions using them.

Example 1 *Predicting the Life Expectancy of Women.*   Consider the data just presented on the life expectancy of women.

a) Fit a regression line to the data using a REGRESSION feature on a grapher.

b) Use the linear model to predict the life expectancy of women in the year 2000.

SOLUTION

a) Using linear regression, we fit a linear equation to the data. We use a grapher and enter the data into a list, like the following. We select the L_1 list to be the independent variable (x) and the L_2 list to be the dependent variable (y). The grapher can then create a scatterplot of the data, as shown on the right.

L1	L2	L3
0	49.1	------
1	53.7	
2	56.3	
3	61.4	
4	65.3	
5	70.9	
6	**73.2**	

L2(7) = 73.2

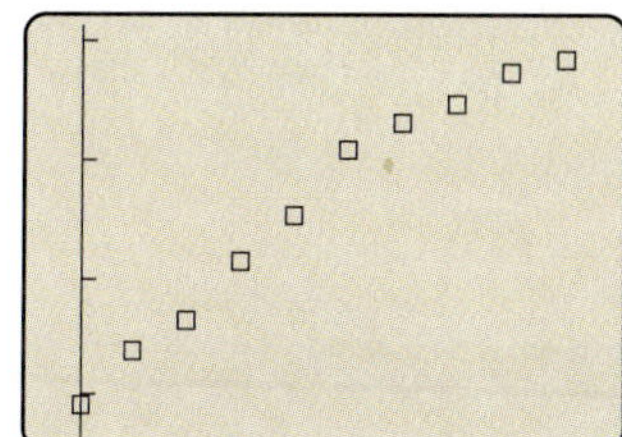

Consider the data points $(0, 49.1)$, $(1, 53.7)$, $(2, 56.3)$, ..., $(9, 78.6)$ on the graph. We want to fit a **regression line**,

$$y = mx + b,$$

to these data points. We use the REGRESSION feature, LINREG(AX + B), to find the regression line. The result is

$$y = 3.4x + 50.7. \qquad \text{Regression line}$$

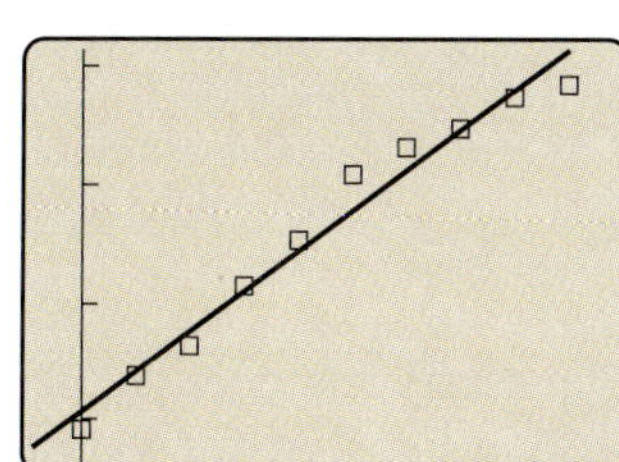

We can then graph the regression line on the same graph as the scatterplot.

b) To predict the life expectancy of women in the year 2000, we substitute the corresponding x-value, $x = 10$, into the formula for the regression function:

$$y = 3.4(10) + 50.7 = 84.7. \qquad \text{In 1980, } x = 8; \text{ in 1990, } x = 9. \\ \text{Thus in 2000, } x = 10.$$

We can also use the grapher to do the calculation. Thus we estimate the life expectancy of women in the year 2000 to be 84.7 yr.

The understanding behind the development of the regression line is best left to a course in calculus and/or statistics.

The Correlation Coefficient

On some graphers, a constant r in $[-1, 1]$, called the **coefficient of linear correlation**, appears with the equation of the regression line. Though we cannot develop a formula for calculating r in this text, keep in mind that r is a number for which $-1 \leq r \leq 1$. It is used to describe the strength of the linear relationship between x and y. The closer $|r|$ is to 1, the better the correlation. A positive value of r also indicates that a positive slope and an increasing function exist. A negative value of r indicates a negative slope and a decreasing function. For the life expectancy data just discussed, $r = 0.9861$, which indicates a very good linear correlation. The following scatterplots summarize the interpretation of a correlation coefficient.

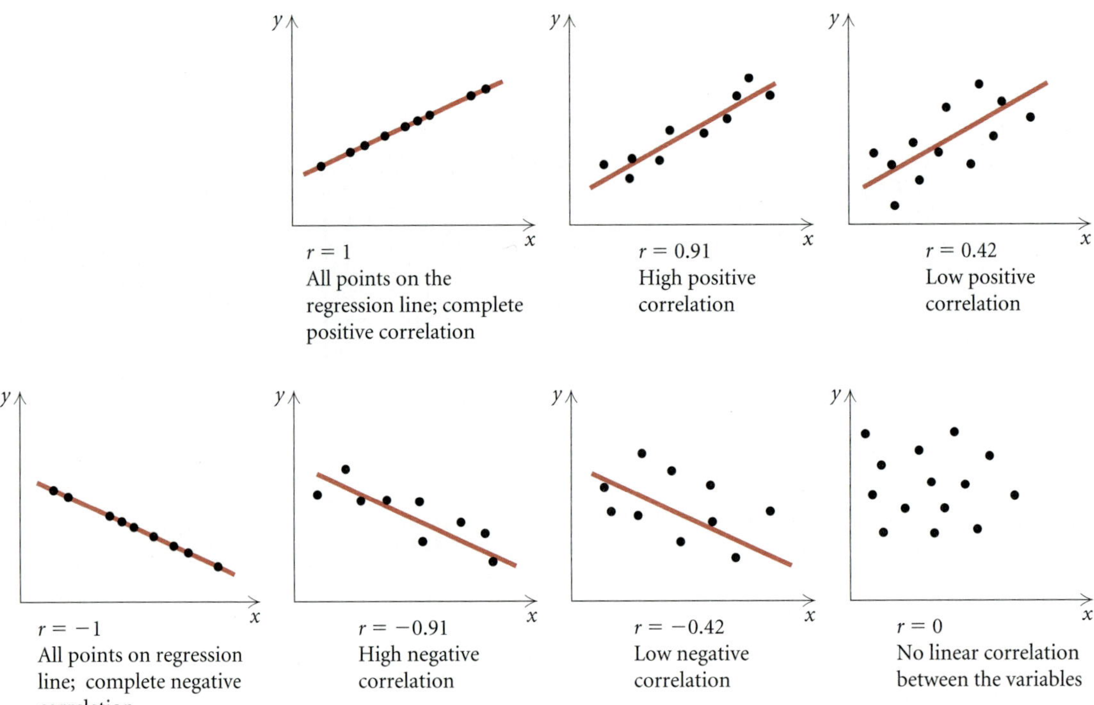

Keep in mind that a high linear correlation coefficient does not necessarily indicate a "cause-and-effect" connection between the variables. We might be able to calculate a high positive correlation between stock prices and rainfall in India, but before we place our life savings in the market, further analysis and common sense should be applied!

1.4 | *Exercise Set*

In Exercises 1–4, determine whether the graph might be modeled by a linear function.

1.

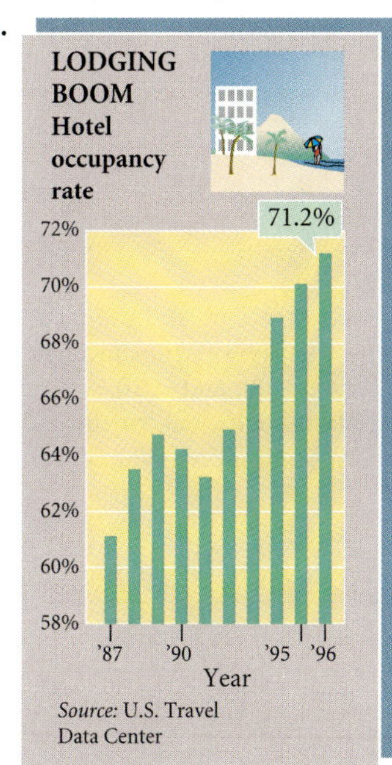

Source: U.S. Travel Data Center

2.

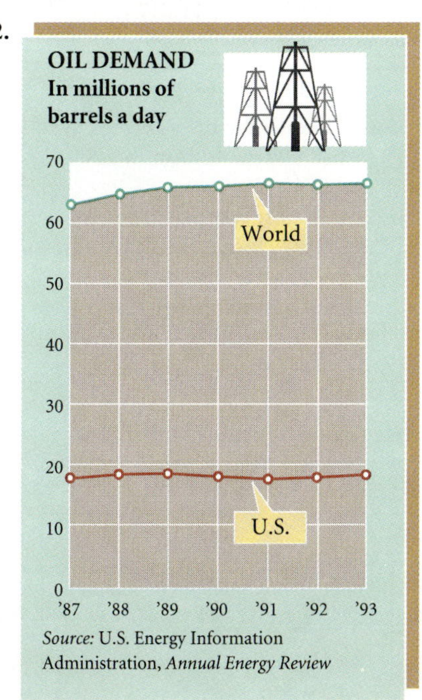

Source: U.S. Energy Information Administration, *Annual Energy Review*

3.

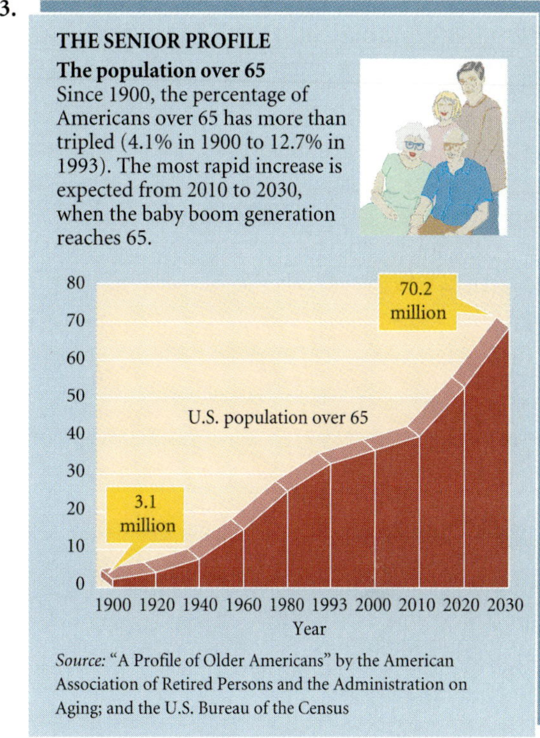

Source: "A Profile of Older Americans" by the American Association of Retired Persons and the Administration on Aging; and the U.S. Bureau of the Census

4.

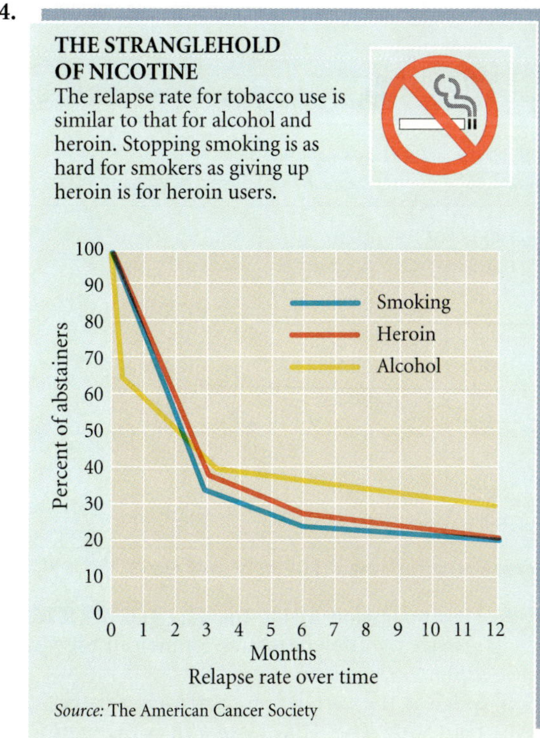

Source: The American Cancer Society

In Exercises 5–8, determine whether the scatterplot might be fit by a linear model.

5.

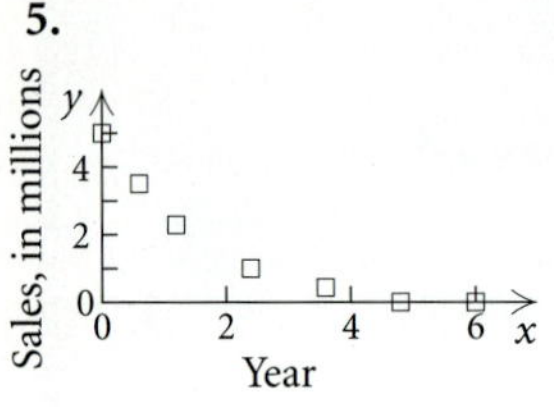

6.

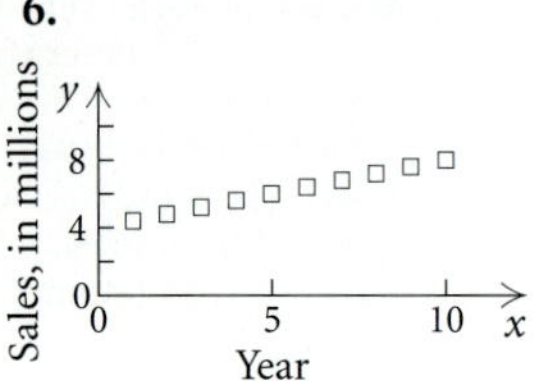

7.

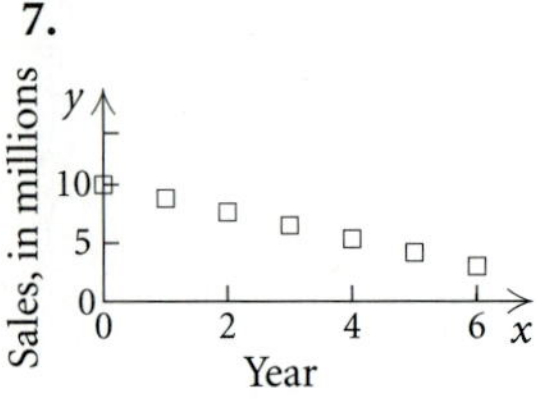

8.

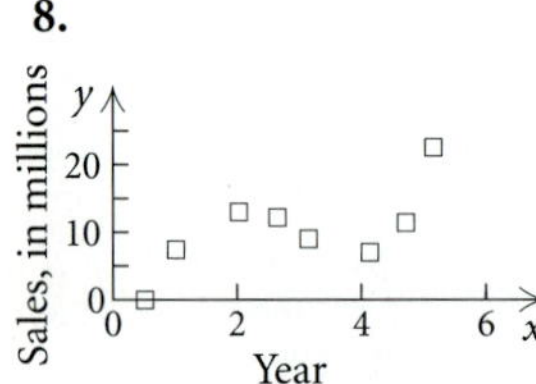

Exercises 9–14 involve creating regression lines on a grapher.

9.

Life Expectancy of Men

YEAR, *x*	LIFE EXPECTANCY, *y* (IN YEARS)	SCATTERPLOT
0. 1900	49.6	
1. 1910	50.2	
2. 1920	54.6	Life expectancy of men
3. 1930	58.0	
4. 1940	60.9	
5. 1950	65.3	
6. 1960	66.6	
7. 1970	67.1	
8. 1980	69.9	
9. 1990	71.8	

Source: Statistical Abstract of the United States.

a) Fit a regression line to the data and use it to predict the life expectancy of men in the year 2000.

b) What is the correlation coefficient for the regression line? How close a fit is the regression line?

10.

Cases of Skin Cancers in the United States

YEAR, *x*	NUMBER OF CASES, *y* (IN THOUSANDS)	SCATTERPLOT
0. 1988	5800	Skin cancer on the rise
1. 1989	6000	
2. 1990	6300	
3. 1991	6500	
4. 1992	6700	

Source: American Cancer Society.

a) Fit a regression line to the data and use it to predict the number of cases of skin cancer in the United States in 1998, 2000, and 2010.

b) What is the correlation coefficient for the regression line? How good a predictor is the regression line?

11.

The Cost of Tuition at State Universities

COLLEGE YEAR	ESTIMATED TUITION
0. (1997–1998)	$ 9,838
1. (1998–1999)	10,424
2. (1999–2000)	11,054
3. (2000–2001)	11,717
4. (2001–2002)	12,420

Source: College Board, Senate Labor Committee.

a) Fit a regression line to the data and use it to predict the cost of college tuition in 2004–2005, 2006–2007, and 2010–2011.

b) What is the correlation coefficient for the regression line? How close a fit is the regression line?

c) A college president asserts that in the college year 2006–2007, the cost of tuition will be $16,619. How well does your estimate compare?

12.

Book Buying Growth in the United States

YEAR, *x*	BOOK SALES, *y* (IN BILLIONS)
0. 1992	$21
1. 1993	23
2. 1994	24
3. 1995	25
4. 1996	26

Source: Book Industry Trends 1995.

a) Fit a regression line to the data and use it to predict book sales in 1998, 2000, and 2010.

b) What is the correlation coefficient for the regression line? How close a fit is the regression line?

c) A bookstore owner asserts that in the year 2000, book sales will be $45 billion. How well does your estimate stack up to this?

13. *Maximum Heart Rate.*   A person who is exercising should not exceed his or her maximum heart rate, which is determined on the basis of that person's sex, age, and resting heart rate. The following table relates resting heart rate and maximum heart rate for a 20-yr-old man.

RESTING HEART RATE, H (IN BEATS PER MINUTE)	MAXIMUM HEART RATE, M (IN BEATS PER MINUTE)
50	166
60	168
70	170
80	172

Source: American Heart Association.

a) Fit a regression line $M = mH + b$ to the data.

b) Predict a maximum heart rate if the resting heart rate is 40, 65, 76, and 84.

c) What is the correlation coefficient? How confident are you about using the regression line as a predictor?

14. *Study Time versus Grades.*   A math instructor asked her students to keep track of how much time each spent studying a chapter on functions in her algebra–trigonometry course. She collected the information together with test scores from that chapter's test. The data are listed in the table below.

STUDY TIME, x (IN HOURS)	TEST GRADE, y (IN PERCENT)
23	81
15	85
17	80
9	75
21	86
13	80
16	85
11	93

a) Fit a regression line to the data.

b) Predict a student's score if he or she studies 24 hr, 6 hr, and 18 hr.

c) What is the correlation coefficient? How confident are you about using the regression line as a predictor?

Skill Maintenance

15. Given $f(x) = 4x^3 - 5x$, find each of the following.

   a) $f(2)$ **b)** $f(-2)$ **c)** $f(a)$ **d)** $f(-a)$

16. Given $f(x) = 5x^2 - 7$, find each of the following.

   a) $f(-3)$ **b)** $f(3)$ **c)** $f(a)$ **d)** $f(-a)$

Synthesis

17. ◆ Why does a correlation coefficient of 1 not guarantee the existence of a cause-and-effect relationship?

18. ◆ Conduct your own experiment. Use some kind of measuring device and gather data relating body height to arm length. Fit a regression line to the data and find the correlation coefficient. How confident would you be about using these data to predict Shaquille O'Neal's arm length, knowing that his height is 7'0"?

19. *Planet Diameters versus Distance from the Sun.* The following table lists the diameters of the planets and their distances from the sun.

PLANET	DIAMETER, d (IN KILOMETERS)	DISTANCE FROM SUN, D (IN ASTRONOMICAL UNITS)*
Mercury	5,000	0.4
Venus	12,000	0.7
Earth	13,000	1.0
Mars	7,000	1.6
Jupiter	143,000	5.2
Saturn	120,000	10.0
Uranus	51,000	19.6
Pluto	5,000	30.8
Neptune	49,000	40.8

*1 astronomical unit (A.U.) = 150 million km.
Source: Encyclopedia Britannica.

a) Find a regression line $D = md + b$, expressing D as a function of d. What is the correlation coefficient?

b) Using a space probe, astronomers discover a new planet that has a diameter of 180,000 km. How confident would you be of using the function of part (a) to predict the distance of this planet from the sun?

20. *Ted Williams and the War Years.* Ted Williams played for the Boston Red Sox from 1939–1960. Many credit him with being the greatest hitter of all time. Unfortunately, his career totals are somewhat less impressive than others because his career was interrupted from 1943–1945 for World War II and from 1952–1953 for the Korean War. Some assert that if he had been playing during the war years, he would have broken Hank Aaron's home run record of 755 and runs batted in (RBI) record of 2297. Below are Williams' statistics.

YEAR, x	HOME RUNS, H	RBIs, R
1939	31	145
1940	23	113
1941	37	120
1942	36	137
1943	0*	0*
1944	0*	0*
1945	0*	0*
1946	38	123
1947	32	114
1948	25	127
1949	43	159
1950	28	97
1951	30	126
1952	1[†]	3[†]
1953	13[†]	34[†]
1954	29	89
1955	28	83
1956	24	82
1957	38	87
1958	26	85
1959	10	43
1960	29	72

*World War II
[†]Korean War

a) Excluding all data from the war years, fit a linear regression equation $H = mx + b$ to the data regarding the number of home runs. Then use the equation to predict how many home runs Williams would have hit in each of the war years. What is the correlation coefficient? How confident are you of your prediction?

b) Would Williams have broken Aaron's home run record?

c) Excluding all data from the war years, fit a linear regression equation $R = mx + b$ to the data regarding the number of RBIs. Then use the equation to predict how many RBIs Williams would have had in each of the war years. What is the correlation coefficient? How confident are you of your prediction?

d) Would Williams have broken Aaron's RBI record?

1.5
Distance, Midpoints, and Circles

- *Find the distance between two points in the plane and the midpoint of a segment.*
- *Find an equation of a circle with a given center and radius, and given an equation of a circle, find the center and the radius.*
- *Graph equations of circles.*

We have seen that graphs can provide a useful way of modeling real-world situations. In carpentry, surveying, engineering, and other fields, it is often necessary to determine distances and midpoints and to produce accurately drawn circles.

The Distance Formula

Suppose that a conservationist needs to determine the distance across an irregularly shaped pond. One way in which the conservationist might proceed is to measure two legs of a right triangle that is situated as shown below. The Pythagorean theorem, $a^2 + b^2 = c^2$, can then be used to find the length of the hypotenuse.

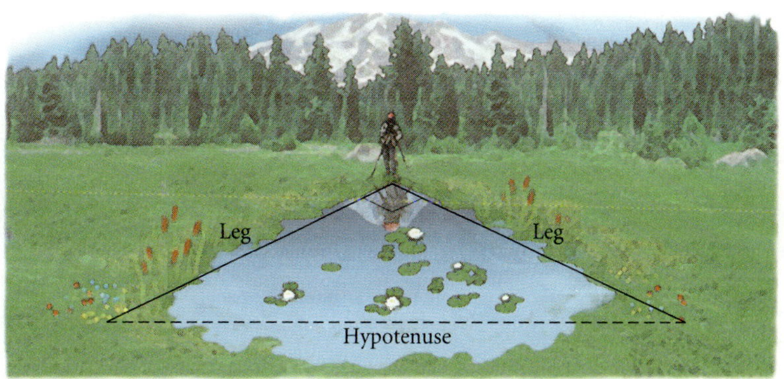

A similar strategy is used to find the distance between two points in a plane. For two points (x_1, y_1) and (x_2, y_2), we can draw a right triangle in which the legs have lengths $|x_2 - x_1|$ and $|y_2 - y_1|$.

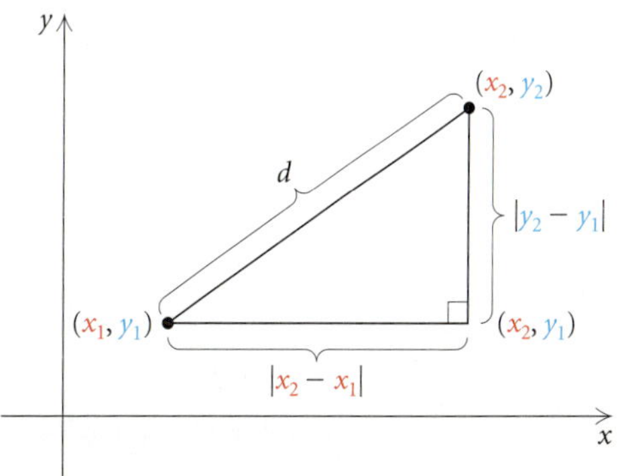

Using the Pythagorean theorem, we have

$$d^2 = |x_2 - x_1|^2 + |y_2 - y_1|^2.$$

Because we are squaring, parentheses can replace the absolute value symbols:

$$d^2 = (x_2 - x_1)^2 + (y_2 - y_1)^2.$$

Taking the principal square root, we obtain the distance formula.

> **The Distance Formula**
>
> The *distance d* between any two points (x_1, y_1) and (x_2, y_2) is given by
> $$d = \sqrt{(x_2 - x_1)^2 + (y_2 - y_1)^2}.$$

The subtraction of the x-coordinates can be done in any order, as can the subtraction of the y-coordinates. Although we derived the distance formula by considering two points not on a horizontal or a vertical line, the distance formula holds for *any* two points.

Example 1   The point $(-2, 5)$ is on a circle that has $(3, -1)$ as its center. Find the length of the radius of the circle.

SOLUTION   Since the length of the radius is the distance from the center to a point on the circle, we substitute into the distance formula:

$$r = \sqrt{[3 - (-2)]^2 + [-1 - 5]^2} \qquad \text{Either point can serve as } (x_1, y_1).$$
$$= \sqrt{5^2 + (-6)^2} = \sqrt{61} \approx 7.810. \qquad \text{Rounded to the nearest thousandth}$$

The circle's radius is approximately 7.8.

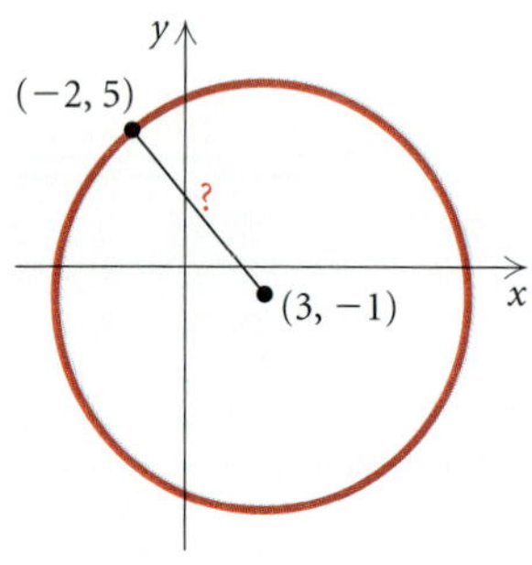

Midpoints of Segments

The distance formula can be used to develop a way of determining the *midpoint* of a segment when the endpoints are known. We state the formula and leave its proof to the exercises.

> **The Midpoint Formula**
>
> If the endpoints of a segment are (x_1, y_1) and (x_2, y_2), then the coordinates of the *midpoint* are
> $$\left(\frac{x_1 + x_2}{2}, \frac{y_1 + y_2}{2} \right).$$

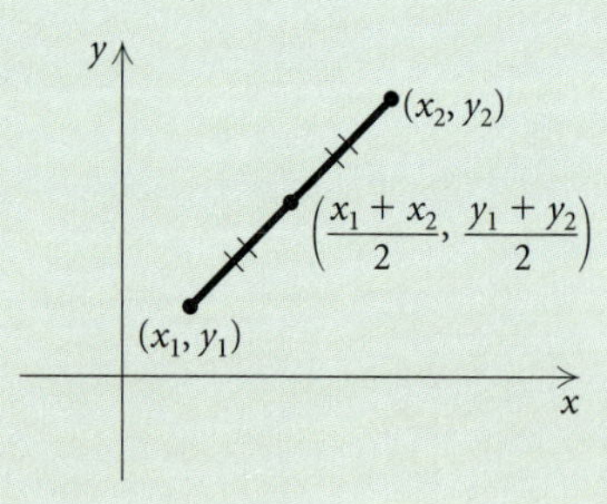

Note that we obtain the coordinates of the midpoint by averaging the coordinates of the endpoints. This is an easy way to remember the midpoint formula.

Example 2   The diameter of a circle connects two points $(2, -3)$ and $(6, 4)$ on the circle. Find the coordinates of the center of the circle.

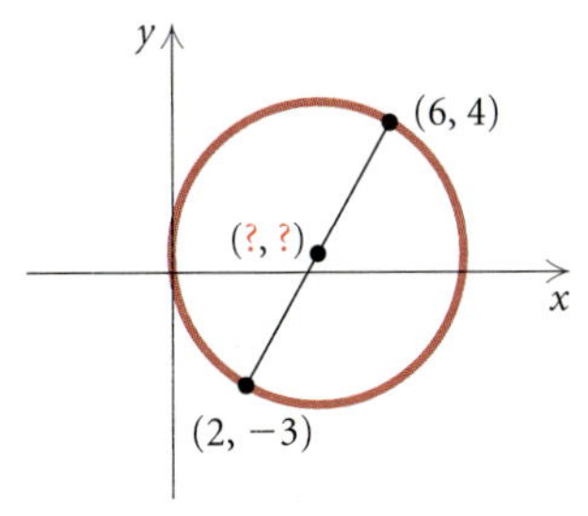

SOLUTION   Using the midpoint formula, we obtain

$$\left(\frac{2+6}{2}, \frac{-3+4}{2}\right), \quad \text{or} \quad \left(\frac{8}{2}, \frac{1}{2}\right), \quad \text{or} \quad \left(4, \frac{1}{2}\right).$$

The coordinates of the center are $\left(4, \frac{1}{2}\right)$.

Circles

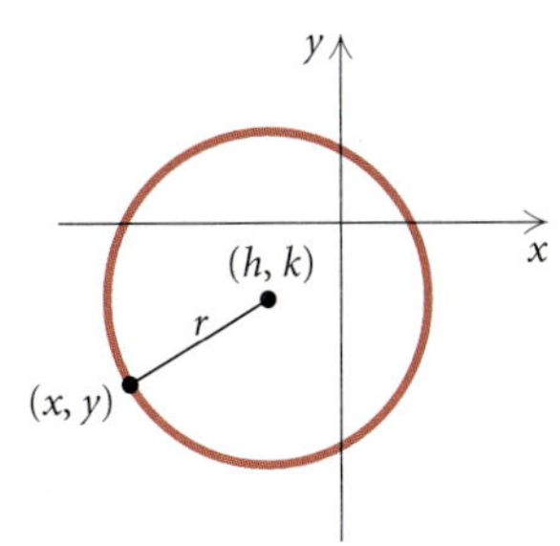

A **circle** is the set of all points in a plane that are a fixed distance r, the radius, from a center (h, k). Thus if a point (x, y) is to be r units from the center, we must have

$$r = \sqrt{(x-h)^2 + (y-k)^2}. \qquad \text{Using the distance formula}$$

Squaring both sides gives an equation of a circle.

The Equation of a Circle

The equation, in standard form, of a circle with center (h, k) and radius r is

$$(x-h)^2 + (y-k)^2 = r^2.$$

Example 3   Find an equation of the circle having radius 5 and center $(3, -7)$.

SOLUTION   Using the standard form, we have

$$[x-3]^2 + [y-(-7)]^2 = 5^2 \qquad \text{Substituting}$$
$$(x-3)^2 + (y+7)^2 = 25.$$

Example 4   Graph the circle: $(x+5)^2 + (y-2)^2 = 16$.

SOLUTION   We write the equation in standard form to determine the center and the radius:

$$[x-(-5)]^2 + [y-2]^2 = 4^2.$$

The center is $(-5, 2)$ and the radius is 4. We locate the center and draw the circle using a compass.

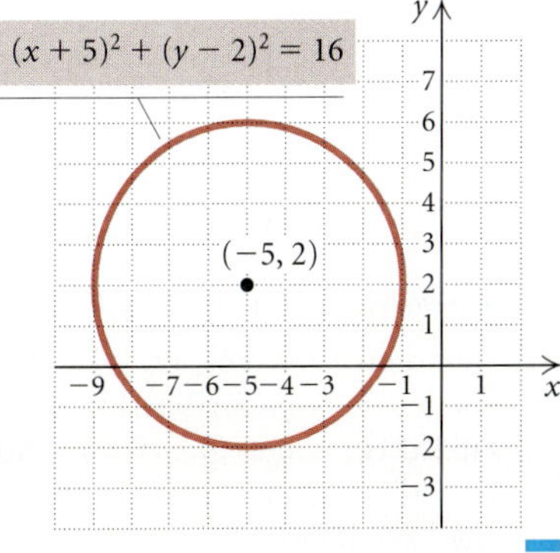

Circles can also be graphed on a grapher.

Example 5   Graph the circle: $x^2 + y^2 = 16$.

SOLUTION   We first solve $x^2 + y^2 = 16$ for y:

$$x^2 + y^2 = 16$$
$$y^2 = 16 - x^2 \qquad \text{Solving for } y^2$$
$$y = \pm\sqrt{16 - x^2}. \qquad \text{Solving for } y$$

We graph both equations $y_1 = \sqrt{16 - x^2}$ and $y_2 = -\sqrt{16 - x^2}$ using the same set of axes and a square viewing window. There are graphers that can draw graphs with a given center and radius directly from the DRAW menu. A square window will still be necessary. Some graphers have the ability to graph an equation like $x^2 + y^2 = 16$ without first solving for y.

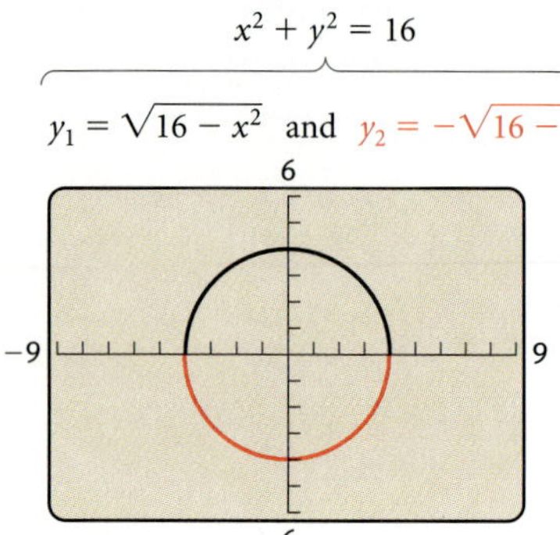

<table>
<tr><td valign="top">

1.5 | *Exercise Set*

Find the distance between each pair of points. Give an exact answer and, where appropriate, an approximation to three decimal places.

1. $(4, 6)$ and $(5, 9)$
2. $(-3, 7)$ and $(2, 11)$
3. $(6, -1)$ and $(9, 5)$
4. $(-4, -7)$ and $(-1, 3)$
5. $(\sqrt{3}, -\sqrt{5})$ and $(-\sqrt{6}, 0)$
6. $(-\sqrt{2}, 1)$ and $(0, \sqrt{7})$

7. The points $(-3, -1)$ and $(9, 4)$ are the endpoints of the diameter of a circle. Find the length of the radius of the circle.

8. The point $(0, 1)$ is on a circle that has center $(-3, 5)$. Find the length of the diameter of the circle.

Use the distance formula and the Pythagorean theorem to determine whether each set of points could be vertices of a right triangle.

9. $(-4, 5)$, $(6, 1)$, and $(-8, -5)$

10. $(-3, 1)$, $(2, -1)$, and $(6, 9)$

11. The points $(-3, 4)$, $(0, 5)$, and $(3, -4)$ are all on the circle $x^2 + y^2 = 25$. Show that these three points are vertices of a right triangle.

12. The points $(-3, 4)$, $(2, -1)$, $(5, 2)$, and $(0, 7)$ are vertices of a quadrilateral. Show that the quadrilateral is a rectangle. (*Hint:* Show that the quadrilateral's opposite sides are the same length and that the two diagonals are the same length.)

</td><td valign="top">

13–18. Find the midpoint of each segment having the endpoints given in Exercises 1–6, respectively.

19. Graph the rectangle described in Exercise 12. Then determine the coordinates of the midpoint for each of the four sides. Are the midpoints vertices of a rectangle?

20. Graph the square with vertices $(-5, -1)$, $(7, -6)$, $(12, 6)$, and $(0, 11)$. Then determine the midpoint for each of the four sides. Are the midpoints vertices of a square?

21. The points $(\sqrt{7}, -4)$ and $(\sqrt{2}, 3)$ are endpoints of the diameter of a circle. Determine the center of the circle.

22. The points $(-3, \sqrt{5})$ and $(1, \sqrt{2})$ are endpoints of the diagonal of a square. Determine the center of the square.

In Exercises 23 and 24, how would you change the window so the circle is not distorted? Answers may vary.

23.

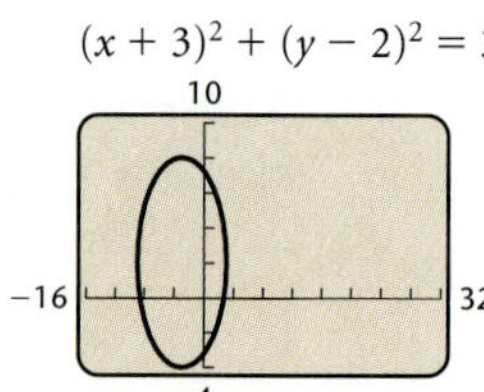

24.

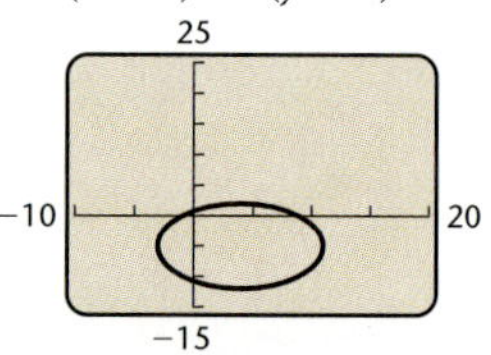

Find an equation for a circle satisfying the given conditions.

25. Center $(2, 3)$, radius of length $\frac{5}{3}$

26. Center $(4, 5)$, diameter of length 8.2

27. Center $(-1, 4)$, passes through $(3, 7)$

</td></tr>
</table>

28. Center $(6, -5)$, passes through $(1, 7)$

29. The points $(7, 13)$ and $(-3, -11)$ are at either end of a diameter.

30. The points $(-9, 4)$, $(-2, 5)$, $(-8, -3)$, and $(-1, -2)$ are vertices of an inscribed square.

31. Center $(-2, 3)$, tangent (touching at one point) to the y-axis

32. Center $(4, -5)$, tangent to the x-axis

Find the center and the radius of each circle. Then graph each circle using a square viewing rectangle.

33. $x^2 + y^2 = 4$

34. $x^2 + y^2 = 81$

35. $x^2 + (y - 3)^2 = 16$
(*Hint:* Solving for y, we get $y = 3 \pm \sqrt{16 - x^2}$.)

36. $(x + 2)^2 + y^2 = 100$

37. $(x - 1)^2 + (y - 5)^2 = 36$

38. $(x - 7)^2 + (y + 2)^2 = 25$

39. $(x + 4)^2 + (y + 5)^2 = 9$

40. $(x + 1)^2 + (y - 2)^2 = 64$

Skill Maintenance

41. Evaluate $x - xy^2 \div 2$ for $x = 5$ and $y = -4$.

42. Evaluate $x^3 - 5x$ for $x = -2$.

43. Simplify: $(a + h)^2 - a^2$.

44. Simplify: $(2a)^3 - 5(2a) - 7$.

Synthesis

Find the distances between each pair of points and find the midpoint of the segment having the given points as endpoints.

45. $(a, \sqrt{a})$ and $(a + h, \sqrt{a + h})$

46. $\left(a, \dfrac{1}{a}\right)$ and $\left(a + h, \dfrac{1}{a + h}\right)$

47. ◆ Explain how the Pythagorean theorem is used to develop the equation of a circle in standard form.

48. ◆ Explain how you could find the coordinates of a point $\frac{7}{8}$ of the way from point A to point B.

Find an equation of a circle satisfying the given conditions.

49. Center $(2, -7)$ with an area of 36π square units

50. Center $(-5, 8)$ with a circumference of 10π units

51. *Swimming Pool.* A swimming pool is being constructed in the corner of a yard, as shown. Before installation, the contractor needs to know

measurements a_1 and a_2. Find them.

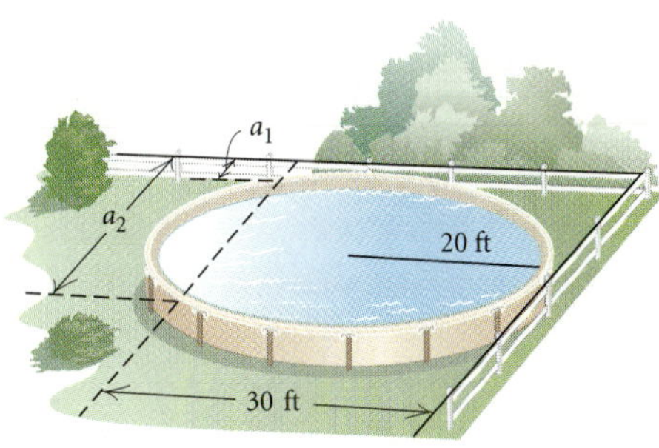

52. *An Arch of a Circle in Carpentry.* Ace Carpentry needs to cut an arch for the top of an entranceway. The arch needs to be 8 ft wide and 2 ft high. To draw the arch, the carpenters will use a stretched string with chalk attached at an end as a compass.

a) Using a coordinate system, locate the center of the circle.

b) What radius should the carpenters use to draw the arch?

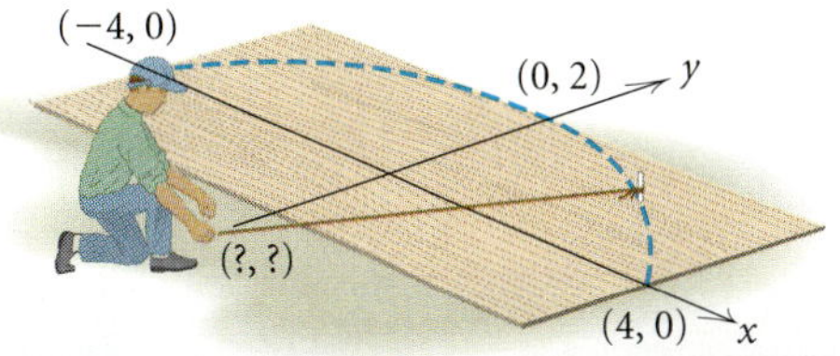

*Determine whether each of the following points lies on the **unit circle**, $x^2 + y^2 = 1$.*

53. $\left(\dfrac{\sqrt{3}}{2}, -\dfrac{1}{2}\right)$

54. $(0, -1)$

55. $\left(-\dfrac{\sqrt{2}}{2}, \dfrac{\sqrt{2}}{2}\right)$

56. $\left(\dfrac{1}{2}, -\dfrac{\sqrt{3}}{2}\right)$

57. Find the point on the y-axis that is equidistant from the points $(-2, 0)$ and $(4, 6)$.

58. Consider any right triangle with base b and height h, situated as shown. Show that the midpoint of the hypotenuse P is equidistant from the three vertices of the triangle.

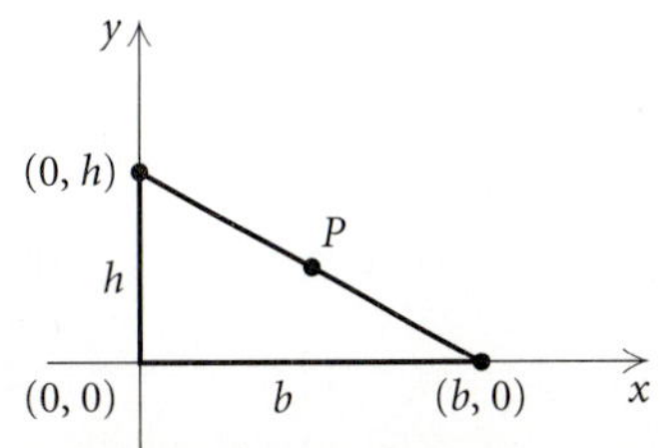

59. Prove the midpoint formula by showing that:

a) $\left(\dfrac{x_1 + x_2}{2}, \dfrac{y_1 + y_2}{2}\right)$ is equidistant from the points (x_1, y_1) and (x_2, y_2); and

b) the distance from (x_1, y_1) to the midpoint plus the distance from (x_2, y_2) to the midpoint equals the distance from (x_1, y_1) to (x_2, y_2).

1.6
Symmetry

- *Determine whether a graph is symmetric with respect to the x-axis, the y-axis, and the origin.*
- *Determine whether a function is even, odd, or neither even nor odd.*

Symmetry

Symmetry occurs often in nature and in art. For example, when viewed from the front, the bodies of most animals are at least approximately symmetric. This means that each eye is the same distance from the center of the bridge of the nose, each shoulder is the same distance from the center of the chest, and so on. Architects have used symmetry for thousands of years to enhance the beauty of buildings.

M. Hyett/VIREO

Although symmetry has important uses throughout mathematics, our present discussion is included to better enable us to graph and analyze equations and functions.

Consider the points $(4, 2)$ and $(4, -2)$ that appear in the graph of $x = y^2$.

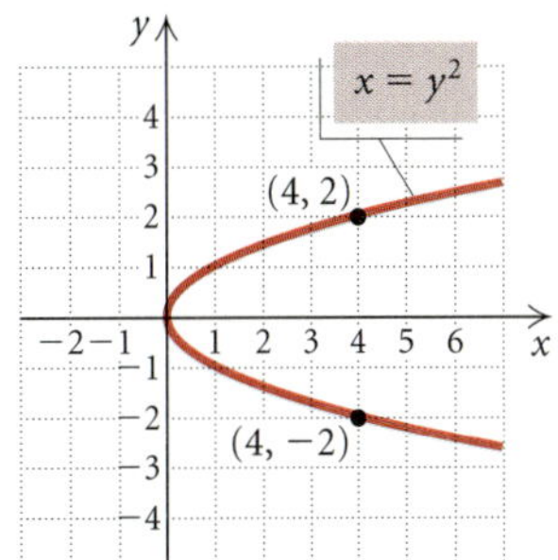

Points like these have the same x-value but opposite y-values, and are **reflections** of each other across the x-axis. Any graph that contains the reflection of every point across the x-axis is said to be **symmetric with respect to the x-axis**. If we fold the graph on the x-axis, the parts above and below the x-axis will coincide.

Consider the points $(3, 4)$ and $(-3, 4)$ that appear in the graph of $y = x^2 - 5$.

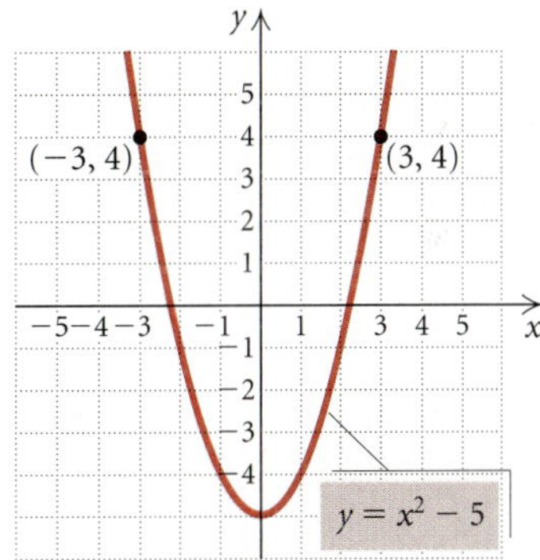

Points like these have the same y-value but opposite x-values, and are **reflections** of each other across the y-axis. Any graph that contains the reflection of every point across the y-axis is said to be **symmetric with respect to the y-axis**. If we fold the graph on the y-axis, the parts to the left and right of the y-axis will coincide.

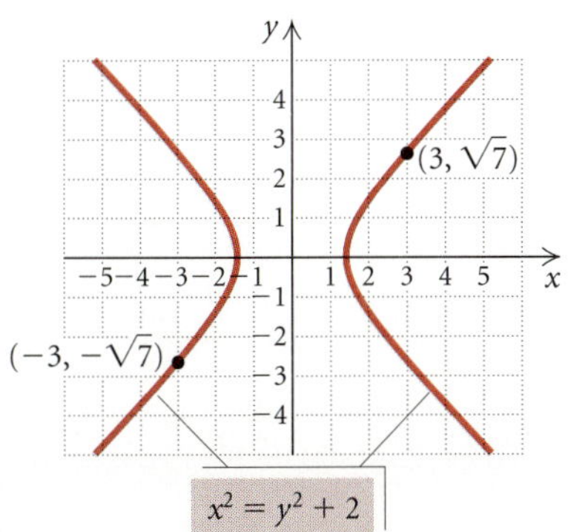

Consider the points $(3, \sqrt{7})$ and $(-3, -\sqrt{7})$ that appear in the graph of $x^2 = y^2 + 2$. (You can create such a graph on a grapher by solving for y, obtaining $y_1 = \sqrt{x^2 - 2}$ and $y_2 = -\sqrt{x^2 - 2}$.) Note that if we take the opposites of the coordinates of one pair, we get the other pair. If for any point (x, y) on a graph the point $(-x, -y)$ is also on the graph, then the graph is said to be **symmetric with respect to the origin**. Visually, if we rotate the graph $180°$ about the origin, the resulting figure coincides with the original.

> ### Algebraic Tests of Symmetry
>
> *x-axis*: If replacing y by $-y$ produces an equivalent equation, then the graph is *symmetric with respect to the x-axis*.
>
> *y-axis*: If replacing x by $-x$ produces an equivalent equation, then the graph is *symmetric with respect to the y-axis*.
>
> *Origin*: If replacing x by $-x$ and y by $-y$ produces an equivalent equation, then the graph is *symmetric with respect to the origin*.

Example 1   Test $y = x^2 + 2$ for symmetry with respect to the x-axis, the y-axis, and the origin.

ALGEBRAIC SOLUTION

X-AXIS

Replace y by $-y$:

$$y = x^2 + 2$$
$$-y = x^2 + 2$$
$$y = -x^2 - 2. \qquad \text{Multiplying both sides by } -1$$

Is the resulting equation equivalent to the original? The answer is no, so the graph *is not* symmetric with respect to the x-axis.

Y-AXIS

Replace x by $-x$:

$$y = x^2 + 2$$
$$y = (-x)^2 + 2$$
$$y = x^2 + 2. \qquad \text{Simplifying}$$

Is the resulting equation equivalent to the original? The answer is yes, so the graph *is* symmetric with respect to the y-axis.

ORIGIN

Replace x by $-x$ and y by $-y$:

$$y = x^2 + 2$$
$$-y = (-x)^2 + 2$$
$$-y = x^2 + 2 \qquad \text{Simplifying}$$
$$y = -x^2 - 2.$$

Is the resulting equation equivalent to the original? The answer is no, so the graph *is not* symmetric with respect to the origin.

GRAPHICAL SOLUTION

We use a grapher to graph the equation. Note that if the graph were folded on the x-axis, the parts above and below the x-axis would not coincide. If it were folded on the y-axis, the parts to the left and right of the y-axis would coincide. If we rotated it 180°, the resulting graph would not coincide with the original graph.

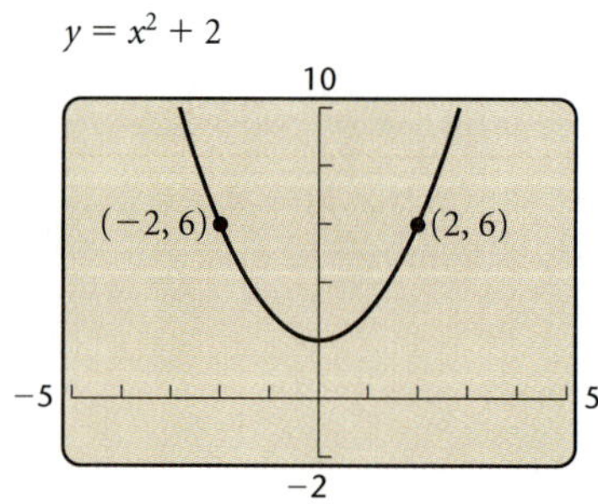

Thus we see that the graph *is not* symmetric with respect to the x-axis or the origin. The graph *is* symmetric with respect to the y-axis.

The algebraic method is often easier to apply than the graphical, especially with equations that we may not be able to graph easily.

Example 2 Test $x^2 + y^4 = 5$ for symmetry with respect to the x-axis, the y-axis, and the origin.

ALGEBRAIC SOLUTION

X-AXIS

Replace y by $-y$:

$$x^2 + y^4 = 5$$

$$x^2 + (-y)^4 = 5$$
$$x^2 + y^4 = 5.$$

The resulting equation is equivalent to the original equation. Thus the graph is symmetric with respect to the x-axis.

  We leave it to the student to verify algebraically that the graph is also symmetric with respect to the y-axis and the origin.

GRAPHICAL SOLUTION

With a grapher, we see symmetry with respect to both axes and with respect to the origin.

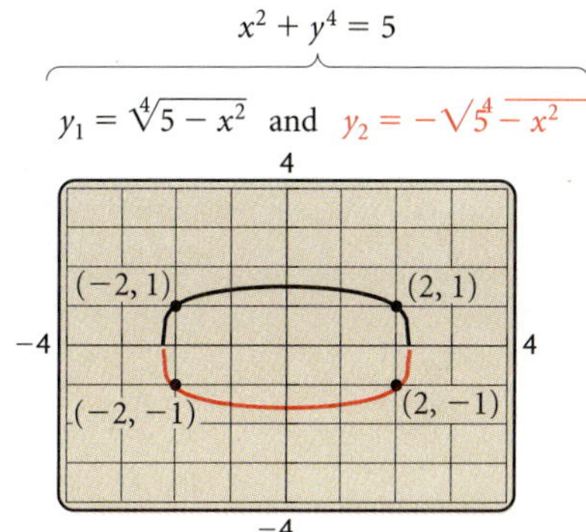

Even and Odd Functions

Now we relate symmetry to graphs of functions.

Interactive Discovery

Consider the function f given by $f(x) = x^2$. Find and graph

$$y_1 = f(x) = x^2, \qquad y_2 = f(-x) = (-x)^2, \quad \text{and} \quad y_3 = -f(x) = -x^2.$$

Use the graphs to determine whether $f(x) = f(-x)$ is an identity. Use the graphs to determine whether $f(x) = -f(x)$ is an identity.

  Finding $f(-x)$ is the same as replacing x with $-x$ in its equation, and finding $-f(x)$ is the same as replacing y with $-y$ in its equation. Can you make a connection between symmetry and the properties $f(x) = f(-x)$ and $f(-x) = -f(x)$?

  Repeat this procedure for $g(x) = x^3$.

The preceding Interactive Discovery leads us to the following definitions and procedure.

Even and Odd Functions

If the graph of a function f is symmetric with respect to the y-axis, we say that it is an *even function*. That is, for each x in the domain of f, $f(x) = f(-x)$.

If the graph of a function f is symmetric with respect to the origin, we say that it is an *odd function*. That is, for each x in the domain of f, $f(-x) = -f(x)$.

> **Algebraic Procedure for Determining Even and Odd Functions**
>
> Given the function $f(x)$:
>
> **1.** Find $f(-x)$ and simplify.
>
> **2.** Find $-f(x)$ and simplify.
>
> **3.** If $f(x) = f(-x)$, then f is even. If $f(-x) = -f(x)$, then f is odd.

Except for the function $f(x) = 0$, a function cannot be *both* even and odd. Thus if we see in step (1) that $f(x) = f(-x)$ (that is, f is even), we need not complete the process.

Example 3   Determine whether each of the following functions is even, odd, or neither.

a) $f(x) = 5x^7 - 6x^3 - 2x$ b) $g(x) = x^4 - 2x$

c) $h(x) = 5x^6 - 3x^2 - 7$

a)

ALGEBRAIC SOLUTION

$$f(x) = 5x^7 - 6x^3 - 2x$$

1. $f(-x) = 5(-x)^7 - 6(-x)^3 - 2(-x)$

$\qquad = 5(-x^7) - 6(-x^3) + 2x$

$\qquad\qquad (-x)^7 = (-1 \cdot x)^7 = (-1)^7 x^7 = -x^7$

$\qquad = -5x^7 + 6x^3 + 2x$

2. $-f(x) = -(5x^7 - 6x^3 - 2x)$

$\qquad\quad = -5x^7 + 6x^3 + 2x$

3. We see that $f(x) \neq f(-x)$. Thus, f is *not* even. We see that $f(-x) = -f(x)$. Thus, f is odd.

GRAPHICAL SOLUTION

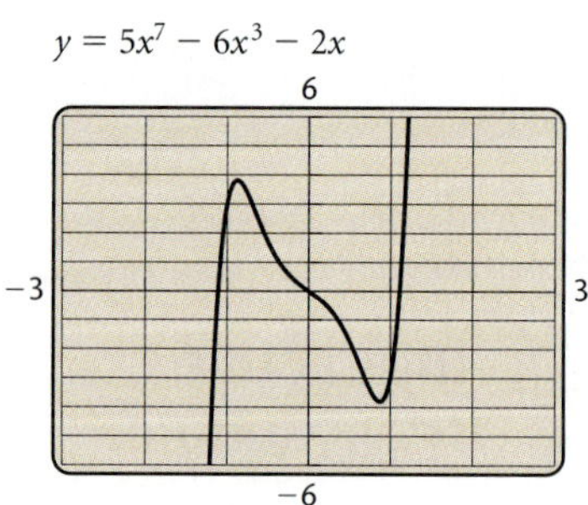

We see visually that the graph is symmetric with respect to the origin. Thus the function is odd.

b)

ALGEBRAIC SOLUTION

$$g(x) = x^4 - 2x$$

1. $g(-x) = (-x)^4 - 2(-x)$

$\qquad = x^4 + 2x$

2. $-g(x) = -(x^4 - 2x)$

$\qquad\quad = -x^4 + 2x$

3. We see that $g(x) \neq g(-x)$. The function is *not* even. We see that $g(-x) \neq -g(x)$. The function is *not* odd. Thus g is neither even nor odd.

GRAPHICAL SOLUTION

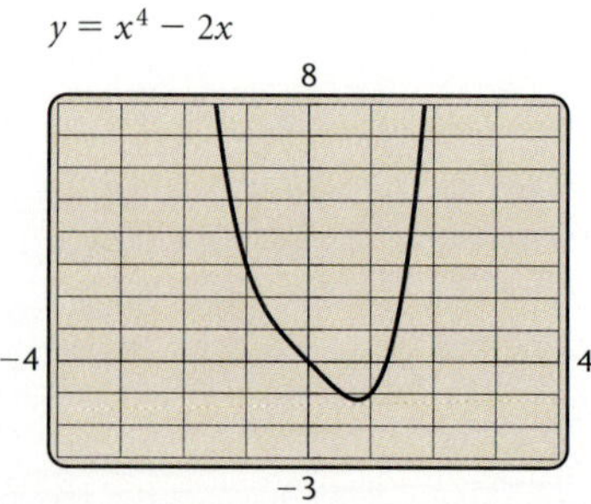

We see visually that the graph is symmetric with respect to neither the y-axis nor the origin. Thus the function is neither even nor odd.

c)

$h(x) = 5x^6 - 3x^2 - 7$

1. $h(-x) = 5(-x)^6 - 3(-x)^2 - 7$
$= 5x^6 - 3x^2 - 7$

We see that $h(x) = h(-x)$. Thus the function is even.

$y = 5x^6 - 3x^2 - 7$

We see visually that the graph is symmetric with respect to the y-axis. Thus the function is even.

1.6 Exercise Set

Determine visually whether each graph is symmetric with respect to the x-axis, the y-axis, and the origin.

1. 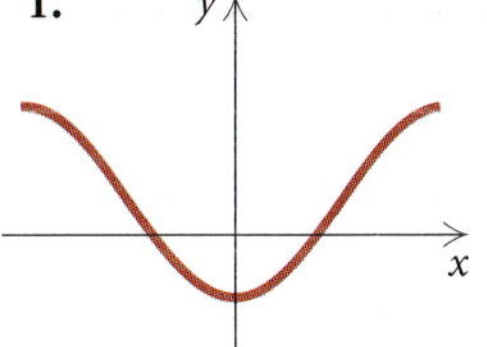

2. 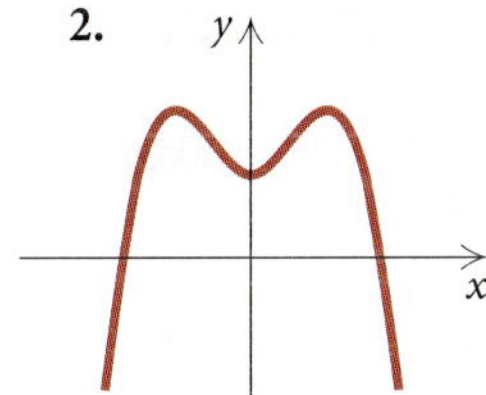

3. 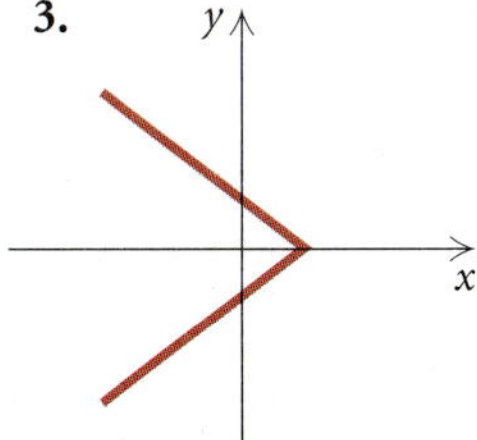

4. 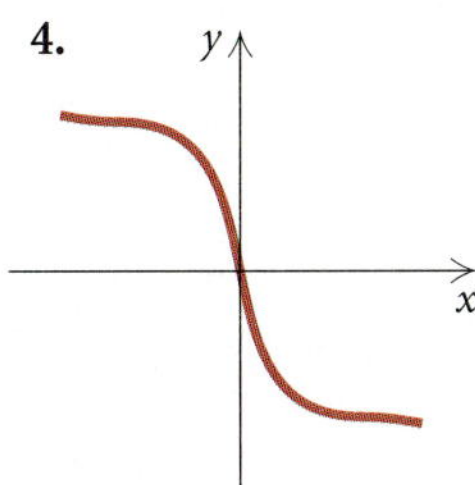

5. 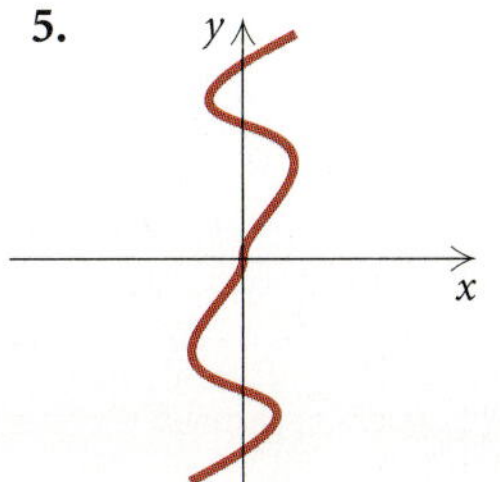

6. 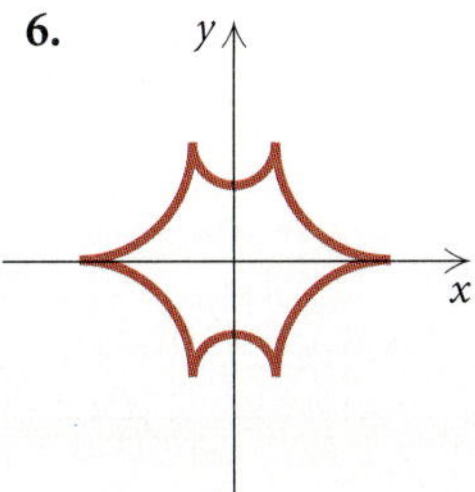

First, graph each equation and determine visually whether it is symmetric with respect to the x-axis, the y-axis, and the origin. Then verify your assertion algebraically.

7. $y = |x| - 2$

8. $y = |x + 5|$

9. $5y = 4x + 5$

10. $2x - 5 = 3y$

11. $5y = 2x^2 - 3$

12. $x^2 + 4 = 3y$

13. $y = \dfrac{1}{x}$

14. $y = -\dfrac{4}{x}$

Test algebraically whether each graph is symmetric with respect to the x-axis, the y-axis, and the origin. Then check your work graphically, if possible, using a grapher.

15. $5x - 5y = 0$

16. $6x + 7y = 0$

17. $3x^2 - 2y^2 = 3$

18. $5y = 7x^2 - 2x$

19. $y = |2x|$

20. $y^3 = 2x^2$

21. $2x^4 + 3 = y^2$

22. $2y^2 = 5x^2 + 12$

23. $3y^3 = 4x^3 + 2$

24. $3x = |y|$

25. $xy = 12$

26. $xy - x^2 = 3$

Determine visually whether each function is even, odd, or neither even nor odd.

27. 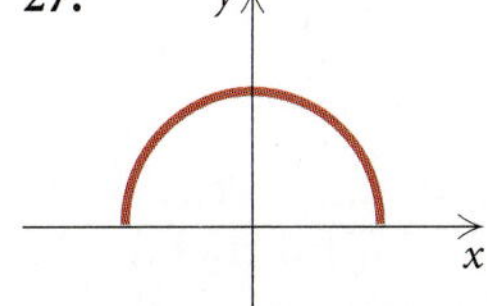

28. 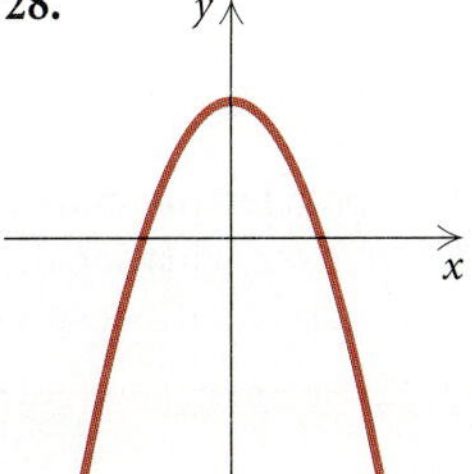

29.
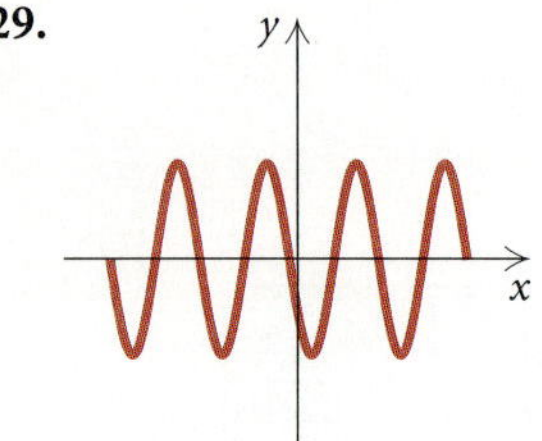

30.
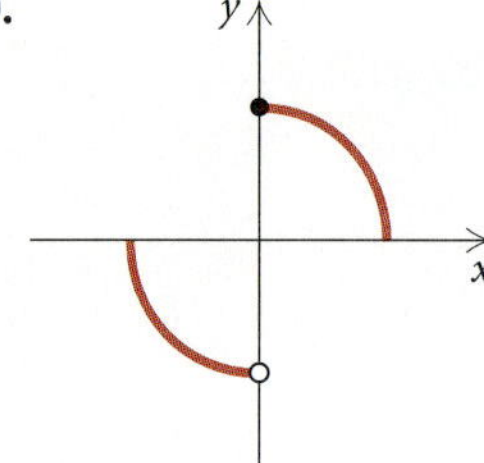

31.
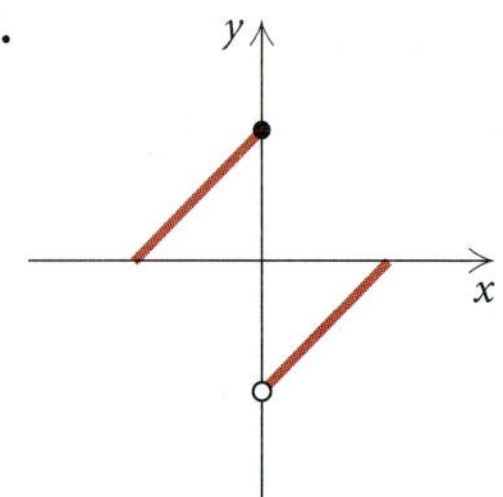

32.
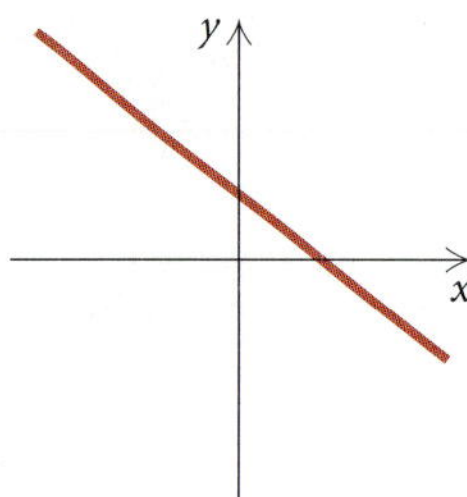

Test algebraically whether each function is even, odd, or neither even nor odd. Then check your work graphically, where possible, using a grapher.

33. $f(x) = -3x^3 + 2x$

34. $f(x) = 2x^2 + 4x$

35. $f(x) = 5x^2 + 2x^4 - 1$

36. $f(x) = 3x^4 - 4x^2$

37. $f(x) = 4x$

38. $f(x) = |3x|$

39. $f(x) = x^{24}$

40. $f(x) = 7x^3 + 4x - 2$

41. $f(x) = x^{17}$

42. $f(x) = x + \dfrac{1}{x}$

43. $f(x) = x - |x|$

44. $f(x) = \sqrt{x}$

45. $f(x) = \sqrt[3]{x}$

46. $f(x) = 8$

47. $f(x) = \sqrt[3]{x - 2}$

48. $f(x) = -\dfrac{2}{x}$

49. $f(x) = \dfrac{1}{x^2}$

50. $f(x) = \sqrt{x^2 + 1}$

Skill Maintenance

Given that $f(x) = 3x^2 - 2x$, find and simplify each of the following.

51. $f(a - 5)$

52. $f(a + 1)$

53. $f(a + 4)$

54. $f(a - 1)$

Synthesis

55. ◆ Consider the constant function $f(x) = 0$. Determine whether the graph of this function is symmetric with respect to the x-axis, the y-axis, and the origin. Determine whether this function is even or odd. In general, can a function be symmetric with respect to the x-axis? Explain.

56. ◆ Describe conditions under which you would know whether a polynomial function

$$f(x) = a_n x^n + a_{n-1} x^{n-1} + \cdots + a_2 x^2 + a_1 x + a_0$$

is even or odd without using an algebraic or graphical procedure. Explain.

Determine whether each function is even, odd, or neither even nor odd. Use any method.

57. $f(x) = x|x^3|$

58. $f(x) = x^2(5 - |x|)$

59. $f(x) = \dfrac{1 - x^4}{x^3 + 1}$

60. $f(x) = \dfrac{x^2 + 1}{x^3 - 1}$

61. $f(x) = x\sqrt{10 - x^2}$

62. $f(x) = \dfrac{-8x}{x^2 + 1}$

Determine whether each graph is symmetric with respect to the x-axis, the y-axis, and the origin. Use any method.

63.
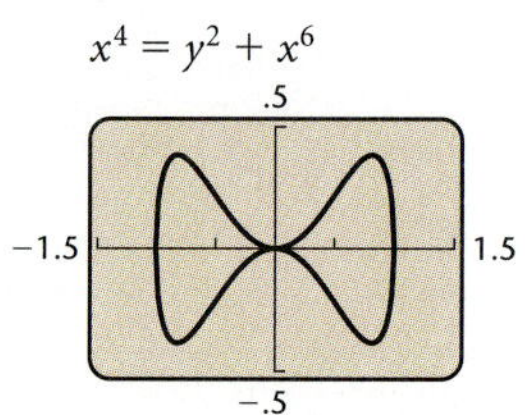

64.
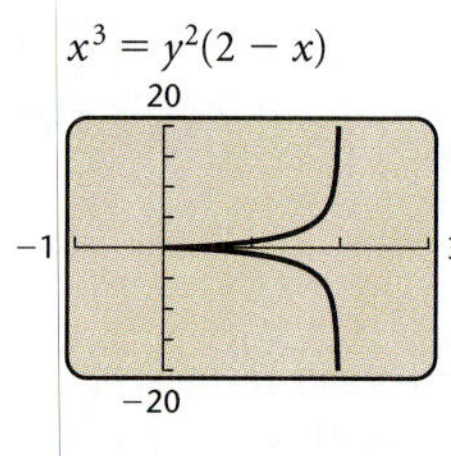

65.
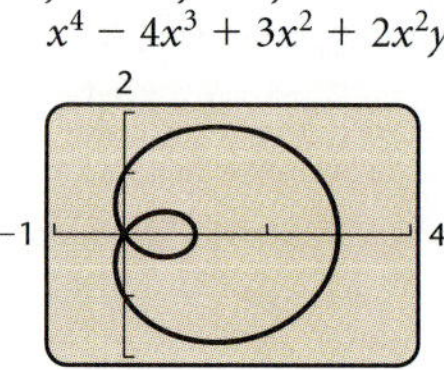

66.
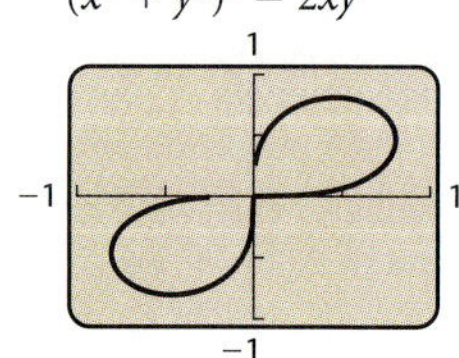

67. Given $f(x) = \frac{1}{2}x^4 - 5x^3 + 2$, let $g(x) = f(-x)$. Graph both f and g using the same set of axes. How do the graphs compare?

68. Consider symmetries with respect to the x-axis, the y-axis, and the origin. Prove that symmetry with respect to any two of these implies symmetry with respect to the other.

1.7
Transformations of Functions

- *Given the graph of a function, graph its transformation under translations, reflections, stretchings, and shrinkings.*

Throughout this chapter, we have considered many kinds of functions. Let's review some of them below.

Constant function:
$y_1 = b$

Linear function:
$y_2 = mx + b$

Squaring function:
$y_3 = x^2$

Quadratic function:
$y_4 = ax^2 + bx + c, a > 0$

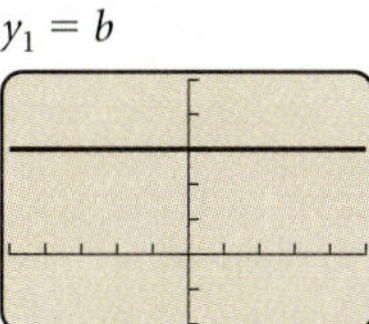 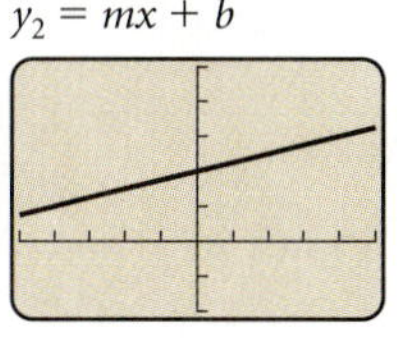 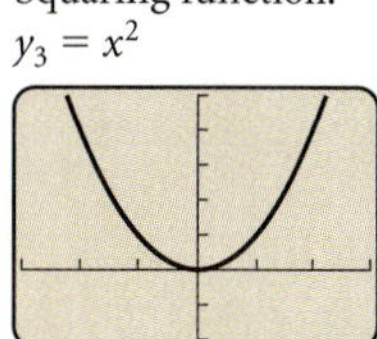 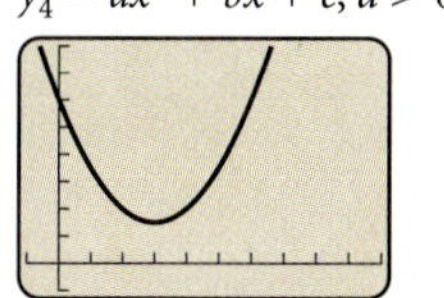

Square root function:
$y_5 = \sqrt{x}$

Cubing function:
$y_6 = x^3$

Cube root function:
$y_7 = \sqrt[3]{x}$

Reciprocal function:
$y_8 = \dfrac{1}{x}$

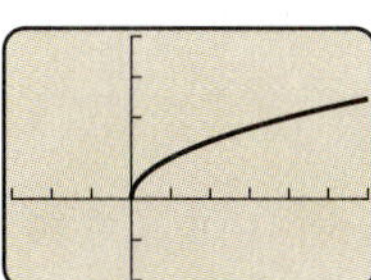 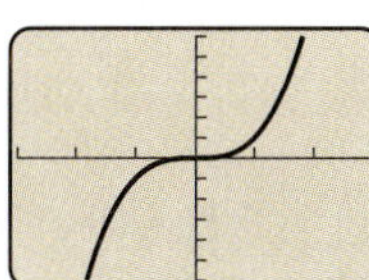 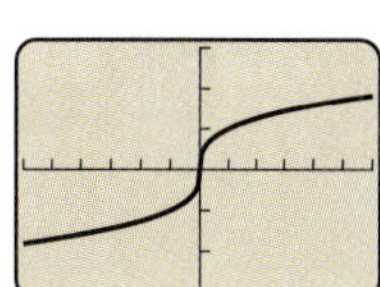 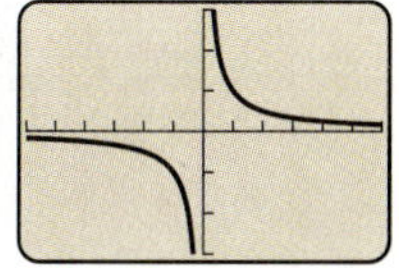

Greatest integer function:
$y_9 = \text{INT}(x)$

Absolute value function:
$y_{10} = |x|$

Upper half-circle:
$y_{11} = \sqrt{a^2 - x^2}$

Lower half-circle:
$y_{12} = -\sqrt{a^2 - x^2}$

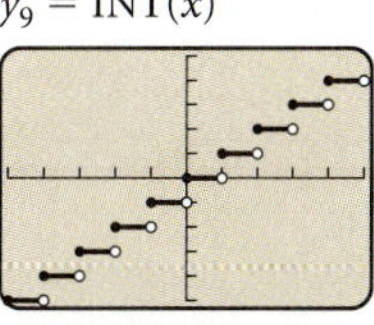 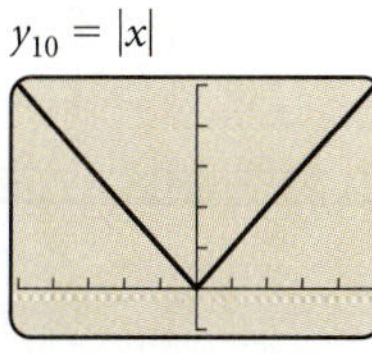 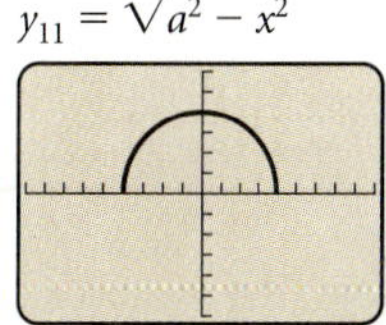 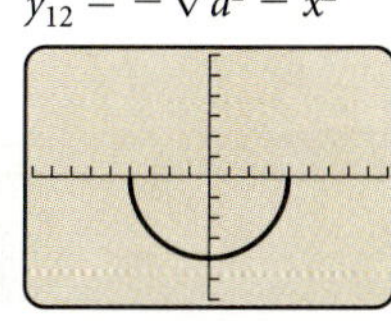

These functions can be considered building blocks for many other functions. We can create graphs of new functions by shifting them horizontally or vertically, stretching or shrinking them, and reflecting them across an axis. In this section, we consider these *transformations*.

Vertical and Horizontal Translations

Suppose we have a function given by $y_1 = f(x)$. Let's explore the graph of a new function $y_2 = f(x) + b$ or $y_3 = f(x) - b$, for $b > 0$.

Interactive Discovery

Consider the function $y_1 = \frac{1}{5}x^4$. Graph it and the functions $y_2 = \frac{1}{5}x^4 - 5$ and $y_3 = \frac{1}{5}x^4 + 3$ using the same viewing window, $[-10, 10, -8, 10]$. What pattern do you see? Test it with some other graphs.

The effect is a shift of $f(x)$ up or down. Such a shift is called a **vertical translation**.

> **Vertical Translation**
>
> For $b > 0$,
>
> the graph of $y_2 = f(x) + b$ is the graph of $y_1 = f(x)$ shifted *up b* units;
>
> the graph of $y_3 = f(x) - b$ is the graph of $y_1 = f(x)$ shifted *down b* units.

Suppose we have a function given by $y_1 = f(x)$. Let's explore the graph of a new function $y_2 = f(x + d)$ or $y_3 = f(x - d)$, for $d > 0$.

Interactive Discovery

Consider the function $y_1 = \frac{1}{5}x^4$. Graph it and the functions $y_2 = \frac{1}{5}(x - 3)^4$ and $y_3 = \frac{1}{5}(x + 7)^4$ using the same viewing window, $[-10, 10, -2, 10]$. What pattern do you see? Test it with some other graphs.

The effect is a shift of $f(x)$ to the right or left. Such a shift is called a **horizontal translation**.

> **Horizontal Translation**
>
> For $d > 0$;
>
> the graph of $y_2 = f(x - d)$ is the graph of $y_1 = f(x)$ shifted *right d* units;
>
> the graph of $y_3 = f(x + d)$ is the graph of $y_1 = f(x)$ shifted *left d* units.

Example 1   Graph each of the following. Before doing so, describe how each graph can be obtained from one of the basic graphs shown on the preceding page.

a) $g(x) = x^2 - 6$

b) $g(x) = \sqrt{x + 2}$

c) $g(x) = \dfrac{1}{x} + 2$

d) $g(x) = |x - 4|$

e) $h(x) = \sqrt{x + 2} - 3$

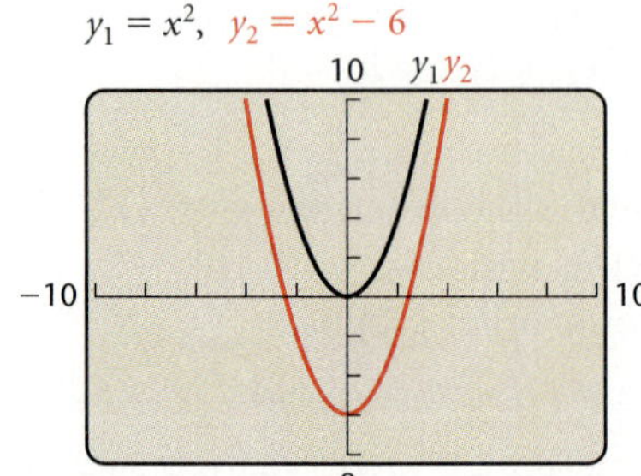

SOLUTION

a) To graph $g(x) = x^2 - 6$, think of the graph of $f(x) = x^2$. Since $g(x) = f(x) - 6$, the graph of $g(x) = x^2 - 6$ is the graph of $f(x) = x^2$, shifted, or translated, *down* 6 units. (See the figure at left.)

b) To graph $g(x) = \sqrt{x + 2}$, think of the graph of $f(x) = \sqrt{x}$. Since $g(x) = f(x + 2)$, the graph of $g(x) = \sqrt{x + 2}$ is the graph of $f(x) = \sqrt{x}$, shifted *left* 2 units.

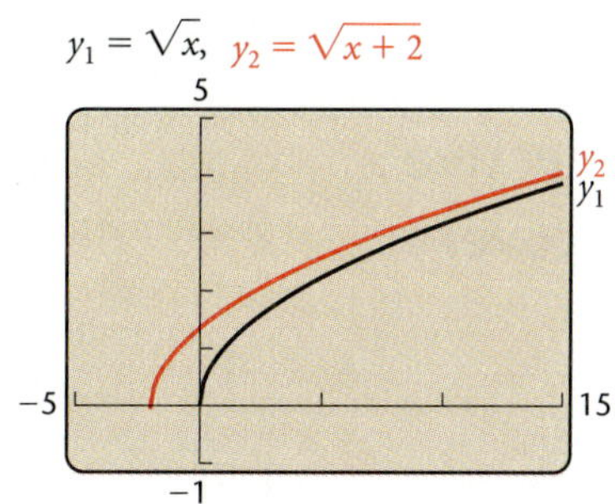

c) To graph $g(x) = 1/x + 2$, think of the graph of $f(x) = 1/x$. Since $g(x) = f(x) + 2$, the graph of $g(x) = 1/x + 2$ is the graph of $f(x) = 1/x$, shifted *up* 2 units. (See the figure on the left below.)

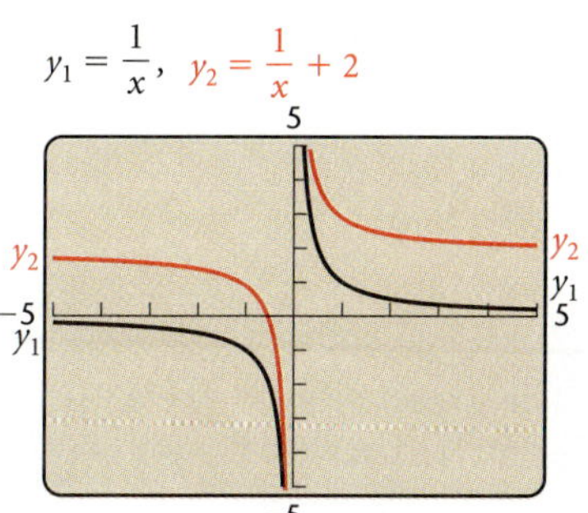

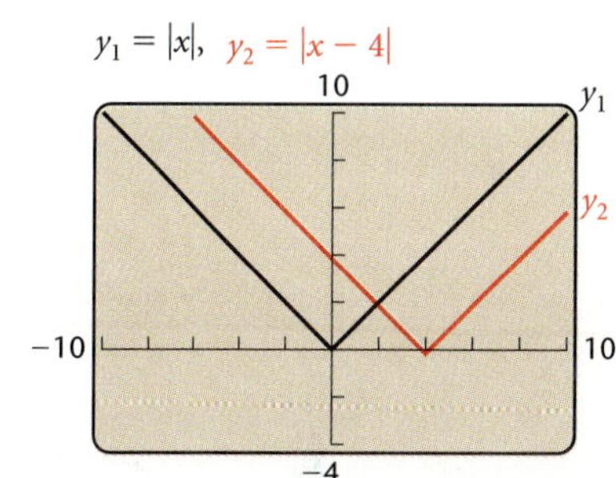

d) To graph $g(x) = |x - 4|$, think of the graph of $f(x) = |x|$. Since $g(x) = f(x - 4)$, the graph of $g(x) = |x - 4|$ is the graph of $f(x) = |x|$ shifted *right* 4 units. (See the figure on the right above.)

e) To graph $h(x) = \sqrt{x + 2} - 3$, think of the graph of $f(x) = \sqrt{x}$. In part (b), we found that the graph of $g(x) = \sqrt{x + 2}$ is the graph of $f(x) = \sqrt{x}$ shifted left 2 units. Since $h(x) = g(x) - 3$, we shift the graph of $g(x) = \sqrt{x + 2}$ *down* 3 units. Together, the graph of $f(x) = \sqrt{x}$ is shifted *left* 2 units and *down* 3 units.

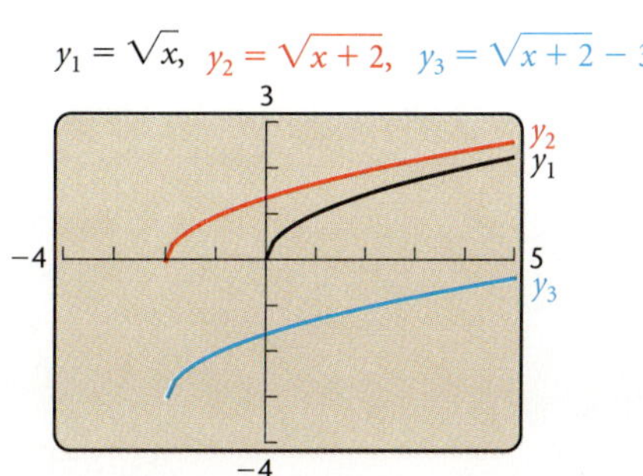

Reflections

Suppose we have a function given by $y = f(x)$. Let's explore the graph of a new function $g(x) = -f(x)$ or $g(x) = f(-x)$.

Interactive Discovery

Consider the functions $y_1 = \frac{1}{5}x^4$ and $y_2 = -\frac{1}{5}x^4$. Graph y_1 in the viewing window $[-10, 10, -10, 10]$. Then graph y_2 in the same viewing window. What pattern do you see? Test it with some other functions in which $y_2 = -y_1$.

Consider the functions $y_1 = 2x^3 - x^4 + 5$ and $y_2 = 2(-x)^3 - (-x)^4 + 5$. Graph y_1 in the viewing window $[-4, 4, -10, 10]$. Then graph y_2 in the same viewing window. What pattern do you see? Test it with some other functions in which x is replaced with $-x$.

Given the graph of $y_1 = f(x)$, we can reflect each point across the x-axis to obtain the graph of $y_2 = -f(x)$. We can reflect each point of y_1 across the y-axis to obtain the graph of $y_2 = f(-x)$. The new graphs are called **reflections** of $f(x)$.

Reflections

The graph of $y_2 = -f(x)$ is the *reflection* of the graph of $y_1 = f(x)$ across the x-axis.

The graph of $y_2 = f(-x)$ is the *reflection* of the graph of $y_1 = f(x)$ across the y-axis.

Example 2   Graph each of the following. Before doing so, describe how each graph can be obtained from the graph of $f(x) = x^3 - 4x^2$.

a) $g(x) = (-x)^3 - 4(-x)^2$ **b)** $g(x) = 4x^2 - x^3$

SOLUTION

a) We first note that

$$f(-x) = (-x)^3 - 4(-x)^2 = g(x).$$

Thus the graph of g is a reflection of the graph of f across the y-axis.

$y_1 = x^3 - 4x^2, \quad y_2 = (-x)^3 - 4(-x)^2$

b) We first note that

$$-f(x) = -(x^3 - 4x^2)$$
$$= -x^3 + 4x^2$$
$$= 4x^2 - x^3$$
$$= g(x).$$

Thus the graph of g is a reflection of the graph of f across the x-axis.

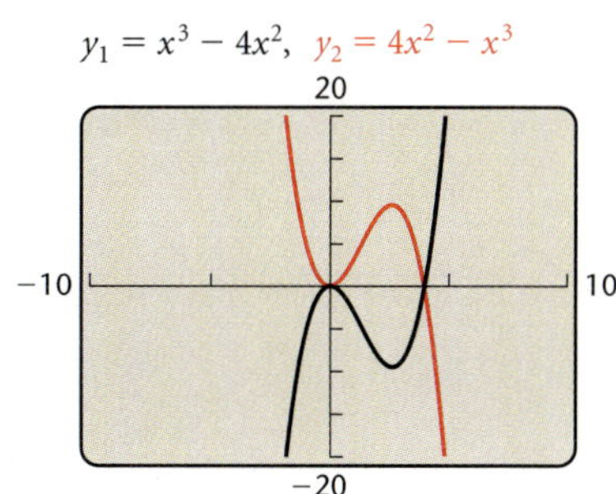

Vertical and Horizontal Stretchings and Shrinkings

Suppose we have a function given by $y_1 = f(x)$. Let's explore the graph of a new function $y_2 = af(x)$ or $y_3 = f(cx)$.

Interactive Discovery

Graph the function $y_1 = x^3 - x$ using the viewing window $[-3, 3, -1, 1]$. Then graph $y_2 = \frac{1}{10}(x^3 - x)$ using the same viewing window. Clear y_2 and graph $y_3 = 2(x^3 - x)$. Then clear y_3 and graph $y_4 = -2(x^3 - x)$. What pattern do you see? Test it with some other graphs.

Now graph y_1 and $y_2 = (2x)^3 - (2x)$ using the same viewing window. Clear y_2 and graph $y_3 = \left(\frac{1}{2}x\right)^3 - \left(\frac{1}{2}x\right)$. Then clear y_3 and graph $y_4 = \left(-\frac{1}{2}x\right)^3 - \left(-\frac{1}{2}x\right)$. What pattern do you see? Test it with some other graphs.

Consider any function f given by $y = f(x)$. Multiplying on the right by any constant a, where $|a| > 1$, to obtain $g(x) = af(x)$ will *stretch* the graph vertically away from the x-axis. If $0 < |a| < 1$, then the graph will be flattened or *shrunk* vertically toward the x-axis. If $a < 0$, the graph is also reflected across the x-axis.

> ### *Vertical Stretching and Shrinking*
>
> The graph of $y_2 = af(x)$ can be obtained from the graph of $y_1 = f(x)$ by
>
> > stretching vertically for $|a| > 1$, or
> >
> > shrinking vertically for $0 < |a| < 1$.
>
> For $a < 0$, the graph is also reflected across the x-axis.

The constant c in the equation $g(x) = f(cx)$ will *stretch* the graph of $y = f(x)$ horizontally away from the y-axis if $0 < |c| < 1$. If $|c| > 1$, the graph will be *shrunk* horizontally toward the y-axis. If $c < 0$, the graph is also reflected across the y-axis.

Horizontal Stretching and Shrinking

The graph of $y_2 = f(cx)$ can be obtained from the graph of $y_1 = f(x)$ by

   shrinking horizontally for $|c| > 1$, or

   stretching horizontally for $0 < |c| < 1$.

For $c < 0$, the graph is also reflected across the y-axis.

It is instructive to use these concepts now to create hand-drawn graphs from a given graph.

Example 3 Shown below is a graph of $y = f(x)$ for some function f. No formula for f is given. Make a hand-drawn graph of each of the following.

a) $g(x) = f(2x)$
b) $h(x) = f\left(\frac{1}{2}x\right)$
c) $t(x) = f\left(-\frac{1}{2}x\right)$

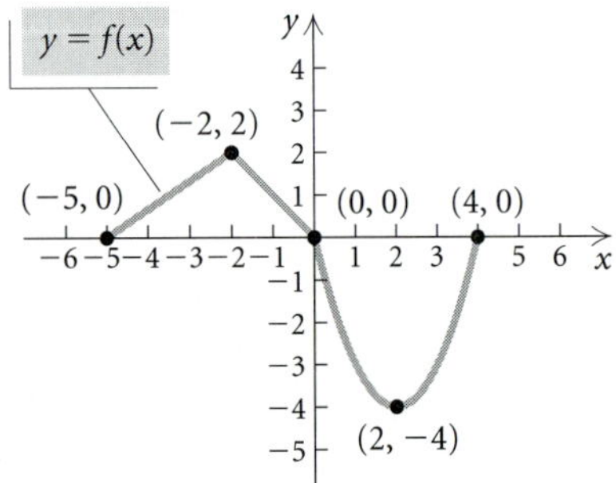

a) Since $|2| > 1$, the graph of $g(x) = f(2x)$ is a horizontal shrinking of the graph of $y = f(x)$. We can consider the key points $(-5, 0)$, $(-2, 2)$, $(0, 0)$, $(2, -4)$, and $(4, 0)$. The transformation divides each x-coordinate by 2 to obtain the key points $(-2.5, 0)$, $(-1, 2)$, $(0, 0)$, $(1, -4)$, and $(2, 0)$ of the graph of $g(x) = f(2x)$. The graph is shown at the top of the following page.

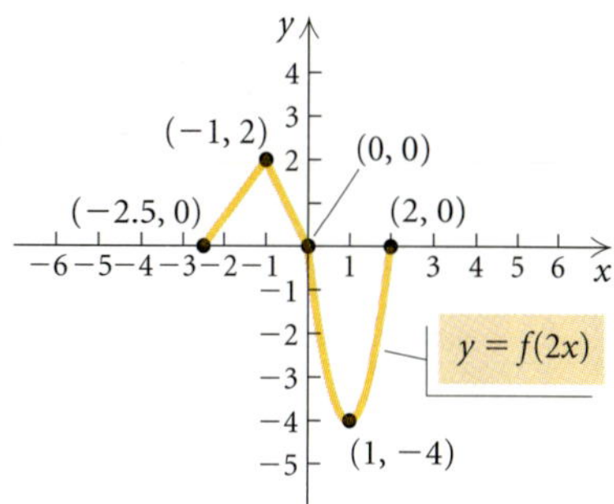

b) Since $\left|\frac{1}{2}\right| < 1$, the graph of $h(x) = f\left(\frac{1}{2}x\right)$ is a horizontal stretching of the graph of $y = f(x)$. We can consider the key points $(-5, 0)$, $(-2, 2)$, $(0, 0)$, $(2, -4)$, and $(4, 0)$. The transformation divides each x-coordinate by $\frac{1}{2}$ (which is the same as multiplying by 2) to obtain the key points $(-10, 0)$, $(-4, 2)$, $(0, 0)$, $(4, -4)$, and $(8, 0)$ of the graph of $h(x) = f\left(\frac{1}{2}x\right)$. The graph is shown below.

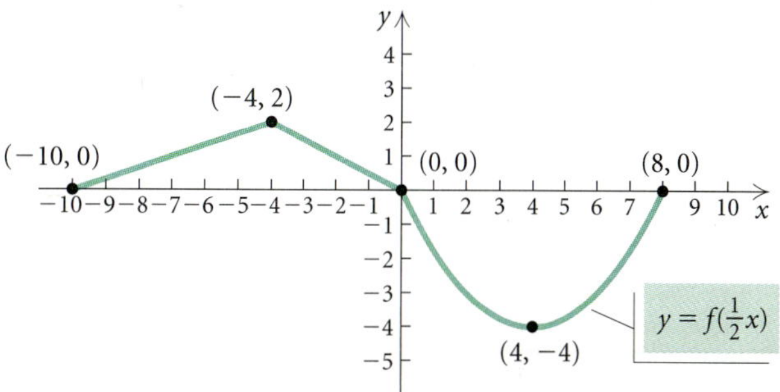

c) The graph of $t(x) = f\left(-\frac{1}{2}x\right)$ can be obtained by reflecting the graph in part (b) across the y-axis. It is shown below.

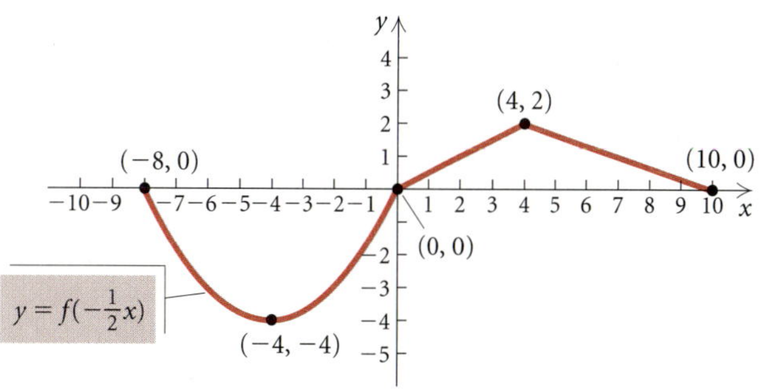

Example 4 Use the graph of $y = f(x)$ given in Example 3. Make a hand-drawn graph of

$$g(x) = -2f(x - 3) + 1.$$

SOLUTION

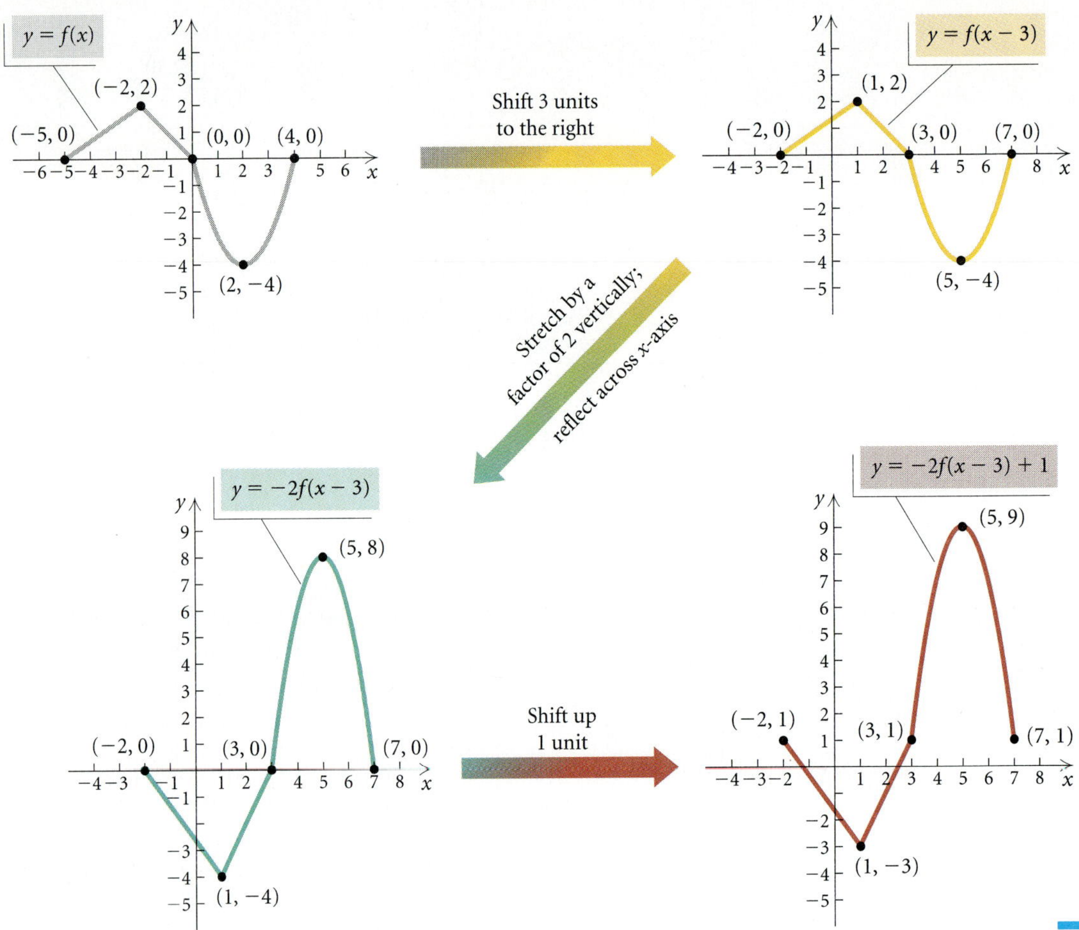

> ### *Vertical or Horizontal Translation*
>
> For $b > 0$,
>
> > the graph of $y_2 = f(x) + b$ is the graph of $y_1 = f(x)$ shifted *up* b units;
> >
> > the graph of $y_2 = f(x) - b$ is the graph of $y_1 = f(x)$ shifted *down* b units.
>
> For $d > 0$,
>
> > the graph of $y_2 = f(x - d)$ is the graph of $y_1 = f(x)$ shifted *right* d units;
> >
> > the graph of $y_2 = f(x + d)$ is the graph of $y_1 = f(x)$ shifted *left* d units.

Reflections

The graph of $y_2 = -f(x)$ is the reflection of the graph of $y_1 = f(x)$ across the x-axis.

The graph of $y_2 = f(-x)$ is the reflection of the graph of $y_1 = f(x)$ across the y-axis.

Vertical Stretching or Shrinking

The graph of $y_2 = af(x)$ can be obtained from the graph of $y_1 = f(x)$ by

stretching vertically for $|a| > 1$, or

shrinking vertically for $0 < |a| < 1$.

For $a < 0$, the graph is also reflected across the x-axis.

Horizontal Stretching or Shrinking

The graph of $y_2 = f(cx)$ can be obtained from the graph of $y_1 = f(x)$ by

shrinking horizontally for $|c| > 1$, or

stretching horizontally for $0 < |c| < 1$.

For $c < 0$, the graph is also reflected across the y-axis.

1.7 Exercise Set

Graph each of the following on a grapher. Before doing so, describe how each graph can be obtained from one of the basic graphs at the beginning of this section.

1. $f(x) = x^2 + 1$

2. $f(x) = x^3 - 4$

3. $f(x) = \dfrac{1}{x} + 3$

4. $f(x) = \sqrt{x} + 2$

5. $f(x) = \sqrt{x} - 5$

6. $f(x) = -\sqrt{x}$

7. $f(x) = -x^2$

8. $f(x) = (x - 4)^2$

9. $f(x) = |x - 3|$

10. $g(x) = |3x|$

11. $g(x) = (x + 5)^3$

12. $g(x) = \dfrac{1}{x - 5}$

13. $g(x) = (x + 1)^2$

14. $f(x) = 2x^2$

15. $f(x) = \frac{1}{2}\sqrt{x}$

16. $f(x) = \frac{1}{2}x^3$

17. $g(x) = \dfrac{2}{x}$

18. $f(x) = |x - 3| - 4$

19. $f(x) = 3\sqrt{x} - 5$

20. $f(x) = 5 - \dfrac{1}{x}$

21. $f(x) = \frac{1}{2}(x - 3)^2$

22. $f(x) = \sqrt{x - 3} + 2$

23. $g(x) = \left|\frac{1}{3}x\right| - 4$

24. $f(x) = \frac{2}{3}x^3 - 4$

25. $f(x) = (x + 5)^2 - 4$

26. $f(x) = (-x)^3 - 5$

27. $f(x) = -\frac{1}{4}(x - 5)^2$

28. $g(x) = \sqrt{-x} - 2$

29. $f(x) = \dfrac{1}{x + 3} + 2$

30. $g(x) = (x - 2)^3 - 5$

31. $f(x) = (x + 4)^3 + 3$

32. $f(x) = \dfrac{1}{-x} + 2$

33. $f(x) = \sqrt{-x} + 5$

34. $f(x) = \frac{4}{5}(x - 4)^2 - 5$

35. $f(x) = 3(x + 4)^2 - 3$

36. $g(x) = 2(x + 1)^3 + 4$

37. $f(x) = 0.43(x - 3)^3 + 2.4$

38. $f(x) = 0.3|x - 5.2| + 2.8$

39. $g(x) = 2.8(x + 5.2)^2 + 1.1$

40. $f(x) = 1.8\sqrt{x - 3.4} - 4.8$

Write an equation for a function that has a graph with the given characteristics. Check your answer on a grapher.

41. The shape of $y = x^2$, but upside-down and shifted right 8 units

42. The shape of $y = \sqrt{x}$, but shifted left 6 units and down 5 units

43. The shape of $y = |x|$, but shifted left 7 units and up 2 units

44. The shape of $y = x^3$, but upside-down and shifted right 5 units

45. The shape of $y = 1/x$, but shrunk vertically by a factor of $\frac{1}{2}$ and shifted down 3 units

46. The shape of $y = x^2$, but shifted right 6 units and up 2 units

47. The shape of $y = x^2$, but upside-down and shifted right 3 units and up 4 units

48. The shape of $y = |x|$, but stretched horizontally by a factor of 2 and shifted down 5 units

49. The shape of $y = \sqrt{x}$, but reflected across the y-axis and shifted left 2 units and down 1 unit

50. The shape of $y = 1/x$, but reflected across the x-axis and shifted up 1 unit

51. The shape of $y = x^3$, but shifted left 4 units and shrunk vertically by a factor of 0.83.

A graph of $y = f(x)$ follows. No formula for f is given. Make a hand-drawn graph of each of the following.

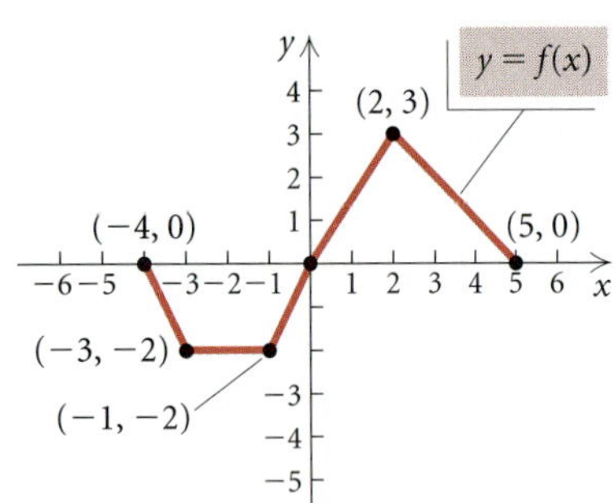

52. $g(x) = \frac{1}{2}f(x)$

53. $g(x) = -2f(x)$

54. $g(x) = f(2x)$

55. $g(x) = f\left(-\frac{1}{2}x\right)$

56. $g(x) = -3f(x + 1) - 4$

57. $g(x) = -\frac{1}{2}f(x - 1) + 3$

The graph of the function f is shown in figure (a). Match each function g in Exercises 58–65 with the appropriate graph from (a)–(h). Some graphs may be used more than once.

a)

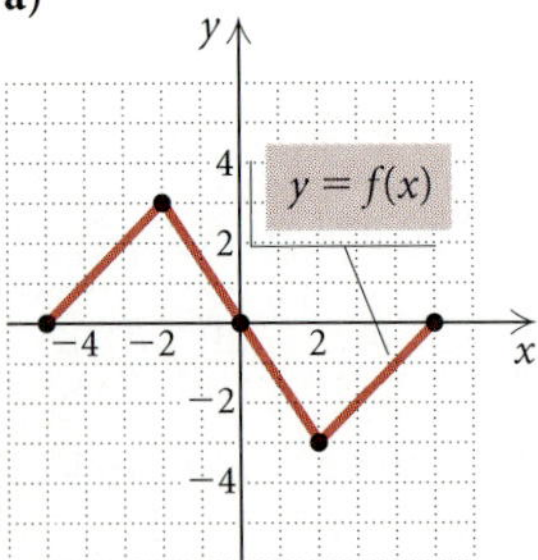

b)

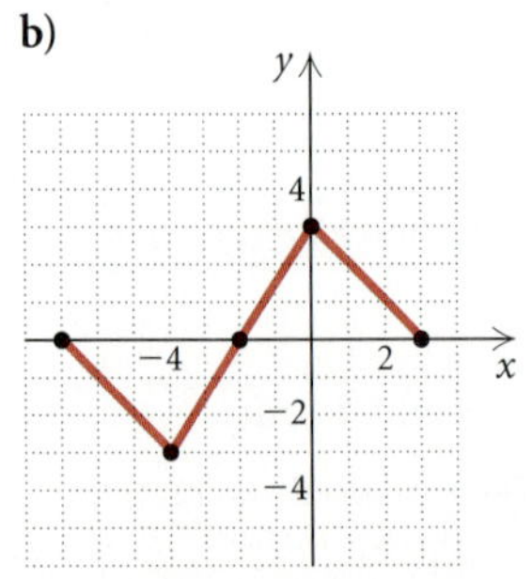

c)

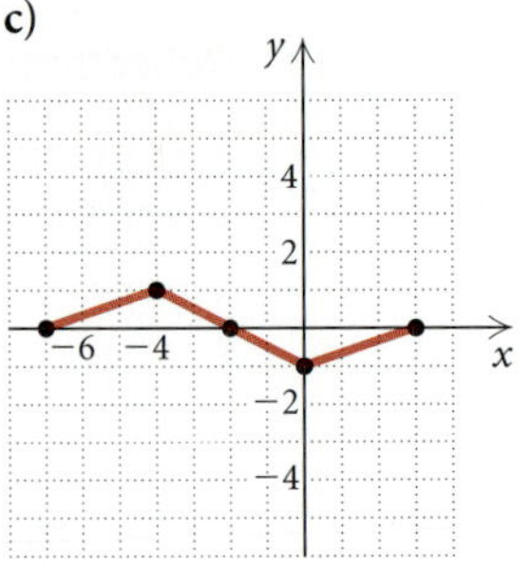

d)

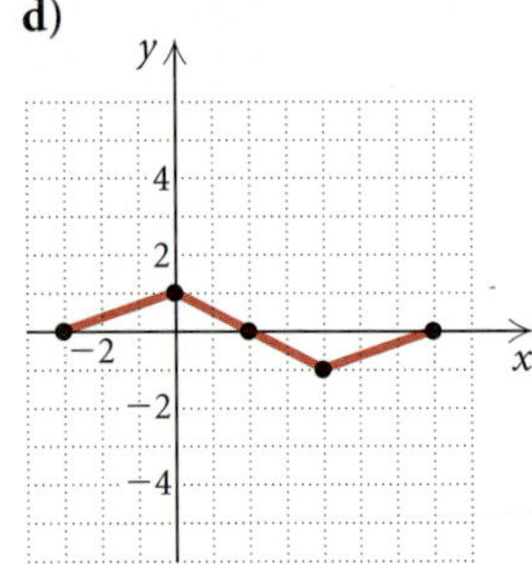

e)

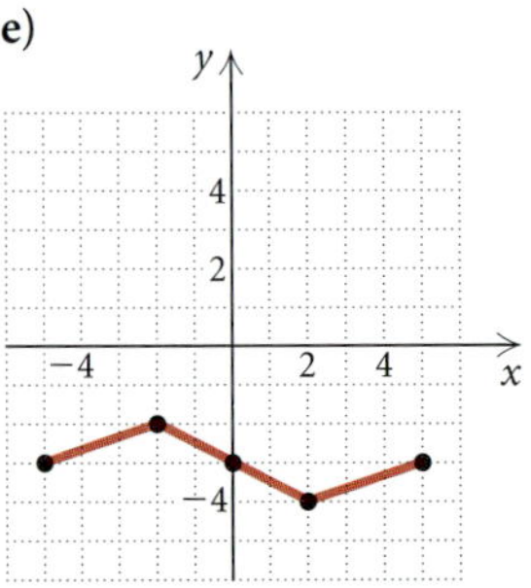

f)

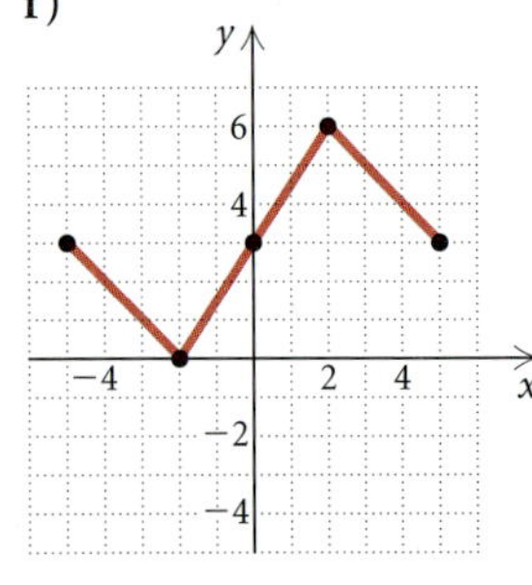

g)

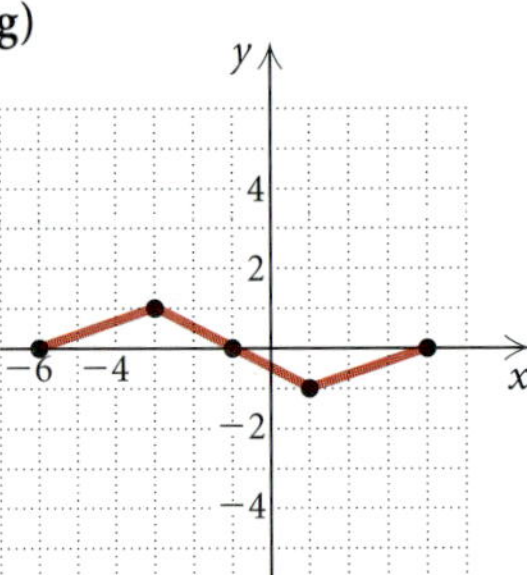

h)

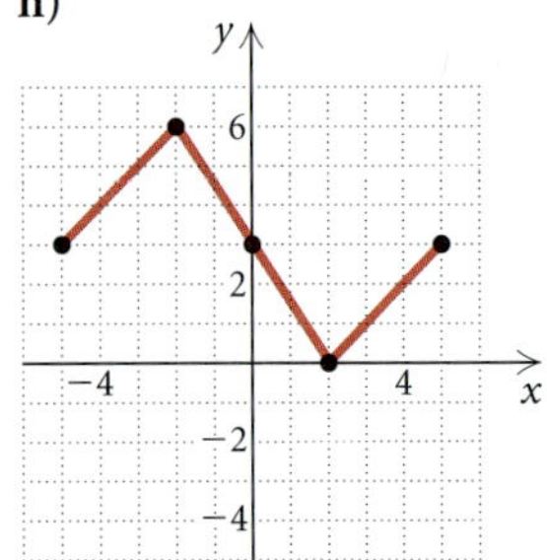

58. $g(x) = f(x) + 3$

59. $g(x) = f(-x) + 3$

60. $g(x) = -f(-x)$

61. $g(x) = -f(x) + 3$

62. $g(x) = \dfrac{1}{3}f(x) - 3$

63. $g(x) = \dfrac{1}{3}f(x - 2)$

64. $g(x) = -f(x + 2)$

65. $g(x) = \dfrac{1}{3}f(x + 2)$

Use a grapher to graph each set of functions using a SIMULTANEOUS *feature. (With this feature, you cannot tell which function is being graphed first.) After the curves are displayed, match each curve with the appropriate function. Check your answers by setting the* WINDOW *format to LabelOn. Then when you use the* TRACE *feature, the label of the curve being traced will appear.*

66. $y_1 = (x + 1)^2,$
$y_2 = (x - 3)^2 + 5,$
$y_3 = (x + 3)^2 - 5,$
$y_4 = x^2 + 1$

67. $y_1 = (x - 5)^3 + 4,$
$y_2 = (x + 5)^3 - 4,$
$y_3 = (x - 2)^3,$
$y_4 = x^3 - 2$

68. $y_1 = \dfrac{1}{x + 5},$

$y_2 = \dfrac{1}{x - 3} + 4,$

$y_3 = \dfrac{1}{x + 3} - 4,$

$y_4 = \dfrac{1}{x} + 5$

69. $y_1 = \sqrt{9 - x^2},$

$y_2 = -\sqrt{9 - x^2},$

$y_3 = \dfrac{2}{3}\sqrt{9 - x^2},$

$y_4 = 4 - \sqrt{9 - x^2}$

70. Using a $[-10, 10, -200, 800]$ window with Yscl $= 100$, show that

$$y_1 = x^4 - 12x^3 + 34x^2 + 12x - 35$$

and

$$y_2 = x^4 + 12x^3 + 34x^2 - 12x - 35$$

are reflections of each other across the y-axis. Verify this algebraically.

For each pair of functions, determine if $g(x) = f(-x)$ using algebra. Then, using the TABLE *feature on a grapher, check your answers by looking at y_1 and y_2 for x-values near 0. You can then check graphically, but be careful to use a suitable window.*

71. $f(x) = 2x^4 - 35x^3 + 3x - 5,$
$\quad g(x) = 2x^4 + 35x^3 - 3x - 5$

72. $f(x) = \frac{1}{4}x^4 + \frac{1}{5}x^3 - 81x^2 - 17,$
$\quad g(x) = \frac{1}{4}x^4 + \frac{1}{5}x^3 + 81x^2 - 17$

A graph of the function $f(x) = x^3 - 3x^2$ is shown below. Exercises 73–76 show graphs of functions transformed from this one. Find a formula for each function.

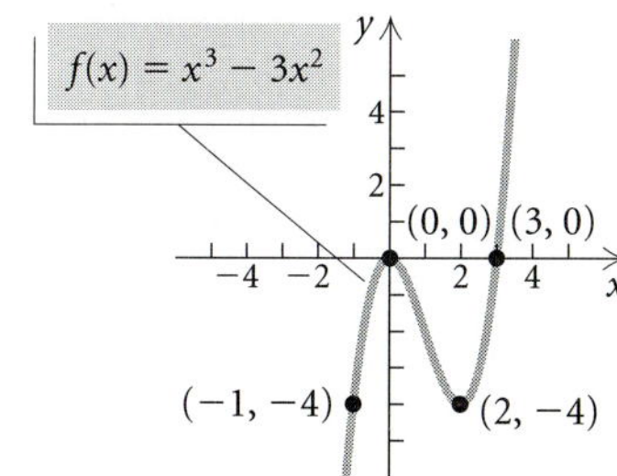

73.

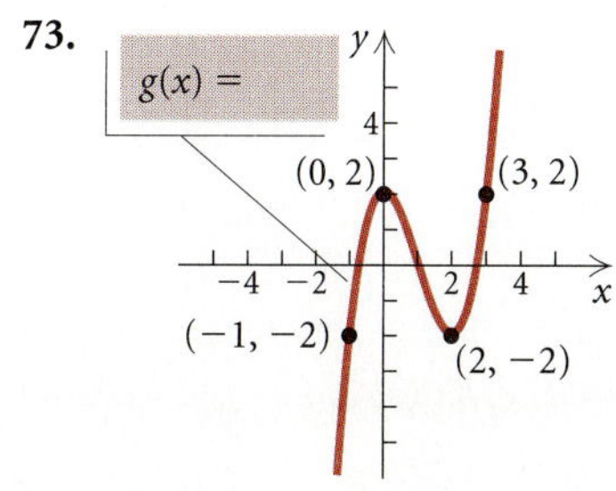

74.

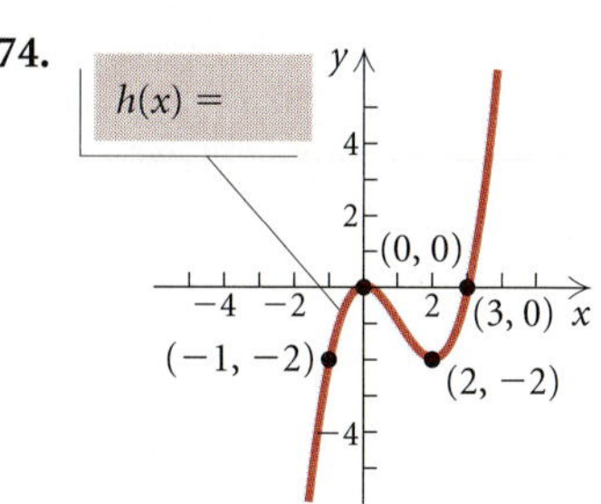

75.

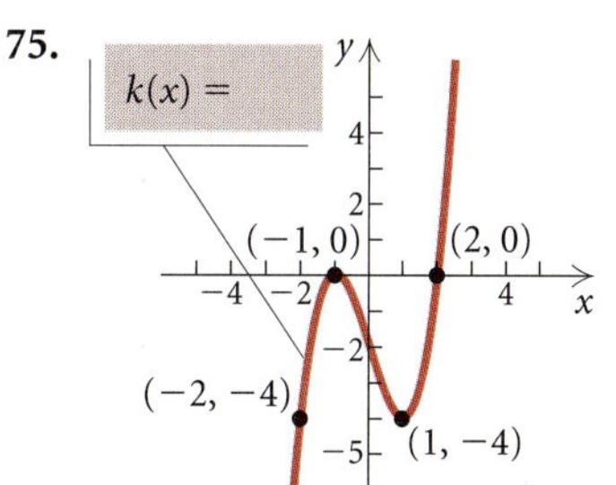

76.

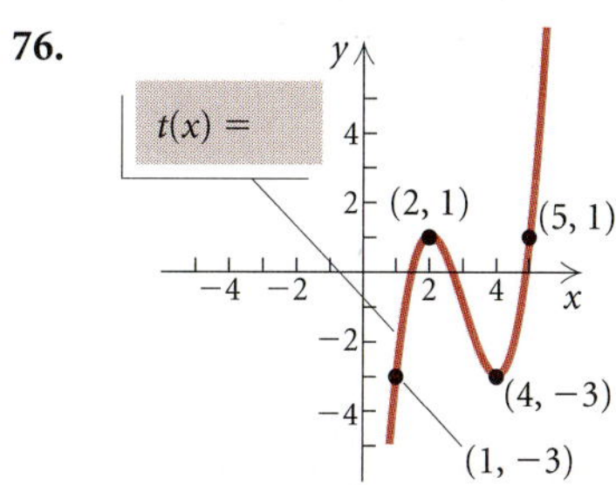

Skill Maintenance

Given that $f(x) = x^3 - 7x$ and $g(x) = x^2 + 1$, find each of the following.

77. $f(3)$

78. $g(3)$

79. $f(3) + g(3)$

80. $f(4) - g(4)$

Synthesis

81. ◈ Explain in your own words why the graph of $y = f(-x)$ is a reflection of the graph of $y = f(x)$ across the y-axis.

82. ◈ Without drawing the graph, describe what the graph of $f(x) = |x^2 - 9|$ looks like.

83. The graph of $f(x) = \sqrt{x}$ passes through the points $(0, 0)$, $(1, 1)$, and $(4, 2)$. Transform this function to one whose graph passes through the points $(4, 7)$, $(5, 6)$, and $(8, 5)$. Check your answer on a grapher.

84. The graph of $f(x) = |x|$ passes through the points $(-3, 3)$, $(0, 0)$, and $(3, 3)$. Transform this function to one whose graph passes through the points $(5, 1)$, $(8, 4)$, and $(11, 1)$. Check your answer on a grapher.

Graph each of the following on a grapher. Before doing so, describe how each graph can be obtained from a more basic graph. Give the domain and the range of each function.

85. $f(x) = \text{INT}\left(x - \frac{1}{2}\right)$

86. $g(x) = \sqrt{7 - x}$

87. $f(x) = 3 \cdot \text{INT}(x + 2) + 1$

88. $f(x) = |x^2 - 4|$ **89.** $g(x) = \left|\dfrac{1}{x}\right|$

90. $f(x) = |\sqrt{x} - 1|$ **91.** $f(x) = \big||x| - 5\big|$

92. Find the roots, or zeros, of $f(x) = 3x^5 - 20x^3$. Then without using a grapher, state the roots of $f(x - 3)$ and $f(x + 8)$.

93. If $(3, 4)$ is a point on the graph of $y = f(x)$, what point do you know is on the graph of $y = 2f(x)$? of $y = 2 + f(x)$? of $y = f(2x)$?

94. If $(-1, 5)$ is a point on the graph of $y = f(x)$, find b such that $(2, b)$ is on the graph of $y = f(x - 3)$.

1.8 The Algebra of Functions

- *Compute function values for the sum, the difference, the product, and the quotient of two functions, and determine the domains.*
- *Find the composition of two functions and the domain of the composition; and decompose a function as a composition of two functions.*

We now consider five methods of combining two functions to obtain a new function. Among these are addition, subtraction, multiplication, and division.

Sums, Differences, Products, and Quotients

Consider the following two functions f and g:

$$f(x) = x + 2 \quad \text{and} \quad g(x) = x^2 + 1.$$

Since $f(3) = 3 + 2 = 5$ and $g(3) = 3^2 + 1 = 10$, we have

$$f(3) + g(3) = 5 + 10 = 15, \qquad f(3) - g(3) = 5 - 10 = -5,$$

$$f(3) \cdot g(3) = 5 \cdot 10 = 50, \quad \text{and} \quad \frac{f(3)}{g(3)} = \frac{5}{10} = \frac{1}{2}.$$

In fact, so long as x is in the domain of both f and g, we can easily compute $f(x) + g(x)$, $f(x) - g(x)$, $f(x) \cdot g(x)$, and, assuming $g(x) \neq 0$, $f(x)/g(x)$. Notation has been developed to facilitate this work.

Sums, Differences, Products, and Quotients of Functions

If f and g are functions and x is in the domain of each function, then

$$(f + g)(x) = f(x) + g(x), \qquad (f - g)(x) = f(x) - g(x),$$
$$(fg)(x) = f(x) \cdot g(x), \qquad (f/g)(x) = f(x)/g(x),$$
$$\text{provided } g(x) \neq 0.$$

Example 1   Given that $f(x) = x + 1$ and $g(x) = \sqrt{x + 3}$, find each of the following.

a) $(f + g)(x)$ **b)** $(f + g)(6)$ **c)** $(f + g)(-4)$

SOLUTION

a) $(f + g)(x) = f(x) + g(x)$
$$= (x + 1) + \sqrt{x + 3} \qquad \text{This cannot be simplified.}$$

b) We can find $(f + g)(6)$ provided 6 is in the domain of each function. The domain of f is all real numbers. The domain of g is all real numbers x for which $x + 3 \geq 0$, or $x \geq -3$. This is the interval $[-3, \infty)$. Because 6 is in both domains, we have

$$f(6) = 6 + 1 = 7, \qquad g(6) = \sqrt{6 + 3} = \sqrt{9} = 3,$$
$$(f + g)(6) = f(6) + g(6) = 7 + 3 = 10.$$

Another method is to use the formula found in part (a):

$$(f + g)(6) = (6 + 1) + \sqrt{6 + 3} = 7 + \sqrt{9} = 7 + 3 = 10.$$

c) To find $(f + g)(-4)$, we must first determine whether -4 is in the domain of each function. We note that -4 is not in the domain of g, $[-3, \infty)$. That is, $\sqrt{-4 + 3}$ is not a real number. Thus, $(f + g)(-4)$ does not exist.   ▬

It is useful to view the concept of the sum of two functions graphically. In the graph on the left below, we see the graphs of two functions f and g and their sum, $f + g$. Consider finding $(f + g)(4) = f(4) + g(4)$. We can locate $g(4)$ on the graph of g and use a compass to measure it. Then we move that setting on top of $f(4)$ and add. The sum gives us $(f + g)(4)$.

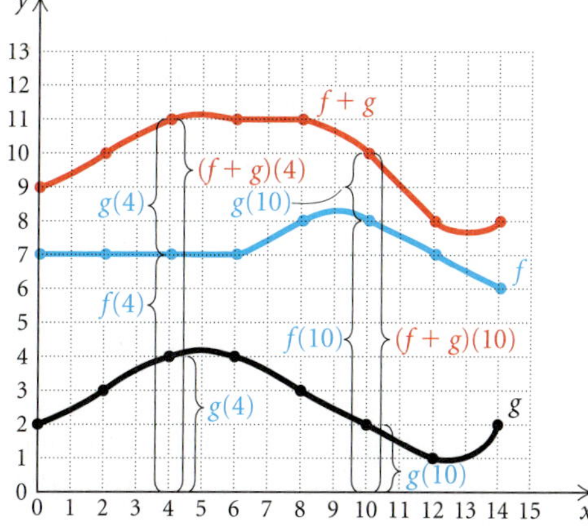

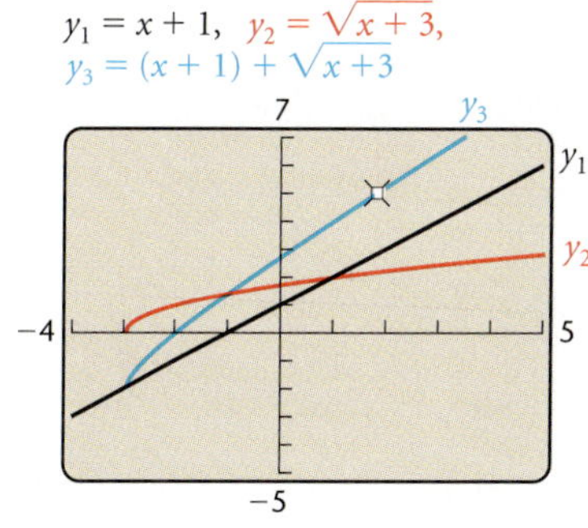

With this in mind, let's view Example 1 from a graphical perspective. See the graph on the right above. Use a grapher to graph

$$y_1 = x + 1, \qquad y_2 = \sqrt{x + 3}, \quad \text{and} \quad y_3 = (x + 1) + \sqrt{x + 3}.$$

(On some graphers, you can enter y_1 and y_2 and then key the grapher to enter $y_3 = y_1 + y_2$ directly.) Note that because the domain of y_2 is $[-3, \infty)$, the graph of the sum exists only over $[-3, \infty)$. By using the TRACE feature and jumping the cursor vertically among the three graphs, we can confirm that the y-coordinates of the graph of $(f + g)(x)$ are the sums of the corresponding y-coordinates of the graphs of $f(x)$ and $g(x)$.

Example 2   Given that $f(x) = x^2 - 4$ and $g(x) = x + 2$, find each of the following.

a) $(f + g)(x)$ b) $(f - g)(x)$ c) $(fg)(x)$

d) $(f/g)(x)$ e) $(gg)(x)$

f) The domain of $f + g, f - g, fg,$ and f/g

SOLUTION

a) $(f + g)(x) = f(x) + g(x) = (x^2 - 4) + (x + 2) = x^2 + x - 2$

b) $(f - g)(x) = f(x) - g(x) = (x^2 - 4) - (x + 2) = x^2 - x - 6$

c) $(fg)(x) = f(x) \cdot g(x) = (x^2 - 4)(x + 2) = x^3 + 2x^2 - 4x - 8$

d) $(f/g)(x) = \dfrac{f(x)}{g(x)}$

$\qquad = \dfrac{x^2 - 4}{x + 2}$ Note that $x \neq -2$.

$\qquad = \dfrac{(x + 2)(x - 2)}{x + 2}$ Factoring

$\qquad = x - 2$ Removing a factor of 1: $\dfrac{x + 2}{x + 2} = 1$

Thus, $(f/g)(x) = x - 2$ with the added stipulation that $x \neq -2$.

e) $(gg)(x) = [g(x)]^2 = (x + 2)^2 = x^2 + 4x + 4$

f) The domain of f is the set of all real numbers. The domain of g is the set of all real numbers. The domain of $f + g, f - g,$ and fg is the set of numbers in the intersection of the domains—that is, the set of numbers in both domains, which is again the set of real numbers. For f/g, we must exclude -2, since $g(-2) = 0$. Thus the domain of f/g is the set of real numbers excluding -2, or $(-\infty, -2) \cup (-2, \infty)$.

The Composition of Functions

In the real world, it is not uncommon for a function's output to depend on some input that is itself an output of some function. For instance, the amount a person pays as state income tax usually depends on the amount of adjusted gross income on the person's federal tax return, which, in turn, depends on his or her annual earnings. Such functions are called *composite functions*.

To illustrate how composite functions work, suppose a chemistry student needs a formula to convert Fahrenheit temperatures to Kelvin units. The formula $c(t) = \frac{5}{9}(t - 32)$ gives the Celsius temperature $c(t)$ that corresponds to the Fahrenheit temperature t. The formula $k(c) = c + 273$ gives the Kelvin temperature $k(c)$ that corresponds to the Celsius temperature c. Thus, 50° Fahrenheit corresponds to

$$c(50) = \tfrac{5}{9}(50 - 32) = \tfrac{5}{9}(18) = 10° \text{ Celsius}$$

and since 10° Celsius corresponds to

$$k(10) = 10 + 273 = 283° \text{ Kelvin},$$

we see that 50° Fahrenheit is the same as 283° Kelvin. This two-step procedure can be used to convert any Fahrenheit temperature to Kelvin units.

In the table shown at left, we use a grapher to convert Fahrenheit temperatures, x, to Celsius temperatures, y_1, using $y_1 = \tfrac{5}{9}(x - 32)$. We also convert Celsius temperatures to Kelvin units, y_2, using $y_2 = y_1 + 273$.

A student making numerous conversions might look for a formula that converts directly from Fahrenheit to Kelvin. Such a formula can be found by substitution:

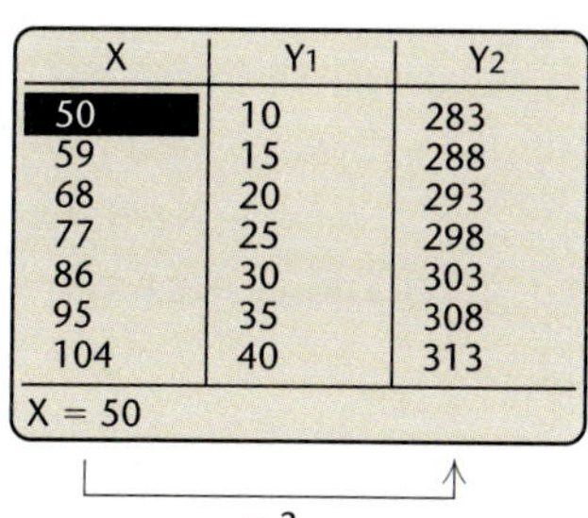

X	Y₁	Y₂
50	10	283
59	15	288
68	20	293
77	25	298
86	30	303
95	35	308
104	40	313

X = 50

= ?

$$y_2 = y_1 + 273$$
$$= \left[\frac{5}{9}(x - 32)\right] + 273 \qquad \textbf{Substituting}$$
$$= \frac{5}{9}x - \frac{160}{9} + 273$$
$$= \frac{5}{9}x - \frac{160}{9} + \frac{2457}{9}$$
$$= \frac{5x + 2297}{9}. \qquad \textbf{Simplifying}$$

We can show on a grapher that the same values that appear in the table for y_2 will appear when y_2 is entered as

$$y_2 = \frac{5x + 2297}{9}.$$

In the more commonly used function notation, we have

$$k(c(t)) = c(t) + 273$$
$$= \frac{5}{9}(t - 32) + 273 \qquad \textbf{Substituting}$$
$$= \frac{5t + 2297}{9}. \qquad \textbf{Simplifying as above}$$

Since the last equation expresses the Kelvin temperature as a new function, K, of the Fahrenheit temperature, t, we can write

$$K(t) = \frac{5t + 2297}{9},$$

where $K(t) = $ the Kelvin temperature corresponding to the Fahrenheit temperature, t. Here we have $K(t) = k(c(t))$. The new function K is called the **composition** of k and c and can be denoted $k \circ c$ (read "k composed with c," "the composition of k and c," or "k circle c").

Composition of Functions

The *composite function* $f \circ g$, the *composition* of f and g, is defined as

$$(f \circ g)(x) = f(g(x)),$$

where x is in the domain of g and $g(x)$ is in the domain of f.

Example 3   Given that $f(x) = 2x - 5$ and $g(x) = x^2 - 3x + 8$, find each of the following.

a) $(f \circ g)(7)$ and $(g \circ f)(7)$ **b)** $(f \circ g)(x)$ and $(g \circ f)(x)$

SOLUTION   Consider each function separately:

$$f(x) = 2x - 5 \qquad \text{This function multiplies each input by 2 and subtracts 5.}$$

and

$$g(x) = x^2 - 3x + 8. \qquad \text{This function squares an input, subtracts 3 times the input from the result, and then adds 8.}$$

a) To find $(f \circ g)(7)$, we first find $g(7)$. Then we use $g(7)$ as an input for f:

$$(f \circ g)(7) = f(g(7)) = f(7^2 - 3 \cdot 7 + 8)$$
$$= f(36) = 2 \cdot 36 - 5$$
$$= 67.$$

To find $(g \circ f)(7)$, we first find $f(7)$. Then we use $f(7)$ as an input for g:

$$(g \circ f)(7) = g(f(7)) = g(2 \cdot 7 - 5)$$
$$= g(9) = 9^2 - 3 \cdot 9 + 8$$
$$= 62.$$

b) To find $(f \circ g)(x)$, we substitute $g(x)$ for x in the equation for $f(x)$:

$$(f \circ g)(x) = f(g(x)) = f(x^2 - 3x + 8)$$
$$\text{Substituting } x^2 - 3x + 8 \text{ for } g(x)$$
$$= 2(x^2 - 3x + 8) - 5$$
$$= 2x^2 - 6x + 16 - 5$$
$$= 2x^2 - 6x + 11.$$

To find $(g \circ f)(x)$, we substitute $f(x)$ for x in the equation for $g(x)$:

$$(g \circ f)(x) = g(f(x)) = g(2x - 5) \qquad \text{Substituting } 2x - 5 \text{ for } f(x)$$
$$= (2x - 5)^2 - 3(2x - 5) + 8$$
$$= 4x^2 - 20x + 25 - 6x + 15 + 8$$
$$= 4x^2 - 26x + 48.$$

Note in Example 3 that, as a rule, $(f \circ g)(x) \neq (g \circ f)(x)$. We can check this graphically.

Example 4 Given that $f(x) = \sqrt{x}$ and $g(x) = x - 3$:

a) Find $h(x)$ and $k(x)$ if $h = f \circ g$ and $k = g \circ f$.

b) Graph h and k.

c) Find the domains of h and k.

SOLUTION

a) $h(x) = (f \circ g)(x) = f(g(x)) = f(x - 3) = \sqrt{x - 3}$

$k(x) = (g \circ f)(x) = g(f(x)) = g(\sqrt{x}) = \sqrt{x} - 3$

b) We actually used function composition when we worked with transformations in Section 1.7. For example, we know that the graph of $h(x) = \sqrt{x - 3}$ has the same shape as $y = \sqrt{x}$ shifted right 3 units. This occurs because g subtracts 3 units from each input before f takes the square root. When this sequence is reversed, as in the graph of $k(x) = \sqrt{x} - 3$, the subtraction of 3 occurs *after* the square root is taken. Thus the graph of k has the same shape as the graph of $y = \sqrt{x}$ shifted *down* 3 units.

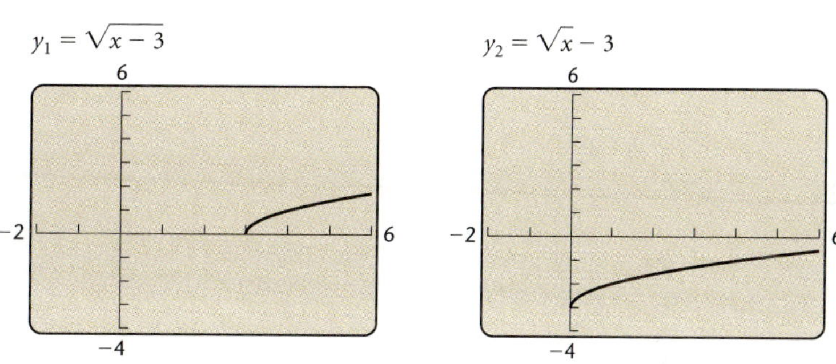

c) The domain of f is $\{x | x \geq 0\}$, or the interval $[0, \infty)$. The domain of g is $\mathbb{R}$.

To find the domain of $h = f \circ g$, we must consider that the outputs for g will serve as inputs for f. Since the inputs for f cannot be negative, we must have

$$g(x) = x - 3, \quad \text{where } x - 3 \geq 0,$$

$$\text{or} \quad g(x) = x - 3, \quad \text{where } x \geq 3.$$

Thus the domain of $h = \{x | x \geq 3\}$, or the interval $[3, \infty)$, as the graph confirms.

To find the domain of $k = g \circ f$, we must consider that the outputs for f will serve as inputs for g. Since g can accept *any* real number as an input, any output from f is acceptable. Thus the domain of $k = \{x | x \geq 0\}$, as the graph confirms.

Decomposing a Function as a Composition

In calculus, one needs to recognize how a function can be expressed as a composition. In this way, we are "decomposing" the function.

Example 5 Find $f(x)$ and $g(x)$ such that $h(x) = (f \circ g)(x)$: $h(x) = (2x - 3)^5$.

SOLUTION We have expressed $h(x)$ as $(2x - 3)$ to the 5th power. Two functions that can be used for the composition are

$$f(x) = x^5 \quad \text{and} \quad g(x) = 2x - 3.$$

We can check by forming the composition:

$$h(x) = (f \circ g)(x) = f(g(x)) = f(2x - 3) = (2x - 3)^5.$$

This is the most "obvious" answer to the question. There can be other less obvious answers. For example, if

$$f(x) = (x + 7)^5 \quad \text{and} \quad g(x) = 2x - 10,$$

then

$$h(x) = (f \circ g)(x) = f(g(x))$$
$$= f(2x - 10) = [(2x - 10) + 7]^5 = (2x - 3)^5.$$

1.8 Exercise Set

Given that $f(x) = x^2 - 3$ and $g(x) = 2x + 1$, find each of the following, if it exists.

1. $(f + g)(5)$ **2.** $(f - g)(3)$

3. $(f \circ g)(4)$ **4.** $(g \circ f)(4)$

5. $(fg)(2)$ **6.** $(fg)(-2)$

7. $(f/g)(-1)$ **8.** $(f/g)\left(-\frac{1}{2}\right)$

9. $(fg)\left(-\frac{1}{2}\right)$ **10.** $(f/g)(-\sqrt{3})$

11. $(f/g)(\sqrt{3})$ **12.** $(f - g)(0)$

For each pair of functions in Exercises 13–16:

a) Find the domain of f, g, $f + g$, $f - g$, fg, ff, f/g, g/f, $f \circ g$, and $g \circ f$.

b) Find $(f + g)(x)$, $(f - g)(x)$, $(fg)(x)$, $(ff)(x)$, $(f/g)(x)$, $(g/f)(x)$, $(f \circ g)(x)$, and $(g \circ f)(x)$.

13. $f(x) = x - 3$, $g(x) = \sqrt{x + 4}$

14. $f(x) = x^2 - 1$, $g(x) = 2x + 5$

15. $f(x) = x^3$, $g(x) = 2x^2 + 5x - 3$

16. $f(x) = x^2$, $g(x) = \sqrt{x}$

For each pair of functions in Exercises 17–20, find $(f + g)(x)$, $(f - g)(x)$, $(fg)(x)$, $(f/g)(x)$, and $(f \circ g)(x)$.

17. $f(x) = x^2 - 4$, $g(x) = x^2 + 2$

18. $f(x) = x^2 + 2$, $g(x) = 4x - 7$

19. $f(x) = \sqrt{x - 7}$, $g(x) = x^2 - 25$

20. $f(x) = \sqrt{x - 1}$, $g(x) = \sqrt{3x - 8}$

For Exercises 21–26, consider the functions F and G as shown in the following graph.

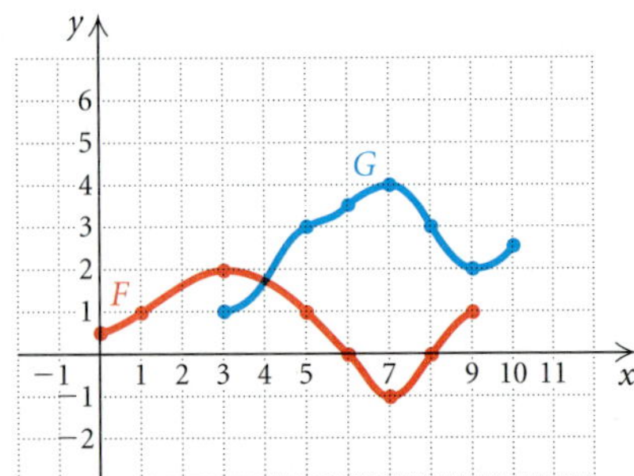

21. Find the domain of F, the domain of G, and the domain of $F + G$.

22. Find the domain of $F - G$, FG, and F/G.

23. Find the domain of G/F.

24. Graph $F + G$.

25. Graph $G - F$.

26. Graph $F - G$.

Find $(f \circ g)(x)$ and $(g \circ f)(x)$.

27. $f(x) = x + 3$, $g(x) = x - 3$

28. $f(x) = \frac{4}{5}x$, $g(x) = \frac{5}{4}x$

29. $f(x) = 3x - 7$, $g(x) = \dfrac{x + 7}{3}$

30. $f(x) = \frac{2}{3}x - \frac{4}{5}$, $g(x) = 1.5x + 1.2$

31. $f(x) = 20$, $g(x) = 0.05$

32. $f(x) = x^4$, $g(x) = \sqrt[4]{x}$

33. $f(x) = \sqrt{x + 5}$, $g(x) = x^2 - 5$

34. $f(x) = x^5 - 2$, $g(x) = \sqrt[5]{x + 2}$

35. $f(x) = \dfrac{1 - x}{x}$, $g(x) = \dfrac{1}{1 + x}$

36. $f(x) = \dfrac{x^2 - 1}{x^2 + 1}$, $g(x) = \dfrac{3x - 4}{5x - 2}$

37. $f(x) = x^3 - 5x^2 + 3x + 7$, $g(x) = x + 1$

38. $f(x) = x - 1$, $g(x) = x^3 + 2x^2 - 3x - 9$

Find $f(x)$ and $g(x)$ such that $h(x) = (f \circ g)(x)$. Answers may vary, but try to select the most obvious answer.

39. $h(x) = (4 + 3x)^5$

40. $h(x) = \sqrt[3]{x^2 - 8}$

41. $h(x) = \dfrac{1}{(x - 2)^4}$

42. $h(x) = \dfrac{1}{\sqrt{3x + 7}}$

43. $h(x) = \dfrac{x^3 - 1}{x^3 + 1}$

44. $h(x) = \left|9x^2 - 4\right|$

45. $h(x) = \left(\dfrac{2 + x^3}{2 - x^3}\right)^6$

46. $h(x) = (\sqrt{x} - 3)^4$

47. $h(x) = \sqrt{\dfrac{x - 5}{x + 2}}$

48. $h(x) = \sqrt{1 + \sqrt{1 + x}}$

49. $h(x) = (x + 2)^3 - 5(x + 2)^2 + 3(x + 2) - 1$

50. $h(x) = 2(x - 1)^{5/3} + 5(x - 1)^{2/3}$

51. *Total Cost, Revenue, and Profit.* In economics, functions that involve revenue, cost, and profit are used. For example, suppose that $R(x)$ and $C(x)$ denote the total revenue and the total cost, respectively, of producing a new kind of tool for King Hardware Stores. Then the difference

$$P(x) = R(x) - C(x)$$

represents the total profit for producing x tools. Given

$$R(x) = 60x - 0.4x^2 \quad \text{and} \quad C(x) = 3x + 13,$$

find each of the following.

a) $P(x)$

b) $R(100)$, $C(100)$, and $P(100)$

c) Use a grapher and graph the three functions using the viewing window $[0, 160, 0, 3000]$.

52. *Dress Sizes.* A dress that is size x in France is size $s(x)$ in the United States, where $s(x) = x - 32$. A dress that is size x in the United States is size $y(x)$

in Italy, where $y(x) = 2(x + 12)$. Find a function that will convert French dress sizes to Italian dress sizes.

53. *Ripple Spread.* A stone is thrown into a pond. A circular ripple is spreading over the pond in such a way that the radius is increasing at the rate of 3 ft/sec.

a) Find a function $r(t)$ for the radius in terms of t.

b) Find a function $A(r)$ for the area of the ripple in terms of the radius r.

c) Find $(A \circ r)(t)$. Explain the meaning of this function.

54. *Airplane Distance.* An airplane is 300 ft from the control tower at the end of the runway. It takes off at a speed of 250 mph.

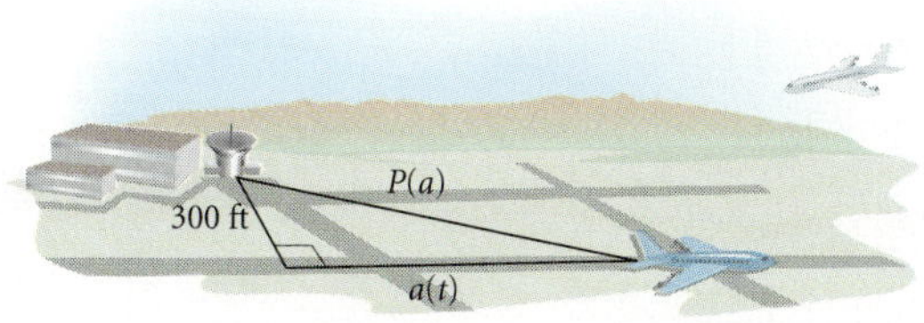

a) Let $a(t)$ = the distance that the airplane travels down the runway in time t. Find a formula for the function.

b) Let P = the distance of the airplane from the control tower. Find a formula for P in terms of the distance a—that is, find an expression for $P(a)$.

c) Find $(P \circ a)(t)$. Explain the meaning of this function.

Skill Maintenance

Refer to the graph accompanying Exercises 21–26.

55. Find the range of F.

56. Find the range of G.

57. Evaluate $5x^3 - 7x^2 + 2x - 7$ for $x = -1$.

58. Factor: $9x^5 - 64x^4 + 72x^3$.

Synthesis

59. ◆ Does $(f + g)(x) = (f \circ g)(x)$ for any functions f and g? Why or why not?

60. ◆ Explain how the function f given by $f(x) = (x - 5)^2 + 3$ can be regarded as a composition of two functions, each of which is a translation.

In Exercises 61–64, use a grapher to graph f, g, and $f + g$ using the same viewing window.

61. $f(x) = x^2$, $g(x) = 3 - 2x$

62. $f(x) = \dfrac{2}{x}$, $g(x) = x$

63. $f(x) = 5 - x^2$, $g(x) = x^2 - 5$

64. $f(x) = |x|$, $g(x) = \text{INT}(x)$

Refer to the graph accompanying Exercises 21–26.

65. Find the domain of $F \circ G$, $G \circ G$, and $F \circ F$.

66. Find the range of $F \circ G$.

67. Consider $f(x) = 3x + b$ and $g(x) = 2x - 1$. Find b such that $(f \circ g)(x) = (g \circ f)(x)$ for all real numbers x.

68. Consider $f(x) = 3x - 4$ and $g(x) = mx + b$. Find m and b such that

$$(f \circ g)(x) = (g \circ f)(x) = x$$

for all real numbers x.

69. For $f(x) = 1/(1 - x)$, find $(f \circ f)(x)$ and $(f \circ (f \circ f))(x)$.

State whether each of the following is true or false.

70. The composition of two even functions is even.

71. The sum of two even functions is even.

72. The product of two odd functions is odd.

73. The composition of two odd functions is odd.

74. The product of an even function and an odd function is odd.

75. Show that if f is *any* function, then the function E defined by

$$E(x) = \frac{f(x) + f(-x)}{2}$$

is even.

76. Show that if f is *any* function, then the function O defined by

$$O(x) = \frac{f(x) - f(-x)}{2}$$

is odd.

77. Consider the functions E and O of Exercises 75 and 76.

a) Show that $f(x) = E(x) + O(x)$. This means that every function can be expressed as the sum of an even and an odd function.

b) Let $f(x) = 4x^3 - 11x^2 + \sqrt{x} - 10$. Express f as a sum of an even function and an odd function.

CHAPTER

1 Summary and Review

Important Properties and Formulas

Terminology about Lines

Slope: $\quad m = \dfrac{y_2 - y_1}{x_2 - x_1}$

The Slope–intercept Equation:
$$y = mx + b$$

The Point–slope Equation:
$$y - y_1 = m(x - x_1)$$

The Two-point Equation:
$$y - y_1 = \frac{y_2 - y_1}{x_2 - x_1}(x - x_1)$$

Horizontal Lines: $\quad y = b$

Vertical Lines: $\quad x = a$

Parallel Lines: $\quad m_1 = m_2, \ b_1 \neq b_2$

Perpendicular Lines:
$$m_1 m_2 = -1, \text{ or}$$
$$x = a, y = b$$

The Distance Formula
$$d = \sqrt{(x_2 - x_1)^2 + (y_2 - y_1)^2}$$

The Midpoint Formula
$$\left(\frac{x_1 + x_2}{2}, \frac{y_1 + y_2}{2} \right)$$

Equation of a Circle
$$(x - h)^2 + (y - k)^2 = r^2$$

Tests for Symmetry

x-axis: If replacing y by $-y$ produces an equivalent equation, then the graph is symmetric with respect to the *x*-axis.

y-axis: If replacing x by $-x$ produces an equivalent equation, then the graph is symmetric with respect to the *y*-axis.

Origin: If replacing x by $-x$ and y by $-y$ produces an equivalent equation, then the graph is symmetric with respect to the origin.

Even Function: $\quad f(-x) = f(x)$

Odd Function: $\quad f(-x) = -f(x)$

Transformations

Vertical Translation: $\qquad y = f(x) \pm b$

Horizontal Translation: $\qquad y = f(x \pm d)$

Reflection across the *x*-axis: $\quad y = -f(x)$

Reflection across the *y*-axis: $\quad y = f(-x)$

Vertical Stretching or Shrinking:
$$y = af(x)$$

Horizontal Stretching or Shrinking:
$$y = f(ax)$$

The Algebra of Functions

The Sum of Two Functions:
$$(f + g)(x) = f(x) + g(x)$$

The Difference of Two Functions:
$$(f - g)(x) = f(x) - g(x)$$

The Product of Two Functions:
$$(fg)(x) = f(x) \cdot g(x)$$

The Quotient of Two Functions:
$$(f/g)(x) = f(x)/g(x), \ g(x) \neq 0$$

The Composition of Two Functions:
$$(f \circ g)(x) = f(g(x))$$

REVIEW EXERCISES

Determine whether each relation is a function. Find the domain and the range.

1. $\{(3, 1), (5, 3), (7, 7), (3, 5)\}$

2. $\{(2, 7), (-2, -7), (7, -2), (0, 2), (1, -4)\}$

Determine whether each graph is that of a function

3. 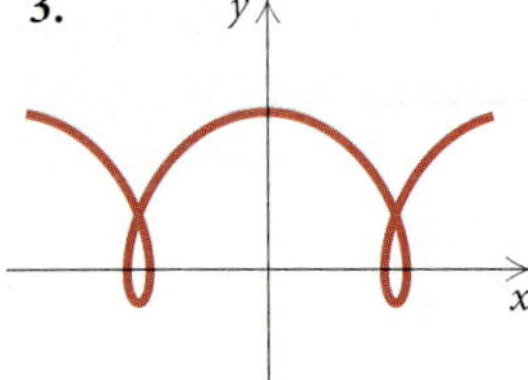

4. 

5. 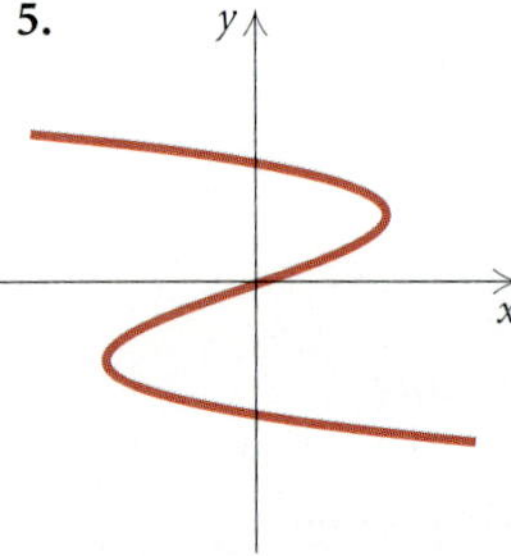

6. 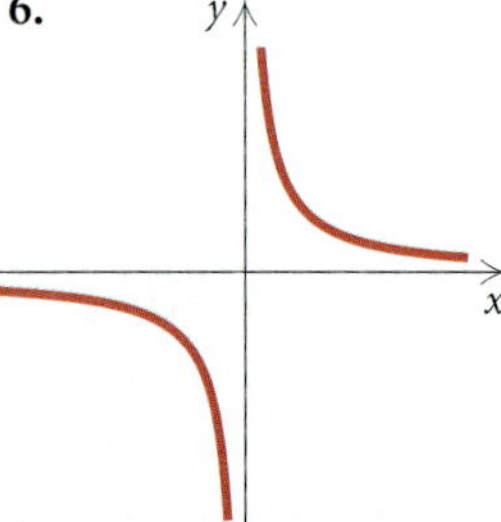

Find the domain of each function. Do not use a grapher.

7. $f(x) = 4 - 5x + x^2$

8. $f(x) = \sqrt{7 - 3x}$

9. $f(x) = \dfrac{1}{x^2 - 6x + 5}$

10. $f(x) = \dfrac{-5x}{|16 - x^2|}$

Use a grapher to graph each function. Then visually estimate the domain and the range.

11. $f(x) = \sqrt{16 - (x + 1)^2}$

12. $f(x) = |x - 3| + |x + 4| - 12$

13. $f(x) = x^3 - 5x^2 + 2x - 7$

14. $f(x) = x^4 - 8x^2 - 3$

15. Given that $f(x) = x^2 - x - 3$, find:

    a) $f(0)$;   **b)** $f(-3)$;

    c) $f(a - 1)$;   **d)** $\dfrac{f(x + h) - f(x)}{h}$.

Use a grapher to determine graphically whether each of the following seems to be an identity.

16. $\sqrt{x + 9} = \sqrt{x} + 3$

17. $x^2 \cdot x^3 = x^5$

18. $x^2 + x^3 = x^5$

19. $x + 3 = 3 + x$

Make a hand-drawn graph of each of the following. Check your results on a grapher.

20. $f(x) = \begin{cases} x^3, & \text{for } x < -2, \\ |x|, & \text{for } -2 \le x \le 2, \\ \sqrt{x - 1}, & \text{for } x > 2 \end{cases}$

21. $f(x) = \begin{cases} \dfrac{x^2 - 1}{x + 1}, & \text{for } x \ne -1, \\ 3, & \text{for } x = -1 \end{cases}$

22. $f(x) = \text{INT}(x)$

23. $f(x) = \text{INT}(x - 3)$

24. Use the TABLE feature to complete the following input–output table for the equations

$$y_1 = \sqrt{16 - (x + 1)^2} \quad \text{and} \quad y_2 = \frac{10x}{x^2 + 1}.$$

X	Y1	Y2
2.7		
2.8		
2.9		
3.0		
3.1		
3.2		
3.3		

X = 2.7

Then use the TABLE feature to consider inputs from -5.2 to -4.6.

    a) What does the table tell you about the domain of each function? the range?

    b) What seems to be the largest output of each function? the smallest?

Use a grapher. For each function in Exercises 25–27, estimate:

a) *the roots, or zeros;*

b) *where the function is increasing or decreasing;*

c) *any relative maxima or minima.*

25. $f(x) = 1.1x^3 - 9.04x^2 - 18x + 702$

26. $f(x) = \dfrac{20}{(x + 4)^2 + 1} - 5$

27. $f(x) = 0.8x\sqrt{16 - (x + 1)^2}$

28. *Inscribed Rectangle.* A rectangle is inscribed in a semicircle of radius 2, as shown. The variable $x =$ half the length of the rectangle. Express the area of the rectangle as a function of x.

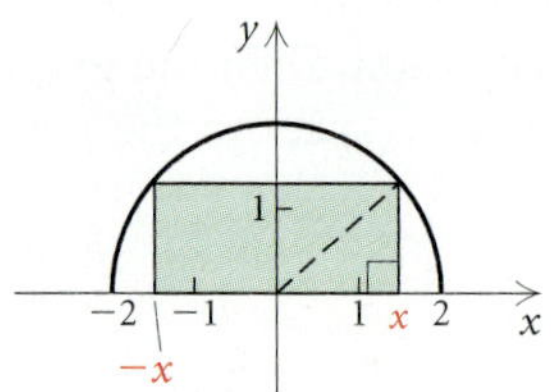

29. *Minimizing Surface Area.*   A container firm is designing an open-top rectangular box, with a square base, that will hold 108 in³. Let $x =$ the length of a side of the base.

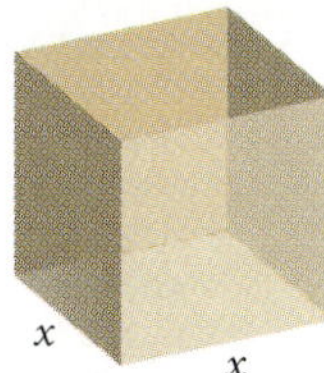

a) Express the surface area as a function of x.
b) Find the domain of the function.
c) Graph the function.
d) What dimensions minimize the surface area of the box?

Each table of data contains input–output values for a function. Answer the following questions:

a) *Is the change in the inputs, x, the same?*
b) *Is the change in the outputs, y, the same?*
c) *Is the function linear?*

30.

x	y
−3	8
−2	11
−1	14
0	17
1	20
2	22
3	26

31.

x	y
20	11.8
30	24.2
40	36.6
50	49.0
60	61.4
70	73.8
80	86.2

32. Find the slope and the y-intercept of the graph of $-2x - y = 7$.

Write a point–slope equation for a line with the following characteristics.

33. $m = -\frac{2}{3}$, y-intercept $(0, -4)$

34. $m = 3$, passes through $(-2, -1)$

35. Passes through $(4, 1)$ and $(-2, -1)$

Determine whether the lines are parallel, perpendicular, or neither.

36. $3x - 2y = 8,$
  $6x - 4y = 2$

37. $y - 2x = 4,$
  $2y - 3x = -7$

38. $y = \frac{3}{2}x + 7,$
  $y = -\frac{2}{3}x - 4$

Given the point $(1, -1)$ and the line $2x + 3y = 4$:

39. Find an equation of the line containing the given point and parallel to the given line.

40. Find an equation of the line containing the given point and perpendicular to the given line.

41. *Total Cost.*   Clear County Cable Television charges a \$25 installation fee and \$20 per month for "basic" service. Write and graph an equation that can be used to determine the total cost, $C(t)$, of t months of basic cable television service. Find the total cost of 6 months of service.

42. *Temperature and Depth of the Earth.*   The function T given by

$$T(d) = 10d + 20$$

can be used to determine the temperature T, in degrees Celsius, at a depth d, in kilometers, inside the earth.

a) Find $T(5)$, $T(20)$, and $T(1000)$.
b) Graph T.
c) The radius of the earth is about 5600 km. Use this fact to determine the domain of the function.

For Exercise 43, use a grapher with a regression feature.

43. *Total Revenue of Pepsico.*   Pepsico is a company that owns many fast-food chains, such as Taco Bell and Pizza Hut. It sells other food items through subsidiaries such as Frito Lay and Pepsi-Cola. Pepsico has experienced great sales growth during the past few years, as shown in the following table.

YEAR, x	TOTAL REVENUE, y (IN BILLIONS)
0. 1990	\$17.8
1. 1991	19.6
2. 1992	22.0
3. 1993	25.0
4. 1994	??

Source: Pepsico Annual Report

a) Fit a regression line to the data and use it to predict total revenue in 1998, 2000, and 2010.
b) What is the correlation coefficient for the regression line? How close a fit is the regression line?

44. Find the distance between $(3, 7)$ and $(-2, 4)$.

45. Find the midpoint of the segment with endpoints $(3, 7)$ and $(-2, 4)$.

46. Find an equation of the circle with center $(-2, 6)$ and radius $\sqrt{13}$.

47. Find the center and radius of the circle
$$(x + 1)^2 + (y - 3)^2 = 16.$$

48. Find an equation of the circle having a diameter with endpoints $(-3, 5)$ and $(7, 3)$.

First, graph each equation and determine visually whether it is symmetric with respect to the x-axis, the y-axis, and the origin. Then verify your assertion algebraically.

49. $x^2 + y^2 = 4$ **50.** $y^2 = x^2 + 3$

51. $x + y = 3$ **52.** $y = x^2$

53. $y = x^3$ **54.** $y = x^4 - x^2$

Determine visually whether each function is even, odd, or neither even nor odd.

55.   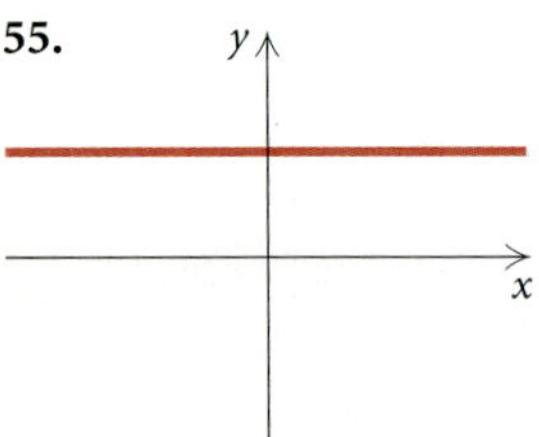

56.   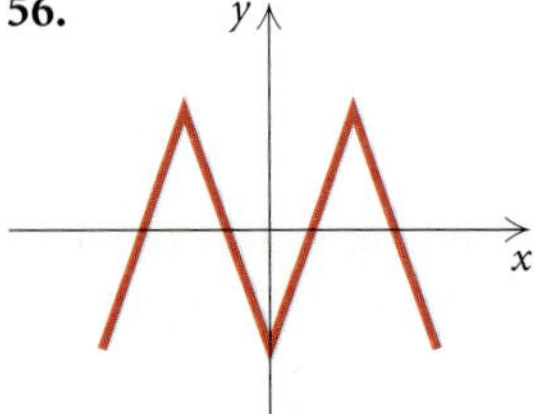

57.   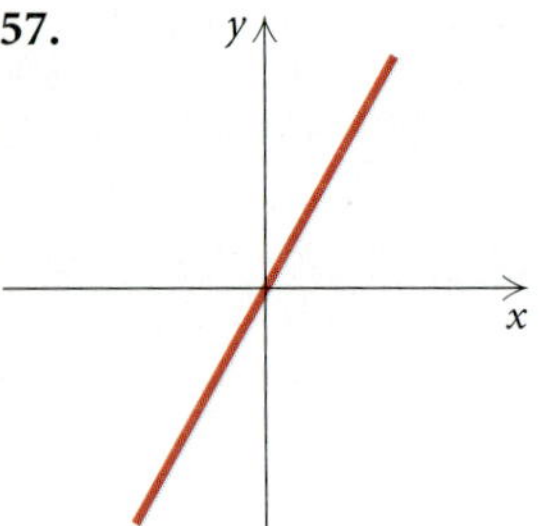

58.   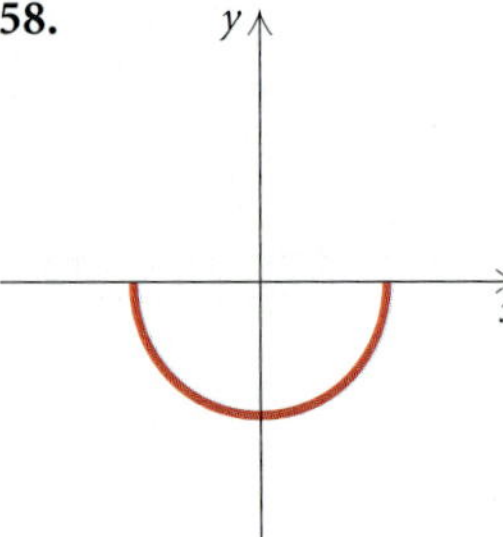

Test algebraically whether each function is even, odd, or neither even nor odd. Then check your work graphically, where possible, using a grapher.

59. $f(x) = 9 - x^2$ **60.** $f(x) = x^3 - 2x + 4$

61. $f(x) = x^7 - x^5$ **62.** $f(x) = |x|$

63. $f(x) = \sqrt{16 - x^2}$ **64.** $f(x) = \dfrac{10x}{x^2 + 1}$

Write an equation for a function that has a graph with the given characteristics. Check your answer on a grapher.

65. The shape of $y = x^2$, but shifted left 3 units

66. The shape of $y = \sqrt{x}$, but upside down and shifted right 3 units and up 4 units

67. The shape of $y = |x|$, but stretched vertically by a factor of 2 and shifted right 3 units

A graph of $y = f(x)$ is shown. No formula for f is given. Make a hand-drawn graph of each of the following.

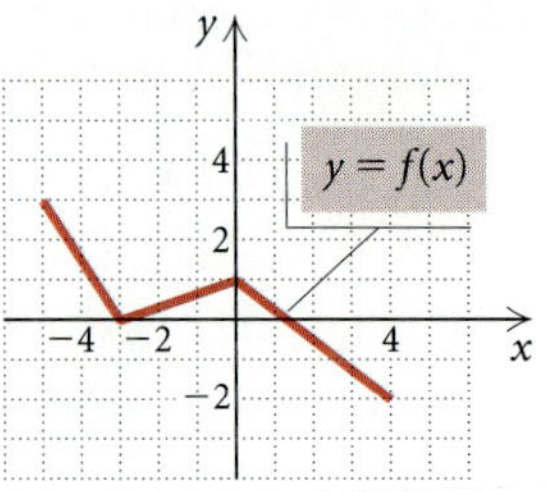

68. $y = f(x - 1)$ **69.** $y = f(2x)$

70. $y = -2f(x)$ **71.** $y = 3 + f(x)$

For each pair of functions:

a) *Find the domain of f, g, f + g, f − g, fg, f/g, f ∘ g, and g ∘ f.*

b) *Find $(f + g)(x)$, $(f - g)(x)$, $fg(x)$, $(f/g)(x)$, $(f \circ g)(x)$, and $(g \circ f)(x)$.*

72. $f(x) = \dfrac{4}{x^2}$; $g(x) = 3 - 2x$

73. $f(x) = 3x^2 + 4x$; $g(x) = 2x - 1$

74. Given the total-revenue and total-cost functions
$$R(x) = 120x - 0.5x^2 \quad \text{and} \quad C(x) = 15x + 6,$$
find total profit $P(x)$.

Find $f(x)$ and $g(x)$ such that $h(x) = (f \circ g)x$:

75. $h(x) = \sqrt{5x + 2}$ **76.** $h(x) = 4(5x - 1)^2 + 9$

Synthesis

77. ◈ Given that $f(x) = 4x^3 - 2x + 7$, find

a) $f(x) + 2$;   **b)** $f(x + 2)$;   **c)** $f(x) + f(2)$.

Then discuss how each expression differs from the other.

78. ◈ Given the graph of $y = f(x)$, explain and contrast the effect of the constant c on the graphs of $y = f(cx)$ and $y = cf(x)$.

79. ◈ **a)** Graph several functions of the type $y_1 = f(x)$ and $y_2 = |f(x)|$ on a grapher. Describe a procedure, involving transformations, for creating hand-drawn graphs of y_2 from y_1.

b) Describe a procedure, involving transformations, for creating hand-drawn graphs of $y_2 = f(|x|)$ from $y_1 = f(x)$.

Find the domain.

80. $f(x) = \dfrac{\sqrt{1 - x}}{x - |x|}$ **81.** $f(x) = (x - 9x^{-1})^{-1}$

82. Prove that the sum of two odd functions is odd.

83. Describe how the graph of $y = -f(-x)$ is obtained from the graph of $y = f(x)$.

Polynomial and Rational Functions 2

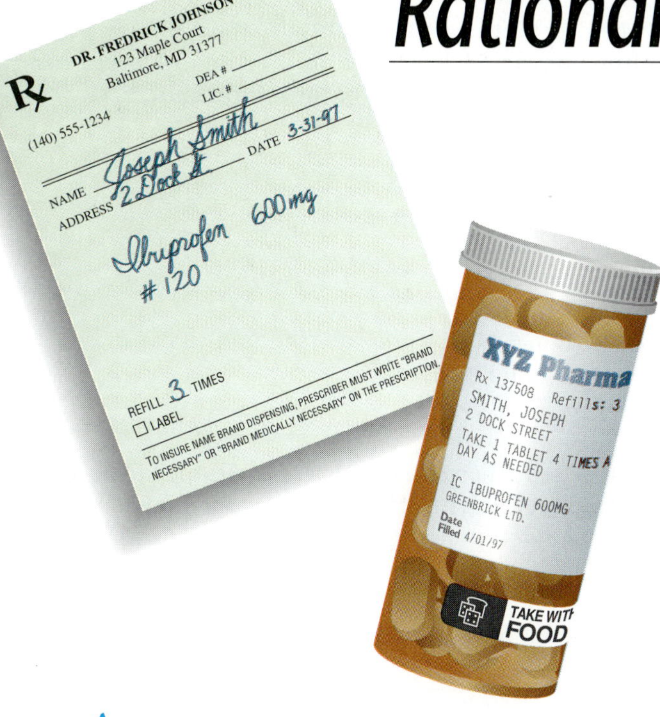

A *polynomial function* is a function that can be defined by a polynomial expression. A *rational function* is a function that can be defined as a quotient of two polynomials. We will study both kinds of functions and examine zeros of polynomials in greater depth. We will also model real data with quadratic, cubic, and quartic polynomial functions and use these functions to make predictions. In addition, we will study the graphs of rational functions and use the graphs of both polynomial and rational functions to solve related inequalities.

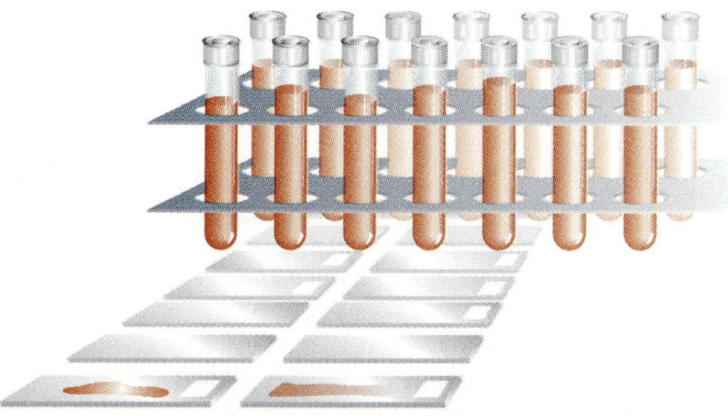

Ibuprofen is a medication used to relieve pain. The fourth-degree, or quartic, polynomial function
$$M(t) = 0.5t^4 + 3.45t^3 - 96.65t^2 + 347.7t, \quad 0 \le t \le 6$$
can be used to estimate the number of milligrams of ibuprofen in the bloodstream t hours after 400 mg of the medication has been swallowed.

X	Y₁	
0	0	
1	255	
2	344.4	
3	306.9	
4	193.2	
5	66	
6	0	
X = 2		

$$y = 0.5x^4 + 3.45x^3 - 96.65x^2 + 347.7x$$

171

2.1

Introduction to Polynomial Functions and Complex Numbers

- *Use a grapher to graph a polynomial function and find its real-number zeros, relative maximum and minimum values, and domain and range.*
- *Perform computations involving complex numbers.*

Because of the grapher's capabilities, we have already studied many aspects of polynomial functions in Chapter 1. In this section, we begin a more rigorous study of how to find the zeros of these functions.

Polynomial Function

A *polynomial function P* is given by

$$P(x) = a_n x^n + a_{n-1}x^{n-1} + a_{n-2}x^{n-2} + \cdots + a_1 x + a_0,$$

where the coefficients $a_n, a_{n-1}, \ldots, a_1, a_0$ are real numbers and the exponents are whole numbers.

In descending order, the first nonzero coefficient is assumed to be a_n and is called the **leading coefficient**. The **degree** of the polynomial function is n. Some examples of types of polynomial functions and their names are as follows.

POLYNOMIAL FUNCTION	DEGREE	EXAMPLE
Constant	0	$f(x) = 3$
Linear	1	$f(x) = \frac{2}{3}x + 1$
Quadratic	2	$f(x) = 2x^2 - x + 3$
Cubic	3	$f(x) = x^3 + 2x^2 + x - 5$
Quartic	4	$f(x) = -x^4 - 1.1x^3 + 0.3x^2 - 2.8x - 1.7$

The constant function $f(x) = 0$ is said to have no degree because there is no nonzero coefficient. This function can be described in many ways: $f(x) = 0 = 0x^2 = 0x^3 = 0x^{53}$, and so on.

From our study of functions in Chapter 1, we know how to use a grapher to at least estimate many characteristics of a polynomial function. Let's consider an example for review.

Function: $f(x) = x^4 - 6x^3 - 4x^2 + 53.2x - 42.6$

Zeros: $-2.975, 0.950, 3, 5.025$

Relative maximum: 15.921 at $x = 1.914$

Relative minima: -106.907 at $x = -1.643$, -23.101 at $x = 4.229$

Domain: All real numbers, $\mathbb{R}$

Range: $[-106.907, \infty)$

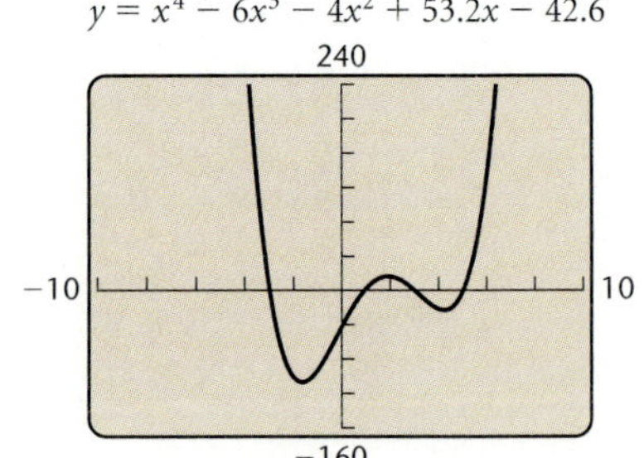

With the techniques presented in this chapter, we will be able to predict certain characteristics of the graph of a polynomial by examining its formula. This will help us to find exact, rather than approximate, values in certain cases and to have greater insight into the creation of graphs. Before proceeding, we need to expand the real-number system to one called the **complex-number system** in which certain polynomial functions have zeros that may not be real numbers.

The Complex Numbers

Let's begin by considering the polynomial function $f(x) = x^2 - 5$.

To find the zeros of the function, we solve the equation $x^2 - 5 = 0$.

Example 1 Solve: $x^2 - 5 = 0$.

ALGEBRAIC SOLUTION

We use the principle of square roots:

$$x^2 - 5 = 0$$
$$x^2 = 5 \qquad \text{Adding 5}$$
$$x = \pm\sqrt{5}. \qquad \text{Using the principle of square roots}$$

The solutions of the equation are $-\sqrt{5}$ and $\sqrt{5}$. The solution set is $\{-\sqrt{5}, \sqrt{5}\}$. Note that $\pm\sqrt{5} \approx \pm 2.236$.

GRAPHICAL SOLUTION

To solve $x^2 - 5 = 0$, we first graph the function $f(x) = x^2 - 5$ and estimate its zeros.

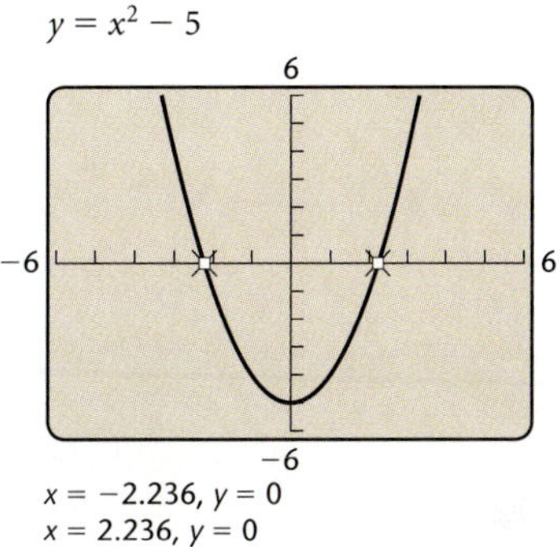

We see that there are zeros near -2 and 2. Using TRACE and ZOOM to approximate the zeros, we get the approximations -2.236 and 2.236. The zeros of the function are the solutions of the equation $x^2 - 5 = 0$. The solution set is $\{-2.236, 2.236\}$.

A faster method might be to use SOLVE. On some graphers, a POLY feature allows you to enter a polynomial and gives all the zeros directly.

Note that the real solutions found with the grapher were approximations. Now let's consider an equation that has no real-number solutions.

Example 2 Solve: $x^2 + 1 = 0$.

$$x^2 + 1 = 0$$

$$x^2 = -1 \qquad \text{Subtracting 1}$$

$$x = \pm\sqrt{-1} \qquad \text{Using the principle of square roots}$$

There are no real-number square roots of negative numbers. Thus the equation $x^2 + 1 = 0$ has no real-number solutions.

We first graph the function $f(x) = x^2 + 1$.

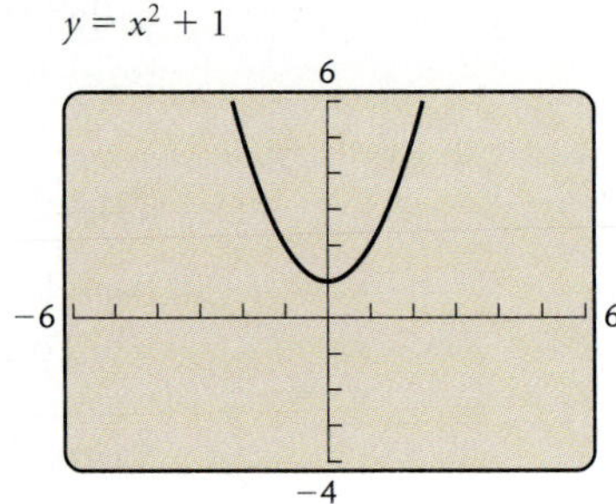

We see that since the graph does not cross the x-axis, it has no x-intercepts. This means that the function $f(x) = x^2 + 1$ has no real-number zeros. Thus there are no real-number solutions of the equation $x^2 + 1 = 0$.

A new kind of number, called *imaginary*, was defined (invented) so that negative numbers would have square roots and equations like the one in Example 2 would have solutions. These new numbers make use of a number named i, defined so that $i = \sqrt{-1}$ and $i^2 = -1$.

> **The Number i**
>
> The number i is defined such that
>
> $$i = \sqrt{-1} \quad \text{and} \quad i^2 = -1.$$

To express roots of negative numbers in terms of i, we can use the fact that $\sqrt{-p} = \sqrt{-1}\sqrt{p}$ when p is a positive real number.

Example 3 Express each number in terms of i.

a) $\sqrt{-7}$ b) $\sqrt{-16}$

c) $-\sqrt{-13}$ d) $-\sqrt{-64}$

e) $\sqrt{-48}$

SOLUTION

i is *not* under the radical.

a) $\sqrt{-7} = \sqrt{-1 \cdot 7} = \sqrt{-1} \cdot \sqrt{7} = i\sqrt{7}$, or $\sqrt{7}i$

b) $\sqrt{-16} = \sqrt{-1 \cdot 16} = \sqrt{-1} \cdot \sqrt{16} = i \cdot 4 = 4i$

c) $-\sqrt{-13} = -\sqrt{-1 \cdot 13} = -\sqrt{-1} \cdot \sqrt{13} = -i\sqrt{13}$, or $-\sqrt{13}i$

d) $-\sqrt{-64} = -\sqrt{-1 \cdot 64} = -\sqrt{-1} \cdot \sqrt{64} = -i \cdot 8 = -8i$

e) $\sqrt{-48} = \sqrt{-1 \cdot 48} = \sqrt{-1} \cdot \sqrt{48}$
$$= i\sqrt{48} = i \cdot 4\sqrt{3} = 4\sqrt{3}i, \quad \text{or} \quad 4i\sqrt{3}$$

Imaginary Number

An *imaginary number* is a number that can be written $a + bi$, where a and b are real numbers and $b \neq 0$.

Don't let the word "imaginary" mislead you. Imaginary numbers have very "real" uses in engineering and science. The following are examples of imaginary numbers:

$$\left. \begin{array}{l} -6 + 2i \\ \frac{3}{5} - \frac{1}{4}i \\ \pi - 3.7i \\ 3 + i\sqrt{2} \end{array} \right\} \quad \text{(here } a \neq 0,\ b \neq 0\text{);}$$

$$17i \qquad \text{(here } a = 0,\ b \neq 0\text{).}$$

The number i and the numbers in Example 3 are also examples of *imaginary numbers* with $a = 0$.

When a and b are real numbers and b could be 0, the number $a + bi$ is said to be **complex**. The real number a is said to be the **real part** of $a + bi$, and the real number b is said to be the **imaginary part**.

Complex Number

A *complex number* is any number that can be written $a + bi$, where a and b are any real numbers. (Note that a and b both can be 0.)

The following are examples of complex numbers:

$$7 + 3i \qquad \text{(here } a \neq 0,\ b \neq 0\text{);}$$
$$8 \qquad \text{(here } a \neq 0,\ b = 0\text{);}$$
$$4i \qquad \text{(here } a = 0,\ b \neq 0\text{);}$$
$$0 \qquad \text{(here } a = 0,\ b = 0\text{).}$$

Note that when $b = 0$, $a + 0i = a$, so every real number is a complex number. Complex numbers like $17i$ or $4i$, in which $a = 0$ and $b \neq 0$, are imaginary numbers with no real part. Such numbers are called **pure imaginary numbers**. The relationships among various types of complex numbers are shown below.

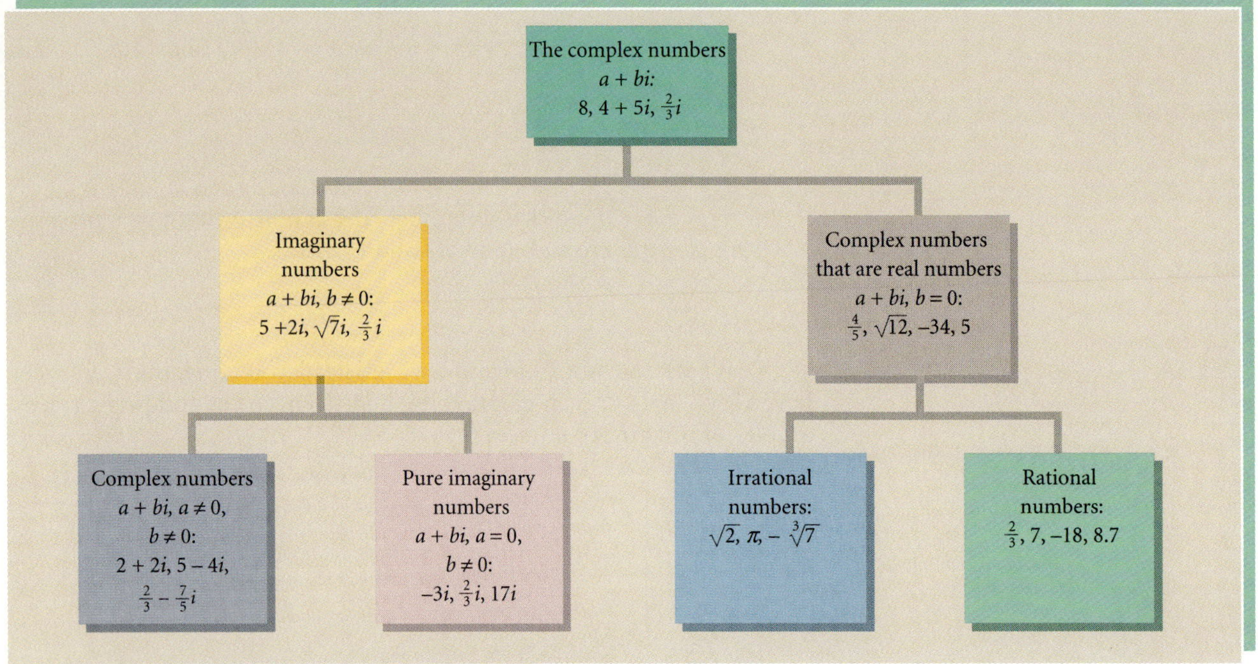

Addition and Subtraction

The complex numbers obey the commutative, associative, and distributive laws. Thus we can add and subtract them as we do binomials.

Example 4 Add or subtract and simplify each of the following.

a) $(8 + 6i) + (3 + 2i)$ 　　　　　　　**b)** $(4 + 5i) - (6 - 3i)$

SOLUTION

a) $(8 + 6i) + (3 + 2i) = (8 + 3) + (6i + 2i)$

        Collecting the real parts and the imaginary parts

$$= 11 + (6 + 2)i = 11 + 8i$$

b) $(4 + 5i) - (6 - 3i) = (4 - 6) + [5i - (-3i)]$

       Note that the 6 and the −3i are both being subtracted.

$$= -2 + 8i$$

Multiplication

For complex numbers, the property $\sqrt{a}\sqrt{b} = \sqrt{ab}$ does *not* hold in general, but it does hold when $a = -1$ and b is a nonnegative number. To multiply square roots of negative real numbers, we first express them in terms of i. For example,

$$\sqrt{-2} \cdot \sqrt{-5} = \sqrt{-1} \cdot \sqrt{2} \cdot \sqrt{-1} \cdot \sqrt{5} = i\sqrt{2} \cdot i\sqrt{5}$$
$$= i^2\sqrt{10} = -1\sqrt{10} = -\sqrt{10} \quad \text{is correct!}$$

But

$$\sqrt{-2} \cdot \sqrt{-5} = \sqrt{(-2)(-5)} = \sqrt{10} \quad \text{is wrong!}$$

Keeping this and the fact that $i^2 = -1$ in mind, we multiply in much the same way that we do with real numbers.

Example 5   Multiply and simplify each of the following.

a) $\sqrt{-16} \cdot \sqrt{-25}$ **b)** $\sqrt{-5} \cdot \sqrt{-7}$ **c)** $-4i(3 - 5i)$

d) $(1 + 2i)(1 + 3i)$ **e)** $(3 - 7i)^2$

SOLUTION

a) $\sqrt{-16} \cdot \sqrt{-25} = \sqrt{-1} \cdot \sqrt{16} \cdot \sqrt{-1} \cdot \sqrt{25}$

$$\begin{aligned}
&= i \cdot 4 \cdot i \cdot 5 \\
&= i^2 \cdot 20 \\
&= -1 \cdot 20 \qquad i^2 = -1 \\
&= -20
\end{aligned}$$

b) $\sqrt{-5} \cdot \sqrt{-7} = \sqrt{-1} \cdot \sqrt{5} \cdot \sqrt{-1} \cdot \sqrt{7}$

$$\begin{aligned}
&= i \cdot \sqrt{5} \cdot i \cdot \sqrt{7} \\
&= i^2 \cdot \sqrt{35} \\
&= -1 \cdot \sqrt{35} \qquad i^2 = -1 \\
&= -\sqrt{35}
\end{aligned}$$

c) $-4i(3 - 5i) = -4i \cdot 3 + (-4i)(-5i)$ **Using the distributive law**

$$\begin{aligned}
&= -12i + 20i^2 \\
&= -12i - 20 \qquad && i^2 = -1 \\
&= 20 \quad 12i \qquad && \textbf{Writing in the form } a + bi
\end{aligned}$$

d) $(1 + 2i)(1 + 3i) = 1 + 3i + 2i + 6i^2$ **Multiplying each term of one number by every term of the other (FOIL)**

$$\begin{aligned}
&= 1 + 3i + 2i - 6 \qquad && i^2 = -1 \\
&= -5 + 5i \qquad && \textbf{Collecting like terms}
\end{aligned}$$

e) $(3 - 7i)^2 = 3^2 - 2 \cdot 3 \cdot 7i + (7i)^2$ **Recall that** $(A - B)^2 = A^2 - 2AB + B^2$

$$\begin{aligned}
&= 9 - 42i + 49i^2 \\
&= 9 - 42i - 49 \qquad && i^2 = -1 \\
&= -40 - 42i
\end{aligned}$$

Recall that -1 raised to an *even* power is 1, and -1 raised to an *odd* power is -1. Simplifying powers of i can then be done by using the fact that $i^2 = -1$ and expressing the given power of i in terms of i^2. Consider the following:

$$\begin{aligned}
i &= \sqrt{-1}, & i^4 &= (i^2)^2 = (-1)^2 = 1, \\
i^2 &= -1, & i^5 &= i^4 \cdot i = (i^2)^2 \cdot i = (-1)^2 \cdot i = i, \\
i^3 &= i^2 \cdot i = (-1)i = -i, & i^6 &= (i^2)^3 = (-1)^3 = -1.
\end{aligned}$$

Note that the powers of i cycle themselves through the values i, -1, $-i$, and 1.

Example 6   Simplify each of the following.

a) i^{37}　　　　**b)** i^{58}　　　　**c)** i^{75}　　　　**d)** i^{80}

SOLUTION

a) $i^{37} = i^{36} \cdot i = (i^2)^{18} \cdot i = (-1)^{18} \cdot i = 1 \cdot i = i$

b) $i^{58} = (i^2)^{29} = (-1)^{29} = -1$

c) $i^{75} = i^{74} \cdot i = (i^2)^{37} \cdot i = (-1)^{37} \cdot i = -1 \cdot i = -i$

d) $i^{80} = (i^2)^{40} = (-1)^{40} = 1$

Conjugates and Division

Conjugates of complex numbers are defined as follows.

> **Conjugate of a Complex Number**
>
> The *conjugate* of a complex number $a + bi$ is $a - bi$, and the *conjugate* of $a - bi$ is $a + bi$.

Example 7   Find the conjugate of each of the following.

a) $-3 + 7i$　　　　**b)** $14 - 5i$　　　　**c)** $4i$

SOLUTION

a) The conjugate of $-3 + 7i$ is $-3 - 7i$.

b) The conjugate of $14 - 5i$ is $14 + 5i$.

c) The conjugate of $4i$ is $-4i$.

The product of a complex number and its conjugate is a real number.

Example 8   Multiply each of the following.

a) $(5 + 7i)(5 - 7i)$　　　　　　**b)** $(8i)(-8i)$

SOLUTION

a) $(5 + 7i)(5 - 7i) = 5^2 - (7i)^2$　　　Using $(A + B)(A - B) = A^2 - B^2$

$$= 25 - 49i^2$$
$$= 25 - 49(-1)$$
$$= 74$$

b) $(8i)(-8i) = -64i^2$
$$= -64(-1)$$
$$= 64$$

When we are dividing complex numbers, conjugates are used.

Example 9   Divide $2 - 5i$ by $1 - 6i$.

SOLUTION   We write fractional notation and then multiply by 1:

$$\frac{2 - 5i}{1 - 6i} = \frac{2 - 5i}{1 - 6i} \cdot \frac{1 + 6i}{1 + 6i}$$

Note that $1 + 6i$ is the conjugate of the divisor.

$$= \frac{(2 - 5i)(1 + 6i)}{(1 - 6i)(1 + 6i)}$$

$$= \frac{2 + 7i - 30i^2}{1 - 36i^2}$$

$$= \frac{32 + 7i}{1 + 36} \qquad i^2 = -1$$

$$= \frac{32}{37} + \frac{7}{37}i.$$

2.1 | *Exercise Set*

Use a grapher to graph each of the following polynomial functions. Then estimate the function's (a) real-number zeros; (b) relative maxima; (c) relative minima; (d) domain and range. (Review Sections 1.1 and 1.2, if needed.)

1. $f(x) = x^2 - x - 6$

2. $f(x) = 20 - x - x^2$

3. $f(x) = x^3 + 5x^2 - x - 5$

4. $f(x) = x^4 - 29x + 100$

5. $f(x) = -2x^3 + x^2 + 14x + 10$

6. $f(x) = 1.1x^4 + 4.2x^3 - 35.08x^2 - 158.8x + 294.7$

7. $f(x) = 0.001x^3 - 0.023x$

8. $f(x) = -215.3x^4 + 1647.1x^2 - 2800$

Use a grapher to show that there are no real-number zeros for each of the following polynomial functions.

9. $f(x) = x^2 + 7$

10. $f(x) = x^2 + 2$

11. $f(x) = x^2 - 6x + 10$

12. $f(x) = x^2 + 8x + 17$

13. $f(x) = x^4 - 4x^3 + 6x^2 - 4x + 3$

14. $f(x) = \frac{2}{3}(x^4 + 4x^3 + 6x^2 + 4x + 3)$

Express in terms of i.

15. $\sqrt{-17}$

16. $\sqrt{-25}$

17. $\sqrt{-49}$

18. $\sqrt{-13}$

19. $-\sqrt{-81}$

20. $-\sqrt{-20}$

21. $6 - \sqrt{-84}$

22. $7 - \sqrt{-60}$

23. $-\sqrt{76} + \sqrt{-125}$

24. $\sqrt{-4} + \sqrt{-12}$

25. $\sqrt{-5}\sqrt{-11}$

26. $-\sqrt{-9}\sqrt{-7}$

27. $\dfrac{-\sqrt{25}}{\sqrt{-16}}$

28. $\dfrac{-\sqrt{-36}}{\sqrt{-9}}$

Simplify. Write answers in the form $a + bi$, where a and b are real numbers.

29. $(-5 + 3i) + (7 + 8i)$

30. $(-6 - 5i) + (9 + 2i)$

31. $(4 - 9i) + (1 - 3i)$

32. $(7 - 2i) + (4 - 5i)$

33. $(3 + \sqrt{-16}) + (2 + \sqrt{-25})$

34. $(7 - \sqrt{-36}) + (2 + \sqrt{-9})$

35. $(10 + 7i) - (5 + 3i)$

36. $(-3 - 4i) - (8 - i)$

37. $(13 + 9i) - (8 + 2i)$

38. $(-7 + 12i) - (3 - 6i)$

39. $(1 + 3i)(1 - 4i)$

40. $(1 - 2i)(1 + 3i)$

41. $(2 + 3i)(2 + 5i)$

42. $(3 - 5i)(8 - 2i)$

43. $7i(2 - 5i)$

44. $3i(6 + 4i)$

45. $(3 + \sqrt{-16})(2 + \sqrt{-25})$

46. $(7 - \sqrt{-16})(2 + \sqrt{-9})$

47. $(5 - 4i)(5 + 4i)$

48. $(5 + 9i)(5 - 9i)$

49. $(4 + 2i)^2$

50. $(5 - 4i)^2$

51. $(-2 + 7i)^2$

52. $(-3 + 2i)^2$

53. $\dfrac{2 + \sqrt{3}i}{5 - 4i}$

54. $\dfrac{\sqrt{5} + 3i}{1 - i}$

55. $\dfrac{4 + i}{-3 - 2i}$

56. $\dfrac{5 - i}{-7 + 2i}$

57. $\dfrac{3}{5 - 11i}$

58. $\dfrac{i}{2 + i}$

59. $\dfrac{1 + i}{(1 - i)^2}$

60. $\dfrac{1 - i}{(1 + i)^2}$

61. $\dfrac{4 - 2i}{1 + i} + \dfrac{2 - 5i}{1 + i}$

62. $\dfrac{3 + 2i}{1 - i} + \dfrac{6 + 2i}{1 - i}$

Simplify.

63. i^{11} **64.** i^7 **65.** i^{35} **66.** i^{24}

67. i^{64} **68.** i^{42} **69.** $(-i)^{71}$ **70.** $(-i)^6$

71. $(5i)^4$ **72.** $(2i)^5$

Skill Maintenance

Factor.

73. $x^2 + 8x + 16$

74. $y^2 + 20y + 100$

75. $x^2 - 10x + 25$

76. $x^2 - 18x + 81$

Synthesis

77. ◈ Is the sum of two imaginary numbers always an imaginary number? Why or why not? What about the product?

78. ◈ Is the domain of every polynomial function $\mathbb{R}$? Why or why not?

Solve.

79. $(2 + i)x - i = 5 + i$

80. $5i = (3 + i)x + i$

81. $5ix + 3 + 2i = (3 - 2i)x + 3i$

82. $(2 + 3i)x - 2i = 2ix + 5 - 4i$

Determine whether each of the following is true or false.

83. The sum of two numbers that are conjugates of each other is always a real number.

84. The conjugate of a sum is the sum of the conjugates of the individual complex numbers.

85. The conjugate of a product is the product of the conjugates of the individual complex numbers.

Let $z = a + bi$ and $\bar{z} = a - bi$.

86. Find a general expression for $1/z$.

87. Find a general expression for $z\bar{z}$.

88. Solve $z + 6\bar{z} = 7$ for z.

2.2
Quadratic Equations and Functions

- *Solve quadratic equations by completing the square and using the quadratic formula.*
- *Solve equations that are reducible to quadratic equations.*
- *Find the vertex, the line of symmetry, and the maxima or minima of the graph of a quadratic function using the method of completing the square.*
- *Graph quadratic functions.*

In this section, we develop a method for solving any quadratic equation as well as a method for graphing quadratic functions. We define quadratic equations and functions as follows.

Quadratic Equations

A *quadratic equation* is an equation equivalent to

$$ax^2 + bx + c = 0, \quad a \neq 0,$$

where the coefficients a, b, and c are real numbers.

> **Quadratic Functions**
>
> A *quadratic function f* is a second-degree polynomial function
>
> $$f(x) = ax^2 + bx + c, \quad a \neq 0,$$
>
> where a, b, and c are real numbers.

The zeros of a quadratic function $f(x) = ax^2 + bx + c$ are the solutions of the quadratic equation $ax^2 + bx + c = 0$. We have already solved certain quadratic equations and graphed some quadratic functions. In this section, we develop methods for finding exact solutions of *any* quadratic equation and for analyzing the graph of any quadratic function. These techniques will allow us to find the exact value of a maximum or a minimum.

Completing the Square

Quadratic equations like $x^2 - 5x - 6 = 0$ and $6x^2 + 19x - 7 = 0$ can be solved by factoring (see Section R.7):

$$x^2 - 5x - 6 = 0 \qquad\qquad 6x^2 + 19x - 7 = 0$$
$$(x - 6)(x + 1) = 0 \qquad\qquad (3x - 1)(2x + 7) = 0$$
$$x - 6 = 0 \ or \ x + 1 = 0 \qquad 3x - 1 = 0 \ or \ 2x + 7 = 0$$
$$x = 6 \ or \quad x = -1 \qquad\qquad x = \tfrac{1}{3} \ or \quad x = -\tfrac{7}{2}$$

The solutions are -1 and 6. $\qquad$ The solutions are $\tfrac{1}{3}$ and $-\tfrac{7}{2}$.

The principle of square roots can be used to solve quadratic equations like $x^2 - 9 = 0$ and $x^2 + 13 = 0$:

$$x^2 - 9 = 0 \qquad\qquad\qquad x^2 + 13 = 0$$
$$x^2 = 9 \qquad\qquad\qquad\quad x^2 = -13$$
$$x = \pm 3 \qquad\qquad\qquad\quad x = \pm\sqrt{13}\,i$$

The solutions are -3 and 3. $\qquad$ The solutions are $-\sqrt{13}\,i$ and $\sqrt{13}\,i$.

We can find solutions for the first three equations graphically. A graphical approach would not yield the solution to $x^2 + 13 = 0$. Some graphers can find complex solutions, but the procedures are not graphical. None of the preceding methods would yield the exact solutions to an equation like $x^2 - 6x - 10 = 0$. If we wish to do that, we can use a procedure called *completing the square* and then use the principle of square roots.

Example 1 Solve

$$x^2 - 6x - 10 = 0$$

by completing the square.

SOLUTION Our goal is to find an equivalent polynomial equation of the form $x^2 + bx + c = d$ in which $x^2 + bx + c$ is a perfect square. This is accomplished as follows:

$$x^2 - 6x - 10 = 0$$
$$x^2 - 6x = 10 \qquad \text{Adding 10}$$
$$x^2 - 6x + 9 = 10 + 9 \qquad \text{Adding 9 to complete the square: } \tfrac{1}{2}(-6) = -3 \text{ and } (-3)^2 = 9$$
$$x^2 - 6x + 9 = 19.$$

Because $x^2 - 6x + 9$ is a perfect square, we are able to write it as the square of a binomial. We can then use the principle of square roots to finish the solution:

$$(x - 3)^2 = 19 \qquad \text{Factoring}$$
$$x - 3 = \pm\sqrt{19} \qquad \text{Using the principle of square roots}$$
$$x = 3 \pm \sqrt{19}. \qquad \text{Adding 3}$$

The solutions are $3 + \sqrt{19}$ and $3 - \sqrt{19}$, or simply $3 \pm \sqrt{19}$ (read "three plus or minus $\sqrt{19}$"). The solution set is $\{3 + \sqrt{19}, 3 - \sqrt{19}\}$, or simply $\{3 \pm \sqrt{19}\}$.

Decimal approximations for $3 \pm \sqrt{19}$ could be found on a grapher, that is, $3 + \sqrt{19} \approx 7.359$ and $3 - \sqrt{19} \approx -1.359$. The solutions could also be found using a graphical approach.

Example 2   Solve: $2x^2 - 1 = 3x$.

SOLUTION

$$2x^2 - 1 = 3x$$
$$2x^2 - 3x - 1 = 0 \qquad \text{Subtracting } 3x; \text{ we are unable to factor the result.}$$
$$2x^2 - 3x = 1 \qquad \text{Adding 1}$$
$$x^2 - \frac{3}{2}x = \frac{1}{2} \qquad \text{Dividing by 2}$$
$$x^2 - \frac{3}{2}x + \frac{9}{16} = \frac{1}{2} + \frac{9}{16} \qquad \text{Completing the square: } \tfrac{1}{2}\left(-\tfrac{3}{2}\right) = -\tfrac{3}{4} \text{ and } \left(-\tfrac{3}{4}\right)^2 = \tfrac{9}{16}$$
$$\left(x - \frac{3}{4}\right)^2 = \frac{17}{16} \qquad \text{Factoring and simplifying}$$
$$x - \frac{3}{4} = \pm\frac{\sqrt{17}}{4} \qquad \text{Using the principle of square roots and the quotient rule for radicals}$$
$$x = \frac{3 \pm \sqrt{17}}{4} \qquad \text{Adding } \tfrac{3}{4}$$

The solution set is $\left\{\dfrac{3 \pm \sqrt{17}}{4}\right\}$.

You should find approximate solutions using scientific keys on a grapher and check the solutions graphically.

> To solve a quadratic equation by completing the square:
>
> 1. Isolate the terms with variables on one side of the equation and arrange them in descending order.
> 2. Divide by the coefficient of the squared term if that coefficient is not 1.
> 3. Complete the square by taking half the coefficient of the first-degree term and adding its square on both sides of the equation.
> 4. Express one side of the equation as the square of a binomial.
> 5. Use the principle of square roots.
> 6. Solve for the variable.

Using the Quadratic Formula

Because completing the square works for *any* quadratic equation, it can be used to solve the general quadratic equation $ax^2 + bx + c = 0$ for x. The result will be a formula that can be used to solve any quadratic equation quickly.

Consider any quadratic equation in standard form:

$$ax^2 + bx + c = 0, \quad a > 0.$$

If a is negative, we first multiply on both sides by -1. Then we solve by completing the square. As we carry out the steps, compare them with those of Example 2.

$$ax^2 + bx = -c \qquad \text{Adding } -c$$

$$x^2 + \frac{b}{a}x = -\frac{c}{a} \qquad \text{Dividing by } a \text{, since } a > 0$$

Half of $\frac{b}{a}$ is $\frac{b}{2a}$ and $\left(\frac{b}{2a}\right)^2 = \frac{b^2}{4a^2}$. Thus we add $\frac{b^2}{4a^2}$:

$$x^2 + \frac{b}{a}x + \frac{b^2}{4a^2} = -\frac{c}{a} + \frac{b^2}{4a^2} \qquad \text{Adding } \frac{b^2}{4a^2} \text{ to complete the square}$$

$$\left(x + \frac{b}{2a}\right)^2 = -\frac{4ac}{4a^2} + \frac{b^2}{4a^2} \qquad \text{Factoring and finding a common denominator:}$$
$$-\frac{c}{a} = -\frac{4a}{4a} \cdot \frac{c}{a} = -\frac{4ac}{4a^2}$$

$$\left(x + \frac{b}{2a}\right)^2 = \frac{b^2 - 4ac}{4a^2}$$

$$x + \frac{b}{2a} = \pm\frac{\sqrt{b^2 - 4ac}}{2a} \qquad \text{Using the principle of square roots and the quotient rule for radicals; since } a > 0, \sqrt{4a^2} = 2a.$$

$$x = \frac{-b \pm \sqrt{b^2 - 4ac}}{2a}. \qquad \text{Adding } -\frac{b}{2a}$$

Program

QUADFORM: This program solves a quadratic equation, giving real and complex solutions. (See the Graphing Calculator Manual that accompanies this text.)

The Quadratic Formula

The solutions of $ax^2 + bx + c = 0$, $a \neq 0$, are given by

$$x = \frac{-b \pm \sqrt{b^2 - 4ac}}{2a}.$$

Example 3   Solve $3x^2 + 2x = 7$. Find exact solutions and approximate solutions rounded to the nearest thousandth.

We show both algebraic and graphical solutions. Note that only the algebraic approach yields the exact solutions.

ALGEBRAIC SOLUTION

After finding standard form, we are unable to factor, so we identify a, b, and c in order to use the quadratic formula:

$$3x^2 + 2x - 7 = 0;$$
$$a = 3, \quad b = 2, \quad c = -7.$$

We then use the quadratic formula:

$$x = \frac{-b \pm \sqrt{b^2 - 4ac}}{2a}$$

$$= \frac{-2 \pm \sqrt{2^2 - 4(3)(-7)}}{2(3)} \quad \text{Substituting}$$

$$= \frac{-2 \pm \sqrt{4 + 84}}{6} = \frac{-2 \pm \sqrt{88}}{6}$$

$$= \frac{-2 \pm \sqrt{4 \cdot 22}}{6} = \frac{-2 \pm 2\sqrt{22}}{6}$$

$$= \frac{2}{2} \cdot \frac{-1 \pm \sqrt{22}}{3} = \frac{-1 \pm \sqrt{22}}{3}.$$

The exact solutions are

$$\frac{-1 - \sqrt{22}}{3} \quad \text{and} \quad \frac{-1 + \sqrt{22}}{3}.$$

Using the scientific keys on a grapher, we approximate the solutions to be -1.897 and 1.230.

GRAPHICAL SOLUTION

To solve $3x^2 + 2x = 7$, or $3x^2 + 2x - 7 = 0$, we first graph the function $f(x) = 3x^2 + 2x - 7$. Then we look for points where the graph crosses the x-axis. It appears that there are two possible zeros, one near -2 and one near 1. We can use TRACE and ZOOM to approximate these zeros, or we can use a SOLVE or POLY feature.

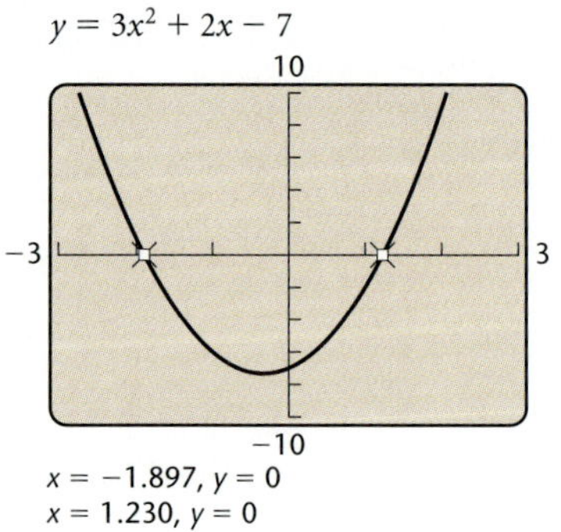

We get the approximate zeros -1.896805 and 1.2301386, or -1.897 and 1.230, rounded to three decimal places. The zeros of the function are the solutions of the equation $3x^2 + 2x = 7$.

Not all quadratic equations can be solved graphically.

Example 4 Solve: $x^2 + 5x + 8 = 0$.

ALGEBRAIC SOLUTION

To find the solutions, we use the quadratic formula. Here

$$a = 1, \quad b = 5, \quad c = 8;$$

$$x = \frac{-b \pm \sqrt{b^2 - 4ac}}{2a}$$

$$= \frac{-5 \pm \sqrt{5^2 - 4(1)(8)}}{2 \cdot 1} \qquad \textbf{Substituting}$$

$$= \frac{-5 \pm \sqrt{-7}}{2} \qquad \textbf{Simplifying}$$

$$= \frac{-5 \pm \sqrt{7}i}{2}.$$

The solutions are $-\dfrac{5}{2} - \dfrac{\sqrt{7}}{2}i$ and $-\dfrac{5}{2} + \dfrac{\sqrt{7}}{2}i$.

GRAPHICAL SOLUTION

The graph of the function $f(x) = x^2 + 5x + 8$ shows no x-intercepts.

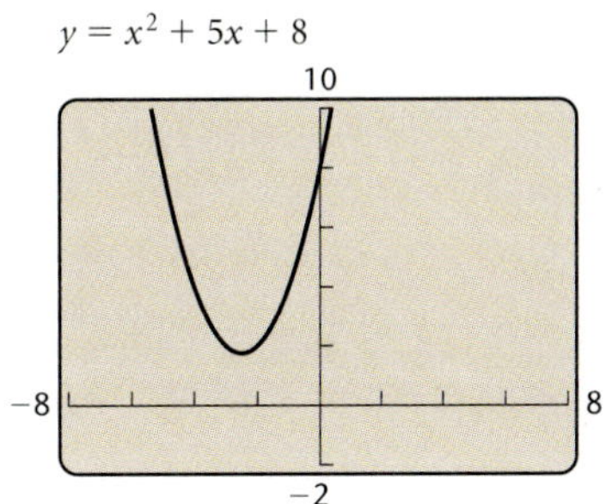

Thus there are no real-number solutions of the equation $x^2 + 5x + 8 = 0$. This is a quadratic equation that cannot be solved graphically.

The Discriminant

From the quadratic formula, we know that the solutions x_1 and x_2 of a quadratic equation are given by

$$x_1 = \frac{-b + \sqrt{b^2 - 4ac}}{2a} \quad \text{and} \quad x_2 = \frac{-b - \sqrt{b^2 - 4ac}}{2a}.$$

The expression $b^2 - 4ac$ shows the nature of the solutions. This expression is called the **discriminant**. If it is 0, then it doesn't matter whether we choose the plus or the minus sign in the formula. There is just one solution, and we sometimes say that there is one repeated real solution. If the discriminant is positive, there will be two real solutions. If it is negative, we will be taking the square root of a negative number; hence there will be two nonreal solutions, and they will be complex conjugates.

Discriminant

For $ax^2 + bx + c = 0$:

$b^2 - 4ac = 0 \longrightarrow$ One real-number solution;

$b^2 - 4ac > 0 \longrightarrow$ Two different real-number solutions;

$b^2 - 4ac < 0 \longrightarrow$ Two different imaginary-number solutions, complex conjugates.

In Example 3, the discriminant, 88, is positive, indicating that there are two different real-number solutions. If the discriminant is negative, as it is in Example 4, we know that there are two different imaginary-number solutions.

Equations Reducible to Quadratic

Some equations can be treated as quadratic, provided that we make a suitable substitution. For example, consider the following:

$$(x^2 - x)^2 - 14(x^2 - x) + 24 = 0$$

$$u^2 - 14u + 24 = 0. \qquad \text{Substituting } u \text{ for } x^2 - x$$

The equation $u^2 - 14u + 24 = 0$ can be solved by factoring or using the quadratic formula. Once we have solved for u, we can reverse our substitution and solve for x. Equations that can be solved in this manner are said to be **reducible to quadratic**, or in **quadratic form**.

Example 5   Solve: $(x^2 - x)^2 - 14(x^2 - x) + 24 = 0$.

ALGEBRAIC SOLUTION

We let $u = x^2 - x$ and substitute:

$$u^2 - 14u + 24 = 0 \qquad \text{Substituting } u \text{ for } x^2 - x$$

$$(u - 12)(u - 2) = 0 \qquad \text{Factoring}$$

$$u - 12 = 0 \quad or \quad u - 2 = 0 \qquad \text{Principle of zero products}$$

$$u = 12 \quad or \quad u = 2.$$

A common error is to stop here, having solved for u, and forget to solve for x. Remember that you must find values for the *original* variable! We now substitute $x^2 - x$ for u and solve for x:

$$x^2 - x = 12 \quad or \quad x^2 - x = 2$$

$$\text{Substituting } x^2 - x \text{ for } u$$

$$x^2 - x - 12 = 0 \quad or \quad x^2 - x - 2 = 0$$

$$(x + 3)(x - 4) = 0 \quad or \quad (x + 1)(x - 2) = 0$$

$$\text{Factoring}$$

$$x + 3 = 0 \text{ or } x - 4 = 0 \text{ or } x + 1 = 0 \text{ or } x - 2 = 0$$

$$x = -3 \text{ or} \quad x = 4 \text{ or} \quad x = -1 \text{ or} \quad x = 2.$$

The solutions are -3, 4, -1, and 2. The solution set is $\{-3, -1, 2, 4\}$.

GRAPHICAL SOLUTION

We can obtain a graphical solution using procedures discussed earlier. We need not substitute or simplify.

$$y = (x^2 - x)^2 - 14(x^2 - x) + 24$$

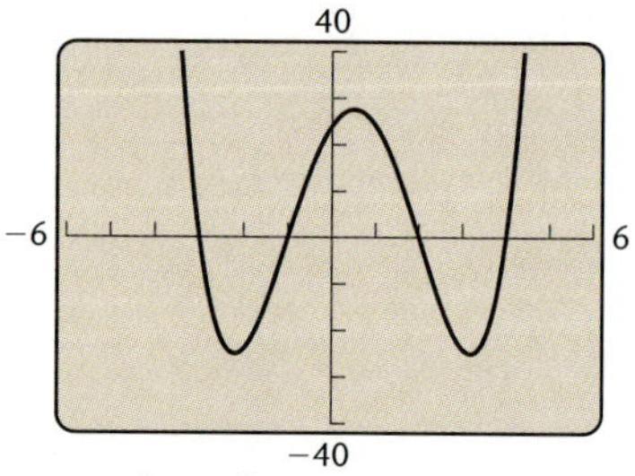

$x = -3, y = 0$
$x = -1, y = 0$
$x = 2, y = 0$
$x = 4, y = 0$

The solutions appear to be -3, -1, 2, and 4. The grapher also provides a quick visual check of the algebraic solution.

Interactive Discovery

See if you can find a way to use the TABLE feature to check the solutions in Example 5.

Example 6 Solve: $t^{2/5} - t^{1/5} - 2 = 0$.

ALGEBRAIC SOLUTION

Note that since $t^{2/5}$ can be rewritten as $(t^{1/5})^2$, the equation can be rewritten as $(t^{1/5})^2 - t^{1/5} - 2 = 0$. We let $u = t^{1/5}$ and solve the following equation:

$$u^2 - u - 2 = 0 \qquad \text{Substituting } u \text{ for } t^{1/5}$$
$$(u - 2)(u + 1) = 0$$
$$u - 2 = 0 \quad \text{or} \quad u + 1 = 0$$
$$u = 2 \quad \text{or} \quad u = -1.$$

Now we substitute $t^{1/5}$ for u and solve:

$$t^{1/5} = 2 \quad \text{or} \quad t^{1/5} = -1$$
$$(t^{1/5})^5 = 2^5 \quad \text{or} \quad (t^{1/5})^5 = (-1)^5 \qquad \text{Principle of powers;}$$
$$\text{raising to the fifth power}$$
$$t = 32 \quad \text{or} \quad t = -1.$$

CHECK: For 32:

$$t^{2/5} - t^{1/5} - 2 = 0$$
$$\overline{32^{2/5} - 32^{1/5} - 2 \ ? \ 0}$$
$$(32^{1/5})^2 - 32^{1/5} - 2$$
$$2^2 - 2 - 2$$
$$0 \ \vert \ 0 \quad \text{TRUE}$$

For -1:

$$t^{2/5} - t^{1/5} - 2 = 0$$
$$\overline{(-1)^{2/5} - (-1)^{1/5} - 2 \ ? \ 0}$$
$$[(-1)^{1/5}]^2 - (-1)^{1/5} - 2$$
$$(-1)^2 - (-1) - 2$$
$$0 \ \vert \ 0 \quad \text{TRUE}$$

Both numbers check. The solution set is $\{-1, 32\}$.

GRAPHICAL SOLUTION

We graph the equation $y = x^{2/5} - x^{1/5} - 2$ using a grapher, entering the equation as $y = (x^2)^\wedge(1/5) - x^\wedge(1/5) - 2$, or as $(x^\wedge(1/5))^2 - x^\wedge(1/5) - 2$.

If entered as $y = x^\wedge(2/5) - x^\wedge(1/5) - 2$, most graphers will omit the portion of the graph to the left of the y-axis. Those graphers reject negative inputs even though they are in the domain of the function.

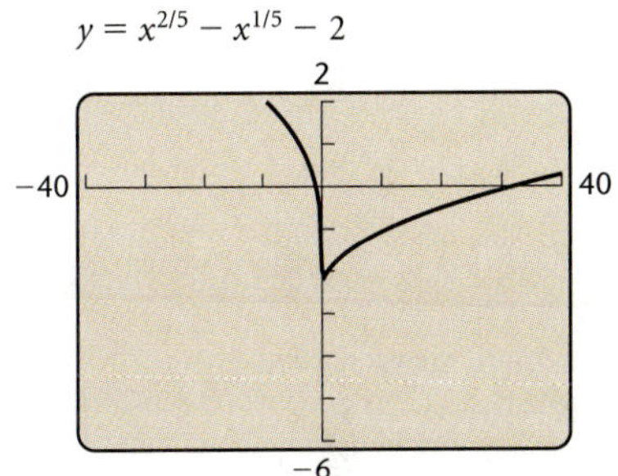

Then using graphical solving procedures as we have before, we get the solutions -1 and 32.

Example 7 Solve: $x^4 - 6x^2 + 7 = 0$.

In this case, we can find an approximate solution using a grapher, but the algebraic method yields exact solutions.

ALGEBRAIC SOLUTION

Letting $u = x^2$, we substitute u for x^2 and u^2 for x^4:

$$x^4 - 6x^2 + 7 = 0$$
$$(x^2)^2 - 6x^2 + 7 = 0$$
$$u^2 - 6u + 7 = 0.$$

To solve for u, we use the quadratic formula with $a = 1$, $b = -6$, and $c = 7$:

$$u = \frac{-b \pm \sqrt{b^2 - 4ac}}{2a}$$

$$= \frac{-(-6) \pm \sqrt{(-6)^2 - 4 \cdot 1 \cdot 7}}{2 \cdot 1}$$

$$= \frac{6 \pm \sqrt{8}}{2}$$

$$= \frac{2 \cdot 3 \pm 2\sqrt{2}}{2 \cdot 1}$$

$$= \frac{2}{2} \cdot \frac{3 \pm \sqrt{2}}{1}$$

$$= 3 \pm \sqrt{2}.$$

We have now solved for u. Next, we substitute x^2 for u and solve for x:

$$x^2 = 3 + \sqrt{2} \quad or \quad x^2 = 3 - \sqrt{2}$$
$$x = \pm\sqrt{3 + \sqrt{2}} \quad or \quad x = \pm\sqrt{3 - \sqrt{2}}.$$

Thus the solutions are

$$-\sqrt{3 + \sqrt{2}}, \quad \sqrt{3 + \sqrt{2}}, \quad -\sqrt{3 - \sqrt{2}}, \quad \text{and} \quad \sqrt{3 - \sqrt{2}},$$

or approximately

$$-2.101, \quad 2.101, \quad -1.259, \quad \text{and} \quad 1.259.$$

GRAPHICAL SOLUTION

We graph $y = x^4 - 6x^2 + 7$.

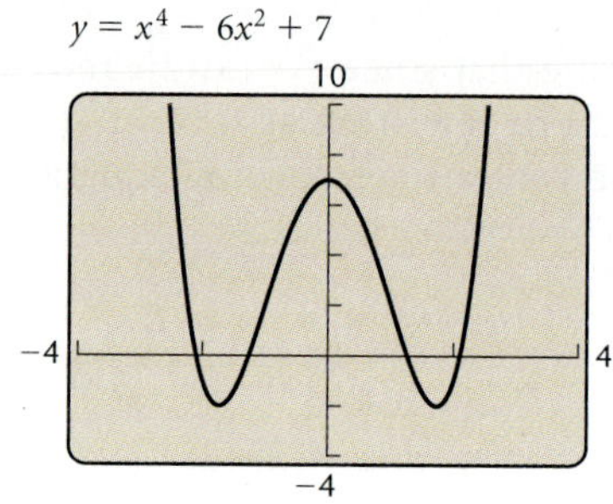

Then using graphical solving procedures as we have before, we get the approximate solutions -2.101, -1.259, 1.259, and 2.101.

Graphing Quadratic Functions of the Type $f(x) = a(x - h)^2 + k$

The graph of a quadratic function is called a **parabola**. We will see that the graph of every parabola evolves from the graph of the squaring function $f(x) = x^2$ using the transformations that we discussed in Section 1.7.

<table>
<tr><td>

Interactive Discovery

</td><td>

Think of transformations and look for patterns. Consider the following functions:

$$y_1 = x^2, \qquad y_2 = -0.4x^2,$$
$$y_3 = -0.4(x - 2)^2, \qquad y_4 = -0.4(x - 2)^2 + 3.$$

Graph y_1 and y_2. How do you get from the graph of y_1 to y_2?
Graph y_2 and y_3. How do you get from the graph of y_2 to y_3?
Graph y_3 and y_4. How do you get from the graph of y_3 to y_4?
  Consider the following functions:

$$y_1 = x^2, \qquad y_2 = 2x^2,$$
$$y_3 = 2(x + 3)^2, \qquad y_4 = 2(x + 3)^2 - 5.$$

Graph y_1 and y_2. How do you get from the graph of y_1 to y_2?
Graph y_2 and y_3. How do you get from the graph of y_2 to y_3?
Graph y_3 and y_4. How do you get from the graph of y_3 to y_4?

</td></tr>
</table>

We get the graph of $f(x) = a(x - h)^2 + k$ from the graph of $f(x) = x^2$ as follows:

$$f(x) = x^2$$
$$\downarrow$$
$$f(x) = ax^2 \qquad$$ **Vertical stretching or shrinking with a reflection across the x-axis if $a < 0$**
$$\downarrow$$
$$f(x) = a(x - h)^2 \qquad$$ **Horizontal translation**
$$\downarrow$$
$$f(x) = a(x - h)^2 + k. \qquad$$ **Vertical translation**

Consider the following graphs of the form $f(x) = a(x - h)^2 + k$. The point at which the graph turns is called the **vertex**. If $f(x)$ has a maximum or a minimum value, it occurs at the vertex. Each graph has a line $x = h$ that is called the **line of symmetry**.

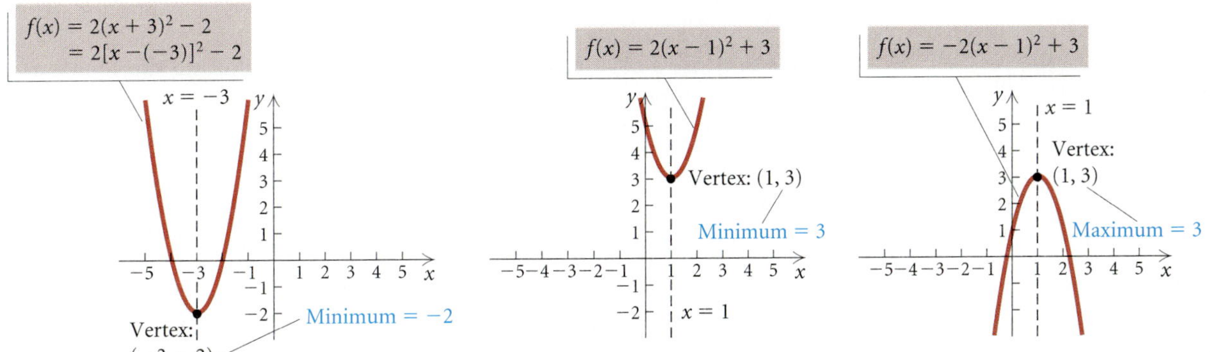

We summarize as follows.

The graph of $f(x) = a(x - h)^2 + k$:

    opens up if $a > 0$ and down if $a < 0$;

    has (h, k) as a vertex;

    has $x = h$ as the line of symmetry;

    has k as a minimum value (output) if $a > 0$ and has k as a maximum value if $a < 0$.

If the equation is in the form $f(x) = a(x - h)^2 + k$, we can determine a great deal of information about the graph without graphing.

FUNCTION	$f(x) = 3\left(x - \frac{1}{4}\right)^2 - 2$ $= 3\left(x - \frac{1}{4}\right)^2 + (-2)$	$g(x) = -3(x + 5)^2 + 7$ $= -3[x - (-5)]^2 + 7$
VERTEX	$\left(\frac{1}{4}, -2\right)$	$(-5, 7)$
LINE OF SYMMETRY	$x = \frac{1}{4}$	$x = -5$
MAXIMUM	No: $3 > 0$, graph opens up.	Yes, 7: $-3 < 0$, graph opens down.
MINIMUM	Yes, -2; $3 > 0$, graph opens up.	No: $-3 < 0$, graph opens down.

Note that the vertex (h, k) is used to find the maximum or the minimum value of the function. The maximum or minimum is the number k, *not* the ordered pair (h, k).

Graphing Quadratic Functions of the Type $f(x) = ax^2 + bx + c, a \neq 0$

We now use a modification of the method of completing the square as an aid in graphing and analyzing quadratic functions of the form $f(x) = ax^2 + bx + c, a \neq 0$.

Example 8   Complete the square to find the vertex, the line of symmetry, and the maximum or minimum value of $f(x) = x^2 + 10x + 23$. Then graph the function.

SOLUTION   To express $f(x) = x^2 + 10x + 23$ in the form $f(x) = a(x - h)^2 + k$, we proceed as follows. We construct a trinomial square. To do so, we take half the coefficient of x and square it. The number is $(10/2)^2$, or 25. We now add and subtract that number on the right-hand side. We can think of this as adding $25 - 25$, which is 0.

$$f(x) = x^2 + 10x + 23$$

Note that 25 completes the square for $x^2 + 10x$.

$$= x^2 + 10x + 25 - 25 + 23$$

Adding $25 - 25$, or 0, to the right side

$$= (x^2 + 10x + 25) - 25 + 23$$

$$= (x + 5)^2 - 2$$

Factoring and simplifying

$$= [x - (-5)]^2 + (-2)$$

Writing in the form $f(x) = a(x - h)^2 + k$

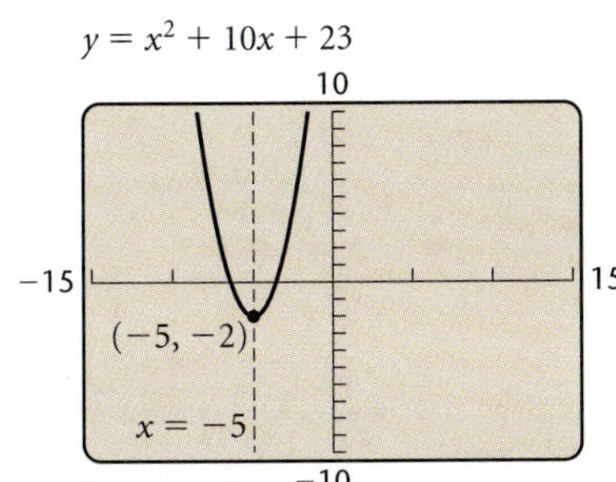

From this form of the function, we know the following:

Vertex: $(-5, -2)$;

Line of symmetry: $x = -5$;

Minimum $= -2$.

The graph of f is a shift of the graph of $y = x^2$ left 5 units and down 2 units.

Keep in mind that a line of symmetry is not part of the graph; it is a characteristic of the graph. If you fold the graph on its line of symmetry, the two halves of the graph will coincide.

Example 9 Complete the square to find the vertex, the line of symmetry, and the maximum or minimum value of $g(x) = x^2/2 - 4x + 8$. Then graph the function.

SOLUTION To complete the square, we factor $\frac{1}{2}$ out of the first two terms. This makes the coefficient of x^2 inside the parentheses 1:

$$g(x) = \frac{x^2}{2} - 4x + 8$$

$$= \frac{1}{2}(x^2 - 8x) + 8.$$

Factoring $\frac{1}{2}$ out of the first two terms

Now we complete the square inside the parentheses: Half of -8 is -4, and $(-4)^2 = 16$. We add and subtract 16 inside the parentheses:

$$g(x) = \tfrac{1}{2}(x^2 - 8x + 16 - 16) + 8$$

$$= \tfrac{1}{2}(x^2 - 8x + 16) - \tfrac{1}{2} \cdot 16 + 8$$

Using the distributive law to remove -16 from within the parentheses

$$= \tfrac{1}{2}(x - 4)^2.$$

Factoring and simplifying

We know the following:

Vertex: $(4, 0)$;

Line of symmetry: $x = 4$;

Minimum $= 0$.

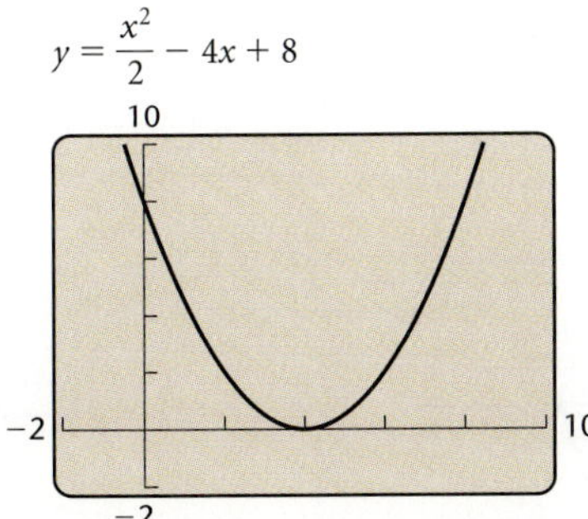

The graph of g is a shrinking of the graph of $y = x^2$ vertically followed by a shift to the right 4 units.

Example 10 Complete the square to find the vertex, the line of symmetry, and the maximum or minimum value of $f(x) = -2x^2 + 10x - \frac{23}{2}$. Then graph the function.

SOLUTION

$$f(x) = -2x^2 + 10x - \frac{23}{2}$$

$$= -2(x^2 - 5x) - \frac{23}{2} \qquad \text{Factoring } -2 \text{ out of the first two terms}$$

$$= -2\left(x^2 - 5x + \frac{25}{4} - \frac{25}{4}\right) - \frac{23}{2} \qquad \text{Completing the square inside the parentheses}$$

$$= -2\left(x^2 - 5x + \frac{25}{4}\right) - 2\left(-\frac{25}{4}\right) - \frac{23}{2} \qquad \text{Removing } -\frac{25}{4} \text{ from within the parentheses}$$

$$= -2\left(x - \frac{5}{2}\right)^2 + \frac{25}{2} - \frac{23}{2}$$

$$= -2\left(x - \frac{5}{2}\right)^2 + 1.$$

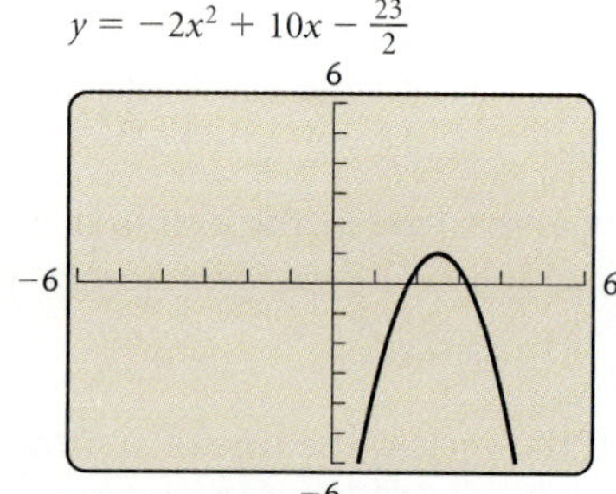

This form of the function yields the following:

Vertex: $\left(\frac{5}{2}, 1\right)$:

Line of symmetry: $x = \frac{5}{2}$;

Maximum $= 1$.

The graph is found by stretching the graph of $f(x) = x^2$ vertically, reflecting the graph across the x-axis, shifting the graph to the right $\frac{5}{2}$ units, and shifting that curve up 1 unit.

In many situations, we want to find the coordinates of the vertex directly from the equation $f(x) = ax^2 + bx + c$ using a formula. One way to develop such a formula is to consider the x-coordinate of the vertex as centered between the x-intercepts, or zeros, of the function. By averaging the two solutions of $ax^2 + bx + c = 0$, we find a formula for the x-coordinate of the vertex:

$$x\text{-coordinate of vertex} = \frac{\dfrac{-b - \sqrt{b^2 - 4ac}}{2a} + \dfrac{-b + \sqrt{b^2 - 4ac}}{2a}}{2}$$

$$= \frac{\dfrac{-2b}{2a}}{2} = \frac{-\dfrac{b}{a}}{2} = -\frac{b}{2a}.$$

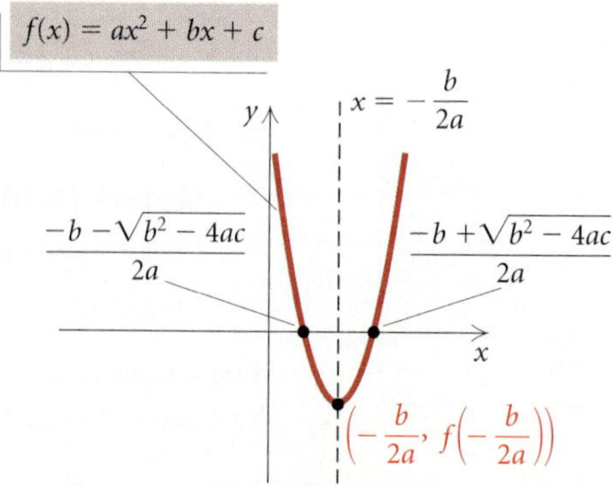

We use this value of x to find the y-coordinate of the vertex, $f\left(-\dfrac{b}{2a}\right)$.

The Vertex of a Parabola

The *vertex* of the graph of $f(x) = ax^2 + bx + c$ is

$$\left(-\frac{b}{2a},\, f\left(-\frac{b}{2a}\right)\right).$$

We calculate the   We substitute to
x-coordinate.   find the y-coordinate.

Example 11   For the function $f(x) = -x^2 + 14x - 47$:

a) Find the vertex.

b) Determine whether there is a maximum or minimum value and find that value.

c) Find the range.

d) On what intervals is the function increasing? decreasing?

SOLUTION   There is no need to graph the function.

a) The x-coordinate of the vertex is

$$\frac{-b}{2a}, \quad \text{or} \quad \frac{-14}{2(-1)}, \quad \text{or} \quad 7.$$

Since

$$f(7) = -7^2 + 14 \cdot 7 - 47 = 2,$$

the vertex is $(7, 2)$.

b) Since $a = -1$ and is negative, the graph opens down so the second coordinate of the vertex, 2, is the maximum value of the function.

c) The range is $(-\infty, 2]$.

d) Since the graph opens down, function values increase to the left of the vertex and decrease to the right of the vertex. Thus the function is increasing on the interval $(-\infty, 7]$ and decreasing on $[7, \infty)$.

You should visually check the results of Example 11 with a grapher.

2.2 Exercise Set

Solve by factoring.

1. $x^2 - 8x + 12 = 0$

2. $5x^2 + 42x + 16 = 0$

3. $7x^2 + 6 = 43x$

4. $23x = 12 + 10x^2$

Solve using the principle of square roots.

5. $x^2 = 81$

6. $x^2 - 100 = 0$

7. $9x^2 - 100 = 0$

8. $4x^2 - 25 = 0$

9. $x^2 + 10 = 0$

10. $x^2 = -14$

11. $(x + 3)^2 = 81$

12. $(x - 5)^2 = 64$

13. $x^2 - 6x + 9 = 1$

14. $x^2 + 8x + 16 = 9$

Solve by completing the square to obtain exact solutions. Check your work on a grapher, if appropriate.

15. $x^2 + 6x = 7$
16. $x^2 + 8x = -15$
17. $x^2 = 8x - 9$
18. $x^2 = 22 + 10x$
19. $x^2 + 8x + 25 = 0$
20. $x^2 + 6x + 13 = 0$
21. $3x^2 + 5x - 2 = 0$
22. $2x^2 - 5x - 3 = 0$
23. $6x + 1 = 4x^2$
24. $3x^2 + 5x = 3$
25. $2x^2 - 4 = 5x$
26. $4x^2 - 2 = 3x$

Solve. Use any method, but obtain exact solutions. Check your work on a grapher, if appropriate.

27. $x^2 - 2x = 15$
28. $x^2 + 4x = 5$
29. $5m^2 + 3m = 2$
30. $2y^2 - 3y - 2 = 0$
31. $3x^2 + 6 = 10x$
32. $3t^2 + 8t + 3 = 0$
33. $x^2 + x + 2 = 0$
34. $x^2 + 1 = x$
35. $5t^2 - 8t = 3$
36. $5x^2 + 2 = x$
37. $3x^2 + 4 = 5x$
38. $2t^2 - 5t = 1$

For each of the following, consider only $b^2 - 4ac$, the discriminant of the quadratic formula, to determine whether imaginary solutions exist.

39. $4x^2 = 8x + 5$
40. $4x^2 - 12x + 9 = 0$
41. $x^2 + 3x + 4 = 0$
42. $x^2 - 2x + 4 = 0$
43. $5t^2 - 7t = 0$
44. $5t^2 - 4t = 11$

Use a grapher to solve graphically. Find solutions to the nearest thousandth.

45. $x^2 - 5x - 4 = 0$
46. $x^2 + 7x - 6 = 0$
47. $3x^2 + 2x - 4 = 0$
48. $5.02x^2 - 4.19x - 2.057 = 0$

Solve. Find exact solutions. Check your work on a grapher, if possible.

49. $(2x - 3)^2 - 5(2x - 3) + 6 = 0$
50. $(3x + 2)^2 + 7(3x + 2) - 8 = 0$
51. $m^{2/3} - 2m^{1/3} - 8 = 0$
52. $t^{2/3} + t^{1/3} - 6 = 0$
53. $x^4 - 5x^2 + 4 = 0$
54. $x^4 + 3 = 4x^2$
55. $(2t^2 + t)^2 - 4(2t^2 + t) + 3 = 0$
56. $12 = (m^2 - 5m)^2 + (m^2 - 5m)$
57. $x - 3\sqrt{x} - 4 = 0$
58. $2x - 9\sqrt{x} + 4 = 0$

Complete the square to:

a) *find the vertex;*
b) *find the line of symmetry;*

c) *determine whether there is a maximum or minimum value and find that value.*

Then check your answers with a grapher.

59. $f(x) = x^2 - 8x + 12$
60. $g(x) = x^2 + 7x - 8$
61. $f(x) = x^2 - 7x + 12$
62. $g(x) = x^2 - 5x + 6$
63. $g(x) = 2x^2 + 6x + 8$
64. $f(x) = 2x^2 - 10x + 14$
65. $g(x) = -2x^2 + 2x + 1$
66. $f(x) = -3x^2 - 3x + 1$

In Exercises 67–74, match the equation with figures (a)–(h). Use a grapher only as a check.

a) 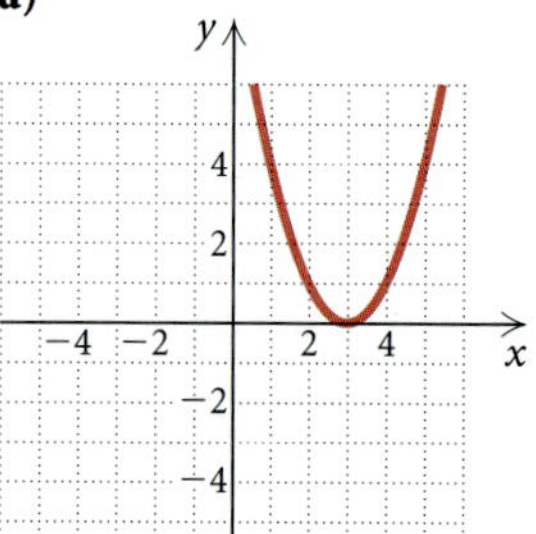

b) 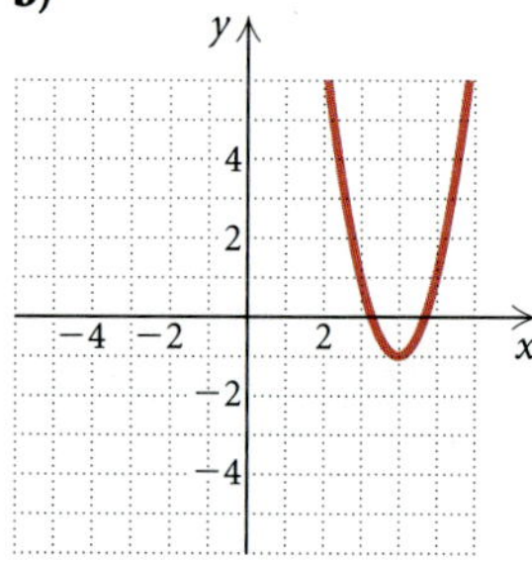

c) 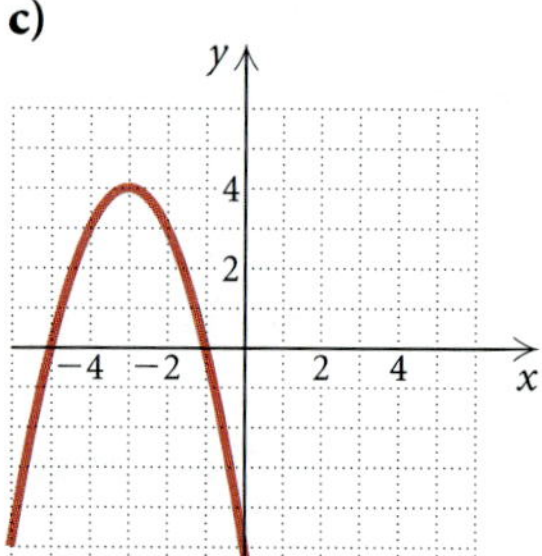

d) 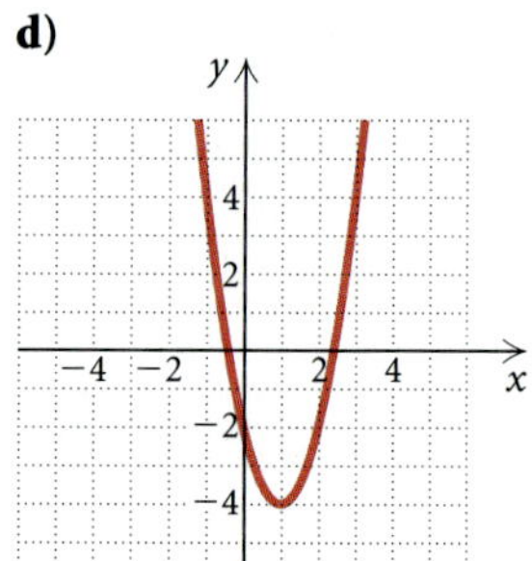

e) 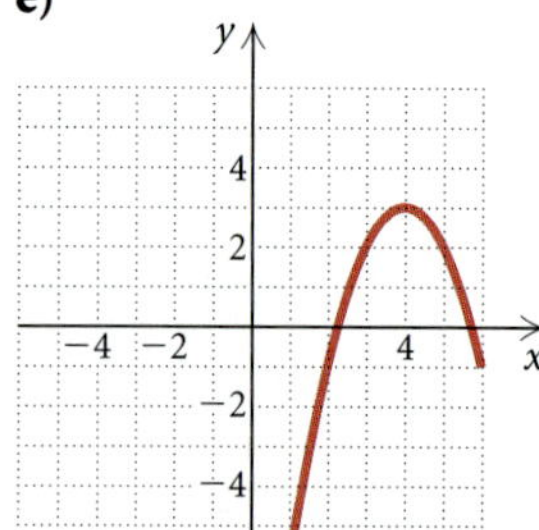

f) 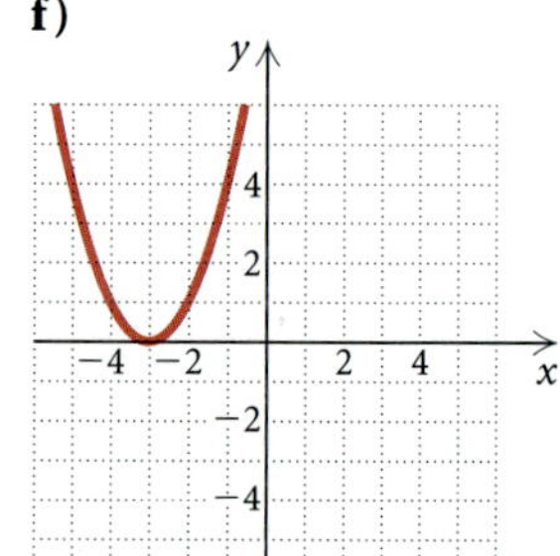

g)

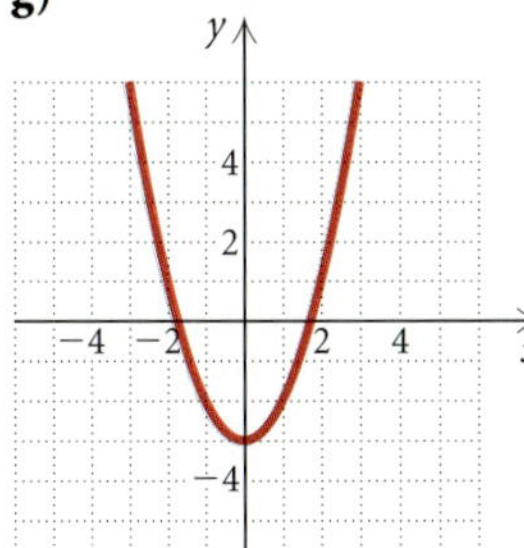

h)

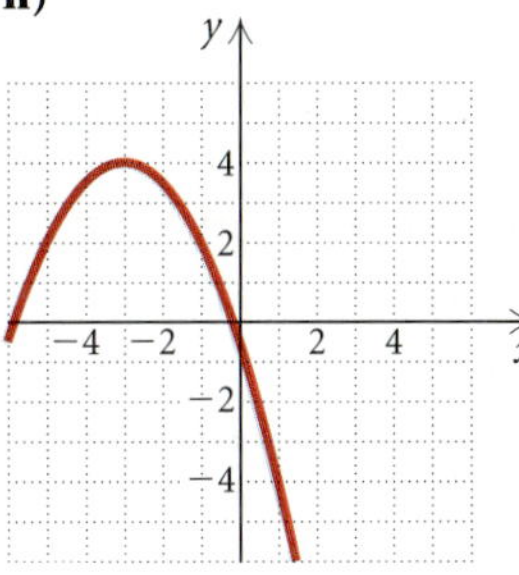

67. $y = (x + 3)^2$

68. $y = -(x - 4)^2 + 3$

69. $y = 2(x - 4)^2 - 1$

70. $y = x^2 - 3$

71. $y = -\frac{1}{2}(x + 3)^2 + 4$

72. $y = (x - 3)^2$

73. $y = -(x + 3)^2 + 4$

74. $y = 2(x - 1)^2 - 4$

In Exercises 75–82:

a) *Find the vertex.*

b) *Determine whether there is a maximum or minimum value and find that value.*

c) *Find the range.*

d) *Find the intervals on which the function is increasing and the function is decreasing.*

Then check your answers with a grapher.

75. $f(x) = x^2 - 6x + 5$

76. $f(x) = x^2 + 4x - 5$

77. $f(x) = 2x^2 + 4x - 16$

78. $f(x) = \frac{1}{2}x^2 - 3x + \frac{5}{2}$

79. $f(x) = -\frac{1}{2}x^2 + 5x - 8$

80. $f(x) = -2x^2 - 24x - 64$

81. $f(x) = 3x^2 + 6x + 5$

82. $f(x) = -3x^2 + 24x - 49$

Skill Maintenance

Solve.

83. $\dfrac{5}{x + 4} + \dfrac{14}{x + 7} = 4$

84. $\dfrac{5}{x + 2} + \dfrac{3x}{x + 6} = 2$

85. Claudia jogs at a rate of 8 mph for 45 min. How far does she run?

Synthesis

86. ◆ Is it possible for a quadratic function to have one real zero and one imaginary zero? Why or why not?

87. ◆ The graph of a quadratic function can have 0, 1, or 2 *x*-intercepts. How can you predict the number of *x*-intercepts without drawing the graph or (completely) solving an equation?

Solve.

88. $x^2 + \sqrt{5}\,x - \sqrt{3} = 0$

89. $x^2 + x - \sqrt{2} = 0$

90. $9t(t + 2) - 3t(t - 2) = 2(t + 4)(t + 6)$

91. $2t^2 + (t - 4)^2 = 5t(t - 4) + 24$

92. $(x - 3)^{2/3} = 2$

93. $\sqrt[4]{x + 2} = \sqrt{3x + 1}$

94. $x^6 - 28x^3 + 27 = 0$

95. $\sqrt{x - 3} - \sqrt[4]{x - 3} = 2$

96. $x^2 + 3x + 1 - \sqrt{x^2 + 3x + 1} = 8$

97. $\left(y + \dfrac{2}{y}\right)^2 + 3y + \dfrac{6}{y} = 4$

For each equation in Exercises 98–101, under the given condition:

a) *Find k.*

b) *Find a second solution.*

98. $kx^2 - 2x + k = 0$; one solution is -3

99. $kx^2 - 17x + 33 = 0$; one solution is 3

100. $x^2 - (6 + 3i)x + k = 0$; one solution is 3

101. $x^2 - kx + 2 = 0$; one solution is $1 + i$

102. Find b such that $f(x) = -4x^2 + bx + 3$ has a maximum value of 50.

103. Find c such that $f(x) = -0.2x^2 - 3x + c$ has a maximum value of -225.

104. Find a quadratic function that has $(4, -5)$ as a vertex and contains the point $(-3, 1)$.

105. Graph: $f(x) = (|x| - 5)^2 - 3$.

106. Solve $\frac{1}{2}at + v_0 t + x_0 = 0$ for t.

2.3
Polynomial Models and Curve Fitting

- *Solve applications using polynomial models.*
- *Fit quadratic, cubic, and quartic polynomial functions to data and make predictions.*

Polynomial functions have many uses as models in science, engineering, and business. In this section, we examine some of these models. Keep in mind that almost any model has its limitations.

The simplest use of polynomial functions in applications occurs when we merely evaluate a polynomial function. In such cases, a model has already been developed.

Example 1   *Threshold Weight.* In a study performed by Alvin Shemesh, it was found that the **threshold weight** *W*, defined as the weight above which the risk of death rises dramatically, is given by

$$W(h) = \left(\frac{h}{12.3}\right)^3,$$

where *W* is in pounds and *h* is a person's height, in inches. Find the threshold weight of a person who is 5 ft, 7 in. tall.

SOLUTION   Since the model has already been developed, we need only evaluate. Note that the function *W* expresses weight as a function of height, in inches. The domain is determined from the context of the application to be $(0, \infty)$. We convert 5 ft, 7 in. to inches,

$$5 \text{ ft, } 7 \text{ in.} = 5(12 \text{ in.}) + 7 \text{ in.} = 67 \text{ in.,}$$

and evaluate the function:

$$W(67) = \left(\frac{67}{12.3}\right)^3 \approx 162.$$

A 5 ft, 7 in. person has a threshold weight of about 162 lb.

Example 2   *Ibuprofen in the Bloodstream.* Ibuprofen is a pain relief medication. The polynomial function

$$M(t) = 0.5t^4 + 3.45t^3 - 96.65t^2 + 347.7t, \quad 0 \le t \le 6,$$

can be used to estimate the number of milligrams of ibuprofen in the bloodstream *t* hours after 400 mg of the medication has been swallowed.

a) Graph the function using the viewing windows $[-24, 15, -14{,}400, 8000]$ with Xscl = 3 and Yscl = 800 and $[0, 6, -200, 500]$ with Xscl = 1 and Yscl = 100.

b) Discuss the limitations of the model.

c) Find the domain, the relative maximum, and the range.

d) Find the number of milligrams in the bloodstream at $t = 0, 0.5, 1, 1.5,$ and so on, up to 3 hr.

SOLUTION

a) The graphs using the two viewing windows are shown at the top of the following page.

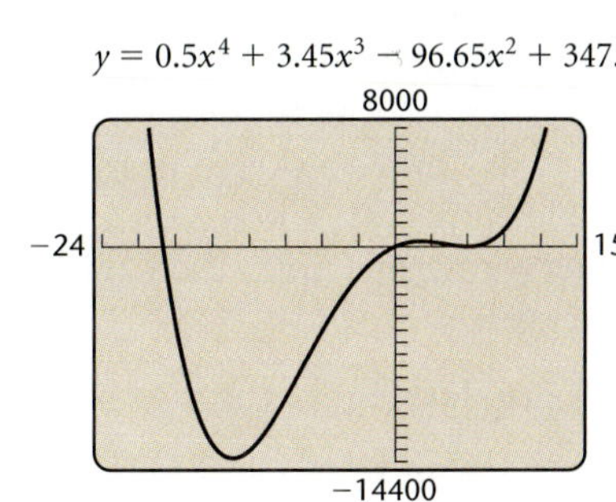

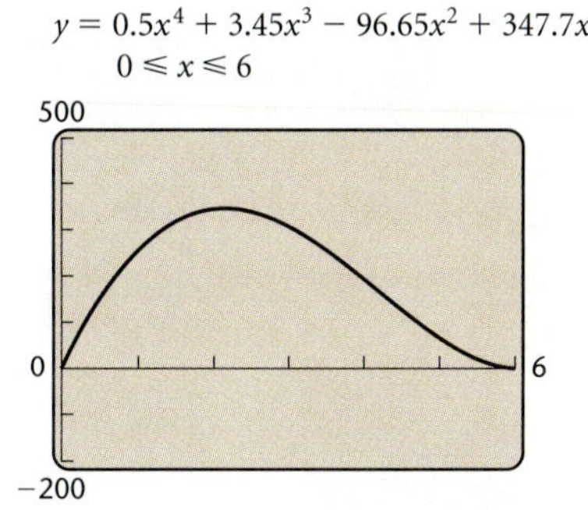

b) The domain of a polynomial function, unless restricted by a statement of the function, is $\mathbb{R}$. The graph on the left above gives a fairly complete picture of the curvature of the polynomial M. The implications of the application restrict the domain of the function. Presuming a patient had not taken any of the medication before, it seems reasonable that $M(0) = 0$; that is, at time 0, there are 0 mg of the medication in the bloodstream. After the medication has been taken, $M(t)$ will be positive for a period of time and eventually dissipate back to 0 and certainly not increase (unless another dose is taken). Thus the graph on the right above suggests that a better model includes a restriction on t,

$$M(t) = 0.5t^4 + 3.45t^3 - 96.65t^2 + 347.7t, \quad 0 \leq t \leq 6.$$

c) Given the restriction, the domain is $[0, 6]$. To find the range, we need to find the relative maximum. Some graphers can do this directly if we first enter an interval and a guess. For others, some use of TRACE and ZOOM will yield an estimate of the relative maximum (see Section 1.4). The maximum is about 345.76. Thus the range is about $[0, 345.76]$.

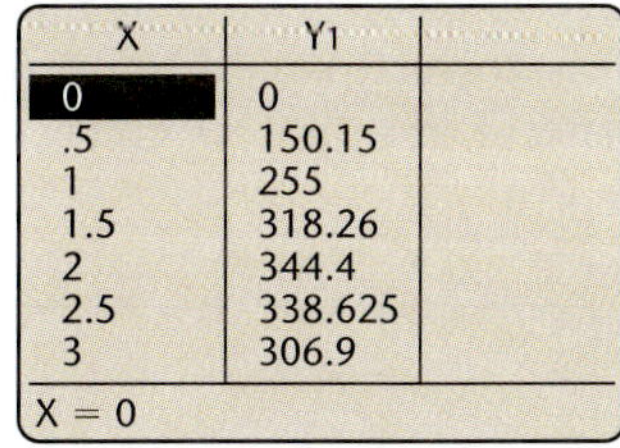

d) We can evaluate the function in a number of ways. To use a grapher, see your manual. If the grapher has a TABLE feature, you can start at 0 and use a step-value of 0.5.   ▬

We considered problems involving maxima or minima in Section 1.2. In those problems and in Example 2 above, the results were usually approximations. With the concepts developed in Section 2.2, we can get exact results when the function is quadratic. We can then check these results on a grapher.

Example 3 *Maximizing Area.* A stone mason has enough stones to enclose a rectangular patio with 60 ft of stone wall. If the attached house forms one side of the rectangle, what is the maximum area that the mason can enclose? What should the dimensions of the patio be in order to yield this area?

SOLUTION   In this case, it is helpful to make use of the five-step problem-solving strategy that we introduced in Section R.9.

1. Familiarize. Suppose the patio were 10 ft wide. It would then be $60 - 2 \cdot 10 = 40$ ft long. The area would be $(10 \text{ ft})(40 \text{ ft}) = 400 \text{ ft}^2$.

To see if this is the maximum area possible, we would need to check other possible combinations of length and width. Instead, we will try to find an area function and determine where a maximum occurs.

We make a sketch of the situation, using w to represent the patio's width, in feet. Since only 60 ft of stone wall is available and the house serves as one side, then $(60 - 2w)$ feet of stone is available for the length.

2. **Translate.** Since the area of a rectangle is length times width, we have

$$A(w) = (60 - 2w)w$$
$$= -2w^2 + 60w,$$

where $A(w) = $ the area of the patio, in square feet, as a function of the width, w.

3. **Carry out.** To solve this problem, we need to determine the maximum value of $A(w)$ and find the dimensions for which that maximum occurs. Since A is a quadratic function and w^2 has a negative coefficient, we know that the function has a maximum value that occurs at the vertex of the graph of the function. The first coordinate of the vertex is

$$w = -\frac{b}{2a} = -\frac{60}{2(-2)} = 15 \text{ ft.}$$

Thus, if $w = 15$ ft, then the length $l = 60 - 2 \cdot 15 = 30$ ft; and the area is $15 \cdot 30$, or 450 ft^2.

4. **Check.** As a partial check, we note that 450 ft$^2 >$ 400 ft^2, which is the area we found in a guess in the familiarize step. A more complete check, assuming that the function $A(w)$ is correct, would examine a table of values for $A(w) = (60 - 2w)w$ and/or examine its graph.

X	Y₁
14.7	449.82
14.8	449.92
14.9	449.98
15	450
15.1	449.98
15.2	449.92
15.3	449.82

X = 15

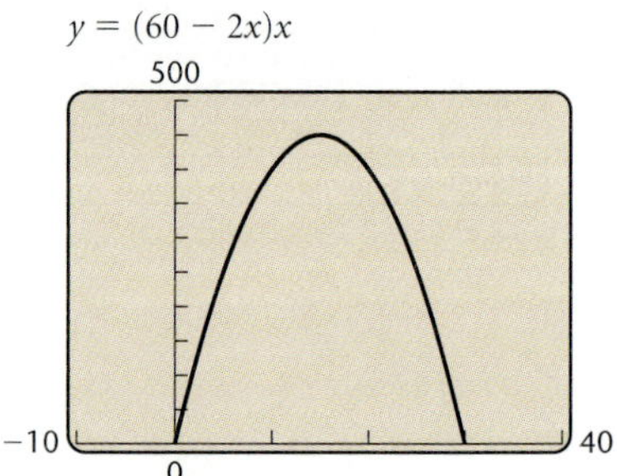

5. **State.** A maximum area of 450 ft^2 will occur if the patio is 15 ft wide and 30 ft long.

Example 4 *Finding the Depth of a Well.* A chlorine tablet is dropped into a well. Two seconds later, the sound of the splash is heard at the top of the well. The speed of the sound is 1100 ft/sec. How far is the top of the well from the water?

SOLUTION

1. **Familiarize.** We first make a drawing and label it with known and

unknown information. We let s = the depth of the well, in feet, t_1 = the time, in seconds, it takes for the tablet to hit the water, and t_2 = the time, in seconds, it takes for the sound to reach the top of the well.

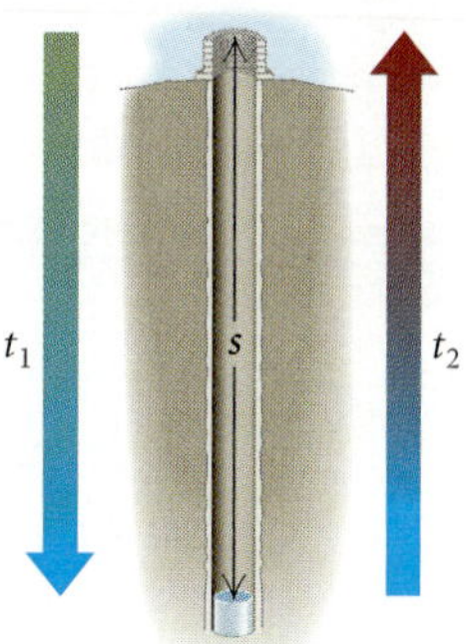

This gives us the equation

$$t_1 + t_2 = 2. \tag{1}$$

2. **Translate.** Can we find any relationship between the two times and the distance s? Often in problem solving you may need to look up related formulas in a physics book, another mathematics book, or maybe an encyclopedia. It turns out that the formula

$$s = 16t^2$$

gives the distance, in feet, that a dropped object falls in t seconds. The time t_1 that it takes the tablet to hit the water can be found as follows:

$$s = 16t_1^2, \quad \text{or} \quad t_1 = \frac{\sqrt{s}}{4}. \tag{2}$$

To find an expression for t_2, the time it takes the sound to travel to the top of the well, recall that *Distance* = *Rate · Time*. Thus,

$$s = 1100t_2, \quad \text{or} \quad t_2 = \frac{s}{1100}. \tag{3}$$

We now have expressions for t_1 and t_2, both in terms of s. Substituting into Equation (1), we obtain

$$t_1 + t_2 = 2, \quad \text{or} \quad \frac{\sqrt{s}}{4} + \frac{s}{1100} = 2. \tag{4}$$

3. **Carry out.** We solve Equation (4) for s. Multiplying by 1100, we get

$$275\sqrt{s} + s = 2200,$$

or

$$s + 275\sqrt{s} - 2200 = 0.$$

This equation is reducible to quadratic with $u = \sqrt{s}$. Substituting, we get

$$u^2 + 275u - 2200 = 0.$$

Using the quadratic formula, we can solve for u:

$$u = \frac{-b \pm \sqrt{b^2 - 4ac}}{2a}$$

$$= \frac{-275 + \sqrt{275^2 - 4 \cdot 1 \cdot (-2200)}}{2 \cdot 1}$$

We want only the positive solution.

$$= \frac{-275 + \sqrt{84{,}425}}{2}$$

$$\approx 7.78.$$

Since $u \approx 7.78$, we have $\sqrt{s} \approx 7.78$, so $s \approx 60.5$.

4. **Check**. To check, we can substitute 60.5 for s in Equations (2) and (3) and see that $t_1 + t_2 \approx 2$. We leave the mathematics for the student.

5. **State**. The top of the well is about 60.5 ft above the water.

Interactive Discovery

Check the solution of Equation (4) in Example 4 with a grapher. Find the intersection of the curves $y_1 = \sqrt{x}/4 + x/1100$ and $y_2 = 2$. If necessary, review the Introduction to Graphs and Graphers for details.

Polynomial Curve Fitting

By looking at an input–output table, we can tell whether the data fit a polynomial function. Let's consider an example:

x	y	FIRST DIFFERENCE	SECOND DIFFERENCE	THIRD DIFFERENCE
−11	−1194			
		912		
−7	−282		−640	
		272		384
−3	−10		−256	
		16		384
1	6		128	
		144		384
5	150		512	
		656		384
9	806		896	
		1552		
13	2358			

In the first column, we see that the differences of the x-values are always the same, 4. Next, we take the first differences of the y-values, but these are not constant. We continue, taking the second differences; these are still not constant. The third differences, however, are all constant—the number 384. The following theorem tells us that the data in this table fit a third-degree, or cubic, polynomial.

> **The Polynomial Difference Theorem**
>
> A function f is a polynomial function of degree n if and only if for any set of x-values that differ by the same number, the nth differences of the corresponding y-values are the same nonzero constant.

Thus we can look at an input–output table and know for sure whether the data fit a polynomial function. Unfortunately, real-world data do not always make such a perfect fit. We can still find quadratic, cubic, and quartic polynomials that fit the data approximately. The remainder of this section presumes the use of a grapher that does quadratic, cubic, and quartic regression.

As we move through this text, we develop a "stable" of functions. These can serve as models for many applications. Let's look at some that we have considered thus far.

Linear Function:
$$y_1 = mx + b$$

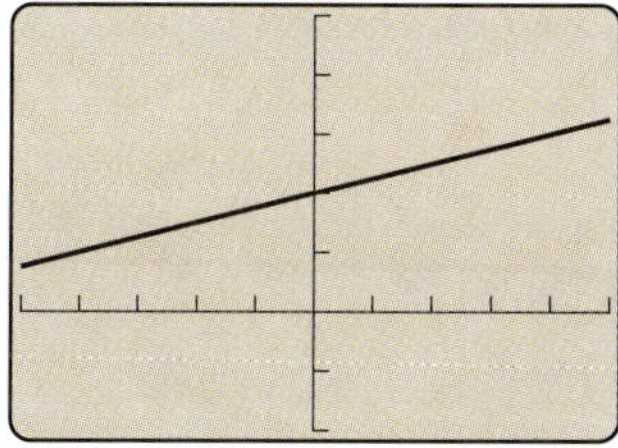

Quadratic Function:
$$y_2 = ax^2 + bx + c, a > 0$$

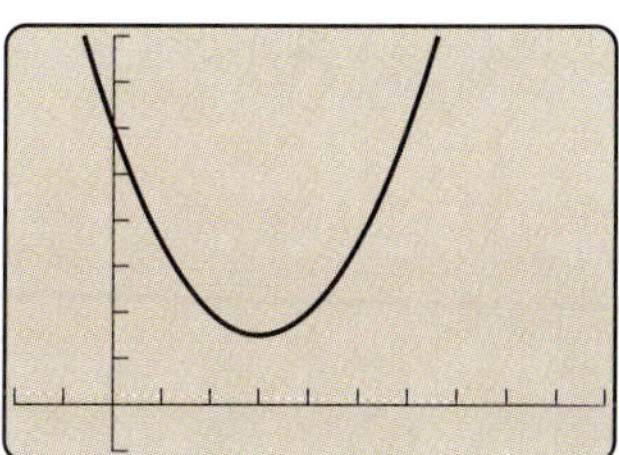

Quadratic Function:
$$y_3 = ax^2 + bx + c, a < 0$$

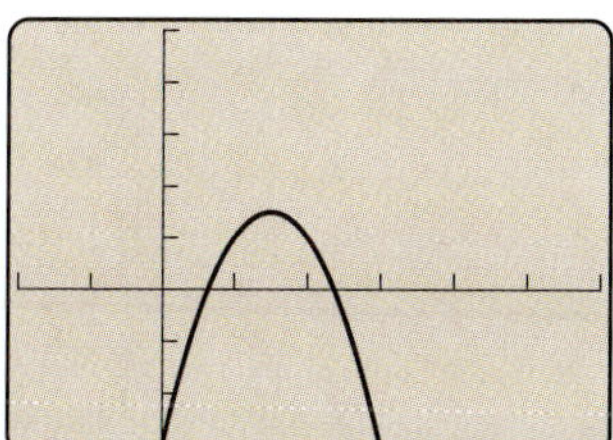

Cubic Function:
$$y_4 = ax^3 + bx^2 + cx + d$$

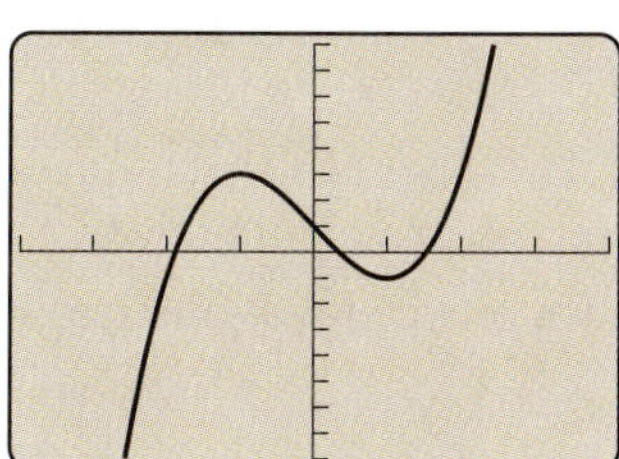

Quartic Function:
$$y_5 = ax^4 + bx^3 + cx^2 + dx + e$$

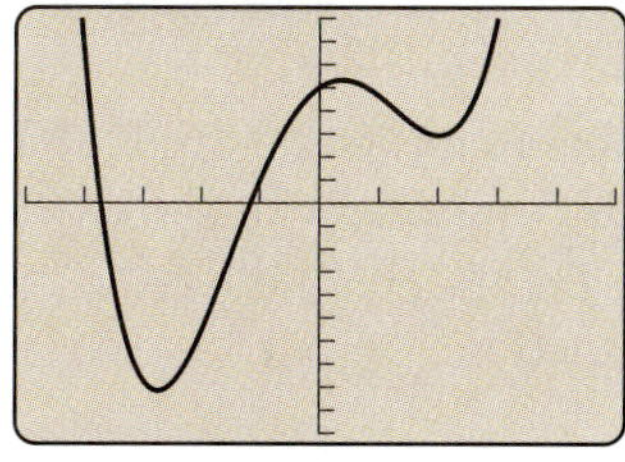

Now let's consider some real-world data. How can we decide which function might fit the data? Our simple way is to graph the data and look over the result. Simply stated, we "eyeball" the situation. Then we use a grapher to find a regression equation and make predictions.

Example 5 *Hours of Sleep versus Death Rate.* In a study by Dr. Harold J. Morowitz of Yale University, data were gathered that showed the relationship between the death rate of men and the average number of hours per day that the men slept.

Death Rate Related to Sleep

AVERAGE NUMBER OF HOURS OF SLEEP, x	DEATH RATE PER 100,000 MALES, y
5	1121
6	805
7	626
8	813
9	967

Source: "Hiding in the Hammond Report," *Hospital Practice,* by Harold J. Morowitz.

a) Make a scatterplot of the data.

b) Determine which, if any, of the functions seems to fit the data.

c) Use a grapher to fit the function to the data. Graph the equation using the same axes as the scatterplot.

d) Predict the death rate of males who sleep 4 hr, 5.5 hr, 7.5 hr, and 10 hr.

SOLUTION

a) We make a scatterplot of the data as shown on the left below.

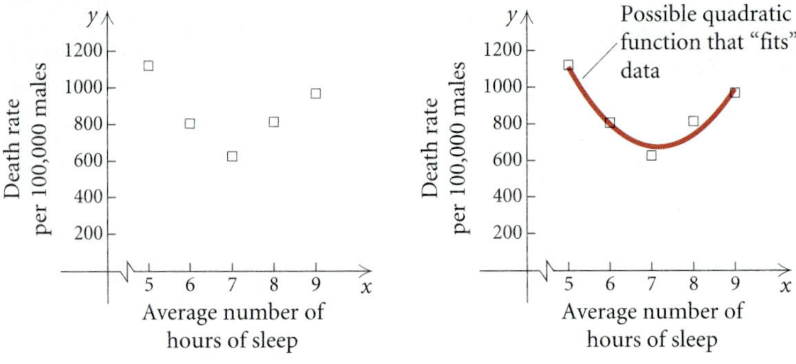

b) Note that the rate drops and then rises. This suggests that a quadratic function might fit the data. See the graph on the right above.

c) Using the quadratic REGRESSION feature, we get the following quadratic function:

$$f(x) = y = 93.28571429x^2 - 1336x + 5460.828571.$$

Depending on the accuracy required, we might round these coefficients, though it is easy to keep this equation in the grapher.

d) We then compute the outputs:

$$f(4) \approx 1609.4, \qquad f(7.5) \approx 688.1,$$
$$f(5.5) \approx 934.7, \qquad f(10) \approx 1429.4.$$

These can also be found with the TABLE feature set in ASK mode, after copying the quadratic function to the $y=$ screen. The results assert, for instance, that the death rate is about 688 per 100,000 for males who sleep 7.5 hr per night.

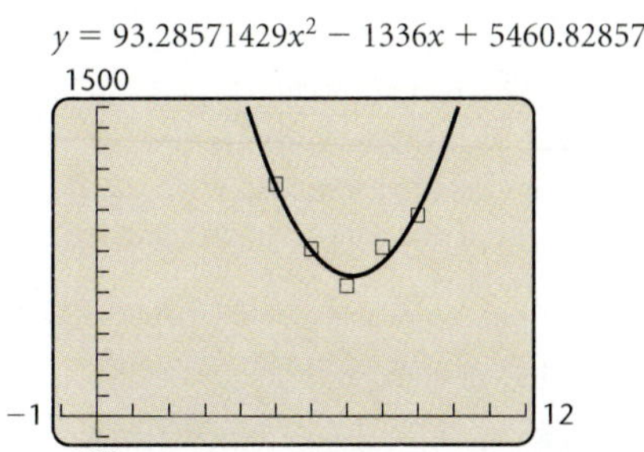

One must always be aware that a model such as the one in Example 5 has its limitations.

Interactive Discovery

In the model of Example 5, find and interpret $f(0)$ and $f(24)$. Argue whether such computations have meaning in the real world. What about $f(-1)$ and $f(25)$?

2.3   *Exercise Set*

1. *Vertical Leap.* A formula relating an athlete's vertical leap V, in inches, to hang time T, in seconds, is

$$V = 48T^2.$$

Anfernee Hardaway of the Orlando Magic has a vertical leap of 36 in. What is his hang time? (*Source:* National Basketball Association)

2. *Projectile Motion.* A stone thrown downward with an initial velocity of 34.3 m/sec will travel a distance of s meters, where

$$s(t) = 4.9t^2 + 34.3t$$

and t is in seconds. If a stone is thrown downward at 34.3 m/sec from a height of 294 m, how long will it take the stone to hit the ground?

3. *Games in a Sports League.* If there are x teams in a sports league and all the teams play each other twice, a total of $N(x)$ games are played, where

$$N(x) = x^2 - x.$$

A softball league has 9 teams, each of which plays the others twice. If the league pays \$45 per game for the field and umpires, how much will it cost to play the entire schedule?

4. *Windmill Power.* Under certain conditions, the power P, in watts per hour, generated by a windmill with winds blowing v miles per hour is given by

$$P(v) = 0.015v^3.$$

a) Find the power generated by 15-mph winds.
b) How fast must the wind blow in order to generate 120 watts of power in 1 hr?

5. *Interest Compounded Annually.* When P dollars is invested at interest rate i, compounded annually, for t years, the investment grows to A dollars, where

$$A = P(1 + i)^t.$$

a) Find the interest rate i if \$2560 grows to \$3610 in 2 yr.

b) Find the interest rate i if \$10,000 grows to \$13,310 in 3 yr.

6. *Maximizing Area.*   A fourth-grade class decides to enclose a rectangular garden, using the side of the school as one side of the rectangle. What is the maximum area that the class can enclose with 32 ft of fence? What should the dimensions of the garden be in order to yield this area?

7. *Maximizing Volume.*   A plastics manufacturer plans to produce a one-compartment vertical file by bending the long side of an 8-in. by 14-in. sheet of plastic along two lines to form a U-shape. How tall should the file be in order to maximize the volume that the file can hold?

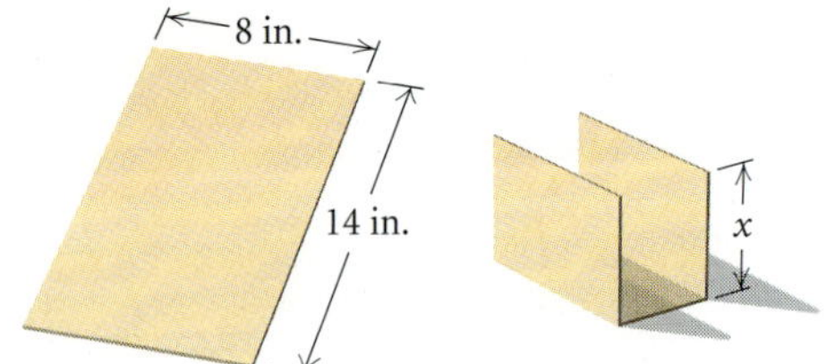

8. *Maximizing Area.*   The sum of the base and the height of a parallelogram is 69 cm. Find the dimensions for which the area is a maximum.

9. *Maximizing Area.*   The sum of the base and the height of a triangle is 20 cm. Find the dimensions for which the area is a maximum.

10. *Finding the Height of a Cliff.*   A water balloon is dropped from a cliff. Exactly 3 sec later, the sound of the balloon hitting the ground reaches the top of the cliff. How high is the cliff? (*Hint:* See Example 4.)

11. *Finding the Height of an Elevator Shaft.*   Jenelle dropped a screwdriver from the top of an elevator shaft. Exactly 5 sec later, she hears the sound of the screwdriver hitting the bottom of the shaft. How tall is the elevator shaft? (*Hint:* See Example 4.)

12. *Minimizing Cost.*   Aki's Bicycle Designs has determined that when x hundred bicycles are built, the average cost per bicycle is given by

$$C(x) = 0.1x^2 - 0.7x + 2.425,$$

where $C(x)$ is in hundreds of dollars. How many bicycles should be built in order to minimize the average cost per bicycle?

Maximizing Profit.   In business, profit is the difference between revenue and cost, that is,

$$Total\ profit = Total\ revenue - Total\ cost,$$
$$P(x) = R(x) - C(x),$$

where x is the number of units. Find the maximum profit and the number of units that must be sold in order to yield the maximum profit for each of the following.

13. $R(x) = 50x - 0.5x^2,\ C(x) = 10x + 3$

14. $R(x) = 5x,\ C(x) = 0.001x^2 + 1.2x + 60$

15. *Maximizing Area.*   A rancher needs to enclose two adjacent rectangular corrals, one for sheep and one for cattle. If a river forms one side of the corrals and 240 yd of fencing is available, what is the largest total area that can be enclosed?

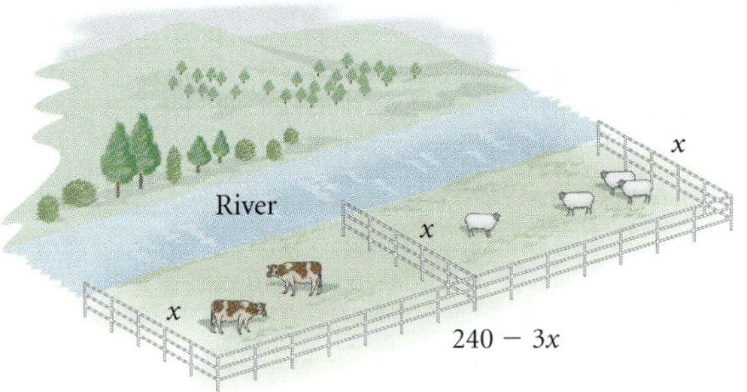

16. *Norman Window.*   A Norman window is a rectangle with a semicircle on top. Sky Blue Windows is designing a Norman window that will require 24 ft of trim. What dimensions will allow the maximum amount of light to enter a house?

For the scatterplots and graphs in Exercises 17–25, determine which, if any, of the following functions might be used as a model for the data.

a) *Linear,* $f(x) = mx + b$
b) *Quadratic,* $f(x) = ax^2 + bx + c,\ a > 0$
c) *Quadratic,* $f(x) = ax^2 + bx + c,\ a < 0$
d) *Polynomial, not quadratic or linear*

17.

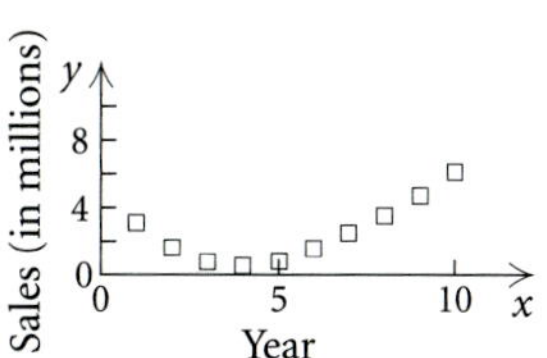

18.

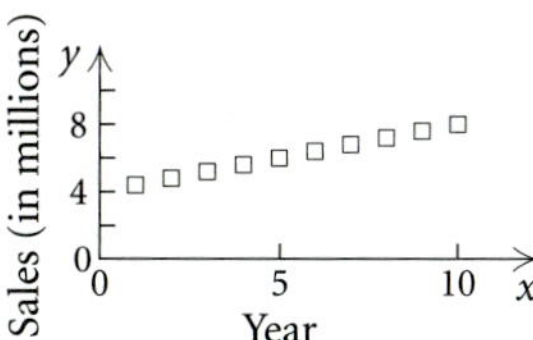

19.

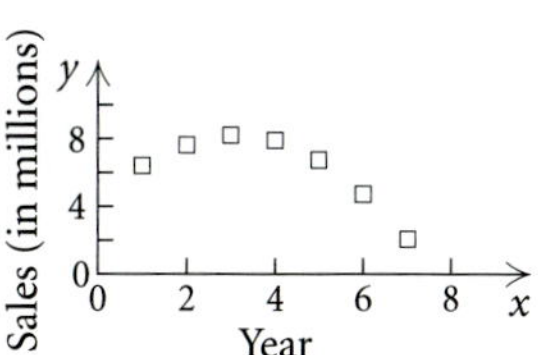

20.

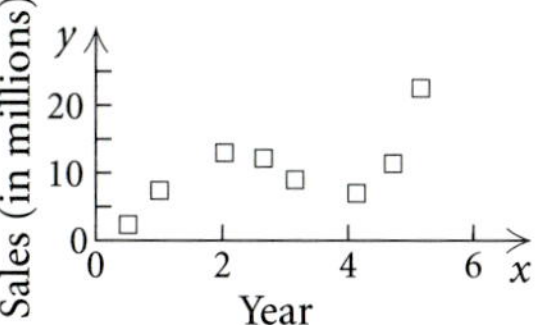

21.

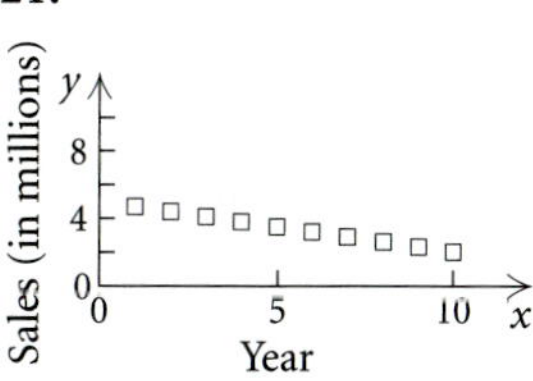

22.

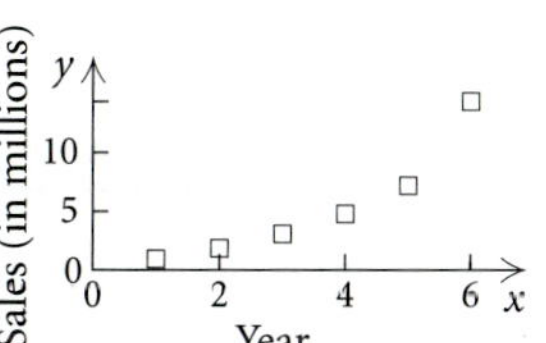

24.

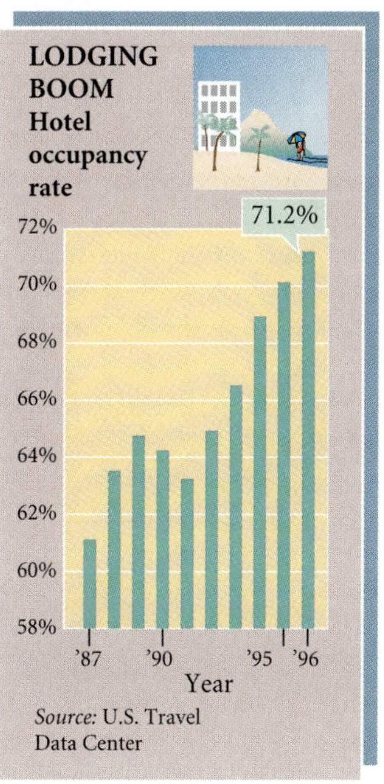

23.

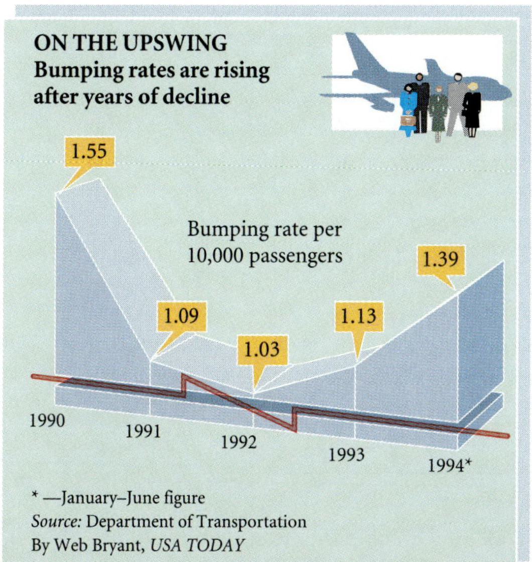

25.

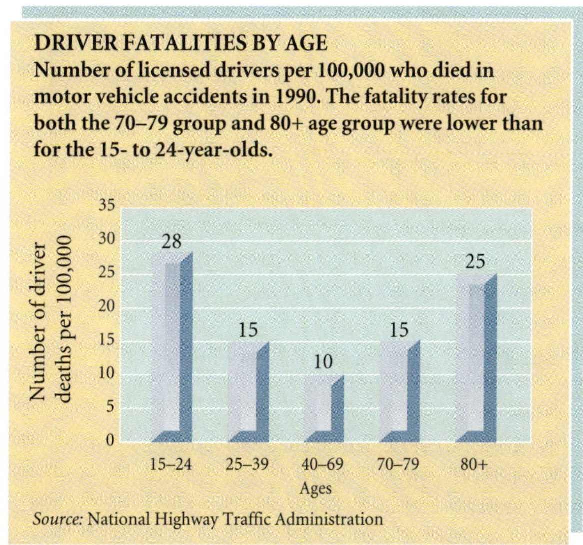

For each table of data in Exercises 26–29, answer the following questions.

a) *Is the change in the inputs x the same?*
b) *Find the first differences of the y-values. Is the change in the outputs y the same?*
c) *Find the second, third, and fourth differences, as needed. Is any set of differences nonzero and constant?*
d) *Does a quadratic, cubic, or quartic function fit the data?*
e) *If so, fit a function to the data using the* REGRESSION *feature of a grapher.*
f) *Then find the missing data.*

26.

x	y
1	7
6	47
11	137
16	277
21	467
26	707
31	997
36	?
?	2657

27.

x	y
−11	1204
−7	292
−3	20
1	4
5	−140
9	−796
13	−2348
17	?
21	?

28.

x	y
10	37
12	42.6
14	47.4
16	51.4
18	54.6
20	57
22	58.6
24	?
?	45.1

29.

x	y
−8	3447
−5	369
−2	−27
1	−9
4	99
7	1917
10	9009
12	?
?	−19.94

30. *Airline Passenger Bumping.*   Use the data in the graph of Exercise 23.

a) Use the REGRESSION feature of a grapher to fit a quadratic function to the data.
b) Predict the rate of passenger bumping in 1998, 2000, and 2005.

Shoe Size and Life Expectancy.   The data in the following table are the result of a Swedish study relating one's life expectancy to the size of one's feet. Use these data in Exercises 31 and 32.

SHOE SIZE	LIFE EXPECTANCY FOR WOMEN (IN YEARS)	SHOE SIZE	LIFE EXPECTANCY FOR MEN (IN YEARS)
4	69	5	66
5	76	6	69
6	82	7	69
7	84	8	72
8	75	9	75
9	72	10	77
10	70	11	82
11	69	12	79
		13	72
		14	69

Source: Orthopedic Quarterly.

31. a) Use the REGRESSION feature of a grapher to fit a quadratic function to the data for women.
b) Estimate the life expectancy of women with shoe sizes of $5\frac{1}{2}$ and $8\frac{1}{2}$.

32. a) Use the REGRESSION feature of a grapher and fit a quadratic function to the data for men.
b) Estimate the life expectancy of men with shoe sizes of $10\frac{1}{2}$ and $15\frac{1}{2}$.

33. *Prices of Personal Computers.*   The price P of a personal computer has varied greatly in recent years. The data in the table relate price P, in dollars, to time t, in years, where $t = 1$ corresponds to 1981.

YEAR, t	AVERAGE PRICE P OF A PERSONAL COMPUTER
1981 ($t = 1$)	$2290
1982 ($t = 2$)	1500
1983 ($t = 3$)	1400
1984 ($t = 4$)	1850
1985 ($t = 5$)	2540
1986 ($t = 6$)	2400
1987 ($t = 7$)	1720
1988 ($t = 8$)	1860
1989 ($t = 9$)	1930
1990 ($t = 10$)	2100
1991 ($t = 11$)	1820
1992 ($t = 12$)	1640

a) Find a cubic function $y = ax^3 + bx^2 + cx + d$ that fits the data.

b) Graph the cubic function of part (a).
c) Use the cubic function to predict the price of a personal computer in 1998.
d) Find a quartic polynomial function
$y = ax^4 + bx^3 + bx^2 + cx + d$ that fits the data.
e) Graph the quartic function of part (d).
f) Use the quartic function to predict the price of a personal computer in 1998.

34. *Damage to the Ozone Layer.*   The concentration of chlorine compounds in the stratosphere serves as an indicator of damage to the ozone layer. The data in the following table show the relationship of estimated chlorine concentration in the atmosphere, in parts per billion (ppb), to the year.

YEAR	CHLORINE CONCENTRATION (IN PARTS PER BILLION)
1985	2.5
1995	3.3
2010	3.9
2035	3.7

Source: Adapted by U.S. EPA by NRDC Earth Action Guide, "Saving the Ozone."

a) Use the REGRESSION feature on a grapher to find a linear, a quadratic, a cubic, and a quartic function to fit the data.
b) It is estimated that in 2060 the chlorine concentration will be 3.3 ppb. Which function in part (a) would best make this prediction?
c) Use the answer to part (b) to predict the chlorine concentration in 2085.

Skill Maintenance

Simplify.

35. $\dfrac{5x^7 + 25x^6 + 10x^4 - 20x^3}{5x^3}$

36. $\dfrac{8a^5 - 10a^4 + 12a^3}{2a^3}$

37. Multiply: $(5x^3 - 2x^2 + 5x - 1)(x - 3)$.

38. Subtract: $(8x^2 + 5x - 7) - (3x^2 - 6x + 9)$.

Synthesis

39. ◆ Regarding Exercise 33, discuss the merits of which function to use to predict the price of a personal computer in 1998. How could you test your assertions?

40. ◆ Write a problem for a classmate to solve. Design the problem so that it is a maximum or minimum problem using a quadratic function.

41. In early 1995, $2000 was deposited at a certain interest rate. One year later, $1200 was deposited in another account at the same rate. At the end of that year, there was a total of $3573.80 in both accounts. What is the annual interest rate?

42. *Minimizing Area.*   A 24-in. piece of string is cut into two pieces. One piece is used to form a circle while the other is used to form a square. How should the string be cut so that the sum of the areas is a minimum?

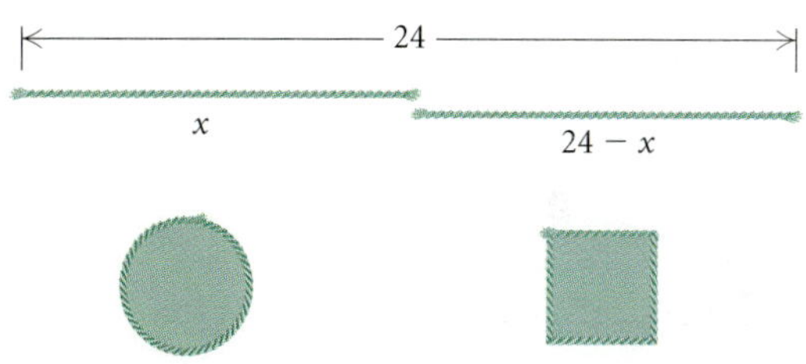

For each of the two input–output tables in Exercises 43 and 44:

a) *Fit a cubic function to the existing data. Then use that equation to find the missing values.*
b) *Fit a quartic function to the existing data. Then use that equation to find the missing values.*

43.

x	y
−4	?
−3	7
−2	11
−1	33
0	45
1	19
?	23
3	46
4	?

44.

x	y
20	12.4
?	24.3
30	28.5
40	19.6
?	39.2
50	78.4
60	196.8
70	793.6
80	?

45. *The Effect of Advertising on Movie Revenue.*   The following table contains actual data pertaining to the box office revenue of certain movies together with the amount of money that was spent to advertise each movie. We want to explore whether box office revenue is directly related to advertising expenditures.

MOVIE	ADVERTISING BUDGET (IN MILLIONS)	BOX OFFICE REVENUE (IN MILLIONS)
The Lion King	$23.3	$300.4
Forrest Gump	25.0	298.5
True Lies	20.5	146.3
The Santa Clause	19.4	137.8
The Flintstones	10.6	130.6
Clear & Present Danger	17.9	121.8
Speed	17.2	121.2
The Mask	11.2	118.8
Maverick	13.8	101.6
Interview with the Vampire	15.4	100.7

Source: The Hollywood Reporter.

a) Using a ZOOMSTAT feature, make a scatterplot of the data.

b) Analyze the scatterplot and try to decide whether a linear, quadratic, cubic, or quartic equation might fit the data.

c) Use the REGRESSION feature on a grapher to find a quadratic, a cubic, and a quartic function to fit the data. Assume that advertising is the independent variable.

d) Plot each regression equation with the scatterplot. Examine the results and try to decide which function fits best.

e) You are about to make a blockbuster movie called *A Day in the Life of a Math Professor*. You decide to spend $30 million to advertise the film. Use each of the three functions found in part (c) to predict the box office revenue from the movie.

2.4

Polynomial Division; The Remainder and Factor Theorems

- *Do long division with polynomials and determine whether one polynomial is a factor of another.*
- *Use synthetic division to divide a polynomial by $x - c$.*
- *Use the remainder theorem to find a function value $f(c)$.*
- *Use the factor theorem to determine whether $x - c$ is a factor of $f(x)$.*

In general, finding exact zeros of polynomial functions is neither easy nor straightforward. In this section and the one that follows, we develop concepts that help us find exact zeros of polynomial functions with degree 3 or greater.

We will now allow the coefficients of a polynomial to be complex numbers. In certain cases, we will restrict the coefficients to be real numbers, rational numbers, or integers, as shown in the following examples.

POLYNOMIAL	TYPE OF COEFFICIENT
$5x^3 - 3x^2 + (2 + 4i)x + i$	Complex
$5x^3 - 3x^2 + \sqrt{2}x - \pi$	Real
$5x^3 - 3x^2 + \frac{2}{3}x - \frac{7}{4}$	Rational
$5x^3 - 3x^2 + 8x - 11$	Integer

Let's review the meaning of a zero of a function.

Example 1 Consider $P(x) = x^3 + 2x^2 - 5x - 6$. Determine whether each of the numbers 3 and -1 is a zero of $P(x)$.

SOLUTION We have

$$P(3) = (3)^3 + 2(3)^2 - 5(3) - 6 = 24.$$ **Substituting 3 into the polynomial**

Since $P(3) \neq 0$, 3 is *not* a zero of the polynomial:

$$P(-1) = (-1)^3 + 2(-1)^2 - 5(-1) - 6 = 0.$$ **Substituting -1 into the polynomial**

Since $P(-1) = 0$, we know that -1 is a zero of $P(x)$.

Division and Factors

When we divide one polynomial by another, we obtain a quotient and a remainder. If the remainder is 0, then the divisor is a **factor** of the dividend.

Example 2 Divide to determine whether $x + 1$ is a factor of

$$x^3 + 2x^2 - 5x - 6.$$

SOLUTION

$$
\begin{array}{r}
\text{Quotient} \\
x^2 + x - 6 \\
x + 1 \,)\overline{x^3 + 2x^2 - 5x - 6} \\
\underline{x^3 + x^2} \\
x^2 - 5x \\
\underline{x^2 + x} \\
-6x - 6 \\
\underline{-6x - 6} \\
0 \longleftarrow \text{Remainder}
\end{array}
$$

Divisor → ($x + 1$)

Since the remainder is 0, we know that $x + 1$ is a factor of $x^3 + 2x^2 - 5x - 6$.

Example 3 Divide to determine whether $x^2 + 3x - 1$ is a factor of $x^4 - 81$.

SOLUTION

$$
\begin{array}{r}
x^2 - 3x + 10 \\
x^2 + 3x - 1 \,)\overline{x^4 \qquad\qquad\qquad - 81} \\
\underline{x^4 + 3x^3 - x^2} \\
-3x^3 + x^2 \\
\underline{-3x^3 - 9x^2 + 3x} \\
10x^2 - 3x - 81 \\
\underline{10x^2 + 30x - 10} \\
-33x - 71
\end{array}
$$

Note that spaces have been left for missing terms.

Since the remainder is not 0, we know that $x^2 + 3x - 1$ is not a factor of $x^4 - 81$.

When we divide a polynomial $P(x)$ by a divisor $d(x)$, a polynomial $Q(x)$ is the quotient and a polynomial $R(x)$ is the remainder. The remainder must either be 0 or have degree less than that of $d(x)$. To check a division, we multiply the quotient by the divisor and add the remainder, to see if we get the dividend. Thus these polynomials are related as follows:

$$P(x) = d(x) \cdot Q(x) + R(x)$$

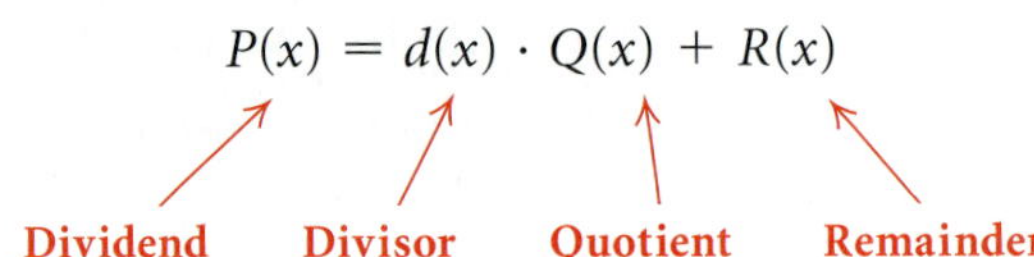

For example, if $P(x) = x^3 + 2x^2 - 5x - 6$ and $d(x) = x + 1$, then $Q(x) = x^2 + x - 6$ and $R(x) = 0$, and

$$\underbrace{x^3 + 2x^2 - 5x - 6}_{P(x)} = \underbrace{(x + 1)}_{d(x)} \cdot \underbrace{(x^2 + x - 6)}_{Q(x)} + \underbrace{0}_{R(x)}.$$

If $P(x) = x^4 - 81$ and $d(x) = x^2 + 3x - 1$, then $Q(x) = x^2 - 3x + 10$ and $R(x) = -33x - 71$, and

$$\underbrace{x^4 - 81}_{P(x)} = \underbrace{(x^2 + 3x - 1)}_{d(x)} \cdot \underbrace{(x^2 - 3x + 10)}_{Q(x)} + \underbrace{(-33x - 71)}_{R(x)}.$$

The Remainder Theorem and Synthetic Division

The Remainder Theorem

If a number c is substituted for x in the polynomial $f(x)$, then the result $f(c)$ is the remainder that would be obtained by dividing $f(x)$ by $x - c$. That is, if $f(x) = (x - c) \cdot Q(x) + R$, then $f(c) = R$.

Proof: The equation $f(x) = d(x) \cdot Q(x) + R(x)$, where $d(x) = x - c$, is the basis of this proof. If we divide $f(x)$ by $x - c$, we obtain a quotient $Q(x)$ and a remainder $R(x)$ related as follows:

$$f(x) = (x - c) \cdot Q(x) + R(x).$$

The remainder $R(x)$ must either be 0 or have degree less than $x - c$. Thus, $R(x)$ must be a constant. Let's call this constant R. In the expression above, we get a true sentence whenever we replace x with any number. Let's replace x with c. We get

$$\begin{aligned} f(c) &= (c - c) \cdot Q(c) + R \\ &= 0 \cdot Q(c) + R \\ &= R. \end{aligned}$$

This tells us that the function value $f(c)$ is the remainder obtained when we divide $f(x)$ by $x - c$.

The remainder theorem motivates us to find a rapid way of dividing by $x - c$, in order to find function values. To streamline division, we can arrange the work so that duplicate and unnecessary writing is avoided. Consider the following.

A.
$$
\begin{array}{r}
4x^2 + 5x + 11 \\
x - 2\overline{)4x^3 - 3x^2 + x + 7} \\
\underline{4x^3 - 8x^2} \\
5x^2 + x \\
\underline{5x^2 - 10x} \\
11x + 7 \\
\underline{11x - 22} \\
29
\end{array}
$$

B.
$$
\begin{array}{r}
4 \quad 5 \quad 11 \\
1 - 2\overline{)4 - 3 + 1 + 7} \\
\underline{4 - 8} \\
5 + 1 \\
\underline{5 - 10} \\
11 + 7 \\
\underline{11 - 22} \\
29
\end{array}
$$

The division in (B) is the same as that in (A), but we wrote only the coefficients. The color numerals are duplicated, so we look for an arrangement in which they are not duplicated. We can also simplify things by using the opposite of -2 and then adding instead of subtracting. When things are thus "collapsed," we have the algorithm known as **synthetic division**.

C. *Synthetic Division*

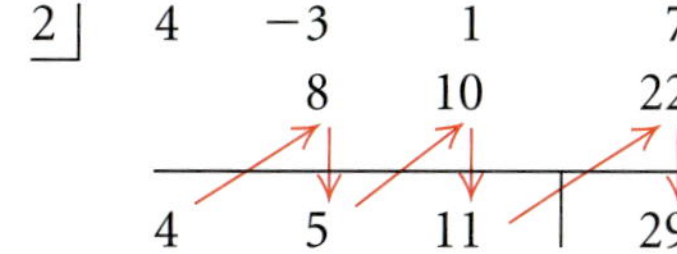

We "bring down" the 4. Then we multiply it by the 2 to get 8 and add to get 5. We then multiply 5 by 2 to get 10, add, and so on. The last number, 29, is the remainder. The others, 4, 5, and 11, are the coefficients of the quotient.

We write a 0 in the synthetic division for a missing term in the dividend.

Program

SYNDIV: This program does synthetic division.

Example 4 Use synthetic division to find the quotient and the remainder:

$$(2x^3 + 7x^2 - 5) \div (x + 3).$$

SOLUTION First we note that $x + 3 = x - (-3)$.

$$
\begin{array}{r}
-3 \,\big|\; 2 \quad 7 \quad 0 \quad -5 \\
\underline{-6 \quad -3 \quad 9} \\
2 \quad 1 \quad -3 \quad 4
\end{array}
$$

Note: We must write 0's for missing terms.

The quotient is $2x^2 + x - 3$. The remainder is 4.

Values of Polynomial Functions

We can now apply synthetic division to find polynomial function values.

Example 5 Given that $f(x) = 2x^5 - 3x^4 + x^3 - 2x^2 + x - 8$, find $f(10)$.

SOLUTION By the remainder theorem, $f(10)$ is the remainder when $f(x)$ is divided by $x - 10$. We use synthetic division to find that remainder.

$$\begin{array}{r|rrrrrr} 10 & 2 & -3 & 1 & -2 & 1 & -8 \\ & & 20 & 170 & 1710 & 17{,}080 & 170{,}810 \\ \hline & 2 & 17 & 171 & 1708 & 17{,}081 & 170{,}802 \end{array}$$

Thus, $f(10) = 170{,}802$.

Compare the computations in Example 5 with those in a direct substitution:

$$f(10) = 2(10)^5 - 3(10)^4 + (10)^3 - 2(10)^2 + 10 - 8.$$

The computations in synthetic division are less complicated.

Example 6   Determine whether -4 is a zero of $f(x)$, where $f(x) = x^3 + 8x^2 + 8x - 32$.

SOLUTION   We use synthetic division and the remainder theorem to find $f(-4)$.

$$\begin{array}{r|rrrr} -4 & 1 & 8 & 8 & -32 \\ & & -4 & -16 & 32 \\ \hline & 1 & 4 & -8 & 0 \end{array}$$

Since $f(-4) = 0$, the number -4 is a zero of $f(x)$.

Finding Factors of Polynomials

We now consider the following useful corollary of the remainder theorem.

> **The Factor Theorem**
>
> For a polynomial $f(x)$, if $f(c) = 0$, then $x - c$ is a factor of $f(x)$.

Proof: If we divide $f(x)$ by $x - c$, we obtain a quotient and a remainder, related as follows:

$$f(x) = (x - c) \cdot Q(x) + f(c).$$

Then if $f(c) = 0$, we have

$$f(x) = (x - c) \cdot Q(x),$$

so $x - c$ is a factor of $f(x)$.

The factor theorem is very useful in factoring polynomials, and hence in solving equations.

Example 7   Let $f(x) = x^3 + 2x^2 - 5x - 6$. Factor $f(x)$ and solve the equation $f(x) = 0$.

SOLUTION   We look for linear factors of the form $x - c$. Let's try $x - 1$. We use synthetic division to see whether $f(1) = 0$.

$$
\begin{array}{r|rrrr}
1 & 1 & 2 & -5 & -6 \\
  &   & 1 & 3 & -2 \\
\hline
  & 1 & 3 & -2 & \;-8
\end{array}
$$

Since $f(1) \neq 0$, we know that $x - 1$ is not a factor of $f(x)$. We try $x + 1$ or $x - (-1)$ in the form $x - c$.

$$
\begin{array}{r|rrrr}
-1 & 1 & 2 & -5 & -6 \\
   &   & -1 & -1 & 6 \\
\hline
   & 1 & 1 & -6 & \;0
\end{array}
$$

Since $f(-1) = 0$, we know that $x + 1$ is one factor and the quotient, $x^2 + x - 6$, is another. Thus,

$$f(x) = (x + 1)(x^2 + x - 6).$$

The trinomial is easily factored, so we have

$$f(x) = (x + 1)(x + 3)(x - 2).$$

Our goal is to solve the equation $f(x) = 0$. To do so, we use the principle of zero products. The solutions are -1, -3, and 2. Thus the solution set is $\{-1, -3, 2\}$. Check these solutions on a grapher.   ▬

2.4 Exercise Set

1. Determine whether 4, 5, and -2 are zeros of
$$f(x) = x^3 - 9x^2 + 14x + 24.$$

2. Determine whether 2, 3, and -1 are zeros of
$$f(x) = 2x^3 - 3x^2 + x - 1.$$

3. For $f(x)$ in Exercise 1, use long division to determine which of the following are factors of $f(x)$.

   a) $x - 4$ **b)** $x - 5$ **c)** $x + 2$

4. For $f(x)$ in Exercise 2, use long division to determine which of the following are factors of $f(x)$.

   a) $x - 2$ **b)** $x - 3$ **c)** $x + 1$

In each of the following, a polynomial $P(x)$ and a divisor $d(x)$ are given. Find the quotient $Q(x)$ and the remainder $R(x)$ when $P(x)$ is divided by $d(x)$, and express $P(x)$ in the form $d(x) \cdot Q(x) + R(x)$.

5. $P(x) = x^3 + 6x^2 - x - 30,$
   $d(x) = x - 2$

6. $P(x) = 2x^3 - 3x^2 + x - 1,$
   $d(x) = x - 2$

7. $P(x) = x^3 + 6x^2 - x - 30,$
   $d(x) = x - 3$

8. $P(x) = 2x^3 - 3x^2 + x - 1,$
   $d(x) = x - 3$

9. $P(x) = x^3 - 8,$
   $d(x) = x + 2$

10. $P(x) = x^3 + 27,$
   $d(x) = x + 1$

11. $P(x) = x^4 + 9x^2 + 20,$
   $d(x) = x^2 + 4$

12. $P(x) = x^4 + x^2 + 2,$
   $d(x) = x^2 + x + 1$

13. $P(x) = 5x^7 - 3x^4 + 2x^2 - 3,$
   $d(x) = 2x^2 - x + 1$

14. $P(x) = 6x^5 + 4x^4 - 3x^2 + x - 2,$
   $d(x) = 3x^2 + 2x - 1$

Use synthetic division to find the quotient and the remainder.

15. $(2x^4 + 7x^3 + x - 12) \div (x + 3)$

16. $(x^3 - 7x^2 + 13x + 3) \div (x - 2)$

17. $(x^3 - 2x^2 - 8) \div (x + 2)$

18. $(x^3 - 3x + 10) \div (x - 2)$

19. $(x^4 - 1) \div (x - 1)$

20. $(x^5 + 32) \div (x + 2)$

21. $(2x^4 + 3x^2 - 1) \div \left(x - \frac{1}{2}\right)$

22. $(3x^4 - 2x^2 + 2) \div \left(x - \frac{1}{4}\right)$

23. $(x^4 - y^4) \div (x - y)$

24. $(x^3 + 3ix^2 - 4ix - 2) \div (x + i)$

Use synthetic division to find the function values.

25. $f(x) = x^3 - 6x^2 + 11x - 6$; find $f(1)$, $f(-2)$, and $f(3)$.

26. $f(x) = x^3 + 7x^2 - 12x - 3$; find $f(-3)$, $f(-2)$, and $f(1)$.

27. $f(x) = 2x^5 - 3x^4 + 2x^3 - x + 8$; find $f(20)$ and $f(-3)$.

28. $f(x) = x^5 - 10x^4 + 20x^3 - 5x - 100$; find $f(-10)$ and $f(5)$.

29. $f(x) = x^4 - 16$; find $f(2)$, $f(-2)$, $f(3)$, and $f(1 - \sqrt{2})$.

30. $f(x) = x^5 + 32$; find $f(2)$, $f(-2)$, $f(3)$, and $f(2 + 3i)$.

Using synthetic division, determine whether the numbers are zeros of the polynomials.

31. $-3, 2$; $f(x) = 3x^3 + 5x^2 - 6x + 18$

32. $-4, 2$; $f(x) = 3x^3 + 11x^2 - 2x + 8$

33. $-3, \frac{1}{2}$; $f(x) = x^3 - \frac{7}{2}x^2 + x - \frac{3}{2}$

34. $i, -i, -2$; $f(x) = x^3 + 2x^2 + x + 2$

Factor the polynomial $f(x)$. Then solve the equation $f(x) = 0$.

35. $f(x) = x^3 + 4x^2 + x - 6$

36. $f(x) = x^3 + 5x^2 - 2x - 24$

37. $f(x) = x^3 - 6x^2 + 3x + 10$

38. $f(x) = x^3 + 2x^2 - 13x + 10$

39. $f(x) = x^3 - x^2 - 14x + 24$

40. $f(x) = x^3 - 3x^2 - 10x + 24$

41. $f(x) = x^4 - x^3 - 19x^2 + 49x - 30$

42. $f(x) = x^4 + 11x^3 + 41x^2 + 61x + 30$

Skill Maintenance

Find an equation for a line with the given slope that passes through the specified point.

43. $m = \frac{3}{4}$; $(-7, 0)$ **44.** $m = -\frac{2}{7}$; $(5, 0)$

Find an equation for a line passing through the given points.

45. $(-1, 5)$, $(6, 3)$ **46.** $(1, -2)$, $(4, 7)$

Synthesis

47. ◆ Is synthetic division always the fastest way to evaluate a polynomial function? What about using a grapher? Why or why not?

48. ◆ Can an nth-degree polynomial have more than n roots? Why or why not?

In Exercises 49 and 50, a graph of a polynomial function is given. On the basis of the graph:

a) *Find as many factors as you can of the polynomial.*
b) *Construct a polynomial function with the zeros shown in the graph.*
c) *Can you find any other polynomial functions with the given zeros?*
d) *Can you find any other polynomial functions with the given zeros and the same graph?*

49. $(-3, 0)$

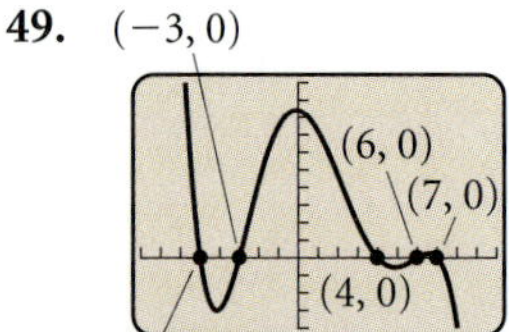

$(-5, 0)$

50. $(-3, 0)$

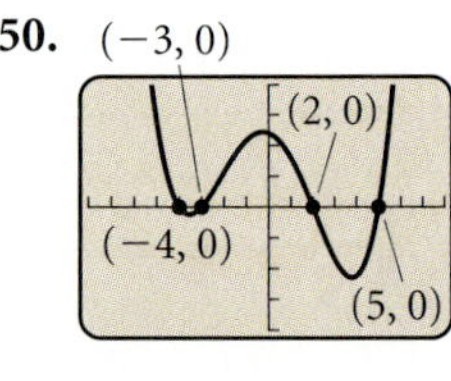

51. Find k so that $x + 2$ is a factor of $x^3 - kx^2 + 3x + 7k$.

52. For what values of k will the remainder be the same when $x^2 + kx + 4$ is divided by $x - 1$ or $x + 1$?

53. *Beam Deflection.*   A beam rests at two points A and B and has a concentrated load applied to its center. Let $y = $ the deflection, in feet, of the beam at a distance of x feet from A. Under certain conditions, this deflection is given by

$$y = \frac{1}{13}x^3 - \frac{1}{14}x.$$

Find the zeros of the polynomial in the interval $[0, 2]$.

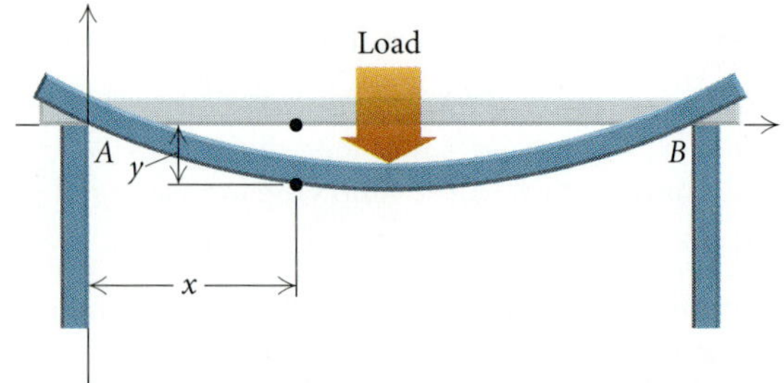

Solve.

54. $\dfrac{6x^2}{x^2 + 11} + \dfrac{60}{x^3 - 7x^2 + 11x - 77} = \dfrac{1}{x - 7}$

55. $\dfrac{2x^2}{x^2 - 1} + \dfrac{4}{x + 3} = \dfrac{32}{x^3 + 3x^2 - x - 3}$

56. Use a grapher to graph $f(x) = x^3 + kx^2 - 19x - 35$ for $k = 15, 19, 20,$ and 24. Compare the graphs. Do the zeros change when you change a coefficient?

57. Find a 15th-degree polynomial for which $x - 1$ is a factor. Answers may vary.

Use synthetic division to divide.

58. $(x^2 - 4x - 2) \div [x - (3 + 2i)]$

59. $(x^2 - 3x + 7) \div (x - i)$

2.5

Theorems about Zeros of Polynomial Functions

- *Factor polynomial functions and find the zeros and their multiplicities.*
- *Find a polynomial with specified zeros.*
- *For a polynomial function with integer coefficients, find the rational zeros and the other zeros, if possible.*
- *For a polynomial function with rational coefficients, find the rational zeros and the other zeros, if possible.*

The Fundamental Theorem of Algebra

A linear, or first-degree, polynomial function $f(x) = mx + b$ (where $m \neq 0$, of course) has just one zero, $-b/m$. It can be shown that any quadratic polynomial function with complex numbers for coefficients has at least one, and at most two, complex zeros. The following theorem is a generalization. No proof is given.

The Fundamental Theorem of Algebra

Every polynomial function of degree n, $n \geq 1$, with complex coefficients, has at least one zero in the system of complex numbers.

Note that although the fundamental theorem of algebra guarantees that a zero exists, it does not tell how to find it. Recall that the zeros of a polynomial function $f(x)$ are the solutions of the polynomial equation $f(x) = 0$. We now develop some concepts that can help in finding zeros. First, we consider a corollary of the fundamental theorem of algebra.

Every polynomial function f of degree n, where $n \geq 1$, with complex coefficients, can be factored into n linear factors (not necessarily unique); that is, $f(x) = a_n(x - c_1)(x - c_2) \cdots (x - c_n)$.

Finding Zeros of Factored Polynomial Functions

When a polynomial function is factored into a product of linear factors, it is easy to find the zeros by solving the equation $f(x) = 0$ using the principle of zero products.

Example 1   Find the zeros of

$$f(x) = (x - 3)(x + 4)(x + 1)(x - 1).$$

SOLUTION   To solve the equation $f(x) = 0$, we use the principle of zero products. The zeros of $f(x)$ are 3, -4, -1, and 1.   ▬

Example 2   Find the zeros of

$$f(x) = 5(x - 2)(x - 2)(x - 2)(x + 1).$$

SOLUTION   To solve the equation $f(x) = 0$, we use the principle of zero products. The zeros of $f(x)$ are 2 and -1.   ▬

In Example 2, the factor $x - 2$ occurs three times. In a case like this, we sometimes say that the zero we obtain from this factor, 2, has a **multiplicity** of 3. If we multiply out the right side, we obtain

$$f(x) = 5x^4 - 25x^3 + 30x^2 + 20x - 40.$$

Had we started with this form of the function, we might have had trouble finding the zeros. Some polynomials, however, can be factored using techniques we already know, such as factoring by grouping.

Example 3   Find the zeros of

$$f(x) = x^3 - 2x^2 - 9x + 18.$$

SOLUTION   We factor by grouping, as follows:

$$
\begin{aligned}
f(x) &= x^3 - 2x^2 - 9x + 18 \\
&= x^2(x - 2) - 9(x - 2) \\
&= (x^2 - 9)(x - 2) \\
&= (x + 3)(x - 3)(x - 2).
\end{aligned}
$$

Then by the principle of zero products, the solutions of the equation $f(x) = 0$ are -3, 3, and 2. These are the zeros of $f(x)$.   ▬

Other factoring techniques can also be used.

Example 4   Find the zeros of

$$f(x) = x^4 + 4x^2 - 45.$$

SOLUTION   We factor as follows:

$$
\begin{aligned}
f(x) &= x^4 + 4x^2 - 45 \\
&= (x^2 - 5)(x^2 + 9) \\
&= (x - \sqrt{5})(x + \sqrt{5})(x - 3i)(x + 3i).
\end{aligned}
$$

Then by the principle of zero products, the solutions of the equation $f(x) = 0$ are $\pm\sqrt{5}$ and $\pm 3i$. These are the zeros of $f(x)$. —

If we tried to use a grapher to find the zeros of f in Example 4, the nonreal solutions $3i$ and $-3i$ could not be seen.

> Every polynomial of degree n, where $n \geq 1$, has at least one zero and at most n zeros.

This is often stated as follows: "Every polynomial of degree n, where $n \geq 1$, has *exactly* n zeros." This statement is not incompatible with the preceding statement, as it would first seem, because one must take multiplicities into account.

Finding Polynomials with Given Zeros

Interactive Discovery

Graph each of the following:

$$y_1 = (x + 2)(x - 1)\left(x - \tfrac{5}{2}\right),$$
$$y_2 = 2(x + 2)(x - 1)\left(x - \tfrac{5}{2}\right),$$
$$y_3 = -\tfrac{1}{2}(x + 2)(x - 1)\left(x - \tfrac{5}{2}\right).$$

How do the graphs compare? How do the zeros compare?

Given several numbers, we can find a polynomial with those numbers as its zeros.

Example 5   Find a polynomial function of degree 3, having the zeros -2, 1, and $3i$.

SOLUTION   Such a polynomial has factors $x + 2$, $x - 1$, and $x - 3i$, so we have

$$f(x) = a_n(x + 2)(x - 1)(x - 3i).$$

The number a_n can be any nonzero number. The simplest polynomial will be obtained if we let it be 1. If we then multiply the factors, we obtain

$$f(x) = x^3 + (1 - 3i)x^2 + (-2 - 3i)x + 6i.$$ —

Example 6   Find a polynomial function of degree 5 with -1 as a zero of multiplicity 3, 4 as a zero of multiplicity 1, and 0 as a zero of multiplicity 1.

SOLUTION   Proceeding as in Example 5, letting $a_n = 1$, we obtain

$$f(x) = (x + 1)^3(x - 4)(x - 0)$$
$$= x^5 - x^4 - 9x^3 - 11x^2 - 4x.$$ —

Zeros of Polynomial Functions with Real Coefficients

Consider the quadratic equation $x^2 - 2x + 2 = 0$, with real coefficients. Its solutions are $1 + i$ and $1 - i$. Note that they are complex conjugates. This generalizes to any polynomial with real coefficients.

> If a complex number $a + bi$, $b \neq 0$, is a zero of a polynomial function $f(x)$ with real coefficients, then its conjugate, $a - bi$, is also a zero. (Nonreal zeros occur in conjugate pairs.)

For the preceding to be true, it is essential that the coefficients be real numbers. We see this in Example 5, where the root $3i$ occurs, but its conjugate does not. This occurs because some of the coefficients of the polynomial are not real.

Rational Coefficients

When a polynomial has rational numbers for coefficients, certain irrational zeros also occur in pairs, as described in the following theorem.

> Suppose that $f(x)$ is a polynomial with rational coefficients. Then if either of the following is a zero, so is the other: $a + c\sqrt{b}$, $a - c\sqrt{b}$, a and c rational, b not a square.

Example 7   Suppose that a polynomial function of degree 6 with rational coefficients has $-2 + 5i$, $-2i$, and $1 - \sqrt{3}$ as some of its zeros. Find the other zeros.

SOLUTION   The other zeros are $-2 - 5i$, $2i$, and $1 + \sqrt{3}$. There are no other zeros since the degree is 6.

Example 8   Find a polynomial function of lowest degree with rational coefficients that has $1 - \sqrt{2}$ and $1 + 2i$ as some of its zeros.

SOLUTION   The function must also have the zeros $1 + \sqrt{2}$ and $1 - 2i$. Thus the polynomial function is

$$f(x) = [x - (1 - \sqrt{2})][x - (1 + \sqrt{2})][x - (1 + 2i)][x - (1 - 2i)]$$
$$= (x^2 - 2x - 1)(x^2 - 2x + 5)$$
$$= x^4 - 4x^3 + 8x^2 - 8x - 5.$$

Integer Coefficients and the Rational Zeros Theorem

It is not always easy to find the zeros of a polynomial function. However, if a polynomial function has integer coefficients, there is a procedure that will yield all the rational zeros.

Example 9   Let $f(x) = x^4 - 5x^3 + 10x^2 - 20x + 24$. Find the other zeros of $f(x)$, given that $2i$ is a zero.

SOLUTION   Since $2i$ is a zero, we know that $-2i$ is also a zero. Thus,

$$f(x) = (x - 2i)(x + 2i) \cdot Q(x)$$

for some $Q(x)$. Since $(x - 2i)(x + 2i) = x^2 + 4$, we know that

$$f(x) = (x^2 + 4) \cdot Q(x).$$

Using division, we find that $Q(x) = x^2 - 5x + 6$, and since we can factor $x^2 - 5x + 6$, we get

$$f(x) = (x^2 + 4)(x - 2)(x - 3).$$

Thus the other zeros are $-2i$, 2, and 3.

The Rational Zeros Theorem

Let

$$P(x) = a_n x^n + a_{n-1} x^{n-1} + \cdots + a_1 x + a_0,$$

where all the coefficients are integers. Consider a rational number denoted by p/q, where p and q are relatively prime (having no common factor besides -1 and 1). If p/q is a zero of $P(x)$, then p is a factor of a_0 and q is a factor of a_n.

Example 10   Given $f(x) = 2x^5 - x^4 - 4x^3 + 2x^2 - 30x + 15$:

a) Find the rational zeros, and then the other zeros.

b) Solve the equation $f(x) = 0$.

c) Factor $f(x)$ into linear factors.

SOLUTION

a) According to the rational zeros theorem, any rational zero of f must be of the form p/q, where p is a factor of 15 and q is a factor of 2. The possibilities are

$$\frac{\text{Possibilities for } p}{\text{Possibilities for } q} : \frac{\pm 1, \pm 3, \pm 5, \pm 15}{\pm 1, \pm 2};$$

$$\text{Possibilities for } p/q: \ 1, -1, 3, -3, 5, -5, 15, -15, \tfrac{1}{2}, -\tfrac{1}{2}, \tfrac{3}{2}, -\tfrac{3}{2},$$
$$\tfrac{5}{2}, -\tfrac{5}{2}, \tfrac{15}{2}, -\tfrac{15}{2}.$$

Rather than use synthetic division to check each of these possibilities, we graph $y_1 = 2x^5 - x^4 - 4x^3 + 2x^2 - 30x + 15$ (see the figure at left). We can then inspect the graph for zeros that appear to be near any of the possible rational zeros.

From the graph, we see that of the possibilities in the list, only the numbers $-\tfrac{5}{2}$, $\tfrac{1}{2}$, and $\tfrac{5}{2}$ might be rational zeros. By synthetic division, we see that only $\tfrac{1}{2}$ is actually a rational zero.

$$\frac{1}{2} \ \big|\ \begin{array}{rrrrrr} 2 & -1 & -4 & 2 & -30 & 15 \\ & 1 & 0 & -2 & 0 & -15 \\ \hline 2 & 0 & -4 & 0 & -30 & 0 \end{array}$$

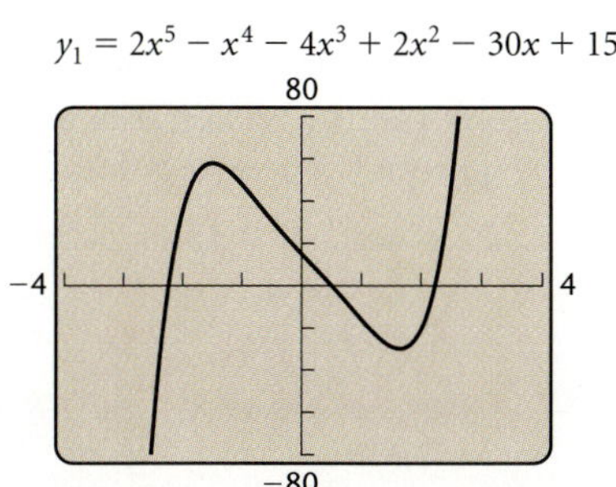

This means that $x - \frac{1}{2}$ is a factor of $f(x)$. To find the other zeros, we write the factorization and try to factor further:

$$\begin{aligned}
f(x) &= \left(x - \tfrac{1}{2}\right)(2x^4 - 4x^2 - 30) \\
&= \left(x - \tfrac{1}{2}\right) \cdot 2 \cdot (x^4 - 2x^2 - 15) \qquad \text{Factoring out the 2} \\
&= \left(x - \tfrac{1}{2}\right) \cdot 2 \cdot (x^2 - 5)(x^2 + 3). \qquad \text{Factoring the trinomial}
\end{aligned}$$

We now use the principle of zero products to determine the zeros:

$$\begin{aligned}
x - \tfrac{1}{2} = 0 \quad &or \quad x^2 - 5 = 0 \quad &&or \quad x^2 + 3 = 0 \\
x = \tfrac{1}{2} \quad &or \quad x^2 = 5 \quad &&or \quad x^2 = -3 \\
x = \tfrac{1}{2} \quad &or \quad x = \pm\sqrt{5} \quad &&or \quad x = \pm\sqrt{3}\,i.
\end{aligned}$$

There is only one rational zero, $\frac{1}{2}$. The other zeros are $\pm\sqrt{5}$ and $\pm\sqrt{3}i$.

b) The solutions of $f(x) = 0$ are $\frac{1}{2}$, $\pm\sqrt{5}$, and $\pm\sqrt{3}\,i$.

c) The factorization into linear factors is

$$f(x) = 2\left(x - \tfrac{1}{2}\right)(x + \sqrt{5})(x - \sqrt{5})(x + \sqrt{3}\,i)(x - \sqrt{3}\,i).$$

Example 11   Given $f(x) = 3x^4 - 11x^3 + 10x - 4$:

a) Find the rational zeros, and then the other zeros.

b) Solve the equation $f(x) = 0$.

c) Factor $f(x)$ into linear factors.

SOLUTION

a) Because the degree of $f(x)$ is 4, there are at most 4 distinct zeros. Although a grapher could be useful, remember that it can only *approximate* real solutions that are not integers.

The rational zeros theorem says that if p/q is a root of $f(x)$, then p must be a factor of -4 and q must be a factor of 3. Thus the possibilities for p/q are

$$\frac{\text{Possibilities for } p}{\text{Possibilities for } q} : \quad \frac{\pm 1, \pm 2, \pm 4}{\pm 1, \pm 3}.$$

If we list all possible quotients, we find the following possibilities for

$$p/q: \quad 1,\ -1,\ 2,\ -2,\ 4,\ -4,\ \tfrac{1}{3},\ -\tfrac{1}{3},\ \tfrac{2}{3},\ -\tfrac{2}{3},\ \tfrac{4}{3},\ -\tfrac{4}{3}.$$

We could substitute each of these possibilities to see if any are zeros, that is, if $f(c) = 0$. We could also use a TABLE feature or some other way to find function values. But, if we use synthetic division, the quotient polynomial becomes a beneficial by-product if a zero is found. We try 1.

$$
\begin{array}{r|rrrrr}
1 & 3 & -11 & 0 & 10 & -4 \\
  &   & 3 & -8 & -8 & 2 \\
\hline
  & 3 & -8 & -8 & 2 & -2
\end{array}
$$

We try -1.

$$\begin{array}{r|rrrrr} -1 & 3 & -11 & 0 & 10 & -4 \\ & & -3 & 14 & -14 & 4 \\ \hline & 3 & -14 & 14 & -4 & 0 \end{array}$$

Since $f(1) = -2$, 1 is not a zero; but $f(-1) = 0$, so -1 is a zero. Using the results of the synthetic division, we can express $f(x)$ as follows:

$$f(x) = (x + 1)(3x^3 - 14x^2 + 14x - 4).$$

We now use $3x^3 - 14x^2 + 14x - 4$ and check the other possible zeros. We use synthetic division again to see whether -1 is a double zero.

$$\begin{array}{r|rrrr} -1 & 3 & -14 & 14 & -4 \\ & & -3 & 17 & -31 \\ \hline & 3 & -17 & 31 & -35 \end{array}$$

It is not. The student can check that none of the remaining integers $(2, -2, 4,$ or $-4)$ are zeros. Let's now try $\frac{2}{3}$.

$$\begin{array}{r|rrrr} 2/3 & 3 & -14 & 14 & -4 \\ & & 2 & -8 & 4 \\ \hline & 3 & -12 & 6 & 0 \end{array}$$

Since $f\left(\frac{2}{3}\right) = 0$, $\frac{2}{3}$ is also a zero.

Using the results of synthetic division, we can factor further:

$$f(x) = (x + 1)\left(x - \tfrac{2}{3}\right)(3x^2 - 12x + 6) \qquad \text{Using the results of the last synthetic division}$$

$$= (x + 1)\left(x - \tfrac{2}{3}\right) \cdot 3 \cdot (x^2 - 4x + 2). \qquad \text{Removing a factor of 3}$$

The quadratic formula can be used to find the zeros of $x^2 - 4x + 2$:

$$x = \frac{-b \pm \sqrt{b^2 - 4ac}}{2a}$$

$$= \frac{-(-4) \pm \sqrt{(-4)^2 - 4 \cdot 1 \cdot 2}}{2 \cdot 1}$$

$$= \frac{4 \pm \sqrt{8}}{2}$$

$$= 2 \pm \sqrt{2}.$$

The rational zeros are -1 and $\frac{2}{3}$. The other zeros are $2 \pm \sqrt{2}$.

b) The solutions of $f(x) = 0$ are $-1, \frac{2}{3},$ and $2 \pm \sqrt{2}$.

c) The complete factorization of $f(x)$ is

$$f(x) = (x + 1)(3x - 2)[x - (2 - \sqrt{2})][x - (2 + \sqrt{2})]$$

$$= (x + 1)(3x - 2)(x - 2 + \sqrt{2})(x - 2 - \sqrt{2}).$$

The Intermediate Value Theorem

Some functions have graphs that are curves with no breaks or holes. Such functions are called **continuous functions**. Loosely speaking, the graph of a continuous function can be drawn without lifting the pencil from the paper. The function f, sketched as follows, is continuous, but g and h are not.

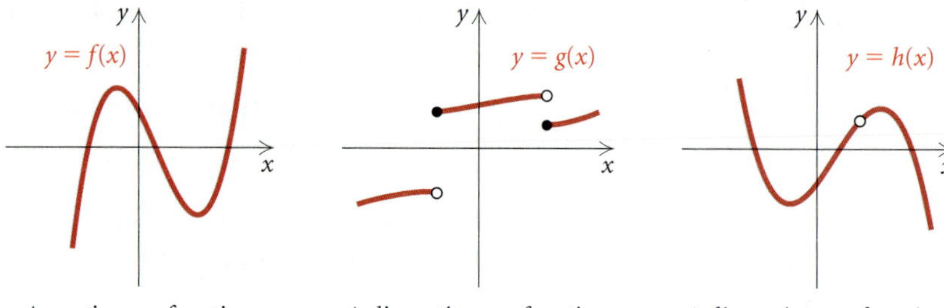

A continuous function   A discontinuous function   A discontinuous function

Interactive Discovery

Consider the following polynomial function, with the two given inputs.

$$P(x) = 12x^3 - 5x^2 - 11x + 6; \quad a = 0.7, b = 0.8$$

Graph this function using the viewing windows $[-1, 2, -2, 10]$ and then $[0.6, 0.92, -0.1, 0.15]$. Does there seem to be a zero between 0.7 and 0.8? Find $P(0.7)$ and $P(0.8)$. Do these function values have opposite signs?

Polynomial functions P are continuous, hence their graphs are unbroken. All polynomial functions have $\mathbb{R}$ as the domain. Suppose we have two function values $P(a)$ and $P(b)$ that have opposite signs. Since P is continuous, its graph must be a curve from $(a, P(a))$ to $(b, P(b))$ without interruption. It follows that the line must cross the x-axis somewhere between a and b.

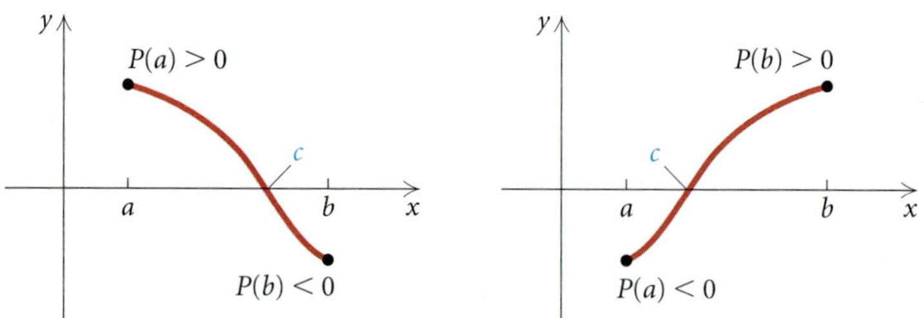

The Intermediate Value Theorem

For any polynomial function $P(x)$ with real coefficients, suppose that for $a \neq b$, $P(a)$ and $P(b)$ are of opposite signs. Then the function has a real zero between a and b.

2.5 | *Exercise Set*

Find the zeros of the polynomial function and state the multiplicity of each.

1. $f(x) = (x + 3)^2(x - 1)$

2. $f(x) = -8(x - 3)^2(x + 4)^3 x^4$

3. $f(x) = x^3(x - 1)^2(x + 4)$

4. $f(x) = (x^2 - 5x + 6)^2$

5. $f(x) = x^4 - 4x^2 + 3$

6. $f(x) = x^4 - 10x^2 + 9$

7. $f(x) = x^3 + 3x^2 - x - 3$

8. $f(x) = x^3 - x^2 - 2x + 2$

Find a polynomial function of degree 3 with the given numbers as zeros.

9. $-2, 3, 5$ **10.** $2, i, -i$

11. $-3, 2i, -2i$ **12.** $1 + 4i, 1 - 4i, -1$

13. $\sqrt{2}, -\sqrt{2}, \sqrt{3}$

14. Find a polynomial function of degree 4 with -2 as a zero of multiplicity 1, 3 as a zero of multiplicity 2, and -1 as a zero of multiplicity 1.

Suppose that a polynomial function of degree 5 with rational coefficients has the given numbers as zeros. Find the other zeros.

15. $6, -3 + 4i, 4 - \sqrt{5}$ **16.** $-2, 3, 4, 1 - i$

Find a polynomial function of lowest degree with rational coefficients that has the given numbers as some of its zeros.

17. $1 + i, 2$ **18.** $2 - i, -1$

19. $-4i, 5$ **20.** $2 - \sqrt{3}, 1 + i$

21. $\sqrt{5}, -3i$ **22.** $-\sqrt{2}, 4i$

Given that the polynomial function has the given zero, find the other zeros.

23. $f(x) = x^4 - 5x^3 + 7x^2 - 5x + 6; \quad -i$

24. $f(x) = x^4 - 16; \quad 2i$

25. $f(x) = x^3 - 6x^2 + 13x - 20; \quad 4$

26. $f(x) = x^3 - 8; \quad 2$

List all possible rational zeros.

27. $f(x) = x^5 - 3x^2 + 1$

28. $f(x) = x^7 + 37x^5 - 6x^2 + 12$

29. $f(x) = 15x^6 + 47x^2 + 2$

30. $f(x) = 10x^{25} + 3x^{17} - 35x + 6$

For each polynomial function:

a) *Find the rational zeros, and then the other zeros.*
b) *Solve the equation $f(x) = 0$.*
c) *Factor $f(x)$ into linear factors.*

31. $f(x) = x^3 + 3x^2 - 2x - 6$

32. $f(x) = x^3 - x^2 - 3x + 3$

33. $f(x) = x^3 - 3x + 2$

34. $f(x) = x^3 - 2x + 4$

35. $f(x) = x^3 - 5x^2 + 11x + 17$

36. $f(x) = 2x^3 + 7x^2 + 2x - 8$

37. $f(x) = 5x^4 - 4x^3 + 19x^2 - 16x - 4$

38. $f(x) = 3x^4 - 4x^3 + x^2 + 6x - 2$

39. $f(x) = x^4 - 3x^3 - 20x^2 - 24x - 8$

40. $f(x) = x^4 + 5x^3 - 27x^2 + 31x - 10$

41. $f(x) = x^3 - 4x^2 + 2x + 4$

42. $f(x) = x^3 - 8x^2 + 17x - 4$

43. $f(x) = x^3 + 8$

44. $f(x) = x^3 - 8$

45. $f(x) = \frac{1}{3}x^3 - \frac{1}{2}x^2 - \frac{1}{6}x + \frac{1}{6}$

46. $f(x) = \frac{2}{3}x^3 - \frac{1}{2}x^2 + \frac{2}{3}x - \frac{1}{2}$

Find only the rational zeros.

47. $f(x) = x^4 + 32$

48. $f(x) = x^6 + 8$

49. $f(x) = x^3 - x^2 - 4x + 3$

50. $f(x) = 2x^3 + 3x^2 + 2x + 3$

51. $f(x) = x^4 + 2x^3 + 2x^2 - 4x - 8$

52. $f(x) = x^4 + 6x^3 + 17x^2 + 36x + 66$

53. $f(x) = x^5 - 5x^4 + 5x^3 + 15x^2 - 36x + 20$

54. $f(x) = x^5 - 3x^4 - 3x^3 + 9x^2 - 4x + 12$

Show that the function f has a zero between a and b using the intermediate value theorem.

55. $f(x) = x^3 + 3x^2 - 9x - 13; \ a = -5, b = -4$

56. $f(x) = x^3 + 3x^2 - 9x - 13; \ a = 2, b = 3$

Skill Maintenance

Solve.

57. $\dfrac{x^2 - 4}{x^2 - 5x + 6} = 0$

58. $\dfrac{2}{x - 2} + \dfrac{1}{x + 2} = \dfrac{3}{x^2 - 4}$

Simplify.

59. $\dfrac{\dfrac{1}{x} + \dfrac{3}{5x}}{\dfrac{2}{x} - \dfrac{4}{15x}}$

60. $\dfrac{\dfrac{5}{x^2} - \dfrac{7}{3x}}{\dfrac{3}{2x^2} + \dfrac{5}{x}}$

Synthesis

61. ◈ Is it possible for a third-degree polynomial with rational coefficients to have no real zeros? Why or why not?

62. ◈ If $Q(x) = -P(x)$, do $P(x)$ and $Q(x)$ have the same zeros? Why or why not?

63. Consider $f(x) = 2x^3 - 5x^2 - 4x + 3$. Find the solutions of each equation.

a) $f(x) = 0$

b) $f(x - 1) = 0$

c) $f(x + 2) = 0$

d) $f(2x) = 0$

64. Use a grapher to find the points of intersection of the graphs of

$$y_1 = 3x^5 + 3x^4 + 54x^3 - 54x^2 + 243x + 243$$
$$\text{and} \quad y_2 = (x + 3)^5.$$

65. Use the rational zeros theorem and the equation $x^2 - 5 = 0$ to show that $\sqrt{5}$ is not a rational number.

66. Use the rational zeros theorem and the equation $x^4 - 12 = 0$ to show that $\sqrt[4]{12}$ is irrational.

Find the rational zeros.

67. $P(x) = 2x^5 - 33x^4 - 84x^3 + 2203x^2 - 3348x - 10{,}080$

68. $P(x) = x^6 - 6x^5 - 72x^4 - 81x^2 + 486x + 5832$

2.6 Rational Functions

• *Graph a rational function, identifying all asymptotes.*
• *Solve applications involving rational functions.*

The sum, difference, or product of two polynomials is a polynomial. Now we turn our attention to functions that represent the quotient of two polynomials. In general, the quotient of two polynomials is *not* itself a polynomial.

The Domain of a Rational Function

Here are some examples of rational functions:

$$f(x) = \frac{1}{x - 3}, \qquad f(x) = \frac{8x^3 - 5x^2 + x - 1}{x^2 - 4},$$

$$f(x) = \frac{2x + 5}{3x^2 - 4x + 7}.$$

Rational Function

A *rational function* is a function f that is a quotient of two polynomials; that is,

$$f(x) = \frac{P(x)}{Q(x)},$$

where $P(x)$ and $Q(x)$ are polynomials with no common factor other than -1 and 1 and where $Q(x)$ is not the zero polynomial. The domain of f consists of all inputs x for which $Q(x) \neq 0$.

Example 1 Consider

$$f(x) = \frac{1}{x-3}.$$

Find the domain and graph f.

SOLUTION   The input 3 results in a denominator of 0. Thus the domain is

$$\{x \mid x \neq 3\}, \text{ or } (-\infty, 3) \cup (3, \infty).$$

The graph of this function is the graph of $y = 1/x$ translated right 3 units. Two versions of the graph are shown below.

CONNECTED **Mode**

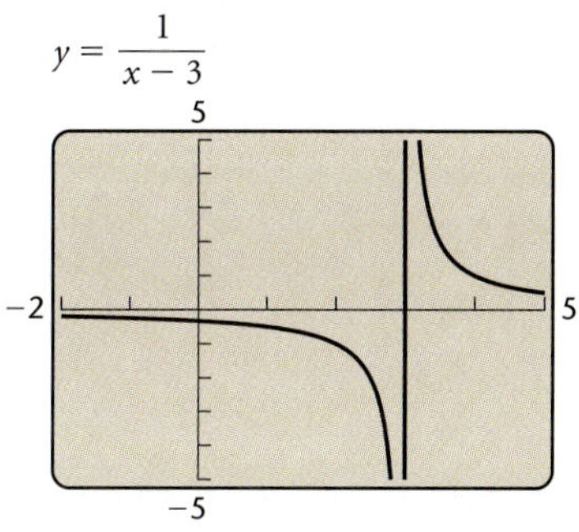

DOT **Mode**

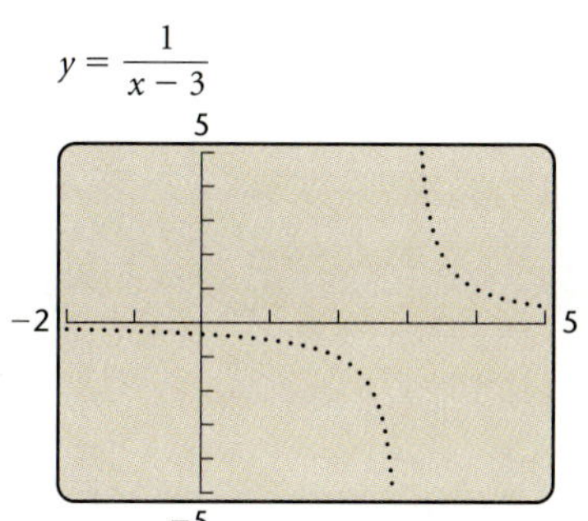

In the Introduction to Graphs and Graphers, we discussed a disadvantage of the grapher that can occur when graphing rational functions. CONNECTED mode can lead to an incorrect graph like the one on the left. Because graphs in CONNECTED mode connect plotted points with line segments, it appears as though the vertical line $x = 3$ is part of the graph. We will see later in this section that a line like $x = 3$, though important in the construction of the graph, is not part of the graph. Graphs in DOT mode, like the one on the right above, simply plot dots representing coordinates of points. If you have a choice when graphing rational functions, use DOT mode.   ▬

Because of the possible deficiencies of graphers when graphing rational functions and because we want you to understand the graphs, we will emphasize the creation of hand-drawn graphs in this section.

Asymptotes

Referring to Example 1, let's explore what happens as x becomes very large positively without bound and becomes very small negatively without bound.

Interactive Discovery

Consider $y = 1/(x - 3)$. Complete the following tables and look for a pattern. What happens to the y-values as the x-values become larger without bound? become smaller without bound? Then do the same for smaller and smaller x-values.

X	Y1		X	Y1
100	.01031		−100	−.0097
200			−400	
1000			−5000	−2E−4
3000			−8000	
10000			−20000	
400000			−6E5	
X = 100			X = −100	

We see from the table on the left that "as x goes to infinity, y goes to 0." We write this as $y \rightarrow 0$ as $x \rightarrow \infty$. Using the table on the right, complete:

$$y \rightarrow \boxed{} \text{ as } x \rightarrow -\infty.$$

We see that

$$\frac{1}{x - 3} \rightarrow 0 \text{ as } x \rightarrow \infty \quad \text{and} \quad \frac{1}{x - 3} \rightarrow 0 \text{ as } x \rightarrow -\infty.$$

From these facts, we say that the curve approaches the x-axis *asymptotically* and that the x-axis is a **horizontal asymptote** for the curve.

Horizontal Asymptote

The line $y = b$ is a *horizontal asymptote* for the graph of f if either or both of the following are true:

$$f(x) \rightarrow b \text{ as } x \rightarrow \infty \quad \text{or} \quad f(x) \rightarrow b \text{ as } x \rightarrow -\infty.$$

The following figures illustrate two ways in which horizontal asymptotes can occur. In each case, the curve gets close to the line $y = b$ either as $x \rightarrow \infty$ or as $x \rightarrow -\infty$. Keep in mind that the symbols ∞ and $-\infty$ convey the idea of increasing positively without bound and decreasing negatively without bound, respectively.

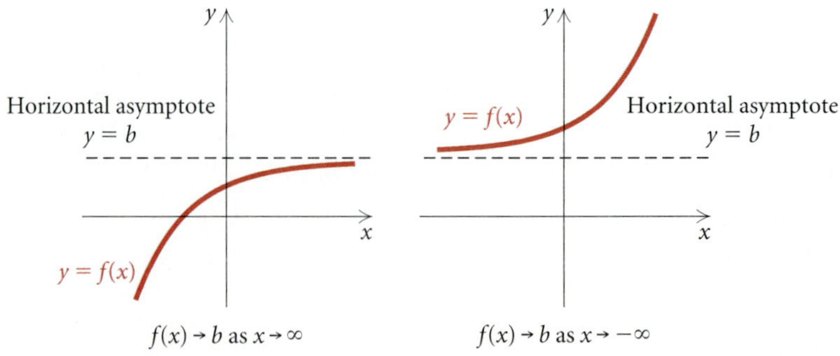

Look again at the graph in Example 1. Let's explore what happens as x-values get closer and closer to 3 from the left. We then explore what happens as x-values get closer and closer to 3 from the right.

Interactive Discovery

Consider $y = 1/(x - 3)$. Using a TABLE feature, complete the following input–output tables and look for a pattern. Trace along the curve from left to right toward an x-value of 3. Then trace from right to left toward 3. What happens to y in each case?

X	Y1			X	Y1	
2.5	−2			**3.6**	1.6667	
2.78				3.1		
2.93				3.01		
2.999				3.001		
2.9999				3.0001		
X = 2.5				X = 3.6		

Complete:

$$y \rightarrow \boxed{} \text{ as } x \rightarrow 3 \text{ from the left;}$$

$$y \rightarrow \boxed{} \text{ as } x \rightarrow 3 \text{ from the right.}$$

We see that as x-values get closer and closer to 3 from the left, the function values (y-values) decrease negatively without bound. In fact, by selecting x close enough to 3, we can get the function values as small negatively as we wish. Similarly, as the x-values approach 3 from the right, the function values increase positively without bound. We write this as

$$f(x) \rightarrow -\infty \text{ as } x \rightarrow 3^- \quad \text{and} \quad f(x) \rightarrow \infty \text{ as } x \rightarrow 3^+.$$

We read "$f(x) \rightarrow -\infty$ as $x \rightarrow 3^-$" as "$f(x)$ decreases negatively without bound as x approaches 3 from the left." We read "$f(x) \rightarrow \infty$ as $x \rightarrow 3^+$" as "$f(x)$ increases positively without bound as x approaches 3 from the right." The vertical line $x = 3$ is said to be a **vertical asymptote** to this curve.

Vertical Asymptote

The line $x = a$ is a *vertical asymptote* for the graph of f if any of the following is true:

$$f(x) \rightarrow \infty \text{ as } x \rightarrow a^- \quad \text{or} \quad f(x) \rightarrow -\infty \text{ as } x \rightarrow a^-, \quad \text{or}$$
$$f(x) \rightarrow \infty \text{ as } x \rightarrow a^+ \quad \text{or} \quad f(x) \rightarrow -\infty \text{ as } x \rightarrow a^+.$$

The following figures show the four ways in which a vertical asymptote can occur.

$f(x) \to \infty$ as $x \to a^-$ $f(x) \to -\infty$ as $x \to a^-$ $f(x) \to \infty$ as $x \to a^+$ $f(x) \to -\infty$ as $x \to a^+$

Determining Vertical Asymptotes

If a is a zero of the denominator of a rational function, then the line $x = a$ is a vertical asymptote.

Example 2   Determine the vertical asymptotes of each of the following functions.

a) $f(x) = \dfrac{3x - 2}{x^3 - 2x^2 - 15x}$

b) $g(x) = \dfrac{x - 2}{x^3 - 5x}$

SOLUTION

a) We factor to find the zeros of the denominator.

$$x^3 - 2x^2 - 15x = x(x - 5)(x + 3)$$

The zeros of the denominator are 0, 5, and -3. Thus the vertical asymptotes are the lines $x = 0$, $x = 5$, and $x = -3$.

b) We factor to find the zeros of the denominator.

$$x^3 - 5x = x(x^2 - 5) = x(x - \sqrt{5})(x + \sqrt{5})$$

The zeros of the denominator are 0, $\sqrt{5}$, and $-\sqrt{5}$. Thus the vertical asymptotes are the lines $x = 0$, $x = \sqrt{5}$, and $x = -\sqrt{5}$.

How can we determine a horizontal asymptote? As x gets very large or very small, the value of the polynomial function $P(x)$ is dominated by the function's leading term. Because of this, if $P(x)$ and $Q(x)$ have the *same* degree, the value of $P(x)/Q(x)$ as $|x| \longrightarrow \infty$ is dominated by the ratio of the numerator's leading coefficient to the denominator's leading coefficient. Let's explore this further.

Interactive Discovery

Consider

$$f(x) = \frac{3x^2 + 2x - 4}{2x^2 - x + 1}.$$

Let $y_1 = 3x^2 + 2x - 4$, $y_2 = 2x^2 - x + 1$, and $y_3 = y_1/y_2$. Using a TABLE feature, complete the following input–output tables and look for a pattern. Check the pattern with Example 2. Trace along the curve for larger and larger values of x. Then trace for smaller and smaller negative values. What happens to y in each case?

X	Y₁	Y₂	Y₃
10	316	191	1.6545
100			
1000			
10000			

X = 10

X	Y₁	Y₂	Y₃
−10	276	211	1.3081
−100			
−1000			
−10000			

X = −10

Complete:

$$y_3 \to \boxed{} \text{ as } x \to \infty; \qquad y_3 \to \boxed{} \text{ as } x \to -\infty.$$

It follows that when the numerator and the denominator of a rational function have the same degree, the line $y = a/b$ is the horizontal asymptote, where a and b are the leading coefficients of the numerator and the denominator, respectively.

Example 3 Find the horizontal asymptote: $f(x) = \dfrac{-7x^4 - 10x^2 + 1}{11x^4 + x - 2}$.

SOLUTION The numerator and the denominator have the same degree, so the line $y = -\frac{7}{11}$ is a horizontal asymptote.　　　　　　　▬

To check Example 3, we could examine a table of values or the graph. A third check, one that is useful in calculus, is to multiply by 1, using $(1/x^4)/(1/x^4)$:

$$f(x) = \frac{-7x^4 - 10x^2 + 1}{11x^4 + x - 2} \cdot \frac{\dfrac{1}{x^4}}{\dfrac{1}{x^4}}$$

$$= \frac{-7 - \dfrac{10}{x^2} + \dfrac{1}{x^4}}{11 + \dfrac{1}{x^3} - \dfrac{2}{x^4}}.$$

As $|x|$ becomes very large, each expression with x in the denominator tends toward zero. Specifically, as $|x| \to \infty$, we have

$$f(x) \to \frac{-7 - 0 + 0}{11 + 0 - 0}, \quad \text{or} \quad f(x) \to -\frac{7}{11}.$$

The horizontal asymptote is $y = -\frac{7}{11}$.

Example 4 Find the horizontal asymptote: $f(x) = \dfrac{2x + 3}{x^3 - 2x^2 + 4}$.

SOLUTION We let $y_1 = 2x + 3$, $y_2 = x^3 - 2x^2 + 4$, and $y_3 = y_1/y_2$, and form a table of values as follows.

X	Y₁	Y₂	Y₃
10	23	804	.02861
100	203	980004	2.1E−4
1000	2003	9.98E8	2E−6
10000	20003	1E12	2E−8

X = 10

X	Y₁	Y₂	Y₃
−10	−17	−1196	.01421
−100	−197	−1E6	1.9E−4
−1000	−1997	−1E9	2E−6
−10000	−19997	−1E12	2E−8

X = −10

Note that as x grows large, the value of y_2 grows much faster than the value of y_1. Because of this, the ratio of y_1/y_2 shrinks toward 0. As $x \longrightarrow -\infty$, the ratio y_1/y_2 behaves in a similar manner. The horizontal asymptote is $y = 0$, that is, the x-axis.

The following statements describe the two ways in which a horizontal asymptote occurs.

Determining a Horizontal Asymptote

When the numerator and the denominator of a rational function have the same degree, the line $y = a/b$ is a horizontal asymptote, where a and b are the leading coefficients of the numerator and the denominator, respectively.

When the degree of the numerator of a rational function is less than the degree of the denominator, the x-axis, or $y = 0$, is a horizontal asymptote.

Remember that the preceding statements on finding horizontal and vertical asymptotes apply to rational functions, which are defined such that the numerator and the denominator have no common factor other than -1 or 1.

The following statements are also true.

> The graph of a rational function never crosses a vertical asymptote.

> The graph of a rational function may or may not cross a horizontal asymptote.

Example 5   Make a hand-drawn graph of

$$g(x) = \frac{2x^2 + 1}{x^2}.$$

Include and label all asymptotes.

SOLUTION Since 0 is a zero of the denominator, the y-axis, $x = 0$, is a vertical asymptote. Note also that the degree of the numerator is the same as the degree of the denominator. Thus, $y = 2/1$, or 2, is the horizontal asymptote.

To complete the graph, we draw the asymptotes with dashed lines. Then we compute some additional ordered pairs and draw the curves.

HAND-DRAWN GRAPH | CHECK ON GRAPHER

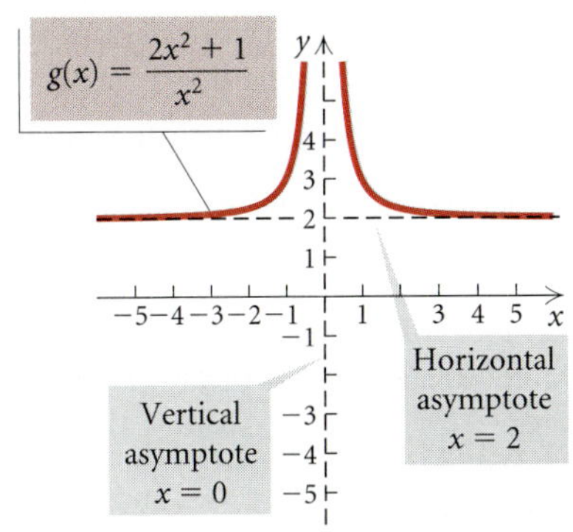

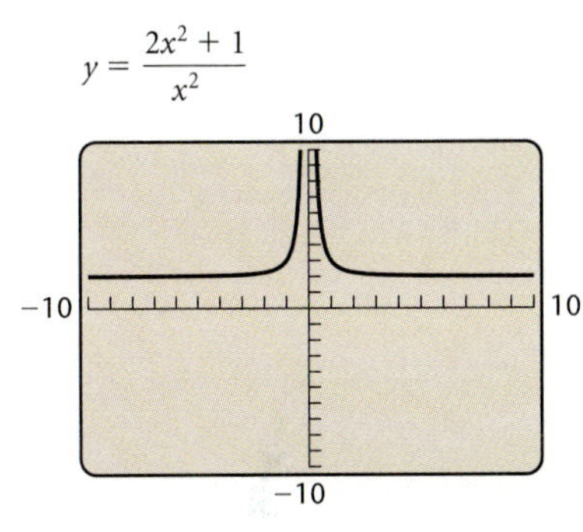

Sometimes a line that is neither horizontal nor vertical is an asymptote. Such a line is called an **oblique asymptote**, or a **slant asymptote**.

Example 6 Find all the asymptotes of

$$f(x) = \frac{2x^2 - 3x - 1}{x - 2}.$$

SOLUTION The line $x = 2$ is a vertical asymptote because 2 is a zero of the denominator. There is no horizontal asymptote. When the degree of the numerator is 1 greater than the degree of the denominator, we divide to find an equivalent expression:

$$\frac{2x^2 - 3x - 1}{x - 2} = (2x + 1) + \frac{1}{x - 2}.$$

$$
\begin{array}{r}
2x + 1 \\
x - 2 \overline{)\,2x^2 - 3x - 1} \\
\underline{2x^2 - 4x} \\
x - 1 \\
\underline{x - 2} \\
1
\end{array}
$$

Now we see that when $|x| \to \infty$, $1/(x - 2) \to 0$ and the value of $f(x) \to 2x + 1$. This means that as $|x|$ becomes very large, the graph of $f(x)$ gets very close to the graph of $y = 2x + 1$. Thus the line $y = 2x + 1$ is the oblique asymptote.

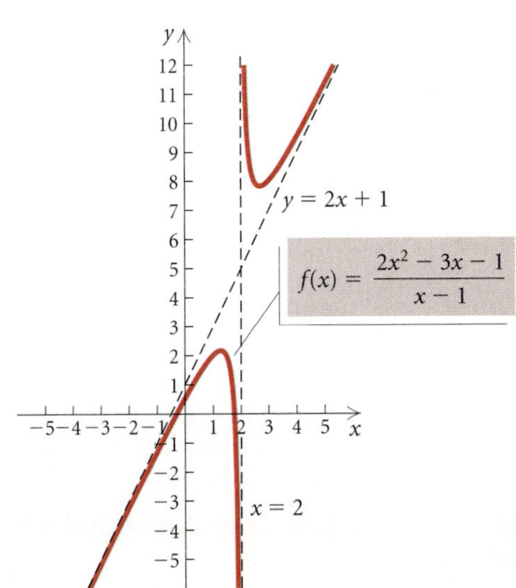

X	Y₁
1	3
2	2.25
3	2.1111
4	2.0625
5	2.04
1/2	6
1/3	11

X = 1/2

Occurrence of Lines as Asymptotes

Vertical asymptotes occur at any x-values that make the denominator 0.

The x-axis is the horizontal asymptote when the degree of the numerator is less than the degree of the denominator.

A horizontal asymptote other than the x-axis occurs when the numerator and the denominator have the same degree.

An oblique asymptote occurs when the degree of the numerator is 1 greater than the degree of the denominator.

There can be only one horizontal or oblique asymptote.

An asymptote is *not* part of the graph.

The following is an outline of a procedure that we can follow to create accurate hand-drawn graphs of rational functions.

To graph a rational function:

1. Find the real zeros of the denominator. Determine the domain of the function and sketch the vertical asymptotes.

2. Find the horizontal or the oblique asymptote, if any, and sketch it.

3. Find the real zeros of the numerator. These are the x-intercepts of the function.

4. Find $f(0)$. This gives the y-intercept of the function.

5. Find other function values to determine the general shape. Then draw the graph.

Example 7   Graph: $f(x) = \dfrac{2x + 3}{3x^2 + 7x - 6}$.

SOLUTION

1. We find the zeros of the denominator by solving $3x^2 + 7x - 6 = 0$. Since

$$3x^2 + 7x - 6 = (3x - 2)(x + 3),$$

the zeros are $\frac{2}{3}$ and -3. Thus the domain excludes $\frac{2}{3}$ and -3 and is $(-\infty, -3) \cup \left(-3, \frac{2}{3}\right) \cup \left(\frac{2}{3}, \infty\right)$. The graph has vertical asymptotes $x = -3$ and $x = \frac{2}{3}$. We sketch these as dashed lines.

2. Because the degree of the numerator is less than the degree of the denominator, the x-axis is the horizontal asymptote. There is no oblique asymptote.

3. To find the zeros of the numerator, we solve $2x + 3 = 0$ and get $x = -\frac{3}{2}$. Thus the pair $\left(-\frac{3}{2}, 0\right)$ is the x-intercept.

4. We find $f(0)$:

$$f(0) = \frac{2 \cdot 0 + 3}{3 \cdot 0^2 + 7 \cdot 0 - 6} = \frac{3}{-6} = -\frac{1}{2}.$$

Thus, $\left(0, -\frac{1}{2}\right)$ is the y-intercept.

5. We find other function values to determine the general shape and then draw the graph. Although vertical asymptotes can never be touched by the graph of a rational function, horizontal asymptotes can be.

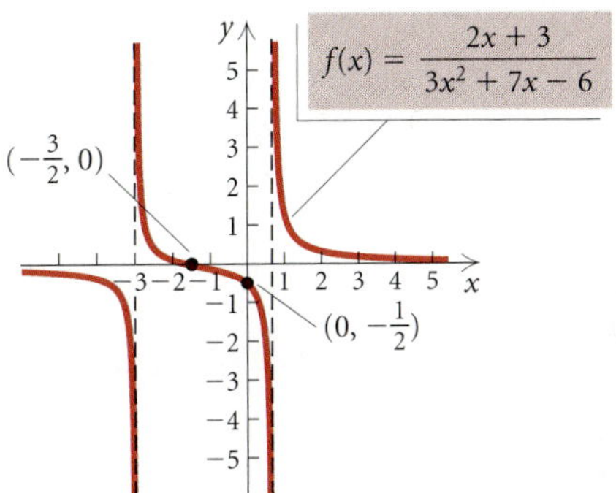

In Example 7, it is somewhat challenging to show on some graphers that $\frac{2}{3}$ is not in the domain of f. We know that the denominator, $3x^2 + 7x - 6$, is 0 when $x = \frac{2}{3}$. However, when $3(2/3)^2 + 7(2/3) - 6$ is entered, the result on a TI-82 is 1 E $^-$13, or 1×10^{-13}. This phenomenon occurs because the grapher approximates $\frac{2}{3}$ as a decimal before squaring or multiplying and this introduces an error. When the factorization $(3x - 2)(x + 3)$ is evaluated for $x = \frac{2}{3}$, the correct result, 0, is displayed. Thus a table of values for

$$y_1 = \frac{2x + 3}{3x^2 + 7x - 6} \quad \text{and} \quad y_2 = \frac{2x + 3}{(3x - 2)(x + 3)}$$

will reveal a discrepancy at $x = \frac{2}{3}$ (listed as .66667 in the x-column). Note that the ERROR message in the y_2-column is correct.

X	Y1	Y2
0	−.5	−.5
.16667	−.7018	−.7018
.33333	−1.1	−1.1
.5	−2.286	−2.286
.66667	4.3E13	ERROR
.83333	2.4348	2.4348
1	1.25	1.25

X = .666666666666...

Example 8   Graph: $g(x) = \dfrac{x^2 - 1}{x^2 + x - 6}$.

SOLUTION

1. We factor the denominator:

$$x^2 + x - 6 = (x + 3)(x - 2).$$

The domain excludes the x-values -3 and 2 and is $(-\infty, -3) \cup (-3, 2) \cup (2, \infty)$. The graph has vertical asymptotes $x = -3$ and $x = 2$. We sketch these as dashed lines.

2. The numerator and the denominator have the same degree, so the horizontal asymptote is determined by the ratio of the leading coefficients: $1/1$, or 1. Thus, $y = 1$ is the horizontal asymptote. We sketch it with a dashed line. There is no oblique asymptote.

3. To find the zeros of the numerator, we solve $x^2 - 1 = 0$. The solutions are -1 and 1. Thus the pairs $(-1, 0)$ and $(1, 0)$ are the x-intercepts.

4. We find $g(0)$:

$$g(0) = \frac{0^2 - 1}{0^2 + 0 - 6} = \frac{-1}{-6} = \frac{1}{6}.$$

Thus, $\left(0, \frac{1}{6}\right)$ is the y-intercept.

5. We find other function values to determine the general shape and then draw the graph.

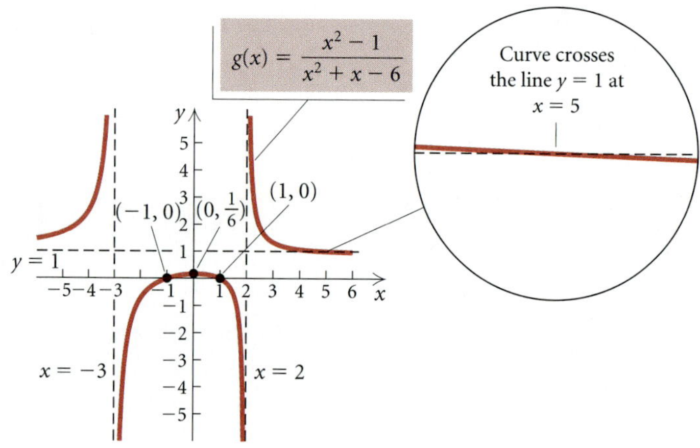

The magnified portion of the graph in Example 8 shows another situation in which a graph can cross its horizontal asymptote.

The graph of

$$f(x) = \frac{2x^3}{x^2 + 1}$$

shown below crosses its oblique asymptote $y = 2x$. Remember, graphs can cross horizontal or oblique asymptotes, but they cannot cross vertical asymptotes.

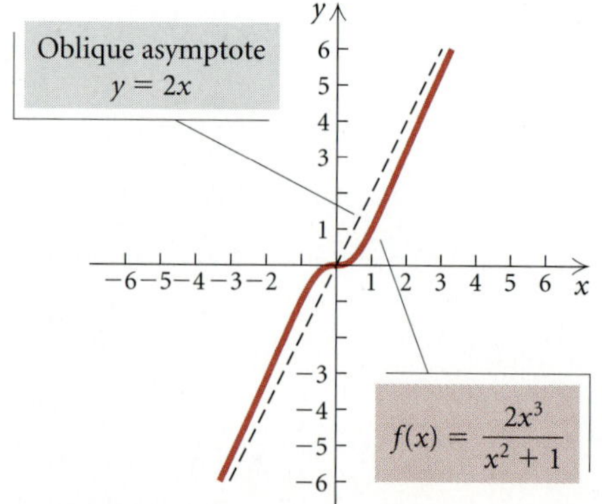

Applications

Example 9   *Temperature During an Illness.* The temperature T, in degrees Fahrenheit, of a person during an illness is given by the function

$$T(t) = \frac{4t}{t^2 + 1} + 98.6,$$

where time t is given in hours since the onset of the illness.

a) Graph the function over the interval $[0, \infty)$.

b) Find the temperature at $t = 0, 1, 2,$ and 5 hr.

c) Find the horizontal asymptote. Complete:

$$T(t) \longrightarrow \boxed{} \text{ as } t \longrightarrow \infty.$$

d) Give the meaning of the answer to part (c) in terms of the application.

e) Find the maximum temperature during the illness.

SOLUTION

a) The graph is shown below.

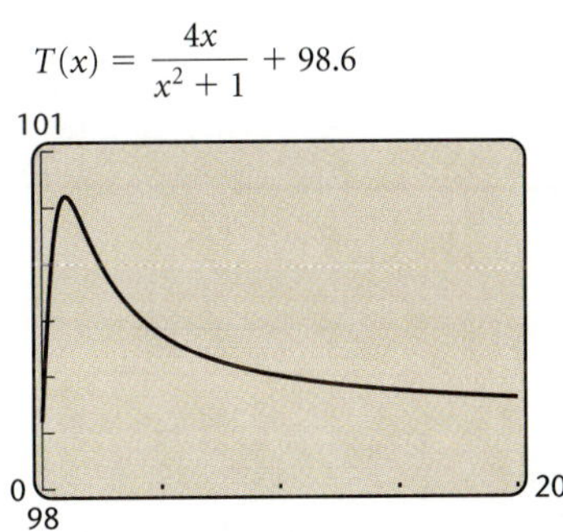

b) We have

$$T(0) = 98.6, \qquad T(1) = 100.6, \qquad T(2) = 100.2, \quad \text{and}$$
$$T(5) = 99.369.$$

c) Since

$$T(t) = \frac{4t}{t^2 + 1} + 98.6$$
$$= \frac{98.6t^2 + 4t + 98.6}{t^2 + 1},$$

the horizontal asymptote is $y = 98.6$. Then it follows that $T(t) \longrightarrow 98.6$ as $t \longrightarrow \infty$. We can check this using a table of values.

d) As time goes on, the temperature returns to "normal," which is $98.6°$.

e) Using a feature on the grapher for finding maximum values or by using TRACE and ZOOM, we find that the maximum temperature is $100.6°$ at $t = 1$ hr.

2.6 Exercise Set

In Exercises 1–6, use a grapher to match the equation with figures (a)–(f), which follow. List all asymptotes.

a)
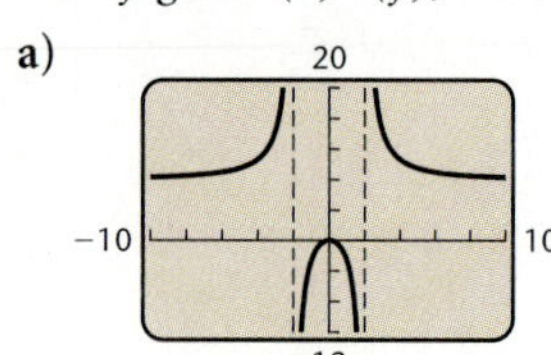

b)
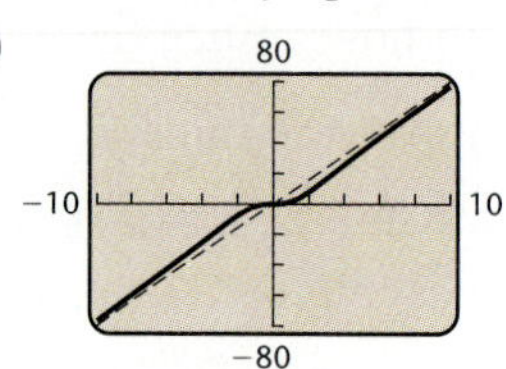

c)
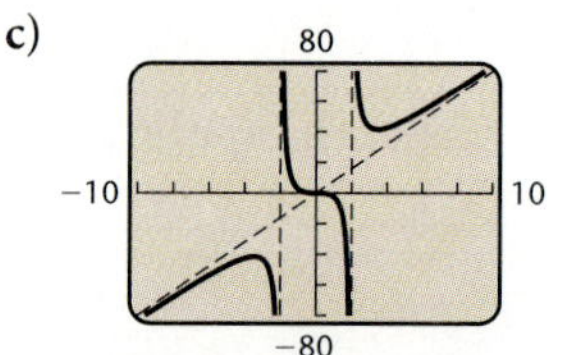

d)
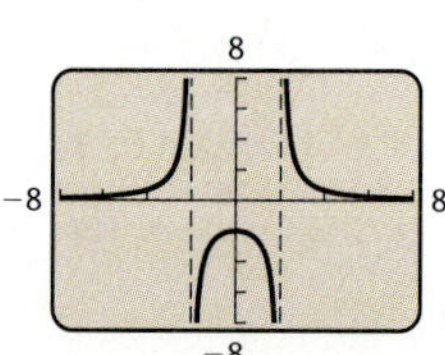

e)
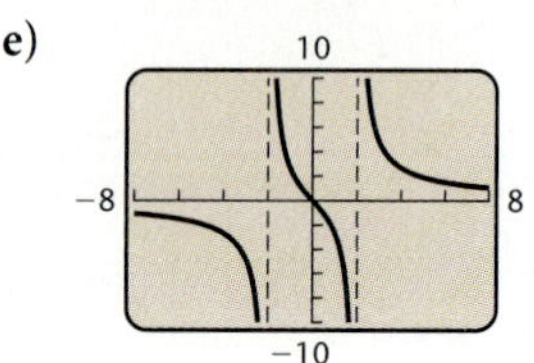

f)
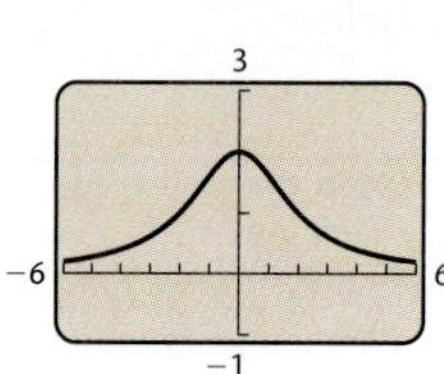

1. $f(x) = \dfrac{8}{x^2 - 4}$

2. $f(x) = \dfrac{8}{x^2 + 4}$

3. $f(x) = \dfrac{8x}{x^2 - 4}$

4. $f(x) = \dfrac{8x^2}{x^2 - 4}$

5. $f(x) = \dfrac{8x^3}{x^2 - 4}$

6. $f(x) = \dfrac{8x^3}{x^2 + 4}$

Make a hand-drawn graph for each of the following. Be sure to label all the asymptotes. Check your work on a grapher.

7. $f(x) = \dfrac{1}{x + 3}$

8. $f(x) = \dfrac{1}{x - 5}$

9. $f(x) = \dfrac{-2}{x - 5}$

10. $f(x) = \dfrac{3}{3 - x}$

11. $f(x) = \dfrac{2x + 1}{x}$

12. $f(x) = \dfrac{3x - 1}{x}$

13. $f(x) = \dfrac{1}{(x - 2)^2}$

14. $f(x) = \dfrac{-2}{(x - 3)^2}$

15. $f(x) = \dfrac{1}{x^2}$

16. $f(x) = \dfrac{1}{3x^2}$

17. $f(x) = \dfrac{1}{x^2 + 3}$

18. $f(x) = \dfrac{-1}{x^2 + 2}$

19. $f(x) = \dfrac{x - 1}{x + 2}$

20. $f(x) = \dfrac{x - 2}{x + 1}$

21. $f(x) = \dfrac{x + 3}{2x^2 - 5x - 3}$

22. $f(x) = \dfrac{3x}{x^2 + 5x + 4}$

23. $f(x) = \dfrac{x^2 - 9}{x + 1}$

24. $f(x) = \dfrac{x^2 - 4}{x - 1}$

25. $f(x) = \dfrac{x^2 + x - 2}{2x^2 + 1}$

26. $f(x) = \dfrac{x^2 - 2x - 3}{3x^2 + 2}$

27. $f(x) = \dfrac{x - 1}{x^2 - 2x - 3}$

28. $f(x) = \dfrac{x + 2}{x^2 + 2x - 15}$

29. $f(x) = \dfrac{x - 3}{(x + 1)^3}$

30. $f(x) = \dfrac{x + 2}{(x - 1)^3}$

31. $f(x) = \dfrac{x^3 + 1}{x}$

32. $f(x) = \dfrac{x^3 - 1}{x}$

33. $f(x) = \dfrac{x^3 + 2x^2 - 15x}{x^2 - 5x - 14}$

34. $f(x) = \dfrac{x^3 + 2x^2 - 3x}{x^2 - 25}$

35. $f(x) = \dfrac{5x^4}{x^4 + 1}$

36. $f(x) = \dfrac{x + 1}{x^2 + x - 6}$

37. $f(x) = \dfrac{x^2}{x^2 - x - 2}$

38. $f(x) = \dfrac{x^2 - x - 2}{x + 2}$

Find a rational function that satisfies the given conditions for each of the following. Answers may vary, but try to give the simplest answer possible.

39. Vertical asymptotes $x = -4$, $x = 5$

40. Vertical asymptotes $x = -4$, $x = 5$; x-intercept $(-2, 0)$

41. Vertical asymptotes $x = -4$, $x = 5$; horizontal asymptote $y = \frac{3}{2}$; x-intercept $(-2, 0)$

42. Oblique asymptote $y = x - 1$

43. *Medical Dosage.* The function

$$N(t) = \frac{0.8t + 1000}{5t + 4}, \quad t \geq 15$$

gives the body concentration $N(t)$, in parts per million, of a certain dosage of medication after time t, in hours.

a) Graph the function on the interval $[15, \infty)$ and complete the following:

$$N(t) \rightarrow \boxed{} \text{ as } t \rightarrow \infty.$$

b) Explain the meaning of the answer to part (a) in terms of the application.

44. *Average Cost.* The average cost per tape, in dollars, for a company to produce x videotapes on exercising is given by the function

$$A(x) = \frac{13x + 100}{x}.$$

a) Graph the function on the interval $(0, \infty)$ and complete the following:

$$A(x) \rightarrow \boxed{} \text{ as } x \rightarrow \infty.$$

b) Explain the meaning of the answer to part (a) in terms of the application.

45. *Population Growth.* The population P, in thousands, of Lordsburg is given by

$$P(t) = \frac{500t}{2t^2 + 9},$$

where $t = $ time, in months.

a) Graph the function over the interval $[0, \infty)$.
b) Find the population at $t = 0, 1, 3,$ and 8 months.
c) Find the horizontal asymptote. Complete:

$$P(t) \rightarrow \boxed{} \text{ as } t \rightarrow \infty.$$

d) Give the meaning of the answer to part (c).
e) Find the maximum population.

46. *Minimizing Surface Area.* The Hold-It Container Co. is designing an open-top rectangular box, with a square base, that will hold 108 cubic centimeters (cc).

a) Express the surface area S as a function of the length x of a side of the base.
b) Graph the function over the interval $(0, \infty)$.
c) Estimate the minimum surface area and the value of x that will yield it.

Inverse Variation. Rational functions like

$$y = \frac{k}{x} \quad and \quad y = \frac{k}{x^2},$$

*where the variables x, y, and k are positive, represent equations of **inverse variation** in applications. The constant k is called the **variation constant**. Note that the function is decreasing over the interval $(0, \infty)$.*

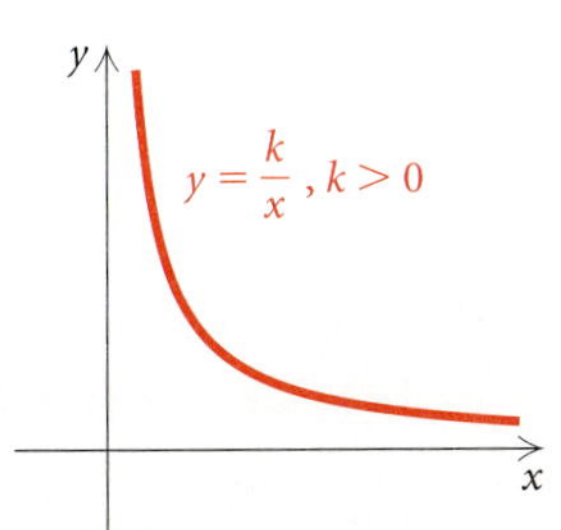

47. *Intensity of Light.* The intensity of light I from a lightbulb varies inversely as the square of the distance d from the bulb. Suppose that $I = 90$ W/m^2 (watts per square meter) when the distance is 5 m. Find the intensity at a distance of 10 m.

48. *Stocks and Gold.* It is theorized by some economists that stock prices S vary inversely as the price G of gold. One day the Dow Jones Industrial Average was 4040 and the price of gold was \$348 per ounce. What will the Dow Jones Industrial Average be if the price of gold rises to \$440 per ounce?

Skill Maintenance

Solve.

49. $3x - 7 > 2$

50. $4x - 9 \leq 6x + 5$

51. $\dfrac{3x - 9}{x^2 + 1} = 0$

52. $\dfrac{1}{x} + \dfrac{3}{2x} = 4$

Synthesis

53. ◈ Under what circumstances will a rational function have a domain consisting of all real numbers?

54. ◈ Explain and contrast the three kinds of asymptotes of rational functions.

55. ◈ Graph

$$y_1 = \frac{x^3 + 4}{x} \quad and \quad y_2 = x^2$$

using the same viewing window. Explain how the parabola $y_2 = x^2$ can be thought of as a nonlinear asymptote for y_1.

Find the nonlinear asymptotes of each function.

56. $f(x) = \dfrac{x^4 + 3x^2}{x^2 + 1}$

57. $f(x) = \dfrac{x^5 + 2x^3 + 4x^2}{x^2 + 2}$

Graph each function. Check on a grapher.

58. $f(x) = \dfrac{x^3 + 4x^2 + x - 6}{x^2 - x - 2}$

59. $f(x) = \dfrac{2x^3 + x^2 - 8x - 4}{x^3 + x^2 - 9x - 9}$

60. $f(x) = \dfrac{x^4 + 3x^3 + 21x^2 - 50x + 80}{x^4 + 8x^3 - x^2 + 20x - 10}$

Holes. Suppose we violate our restriction that rational functions have no common factor in the numerator and the denominator other than -1 or 1. If there is some *other common factor, we can get a so-called hole in the graph where this factor is 0. Make a hand-drawn graph of each of the following, showing any holes.*

61. $f(x) = \dfrac{x^2 - 4}{x - 2}$ **62.** $f(x) = \dfrac{x^2 - 9}{x + 3}$

63. $g(x) = \dfrac{3x^2 - x - 2}{x - 1}$

64. $f(x) = \dfrac{2x + 1}{2x^2 - 5x - 3}$

65. ◆ Which procedure for finding asymptotes can no longer be used if we allow rational functions to have common factors other than -1 or 1 in the numerator and denominator? (See Exercises 61–64.)

Find the domain.

66. $f(x) = \sqrt{x^2 - 4x - 21}$

67. $f(x) = \sqrt{\dfrac{72}{x^2 - 4x - 21}}$

2.7
Polynomial and Rational Inequalities

- *Solve polynomial and rational inequalities.*

We will use a combination of algebraic and graphical methods to solve polynomial and rational inequalities.

Polynomial Inequalities

Just as a quadratic equation can be written in the form $ax^2 + bx + c = 0$, a **quadratic inequality** can be written in the form $ax^2 + bx + c \;\blacksquare\; 0$, where $\blacksquare$ is $<, >, \le,$ or $\ge$. Here are some examples of quadratic inequalities:

$$3x^2 - 2x - 5 > 0, \qquad -\tfrac{1}{2}x^2 + 4x - 7 \le 0.$$

Quadratic inequalities are one type of **polynomial inequality**. Other examples are

$$-2x^4 + x^2 - 3 < 0, \qquad \tfrac{2}{3}x + 4 \ge 0, \quad \text{and} \quad 4x^3 - 2x^2 > 5x + 7.$$

When the inequality symbol in a polynomial inequality is replaced with an equality symbol, a **related equation** is formed. Polynomial inequalities can be easily solved once the related equation has been solved.

Example 1 Solve: $x^3 - x > 0$.

SOLUTION We are asked to find all x-values for which $x^3 - x > 0$. To locate these values, we graph $f(x) = x^3 - x$. Then we note that whenever the function changes sign, its graph passes through an x-intercept. Thus to solve $x^3 - x > 0$, we first solve the related equation to locate any x-intercepts:

$$x^3 - x = 0$$
$$x(x^2 - 1) = 0$$
$$x(x + 1)(x - 1) = 0.$$

The x-intercepts are -1, 0, and 1, as shown on the graph.

The x-intercepts divide the x-axis into four intervals: $(-\infty, -1)$, $(-1, 0)$, $(0, 1)$, and $(1, \infty)$. For x-values within each of these intervals, the sign of $x^3 - x$ must be either uniformly positive or uniformly negative. To determine which, we choose a test value for x from each interval and find $f(x)$. This can be done by substitution or by using the TABLE feature set in ASK mode. We can also simply look at the graph.

INTERVAL	TEST VALUE	SIGN OF $f(x)$
$(-\infty, -1)$	$f(-2) = -6$	Negative
$(-1, 0)$	$f(-0.5) = 0.375$	Positive
$(0, 1)$	$f(0.5) = -0.375$	Negative
$(1, \infty)$	$f(2) = 6$	Positive

Since we are solving $x^3 - x > 0$, the solution set consists of only two of the four intervals, those where the sign of $f(x)$ is *positive*. We see that the solution set is $(-1, 0) \cup (1, \infty)$, or $\{x \mid -1 < x < 0 \text{ or } x > 1\}$.   ▬

The following is a method for solving polynomial inequalities.

To solve a polynomial inequality:

1. Find an equivalent inequality with 0 on one side.

2. Solve the related polynomial equation.

3. Use the solutions to divide the x-axis into intervals. Then select a test value from each interval and determine the polynomial's sign on the interval.

4. Determine the intervals for which the inequality is satisfied and write interval notation or set-builder notation for the solution set. Include the endpoints of the intervals in the solution set if the inequality symbol is $\leq$ or $\geq$.

Example 2   Solve: $3x^4 + 10x \leq 11x^3 + 4$.

By subtracting $11x^3 + 4$, we form the equivalent inequality

$$3x^4 - 11x^3 + 10x - 4 \leq 0.$$

ALGEBRAIC SOLUTION

To solve the related equation

$$3x^4 - 11x^3 + 10x - 4 = 0,$$

we need to use a grapher and/or the theorems of Section 2.5 (see Example 11 in Section 2.5). The solutions are

$$-1, \quad 2 - \sqrt{2}, \quad \tfrac{2}{3}, \quad \text{and} \quad 2 + \sqrt{2}.$$

These numbers divide the x-axis into five intervals.

We then let $f(x) = 3x^4 - 11x^3 + 10x - 4$ and, using test values for $f(x)$, determine the sign of $f(x)$ in each interval:

INTERVAL	TEST VALUE	SIGN OF $f(x)$
$(-\infty, -1)$	$f(-2) = 112$	Positive
$(-1, 2 - \sqrt{2})$	$f(0) = -4$	Negative
$\left(2 - \sqrt{2}, \tfrac{2}{3}\right)$	$f(0.6) = 0.0128$	Positive
$\left(\tfrac{2}{3}, 2 + \sqrt{2}\right)$	$f(1) = -2$	Negative
$(2 + \sqrt{2}, \infty)$	$f(4) = 100$	Positive

Function values are negative in the intervals $(-1, 2 - \sqrt{2})$ and $\left(\tfrac{2}{3}, 2 + \sqrt{2}\right)$. Since the inequality sign is $\leq$, we include the endpoints of the intervals in the solution set. The solution set is

$$[-1, 2 - \sqrt{2}] \cup \left[\tfrac{2}{3}, 2 + \sqrt{2}\right], \quad \text{or}$$
$$\left\{x \mid -1 \leq x \leq 2 - \sqrt{2} \text{ or } \tfrac{2}{3} \leq x \leq 2 + \sqrt{2}\right\}.$$

GRAPHICAL SOLUTION

We graph $y = 3x^4 - 11x^3 + 10x - 4$ using a viewing window that reveals the curvature of the graph.

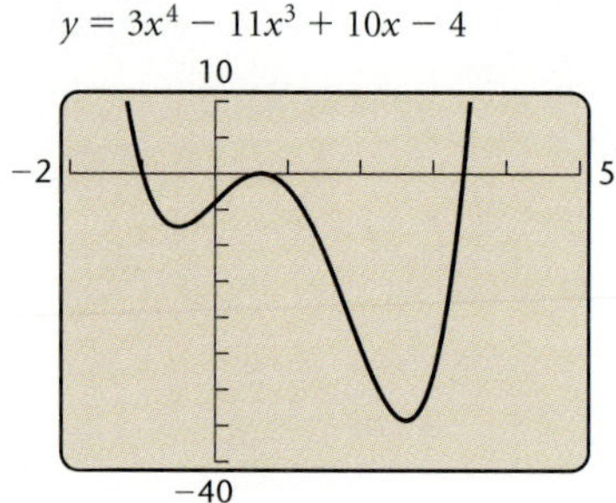

Although this window reveals the curvature, it leaves us uncertain about the number of zeros of the function in the interval $[0, 1]$. By using TRACE and ZOOM or the SOLVE feature (see also Example 10 in the Introduction to Graphs and Graphers), we see that the first zero is -1. The following window shows another view of the zeros in the interval $[0, 1]$.

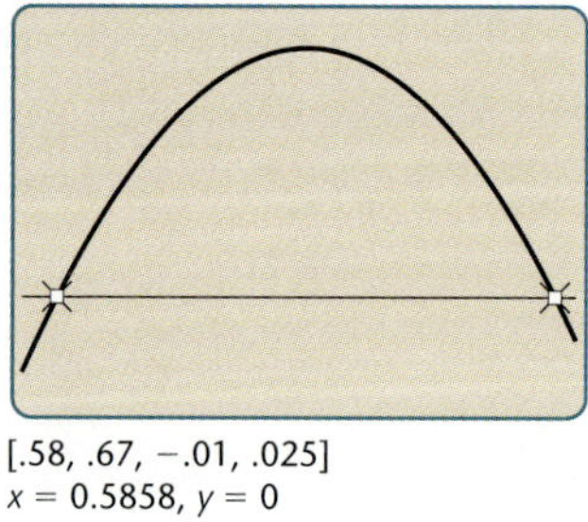

[.58, .67, −.01, .025]
$x = 0.5858, y = 0$
$x = 0.6667, y = 0$

The zeros are about -1, 0.586, 0.667, and 3.414.

The intervals to be considered are $(-\infty, -1)$, $(-1, 0.586)$, $(0.586, 0.667)$, $(0.667, 3.414)$, and $(3.414, \infty)$. We note on the graph where the function is negative. Then including appropriate endpoints, we find that the solution set is approximately

$$[-1, 0.586] \cup [0.667, 3.414], \quad \text{or}$$
$$\{x \mid -1 \leq x \leq 5.586 \text{ or } 0.667 \leq x \leq 3.414\}.$$

Interactive Discovery

Referring to Example 2, use the TRACE feature to confirm that the sign of $3x^4 - 11x^3 + 10x - 4$ changes as the cursor moves through each of the x-intercepts. Then let $y_1 = 3x^4 + 10x$ and $y_2 = 11x^3 + 4$. Show that the graphs of y_1 and y_2 intersect at those values for which $3x^4 - 11x^3 + 10x - 4 = 0$. How can the graphs of y_1 and y_2 be used to solve Example 2? We see from this Interactive Discovery that a grapher can be used to solve a polynomial inequality when one side is not 0.

Rational Inequalities

Some inequalities involve rational expressions and functions. These are called **rational inequalities**. To solve rational inequalities, we need to make some adjustments to the preceding method.

Example 3   Solve: $\dfrac{x - 3}{x + 4} \geq \dfrac{x + 2}{x - 5}$.

We first subtract $(x + 2)/(x - 5)$ in order to find an equivalent inequality with 0 on one side:

$$\frac{x - 3}{x + 4} - \frac{x + 2}{x - 5} \geq 0.$$

ALGEBRAIC SOLUTION

We look for all values of x for which the related function

$$f(x) = \frac{x - 3}{x + 4} - \frac{x + 2}{x - 5}$$

is not defined or is 0. These are called **critical values**.

A look at the denominators shows that $f(x)$ is not defined for $x = -4$ and $x = 5$. Next, we solve $f(x) = 0$:

$$\frac{x - 3}{x + 4} - \frac{x + 2}{x - 5} = 0$$

$$(x + 4)(x - 5)\left(\frac{x - 3}{x + 4} - \frac{x + 2}{x - 5}\right) =$$

$$(x + 4)(x - 5) \cdot 0$$

$$(x - 5)(x - 3) - (x + 4)(x + 2) = 0$$

$$(x^2 - 8x + 15) - (x^2 + 6x + 8) = 0$$

$$-14x + 7 = 0$$

$$x = \tfrac{1}{2}.$$

The critical values are $-4, \tfrac{1}{2}$, and 5. The critical values divide the x-axis into four intervals.

We then use a test value to determine the sign of $f(x)$ in each interval:

GRAPHICAL SOLUTION

We graph

$$y = \frac{x - 3}{x + 4} - \frac{x + 2}{x - 5}$$

in a standard window, which shows its curvature.

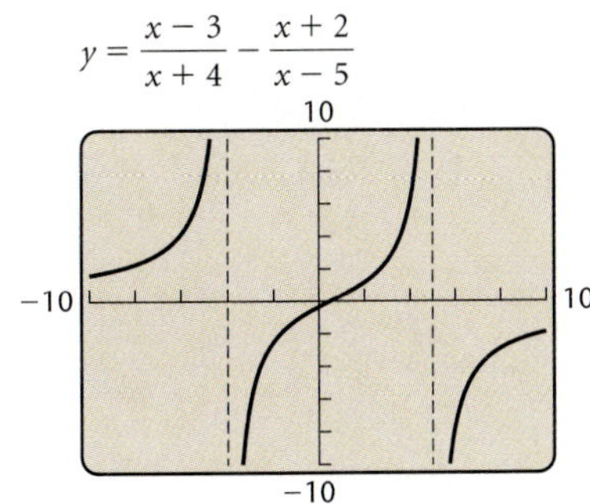

By using TRACE and ZOOM or the SOLVE feature, we find that 0.5 is a zero and that no other numbers are zeros.

We then look for values where the function is not defined. We can determine them quickly by examining the denominators $x + 4$ and $x - 5$, but it is instructive to use CONNECTED

(continued)

INTERVAL	TEST VALUE	SIGN OF $f(x)$
$(-\infty, -4)$	$f(-5) = 7.7$	Positive
$\left(-4, \frac{1}{2}\right)$	$f(-2) = -2.5$	Negative
$\left(\frac{1}{2}, 5\right)$	$f(3) = 2.5$	Positive
$(5, \infty)$	$f(6) = -7.7$	Negative

Function values are positive in the intervals $(-\infty, -4)$ and $\left(\frac{1}{2}, 5\right)$. Note that since $f\left(\frac{1}{2}\right) = 0, \frac{1}{2}$ must be in the solution set. Note that since neither -4 nor 5 is in the domain of f, they cannot be part of the solution set.

The solution set is

$$(-\infty, -4) \cup \left[\tfrac{1}{2}, 5\right).$$

mode and zoom in near the x-value -4. Then use the TRACE feature to observe how the y-values jump from very positive on the left of -4 to very negative on the right of -4. These y-values will actually be offscreen. The same thing occurs near 5.

The **critical values**, where y is either not defined or 0, are -4, 0.5, and 5.

The graph shows where y is positive and where it is negative. Note that the x-values of the asymptotes cannot be in the solution set since y is not defined for these values.

The solution set is

$$(-\infty, -4) \cup \left[\tfrac{1}{2}, 5\right).$$

The following is a method for solving rational inequalities.

To solve a rational inequality:

1. Find an equivalent inequality with 0 on one side.

2. Change the inequality symbol to an equals sign and solve the related equation.

3. Find x-values for which the related rational function is not defined.

4. The numbers found in steps (2) and (3) are called critical values. Use the critical values to divide the x-axis into intervals. Then test an x-value from each interval to determine the function's sign on that interval.

5. Select the intervals for which the inequality is satisfied and write interval notation or set-builder notation for the solution set. If the inequality symbol is $\leq$ or $\geq$, then the solutions to step (2) should be included in the solution set.

It works well to use a combination of algebraic and graphical methods to solve polynomial and rational inequalities. The algebraic methods give exact numbers for the critical values, and the graphical methods allow us to see easily what intervals satisfy the inequality.

2.7 | Exercise Set

In Exercises 1–4, a related function is graphed. Solve the given inequality.

1. $x^3 + 6x^2 < x + 30$

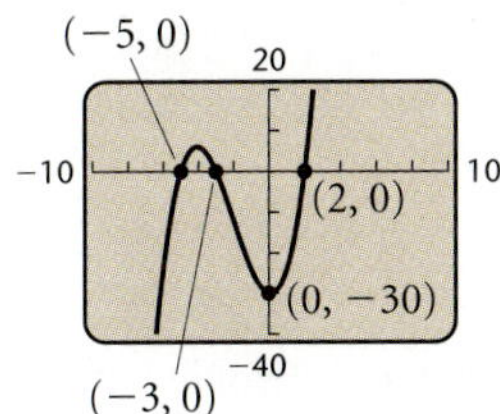

2. $x^4 - 27x^2 - 14x + 120 \geq 0$

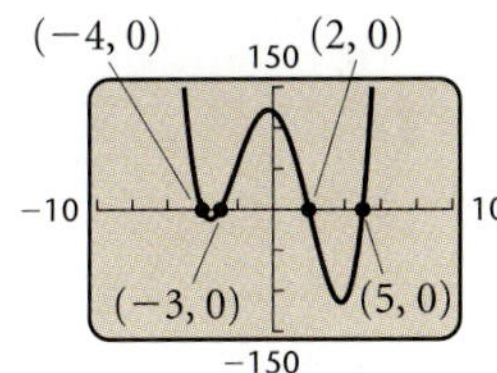

3. $\dfrac{8x}{x^2 - 4} \geq 0$

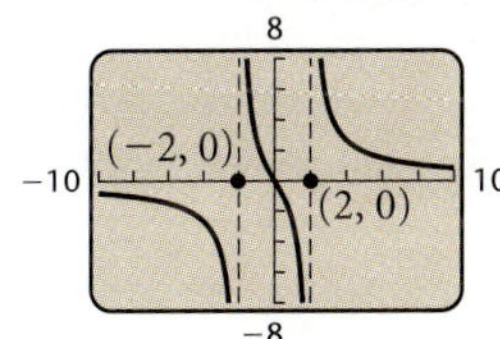

4. $\dfrac{8}{x^2 - 4} < 0$

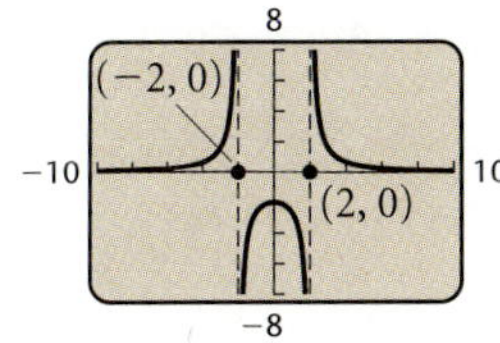

Solve.

5. $(x - 1)(x + 4) < 0$

6. $(x + 3)(x - 5) < 0$

7. $(x - 4)(x + 2) \geq 0$

8. $(x - 2)(x + 1) \geq 0$

9. $x^2 + x - 2 > 0$

10. $x^2 - x - 6 > 0$

11. $x^2 > 25$

12. $x^2 \leq 1$

13. $4 - x^2 \leq 0$

14. $11 - x^2 \geq 0$

15. $6x - 9 - x^2 < 0$

16. $x^2 + 2x + 1 \leq 0$

17. $x^2 + 12 < 4x$

18. $x^2 - 8 > 6x$

19. $4x^3 - 7x^2 \leq 15x$

20. $2x^3 - x^2 < 5$

21. $x^3 + 3x^2 - x - 3 \geq 0$

22. $x^3 + x^2 - 4x - 4 \geq 0$

23. $x^3 - 2x^2 < 5x - 6$

24. $x^3 + x \leq 6 - 4x^2$

25. $x^5 + x^2 \geq 2x^3 + 2$

26. $x^5 + 24 > 3x^3 + 8x^2$

27. $2x^3 + 6 \leq 5x^2 + x$

28. $2x^3 + x^2 < 10 + 11x$

29. $x^3 + 5x^2 - 25x \leq 125$

30. $x^3 - 9x + 27 \geq 3x^2$

31. $0.1x^3 - 0.6x^2 - 0.1x + 2 < 0$

32. $19.2x^3 + 12.8x^2 + 144 \geq 172.8x + 3.2x^4$

33. $\dfrac{1}{x + 4} > 0$

34. $\dfrac{1}{x - 3} \leq 0$

35. $\dfrac{-4}{2x + 5} < 0$

36. $\dfrac{-2}{5 - x} \geq 0$

37. $\dfrac{x - 4}{x + 3} - \dfrac{x + 2}{x - 1} \leq 0$

38. $\dfrac{x + 1}{x - 2} + \dfrac{x - 3}{x - 1} < 0$

39. $\dfrac{2x - 1}{x + 3} \geq \dfrac{x + 1}{3x + 1}$

40. $\dfrac{x + 5}{x - 4} > \dfrac{3x + 2}{2x + 1}$

41. $\dfrac{x + 1}{x - 2} \geq 3$

42. $\dfrac{x}{x - 5} < 2$

43. $x - 2 > \dfrac{1}{x}$

44. $4 \geq \dfrac{4}{x} + x$

45. $\dfrac{2}{x^2 - 4x + 3} \leq \dfrac{5}{x^2 - 9}$

46. $\dfrac{3}{x^2 - 4} \leq \dfrac{5}{x^2 + 7x + 10}$

47. $\dfrac{3}{x^2 + 1} \geq \dfrac{6}{5x^2 + 2}$

48. $\dfrac{4}{x^2 - 9} < \dfrac{3}{x^2 - 25}$

49. $\dfrac{5}{x^2 + 3x} < \dfrac{3}{2x + 1}$

50. $\dfrac{2}{x^2 + 3} > \dfrac{3}{5 + 4x^2}$

51. $\dfrac{5x}{7x - 2} > \dfrac{x}{x + 1}$

52. $\dfrac{x^2 - x - 2}{x^2 + 5x + 6} < 0$

53. $\dfrac{x}{x^2 + 4x - 5} + \dfrac{3}{x^2 - 25} \leq \dfrac{2x}{x^2 - 6x + 5}$

54. $\dfrac{2x}{x^2 - 9} + \dfrac{x}{x^2 + x - 12} \geq \dfrac{3x}{x^2 + 7x + 12}$

55. *Temperature During an Illness.* The temperature T, in degrees Fahrenheit, of a person during an illness is given by the function

$$T(t) = \frac{4t}{t^2 + 1} + 98.6,$$

where t = time, in hours. Find the interval over which the temperature was over $100°$. (See Example 9 in Section 2.6.)

56. *Population Growth.* The population P, in thousands, of Lordsburg is given by

$$P(t) = \frac{500t}{2t^2 + 9},$$

where t = time, in months. Find the interval over which the population was 40 thousand or greater. (See Exercise 45 in Exercise Set 2.6.)

57. *Total Profit.* Flexl, Inc., determines that its total profit is given by the function

$$P(x) = -3x^2 + 630x - 6000.$$

a) Flexl makes a profit for those nonnegative values of x for which $P(x) > 0$. Find the values of x for which Flexl makes a profit.

b) Flexl loses money for those nonnegative values of x for which $P(x) < 0$. Find the values of x for which Flexl loses money.

58. *Height of a Thrown Object.* The function

$$S(t) = -16t^2 + 32t + 1920$$

gives the height S, in feet, of an object thrown from a cliff that is 1920 ft high. Here t is the time, in seconds, that the object is in the air.

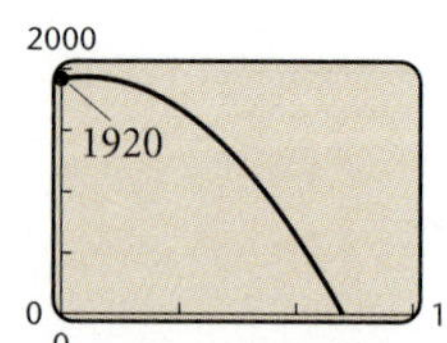

a) For what times is the height greater than 1920 ft?
b) For what times is the height less than 640 ft?

59. *Number of Diagonals.* A polygon with n sides has D diagonals, where D is given by the function

$$D(n) = \frac{n(n - 3)}{2}.$$

Find the number of sides n if

$$27 \leq D \leq 230.$$

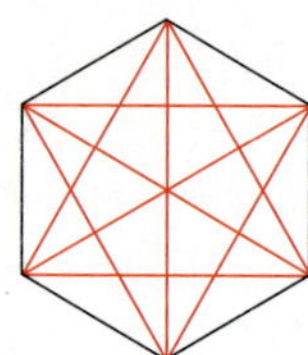

60. *Number of Handshakes.* There are n people in a room. The number N of possible handshakes by all the people in the room is given by the function

$$N(n) = \frac{n(n - 1)}{2}.$$

For what number n of people is

$$66 \leq N \leq 300?$$

Skill Maintenance

61. Solve for b: $a = 3b - 7$.

62. Solve for y: $4x - 5y = 9$.

Find $(f \circ g)(2)$ *for each of the following.*

63. $f(x) = 3x - 7$, $g(x) = x^3 - 1$

64. $f(x) = \sqrt{x} + 1$, $g(x) = 5x^2 - 4$

Synthesis

65. ◈ Under what circumstances would a quadratic inequality have a solution set that is a closed interval?

66. ◈ Under what circumstances would a quadratic inequality have an empty solution set?

Solve.

67. $x^2 + 9 \leq 6x$

68. $x^4 - 6x^2 + 5 > 0$

69. $x^4 + 3x^2 > 4x - 15$

70. $\left| \dfrac{x + 3}{x - 4} \right| < 2$

71. $|x^2 - 5| = 5 - x^2$

72. $(7 - x)^{-2} < 0$

73. $2|x|^2 - |x| + 2 \leq 5$

74. $|x|^2 - 4|x| + 4 \geq 9$

75. $\left| 1 + \dfrac{1}{x} \right| < 3$

76. $\left| 2 - \dfrac{1}{x} \right| \leq 2 + \left| \dfrac{1}{x} \right|$

77. $|x^2 + 3x - 1| < 3$

78. $|1 + 5x - x^2| \geq 5$

79. Write a quadratic inequality for which the solution set is $(-4, 3)$.

80. Write a polynomial inequality for which the solution set is $[-4, 3] \cup [7, \infty)$.

CHAPTER

2 Summary and Review

Important Properties and Formulas

Polynomial Function:

$$P(x) = a_n x^n + a_{n-1} x^{n-1} + a_{n-2} x^{n-2} + \cdots$$
$$+ a_1 x + a_0$$

Complex Number: $a + bi$, a, b real, $i^2 = -1$

Imaginary Number: $a + bi$, $b \neq 0$

Complex Conjugates: $a + bi$, $a - bi$

Quadratic Equation:

$$ax^2 + bx + c = 0, \ a \neq 0, \ a, b, c \text{ real}$$

Quadratic Function:

$$f(x) = ax^2 + bx + c, \ a \neq 0, \ a, b, c \text{ real}$$

Quadratic Formula:

$$x = \frac{-b \pm \sqrt{b^2 - 4ac}}{2a}$$

Polynomial Divison:

$$P(x) = D(x) \cdot Q(x) + R(x)$$

Dividend   Divisor   Quotient   Remainder

The Remainder Theorem: The remainder found by dividing $P(x)$ by $x - c$ is $P(c)$.

The Factor Theorem: $P(c) = 0 \leftrightarrow x - c$ is a factor of $P(x)$.

The Fundamental Theorem of Algebra: Every polynomial of degree n, $n \geq 1$, with complex coefficients has at least one complex-number zero.

The Rational Zeros Theorem: Consider the polynomial equation

$$a_n x^n + a_{n-1} x^{n-1} + a_{n-2} x^{n-2} + \cdots + a_1 x$$
$$+ a_0 = 0,$$

where all the coefficients are integers and $n \geq 1$. Also, consider a rational number p/q, where p and q have no common factor other than -1 and 1. If p/q is a zero of the polynomial equation, then p is a factor of a_0 and q is a factor of a_n.

The Intermediate Value Theorem: For any polynomial P with real coefficients, suppose that $P(a) \neq P(b)$ for $a \neq b$ and that $P(a)$ and $P(b)$ are of opposite signs. Then the polynomial has a real zero between a and b.

Rational Function:

$$f(x) = \frac{P(x)}{Q(x)},$$

where $P(x)$ and $Q(x)$ are polynomials with no common factor other than -1 and 1 and where $Q(x)$ is not the zero polynomial, and the domain consists of all x for which $Q(x) \neq 0$.

Horizontal Asymptote: $y = b$, where

$$f(x) \to b \text{ as } x \to \infty, \quad \text{or} \quad f(x) \to b \text{ as } x \to -\infty.$$

Vertical Asymptote: $x = a$ if any of the following is true:

$$f(x) \to \infty \text{ as } x \to a^-, \quad \text{or}$$
$$f(x) \to -\infty \text{ as } x \to a^-, \quad \text{or}$$
$$f(x) \to \infty \text{ as } x \to a^+, \quad \text{or}$$
$$f(x) \to -\infty \text{ as } x \to a^+.$$

REVIEW EXERCISES

Use a grapher to graph each of the following polynomial functions. Then estimate the function's (a) zeros; (b) relative maxima; (c) relative minima; (d) domain and range.

1. $f(x) = -2x^2 - 3x + 6$

2. $f(x) = x^3 + 3x^2 - 2x - 6$

3. $f(x) = x^4 - 3x^3 + 2x^2$

Express in terms of i.

4. $-\sqrt{-40}$

5. $\sqrt{-12} \cdot \sqrt{-20}$

6. $\dfrac{\sqrt{-49}}{-\sqrt{-64}}$

Simplify each of the following. Leave answers in the form $a + bi$, where a and b are real numbers.

7. $(6 + 2i)(-4 - 3i)$

8. $\dfrac{2 - 3i}{1 - 3i}$

9. $(3 - 5i) - (2 - i)$

10. $(6 + 2i) + (-4 - 3i)$

11. i^{23}

12. $(-3i)^{28}$

Solve by completing the square to obtain exact solutions. Show your work. Check your work on a grapher.

13. $x^2 - 3x = 18$

14. $3x^2 - 12x - 6 = 0$

Solve. Use any method, but obtain exact solutions. Check your work on a grapher, if possible.

15. $3x^2 + 2x = 8$

16. $r^2 - 2r + 10 = 0$

17. $x^2 = 18 + 3x$

18. $x = 2\sqrt{x} - 1$

19. $y^4 - 3y^2 + 1 = 0$

20. $(x^2 - 1)^2 - (x^2 - 1) - 2 = 0$

21. $(p - 3)(3p + 2)(p + 2) = 0$

22. $x^3 + 5x^2 - 4x - 20 = 0$

In Exercises 23 and 24, complete the square to:

a) *find the vertex;*
b) *find the line of symmetry;*
c) *determine whether there is a maximum or minimum value and find that value;*
d) *find the range.*

Then check your answers using a grapher.

23. $f(x) = -4x^2 + 3x - 1$

24. $f(x) = 5x^2 - 10x + 3$

In Exercises 25–28, match the equation with figures (a)–(d), which follow. Do not use a grapher except as a check.

a)

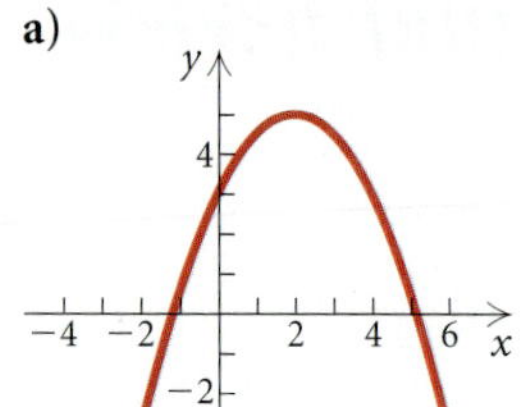

b)

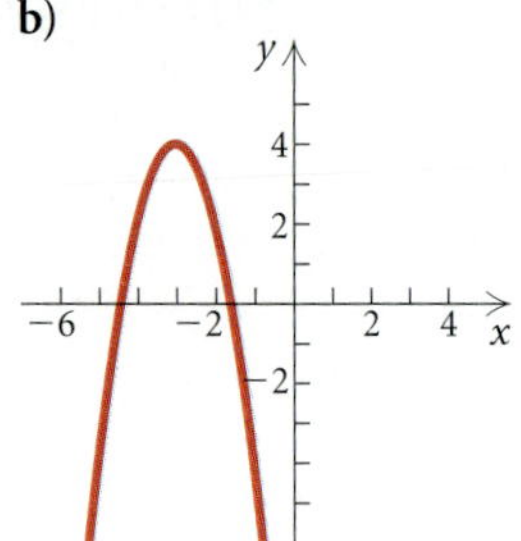

c)

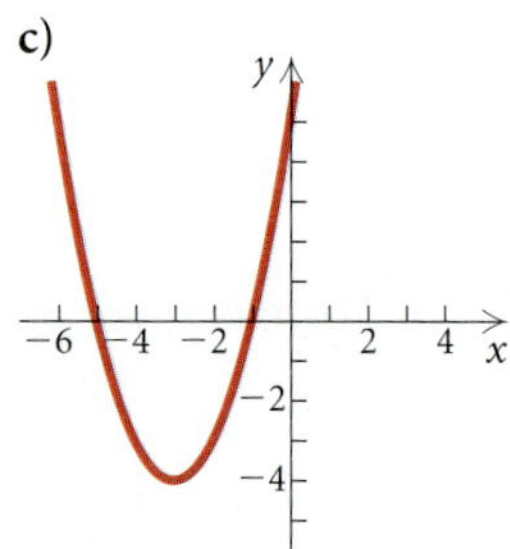

d)

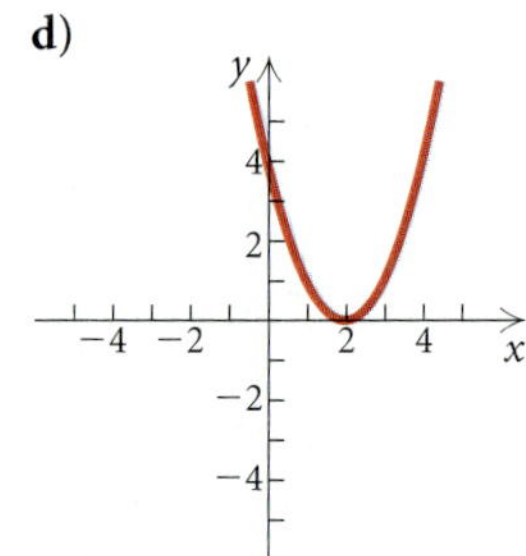

25. $y = (x - 2)^2$

26. $y = (x + 3)^2 - 4$

27. $y = -2(x + 3)^2 + 4$

28. $y = -\frac{1}{2}(x - 2)^2 + 5$

29. *Interest Compounded Annually.* When P dollars is invested at interest rate i, compounded annually, for t years, the investment grows to A dollars, where

$$A = P(1 + i)^t.$$

a) Find the interest rate i if $6250 grows to $6760 in 2 yr.

b) Find the interest rate i if $1,000,000 grows to $1,215,506.25 in 4 yr.

30. *Sidewalk Width.* A 60-ft by 80-ft parking lot is torn up to install a sidewalk of uniform width around its perimeter. The new area of the parking lot is two thirds of the old area. How wide is the sidewalk?

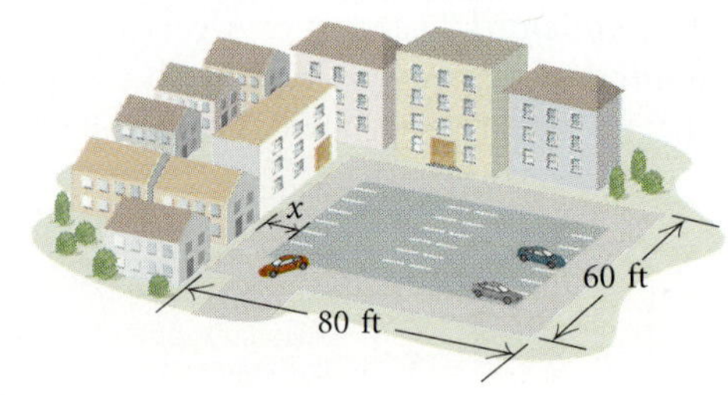

31. *Maximizing Volume.* The Berniers have 24 ft of flexible fencing with which to build a rectangular "toy corral" for a backyard. If the fencing is 2 ft high, what dimensions should the corral have in order to maximize its volume?

32. *Dimensions of a Box.* An open box is made from a 10-cm by 20-cm piece of aluminum by cutting a square from each corner and folding up the edges. The area of the resulting base is 90 cm^2. What is the length of the sides of the squares?

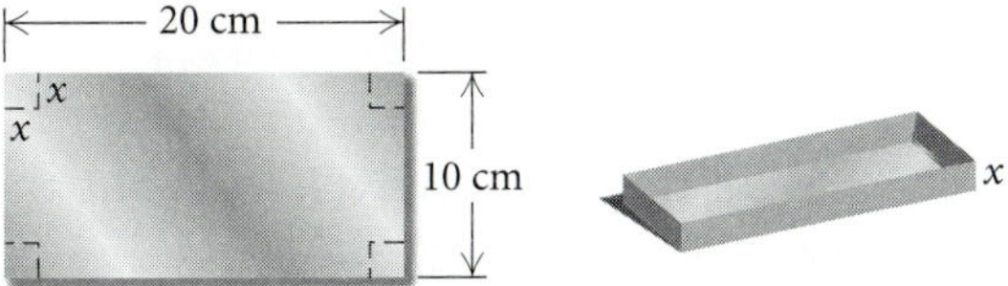

33. *Maximizing Box Dimensions.* An open box is to be made from a 10-cm by 20-cm piece of aluminum by cutting a square from each corner and folding up the edges. What should the length of the sides of the squares be in order to maximize the volume?

34. *Cholesterol Level and the Risk of Heart Attack.* The data in the following table show the relationship of cholesterol level in men to the risk of a heart attack.

CHOLESTEROL LEVEL	NUMBER OF MEN, PER 10,000, WHO SUFFER A HEART ATTACK
100	30
200	65
250	100
275	130

Source: Nutrition Action Healthletter.

  a) Use the REGRESSION feature on a grapher to find a linear, a quadratic, and a cubic function to fit the data.
  b) It is also known that 180 of 10,000 men have a heart attack with a cholesterol level of 300. Which function in part (a) would best make this prediction?
  c) Use the answer to part (b) to predict the heart attack rate for men with a cholesterol level of 350; with one of 400.

Do the long division. Check by multiplying or a grapher.

35. $(20x^3 - 6x^2 - 9x + 10) \div (5x + 2)$

36. $(2x^5 - 3x^3 + x^2 + 4) \div (x^2 - 2)$

In Exercises 37 and 38, a polynomial equation is given along with information about factors. Use the information to solve each equation.

37. $x^3 + 2x^2 - 13x + 10 = 0$;
  $x + 5$ is a factor of $x^3 + 2x^2 - 13x + 10$

38. $x^4 + 5x^3 - 23x^2 - 87x + 140 = 0$;
  $x^2 + x - 20$ is a factor of
  $x^4 + 5x^3 - 23x^2 - 87x + 140$

Use synthetic division to find the quotient and the remainder.

39. $(x^3 + 2x^2 - 13x + 10) \div (x - 5)$

40. $(x^4 + 3x^3 + 3x^2 + 3x + 2) \div (x + 2)$

Use synthetic division to find the indicated function value.

41. $P(x) = x^3 + 2x^2 - 13x + 10$; $P(5)$

42. $P(x) = x^4 - 16$; $P(-2)$

Determine whether the given values are zeros of $P(x)$. Then factor and find any other zeros that exist.

43. $P(x) = x^3 + 7x^2 + 7x - 15$; $-1, 2,$ and -3

44. $P(x) = x^4 + 9x^3 + 19x^2 + 9x - 2$; $1, -4,$ and -2

Find a polynomial equation of lowest degree with integer coefficients and the following as some of its zeros. Use a grapher as a partial check.

45. $4, -1, -2$

46. $-4, -\sqrt{3}, \sqrt{3}, 1$

47. $3, 2 - 3i, 4 + \sqrt{5}, -2$

Find an equation with rational coefficients that has a graph that passes through the given points. Use a grapher as a partial check.

48. $(-2, 0), (-1, 0), (0, 5), (4, 0)$

49. $(-\sqrt{8}, 0), (-3, 0), (0, 12), (1, 0), (\sqrt{8}, 0)$

50. $(-5, 0), (1 - \sqrt{3}, 0), (0, 8), (1 + \sqrt{3}, 0)$

Use the rational zeros theorem and/or factoring to solve each equation.

51. $x^3 - 4x^2 + x + 6 = 0$

52. $x^3 + 7x^2 + 7x = 15$

53. $x^4 + 36x + 63 = 4x^3 + 16x^2$

For each polynomial function:

a) *Solve $P(x) = 0$.*
b) *Graph $y = P(x)$.*
c) *Express $P(x)$ as a product of linear factors.*

54. $P(x) = x^3 + 8x^2 - 19x + 10$

55. $P(x) = x^6 + x^5 - 28x^4 - 16x^3 + 192x^2$

Make a hand-drawn graph for each of the following. Be sure to label all the asymptotes.

56. $f(x) = \dfrac{x^2 - 5}{x + 2}$

57. $f(x) = \dfrac{5}{(x - 2)^2}$

58. $f(x) = \dfrac{x^2 + x - 6}{x^2 - x - 20}$

59. $f(x) = \dfrac{x - 2}{x^2 - 2x - 15}$

In Exercises 60 and 61, find a rational function that satisfies the given conditions. Answers may vary, but try to give the simplest answer possible.

60. Vertical asymptotes $x = -2$, $x = 3$

61. Vertical asymptotes $x = -2$, $x = 3$; horizontal asymptote $y = 4$; x-intercept $(-3, 0)$

62. *Medical Dosage.* The function

$$N(t) = \frac{0.7t + 2000}{8t + 9}, \quad t \geq 5$$

gives the body concentration $N(t)$, in parts per million, of a certain dosage of medication after time t, in hours.

a) Graph the function on the interval $[5, \infty)$ and complete the following:

$$N(t) \longrightarrow \boxed{} \text{ as } t \longrightarrow \infty.$$

b) Explain the meaning of the answer to part (a) in terms of the application.

Solve.

63. $x^2 - 9 < 0$

64. $2x^2 > 3x + 2$

65. $(1 - x)(x + 4)(x - 2) < 0$

66. $\dfrac{x - 2}{x + 3} < 4$

67. $\dfrac{x - 3}{x^2 + x - 20} \geq \dfrac{4}{x^2 - 4}$

68. *Height of a Thrown Object.* The function

$$S(t) = -16t^2 + 80t + 224$$

gives the height S, in feet, of a model rocket launched from a hill that is 224 ft high with a velocity of 80 ft/sec, where $t = $ time, in seconds.

a) Find the maximum height of the rocket and when that height is attained.
b) Determine when the object reaches the ground.
c) Over what interval is the height greater than 320 ft?

69. *Population Growth.*   The population P, in thousands, of Novi is given by

$$P(t) = \frac{8000t}{4t^2 + 10},$$

where $t = $ time, in months. Find the interval over which the population was 400 or greater.

Synthesis

70. ◈ Explain the difference between a polynomial function and a rational function.

71. ◈ Explain and contrast the three types of asymptotes considered for rational functions.

72. Determine whether the following is an identity:

$$4(x^2 + x + 1)^3 - 27x^2(x + 1)^2$$
$$= (x - 1)^2(2x + 1)^2(x + 2)^2.$$

73. *Interest Rate.*   In early 1995, \$3500 was deposited at a certain interest rate. One year later, \$4000 was deposited in another account at the same rate. At the end of that year, there was a total of \$8518.35 in both accounts. What is the annual interest rate?

Solve.

74. $x^2 \geq 5 - 2x$

75. $\left| 1 - \dfrac{1}{x^2} \right| < 3$

76. $x^4 - 2x^3 + 3x^2 - 2x + 2 = 0$

77. $(x - 2)^{-3} < 0$

78. Express $x^3 - 1$ as a product of linear factors.

79. Find k so that $x + 3$ is a factor of $x^3 + kx^2 + kx - 15$.

80. When $x^2 - 4x + 3k$ is divided by $x + 5$, the remainder is 33. Find the value of k.

Find the domain of each function.

81. $f(x) = \sqrt{x^2 + 3x - 10}$

82. $f(x) = \sqrt{x^2 - 3.1x + 2.2} + 1.75$

83. $f(x) = \dfrac{1}{\sqrt{5 - |7x + 2|}}$

Exponential and Logarithmic Functions 3

The number of cellular phones in this country is modeled by an exponential function, where y = the number of telephones, in millions, in the year x. Here $x = 0$ corresponds to 1985. (*Source:* Cellular Telecommunications Industry Association)

X	Y₁	
0	.2	
1	.4	
2	1.0	
3	1.8	
5	4.1	
7	8.6	
9	19.3	

X = 5

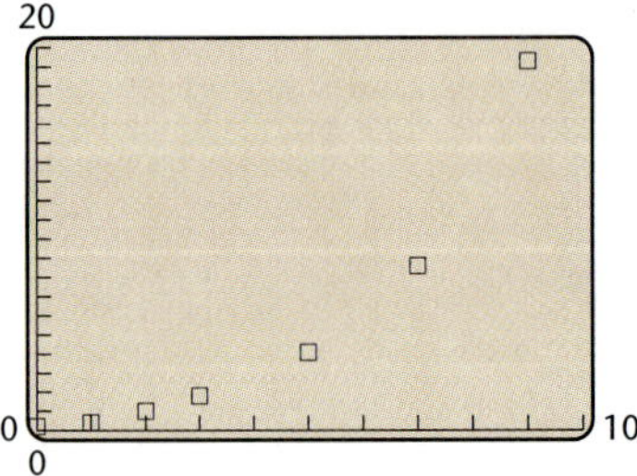

In this chapter, we will consider two kinds of closely related functions. The first, called *exponential functions,* are those that have a variable in the exponent. Such functions have many applications to the growth of populations, commodities, and investments.

Recall that a function takes an input to an output. Suppose we can reverse the process and take the output back to an input. That process produces what we call the *inverse* of the original function. Functions that are inverses of each other are closely related. The inverses of exponential functions, called *logarithmic functions,* or *logarithm functions,* are also important in many applications such as earthquake magnitude, sound level, and chemical pH.

3.1
Inverse Functions

- *Determine whether a function is one-to-one, and if it is, find a formula for its inverse.*
- *Simplify expressions of the type $(f \circ f^{-1})(x)$ and $(f^{-1} \circ f)(x)$.*

When we go from an output of a function back to its input or inputs, we get an inverse relation. When that relation is a function, we have an inverse function. We now study inverse functions and how to find their formulas when the original function has a formula. We do this to understand the relationship between exponential and logarithmic functions.

Inverses

Consider the relation h given as follows:

$$h = \{(-8, 5), (4, -2), (-7, 1), (3.8, 6.2)\}.$$

Suppose we *interchange* the first and second coordinates. The relation we obtain is called the **inverse** of the relation h and is given as follows:

$$\text{Inverse of } h = \{(5, -8), (-2, 4), (1, -7), (6.2, 3.8)\}.$$

> **Inverse Relation**
>
> Interchanging the first and second coordinates of each ordered pair in a relation produces the *inverse relation*.

Example 1   Consider the relation g given by

$$g = \{(2, 4), (-1, 3), (-2, 0)\}.$$

Graph the relation in blue. Find the inverse and graph it in red.

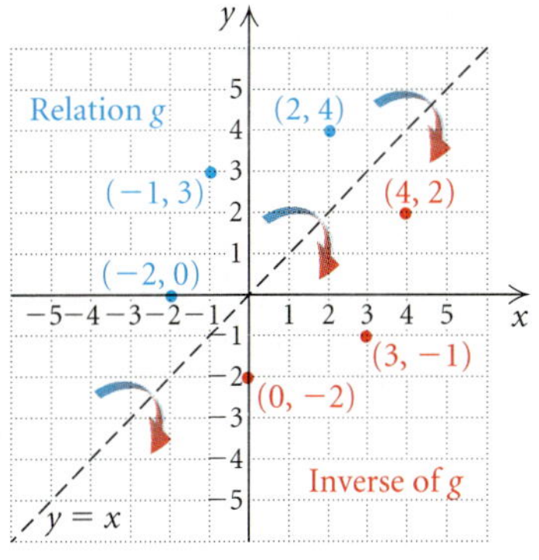

SOLUTION   The relation g is shown in blue in the figure at left. The inverse of the relation is

$$\{(4, 2), (3, -1), (0, -2)\}$$

and is shown in red. The pairs in the inverse are reflections across the line $y = x$.

> **Inverse Relation**
>
> If a relation is defined by an equation, interchanging the variables produces an equation of the *inverse relation*.

Example 2   Find an equation for the inverse of the relation:

$$y = x^2 - 5x.$$

SOLUTION   We interchange x and y and obtain an equation of the inverse:

$$x = y^2 - 5y.$$

Interactive Discovery

Graph each of the following relations:

$$y = 3x + 2, \qquad y = x, \qquad xy = 2,$$
$$x^2 + 3y^2 = 4, \qquad y^2 = 4x - 5.$$

Then find the inverse of each and graph it using the same set of axes. What pattern do you see? Be sure to use a square viewing window.

If a relation is given by an equation, the solutions of the inverse can be found from those of the original equation by interchanging the first and second coordinates of each ordered pair. Thus the graphs of a relation and its inverse are always reflections of each other across the line $y = x$.

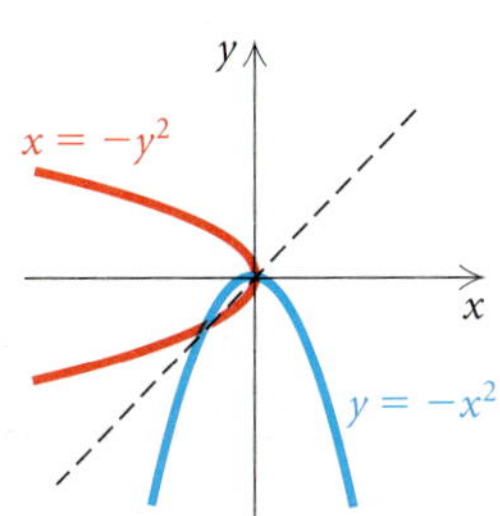

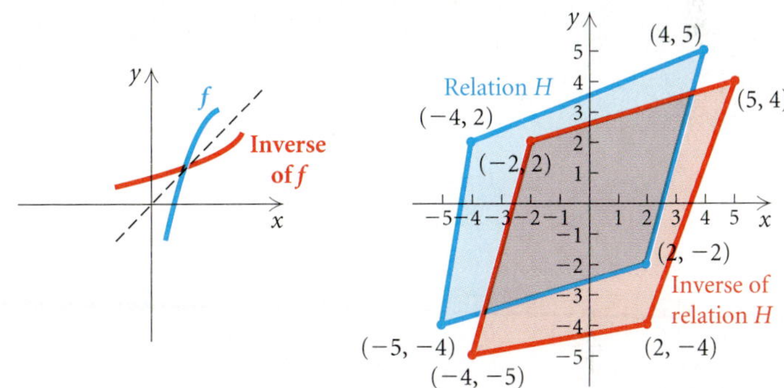

Inverses and One-to-One Functions

Let's consider the following two functions.

NUMBER (DOMAIN)	CUBING (RANGE)
−3	−27
−2	−8
−1	−1
0	0
1	1
2	8
3	27

YEAR (DOMAIN)	FIRST-CLASS POSTAGE COST, IN CENTS (RANGE)
1978	15
1983	20
1984	
1989	25
1991	29
1995	32

Source: U.S. Postal Service.

Suppose we reverse the arrows. Are these inverse relations functions?

NUMBER (RANGE)	CUBING (DOMAIN)
-3 ←	-27
-2 ←	-8
-1 ←	-1
0 ←	0
1 ←	1
2 ←	8
3 ←	27

YEAR (RANGE)	FIRST-CLASS POSTAGE COST, IN CENTS (DOMAIN)
1978 ←	15
1983 ←	20
1984 ←	
1989 ←	25
1991 ←	29
1995 ←	32

We see that the inverse of the cubing function is a function, but the inverse of the postage function is not a function. Like all functions, each input in the postage function has exactly one output. However, the output for both 1983 and 1984 is the same number, 20. Thus in the inverse of the postage function, the input 20 has *two* outputs, 1983 and 1984. When the same output comes from two or more different inputs, the inverse cannot be a function. In the cubing function, each output corresponds to exactly one input, so its inverse is also a function. The cubing function is an example of a **one-to-one function**. If the inverse of a function f is also a function, it is named f^{-1} (read "f-inverse").

The -1 in f^{-1} is *not* an exponent!

One-to-One Function and Inverses

A function f is *one-to-one* if different inputs have different outputs—that is,

$$\text{if} \quad a \neq b, \quad \text{then} \quad f(a) \neq f(b). \quad \text{Or,}$$

A function f is *one-to-one* if when the outputs are the same, the inputs are the same—that is,

$$\text{if} \quad f(a) = f(b), \quad \text{then} \quad a = b.$$

If a function is one-to-one, then its inverse is a function.

The domain of a one-to-one function f is the range of the inverse f^{-1}.

The range of a one-to-one function f is the domain of the inverse f^{-1}.

Example 3   Given the function f described by $f(x) = 2x - 3$, prove that f is one-to-one (that is, it has an inverse that is a function).

SOLUTION   To show that f is one-to-one, we show that if $f(a) = f(b)$, then $a = b$. Assume that $f(a) = f(b)$ for any numbers a and b in the domain of f. Then

$$2a - 3 = 2b - 3$$
$$2a = 2b \qquad \text{Adding 3}$$
$$a = b. \qquad \text{Dividing by 2}$$

Thus, if $f(a) = f(b)$, then $a = b$. This shows that f is one-to-one.

Example 4   Given the function g described by $g(x) = x^2$, prove that g is not one-to-one.

SOLUTION   To prove that g is not one-to-one, we need to find two numbers a and b for which $a \neq b$ and $g(a) = g(b)$. Two such numbers are -3 and 3, because $-3 \neq 3$ and $g(-3) = g(3) = 9$. Thus g is not one-to-one.

The graph below shows a function, in blue, and its inverse, in red. To determine whether the inverse is a function, we can apply the vertical-line test to its graph. By reflecting each such vertical line back across the line $y = x$, we obtain an equivalent **horizontal-line test** for the original function.

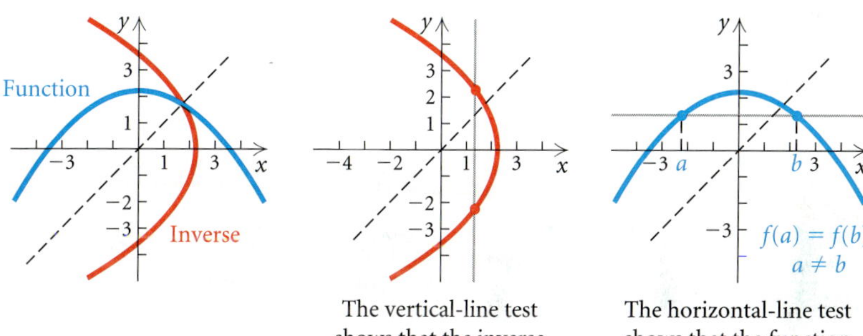

The vertical-line test
shows that the inverse
is not a function.

The horizontal-line test
shows that the function
is not one-to-one.

Horizontal-Line Test

If it is possible for a horizontal line to intersect the graph of a function more than once, then the function is *not* one-to-one and its inverse is *not* a function.

Example 5   Graph each of the following functions. Then determine whether each is one-to-one and thus has an inverse that is a function.

a) $f(x) = 4 - x$

b) $f(x) = x^2$

c) $f(x) = \sqrt[3]{x + 2} + 3$

d) $f(x) = 3x^5 - 20x^3$

SOLUTION   We graph each function using a grapher. Then we apply the horizontal-line test.

a) $y = 4 - x$

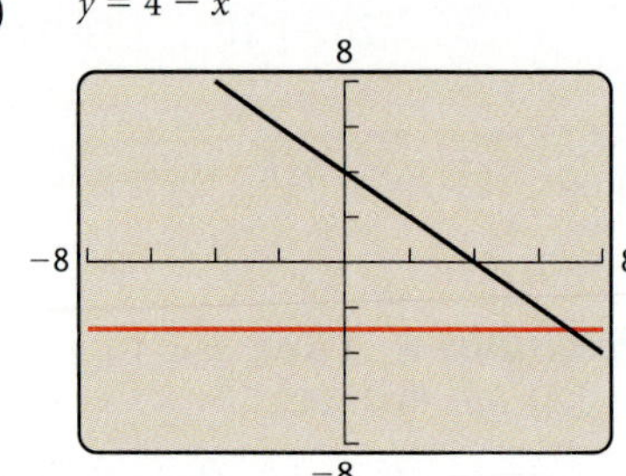

b) $y = x^2$

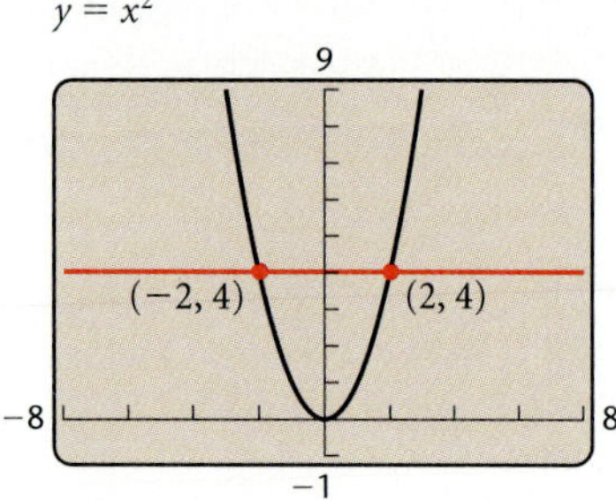

c) $y = \sqrt[3]{x + 2} + 3$

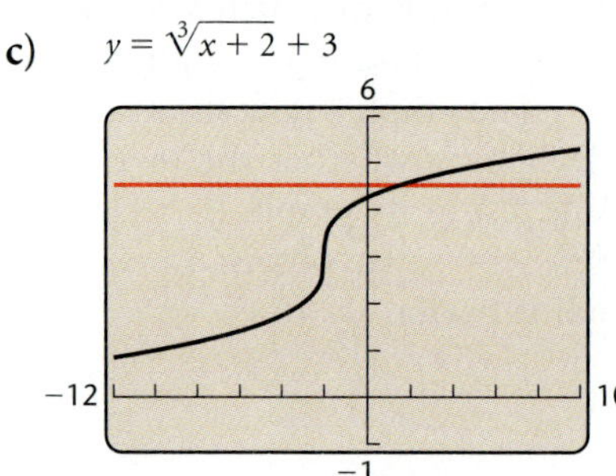

d) $y = 3x^5 - 20x^3$

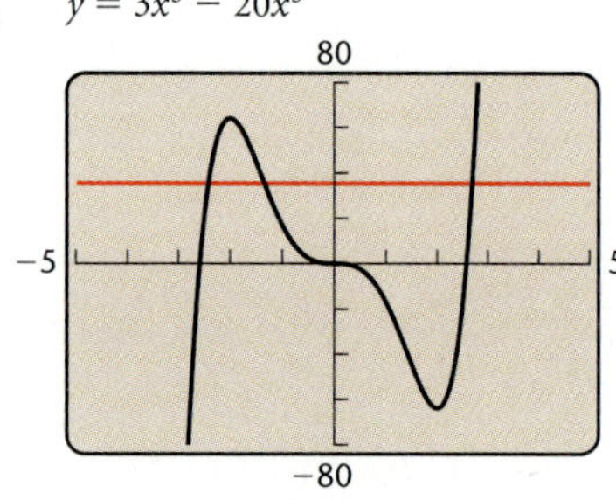

RESULT	**REASON**
a) One-to-one; inverse is a function	No horizontal line crosses the graph more than once.
b) Not one-to-one; inverse is not a function	There are many horizontal lines that cross the graph more than once. In particular, the line $y = 4$ does so. Note that where the line crosses, the first coordinates are -2 and 2. Although these are different inputs, they have the same output—that is, $-2 \neq 2$, but $$f(-2) = (-2)^2 = 4 = 2^2 = f(2).$$
c) One-to-one; inverse is a function	No horizontal line crosses the graph more than once.
d) Not one-to-one; inverse is not a function	There are many horizontal lines that cross the graph more than once.

Finding Formulas for Inverses

Suppose that a function is described by a formula. If it has an inverse that is a function, we proceed as follows to find a formula for f^{-1}.

> **Obtaining a Formula for an Inverse**
>
> If a function f is one-to-one, a formula for its inverse can generally be found as follows:
>
> **1.** Replace $f(x)$ with y.
>
> **2.** Interchange x and y.
>
> **3.** Solve for y.
>
> **4.** Replace y with $f^{-1}(x)$.

Example 6   Determine whether the function $f(x) = 2x - 3$ is one-to-one, and if it is, find a formula for $f^{-1}(x)$.

SOLUTION   The graph of f is shown at left. It passes the horizontal-line test. Thus it is one-to-one and its inverse is a function.

We also proved that f is one-to-one in Example 3.

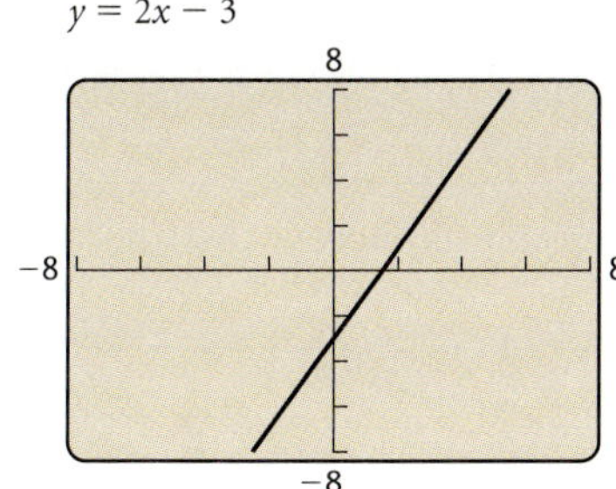

1. Replace $f(x)$ with y: $y = 2x - 3$

2. Interchange x and y: $x = 2y - 3$

3. Solve for y: $x + 3 = 2y$

$$\frac{x + 3}{2} = y$$

4. Replace y with $f^{-1}(x)$: $f^{-1}(x) = \dfrac{x + 3}{2}.$

Consider

$$f(x) = 2x - 3 \quad \text{and} \quad f^{-1}(x) = \frac{x + 3}{2}$$

from Example 6. For the input 5, we have

$$f(5) = 2 \cdot 5 - 3 = 10 - 3 = 7.$$

The output is 7. Now we use 7 for the input in the inverse:

$$f^{-1}(7) = \frac{7 + 3}{2} = \frac{10}{2} = 5.$$

The function f takes the number 5 to 7. The inverse function f^{-1} takes the number 7 back to 5.

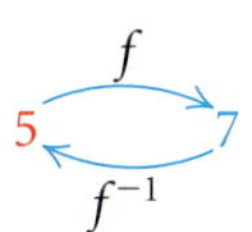

Example 7   Graph

$$f(x) = 2x - 3 \quad \text{and} \quad f^{-1}(x) = \frac{x + 3}{2}$$

using the same set of axes. Then compare the two graphs.

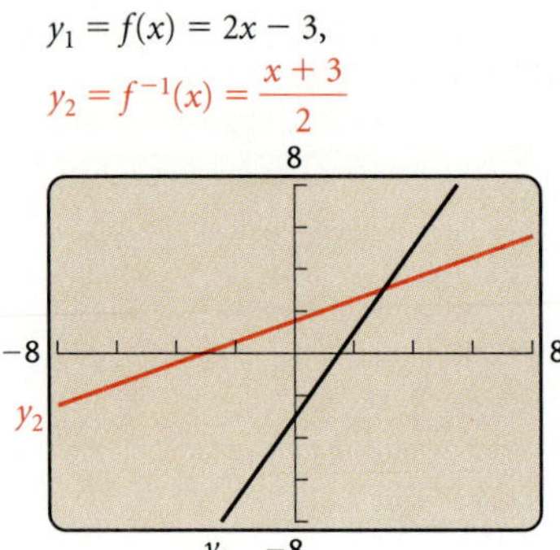

$$y_1 = f(x) = 2x - 3,$$
$$y_2 = f^{-1}(x) = \frac{x + 3}{2}$$

SOLUTION

Method 1. The graphs of f and f^{-1} are shown at left. Note that the graph of f^{-1} can be drawn by reflecting the graph of f across the line $y = x$. That is, if we were to graph $f(x) = 2x - 3$ in wet ink and fold along the line $y = x$, the graph of $f^{-1}(x) = (x + 3)/2$ would be formed by the ink transferred from f.

When we interchange x and y in finding a formula for the inverse of $f(x) = 2x - 3$, we are in effect reflecting the graph of that function across the line $y = x$. For example, when the coordinates of the y-intercept of the graph of f, $(0, -3)$, are reversed, we get the x-intercept of the graph of f^{-1}, $(-3, 0)$.

Method 2. Some graphers have a feature that graphs inverses automatically. If we were to start with $y_1 = 2x - 3$, the graphs of both y_1 and its inverse $y_2 = (x + 3)/2$ would be drawn. Be sure to square the viewing window. Consult your manual.

> The graph of f^{-1} is a reflection of the graph of f across the line $y = x$.

Example 8   Consider $g(x) = x^3 + 2$.

a) Determine whether the function is one-to-one.

b) If it is one-to-one, find a formula for its inverse.

c) Graph the function and its inverse. Use a square viewing window.

SOLUTION

a) The graph of $g(x) = x^3 + 2$ is shown below. It passes the horizontal-line test and thus has an inverse that is a function.

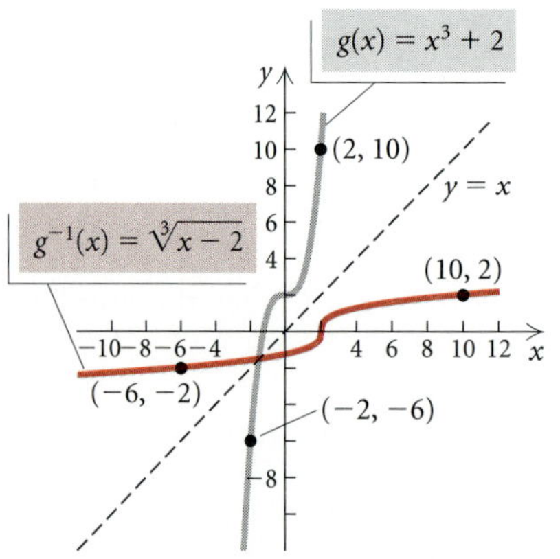

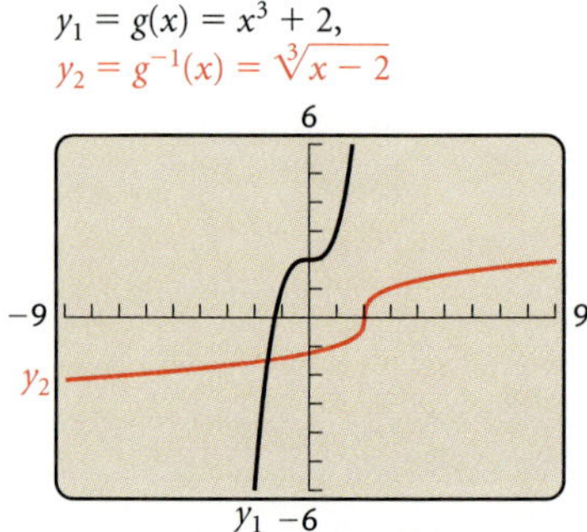

b) We follow the procedure for finding an inverse.

   1. Replace $g(x)$ with y: $y = x^3 + 2$

   2. Interchange x and y: $x = y^3 + 2$

3. Solve for y:
$$x - 2 = y^3$$
$$\sqrt[3]{x - 2} = y$$

4. Replace y with $g^{-1}(x)$: $g^{-1}(x) = \sqrt[3]{x - 2}.$

c) To find the graph, we reflect the graph of $g(x) = x^3 + 2$ across the line $y = x$. This can be done by using a grapher or by plotting points. See the graph on the left above. Be sure to square the viewing window.

Inverse Functions and Composition

Suppose that we were to use some input a for a one-to-one function f and find its output, $f(a)$. The function f^{-1} would then take that output back to a. Similarly, if we began with an input b for the function f^{-1} and found its output, $f^{-1}(b)$, the original function f would then take that output back to b. This is summarized as follows.

If a function f is one-to-one, then f^{-1} is the unique function such that each of the following holds:

$$(f^{-1} \circ f)(x) = f^{-1}(f(x)) = x, \quad \text{for each } x \text{ in the domain of } f$$
$$\text{(range of } f^{-1}), \quad \text{and}$$

$$(f \circ f^{-1})(x) = f(f^{-1}(x)) = x, \quad \text{for each } x \text{ in the domain of}$$
$$f^{-1} \text{ (range of } f).$$

Example 9   Given that $f(x) = 5x + 8$, use composition of functions to show that $f^{-1}(x) = (x - 8)/5$.

SOLUTION   We find $(f^{-1} \circ f)(x)$ and $(f \circ f^{-1})(x)$ and check to see that each is x:

$$(f^{-1} \circ f)(x) = f^{-1}(f(x)) = f^{-1}(5x + 8) = \frac{(5x + 8) - 8}{5} = \frac{5x}{5} = x;$$

$$(f \circ f^{-1})(x) = f(f^{-1}(x))$$
$$= f\left(\frac{x - 8}{5}\right)$$
$$= 5\left(\frac{x - 8}{5}\right) + 8$$
$$= x - 8 + 8 = x.$$

Restricting a Domain

In the case in which the inverse of a function is not a function, the domain can be restricted to allow the inverse to be a function. We saw in Example 5 that $f(x) = x^2$ is not one-to-one. The graph is shown at the top of the following page.

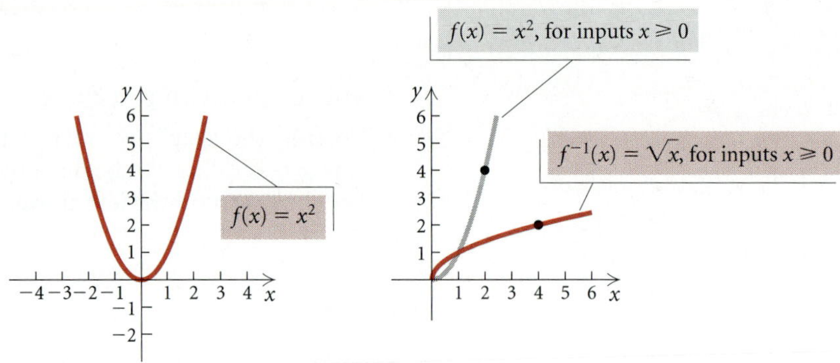

Suppose we had not known this and had tried to find a formula for the inverse as follows:

$$y = x^2 \qquad \text{Replacing } f(x) \text{ with } y$$

$$x = y^2. \qquad \text{Interchanging } x \text{ and } y$$

We cannot solve for y and get only one value, since most real numbers have two square roots:

$$\pm\sqrt{x} = y.$$

This is not the equation of a function. An input of, say, 4 would yield two outputs, -2 and 2. In such cases, it is convenient to consider "part" of the function by restricting the domain of $f(x)$. For example, if we restrict the domain of $f(x) = x^2$ to nonnegative numbers, then its inverse is a function. See the graphs of $f(x) = x^2$, $x \geq 0$, and $f^{-1}(x) = \sqrt{x}$, $x \geq 0$ on the right above.

3.1 Exercise Set

Find the inverse of each relation.

1. $\{(7, 8), (-2, 8), (3, -4), (8, -8)\}$

2. $\{(0, 1), (5, 6), (-2, -4)\}$

3. $\{(-1, -1), (-3, 4)\}$

4. $\{(-1, 3), (2, 5), (-3, 5), (2, 0)\}$

Find an equation of each inverse relation.

5. $y = 4x - 5$

6. $2x^2 + 5y^2 = 4$

7. $x^3y = -5$

8. $y = 3x^2 - 5x + 9$

Graph each equation by substituting and plotting points. Then reflect the graph across the line $y = x$ to obtain the graph of its inverse.

9. $x = y^2 - 3$

10. $y = x^2 + 1$

11. $y = |x|$

12. $x = |y|$

Using the horizontal-line test, determine whether each function is one-to-one.

13. $f(x) = 2.7^x$

14. $f(x) = 2^{-x}$

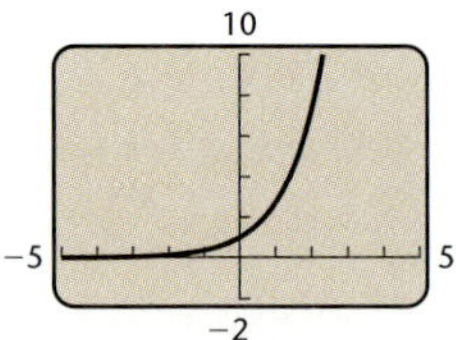

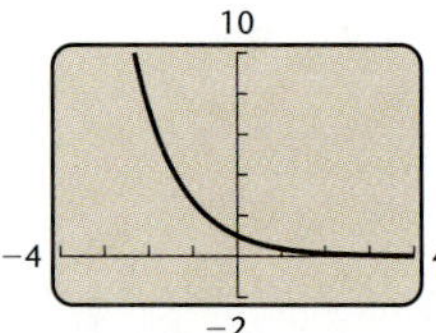

15. $f(x) = 4 - x^2$

16. $f(x) = x^3 - 3x + 1$

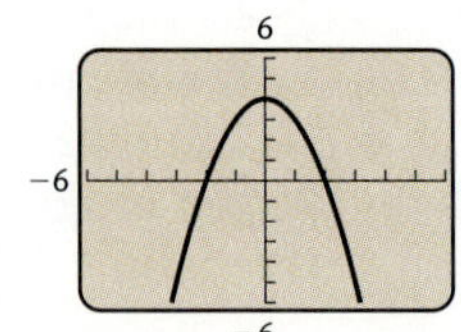

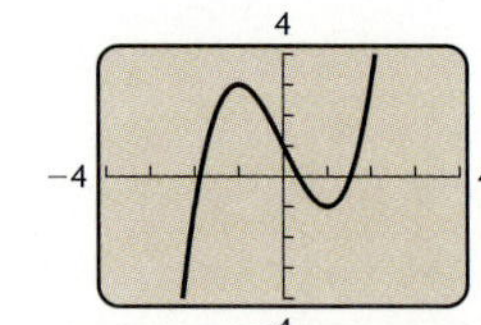

17. $f(x) = \dfrac{8}{x^2 - 4}$ **18.** $f(x) = \sqrt{\dfrac{10}{4 + x}}$

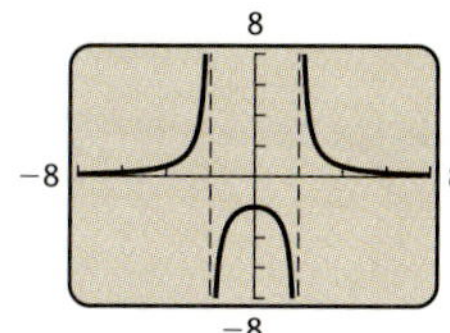

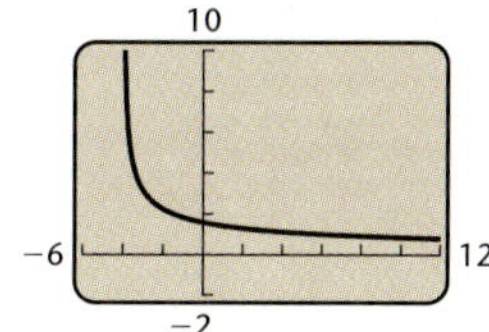

19. $f(x) = \sqrt[3]{x + 2} - 2$ **20.** $f(x) = \dfrac{8}{x}$

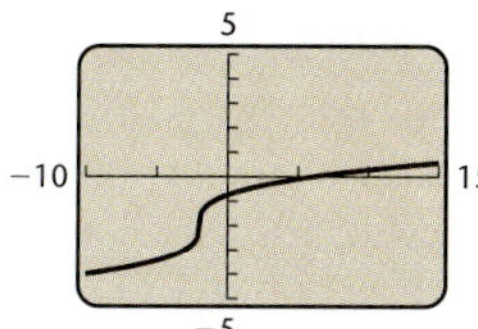

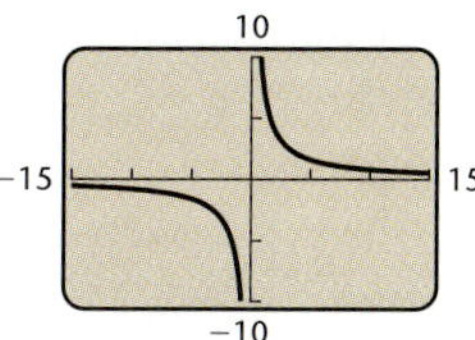

Determine whether each of the following is one-to-one. Use a grapher and the horizontal-line test or use an algebraic procedure.

21. $f(x) = 5x - 8$ **22.** $f(x) = 3 + 4x$

23. $f(x) = 1 - x^2$ **24.** $f(x) = |x| - 2$

25. $f(x) = |x + 2|$ **26.** $f(x) = -0.8$

27. $f(x) = -\dfrac{4}{x}$ **28.** $f(x) = \dfrac{2}{x + 3}$

Graph each function and its inverse using a grapher. Use an inverse drawing feature, if available. Find the domain and the range of f. Find the domain and the range of the inverse f^{-1}.

29. $f(x) = 0.8x + 1.7$ **30.** $f(x) = 2.7 - 1.08x$

31. $f(x) = \tfrac{1}{2}x - 4$ **32.** $f(x) = x^3 - 1$

33. $f(x) = \sqrt{x - 3}$ **34.** $f(x) = -\dfrac{2}{x}$

35. $f(x) = x^2 - 4, \ x \geq 0$

36. $f(x) = 3 - x^2, \ x \geq 0$

37. $f(x) = (3x - 9)^3$

38. $f(x) = \sqrt[3]{\dfrac{x - 3.2}{1.4}}$

Given each function:

a) *Determine whether it is one-to-one, using a grapher if desired.*

b) *If it is one-to-one, find a formula for the inverse.*

39. $f(x) = x + 4$ **40.** $f(x) = 7 - x$

41. $f(x) = 2x$ **42.** $f(x) = 5x + 8$

43. $f(x) = \dfrac{4}{x + 7}$ **44.** $f(x) = -\dfrac{3}{x}$

45. $f(x) = \dfrac{x + 4}{x - 3}$ **46.** $f(x) = \dfrac{5x - 3}{2x + 1}$

47. $f(x) = x^3 - 1$ **48.** $f(x) = (x + 5)^3$

49. $f(x) = x\sqrt{4 - x^2}$

50. $f(x) = 4x^5 - 20x^3 + 2x^2 - 5x + 1$

51. $f(x) = 5x^2 - 2, \ x \geq 2$

52. $f(x) = 4x^2 + 3, \ x \geq 0$

53. $f(x) = \sqrt{x + 1}$

54. $f(x) = \sqrt[3]{x - 8}$

Find each inverse by thinking about the operations of the function and then reversing, or undoing, them. Check your work algebraically.

FUNCTION	INVERSE
55. $f(x) = 3x$	$f^{-1}(x) = $
56. $f(x) = \tfrac{1}{4}x + 7$	$f^{-1}(x) = $
57. $f(x) = -x$	$f^{-1}(x) = $
58. $f(x) = \sqrt[3]{x} - 5$	$f^{-1}(x) = $
59. $f(x) = \sqrt[3]{x - 5}$	$f^{-1}(x) = $
60. $f(x) = x^{-1}$	$f^{-1}(x) = $

For each function f, use composition of functions to show that f^{-1} is as given.

61. $f(x) = \dfrac{7}{8}x, \ f^{-1}(x) = \dfrac{8}{7}x$

62. $f(x) = \dfrac{(x + 5)}{4}, \ f^{-1}(x) = 4x - 5$

63. $f(x) = \dfrac{(1 - x)}{x}, \ f^{-1}(x) = \dfrac{1}{x + 1}$

64. $f(x) = \sqrt[3]{x + 4}, \ f^{-1}(x) = x^3 - 4$

65. Find $f(f^{-1}(5))$ and $f^{-1}(f(a))$:
$$f(x) = x^3 - 4.$$

66. Find $f^{-1}(f(p))$ and $f(f^{-1}(1253))$:
$$f(x) = \sqrt[5]{\dfrac{2x - 7}{3x + 4}}.$$

67. *Dress Sizes in the United States and Italy.* A function that will convert dress sizes in the United States to those in Italy is
$$g(x) = 2(x + 12).$$

a) Find the dress sizes in Italy that correspond to sizes 6, 8, 10, 14, and 18 in the United States.

b) Find a formula for the inverse of the function.

c) Use the inverse function to find the dress sizes in the United States that correspond to 36, 40, 44, 52, and 60 in Italy.

68. *Bus Chartering.* An organization determines that the cost per person of chartering a bus is given by the formula

$$C(x) = \frac{100 + 5x}{x},$$

where x = the number of people in the group and $C(x)$ is in dollars. Determine $C^{-1}(x)$ and explain what it represents.

69. *Reaction Distance.* You are driving a car when a deer suddenly darts across the road in front of you. Your brain registers the emergency and sends a signal to your foot to hit the brake. The car travels a distance D, in feet, during this time, where D is a function of the speed r, in miles per hour, that the car is traveling when you see the deer. That reaction distance D is a linear function given by

$$D(r) = \frac{11r + 5}{10}.$$

a) Find $D(0)$, $D(10)$, $D(20)$, $D(50)$, and $D(65)$.
b) Graph $D(r)$.
c) Find $D^{-1}(r)$ and explain what it represents.
d) Graph the inverse.

70. *Bread Consumption.* The number N of 1-lb loaves of bread consumed per person per year t years after 1995 is given by the function

$$N(t) = 0.6514t + 53.1599.$$

(*Source:* Department of Agriculture)

a) Find the consumption of bread per person in 1998 and 2000.
b) Use a grapher to graph the function and its inverse.
c) Explain what the inverse represents.

Graph each of the following.

71. $y = x^3 - x$ **72.** $x = y^3 - y$

73. $f(x) = \sqrt[3]{x}$ **74.** $f(x) = \dfrac{8}{x^2 - 4}$

75. ◈ Suppose that you have graphed a function using a grapher and you see that it is one-to-one. How could you then use the TRACE feature to make a hand-drawn graph of the inverse?

76. ◈ The following formulas for the conversion between Fahrenheit and Celsius temperatures have been considered several times in this text:

$$C = \tfrac{5}{9}(F - 32)$$

and

$$F = \tfrac{9}{5}C + 32.$$

Discuss these formulas from the standpoint of inverses.

Using only a grapher, determine whether the functions are inverses of each other.

77. $f(x) = \sqrt[3]{\dfrac{x - 3.2}{1.4}},\ g(x) = 1.4x^3 + 3.2$

78. $f(x) = \dfrac{2x - 5}{4x + 7},\ g(x) = \dfrac{7x - 4}{5x + 2}$

79. $f(x) = \dfrac{2}{3},\ g(x) = \dfrac{3}{2}$

80. $f(x) = x^4,\ x \geq 0;\ g(x) = \sqrt[4]{x}$

81. Find three examples of functions that are their own inverses, that is, $f = f^{-1}$.

82. Consider the function f given by

$$f(x) = \begin{cases} x^3 + 2, & x \leq -1, \\ x^2, & -1 < x < 1, \\ x + 1, & x \geq 1. \end{cases}$$

Does f have an inverse that is a function? Why or why not?

3.2 Exponential Functions and Graphs

We now turn our attention to the study of a set of functions very rich in application. Consider the following graphs.

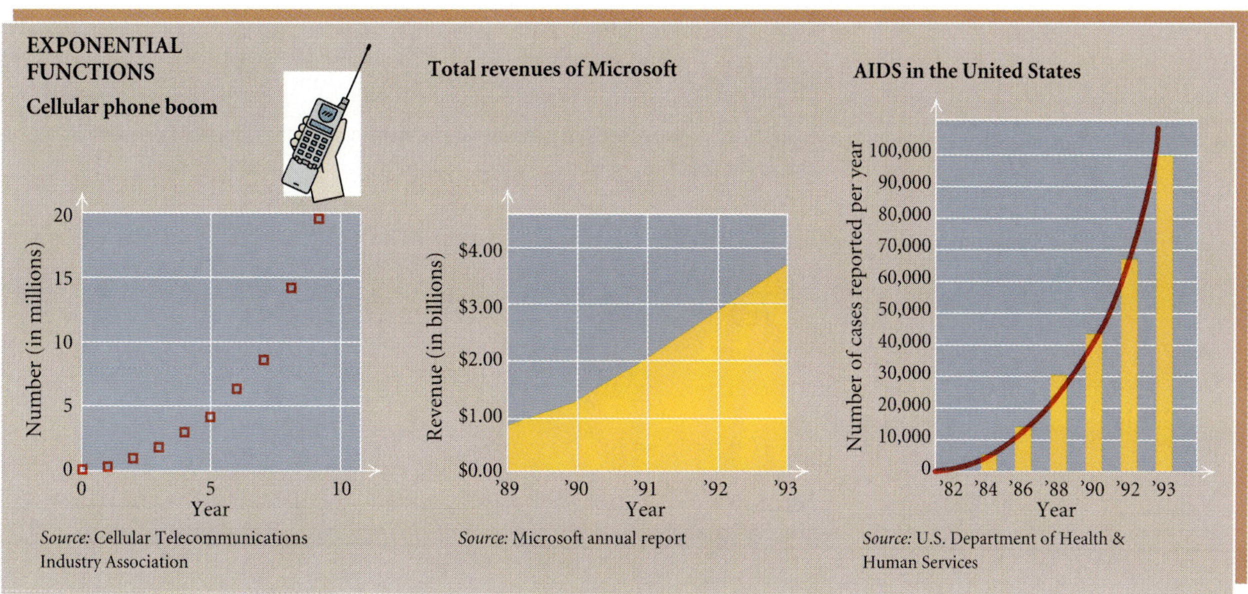

Each of these graphs illustrates the idea of an *exponential function*. The graph on the right shows the numbers of cases of AIDS reported in various years. The curve drawn along the graph approximates an exponential function. In this section, we consider such graphs, their inverses, and some important applications of exponential functions.

Graphing Exponential Functions

We now define exponential functions. We assume that a^x has meaning for any real number x and any positive real number a and that the laws of exponents still hold, though we will not prove them here.

> **Exponential Function**
>
> The function $f(x) = a^x$, where x is a real number, $a > 0$ and $a \neq 1$, is called the *exponential function*, base a.

We require the **base** to be positive in order to avoid the complex numbers that would occur by taking even roots of negative numbers—

an example is the square root of -1, $(-1)^{1/2}$, which is not a real number. The restriction $a \neq 1$ is made to exclude the constant function $f(x) = 1^x = 1$, which does not have an inverse because it is not one-to-one.

The following are examples of exponential functions:

$$f(x) = 2^x, \qquad f(x) = \left(\frac{1}{2}\right)^x, \qquad f(x) = (3.57)^x.$$

Note that, in contrast to functions like $f(x) = x^5$ and $f(x) = x^{1/2}$, the variable in an exponential function is *in the exponent*. Let's now consider graphs of exponential functions.

Example 1 Graph the exponential function: $y = f(x) = 2^x$.

SOLUTION We compute some function values and list the results in a table (at left).

x	$y = f(x) = 2^x$	(x, y)
0	1	$(0, 1)$
1	2	$(1, 2)$
2	4	$(2, 4)$
3	8	$(3, 8)$
-1	$\frac{1}{2}$	$\left(-1, \frac{1}{2}\right)$
-2	$\frac{1}{4}$	$\left(-2, \frac{1}{4}\right)$
-3	$\frac{1}{8}$	$\left(-3, \frac{1}{8}\right)$

$$f(0) = 2^0 = 1; \qquad f(-1) = 2^{-1} = \frac{1}{2^1} = \frac{1}{2};$$

$$f(1) = 2^1 = 2; \qquad f(-2) = 2^{-2} = \frac{1}{2^2} = \frac{1}{4};$$

$$f(2) = 2^2 = 4; \qquad f(-3) = 2^{-3} = \frac{1}{2^3} = \frac{1}{8}$$

$$f(3) = 2^3 = 8;$$

Next, we plot these points and connect them with a smooth curve. Be sure to plot enough points to determine how steeply the curve rises.

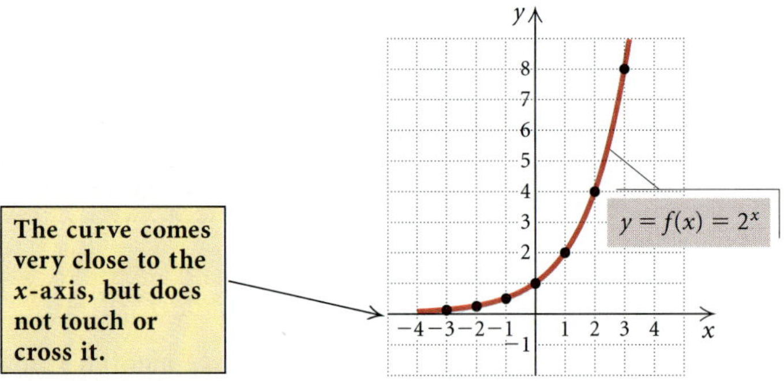

Note that as x increases, the function values increase without bound. As x decreases, the function values decrease, getting close to 0. That is, as $x \rightarrow -\infty$, $y \rightarrow 0$. The x-axis, or the line $y = 0$, is a horizontal asymptote. As the x-inputs decrease, the curve gets closer and closer to this line, but does not cross it.

Check the graph of $y = f(x) = 2^x$ on a grapher. In general, this function is entered as $y = 2\text{\textasciicircum}x$ or $f(x) = 2\text{\textasciicircum}x$. Then graph $y = f(x) = 3^x$. Use the TRACE and ZOOM features to confirm that the graph never touches the x-axis.

Example 2 Graph the exponential function: $y = f(x) = \left(\dfrac{1}{2}\right)^x$.

SOLUTION We compute some function values and list the results in a table (at left). Before we plot these points and draw the curve, note that

$$y = f(x) = \left(\frac{1}{2}\right)^x = (2^{-1})^x = 2^{-x}.$$

This tells us, before we begin graphing, that this graph is a reflection of the graph of $y = 2^x$ across the y-axis.

x	$y = f(x) = 2^{-x}$	(x, y)
0	1	$(0, 1)$
1	$\frac{1}{2}$	$\left(1, \frac{1}{2}\right)$
2	$\frac{1}{4}$	$\left(2, \frac{1}{4}\right)$
3	$\frac{1}{8}$	$\left(3, \frac{1}{8}\right)$
-1	2	$(-1, 2)$
-2	4	$(-2, 4)$
-3	8	$(-3, 8)$

$$f(0) = 2^{-0} = 1; \qquad\qquad f(-1) = 2^{-(-1)} = 2^1 = 2;$$

$$f(1) = 2^{-1} = \frac{1}{2^1} = \frac{1}{2}; \qquad f(-2) = 2^{-(-2)} = 2^2 = 4;$$

$$f(2) = 2^{-2} = \frac{1}{2^2} = \frac{1}{4}; \qquad f(-3) = 2^{-(-3)} = 2^3 = 8$$

$$f(3) = 2^{-3} = \frac{1}{2^3} = \frac{1}{8};$$

Next, we plot these points and connect them with a smooth curve.

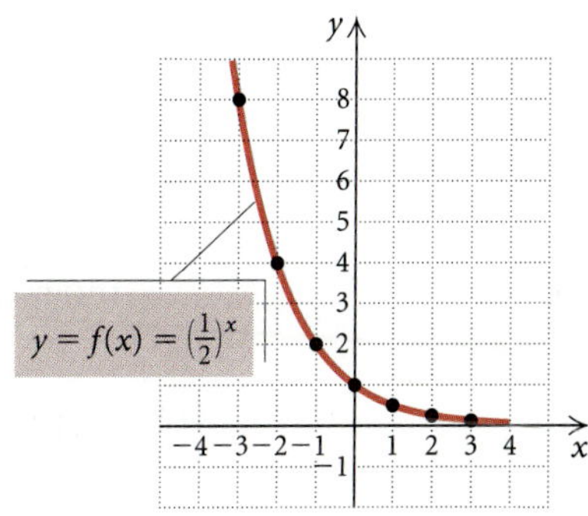

Check the graph in Example 2 on a grapher. Then graph $y = f(x) = \left(\frac{1}{3}\right)^x$. Also, try using the TABLE feature to make a table simultaneously for each of the functions $f(x) = 3^x$ and $g(x) = 3^{-x}$.

Interactive Discovery

Graph each of the following functions using the same set of axes and a viewing window of $[-10, 10, -10, 10]$. Look for patterns in the graphs.

$$f(x) = 1.1^x, \qquad f(x) = 2.7^x, \qquad f(x) = 4^x$$

Next, clear the screen and graph each of the following functions using the same set of axes and viewing window. Again, look for patterns in the graphs.

$$f(x) = 0.1^x, \qquad f(x) = 0.6^x, \qquad f(x) = 0.23^x$$

What relationship do you see between the base a and the shape of the resulting graph of a^x? What do all the graphs have in common? How do they differ?

The preceding examples and interactive discovery illustrate exponential functions with various bases. Let's list some characteristics, keeping in mind that the definition of an exponential function, $f(x) = a^x$, requires that a be positive and different from 1.

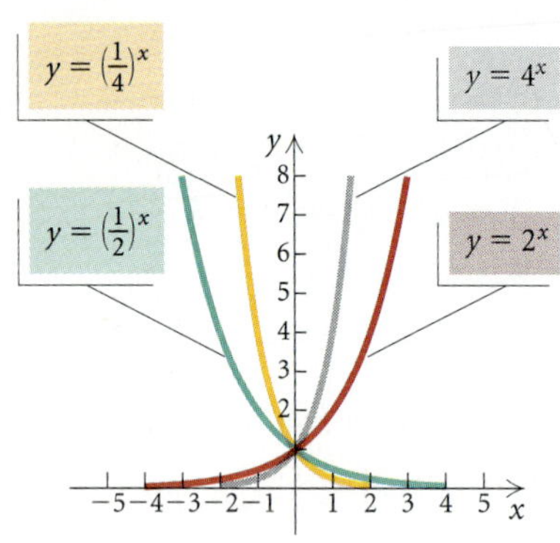

Properties of Exponential Functions

$f(x) = a^x$, $a > 1$:

Continuous

One-to-one

Domain: All real numbers, $\mathbb{R}$

Range: All positive real numbers, $(0, \infty)$

Increasing

Horizontal asymptote is x-axis: $(a^x \to 0$ as $x \to -\infty)$

y-intercept: $(0, 1)$

$f(x) = a^x$ for $0 < a < 1$ or, equivalently, $f(x) = a^{-x}$, $a > 1$:

Continuous

One-to-one

Domain: All real numbers, $\mathbb{R}$

Range: All positive real numbers, $(0, \infty)$

Decreasing

Horizontal asymptote is x-axis: $(a^x \to 0$ as $x \to \infty)$

y-intercept: $(0, 1)$

To graph other types of exponential functions, keep in mind the ideas of translation, stretching, reflection, and combinations of these ideas. All these concepts allow us to visualize the graph before drawing it.

Example 3 Graph each of the following. Before doing so, describe how each graph can be obtained from $f(x) = 2^x$.

a) $f(x) = 2^{x-2}$ **b)** $f(x) = 2^x - 4$ **c)** $f(x) = 5 - 2^{-x}$

SOLUTION

a) The graph of this function is the graph of $y = 2^x$ shifted *right* 2 units.

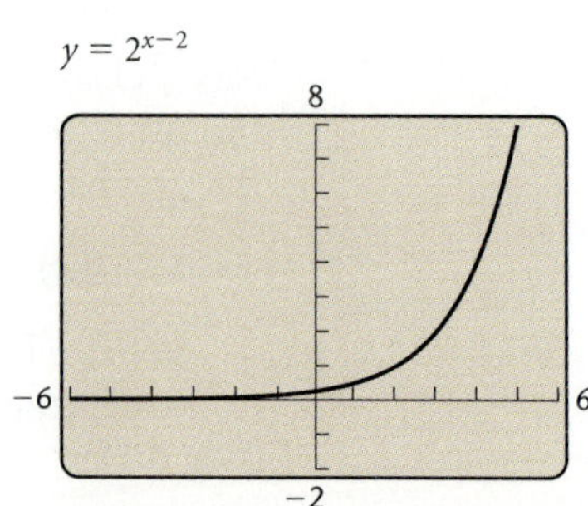

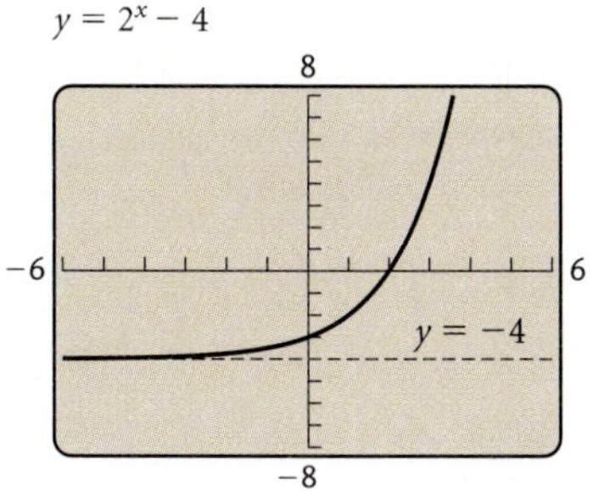

b) The graph is the graph of $y = 2^x$ shifted *down* 4 units (see the graph at left).

c) The graph is a reflection of the graph of $y = 2^x$ across the y-axis, followed by a reflection across the x-axis and then a shift *up* 5 units.

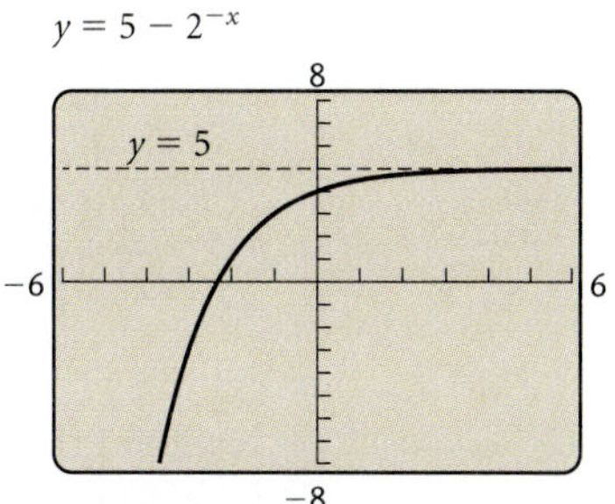

Graphs of Inverses of Exponential Functions

We have noted that every exponential function (with $a > 0$ and $a \neq 1$) is one-to-one. Thus each such function has an inverse that is a function. In the next section, we will name these inverse functions and use them in applications. For now, we draw their graphs by interchanging x and y.

Example 4 Graph: $x = 2^y$.

SOLUTION Note that x is alone on one side of the equation. We can find ordered pairs that are solutions by choosing values for y and then computing the corresponding x-values.

For $y = 0,\ x = 2^0 = 1$.

For $y = 1,\ x = 2^1 = 2$.

For $y = 2,\ x = 2^2 = 4$.

For $y = 3,\ x = 2^3 = 8$.

For $y = -1,\ x = 2^{-1} = \dfrac{1}{2^1} = \dfrac{1}{2}$.

For $y = -2,\ x = 2^{-2} = \dfrac{1}{2^2} = \dfrac{1}{4}$.

For $y = -3,\ x = 2^{-3} = \dfrac{1}{2^3} = \dfrac{1}{8}$.

x		
$x = 2^y$	y	x, y
1	0	$(1, 0)$
2	1	$(2, 1)$
4	2	$(4, 2)$
8	3	$(8, 3)$
$\dfrac{1}{2}$	-1	$\left(\dfrac{1}{2}, -1\right)$
$\dfrac{1}{4}$	-2	$\left(\dfrac{1}{4}, -2\right)$
$\dfrac{1}{8}$	-3	$\left(\dfrac{1}{8}, -3\right)$

(1) Choose values for y.

(2) Compute values for x.

We plot the points and connect them with a smooth curve. Note that the curve does not touch or cross the y-axis. The y-axis is a vertical asymptote.

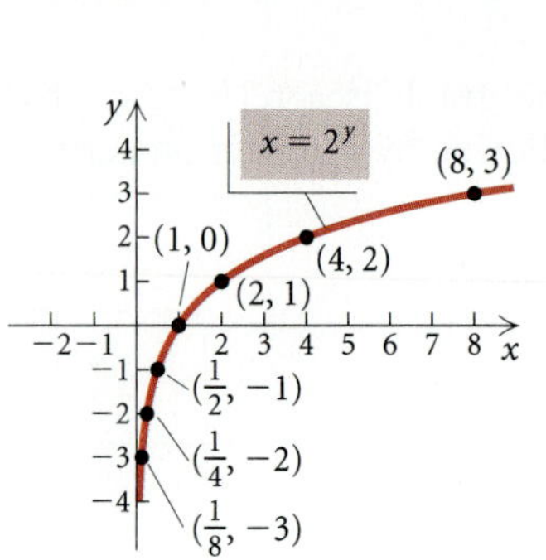

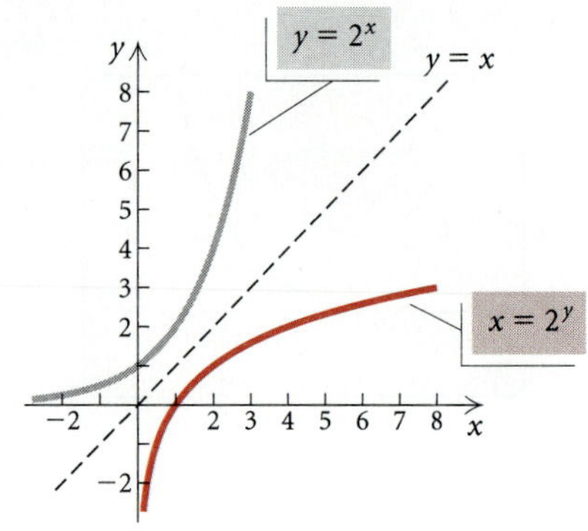

Note too that this curve looks just like the graph of $y = 2^x$, except that it is reflected across the line $y = x$, as we would expect for an inverse. The inverse of $y = 2^x$ is $x = 2^y$. We will explore the inverses of exponential functions further in the next section.

Applications

Graphers are especially helpful when working with exponential functions. They not only facilitate computations but they also allow us to visualize the function. It is worthwhile to get in the habit of creating such a graph, and perhaps looking at a table, even if an exercise or application may not specifically request a graph or table.

Example 5   *Compound Interest.* The amount of money A that a principal P will be worth after t years at interest rate i, compounded n times per year, is given by the formula

$$A = P\left(1 + \frac{i}{n}\right)^{nt}.$$

Suppose that \$100,000 is invested at 8% interest, compounded semi-annually.

a) Find a function for the amount of money after t years.

b) Find the amount of money in the account at $t = 0$, 4, 8, and 10 yr.

c) Graph the function.

d) When will the amount of money in the account reach \$400,000?

SOLUTION

a) Since $P = \$100,000$, $i = 8\% = 0.08$, and $n = 2$, we can substitute these values and form the following function:

$$A(t) = 100,000\left(1 + \frac{0.08}{2}\right)^{2 \cdot t}$$

$$= 100,000(1.04)^{2t}.$$

b) We can enter the function into a grapher, using a TABLE feature set in ASK mode, to compute function values. We can also calculate the values directly on a grapher by substituting in the expression for $A(t)$.

$$A(0) = 100{,}000(1.04)^{2 \cdot 0} = 100{,}000;$$
$$A(4) = 100{,}000(1.04)^{2 \cdot 4} \approx 136{,}856.91;$$
$$A(8) = 100{,}000(1.04)^{2 \cdot 8} \approx 187{,}298.12;$$
$$A(10) = 100{,}000(1.04)^{2 \cdot 10} \approx 219{,}112.31.$$

c) For the graph, we use the viewing window [0, 20, 0, 500,000] because of the large numbers and the fact that negative time values and amounts of money have no meaning in this application.

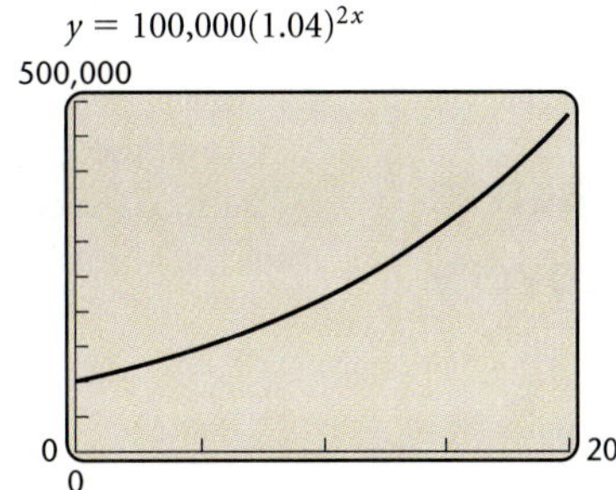

d) To find the amount of time it takes for the account to grow to $400,000, we set

$$100{,}000(1.04)^{2t} = 400{,}000$$

and solve. One way we can do this is by graphing the equations

$$y_1 = 100{,}000(1.04)^{2x} \quad \text{and} \quad y_2 = 400{,}000.$$

Then we can use the TRACE and ZOOM features or the INTERSECT feature as discussed in the Introduction to Graphs and Graphers to estimate the point of intersection.

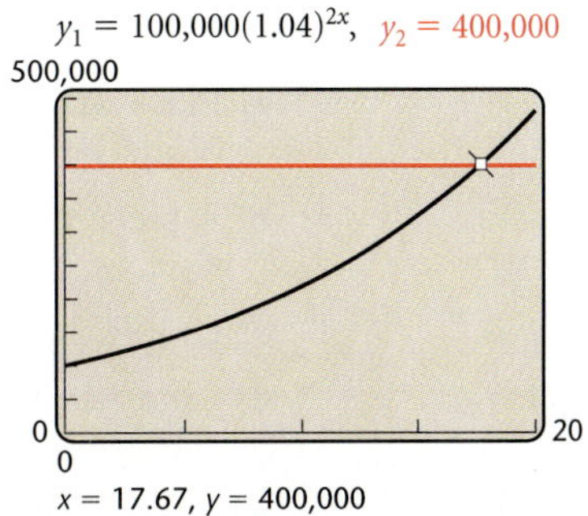

We can also use the SOLVE feature on a grapher. It might be necessary to write the equation as $100{,}000(1.04)^{2x} - 400{,}000 = 0$. In Section 3.5, we will see how to solve equations such as this algebraically.

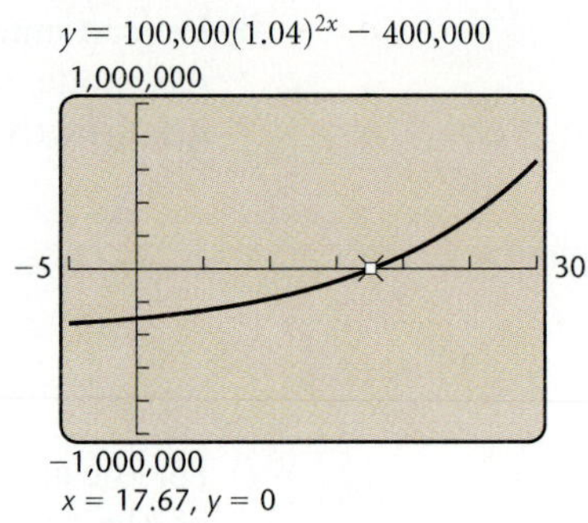

Regardless of the method we use to find the solution, we see that the account grows to $400,000 after about 17.67 yr.

Interactive Discovery

Take a sheet of $8\frac{1}{2}$-in. by 11-in. paper and cut it into two equal pieces. Then cut each of these in half. Next, cut each of the four pieces in half again. Continue this process.

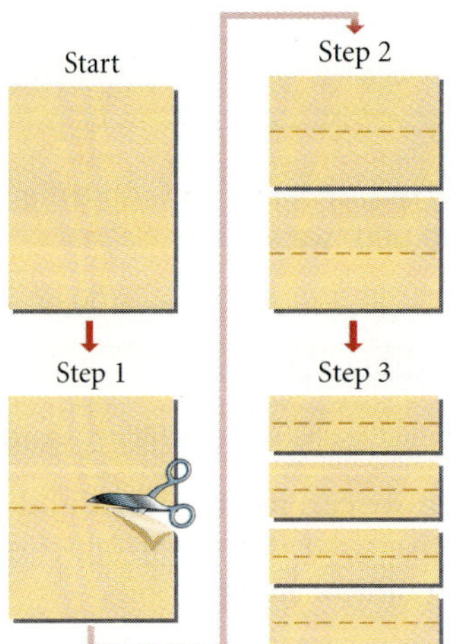

	t	$f(t) = 0.004 \cdot 2^t$
Start	0	$0.004 \cdot 2^0$, or 0.004
Step 1	1	$0.004 \cdot 2^1$, or 0.008
Step 2	2	$0.004 \cdot 2^2$, or 0.016
Step 3	3	
Step 4	4	
Step 5	5	

a) Place all the pieces in a stack and measure the thickness with a micrometer or other measuring device such as a dial caliper.

b) A piece of paper is typically 0.004 in. thick. Check the calculation in part (a) by completing the table shown here.

c) Graph the function $f(t) = 0.004(2)^t$.

d) Compute the thickness of the paper (in miles) after 25 steps.

The Number $e = 2.7182818284\ldots$

We now consider a very special number in mathematics. Though you may not have encountered it before, you will see here and in future mathematics that it has many important applications. To derive this number, we use the compound interest formula $A = P(1 + i/n)^{nt}$ in Example 5. Suppose that $1 is an initial investment at 100% interest for

1 yr (no bank would pay this). The formula above becomes a function A defined in terms of the number of compounding periods n:

$$A(n) = \left(1 + \frac{1}{n}\right)^n.$$

Let's find some function values using the scientific keys on a grapher. Rounding to six decimal places gives us the following table.

n	$A(n) = \left(1 + \dfrac{1}{n}\right)^n$
1 (compounded annually)	\$2.00
2 (compounded semiannually)	\$2.25
3	\$2.370370
4 (compounded quarterly)	\$2.441406
5	\$2.488320
100	\$2.704814
365 (compounded daily)	\$2.714567
8760 (compounded hourly)	\$2.718127

As the values of n get larger and larger, the function values get closer and closer to a number in mathematics, called e. Its decimal representation does not terminate or repeat; it is irrational. In 1741, Leonhard Euler named this number e.

$$e = 2.7182818284\ldots$$

Interactive Discovery

Graph the function $y = (1 + 1/x)^x$. Consider the graph for larger and larger values of x. Does this function have a horizontal asymptote? Explore with a grapher to determine the asymptote, if it exists. You might also try to use the TABLE feature.

We can find values of the exponential function $f(x) = e^x$ using the scientific $\boxed{e^x}$ key on a grapher.

Example 6 Find each value of e^x, to four decimal places, on a grapher.

a) e^3 **b)** $e^{-0.23}$ **c)** e^0 **d)** $100e^{5.8}$ **e)** e^1

SOLUTION

FUNCTION VALUE	READOUT	ROUNDED
a) e^3	20.08553692	20.0855
b) $e^{-0.23}$	0.7945336025	0.7945
c) e^0	1	1
d) $100e^{5.8}$	33029.95599	33029.9560
e) e^1	2.718281828	2.7183

Graphs of Exponential Functions, Base e

We demonstrate ways in which to graph exponential functions.

Example 7 Graph $f(x) = e^x$ and $g(x) = e^{-x}$.

SOLUTION

Method 1. We can compute points for each equation using the $\boxed{e^x}$ key on a grapher. Then we plot these points and draw the graphs of the functions.

x	$f(x) = e^x$	$g(x) = e^{-x}$
-4	0.0183	54.5982
-3	0.0498	20.0855
-2	0.1353	7.3891
-1	0.3679	2.7183
0	1	1
1	2.7183	0.3679
2	7.3891	0.1353
3	20.0855	0.0498
4	54.5982	0.0183

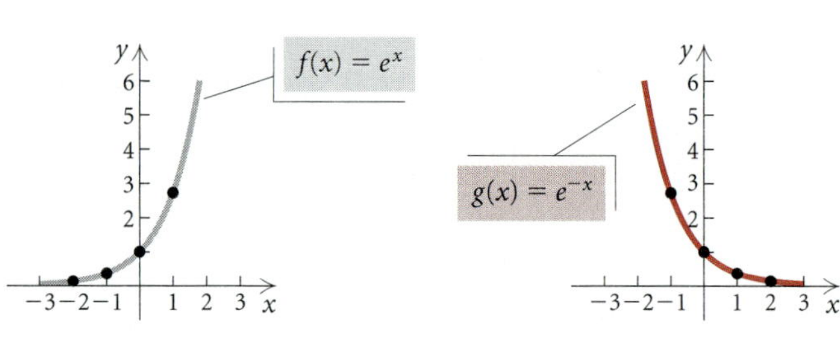

Note that the graph of g is a reflection of the graph of f across the y-axis.

Method 2. On a grapher, we simply enter the equations $y_1 = e^x$ and $y_2 = e^{-x}$. We may need to enter these equations as $y_1 = \exp(x)$ and $y_2 = \exp(-x)$.

Example 8 Graph each of the following using a grapher. Briefly describe how each graph can be obtained from $y = e^x$.

a) $f(x) = e^{-0.5x}$

b) $f(x) = 1 - e^{-2x}$

c) $f(x) = e^{x+3}$

SOLUTION

a) We note that the graph of this function is a horizontal stretching of the graph of $y = e^x$ followed by a reflection across the y-axis.

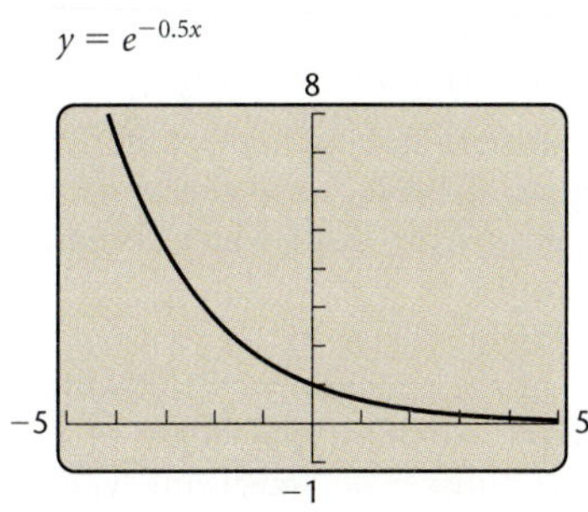

b) The graph is a horizontal shrinking of the graph of $y = e^x$, followed by a reflection across the y-axis, then across the x-axis, followed by a translation up 1 unit. (See the figure on the left below.)

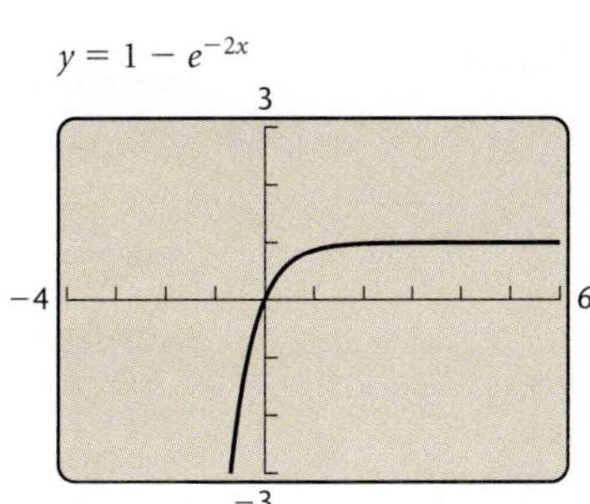

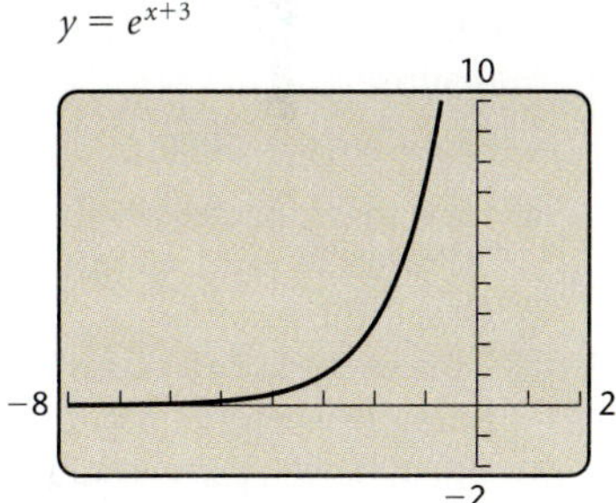

c) The graph is a translation of the graph of $y = e^x$ to the left 3 units. (See the figure on the right above.)

3.2 Exercise Set

Find each of the following, to four decimal places, on a grapher.

1. e^4

2. e^{10}

3. $e^{-2.458}$

4. $e^{1.0345}$

Make a hand-drawn graph of each function. Then check your work using a grapher, if possible.

5. $f(x) = 3^x$

6. $f(x) = 5^x$

7. $f(x) = 6^x$

8. $f(x) = 3^{-x}$

9. $f(x) = \left(\frac{1}{4}\right)^x$

10. $f(x) = \left(\frac{2}{3}\right)^x$

11. $x = 3^y$

12. $x = 4^y$

13. $x = \left(\frac{1}{2}\right)^y$

14. $x = \left(\frac{4}{3}\right)^y$

15. $y = \frac{1}{4}e^x$

16. $y = 2e^{-x}$

17. $f(x) = 1 - e^{-x}$

18. $f(x) = e^x - 2$

Graph each function using a grapher. Describe how each graph can be obtained from a basic exponential function.

19. $f(x) = 2^{x+1}$

20. $f(x) = 2^{x-1}$

21. $f(x) = 2^x - 3$

22. $f(x) = 2^x + 1$

23. $f(x) = 4 - 3^{-x}$

24. $f(x) = 2^{x-1} - 3$

25. $f(x) = \left(\frac{3}{2}\right)^{x-1}$

26. $f(x) = 3^{4-x}$

27. $f(x) = 2^{x+3} - 5$

28. $f(x) = -3^{x-2}$

29. $f(x) = e^{2x}$

30. $f(x) = e^{-0.2x}$

31. $y = e^{-x+1}$

32. $y = e^{2x} + 1$

33. $f(x) = 2(1 - e^{-x})$

34. $f(x) = 1 - e^{-0.01x}$

35. *Growth of AIDS.* The total number of Americans who have contracted AIDS is approximated by the exponential function

$$N(t) = 100,000(1.4)^t,$$

where $t = 0$ corresponds to 1989. (*Source:* U.S. Department of Health & Human Services)

a) According to this function, how many Americans had been infected as of 1997?

b) Predict the number of Americans who will have been infected by 2001.

c) Graph the function.

d) Include the graph of the equation $y = 1,000,000$ with the graph in part (c). Find the length of time it took for the number of Americans who contracted AIDS to reach 1 million.

36. *Growth of Bacteria* Escherichia coli. The bacteria *Escherichia coli* are commonly found in the human bladder. Suppose that 3000 of the bacteria are present at time $t = 0$. Then under certain conditions, t minutes later, the number of bacteria present is

$$N(t) = 3000(2)^{t/20}.$$

a) How many bacteria will be present after 10 min? 20 min? 30 min? 40 min? 60 min?

b) Graph the function.

c) This bacteria can cause bladder infections in humans when the number of bacteria reaches 100,000,000. Use a grapher to include the graph of the equation $y = 100,000,000$ with the graph in part (b). Find the length of time it takes for a bladder infection to be possible.

37. *Recycling Aluminum Cans.* It is estimated that two thirds of all aluminum cans distributed will be recycled each year (*Source:* Alcoa Corporation). A beverage company distributes 350,000 cans. The number still in use after time t, in years, is given by the exponential function

$$N(t) = 350,000\left(\tfrac{2}{3}\right)^t.$$

a) How many cans are still in use after 0 yr? 1 yr? 4 yr? 10 yr?

b) Graph the function.

c) After how long will 2000 of the cans still be in use?

38. *Interest in a College Trust Fund.* Following the birth of a child, Juan deposits $10,000 in a college trust fund where interest is 6.4%, compounded semiannually.

a) Find a function for the amount in the account after t years.

b) Find the amount of money in the account at $t = 0, 4, 8, 10,$ and 18 yr.

c) Graph the function.

d) After how long will the account contain $100,000?

39. *Salvage Value.* A top-quality fax–copying machine is purchased for $5800. Its value each year is about 80% of the value of the preceding year. After t years, its value, in dollars, is given by the exponential function

$$V(t) = 5800(0.8)^t.$$

a) Find the value of the machine after 0 yr, 1 yr, 2 yr, 5 yr, and 10 yr.

b) Graph the function.

c) The company decides to replace the machine when its value has declined to $500. After how long will the machine be replaced?

40. *Revenues of Packard Bell.* Sales of Packard Bell computers have grown exponentially in recent years. The total revenue R, in billions, is given by

$$R(t) = 0.518(1.42)^t,$$

where $t = $ the number of years since 1990.

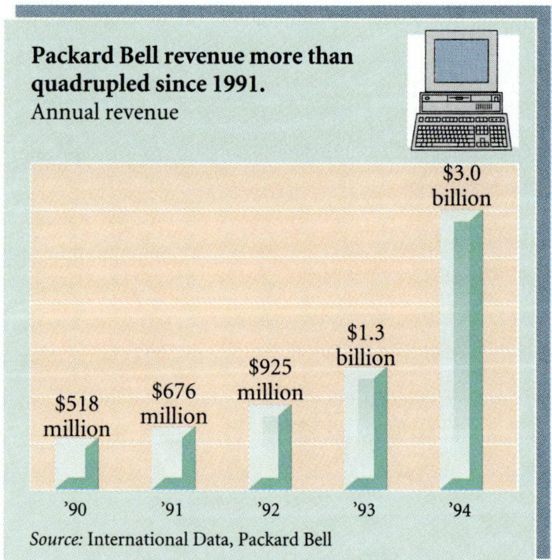

a) Find the total revenue in 1990, 1994, 1998, and 2004.

b) Graph the function.

c) When will revenues be $8 billion?

41. *Timber Demand.* World demand for timber is increasing exponentially. The demand N, in billions of cubic feet, purchased is given by

$$N(t) = 46.6(1.018)^t,$$

where t = the number of years since 1981. (*Source: U.N. Food and Agricultural Organization, American Forest and Paper Association*)

a) Find the demand for timber in 1985, 1997, and 2010.

b) Graph the function.

c) After how many years will the demand for timber be 93.4 billion cubic feet?

42. *Typing Speed.*   Sarah is taking typing in college. After she practices for t hours, her speed, in words per minute, is given by the function

$$S(t) = 200[1 - (0.86)^t].$$

a) What is Sarah's speed after practicing for 10 hr? 20 hr? 40 hr? 100 hr?

b) Graph the function.

c) How much time passes before Sarah's speed is 100 words per minute?

d) Does this graph have an asymptote? If so, what is it, and what is its significance to Sarah's learning?

43. *Advertising.*   A company begins a radio advertising campaign in New York City to market a new CD-ROM video game. The percentage of the target market that buys a game is generally a function of the length of the advertising campaign. The estimated percentage is given by

$$f(t) = 100(1 - e^{-0.04t}),$$

where t = the number of days of the campaign.

a) Find $f(25)$, the percentage of the target market that has bought the product after a 25-day advertising campaign.

b) Graph the function.

c) After how long will 90% of the target market have bought the product?

44. *Growth of a Stock.*   The value of a stock is given by the function

$$V(t) = 58(1 - e^{-1.1t}) + 20,$$

where V = the value of the stock after time t, in months.

a) Find $V(1)$, $V(2)$, $V(4)$, $V(6)$, and $V(12)$.

b) Graph the function.

c) After how long will the value of the stock be \$75?

In Exercises 45–58, use a grapher to match the equation with one of figures (a)–(n), which follow.

a) 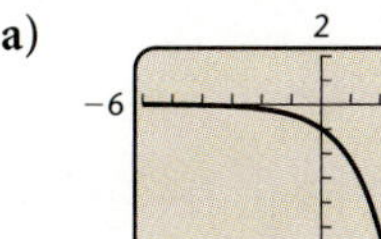

b) 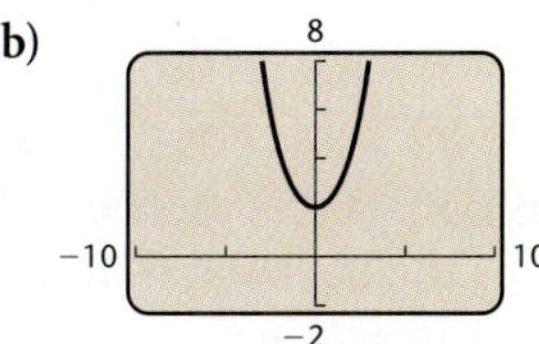

c) 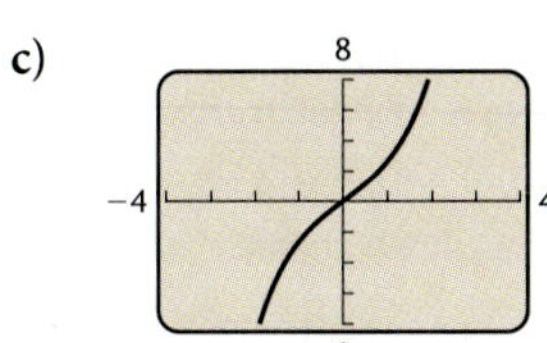

d) 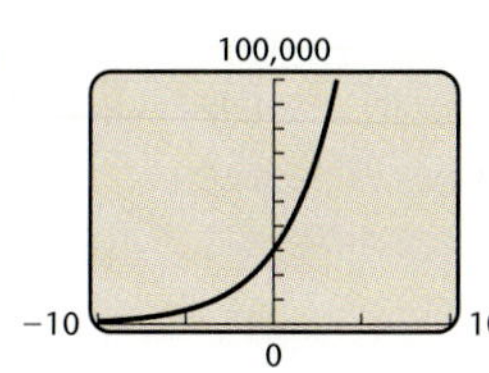

e) 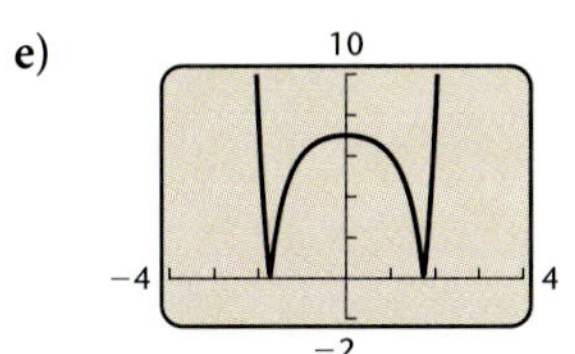

f) 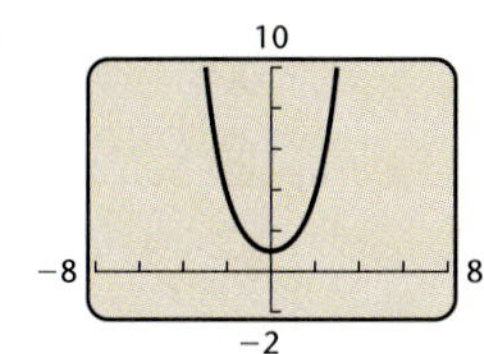

g) 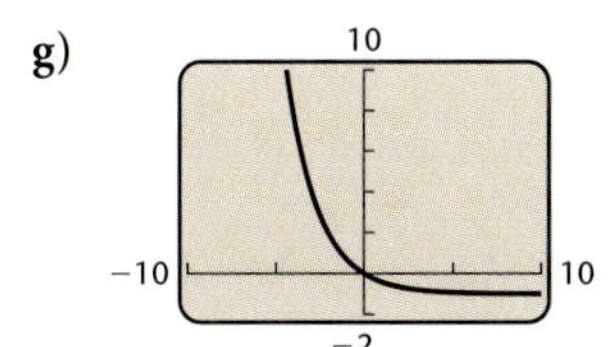

h) 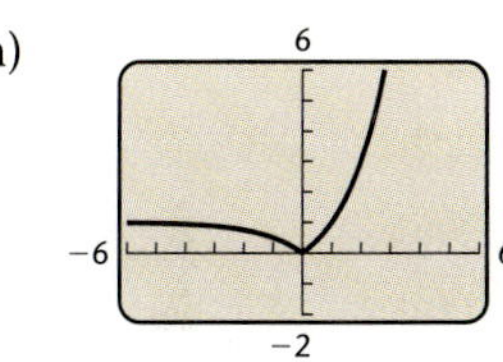

i) 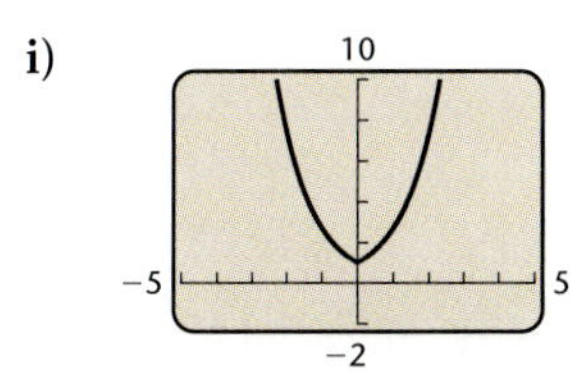

j) 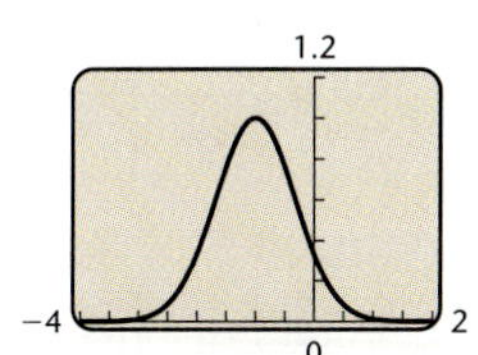

k) 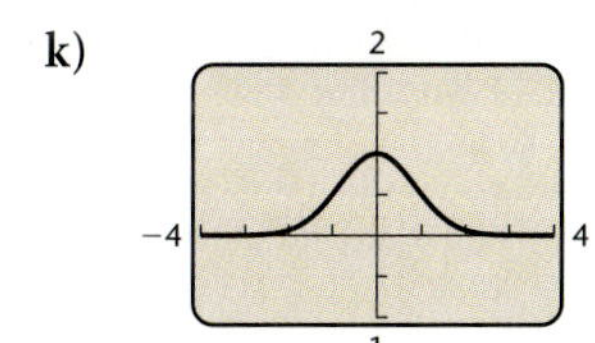

l) 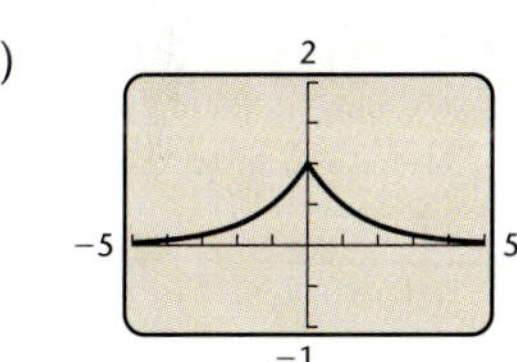

m) 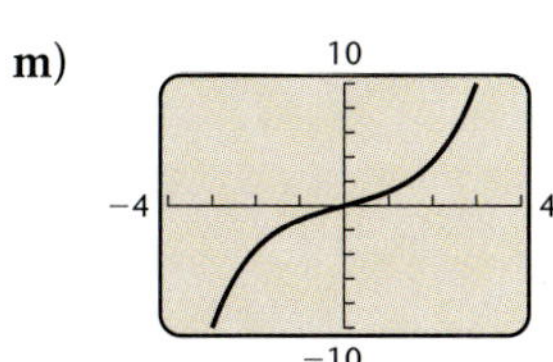

n) 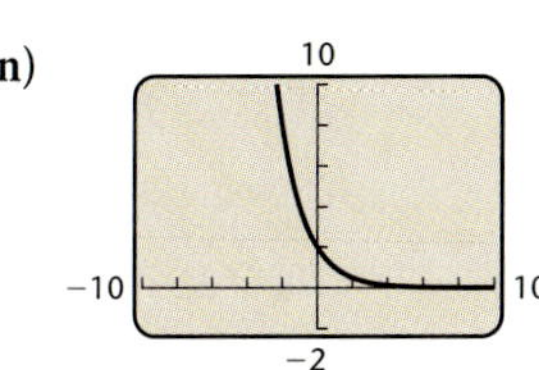

45. $y = 3^x - 3^{-x}$

46. $y = 3^{-(x+1)^2}$

47. $f(x) = -2.3^x$

48. $f(x) = 30{,}000(1.4)^x$

49. $y = 2^{-|x|}$

50. $y = 2^{-(x-1)}$

51. $f(x) = (0.58)^x - 1$

52. $y = 2^x + 2^{-x}$

53. $g(x) = e^{|x|}$

54. $f(x) = |2^x - 1|$

55. $y = 2^{-x^2}$

56. $y = |2^{x^2} - 8|$

57. $g(x) = \dfrac{e^x - e^{-x}}{2}$

58. $f(x) = \dfrac{e^x + e^{-x}}{2}$

Use a grapher to find the point(s) of intersection of the graphs of each of the following pairs of equations.

59. $y = |1 - 3^x|,$
$y = 4 + 3^{-x^2}$

60. $y = 4^x + 4^{-x},$
$y = 8 - 2x - x^2$

61. $y = 2e^x - 3,\ y = \dfrac{e^x}{x}$

62. $y = \dfrac{1}{e^x + 1},\ y = 0.3x + \dfrac{7}{9}$

Use a grapher. Solve graphically.

63. $5.3^x - 4.2^x = 1073$

64. $e^x = x^3$

65. $2^x > 1$

66. $3^x \leq 1$

67. $2^x + 3^x = x^2 + x^3$

68. $31{,}245e^{-3x} = 523{,}467$

Skill Maintenance

Simplify.

69. $a^0,\ a \neq 0$ **70.** a^1 **71.** 10^3 **72.** 10^{-2}

73. To what power must you raise 5 in order to get 25?

74. To what power must you raise 10 in order to get 0.001?

Synthesis

75. ◈ Describe the differences between the graphs of $f(x) = x^3$ and $g(x) = 3^x$.

76. ◈ Suppose that $10,000 is invested for 8 yr at 6.4% interest, compounded annually. In what year will the most interest be earned? Why?

77. ◈ Graph each pair of equations using the same set of axes. Then compare the results in parts (a)

and (b).

a) $y = 3^x,\ x = 3^y$ **b)** $y = 1^x,\ x = 1^y$

78. Which is larger, 7^π or π^7?

79. Graph $f(x) = x^{1/(x-1)}$. Use a grapher and the TABLE feature to identify the horizontal asymptote.

In Exercises 80 and 81:

a) *Graph using a grapher.*
b) *Approximate the zeros.*
c) *Approximate the relative maximum and minimum values. If your grapher has a MAX–MIN feature, use it.*

80. $f(x) = x^2 e^{-x}$ **81.** $f(x) = e^{-x^2}$

82. Consider each of the following functions:

$$y_1 = e^x, \qquad y_2 = 1 + x + \frac{x^2}{2} + \frac{x^3}{6} + \frac{x^4}{24}.$$

a) Use a grapher to graph both functions using the viewing window $[0, 1, -1, 4]$, with Xscl = 1 and Yscl = 1.

b) On the basis of your graphs, would you consider

$$e^x = 1 + x + \frac{x^2}{2} + \frac{x^3}{6} + \frac{x^4}{24}$$

an identity?

c) See if you can prove the equation in part (b) to be an identity. Substitute $x = 1$ in each expression. What is the result?

d) Now go back to the original equations. Change the viewing window, use TRACE and ZOOM and the TABLE feature to examine the graphs in more detail. What do you discover about the graphs?

e) What caution must you be aware of using a grapher to determine whether an equation is an identity?

3.3

Logarithmic Functions and Graphs

• *Graph logarithmic functions.*
• *Convert between exponential and logarithmic equations.*
• *Find common and natural logarithms on a grapher.*

We now consider *logarithmic*, or *logarithm*, *functions*. These functions are inverses of exponential functions and have many applications.

Logarithmic Functions

Consider the function $f(x) = 2^x$ graphed below. We see from the graph that this function passes the horizontal-line test and is one-to-one. Thus f has an inverse that is a function.

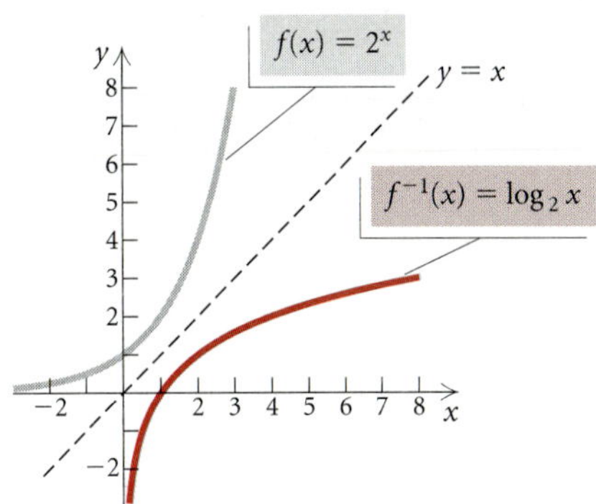

To find a formula for f^{-1} when $f(x) = 2^x$, we try to use the method of Section 3.1:

1. Replace $f(x)$ with y: $y = 2^x$

2. Interchange x and y: $x = 2^y$

3. Solve for y: $y =$ the power to which we raise 2 to get x

4. Replace y with $f^{-1}(x)$: $f^{-1}(x) =$ the power to which we raise 2 to get x.

Mathematicians have defined a new symbol to replace the words "the power to which we raise 2 to get x." That symbol is "$\log_2 x$," read "the logarithm, base 2, of x."

Logarithmic Function, Base 2

"$\log_2 x$," read "the logarithm, base 2, of x," means "the power to which we raise 2 to get x."

Thus if $f(x) = 2^x$, then $f^{-1}(x) = \log_2 x$. For example,

$$f^{-1}(8) = \log_2 8$$
$$= 3,$$

because

3 is the power to which we raise 2 to get 8.

Similarly, $\log_2 13$ is the power to which we raise 2 to get 13. As yet, we have no simpler way to say this other than "$\log_2 13$ is the power to which we raise 2 to get 13." Later, however, we will learn how to approximate this expression using a grapher.

For any exponential function $f(x) = a^x$, its inverse is called a **logarithmic function, base a.** The graph of the inverse can be obtained by reflecting the graph of $y = a^x$ across the line $y = x$, to obtain $x = a^y$. Then $x = a^y$ is equivalent to $y = \log_a x$. We read $\log_a x$ as "the logarithm, base a, of x."

The inverse of $f(x) = a^x$ is given by $f^{-1}(x) = \log_a x$.

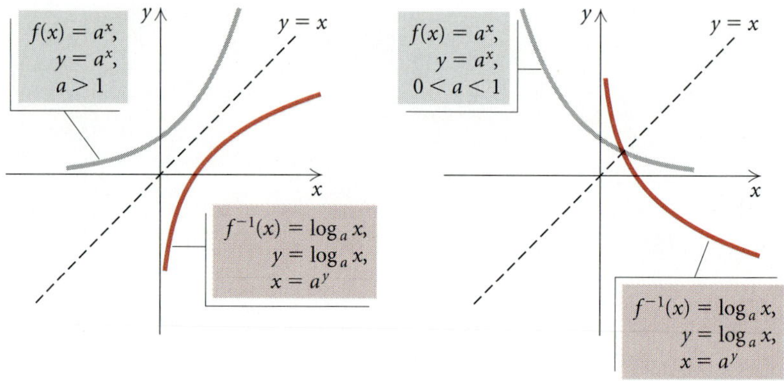

Logarithmic Function, Base a

We define $y = \log_a x$ as that number y such that $x = a^y$, where $x > 0$ and a is a positive constant other than 1.

In the following table, we compare exponential and logarithmic functions. In general, we use a number a that is greater than 1 for the logarithm base.

EXPONENTIAL FUNCTION	LOGARITHMIC FUNCTION
$y = a^x$	$x = a^y$
$f(x) = a^x$	$f^{-1}(x) = \log_a x$
$a > 1$	$a > 1$
Continuous	Continuous
One-to-one	One-to-one
Domain: All real numbers, $\mathbb{R}$	Domain: All positive real numbers, $(0, \infty)$
Range: All positive real numbers, $(0, \infty)$	Range: All real numbers, $\mathbb{R}$
Increasing	Increasing
Horizontal asymptote is x-axis: $(a^x \to 0$ as $x \to -\infty)$	Vertical asymptote is y-axis: $(\log_a x \to -\infty$ as $x \to 0^+)$
y-intercept: $(0, 1)$	x-intercept: $(1, 0)$
There is no x-intercept.	There is no y-intercept.

Converting Between Exponential and Logarithmic Equations

It is helpful in dealing with logarithmic functions to remember that a logarithm of a number is an *exponent*. It is the exponent y in $x = a^y$. You might think to yourself, "the logarithm, base a, of a number x is the power to which a must be raised to get x."

We are led to the following. (The symbol $\longleftrightarrow$ means that the two statements are equivalent; that is, when one is true, the other is true. The words "if and only if" can be used in place of $\longleftrightarrow$.)

$$\log_a x = y \longleftrightarrow x = a^y \qquad \text{A logarithm is an exponent!}$$

Be sure to memorize this relationship! It is probably the most important definition in the chapter. Many times this definition will be justification for a proof or a procedure.

Example 1   Convert each of the following to a logarithmic equation.

a) $16 = 2^x$ **b)** $10^{-3} = 0.001$ **c)** $e^t = 70$

SOLUTION

The exponent is the logarithm.

a) $16 = 2^x \qquad \log_2 16 = x$

The base remains the same.

b) $10^{-3} = 0.001 \longrightarrow \log_{10} 0.001 = -3$

c) $e^t = 70 \longrightarrow \log_e 70 = t$

Example 2   Convert each of the following to an exponential equation.

a) $\log_2 32 = 5$ **b)** $\log_a Q = 8$ **c)** $x = \log_t M$

SOLUTION

The logarithm is the exponent.

a) $\log_2 32 = 5 \qquad 2^5 = 32$

The base remains the same.

b) $\log_a Q = 8 \longrightarrow a^8 = Q$

c) $x = \log_t M \longrightarrow t^x = M$

Finding Certain Logarithms

Let's use the definition of logarithms to find some logarithmic values.

Example 3   Find each of the following logarithms.

a) $\log_{10} 10{,}000$ **b)** $\log_{10} 0.01$
c) $\log_2 8$ **d)** $\log_9 3$
e) $\log_6 1$ **f)** $\log_8 8$

SOLUTION

RESULT	REASON
a) $\log_{10} 10{,}000 = 4$	Think of the meaning of $\log_{10} 10{,}000$. It is the exponent to which we raise 10 to get 10,000. That exponent is 4.

b) $\log_{10} 0.01 = -2$

We have $0.01 = \dfrac{1}{100} = \dfrac{1}{10^2} = 10^{-2}$. The exponent to which we raise 10 to get 0.01 is -2.

c) $\log_2 8 = 3$

$8 = 2^3$. The exponent to which we raise 2 to get 8 is 3.

d) $\log_9 3 = \frac{1}{2}$

$3 = \sqrt{9} = 9^{1/2}$. The exponent to which we raise 9 to get 3 is $\frac{1}{2}$.

e) $\log_6 1 = 0$

$1 = 6^0$. The exponent to which we raise 6 to get 1 is 0.

f) $\log_8 8 = 1$

$8 = 8^1$. The exponent to which we raise 8 to get 8 is 1.

Examples 3(e) and 3(f) illustrate two important properties of logarithms.

$$\log_a 1 = 0 \quad \text{and} \quad \log_a a = 1, \quad \text{for any logarithmic base } a.$$

$\log_a 1 = 0$ follows from the fact that $a^0 = 1$. Thus, $\log_7 1 = 0$, $\log_{10} 1 = 0$, and so on. $\log_a a = 1$ follows from the fact that $a^1 = a$. Thus, $\log_5 5 = 1$, $\log_{10} 10 = 1$, and so on.

Finding Logarithms on a Grapher

Before calculators became so widely available, base-10, or **common logarithms**, were used extensively to simplify complicated calculations. In fact, that is why logarithms were invented. The abbreviation **log**, with no base written, is used to represent the common logarithms, or base 10 logarithms. Thus,

$$\log 29 \quad \text{means} \quad \log_{10} 29.$$

We can approximate $\log 29$ as follows:

$$\log 100 = \log_{10} 100 = 2$$
$$\log 29 = ?$$
$$\log 10 = \log_{10} 10 = 1$$

It seems reasonable that $\log 29$ is between 1 and 2.

On a grapher, the scientific key for common logarithms is generally marked $\boxed{\text{LOG}}$. Using that key, we find that

$$\log 29 \approx 1.462397998 \approx 1.4624$$

rounded to four decimal places. This also tells us that $10^{1.4624} \approx 29$.

Interactive Discovery

Find log 1000 and log 10,000 without using a grapher. Between what two whole numbers is log 9874? Now find log 9874, rounded to four decimal places, on a grapher.

Example 4 Find each of the following common logarithms on a grapher. Round to four decimal places.

a) log 645,778   **b)** log 0.0000239   **c)** log (-3)

SOLUTION

FUNCTION VALUE	READOUT	ROUNDED
a) log 645,778	5.810083246	5.8101
b) log 0.0000239	-4.621602099	-4.6216
c) log (-3)	ERROR*	Does not exist

In Example 4(c), log (-3) does not exist as a real number because there is no real-number power to which we can raise 10 to get -3. The number 10 raised to any real-number power is positive. The common logarithm of a negative number does not exist as a real number.

Natural Logarithms

Logarithms, base e, are called **natural logarithms**. The abbreviation "ln" is generally used for natural logarithms. Thus

$$\ln 53 \quad \text{means} \quad \log_e 53.$$

On a grapher, the scientific key for natural logarithms is generally marked $\boxed{\text{LN}}$. Using that key, we find that

$$\ln 53 \approx 3.970291914 \approx 3.9703$$

rounded to four decimal places.

Example 5 Find each of the following natural logarithms on a grapher. Round to four decimal places.

a) ln 645,778   **b)** ln 0.0000239   **c)** ln (-5)

d) ln e **e)** ln 1

SOLUTION

FUNCTION VALUE	READOUT	ROUNDED
a) ln 645,778	13.37821107	13.3782
b) ln 0.0000239	-10.64163210	-10.6416
c) ln (-5)	ERROR	Does not exist
d) ln e	1	1
e) ln 1	0	0

Note that $\ln e = \log_e e = 1$ and $\ln 1 = \log_e 1 = 0$.

---

*On some graphers, it would be expressed as a complex number—about $0.477 + 1.364i$.

Interactive Discovery

In some textbooks and computer applications, log x is used to represent $\log_e x$ rather than $\log_{10} x$. Explore some methods you might use with a grapher to discover what the base actually is.

Changing Logarithmic Bases

Most graphers give the values of both common logarithms and natural logarithms. To find a logarithm with a base other than 10 or e, we can use the following conversion formula.

> **The Change-of-Base Formula**
>
> For any logarithmic bases a and b, and any positive number M,
>
> $$\log_b M = \frac{\log_a M}{\log_a b}.$$

We will prove this result in the next section.

Program

LOGBASE: This program provides a formula for changing the base of logarithms.

Example 6   Find $\log_5 8$ using common logarithms.

SOLUTION   First we let $a = 10$, $b = 5$, and $M = 8$. Then we substitute into the change-of-base formula:

$$\log_5 8 = \frac{\log_{10} 8}{\log_{10} 5} \qquad \text{Substituting}$$

$$\approx \frac{0.903090}{0.698970}$$

$$\approx 1.2920.$$

To check, we use the power key to verify that $5^{1.2920} \approx 8$.

We can also use base e for a conversion.

Example 7   Find $\log_4 31$ using natural logarithms.

SOLUTION   Substituting e for a, 4 for b, and 31 for M, we have

$$\log_4 31 = \frac{\log_e 31}{\log_e 4} = \frac{\ln 31}{\ln 4} \approx \frac{3.433987}{1.386294} \approx 2.4771.$$

Graphs of Logarithmic Functions

We demonstrate several ways to graph logarithmic functions.

Example 8   Graph: $g(x) = \ln x$.

SOLUTION

Method 1. To graph $y = g(x) = \ln x$, we first write its equivalent exponential equation, $x = e^y$. Then we select values for y and use a grapher to find the corresponding values of e^y. We then plot points, remembering that x still is the first coordinate.

x, or e^y	y
1	0
2.7	1
7.4	2
20	3
0.1	−2
0.4	−1

(1) Select y.
(2) Compute x.

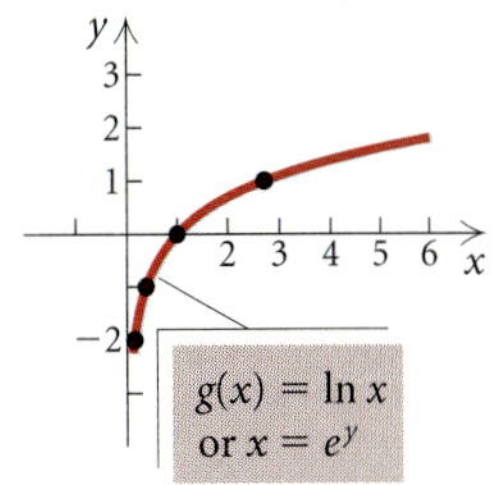

Connect the points with a smooth curve.

Method 2. We can also use the LN key on a grapher to find function values. Then we plot points and draw the graph.

Method 3. We use a grapher to graph $y = \ln x$.

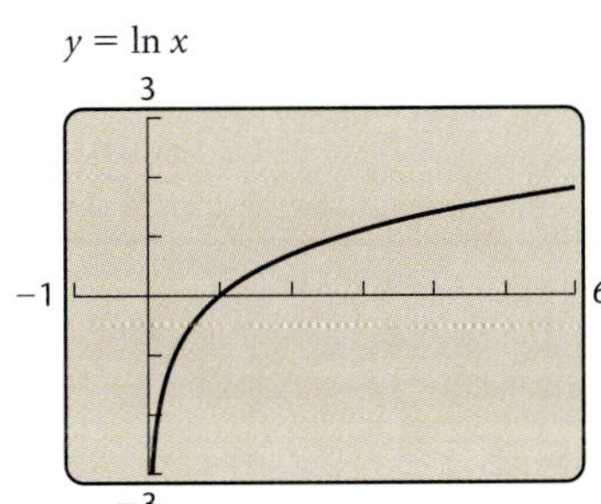

Example 9 Graph $y = f(x) = \log_5 x$.

SOLUTION

Method 1. The equation $y = \log_5 x$ is equivalent to $x = 5^y$. We can find ordered pairs that are solutions by choosing values for y and computing the x-values.

For $y = 0$, $x = 5^0 = 1$.

For $y = 1$, $x = 5^1 = 5$.

For $y = 2$, $x = 5^2 = 25$.

For $y = 3$, $x = 5^3 = 125$.

For $y = -1$, $x = 5^{-1} = \dfrac{1}{5}$.

For $y = -2$, $x = 5^{-2} = \dfrac{1}{25}$.

x, or 5^y	y
1	0
5	1
25	2
125	3
$\dfrac{1}{5}$	−1
$\dfrac{1}{25}$	−2

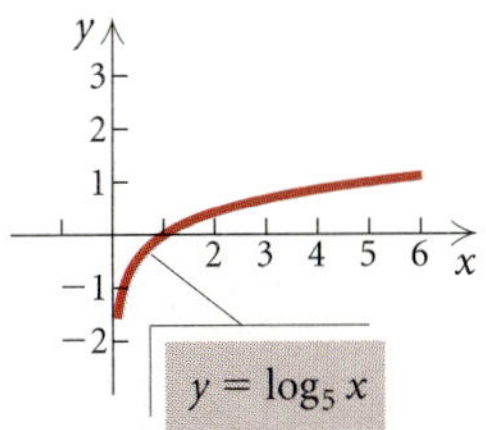

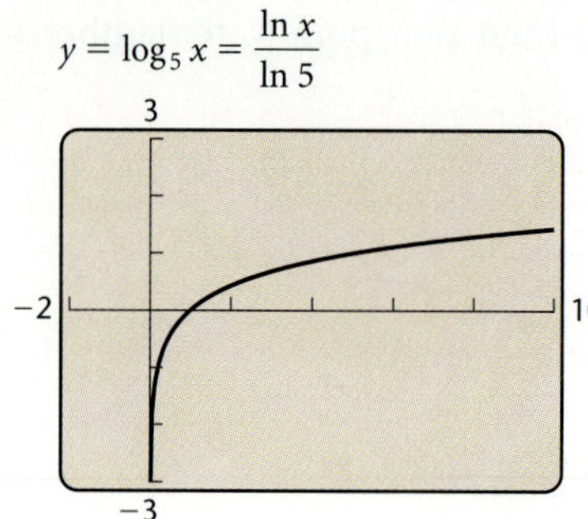

$$y = \log_5 x = \frac{\ln x}{\ln 5}$$

Method 2. To use a grapher, we must first change the base. Here we change from base 5 to base e:

$$y = \log_5 x = \frac{\ln x}{\ln 5}. \qquad \text{Using } \log_b M = \frac{\log_a M}{\log_a b}$$

The graph is as shown at left.

Method 3. Some graphers have a feature that graphs inverses automatically. If we begin with $y_1 = 5^x$, the graphs of both y_1 and its inverse $y_2 = \log_5 x$ will be drawn.

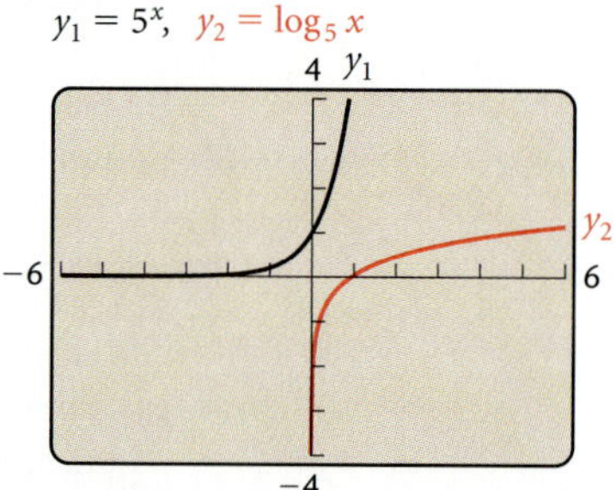

The graph of $f(x) = \log_a x$, for any base a, has the x-intercept $(1, 0)$. The domain is the set of positive real numbers. The range is the set of all real numbers.

Example 10   Graph each of the following using a grapher. Describe how each graph can be obtained from $y = \ln x$. Give the domain of each function and discuss the vertical asymptotes.

a) $f(x) = \ln (x + 3)$

b) $f(x) = 3 - \frac{1}{2} \ln x$

c) $f(x) = |\ln (x - 1)|$

SOLUTION

a) The graph is a shift of the graph of $y = \ln x$ left 3 units.

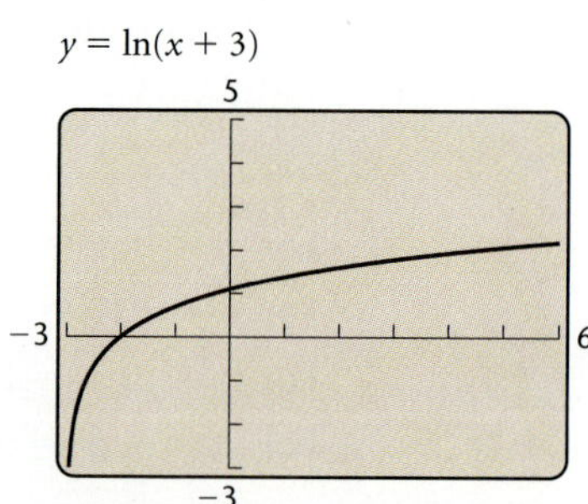

The domain is the set of all real numbers greater than -3, $(-3, \infty)$. The line $x = -3$ is a vertical asymptote.

b) The graph is a shrinking of the graph of $y = \ln x$ in the y-direction, followed by a reflection across the x-axis, and then a translation 3 units up.

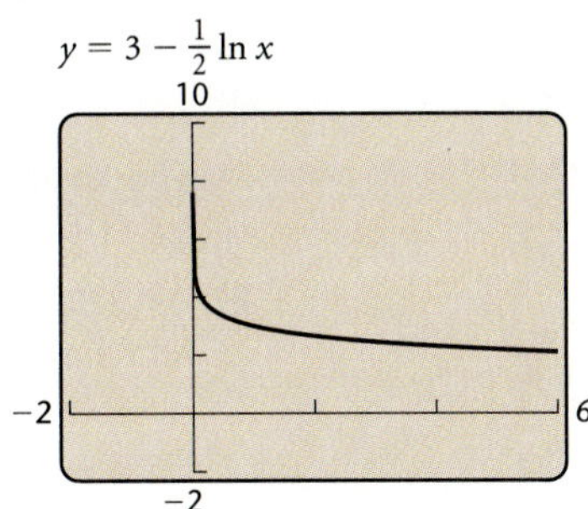

The domain is the set of all positive real numbers, $(0, \infty)$. The y-axis is a vertical asymptote.

c) The graph is a translation of the graph of $y = \ln x$, 1 unit to the right. Then the absolute value has the effect of reflecting negative outputs across the x-axis.

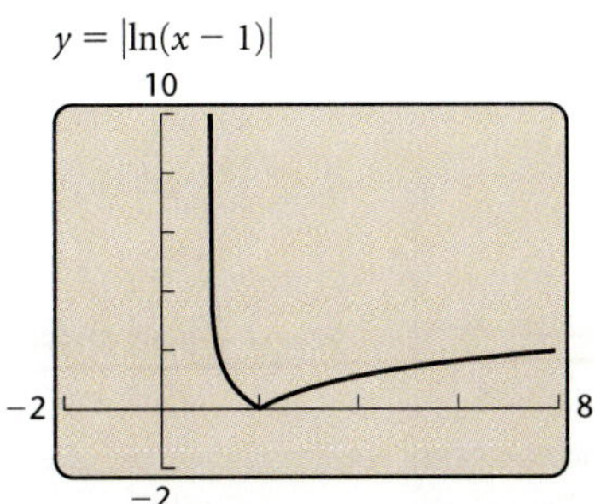

The domain is the set of all real numbers greater than 1, $(1, \infty)$. The line $x = 1$ is a vertical asymptote.

Applications

Example 11 *Walking Speed.* In a study by psychologists Bornstein and Bornstein, it was found that the average walking speed w, in feet per second, of a person living in a city of population P, in thousands, is given by the function

$$w(P) = 0.37 \ln P + 0.05.$$

(Source: International Journal of Psychology)

a) The population of Orlando, Florida, is 1,073,000. Find the average walking speed of people living in Orlando.

b) Graph the equation.

c) A sociologist computes the average walking speed in a city to be about 2.0 ft/sec. Use this information to estimate the population of the city.

SOLUTION

a) We substitute 1073 for P, since P is in thousands:

$$w(1073) = 0.37 \ln 1073 + 0.05 \qquad \text{Substituting}$$
$$\approx 2.6 \text{ ft/sec.} \qquad \text{Finding the natural logarithm and simplifying}$$

The average walking speed of people living in Orlando is about 2.6 ft/sec.

b) We graph with a viewing window of [0, 600, 0, 4] because inputs are very large and outputs are very small by comparison.

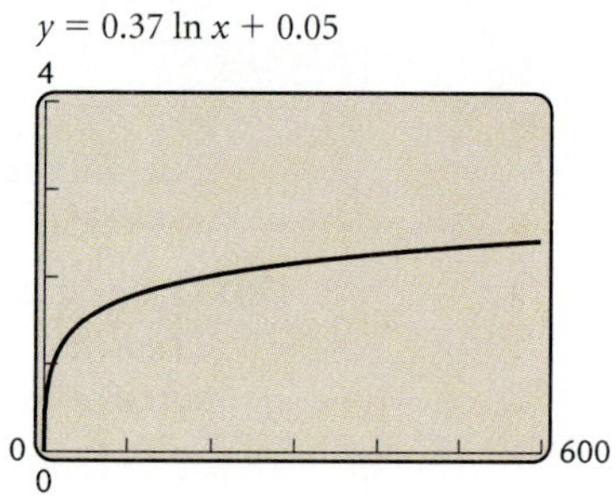

c) To find the population for which the average walking speed is 2.0 ft/sec, we set

$$2.0 = 0.37 \ln P + 0.05$$

and solve. We begin by graphing the equations $y_1 = 0.37 \ln x + 0.05$ and $y_2 = 2$. Then we can use the TRACE and ZOOM features, or the INTERSECT feature, to approximate the point of intersection. See one such TRACE at left. We see that a city with a population of about 194,500 would have an average walking speed of 2.0 ft/sec. Some graphers can solve the equation directly.

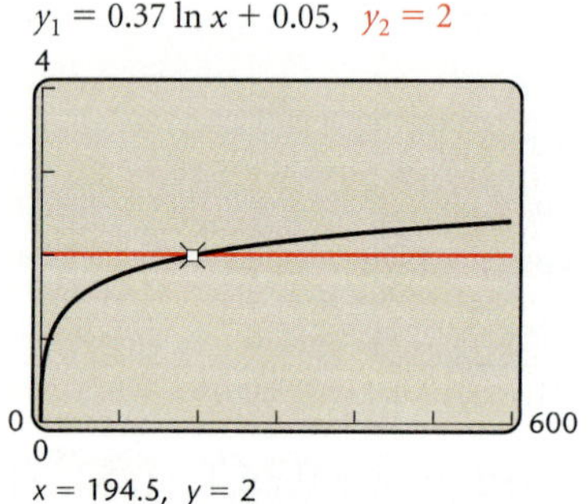

Example 12 *Earthquake Magnitude.* The magnitude R, measured on the Richter scale, of an earthquake of intensity I is defined as

$$R = \log \frac{I}{I_0},$$

where I_0 is a minimum intensity used for comparison. We can think of I_0 as a threshold intensity that is the weakest earthquake that can be recorded on a seismograph. If one earthquake is 10 times as intense as another, its magnitude on the Richter scale is 1 greater than that of the other. If one earthquake is 100 times as intense as another, its magnitude on the Richter scale is 2 higher, and so on. Thus an earthquake whose magnitude is 7 on the Richter scale is 10 times as intense as an earthquake whose magnitude is 6. Earthquakes can be interpreted as multiples of the minimum intensity I_0.

The Kobe, Japan, earthquake of 1995 had an intensity of $10^{6.8} \cdot I_0$. What was its magnitude on the Richter scale?

SOLUTION   We substitute into the formula:

$$R = \log \frac{I}{I_0} = \log \frac{10^{6.8}I_0}{I_0} = \log 10^{6.8} = 6.8.$$

The magnitude of the earthquake was 6.8 on the Richter scale.

3.3 Exercise Set

Make a hand-drawn graph of each of the following. Then check your work using a grapher.

1. $y = \log_3 x$ **2.** $y = \log_4 x$

3. $f(x) = \log x$ **4.** $f(x) = \ln x$

Find each of the following. Do not use a grapher.

5. $\log_2 16$ **6.** $\log_3 9$

7. $\log_5 125$ **8.** $\log_2 64$

9. $\log 0.001$ **10.** $\log 100$

11. $\log_2 \frac{1}{4}$ **12.** $\log_8 2$

13. $\ln 1$ **14.** $\ln e$

15. $\log 10$ **16.** $\log 1$

Convert to a logarithmic equation.

17. $10^3 = 1000$ **18.** $5^{-3} = \frac{1}{125}$

19. $8^{1/3} = 2$ **20.** $10^{0.3010} = 2$

21. $e^3 = t$ **22.** $Q^t = x$

23. $e^2 = 7.3891$ **24.** $e^{-1} = 0.3679$

25. $p^k = 3$ **26.** $e^{-t} = 4000$

Convert to an exponential equation.

27. $\log_5 5 = 1$ **28.** $t = \log_4 7$

29. $\log 0.01 = -2$ **30.** $\log 7 = 0.845$

31. $\ln 30 = 3.4012$ **32.** $\ln 0.38 = -0.9676$

33. $\log_a M = -x$ **34.** $\log_t Q = k$

35. $\log_a T^3 = x$ **36.** $\ln W^5 = t$

Find each of the following on a grapher. Round to four decimal places.

37. $\log 3$ **38.** $\log 8$

39. $\log 532$ **40.** $\log 93,100$

41. $\log 0.57$ **42.** $\log 0.082$

43. $\log (-2)$ **44.** $\ln 50$

45. $\ln 2$ **46.** $\ln (-4)$

47. $\ln 809.3$ **48.** $\ln 0.00037$

49. $\ln (-1.32)$ **50.** $\ln 0$

Find the logarithm using the change-of-base formula.

51. $\log_4 100$ **52.** $\log_3 20$

53. $\log_{100} 0.3$ **54.** $\log_\pi 100$

55. $\log_{200} 50$ **56.** $\log_{5.3} 1700$

For each of the following functions, briefly describe how each graph can be obtained from a basic logarithmic function. Then graph the function using a grapher. Give the domain of each function and discuss the vertical asymptotes.

57. $f(x) = \log_2 (x + 3)$ **58.** $f(x) = \log_3 (x - 2)$

59. $y = \log_3 x - 1$ **60.** $y = 3 + \log_2 x$

61. $f(x) = 4 \ln x$ **62.** $f(x) = \frac{1}{2} \ln x$

63. $y = 2 - \ln x$ **64.** $y = \ln (x + 1)$

Graph each function and its inverse using the same set of axes. Use any method.

65. $f(x) = 3^x,\ f^{-1}(x) = \log_3 x$

66. $f(x) = \log_4 x,\ f^{-1}(x) = 4^x$

67. $f(x) = \log x,\ f^{-1}(x) = 10^x$

68. $f(x) = e^x,\ f^{-1}(x) = \ln x$

69. *Walking Speed.*   Refer to Example 11. Various cities and their populations are given below. Find the average walking speed in each city.

a) Los Angeles, California: 3,486,000
b) Chicago, Illinois: 3,005,000
c) Tucson, Arizona: 406,000
d) Rome, New York: 50,400

70. *Earthquake Magnitude.*   Refer to Example 12. Various locations of earthquakes and their intensities are given below. What was the magnitude on the Richter scale?

a) Mexico City, 1978: $10^{7.85} \cdot I_0$
b) San Francisco, 1906: $10^{8.25} \cdot I_0$
c) Chile, 1960: $10^{9.6} \cdot I_0$
d) Italy, 1980: $10^{7.85} \cdot I_0$
e) San Francisco, 1989: $10^{6.9} \cdot I_0$

71. *Forgetting.*   Students in an accounting class took a final exam. They took equivalent forms of the exam in monthly intervals thereafter. The average score $S(t)$, in percent, after t months was found to be given by the function

$$S(t) = 78 - 15 \log (t + 1), \quad t \geq 0.$$

a) What was the average score when they initially took the test, $t = 0$?
b) What was the average score after 4 months? 24 months?
c) Graph the function.
d) After what time t was the average score 50?

72. *pH of Substances in Chemistry.*   In chemistry, the pH of a substance is defined as

$$pH = -\log [H^+],$$

where H^+ is the hydrogen ion concentration, in moles per liter. Find the pH of each substance.

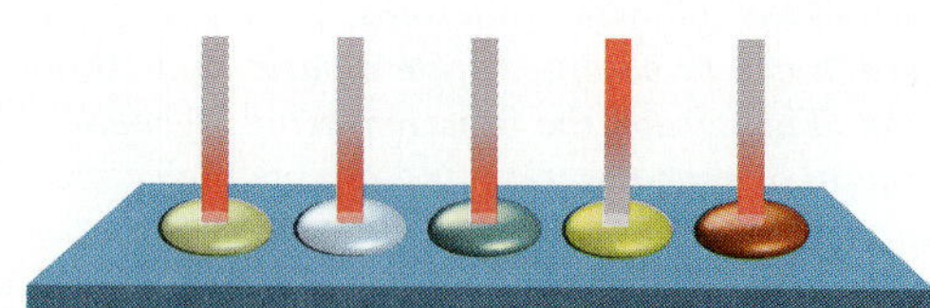

Litmus paper is used to test pH.

SUBSTANCE	HYDROGEN ION CONCENTRATION
a) Pineapple juice	1.6×10^{-4}
b) Hair rinse	0.0013
c) Mouthwash	6.3×10^{-7}
d) Eggs	1.6×10^{-8}
e) Tomatoes	6.3×10^{-5}

73. Find the hydrogen ion concentration of each substance, given the pH. Express the answer in scientific notation.

SUBSTANCE	PH
a) Tap water	7
b) Rainwater	5.4
c) Orange juice	3.2
d) Wine	4.8

74. *Advertising.*   A model for advertising response is given by the function

$$N(a) = 1000 + 200 \ln a, \quad a \geq 1,$$

where $N(a) =$ the number of units sold and $a =$ the amount spent on advertising, in thousands of dollars.

a) How many units were sold after spending $1000 ($a = 1$) on advertising?

b) How many units were sold after spending $5000?
c) Graph the function.
d) How much would have to be spent in order to sell 2000 units?

75. *Loudness of Sound.*   The **loudness L**, in bels (after Alexander Graham Bell), of a sound of intensity I is defined to be

$$L = \log \frac{I}{I_0},$$

where $I_0 =$ the minimum intensity detectable by the human ear (such as the tick of a watch at 20 ft under quiet conditions). If a sound is 10 times as intense as another, its loudness is 1 bel greater than that of the other. If a sound is 100 times as intense as another, its loudness is 2 bels greater, and so on. The bel is a large unit, so a subunit, the **decibel**, is generally used. For L, in decibels, the formula is

$$L = 10 \log \frac{I}{I_0}.$$

Find the loudness, in decibels, of each sound with the given intensity.

SOUND	INTENSITY
a) Library	$2510 \cdot I_0$
b) Dishwater	$2{,}500{,}000 \cdot I_0$
c) Conversational speech	$10^6 \cdot I_0$
d) Heavy truck	$10^9 \cdot I_0$

Skill Maintenance

76. Multiply and simplify: $x^{-5t} \cdot x^{7t}$.

77. Divide and simplify: $\dfrac{y^{-52}}{y^{-76}}$.

Simplify.
78. $(e^{-12})^{-5}$

79. $(10^4)^t$

Synthesis

80. ◆ Explain how the graph of $f(x) = \ln x$ can be used to obtain the graph of $g(x) = e^{x-2}$.

81. ◆ Explain how the graph of $f(x) = e^x$ can be used to obtain the graph of $g(x) = 3 + \ln x$.

Simplify.
82. $\dfrac{\log_3 64}{\log_3 16}$

83. $\dfrac{\log_5 8}{\log_5 2}$

Find the domain of each function.
84. $f(x) = \log_4 x^2$

85. $f(x) = \log_5 x^3$

86. $f(x) = \log (3x - 4)$

87. $f(x) = \ln |x|$

Solve.

88. $\log_2 (x - 3) \geq 4$

89. $\log_2 (2x + 5) < 0$

90. Use a grapher. Find the point(s) of intersection of the graphs.

$$y = 4 \ln x, \qquad y = \frac{4}{e^x + 1}$$

In Exercises 91–94, match the equation with one of figures (a)–(d), which follow. If needed, use a grapher.

a)
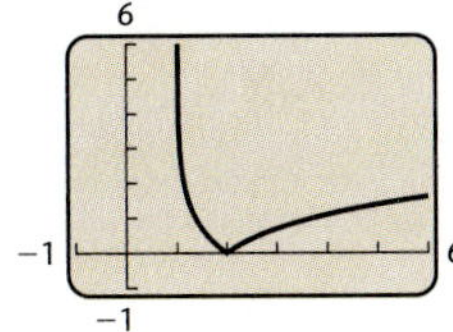

b)
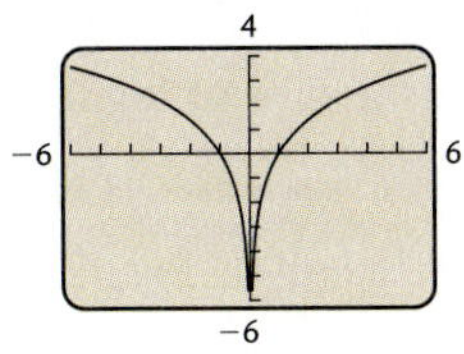

c)
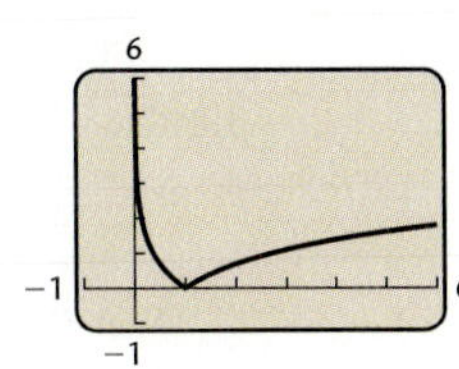

d)
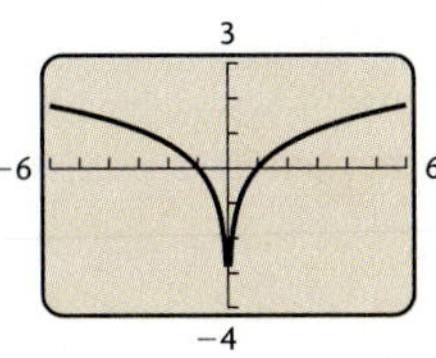

91. $f(x) = \ln |x|$

92. $f(x) = |\ln x|$

93. $f(x) = \ln x^2$

94. $g(x) = |\ln (x - 1)|$

For Exercises 95–98:

a) *Graph the function.*
b) *Estimate the zeros.*
c) *Estimate the relative maximum and the minimum values.*

95. $f(x) = x \ln x$

96. $f(x) = x^2 \ln x$

97. $f(x) = \dfrac{\ln x}{x^2}$

98. $f(x) = e^{-x} \ln x$

3.4
Properties of Logarithmic Functions

- *Convert from logarithms of products, powers, and quotients to expressions in terms of individual logarithms, and conversely.*
- *Simplify expressions of the type $\log_a a^x$ and $a^{\log_a x}$.*

We now establish some properties of logarithmic functions that are fundamental to their use here and in subsequent mathematics. These properties are based on corresponding rules for exponents.

Logarithms of Products

Interactive Discovery

Graph each of the following functions. Use the graphs to discover an identity.

$$y_1 = \log (4x^2 \cdot x^3), \quad y_2 = \log (4x^2) + \log (x^3),$$
$$y_3 = \log (4x^2 + x^3), \quad y_4 = [\log (4x^2)][\log (x^3)]$$

Test your result by using a TABLE feature to complete this table.

X	Y1	Y2	Y3	Y4
1				
10				
20				
300				

X = 1

Graph some other functions to test your discovery.

The first property of logarithms corresponds to the rule of exponents: $a^m \cdot a^n = a^{m+n}$.

The Product Rule

For any positive numbers M and N and any logarithmic base a,

$$\log_a MN = \log_a M + \log_a N.$$

(The logarithm of a product is the sum of the logarithms of the factors.)

Example 1   Express as a sum of logarithms: $\log_3 (9 \cdot 27)$.

SOLUTION   We have

$$\log_3 (9 \cdot 27) = \log_3 9 + \log_3 27. \qquad \text{Using the product rule}$$

As a check, note that

$$\log_3 (9 \cdot 27) = \log_3 243 = 5$$

and that

$$\log_3 9 + \log_3 27 = 2 + 3 = 5.$$

Example 2   Express as a single logarithm: $\log_2 p^3 + \log_2 q$.

SOLUTION   We have

$$\log_2 p^3 + \log_2 q = \log_2 (p^3 q).$$

A Proof of the Product Rule: Let $\log_a M = x$ and $\log_a N = y$. Converting to exponential equations, we have $a^x = M$ and $a^y = N$. Then

$$MN = a^x \cdot a^y$$
$$= a^{x+y}.$$

Converting back to a logarithmic equation, we get

$$\log_a MN = x + y.$$

Remembering what x and y represent, it follows that

$$\log_a MN = \log_a M + \log_a N.$$

Logarithms of Powers

Interactive Discovery

Graph each of the following functions. Use the graphs to discover an identity.

$$y_1 = \log x^5, \qquad y_2 = \log x + \log 5,$$
$$y_3 = 5 \log x, \qquad y_4 = \log (x + 5)$$

Test your result by using a TABLE feature to complete this table.

(continued)

X	Y1	Y2	Y3	Y4
1				
10				
20				
300				

X = 1

Graph some other functions to test your discovery.

The second property of logarithms corresponds to the rule of exponents: $(a^m)^n = a^{mn}$.

The Power Rule

For any positive number M, any logarithmic base a, and any real number p,

$$\log_a M^p = p \log_a M.$$

(The logarithm of a power of M is the exponent times the logarithm of M.)

Example 3 Express each of the following as a product.

a) $\log_a 11^{-3}$ **b)** $\log_a \sqrt[4]{7}$

SOLUTION

a) $\log_a 11^{-3} = -3 \log_a 11$ Using the power rule

b) $\log_a \sqrt[4]{7} = \log_a 7^{1/4}$ Writing exponential notation

$\qquad\qquad = \frac{1}{4} \log_a 7$ Using the power rule

A Proof of the Power Rule: Let $x = \log_a M$. The equivalent exponential equation is $a^x = M$. Raising both sides to the power p, we obtain

$$(a^x)^p = M^p,$$

or

$$a^{xp} = M^p.$$

Converting back to a logarithmic equation, we get

$$\log_a M^p = xp.$$

But $x = \log_a M$, so substituting gives us

$$\log_a M^p = (\log_a M)p$$
$$= p \log_a M.$$

Logarithms of Quotients

Graph each of the following functions. Use the graphs to discover an identity.

$$y_1 = \log\left(\frac{4x^5}{x^2}\right), \qquad y_2 = \log(4x^5) + \log(x^2),$$

$$y_3 = \log(4x^5) - \log(x^2), \qquad y_4 = \log(4x + 5x^2)$$

Test your result by using a TABLE feature to complete this table.

X	Y₁	Y₂	Y₃	Y₄
1				
10				
20				
300				

X = 1

Graph some other functions to test your discovery.

The third property of logarithms corresponds to the rule of exponents:

$$\frac{a^m}{a^n} = a^{m-n}.$$

The Quotient Rule

For any positive numbers M and N, and any logarithmic base a,

$$\log_a \frac{M}{N} = \log_a M - \log_a N.$$

(The logarithm of a quotient is the logarithm of the numerator minus the logarithm of the denominator.)

Example 4   Express as a difference of logarithms: $\log_t \dfrac{8}{w}$.

SOLUTION

$$\log_t \frac{8}{w} = \log_t 8 - \log_t w \qquad \text{Using the quotient rule}$$

Example 5   Express as a single logarithm: $\log_b 64 - \log_b 16$.

SOLUTION

$$\log_b 64 - \log_b 16 = \log_b \frac{64}{16} = \log_b 4$$

A Proof of the Quotient Rule: The proof follows from both the product and the power rules:

$$\log_a \frac{M}{N} = \log_a MN^{-1}$$

$$= \log_a M + \log_a N^{-1} \qquad \text{Using the product rule}$$
$$= \log_a M + (-1)\log_a N \qquad \text{Using the power rule}$$
$$= \log_a M - \log_a N.$$

Note the following.

$\log_a MN \neq (\log_a M)(\log_a N)$	The logarithm of a product is *not* the product of the logarithms.
$\log_a (M + N) \neq \log_a M + \log_a N$	The logarithm of a sum is *not* the sum of the logarithms.
$\log_a \dfrac{M}{N} \neq \dfrac{\log_a M}{\log_a N}$	The logarithm of a quotient is *not* the quotient of the logarithms.

Using the Properties Together

Example 6   Express each of the following in terms of sums and differences of logarithms.

a) $\log_a \dfrac{x^2 y^5}{z^4}$
  b) $\log_a \sqrt[3]{\dfrac{a^2 b}{c^5}}$
  c) $\log_b \dfrac{a y^5}{m^3 n^4}$

SOLUTION

a) $\log_a \dfrac{x^2 y^5}{z^4} = \log_a (x^2 y^5) - \log_a z^4$ 　　Using the quotient rule

$$= \log_a x^2 + \log_a y^5 - \log_a z^4 \qquad \text{Using the product rule}$$
$$= 2\log_a x + 5\log_a y - 4\log_a z \qquad \text{Using the power rule}$$

b) $\log_a \sqrt[3]{\dfrac{a^2 b}{c^5}} = \log_a \left(\dfrac{a^2 b}{c^5}\right)^{1/3}$ 　　Writing exponential notation

$$= \frac{1}{3}\log_a \frac{a^2 b}{c^5} \qquad \text{Using the power rule}$$

$$= \frac{1}{3}(\log_a a^2 b - \log_a c^5) \qquad \text{Using the quotient rule}$$

$$= \frac{1}{3}(2\log_a a + \log_a b - 5\log_a c) \qquad \text{Using the product and power rules. The parentheses are important.}$$

$$= \frac{1}{3}(2 + \log_a b - 5\log_a c) \qquad \log_a a = 1$$

$$= \frac{2}{3} + \frac{1}{3}\log_a b - \frac{5}{3}\log_a c \qquad \text{Multiplying to remove parentheses}$$

c) $\log_b \dfrac{ay^5}{m^3n^4} = \log_b ay^5 - \log_b m^3n^4$ Using the quotient rule

$$= (\log_b a + \log_b y^5) - (\log_b m^3 + \log_b n^4)$$ Using the product rule

$$= \log_b a + \log_b y^5 - \log_b m^3 - \log_b n^4$$ Removing parentheses

$$= \log_b a + 5\log_b y - 3\log_b m - 4\log_b n$$ Using the power rule

Example 7   Express as a single logarithm:

$$5\log_b x - \log_b y + \frac{1}{4}\log_b z.$$

SOLUTION

$$5\log_b x - \log_b y + \frac{1}{4}\log_b z = \log_b x^5 - \log_b y + \log_b z^{1/4}$$

Using the power rule

$$= \log_b \frac{x^5}{y} + \log_b z^{1/4}$$

Using the quotient rule

$$= \log_b \frac{x^5 z^{1/4}}{y}, \text{ or } \log_b \frac{x^5 \sqrt[4]{z}}{y}$$

Using the product rule

Example 8   Given that $\log_a 2 = 0.301$ and $\log_a 3 = 0.477$, find each of the following.

a) $\log_a 6$ b) $\log_a \dfrac{2}{3}$ c) $\log_a 81$

d) $\log_a \sqrt{a}$ e) $\log_a 5$ f) $\dfrac{\log_a 3}{\log_a 2}$

SOLUTION

a) $\log_a 6 = \log_a (2 \cdot 3) = \log_a 2 + \log_a 3$ **Using the product rule**

$$= 0.301 + 0.477$$
$$= 0.778$$

b) $\log_a \frac{2}{3} = \log_a 2 - \log_a 3$ **Using the quotient rule**

$$= 0.301 - 0.477 = -0.176$$

c) $\log_a 81 = \log_a 3^4 = 4\log_a 3$ **Using the power rule**

$$= 4(0.477) = 1.908$$

d) $\log_a \sqrt{a} = \log_a a^{1/2} = \frac{1}{2}\log_a a$ **Using the power rule**

$$= \frac{1}{2} \cdot 1 = \frac{1}{2}$$

e) $\log_a 5$ *cannot be found using these properties and the given information.*

$$(\log_a 5 \neq \log_a 2 + \log_a 3)$$

f) $\dfrac{\log_a 3}{\log_a 2} = \dfrac{0.477}{0.301} \approx 1.585$ We simply divided, not using any of the properties.

In Example 8, a is actually 10 so we have common logarithms. Check as many results as possible using a grapher.

Simplifying Expressions of the Type $\log_a a^x$ and $a^{\log_a x}$

We have two final properties to consider. The first follows from the product rule: Since $\log_a a^x = x \log_a a = x \cdot 1 = x$, we have $\log_a a^x = x$. This property also follows from the definition of a logarithm: x is the power to which we raise a in order to get a^x.

> **The Logarithm of a Base to a Power**
>
> For any base a and any real number x,
>
> $$\log_a a^x = x.$$
>
> (The logarithm, base a, of a to a power is the power.)

Example 9   Simplify each of the following.

a) $\log_a a^8$ **b)** $\ln e^{-t}$ **c)** $\log 10^{3k}$

SOLUTION

a) $\log_a a^8 = 8$ 8 is the power to which we raise a in order to get a^8.
b) $\ln e^{-t} = \log_e e^{-t} = -t$
c) $\log 10^{3k} = \log_{10} 10^{3k} = 3k$

Let $M = \log_a x$. Then $a^M = x$. Substituting $\log_a x$ for M, we obtain $a^{\log_a x} = x$. This also follows from the definition of a logarithm: $\log_a x$ is the power to which a is raised in order to get x.

> **A Base to a Logarithmic Power**
>
> For any base a and any positive real number x,
>
> $$a^{\log_a x} = x.$$
>
> (The number a raised to the power $\log_a x$ is x.)

Example 10   Simplify each of the following.

a) $4^{\log_4 k}$ **b)** $e^{\ln 5}$ **c)** $10^{\log 7t}$

SOLUTION

a) $4^{\log_4 k} = k$

b) $e^{\ln 5} = e^{\log_e 5} = 5$

c) $10^{\log 7t} = 10^{\log_{10} 7t} = 7t$

A Proof of the Change-of-Base Formula: We close this section by proving the change-of-base formula and summarizing the properties of logarithms considered thus far in this chapter. In Section 3.3, we used the change-of-base formula,

$$\log_b M = \frac{\log_a M}{\log_a b},$$

to make base conversions in order to graph logarithmic functions on a grapher. Let $x = \log_b M$. Then by the definition of logarithms, $b^x = M$. We have established that logarithmic functions are one-to-one. This says that $M = N \longleftrightarrow \log_a M = \log_a N$ for any positive numbers M and N. Thus we can take the logarithm base a on both sides of an expression such as $b^x = M$. Then we get $\log_a b^x = \log_a M$. Using the power rule, we obtain $x \log_a b = \log_a M$. Solving for x gives us

$$x = \frac{\log_a M}{\log_a b}, \quad \text{so} \quad x = \log_b M = \frac{\log_a M}{\log_a b}.$$

Following is a summary of the properties of logarithms.

Summary of the Properties of Logarithms

The Product Rule: $\qquad\qquad\qquad \log_a MN = \log_a M + \log_a N$

The Power Rule: $\qquad\qquad\qquad \log_a M^p = p \log_a M$

The Quotient Rule: $\qquad\qquad\qquad \log_a \dfrac{M}{N} = \log_a M - \log_a N$

The Change-of-Base Formula: $\qquad \log_b M = \dfrac{\log_a M}{\log_a b}$

Other Formulas: $\qquad\qquad\qquad \log_a a = 1, \qquad \log_a 1 = 0,$

$\qquad\qquad\qquad\qquad\qquad\qquad\quad \log_a a^x = x, \qquad a^{\log_a x} = x$

3.4 Exercise Set

Express as a sum of logarithms.

1. $\log_3 (81 \cdot 27)$ $\qquad\qquad$ **2.** $\log_2 (8 \cdot 64)$

3. $\log_5 (5 \cdot 125)$ $\qquad\qquad$ **4.** $\log_4 (64 \cdot 32)$

5. $\log_t 8Y$ $\qquad\qquad\qquad$ **6.** $\log_e Qx$

Express as a product.

7. $\log_b t^3$ $\qquad\qquad\qquad$ **8.** $\log_a x^4$

9. $\log y^8$ $\qquad\qquad\qquad$ **10.** $\ln y^5$

11. $\log_c K^{-6}$ $\qquad\qquad\quad$ **12.** $\log_b Q^{-8}$

Express as a difference of logarithms.

13. $\log_t \dfrac{M}{8}$ $\qquad\qquad\quad$ **14.** $\log_a \dfrac{76}{13}$

15. $\log_a \dfrac{x}{y}$

16. $\log_b \dfrac{3}{w}$

Express in terms of sums and differences of logarithms.

17. $\log_a 6xy^5z^4$

18. $\log_a x^3y^2z$

19. $\log_b \dfrac{p^2q^5}{m^4b^9}$

20. $\log_b \dfrac{x^2y}{b^3}$

21. $\log_a \sqrt{\dfrac{x^6}{p^5q^8}}$

22. $\log_c \sqrt[3]{\dfrac{y^3z^2}{x^4}}$

23. $\log_a \sqrt[4]{\dfrac{m^8n^{12}}{a^3b^5}}$

24. $\log_a \sqrt{\dfrac{a^6b^8}{a^2b^5}}$

Express as a single logarithm and simplify, if possible.

25. $\log_a 75 + \log_a 2$

26. $\log 0.01 + \log 1000$

27. $\log 10{,}000 - \log 100$

28. $\ln 54 - \ln 6$

29. $\frac{1}{2}\log_a x + 4\log_a y - 3\log_a x$

30. $\frac{2}{5}\log_a x - \frac{1}{3}\log_a y$

31. $\ln x^2 - 2\ln \sqrt{x}$

32. $\ln 2x + 3(\ln x - \ln y)$

33. $\ln (x^2 - 4) - \ln (x + 2)$

34. $\log_a \dfrac{a}{\sqrt{x}} - \log_a \sqrt{ax}$

35. $\ln x - 3[\ln (x - 5) + \ln (x + 5)]$

36. $\frac{2}{3}[\ln (x^2 - 9) - \ln (x + 3)] + \ln (x + y)$

37. $\frac{3}{2}\ln 4x^6 - \frac{4}{5}\ln 2y^{10}$

38. $120(\ln \sqrt[5]{x^3} + \ln \sqrt[3]{y^2} - \ln \sqrt[4]{16z^5})$

Given that $\log_b 3 = 1.0986$ and $\log_b 5 = 1.6094$, find each of the following.

39. $\log_b \frac{3}{5}$

40. $\log_b 15$

41. $\log_b \frac{1}{5}$

42. $\log_b \frac{5}{3}$

43. $\log_b \sqrt{b}$

44. $\log_b \sqrt{b^3}$

45. $\log_b 5b$

46. $\log_b 9$

47. $\log_b 75$

48. $\log_b \dfrac{1}{b}$

Simplify.

49. $\log_p p^3$

50. $\log_t t^{2713}$

51. $\log_e e^{|x-4|}$

52. $\log_q q^{\sqrt{3}}$

53. $3^{\log_3 4x}$

54. $5^{\log_5 (4x-3)}$

55. $10^{\log w}$

56. $e^{\ln x^3}$

57. $\ln e^{8t}$

58. $\log 10^{-k}$

Solve.

59. $3x - 7 = 5$

60. $x^4 - 6x^2 + 7 = 0$

61. $\dfrac{2x - 1}{x - 4} = 9$

62. $x(x - 3) = 10$

Synthesis

63. ◆ Given that $f(x) = a^x$ and $g(x) = \log_a x$, find $(f \circ g)(x)$ and $(g \circ f)(x)$. These results are alternative proofs of what properties of logarithms already proven in this section? Explain.

64. ◆ Explain the errors, if any, in the following:

$$\log_a ab^3 = (\log_a a)(\log_a b^3)$$
$$= 3\log_a b.$$

Solve for x. Do an algebraic solution.

65. $5^{\log_5 8} = 2x$

66. $\ln e^{3x-5} = -8$

Express as a single logarithm and simplify, if possible.

67. $\log_a (x^2 + xy + y^2) + \log_a (x - y)$

68. $\log_a (a^{10} - b^{10}) - \log_a (a + b)$

Express as a sum or a difference of logarithms.

69. $\log_a \dfrac{x - y}{\sqrt{x^2 - y^2}}$

70. $\log_a \sqrt{9 - x^2}$

71. Given that $\log_a x = 2$, $\log_a y = 3$, and $\log_a z = 4$, find

$$\log_a \dfrac{\sqrt[4]{y^2z^5}}{\sqrt[4]{x^3z^{-2}}}.$$

Determine whether each of the following is true. Assume that a, x, M, and N are positive.

72. $\log_a M + \log_a N = \log_a (M + N)$

73. $\log_a M - \log_a N = \log_a \dfrac{M}{N}$

74. $\dfrac{\log_a M}{\log_a N} = \log_a M - \log_a N$

75. $\dfrac{\log_a M}{x} = \log_a M^{1/x}$

76. $\log_a x^3 = 3\log_a x$

77. $\log_a 8x = \log_a x + \log_a 8$

78. $\log_N (MN)^x = x\log_N M + x$

Suppose that $\log_a x = 2$. Find each of the following.

79. $\log_a \left(\dfrac{1}{x} \right)$

80. $\log_{1/a} x$

81. Simplify:

$\log_{10} 11 \cdot \log_{11} 12 \cdot \log_{12} 13 \cdots \log_{998} 999 \cdot \log_{999} 1000.$

Prove each of the following for any base a and any positive number x.

82. $\log_a \left(\dfrac{1}{x} \right) = -\log_a x = \log_{1/a} x$

83. $\log_a \left(\dfrac{x + \sqrt{x^2 - 5}}{5} \right) = -\log_a \left(x - \sqrt{x^2 - 5} \right)$

3.5
Solving Exponential and Logarithmic Equations

- *Solve exponential and logarithmic equations.*

Solving Exponential Equations

Equations with variables in the exponents, such as $3^x = 20$ and $2^{5x} = 64$, are called **exponential equations**. We now consider solving exponential equations.

Sometimes, as is the case with the equation $2^{5x} = 64$, we can write each side as a power of the same number:

$$2^{5x} = 2^6.$$

We can then set the exponents equal and solve:

$$5x = 6$$
$$x = \tfrac{6}{5}.$$

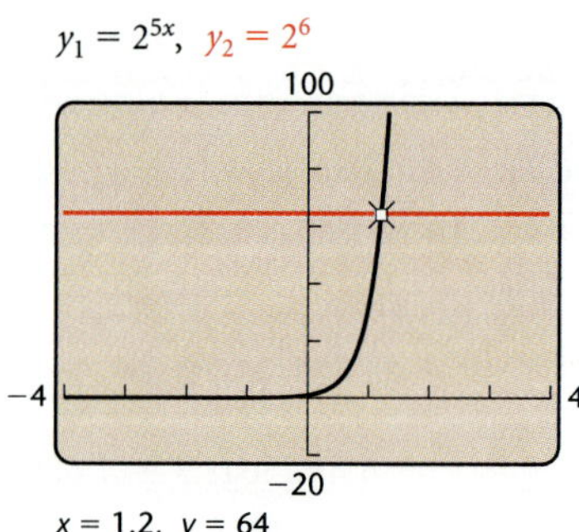

We use the following property.

Base–Exponent Property

For any $a > 0$, $a \neq 1$,

$$a^x = a^y \longleftrightarrow x = y.$$

This property follows from the fact that for any $a > 0$, $a \neq 1$, $f(x) = a^x$ is a one-to-one function. If $a^x = a^y$, then $f(x) = f(y)$. Then since f is one-to-one (see the definition in Section 3.1), it follows that $x = y$. Conversely, if $x = y$, it follows that $a^x = a^y$, since we are raising a to the same power.

Example 1 Solve: $2^{3x-7} = 32$.

ALGEBRAIC SOLUTION

Note that $32 = 2^5$. Thus we can write each side as a power of the same number:

$$2^{3x-7} = 2^5.$$

Since the bases are the same number, 2, we can use the base–exponent property and set the exponents equal:

$$3x - 7 = 5$$
$$3x = 12$$
$$x = 4$$

CHECK:

$$\dfrac{2^{3x-7} = 32}{}$$

$$2^{3(4)-7} \ ? \ 32$$
$$2^{12-7} \ \Big|$$
$$2^5 \ \Big|$$
$$32 \ \Big| \ 32 \quad \text{TRUE}$$

The solution is 4. The solution set is $\{4\}$.

GRAPHICAL SOLUTION

To solve on a grapher, we graph the equations

$$y_1 = 2^{3x-7} \quad \text{and} \quad y_2 = 32.$$

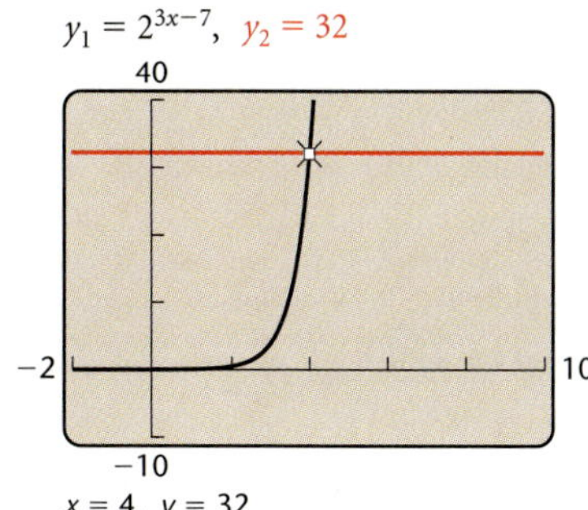

Then we can use the TRACE and ZOOM features or the INTERSECT feature to approximate the point of intersection.

We see that the solution is 4. On some graphers, we can get the solution by solving the equation $2^{3x-7} - 32 = 0$ using a SOLVE feature.

When it does not seem possible to write each side as a power of the same base, we can take the common or natural logarithm on each side and use the power rule for logarithms.

Example 2 Solve: $3^x = 20$.

ALGEBRAIC SOLUTION

We have

$$3^x = 20$$
$$\log 3^x = \log 20 \qquad \text{Taking the common logarithm on both sides}$$
$$x \log 3 = \log 20 \qquad \text{Using the power rule}$$
$$x = \frac{\log 20}{\log 3}. \qquad \text{Dividing by log 3}$$

This is an exact answer. We cannot simplify further, but we can approximate using a grapher:

$$x = \frac{\log 20}{\log 3} \approx \frac{1.3010}{0.4771} \approx 2.7268.$$

We can check this by finding $3^{2.7268}$. The solution is about 2.727.

GRAPHICAL SOLUTION

We graph the equations

$$y_1 = 3^x \quad \text{and} \quad y_2 = 20$$

and use TRACE and ZOOM or INTERSECT to determine a point of intersection.

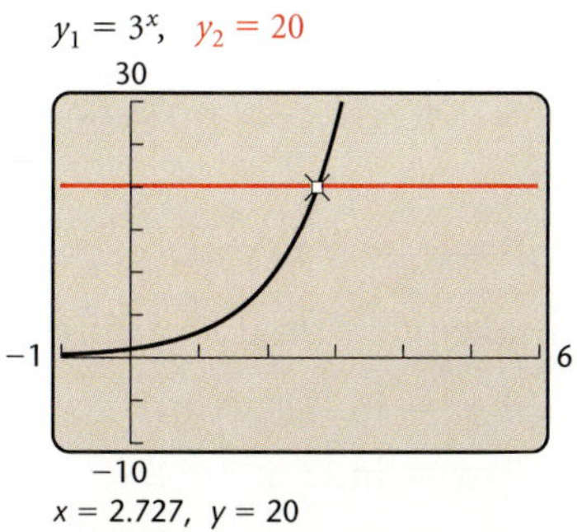

The solution is about 2.727.

It will ease our work if we take the natural logarithm when working with equations that have e as a base.

Example 3 Solve: $e^{0.08t} = 2500$.

ALGEBRAIC SOLUTION

We have

$$e^{0.08t} = 2500$$
$$\ln e^{0.08t} = \ln 2500 \qquad \text{Taking the natural logarithm on both sides}$$
$$0.08t = \ln 2500 \qquad \text{Finding the logarithm of a base to a power: } \log_a a^x = x$$
$$t = \frac{\ln 2500}{0.08} \approx 97.8.$$

The solution is about 97.8.

GRAPHICAL SOLUTION

We graph the equations

$$y_1 = e^{0.08x} \quad \text{and} \quad y_2 = 2500$$

and determine a point of intersection.

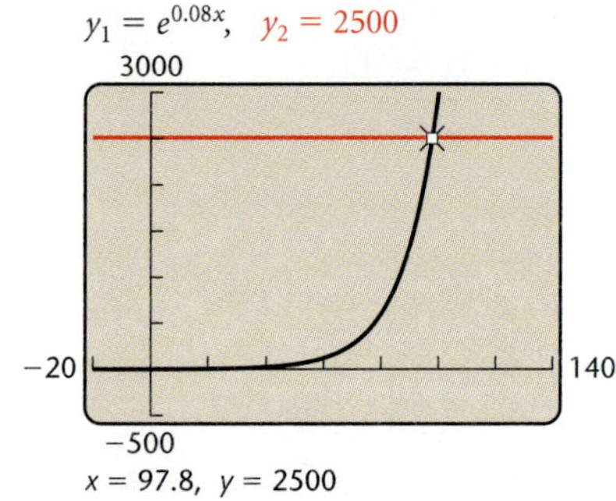

The solution is about 97.8.

Example 4 Solve: $e^x + e^{-x} - 6 = 0$.

ALGEBRAIC SOLUTION

In this case, we have more than one term with x in the exponent. To get a single expression with x in the exponent, we do the following:

$$e^x + e^{-x} - 6 = 0$$
$$e^x + \frac{1}{e^x} - 6 = 0 \qquad \text{Rewriting } e^{-x} \text{ with a positive exponent}$$
$$e^{2x} + 1 - 6e^x = 0. \qquad \text{Multiplying on both sides by } e^x$$

This equation is reducible to quadratic with $u = e^x$:

$$u^2 - 6u + 1 = 0.$$

The coefficients of the reduced quadratic equation are $a = 1$, $b = -6$, and $c = 1$. Using the quadratic formula, we obtain

$$u = e^x = 3 \pm \sqrt{8}.$$

We now take the natural logarithm on both sides:

$$\ln e^x = \ln (3 \pm \sqrt{8})$$
$$x = \ln (3 \pm \sqrt{8}). \qquad \text{Using } \ln e^x = x$$

Approximating each of the solutions, we obtain 1.76 and −1.76.

GRAPHICAL SOLUTION

We begin by graphing $f(x) = e^x + e^{-x} - 6$. We look for the first coordinates of the points where the graph of $y = e^x + e^{-x} - 6$ crosses the x-axis. The zeros of the function are the solutions of the equation.

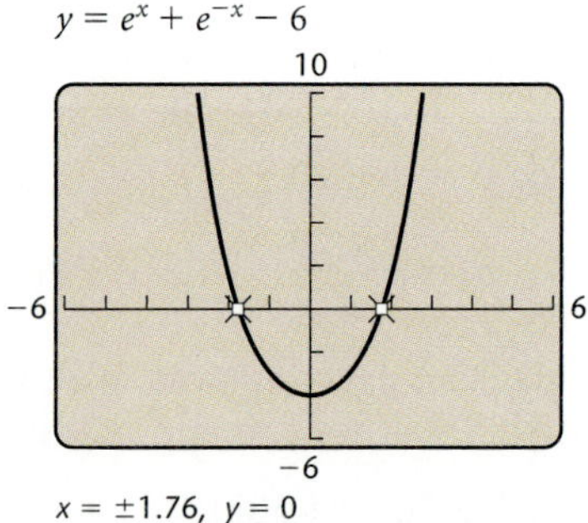

We obtain the approximate solutions −1.76 and 1.76.

It is possible that when encountering an equation like the one in Example 4, you might not recognize that it could be solved in the algebraic manner shown. This points out the value of the graphical solution.

Solving Logarithmic Equations

Equations containing variables in logarithmic expressions, such as $\log_2 x = 4$ and $\log x + \log (x + 3) = 1$, are called **logarithmic equations**.

> To solve logarithmic equations algebraically, first try to obtain a single logarithmic expression on one side and then write an equivalent exponential equation.

Example 5 Solve: $\log_3 x = -2$.

ALGEBRAIC SOLUTION

We have

$$\log_3 x = -2$$
$$3^{-2} = x \qquad \text{Converting to an exponential equation}$$
$$\frac{1}{3^2} = x$$
$$\frac{1}{9} = x.$$

CHECK:
$$\log_3 x = -2$$

$$\log_3 \frac{1}{9} \ ? \ -2$$
$$\log_3 3^{-2}$$
$$-2 \ \bigg| \ -2 \quad \text{TRUE}$$

The solution is $\frac{1}{9}$.

GRAPHICAL SOLUTION

Use the change-of-base formula and graph the equations

$$y_1 = \log_3 x = \frac{\ln x}{\ln 3}$$

and

$$y_2 = -2.$$

Then look for the point(s) of intersection.

The approximate solution is 0.111, which is about $\frac{1}{9}$.

Example 6 Solve: $\log x + \log (x + 3) = 1$.

ALGEBRAIC SOLUTION

In this case, we have common logarithms. Including the base 10's will help us understand the problem:

$$\log_{10} x + \log_{10} (x + 3) = 1$$

$$\log_{10} [x(x + 3)] = 1 \qquad \text{Using the product rule to obtain a single logarithm}$$

$$x(x + 3) = 10^1 \qquad \text{Writing an equivalent exponential equation}$$

$$x^2 + 3x = 10$$

$$x^2 + 3x - 10 = 0$$

$$(x - 2)(x + 5) = 0 \qquad \text{Factoring}$$

$$x - 2 = 0 \quad or \quad x + 5 = 0$$

$$x = 2 \quad or \quad x = -5.$$

CHECK: For 2:

$$\log x + \log (x + 3) = 1$$

$$\log 2 + \log (2 + 3) \;?\; 1$$

$$\log 2 + \log 5$$

$$\log 10$$

$$1 \;\bigg|\; 1 \qquad \text{TRUE}$$

For -5:

$$\log x + \log (x + 3) = 1$$

$$\log (-5) + \log (-5 + 3) \;?\; 1 \qquad \text{FALSE}$$

The number -5 is not a solution because negative numbers do not have real-number logarithms. The solution is 2.

GRAPHICAL SOLUTION

We graph the equations

$$y_1 = \log x + \log (x + 3)$$

and

$$y_2 = 1$$

and find the point(s) of intersection.

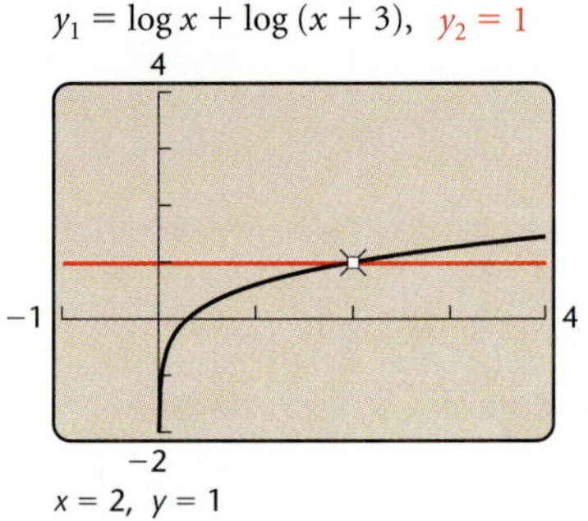

The solution is 2.

Example 7   Solve: $\log_3 (2x - 1) - \log_3 (x - 4) = 2$.

ALGEBRAIC SOLUTION

We have

$$\log_3 (2x - 1) - \log_3 (x - 4) = 2$$

$$\log_3 \frac{2x - 1}{x - 4} = 2 \qquad \textbf{Using the quotient rule}$$

$$\frac{2x - 1}{x - 4} = 3^2 \qquad \textbf{Writing an equivalent exponential equation}$$

$$\frac{2x - 1}{x - 4} = 9$$

$$2x - 1 = 9(x - 4) \qquad \textbf{Multiplying by the LCD, } x - 4$$

$$2x - 1 = 9x - 36$$

$$35 = 7x$$

$$5 = x.$$

CHECK:

$$\log_3 (2x - 1) - \log_3 (x - 4) = 2$$

$$\log_3 (2 \cdot 5 - 1) - \log_3 (5 - 4) \ ? \ 2$$

$$\log_3 9 - \log_3 1$$

$$2 - 0$$

$$2 \ \big| \ 2 \quad \text{TRUE}$$

The solution is 5.

GRAPHICAL SOLUTION

We use the change-of-base formula and graph the equations

$$y_1 = \frac{\ln (2x - 1)}{\ln 3} - \frac{\ln (x - 4)}{\ln 3}$$

and

$$y_2 = 2.$$

Then we find the point(s) of intersection.

$$y_1 = \frac{\ln (2x - 1)}{\ln 3} - \frac{\ln (x - 4)}{\ln 3}, \quad y_2 = 2$$

$x = 5, \ y = 2$

The solution is 5.

Sometimes we encounter equations for which an algebraic solution seems difficult or impossible.

Example 8   Solve: $e^{0.5x} - 7.3 = 2.08x + 6.2$.

ALGEBRAIC SOLUTION

In this case, we have an equation for which an algebraic solution seems difficult or impossible.

GRAPHICAL SOLUTION

We graph the equations

$$y_1 = e^{0.5x} - 7.3 \quad \text{and} \quad y_2 = 2.08x + 6.2$$

and look for points of intersection. (See the figure at left.)
We can also consider the equation

$$y = e^{0.5x} - 7.3 - 2.08x - 6.2, \quad \text{or} \quad y = e^{0.5x} - 2.08x - 13.5,$$

and look for zeros using TRACE and ZOOM or a SOLVE feature. The approximate solutions are -6.471 and 6.610.

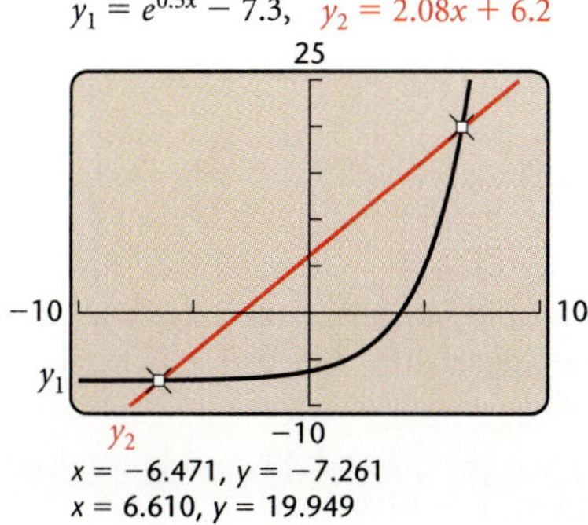

$y_1 = e^{0.5x} - 7.3, \quad y_2 = 2.08x + 6.2$

$x = -6.471, y = -7.261$
$x = 6.610, y = 19.949$

3.5 Exercise Set

Solve each exponential equation algebraically. Then check on a grapher.

1. $3^x = 81$ **2.** $2^x = 32$

3. $2^{2x} = 8$ **4.** $3^{7x} = 27$

5. $2^x = 33$ **6.** $2^x = 40$

7. $5^{4x-7} = 125$ **8.** $4^{3x-5} = 16$

9. $27 = 3^{5x} \cdot 9^{x^2}$ **10.** $3^{x^2+4x} = \frac{1}{27}$

11. $84^x = 70$ **12.** $28^x = 10^{-3x}$

13. $e^t = 1000$ **14.** $e^{-t} = 0.04$

15. $e^{-0.03t} = 0.08$ **16.** $1000e^{0.09t} = 5000$

17. $3^x = 2^{x-1}$ **18.** $5^{x+2} = 4^{1-x}$

19. $(3.9)^x = 48$ **20.** $250 - (1.87)^x = 0$

21. $e^x + e^{-x} = 5$ **22.** $e^x - 6e^{-x} = 1$

23. $\dfrac{e^x + e^{-x}}{e^x - e^{-x}} = 3$ **24.** $\dfrac{5^x - 5^{-x}}{5^x + 5^{-x}} = 8$

Solve each logarithmic equation algebraically. Then check on a grapher.

25. $\log_5 x = 4$ **26.** $\log_2 x = -3$

27. $\log x = -4$ **28.** $\log x = 1$

29. $\ln x = 1$ **30.** $\ln x = -2$

31. $\log_2 (10 + 3x) = 5$ **32.** $\log_5 (8 - 7x) = 3$

33. $\log x + \log (x - 9) = 1$

34. $\log_2 (x + 1) + \log_2 (x - 1) = 3$

35. $\log_8 (x + 1) - \log_8 x = 2$

36. $\log x - \log (x + 3) = -1$

37. $\log_4 (x + 3) + \log_4 (x - 3) = 2$

38. $\ln (x + 1) - \ln x = \ln 4$

39. $\log (2x + 1) - \log (x - 2) = 1$

40. $\log_5 (x + 4) + \log_5 (x - 4) = 2$

Use only a grapher. Find approximate solutions of each equation or approximate the point(s) of intersection of a pair of equations.

41. $e^{7.2x} = 14.009$ **42.** $0.082e^{0.05x} = 0.034$

43. $xe^{3x} - 1 = 3$ **44.** $5e^{5x} + 10 = 3x + 40$

45. $4 \ln (x + 3.4) = 2.5$ **46.** $\ln x^2 = -x^2$

47. $\log_8 x + \log_8 (x + 2) = 2$

48. $\log_3 x + 7 = 4 - \log_5 x$

49. $\log_5 (x + 7) - \log_5 (2x - 3) = 1$

50. $y = \ln 3x, \ y = 3x - 8$

51. $2.3x + 3.8y = 12.4, \ y = 1.1 \ln (x - 2.05)$

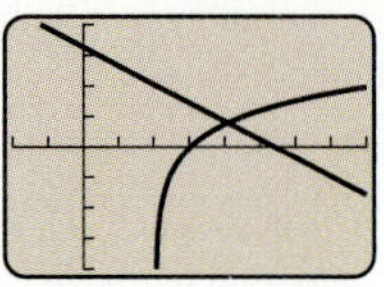

52. $y = 2.3 \ln (x + 10.7), \ y = 10e^{-0.07x^2}$

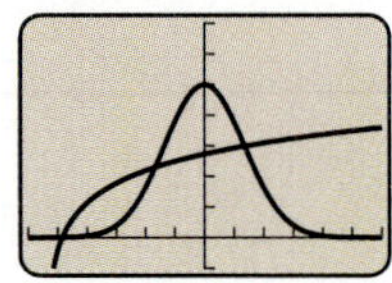

53. $y = 2.3 \ln (x + 10.7), \ y = 10e^{-0.007x^2}$

Skill Maintenance

54. Solve $K = \frac{1}{2}mv^2$ for v.

Solve.

55. $x^4 + 5x^2 = 36$ **56.** $t^{2/3} - 10 = 3t^{1/3}$

57. *Total Sales of Goodyear.* The following table shows factual data regarding total sales of The Goodyear Tire and Rubber Company.

YEAR, x	TOTAL SALES, y (IN MILLIONS)
1. 1991	$10,906.8
2. 1992	11,784.9
3. 1993	11,643.4
4. 1994	12,288.2

Source: The Goodyear Tire and Rubber Company Annual Report.

a) Use linear regression on a grapher to fit an equation $y = mx + b$, where $x = 1$ corresponds to 1991, to the data points. Predict total sales in 1999.

b) Use quadratic regression on a grapher to fit a quadratic equation $y = ax^2 + bx + c$ to the data points. Predict total sales in 1999.

Synthesis

58. ◆ In Example 3, we took the natural logarithm on both sides. What would have happened had we taken the common logarithm? Explain which seems best to you and why.

59. ◆ Explain how Exercises 29 and 30 could be solved using the graph of $f(x) = \ln x$.

Solve using any method.

60. $\ln (\ln x) = 2$

61. $\ln (\log x) = 0$

62. $\ln \sqrt[4]{x} = \sqrt{\ln x}$

63. $\sqrt{\ln x} = \ln \sqrt{x}$

64. $\log_3 (\log_4 x) = 0$

65. $(\log_3 x)^2 - \log_3 x^2 = 3$

66. $(\log x)^2 - \log x^2 = 3$

67. $\ln x^2 = (\ln x)^2$

68. $e^{2x} - 9 \cdot e^x + 14 = 0$

69. $5^{2x} - 3 \cdot 5^x + 2 = 0$

70. $x \left(\ln \frac{1}{6}\right) = \ln 6$

71. $\log_3 |x| = 2$

72. $x^{\log x} = \dfrac{x^3}{100}$

73. $\ln x^{\ln x} = 4$

74. $\dfrac{(e^{3x+1})^2}{e^4} = e^{10x}$

75. $\dfrac{\sqrt{(e^{2x} \cdot e^{-5x})^{-4}}}{e^x \div e^{-x}} = e^7$

76. $|\log_a x| = \log_a |x|$

77. $\ln (x - 2) > 4$

78. $e^x < \dfrac{4}{5}$

79. $|\log_5 x| + 3 \log_5 |x| = 4$

80. $|2^{x^2} - 8| = 3$

81. Given that $a = \log_8 225$ and $b = \log_2 15$, express a as a function of b.

82. Given that $a = (\log_{125} 5)^{\log_5 125}$, find the value of $\log_3 a$.

83. Given that
$$\log_2 [\log_3 (\log_4 x)] = \log_3 [\log_2 (\log_4 y)]$$
$$= \log_4 [\log_3 (\log_2 z)]$$
$$= 0,$$
find $x + y + z$.

84. Given that $f(x) = e^x - e^{-x}$, find $f^{-1}(x)$ if it exists.

3.6

Applications and Models: Growth and Decay

- *Solve applications involving exponential growth and decay.*
- *Find models involving exponential and logarithmic functions.*

Exponential and logarithmic functions with base e are rich in applications to many fields such as business, science, psychology, and sociology. In this section, we consider some basic applications and then use curve fitting to do others.

Population Growth

The function

$$P(t) = P_0 e^{kt}$$

is a model of many kinds of population growth, whether it be a population of people, bacteria, cellular phones, or money. In this function, P_0 is the population at time 0, P is the population after time t, and k is called the **exponential growth rate**. The graph of such an equation is shown at right.

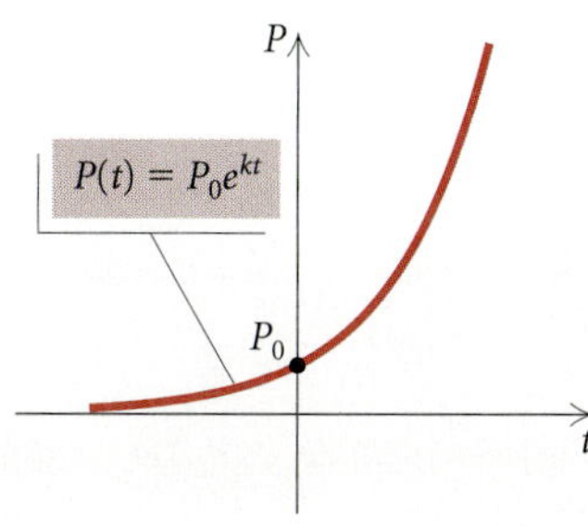

Example 1 *Population Growth of the United States.* In 1997, the population of the United States was about 266 million and the exponential growth rate was 0.9% per year. (*Source:* Statistical Abstract of the United States)

a) Find the exponential growth function.

b) What will the population be in 2002?

c) Graph the exponential growth function.

d) When will the population be double what it was in 1997?

SOLUTION

a) At $t = 0$ (1997), the population was 266 million. We substitute 266 for P_0 and 0.9%, or 0.009, for k to obtain the exponential growth function

$$P(t) = 266e^{0.009t}.$$

b) In 2002, $t = 5$; that is, 5 yr have passed. To find the population in 2002, we substitute 5 for t:

$$P(5) = 266e^{0.009(5)} = 266e^{0.045} \approx 278.$$

In 2002, the population of the United States will be about 278 million.

c) Using a grapher, we obtain the graph (at left) of the exponential growth function.

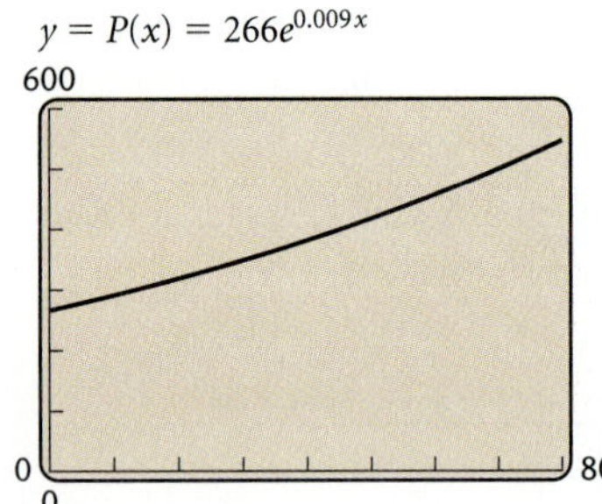

d) We are looking for the time T for which $P(T) = 2 \cdot 266$, or 532. The number T is called the **doubling time**. To find T, we solve the equation $532 = 266e^{0.009T}$.

ALGEBRAIC SOLUTION

We have

$532 = 266e^{0.009T}$	**Substituting 532 for $P(T)$**
$2 = e^{0.009T}$	**Dividing by 266**
$\ln 2 = \ln e^{0.009T}$	**Taking the natural logarithm on both sides**
$\ln 2 = 0.009T$	**$\ln e^x = x$**
$\dfrac{\ln 2}{0.009} = T$	**Dividing by 0.009**
$77 \approx T.$	

The population of the United States will be double what it was in 1997 in about 77 yr.

GRAPHICAL SOLUTION

To solve on a grapher, we graph the equations

$$y_1 = 266e^{0.009x} \quad \text{and} \quad y_2 = 532$$

and find the first coordinate of their point of intersection.

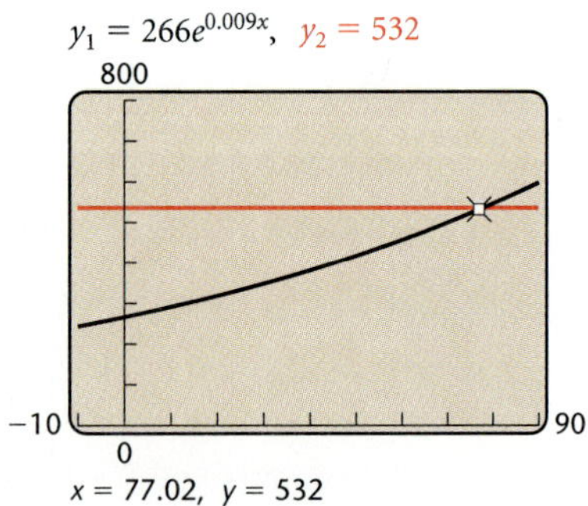

On some graphers, we can get the solution by having the grapher solve the equation directly. We see that the solution is about 77 yr.

No wonder ecologists are so concerned about population growth.

Interest Compounded Continuously

Here we explore the mathematics behind the concept of **interest compounded continuously**. Suppose that an amount P_0 is invested in a savings account at interest rate k *compounded continuously*. The amount $P(t)$ in the account after t years is given by the exponential function

$$P(t) = P_0 e^{kt}.$$

Example 2 *Interest Compounded Continuously.* Suppose $2000 is invested at interest rate k, compounded continuously, and grows to $2983.65 in 5 yr.

a) What is the interest rate?

b) Find the exponential growth function.

c) What will the balance be after 10 yr?

d) After how long will the $2000 have doubled?

SOLUTION

a) At $t = 0$, $P(0) = P_0 = \$2000$. Thus the exponential growth function is

$$P(t) = 2000 e^{kt}.$$

We know that $P(5) = \$2983.65$. We substitute and solve for k, as shown below.

<table>
<tr><td valign="top">

ALGEBRAIC SOLUTION

We have

$2983.65 = 2000e^{k(5)}$ **Substituting 2983.65 for** $P(t)$

$2983.65 = 2000e^{5k}$

$\dfrac{2983.65}{2000} = e^{5k}$ **Dividing by 2000**

$\ln \dfrac{2983.65}{2000} = \ln e^{5k}$ **Taking the natural logarithm**

$\ln \dfrac{2983.65}{2000} = 5k$ **Using** $\ln e^x = x$

$\dfrac{\ln \dfrac{2983.65}{2000}}{5} = k$

$0.08 \approx k.$

The interest rate is about 0.08, or 8%.

</td><td valign="top">

GRAPHICAL SOLUTION

We can solve by graphing the equations

$$y_1 - 2000e^{5x} \quad \text{and} \quad y_2 - 2983.65$$

on a grapher. Then we can use TRACE and ZOOM or other features of a grapher to find an approximation for the first coordinate of the point of intersection.

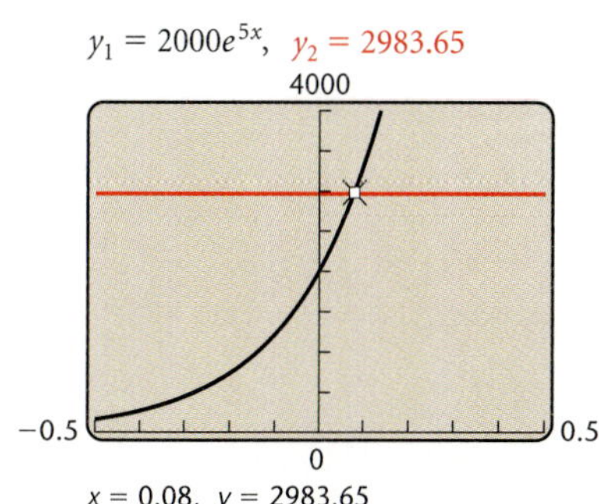

The solution is about 0.08, or 8%.

</td></tr>
</table>

b) The exponential growth function is

$$P(t) = 2000 e^{0.08t}.$$

c) The balance after 10 yr is

$$P(10) = 2000e^{0.08(10)}$$
$$= 2000e^{0.8}$$
$$\approx \$4451.08.$$

d) We solve using both the algebraic method and a grapher.

ALGEBRAIC SOLUTION

To find the doubling time T, we set $P(T) = \$4000$ and solve for T:

$$4000 = 2000e^{0.08T}$$

$2 = e^{0.08T}$ **Dividing by 2000**

$\ln 2 = \ln e^{0.08T}$ **Taking the natural logarithm**

$\ln 2 = 0.08T$ **$\ln e^x = x$**

$\dfrac{\ln 2}{0.08} = T$ **Dividing by 0.08**

$8.7 \approx T.$

Thus the original investment of $2000 will double in about 8.7 yr.

GRAPHICAL SOLUTION

To solve on a grapher, we graph the equations

$$y_1 = 2000e^{0.08x} \quad \text{and} \quad y_2 = 4000$$

and find the first coordinate of their point of intersection.

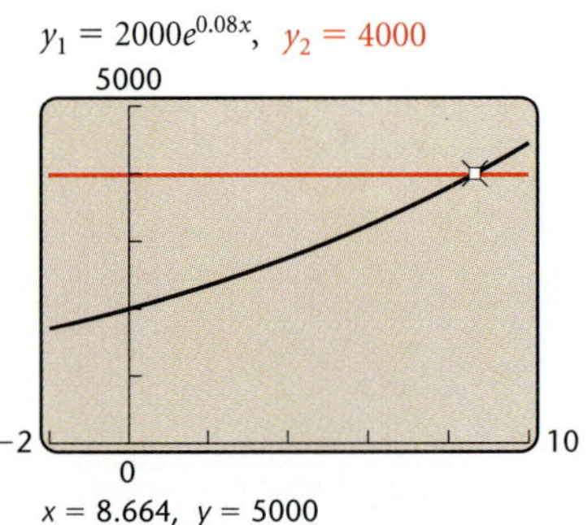

The solution is about 8.7.

We can find a general expression relating the growth rate k and the doubling time T by solving the following equation:

$2P_0 = P_0e^{kT}$ **Substituting $2P_0$ for P and T for t**

$2 = e^{kT}$ **Dividing by P_0**

$\ln 2 = \ln e^{kt}$ **Taking the natural logarithm**

$\ln 2 = kT$ **Using $\ln e^x = x$**

$\dfrac{\ln 2}{k} = T.$

Growth Rate and Doubling Time

The *growth rate* k and the *doubling time* T are related by

$$kT = \ln 2, \quad \text{or} \quad k = \frac{\ln 2}{T}, \quad \text{or} \quad T = \frac{\ln 2}{k}.$$

Note that the relationship between k and T does not depend on P_0.

Example 3 *World Population Growth.* The population of the world is now doubling every 24.8 yr. What is the exponential growth rate?

SOLUTION   We have

$$k = \frac{\ln 2}{T} \approx \frac{0.693147}{24.8} \approx 2.8\%.$$

The growth rate of the world population is about 2.8% per year.

Models of Limited Growth

The model $P(t) = P_0 e^{kt}$ has many applications involving what may seem to be unlimited population growth. There can be factors that prevent a population from exceeding some limiting value—perhaps a limitation on food, living space, or other natural resources. One model of such growth is

$$P(t) = \frac{a}{1 + be^{-kt}},$$

which is called a **logistic equation**. This function increases toward a *limiting value a* as $t \to \infty$.

Example 4   *Limited Population Growth.* A ship carrying 1000 passengers has the misfortune to be shipwrecked on a small island from which the passengers are never rescued. The natural resources of the island limit the population to 5780. The population gets closer and closer to this limiting value, but never reaches it. The population of the island after time t, in years, is given by the logistic equation

$$P(t) = \frac{5780}{1 + 4.78e^{-0.4t}}.$$

a) Find the population after 0, 1, 2, 5, 10, and 20 yr.

b) Graph the function.

SOLUTION

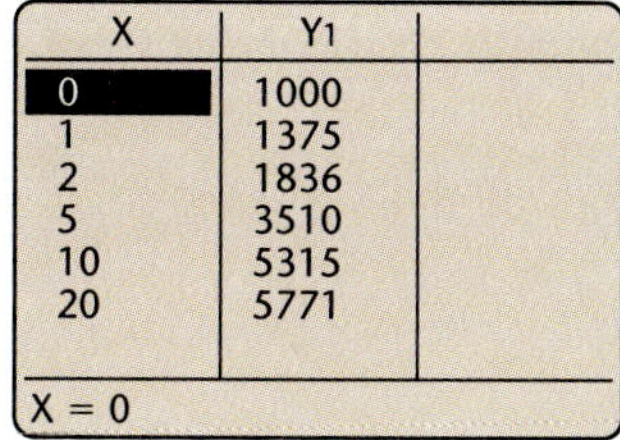

a) We use a grapher to find the function values, listing them in a table as shown at left. (We round to the nearest unit.) A grapher with a TABLE feature set in ASK mode can also be used. Thus the population will be 1000 after 0 yr, 1375 after 1 yr, 1836 after 2 yr, 3510 after 5 yr, 5315 after 10 yr, and 5771 after 20 yr.

b) We use a grapher to graph the function. The graph is the S-shaped curve shown below. Note that this function increases toward a limiting value of 5780.

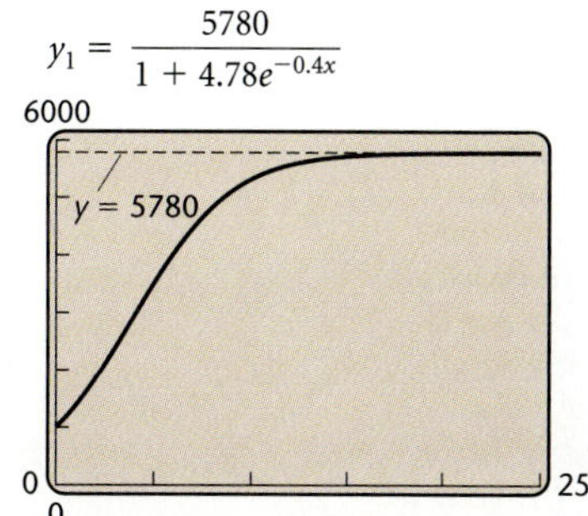

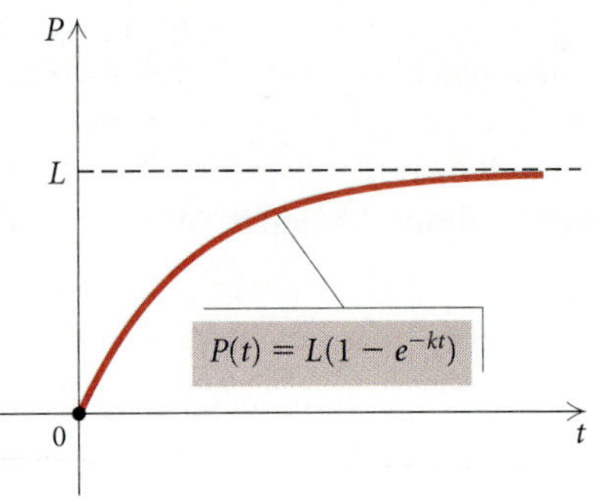

Another model of limited growth is provided by the function

$$P(t) = L(1 - e^{-kt}),$$

which is shown graphed at left. This function also increases toward a limiting value L, as $x \to \infty$.

Exponential Decay

The function

$$P(t) = P_0 e^{-kt}$$

is an effective model of the decline, or decay, of a population. An example is the decay of a radioactive substance. In this case, P_0 is the amount of the substance at time $t = 0$, and $P(t)$ is the amount of the substance left after time t, where k is a positive constant that depends on the situation. The constant k is called the **decay rate**.

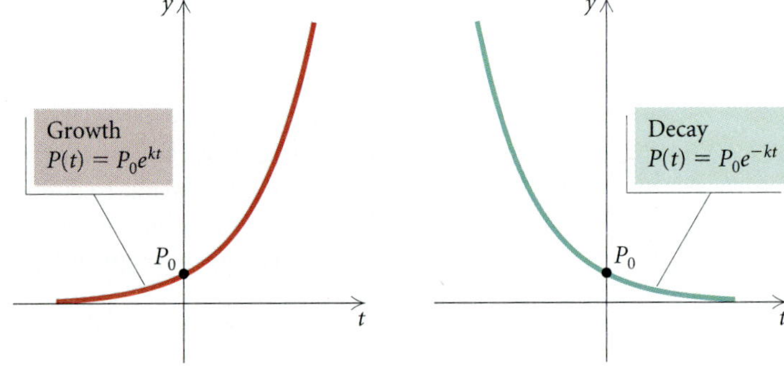

How can scientists determine that an animal bone has lost 30% of its carbon-14? The assumption is that the percentage of carbon-14 in the atmosphere and in living plants and animals is the same. When a plant or an animal dies, the amount of carbon-14 decays exponentially. The scientist burns the animal bone and uses a Geiger counter to determine the percentage of the smoke that is carbon-14. It is the amount that this varies from the percentage in the atmosphere that tells how much carbon-14 has been lost.

The process of carbon-14 dating was developed by the American chemist Willard E. Libby in 1952. It is known that the radioactivity in a living plant is 16 disintegrations per gram per minute. Since the half-life of carbon-14 is 5750 years, an object with an activity of 8 disintegrations per gram per minute is 5750 years old, one with an activity of 4 disintegrations per gram per minute is 11,500 years old, and so on. Carbon-14 dating can be used to measure the age of objects from 30,000 to 40,000 years old. Beyond such an age, it is too difficult to measure the radioactivity and some other method would have to be used.

Carbon-14 was indeed used to find the age of the Dead Sea Scrolls. It was used recently to refute the authenticity of the Shroud of Turin, presumed to have covered the body of Christ.

The **half-life** of bismuth is 5 days. This means that half of an amount of bismuth will cease to be radioactive in 5 days. The effect of half-life T is shown in the graph below for nonnegative inputs. The exponential function gets close to 0, but never reaches 0, as t gets very large. Thus, according to an exponential decay model, a radioactive substance never completely decays.

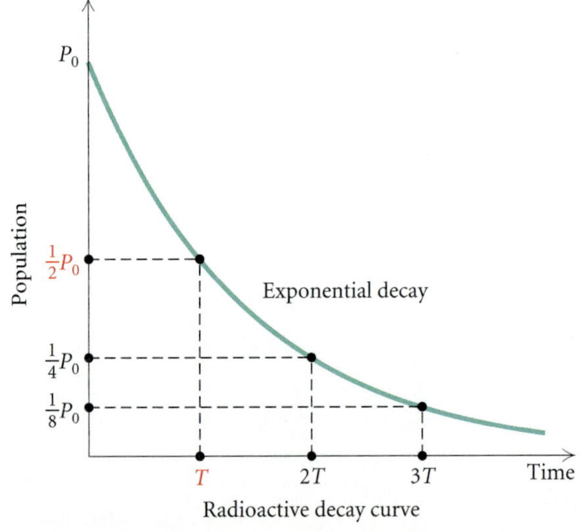

Radioactive decay curve

Example 5   *Carbon Dating.* The radioactive element carbon-14 has a half-life of 5750 yr. The percentage of carbon-14 present in the remains of organic matter can be used to determine the age of that organic matter. Archeologists discovered that the linen wrapping from one of the Dead Sea Scrolls had lost 22.3% of its carbon-14 at the time it was found. How old was the linen wrapping?

SOLUTION   We first find k. When $t = 5750$ (the half-life), $P(t)$ will be half of P_0. We substitute $\frac{1}{2}P_0$ for $P(t)$ and 5750 for t and solve for k. Then

$$\tfrac{1}{2}P_0 = P_0 e^{-k(5750)}$$

or

$$\tfrac{1}{2} = e^{-5750k}.$$

We take the natural logarithm on both sides:

$$\ln \tfrac{1}{2} = \ln e^{-5750k}$$
$$= -5750k.$$

Then

$$k = \frac{\ln 0.5}{-5750}$$
$$\approx 0.00012.$$

We could also solve the equation $\frac{1}{2} = e^{-5750k}$ using a grapher. Now we have the function

$$P(t) = P_0 e^{-0.00012t}.$$

(This equation can be used for any subsequent carbon-dating problem.) If the linen wrapping has lost 22.3% of its carbon-14 from an initial amount P_0, then $77.7\% P_0$ is the amount present. To find the age t of the wrapping, we solve the following equation for t:

$$77.7\% P_0 = P_0 e^{-0.00012t} \qquad \textbf{Substituting 77.7\%\textit{P}}_0 \textbf{ for \textit{P}}$$
$$0.777 = e^{-0.00012t}$$
$$\ln 0.777 = \ln e^{-0.00012t}$$
$$\ln 0.777 = -0.00012t \qquad \textbf{ln } e^x = x$$
$$\frac{\ln 0.777}{-0.00012} = t$$
$$2103 \approx t.$$

Thus the linen wrapping on the Dead Sea Scrolls is about 2103 yr old.

Exponential and Logarithmic Curve Fitting

We have added several new functions to our candidates for curve fitting. Let's review some of them.

In 1947, a Bedouin youth looking for a stray goat climbed into a cave at Kirbet Qumran on the shores of the Dead Sea near Jericho and came upon earthenware jars containing an incalculable treasure of ancient manuscripts. Shown here are fragments of those so-called Dead Sea Scrolls, a portion of some 600 or so texts found so far and which concern the Jewish books of the Bible. Officials date them before 70 A.D., making them the oldest Biblical manuscripts by 1000 years.

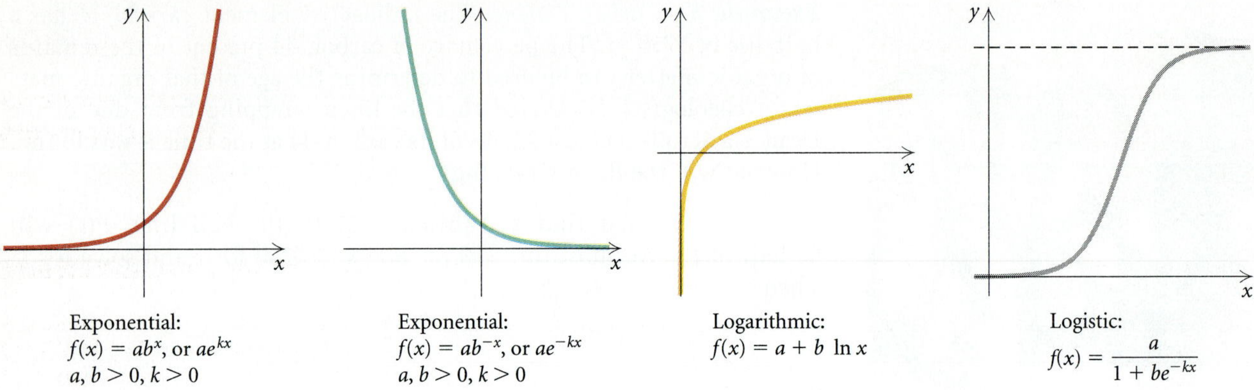

Exponential:
$f(x) = ab^x$, or ae^{kx}
$a, b > 0, k > 0$

Exponential:
$f(x) = ab^{-x}$, or ae^{-kx}
$a, b > 0, k > 0$

Logarithmic:
$f(x) = a + b \ln x$

Logistic:
$$f(x) = \frac{a}{1 + be^{-kx}}$$

Now, when we analyze a set of data for curve fitting, these models can be considered as well as linear, quadratic, polynomial, and rational functions. Indeed, some graphers can use *regression* to fit an exponential, a logarithmic, or a logistic* equation to a set of data.

Example 6 *Cellular Phones.* The number of cellular phones in use has grown dramatically in recent years. Let's examine the following data.

x, YEAR	NUMBER OF CELLULAR PHONES, y (IN MILLIONS)
0. 1985	0.2
1. 1986	0.4
2. 1987	1.0
3. 1988	1.8
4. 1989	2.8
5. 1990	4.1
6. 1991	6.2
7. 1992	8.6
8. 1993	13.0
9. 1994	19.3

Source: Cellular Telecommunications Industry Association

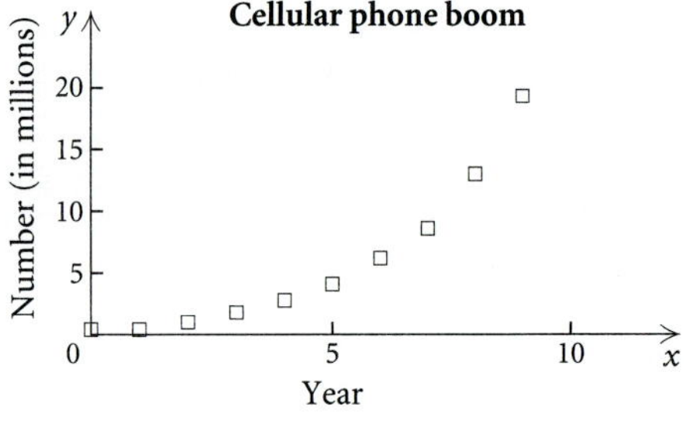

From the scatterplot on the right above, it would appear that we have exponential growth.

a) Use a grapher and regression to fit an exponential function to the data.

b) Graph the function.

c) Predict the number of cellular phones in use in the year 2000.

---

*The TI-83 is an example.

SOLUTION

a) We are fitting the data to an equation of the type $y = a \cdot b^x$. Entering the data into the grapher and carrying out the regression procedure, we find that

$$a = 0.3039703339,$$
$$b = 1.627328317,$$
$$r = 0.9860606791.$$

This tells us that the equation is

$$y = 0.3039703339(1.627328317)^x,$$

or about

$$y = 0.304(1.627)^x.$$

The *correlation coefficient* is very close to 1. This gives us a good indication that the data fit an exponential equation.

b) The graph is shown at left.

c) To predict the number of cellular phones in use in 2000, we substitute 15 for x in the exponential equation:

$$y = 0.304(1.627)^x$$
$$= 0.304(1.627)^{15} \approx 450.$$

Thus according to this model, there will be about 450 million cellular phones in use in 2000.

On some graphers, there may be a REGRESSION feature that yields an exponential function, base e. If not, and you wish to find such a function, a conversion can be done using the following.

Converting from Base b to Base e

$$b^x = e^{x(\ln b)}$$

Then, for the equation in Example 6, we have

$$y = 0.304(1.627)^x$$
$$= 0.304e^{x(\ln 1.627)} = 0.304e^{0.487x}.$$

We can prove this conversion formula using properties of logarithms, as follows:

$$e^{x(\ln b)} = e^{\ln b^x} = b^x.$$

Power Models

There are many situations in which so-called **power models** $y = ax^b$ can be fit to data using regression. Note that the constants to be determined are a and b. The base is the variable x.

Example 7   *Cholesterol Level and the Risk of Heart Attack.* The data in the following table show the relationship of cholesterol level in men to the risk of a heart attack.

CHOLESTEROL LEVEL, x	MEN, PER 10,000, WHO SUFFER A HEART ATTACK, y
100	30
200	65
250	100
275	130
300	180

Source: Nutrition Action Healthletter.

a) Use the REGRESSION feature on a grapher to find a power function to fit the data.

b) Graph the function.

c) Use the answer to part (b) to predict the heart attack rate for men with cholesterol levels of 350 and 400.

SOLUTION

a) We are fitting an equation of the type $y = ax^b$ to the data. Entering the data into the grapher and carrying out the regression procedure, we find that

$$a = 0.0241789574,$$
$$b = 1.527457172,$$
$$r = 0.9739361336.$$

This tells us that the equation is about $y = 0.024x^{1.527}$. The *correlation coefficient* of about 0.974 is very close to 1. This indicates that the data fit a power function fairly well, although an exponential function with $r = 0.996$ is a better fit. For illustrative purposes, though, we will continue with the power model.

b) The graph is shown at left.

c) To predict the heart attack rate for men with cholesterol levels of 350 and 400, we substitute 350 and 400 for x in the power equation:

$$y = 0.024x^{1.527} = 0.024(350)^{1.527} \approx 184,$$
$$y = 0.024x^{1.527} = 0.024(400)^{1.527} \approx 226.$$

Thus the heart attack rate is about 184 out of 10,000 for men with a cholesterol level of 350 and about 226 out of 10,000 for men with a cholesterol level of 400.

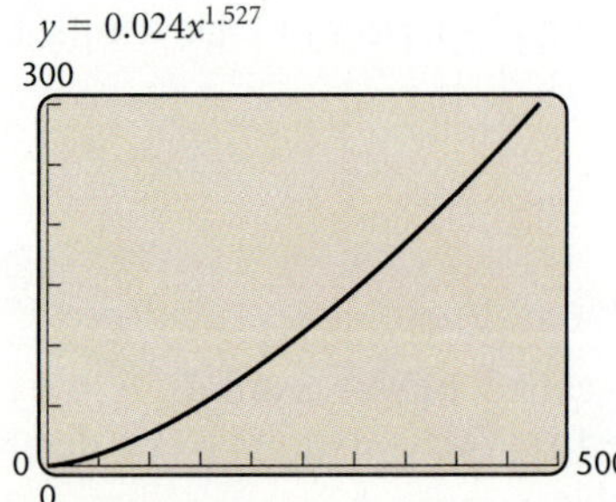

3.6 | Exercise Set

1. *World Population Growth.*   In 1997, the world population was 5.8 billion. The exponential growth rate was 1.5% per year.

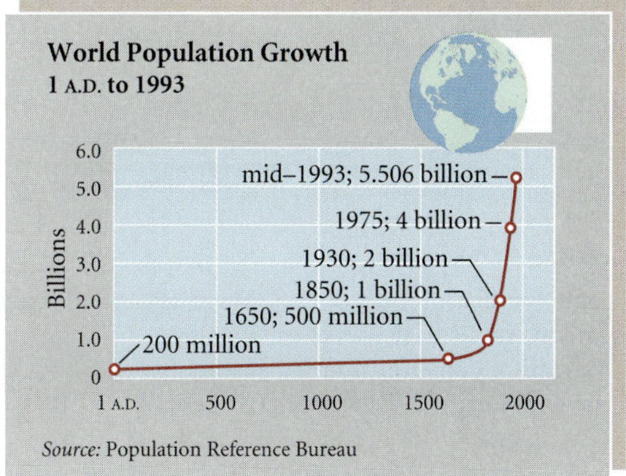

Source: Population Reference Bureau

a) Find the exponential growth function.
b) Predict the population of the world in 2000 and in 2010.
c) When will the world population be 8 billion?
d) Find the doubling time.

2. *Population Growth of Rabbits.*   Under ideal conditions, a population of rabbits has an exponential growth rate of 11.7% per day. Consider an initial population of 100 rabbits.

a) Find the exponential growth function.
b) What will the population be after 7 days?
c) Find the doubling time.
d) Graph the function.

3. *Population Growth.*   Complete the following table.

POPULATION	GROWTH RATE, k	DOUBLING TIME, T
a) Mexico	3.5% per year	
b) Europe		69.31 yr
c) Oil reserves	10% per year	
d) Coal reserves	4% per year	
e) Alaska		24.8 yr
f) Central America		19.8 yr

4. *Female Olympic Athletes.*   In 1985, the number of female athletes participating in Summer Olympic-Type Games was 500. In 1996, about 3600 participated in the Summer Olympics in Atlanta. Assuming the exponential model applies:

a) Find the value of k ($P_0 = 500$), and write the function.
b) Estimate the number of female athletes in the Summer Olympics of 2000.

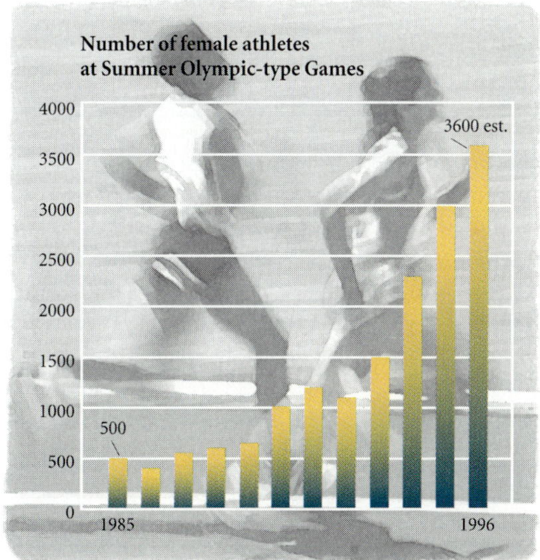

5. *Population Growth of the Virgin Islands.*   The population of the U.S. Virgin Islands has a growth rate of 2.6% per year. In 1990, the population was 512,000. The land area of the Virgin Islands is 3,097,600 square yards. Assuming this growth rate continues and is exponential, after how long will there be one person for every square yard of land? (*Source:* Statistical Abstract of the United States)

6. *Value of Manhattan Island.*   In 1626, Peter Minuit of the Dutch West India Company purchased Manhattan Island from the Indians for $24. Assuming an exponential rate of inflation of 8% per year, how much will Manhattan be worth in 2000?

7. *Interest Compounded Continuously.*   Suppose $10,000 is invested at an interest rate of 5.4% per year, compounded continuously.

a) Find the exponential function that describes the amount in the account after time t, in years.
b) What is the balance after 1 yr? 2 yr? 5 yr? 10 yr?
c) What is the doubling time?

8. *Interest Compounded Continuously.* Complete the following table.

INITIAL INVESTMENT AT $t = 0$, P_0	INTEREST RATE, k	DOUBLING TIME, T	AMOUNT AFTER 5 YR
a) $35,000	6.2%		
b) $5000			$ 7,130.90
c)	8.4%		$11,414.71
d)		11 yr	$17,539.32

9. *Carbon Dating.* A mummy discovered in the pyramid Khufu in Egypt has lost 46% of its carbon-14. Determine its age.

10. *Carbon Dating.* The statue of Zeus at Olympia in Greece is one of the Seven Wonders of the World. It is made of gold and ivory. The ivory was found to have lost 35% of its carbon-14. Determine the age of the statue.

11. *Radioactive Decay.* Complete the following table.

RADIOACTIVE SUBSTANCE	DECAY RATE, k	HALF-LIFE, T
a) Polonium		3 min
b) Lead		22 yr
c) Iodine-131	9.6% per day	
d) Krypton-85	6.3% per year	
e) Strontium-90		25 yr
f) Uranium-238		4560 yr
g) Plutonium		23,105 yr

12. *Decline of Long-Playing Records.* The sales S of long-playing records has declined considerably in the past 10 yr because of the emergence of the cassette tape and compact disc. In 1983, 205 million LP records were sold and in 1993, 1.2 million records were sold. (*Source:* Recording Industry Association of America) Assuming the sales are decreasing according to the exponential-decay model:

a) Find the value k, and write an exponential function that describes the number of long-playing records sold after time t, in years.

b) Estimate the sales of LP records in the year 2000.

c) In what year (theoretically) will only 1 long-playing record be sold?

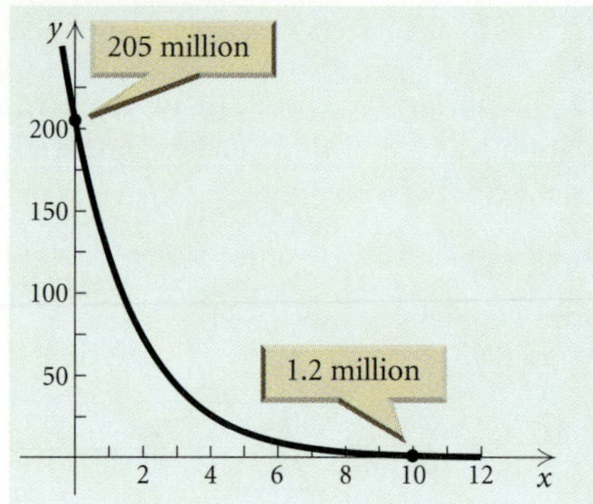

13. *Decline in Beef Consumption.* In 1985, the average annual consumption of beef B was about 80 lb per person. In 1996, it was about 67 lb per person. Assuming consumption is decreasing according to the exponential-decay model:

a) Find the value k, and write an equation that describes beef consumption after time t, in years.

b) Estimate the consumption of beef in the year 2000.

c) After how many years (theoretically) will the average annual consumption of beef be 20 lb per person?

14. *The Value of Eddie Murray's Baseball Card.* The collecting of baseball cards has become a popular hobby. The card shown here is a photograph of Eddie Murray in his rookie season of 1978.

In 1983, the value of the card was $7.75 and in 1987, its value was $27.00. (*Source: Sport Collectors Digest*) Assuming that the value of the card has grown exponentially:

a) Find the value k and determine the exponential growth function, assuming $V_0 = 7.75$.

b) Estimate the value of the card in 1995 and in 2000. Check your answer for 1995 with a baseball card dealer.

c) What is the doubling time for the value of the card?

d) After how long will the value of the card be $2000?

15. *Spread of an Epidemic.* In a town whose population is 2000, a disease creates an epidemic. The number of people N infected t days after the disease has begun is given by the function

$$N(t) = \frac{2000}{1 + 19.9e^{-0.6t}}.$$

a) How many are initially infected with the disease $(t = 0)$?

b) Find the number infected after 2 days, 5 days, 8 days, 12 days, and 16 days.

c) Graph the function.

16. *Acceptance of Seat Belt Laws.* In recent years, many states have passed mandatory seat belt laws. The total number of states N that have passed a seat belt law t years after 1984 is given by the function

$$N(t) = \frac{50}{1 + 22e^{-0.6t}}.$$

(*Source:* National Highway Traffic Safety Administration)

a) How many states had passed the law in 1984? ($t = 0$ corresponds to 1984.)

b) Find the number of states that had passed the law by 1988, 1992, 1996, and 2002.

c) Graph the function.

d) If the function were to continue to be appropriate, would all 50 states ever pass the law? Explain.

17. *Limited Population Growth in a Lake.* A lake is stocked with 400 fish of a new variety. The size of the lake, the availability of food, and the number of other fish restrict growth in the lake to a *limiting value* of 2500. The population of fish in the lake after time t, in months, is given by the function

$$P(t) = \frac{2500}{1 + 5.25e^{-0.32t}}.$$

a) Find the population after 0, 1, 5, 10, 15, and 20 months.

b) Graph the function.

For each of the following scatterplots, determine which, if any, of these functions might be used as a model for the data.

a) *Quadratic, $f(x) = ax^2 + bx + c$*

b) *Polynomial, not quadratic*

c) *Exponential, $f(x) = ab^x$, or Be^{kx}, $k > 0$*

d) *Exponential, $f(x) = ab^x$, or Be^{-kx}, $k > 0$*

e) *Logarithmic, $f(x) = a + b \ln x$*

f) *Logistic, $f(x) = \dfrac{a}{1 + be^{-kx}}$*

18.

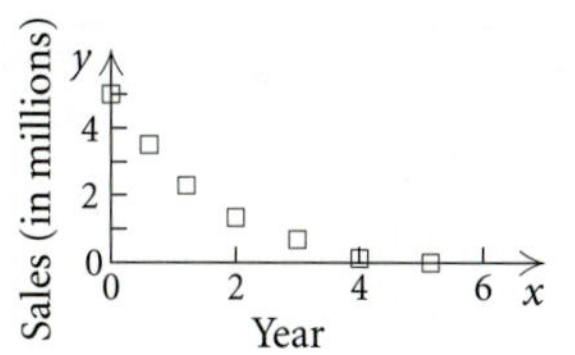

19.

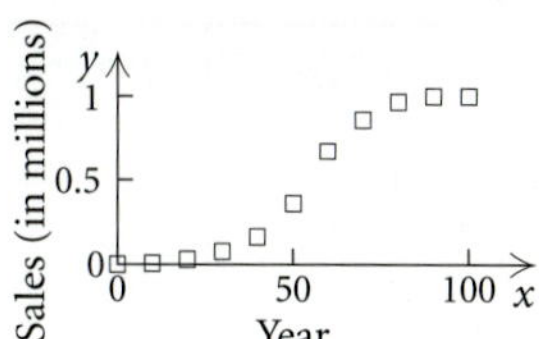

20.

21.

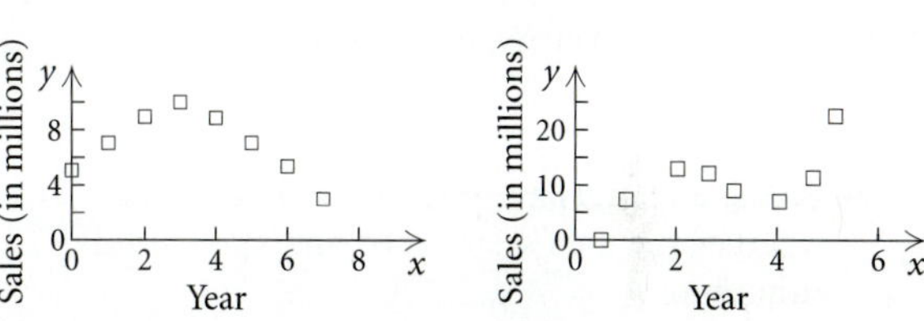

22.

23.

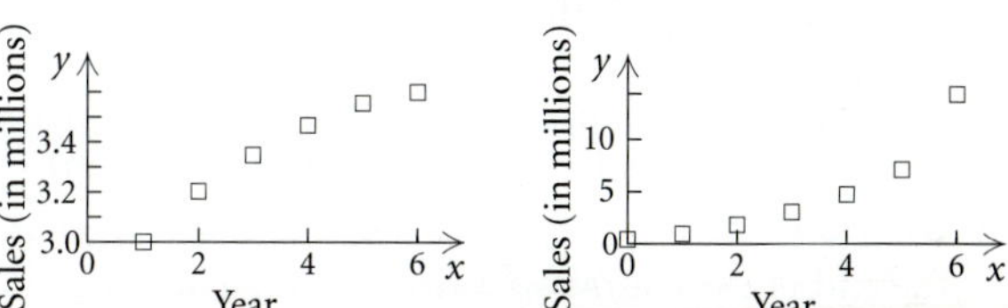

24. *Opinion on Capital Punishment.* The following table contains factual data from a recent survey of college freshmen regarding their opinion of capital punishment.

x, YEAR	PERCENTAGE OF COLLEGE FRESHMEN AGREEING THAT CAPITAL PUNISHMENT SHOULD BE ABOLISHED, y
0. 1971	56
8. 1979	32
13. 1984	24
18. 1989	21
23. 1994	18

Source: UCLA Higher Education Research Institute.

Using a grapher:

a) Create a scatterplot of the data. Determine whether an exponential function appears to fit the data.

b) Use regression to fit an exponential function $y = ab^x$, or ae^{-kx}, to the data, where $x = $ the number of years after 1971.

c) Use the function to predict the percent of college freshmen in 2000 agreeing that capital punishment should be abolished.

d) In what year would only 1% agree that capital punishment should be abolished?

25. *Total Revenue of Microsoft, Inc.*

x, YEAR	TOTAL REVENUE, y (IN BILLIONS)
0. 1989	$0.8
1. 1990	1.1
2. 1991	1.8
3. 1992	2.7
4. 1993	3.75

Source: Microsoft Annual Report.

Using a grapher:

a) Create a scatterplot of the data. Determine whether the data seem to fit an exponential function.

b) Use regression to fit an exponential function $y = ab^x$, or ae^{kx}, to the data, where $x =$ the number of years after 1989.

c) Use the function to predict the total revenue of Microsoft, Inc., in 2000.

d) In what year would the revenue be $20 billion?

26. *Forgetting.* In an art class, students were tested at the end of the course on a final exam. Then they were retested with an equivalent test at subsequent time intervals. Their scores after time t, in months, are given in the following table.

TIME, t (IN MONTHS)	SCORE, y
1	84.9%
2	84.6%
3	84.4%
4	84.2%
5	84.1%
6	83.9%

Using a grapher:

a) Use regression to fit a logarithmic function $y = a + b \ln x$ to the data.

b) Use the function to predict test scores after 8, 10, 24, and 36 months.

c) After how long will the test scores fall below 82%?

27. *Video Rentals.* Video rental spending has been on the increase. The data in the following table shows the average amount of video rental spending per family in recent years.

x, YEAR	VIDEO RENTAL SPENDING PER FAMILY, y
1. 1991	$113
2. 1992	114
3. 1993	119
4. 1994	122
5. 1995	129

Source: Media Group Research.

a) Use the REGRESSION feature on a grapher to fit both an exponential and a power function to the data, where $x = 1$ corresponds to 1991.

b) Graph each function.

c) Use each function in part (a) to predict the video rental spending in 2005.

d) Discuss the relative merits of using each function as a predictor.

28. *Effect of Advertising.* A company introduces a new software product on a trial run in a city. They advertised the product on television and found the following data relating the percentage P of people who bought the product after x ads were run.

NUMBER OF ADS, x	PERCENTAGE WHO BOUGHT, P
0	0.2
10	0.7
20	2.7
30	9.2
40	27
50	57.6
60	83.3
70	94.8
80	98.5
90	99.6

a) Use the REGRESSION feature on a grapher to fit a logistic function

$$P(x) = \frac{a}{1 + be^{-kx}}$$

to the data.

b) What percent will buy the product when 55 ads are run? 100 ads?

c) Find an asymptote for the graph. Interpret the asymptote in terms of the advertising situation.

Skill Maintenance

Find the missing lengths in each right triangle.

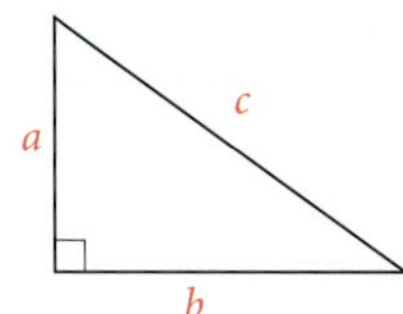

29. $b = 1$, $c = 2$ **30.** $a = 1$, $b = 1$

31. $a = 47$, $b = 34$ **32.** $b = \sqrt{13}$, $c = 200$

Synthesis

33. ◆ Browse through some newspapers or magazines until you find some data and/or a graph that seem as though they can be fit to an exponential function. Make a case for why such a fit is appropriate. Then fit an exponential function to the data and make some predictions.

34. ◆ *Atmospheric Pressure.* Atmospheric pressure P at an altitude a is given by

$$P = P_0 e^{-0.00005a},$$

where $P_0 =$ the pressure at sea level ≈ 14.7 lb/in^2 (pounds per square inch). Explain how a barometer, or some device for measuring atmospheric pressure, can be used to find the height of a skyscraper.

35. *Present Value.* Following the birth of a child, a parent wants to make an initial investment P_0 that will grow to $50,000 for the child's education at age 18. Interest is compounded continuously at 7%. What should the initial investment be? Such an amount is called the **present value** of $50,000 due 18 yr from now.

36. *Present Value.* Referring to Exercise 35:

a) Solve $P = P_0 e^{kt}$ for P_0.
b) Find the present value of $50,000 due 18 yr from now at interest rate 6.4%.

37. *Supply and Demand.* The supply and demand for the sale of a certain type of VCR are given by

$$S(p) = 480e^{-0.003p} \quad \text{and} \quad D(p) = 150e^{0.004p},$$

where $S(p) =$ the number of VCRs that the company is willing to sell at price p and $D(p) =$ the quantity that the public is willing to buy at price p. Find p, called the **equilibrium price**, such that $D(p) = S(p)$.

38. *Carbon Dating.* Recently, while digging in Chaco Canyon, New Mexico, archeologists found corn pollen that was 4000 yr old. This was evidence that Native Americans had been cultivating crops in the Southwest centuries earlier than scientists had thought. What percent of the carbon-14 had been lost from the pollen? (*Source: American Anthropologist*)

39. *Newton's Law of Cooling.* Suppose a body with temperature T_1 is placed in surroundings with temperature T_0 different from that of T_1. The body will either cool or warm to temperature $T(t)$ after time t, in minutes, where

$$T(t) = T_0 + |T_1 - T_0|e^{-kt}.$$

A cup of coffee with temperature 105°F is placed in a freezer with temperature 0°F. After 5 min, the temperature of the coffee is 70°F. What will its temperature be after 10 min?

40. *When Was the Murder Committed?* The police discover the body of a math professor. Critical to solving the crime is determining when the murder was committed. The coroner arrives at the murder scene at 12:00 P.M. She immediately takes the temperature of the body and finds it to be 94.6°. She then takes the temperature 1 hr later and finds it to be 93.4°. The temperature of the room is 70°. When was the murder committed? (Use Newton's law of cooling in Exercise 39.)

41. *Electricity.* The formula

$$i = \frac{V}{R}[1 - e^{-(R/L)t}]$$

occurs in the theory of electricity. Solve for t.

42. *The Beer–Lambert Law.* A beam of light enters a medium such as water or smog with initial intensity I_0. Its intensity decreases depending on the thickness (or concentration) of the medium. The intensity I at a depth (or concentration) of x units is given by

$$I = I_0 e^{-\mu x}.$$

The constant μ (the Greek letter "mu") is called the **coefficient of absorption**, and it varies with the medium. For sea water, $\mu = 1.4$.

a) What percentage of light intensity I_0 remains at a depth of sea water that is 1 m? 3 m? 5 m? 50 m?
b) Plant life cannot exist below 10 m. What percentage of I_0 remains at 10 m?

43. Given that $y = ae^x$, take the natural logarithm on both sides. Let $Y = \ln y$. Consider Y as a function of x. What kind of function is Y?

44. Given that $y = ax^b$, take the natural logarithm on both sides. Let $Y = \ln y$ and $X = \ln x$. Consider Y as a function of X. What kind of function is Y?

CHAPTER 3 Summary and Review

Important Properties and Formulas

One-to-One Function:	$f(a) = f(b) \longrightarrow a = b$
Exponential Function:	$f(x) = a^x$
The Number $e = 2.7182818284\ldots$	
Logarithmic Function:	$f(x) = \log_a x$
A Logarithm is an Exponent:	$\log_a x = y \longleftrightarrow x = a^y$
The Change-of-Base Formula:	$\log_b M = \dfrac{\log_a M}{\log_a b}$
The Product Rule:	$\log_a MN = \log_a M + \log_a N$
The Power Rule:	$\log_a M^p = p \log_a M$
The Quotient Rule:	$\log_a \dfrac{M}{N} = \log_a M - \log_a N$
Other Properties:	$\log_a a = 1, \qquad \log_a 1 = 0,$
	$\log_a a^x = x, \qquad a^{\log_a x} = x$
Base–Exponent Property:	$a^x = a^y \longleftrightarrow x = y$
Exponential Growth Model:	$P(t) = P_0 e^{kt}$
Exponential Decay Model:	$P(t) = P_0 e^{-kt}$
Interest Compounded Continuously:	$P(t) = P_0 e^{kt}$
Limited Growth:	$P(t) = \dfrac{a}{1 + be^{-kt}}$

REVIEW EXERCISES

1. Find the inverse of the relation

$\{(1.3, -2.7), (8, -3), (-5, 3), (6, -3), (7, -5)\}.$

2. Find an equation of the inverse relation.

a) $y = 3x^2 + 2x - 1$
b) $0.8x^3 - 5.4y^2 = 3x$

Given each function:

a) *Determine whether it is one-to-one, using a grapher if desired.*
b) *If it is one-to-one, find a formula for the inverse.*

3. $f(x) = \sqrt{x - 6}$ **4.** $f(x) = x^3 - 8$

5. $f(x) = 3x^2 + 2x - 1$ **6.** $f(x) = e^x$

7. Find $f(f^{-1}(657))$: $f(x) = \dfrac{4x^5 - 16x^{37}}{119x}$, $x > 0$.

In Exercises 8–13, match the equation with one of figures (a)–(f), which follow. If needed, use a grapher.

a) 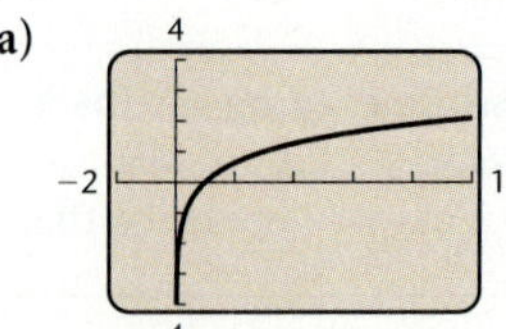

b) 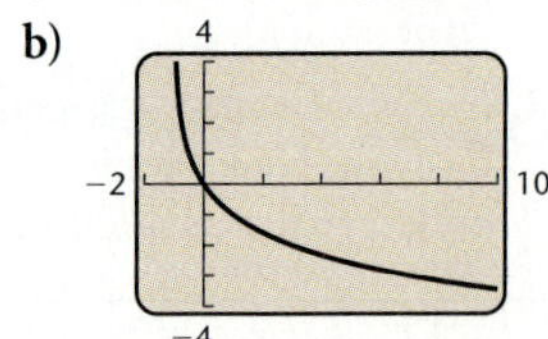

c) 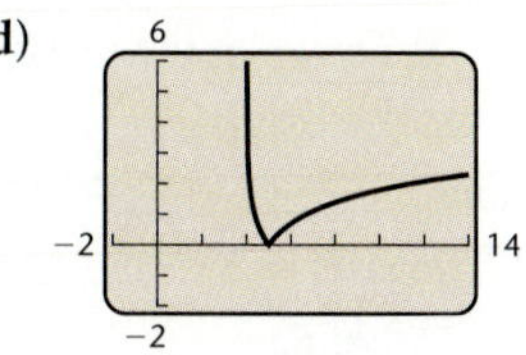

d)

e) 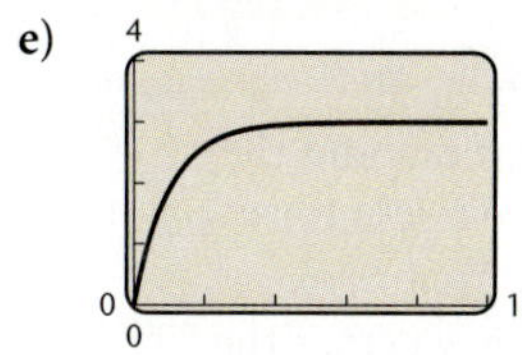

f) 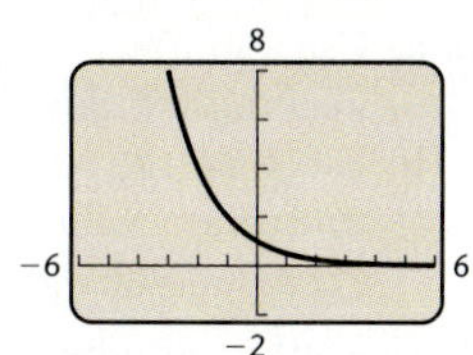

8. $f(x) = e^{x-3}$

9. $f(x) = \log_3 x$

10. $y = -\log_3 (x + 1)$

11. $y = \left(\frac{1}{2}\right)^x$

12. $f(x) = 3(1 - e^{-x})$, $x \geq 0$

13. $f(x) = |\ln (x - 4)|$

14. Convert to an exponential equation:
$$\log_4 x = 2.$$

15. Convert to a logarithmic equation:
$$e^x = 80.$$

Solve. Use any method.

16. $\log_4 x = 2$

17. $3^{1-x} = 9^{2x}$

18. $e^x = 80$

19. $4^{2x-1} - 3 = 61$

20. $\log_{16} 4 = x$

21. $\log_x 125 = 3$

22. $\log_2 x + \log_2 (x - 2) = 3$

23. $\log (x^2 - 1) - \log (x - 1) = 1$

24. $\log x^2 = \log x$

25. $e^{-x} = 0.02$

Express as a single logarithm and simplify if possible.

26. $3 \log_b x - 4 \log_b y + \frac{1}{2} \log_b z$

27. $\ln (x^3 - 8) - \ln (x^2 + 2x + 4) + \ln (x + 2)$

Express in terms of sums and differences of logarithms.

28. $\ln \sqrt[4]{wr^3}$

29. $\log \sqrt[3]{\dfrac{M^2}{N}}$

Given that $\log_a 2 = 0.301$, $\log_a 5 = 0.699$, and $\log_a 6 = 0.778$, find each of the following.

30. $\log_a 3$

31. $\log_a 50$

32. $\log_a \frac{1}{5}$

33. $\log_a \sqrt[3]{5}$

Simplify.

34. $\ln e^{-5k}$

35. $\log_5 5^{-6t}$

36. How long will it take an investment to double itself if it is invested at 5.4%, compounded continuously?

37. The population of a city doubled in 30 yr. What was the exponential growth rate?

38. How old is a skeleton that has lost 27% of its carbon-14?

39. The hydrogen ion concentration of milk is 2.3×10^{-6}. What is the pH?

40. What is the loudness, in decibels, of a sound whose intensity is $1000I_0$?

41. *The Population of Brazil.* The population of Brazil was 52 million in 1959, and the exponential growth rate was 2.8% per year. (*Source:* U.S. Bureau of the Census, World Population Profile)

    a) Find the exponential growth function.
    b) What will the population be in 2000? in 2020?
    c) When will the population be 300 million?
    d) What was the doubling time?

42. *Toll-free 800 Numbers.* The use of toll-free 800 numbers has grown exponentially. In 1967, there were 7 million such calls and in 1991, there were 10.2 billion such calls. (*Source:* Federal Communication Commission)

    a) Find the exponential growth rate k.
    b) Find the exponential growth function.
    c) Graph the exponential growth function.
    d) How many toll-free 800 number calls will be placed in 1998? in 2005?
    e) In what year will 20 billion such calls be placed?

43. *Walking Speed.* The average walking speed w, in feet per second, of a person living in a city of population P, in thousands, is given by the function
$$w(P) = 0.37 \ln P + 0.05.$$

    a) The population of Austin, Texas, is 466,000. Find the average walking speed.
    b) A city's population has an average walking speed of 3.4 ft/sec. Find the population.

44. *Multimedia Personal Computers.* The following table contains estimated data regarding the number of multimedia personal computers installed in homes in various years.

x, YEAR	MULTIMEDIA PERSONAL COMPUTERS, y (IN MILLIONS)
0. 1992	2
1. 1993	4.0
2. 1994	8.0
3. 1995	16.0
4. 1996	22.5
5. 1997	31.4

Source: Piper Joffray Research.

Using a grapher:

a) Create a scatterplot of the data. Determine whether the data seem to fit an exponential function.

b) Use regression to fit an exponential function $y = ab^x$, or ae^{kx}, to the data, where $x =$ the number of years after 1992.

c) Use the function to predict the number of multimedia personal computers in homes in 2000.

d) In how many years will there be 100 million such computers in homes?

Synthesis

45. ◈ Suppose that you were trying to convince a fellow student that

$$\log_2 (x + 5) \neq \log_2 x + \log_2 5.$$

Give as many explanations as you can.

46. ◈ Describe the difference between $f^{-1}(x)$ and $[f(x)]^{-1}$.

Solve.

47. $|\log_4 x| = 3$ **48.** $\log x = \ln x$

49. $5^{\sqrt{x}} = 625$

50. a) Use a grapher to graph $f(x) = 5e^{-x} \ln x$ in the viewing window $[-1, 10, -5, 5]$.

b) Estimate the relative maximum and the minimum values. Use a MAX–MIN feature if such exists on your grapher.

51. Use only a grapher. Determine whether the following functions are inverses of each other:

$$f(x) = \frac{4 + 3x}{x - 2}, \qquad g(x) = \frac{x + 4}{x - 3}.$$

52. Find the domain: $f(x) = \log_3 (\ln x)$.

53. Find the points of intersection of the graphs of the following equations:

$$y = 5x^2 e^{-x}, \qquad y = 2 - e^{-x^2}.$$

Systems and Matrices

4

The equilibrium point (x, y) for a product indicates the price y at which the amount x of the product that the seller willingly supplies is the same as the amount that the consumer willingly demands. The equilibrium point for the Great Tunes portable CD player is the solution of this system of equations, where y is the price, in dollars, and x is the number of units, in millions:

$$y = 200 - 25x,$$
$$y = 90 + 30x.$$

X	Y₁	Y₂
0	200	90
1	175	120
2	150	150
3	125	180
4	100	210
5	75	240
6	50	270

X = 2

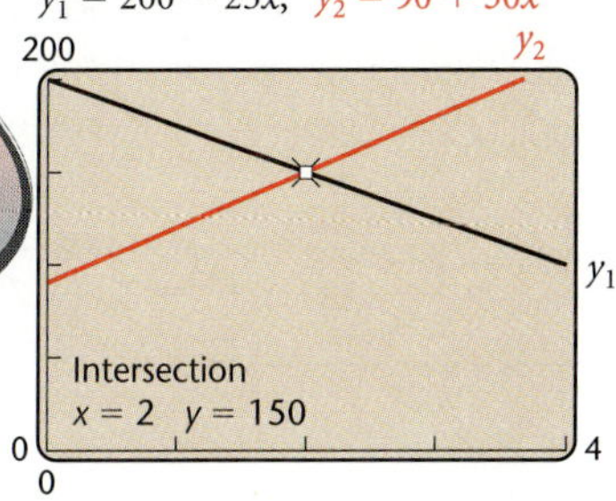

It is often desirable or even necessary to use two or more variables to model a situation in a field such as business, science, psychology, engineering, education, or sociology. When this is the case, we write and solve a *system of equations* in order to answer questions about that situation. In this chapter, we study systems of equations and several methods for solving them. We also use *systems of inequalities*, along with *linear programming*, to model applications and to find the maximum and minimum values of functions subject to a set of restrictions, or constraints.

321

4.1

Systems of Equations in Two Variables

- *Solve a system of two linear equations in two variables by graphing.*
- *Solve a system of two linear equations in two variables using the substitution and the elimination methods.*
- *Use systems of two linear equations to solve applied problems.*

A **system of equations** is composed of two or more equations considered simultaneously. For example,

$$x + y = 11,$$
$$3x - y = 5$$

is a system of two linear equations in two variables. The solution set of this system consists of all ordered pairs that make *both* equations true. Note that $(4, 7)$ is a solution of the system of equations above. We can verify this by substituting 4 for x and 7 for y in *each* equation.

$x + y = 11$	$3x - y = 5$
$4 + 7 \ ? \ 11$	$3 \cdot 4 - 7 \ ? \ 5$
$11 \ | \ 11$ TRUE	$12 - 7$
	$5 \ | \ 5$ TRUE

Interactive Discovery

Graph $y_1 = 2x - 5$ and $y_2 = -x + 1$ in the same viewing window. Use the grapher's TRACE and ZOOM features or the INTERSECT feature to find the coordinates of the point of intersection of the graphs. Do the same for $y_1 = x + 4$ and $y_2 = -1.25x - 5$ and for $y_1 = 3x - 7$ and $y_2 = -2x + \frac{1}{2}$. What does each of these ordered pairs represent?

Solving Systems of Equations Graphically

Recall that the graph of a linear equation is a line that contains all the ordered pairs in the solution set of the equation. When we graph a system of linear equations, each point at which the equations intersect is a solution of *both* equations and therefore a solution of the system of equations.

Consider the system

$$x + y = 11,$$
$$3x - y = 5.$$

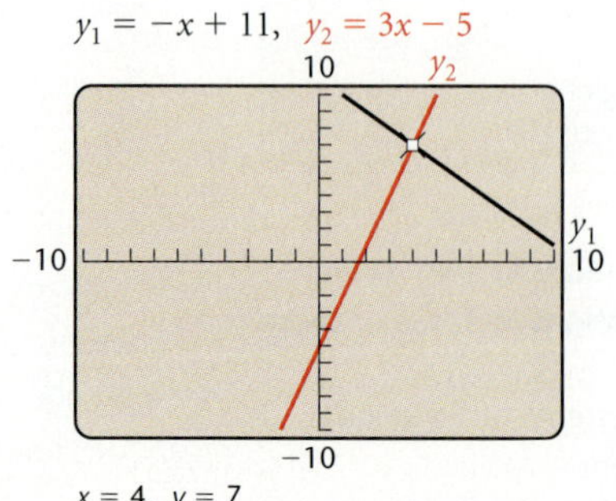

When we graph these equations on the same set of axes, we see that they intersect at a single point, $(4, 7)$, so $(4, 7)$ is a solution of the system of equations. (Note that when a grapher is used, it might be necessary to write each equation in "$Y = \cdots$" form. If so, we would graph $y_1 = -x + 11$ and $y_2 = 3x - 5$.)

To check this result, we can substitute 4 for x and 7 for y in both equations, as we did above. We can also use the TABLE feature on a grapher to do a check, observing that when $x = 4$, $y_1 = 7$ and $y_2 = 7$.

X	Y1	Y2
1	10	−2
2	9	1
3	8	4
4	7	7
5	6	10
6	5	13
7	4	16

X = 4

The graphs of most of the systems of equations that we use to model applications intersect at a single point, like the system above. However, it is possible that the graphs will have no points in common or infinitely many points in common. Each of these possibilities is illustrated below.

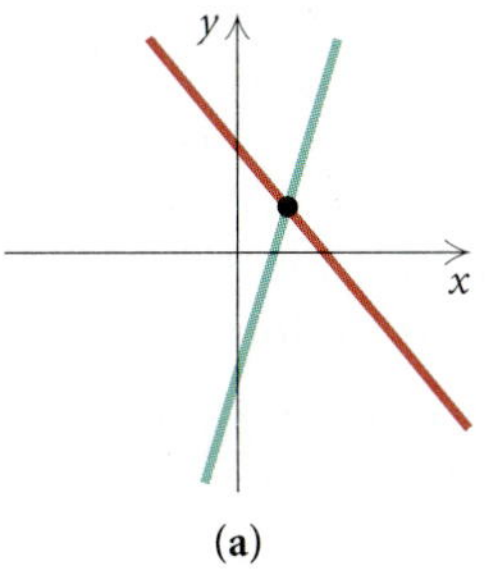

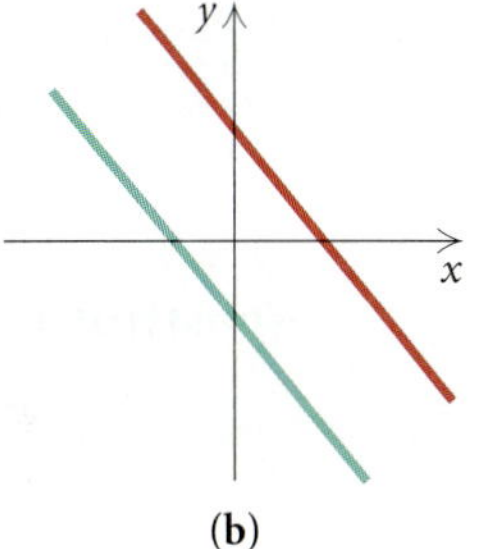

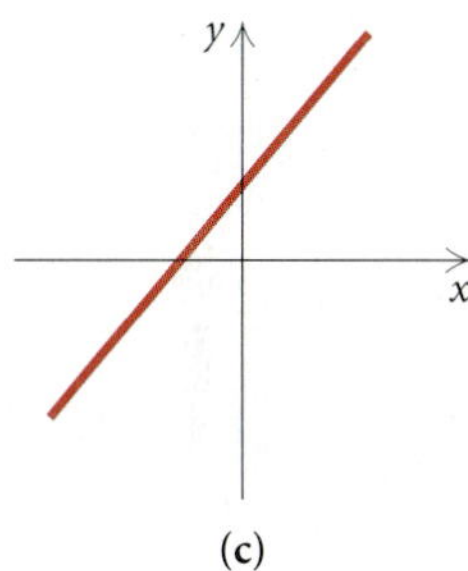

(a)

(b)

(c)

Exactly one common point
One solution

Parallel lines
No common points
No solution

Lines are identical
Infinitely many common points
Infinitely many solutions

If a system of two linear equations in two variables has at least one solution, it is **consistent**. If the system has no solutions, it is **inconsistent**. In addition, if a system of two linear equations in two variables has an infinite number of solutions, it is **dependent**. Otherwise, it is **independent**.

The system graphed in figure (a) above is consistent and independent; the system in (b) is inconsistent and independent; and the system in (c) is consistent and dependent.

The Substitution Method

Graphing helps us picture the solution of a system of equations, but solving by graphing is not always accurate when solutions are not integers. A grapher can improve the accuracy of the graphing method, but when the solution is a pair like $\left(\frac{43}{27}, -\frac{19}{27}\right)$, the grapher generally gives only an approximate solution.

Algebraic methods for solving systems of equations, when used correctly, always give accurate results. One such technique is the **substitution method**. It is used most often when a variable is alone on one side of an equation or when it is easy to solve for a variable. To apply the substitution method, we begin by using one of the equations to express one variable in terms of the other; then we substitute that expression in the other equation of the system.

Example 1   Use the substitution method to solve the system

$$x + y = 11, \qquad (1)$$
$$3x - y = 5. \qquad (2)$$

SOLUTION   First, we solve equation (1) for y. (We could just as well solve for x.)

$$y = 11 - x$$

Then we substitute $11 - x$ for y in equation (2). This gives an equation in one variable, which we know how to solve.

$$3x - (11 - x) = 5$$
$$3x - 11 + x = 5 \qquad \text{\textbf{Removing parentheses}}$$
$$4x - 11 = 5$$
$$4x = 16$$
$$x = 4$$

Now we substitute 4 for x in either equation (1) or (2) (this is called **back-substitution**) and solve for y. We choose equation (1):

$$4 + y = 11$$
$$y = 7.$$

We have previously checked the pair $(4, 7)$ in both equations both algebraically and graphically. The solution of the system of equations is $(4, 7)$.

The Elimination Method

Another algebraic technique for solving systems of equations is the **elimination method**. With this method, we eliminate a variable by adding two equations. Before we add, it might be necessary to multiply one or both equations by suitable constants in order to find two equations in which coefficients differ only by sign.

Example 2   Solve each of the following systems using the elimination method.

a) $2x + y = 2, \qquad (1)$ **b)** $\;\; 4x + 3y = 11, \qquad (1)$

$\quad\;\; x - y = 7 \qquad\;\; (2)$ $\;\; -5x + 2y = 15 \qquad (2)$

SOLUTION

a) Since the y-coefficients differ only by sign, we can eliminate y by adding the equations:

$$\begin{aligned} 2x + y &= 2 \qquad (1)\\ \underline{x - y} &= \underline{7} \qquad (2)\\ 3x \quad\;\; &= 9 \qquad \text{\textbf{Adding}}\\ x &= 3. \end{aligned}$$

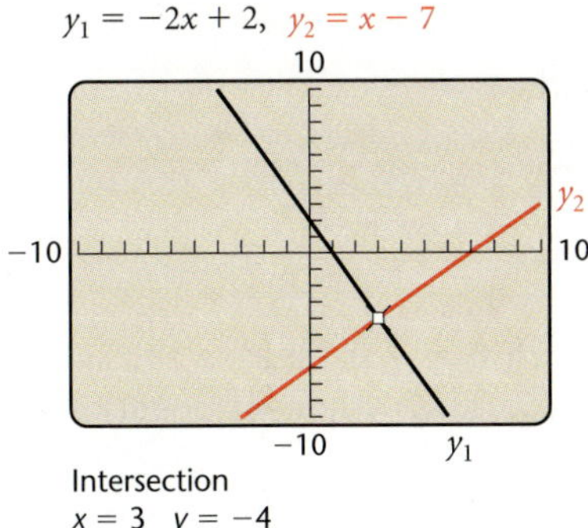

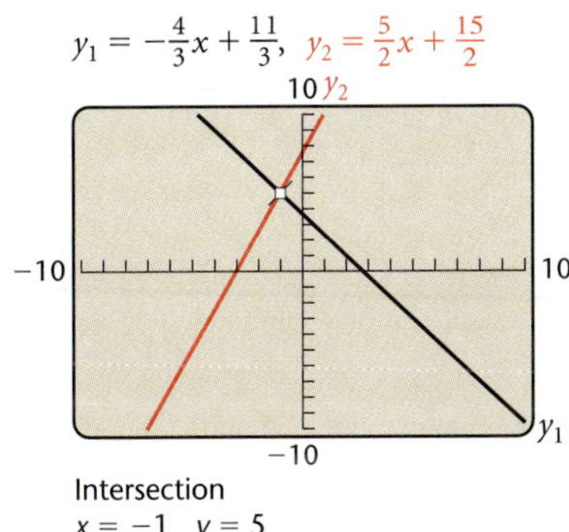

We then back-substitute 3 for x in either equation and solve for y. We choose equation (1):

$$2 \cdot 3 + y = 2$$
$$6 + y = 2$$
$$y = -4.$$

We can check the pair $(3, -4)$ either algebraically by substituting in both equations or graphically using the INTERSECT feature, as shown at left. The solution is $(3, -4)$.

b) We can obtain x-coefficients that differ only by sign by multiplying the first equation by 5 and the second equation by 4:

$$\begin{array}{ll} 20x + 15y = 55 & \text{Multiplying equation (1) by 5} \\ \underline{-20x + 8y = 60} & \text{Multiplying equation (2) by 4} \\ 23y = 115 & \text{Adding} \\ y = 5. \end{array}$$

We then back-substitute 5 for y in either equation (1) or (2) and solve for x. We choose equation (1):

$$\begin{array}{ll} 4x + 3 \cdot 5 = 11 & \text{Substituting 5 for } y \text{ in equation (1)} \\ 4x + 15 = 11 & \\ 4x = -4 & \\ x = -1. & \end{array}$$

We can check the pair $(-1, 5)$ either algebraically by substituting in both equations or graphically, as shown at left. The solution is $(-1, 5)$.

In Example 2(b), the two systems

$$\begin{array}{l} 4x + 3y = 11, \\ -5x + 2y = 15 \end{array} \quad \text{and} \quad \begin{array}{l} 20x + 15y = 55, \\ -20x + 8y = 60 \end{array}$$

are **equivalent** because they have exactly the same solutions. When we use the elimination method, we often multiply one or both equations by constants to find equivalent equations that allow us to eliminate a variable by adding.

Example 3   Solve each of the following systems using the elimination method.

a) $\begin{array}{ll} x - 3y = 1, & (1) \\ -2x + 6y = 5 & (2) \end{array}$

b) $\begin{array}{ll} 2x + 3y = 6, & (1) \\ 4x + 6y = 12 & (2) \end{array}$

SOLUTION

a) We multiply equation (1) by 2 and add:

$$2x - 6y = 2 \qquad \text{Multiplying equation (1) by 2}$$
$$\underline{-2x + 6y = 5} \qquad (2)$$
$$0 = 7. \qquad \text{Adding}$$

There are no values of x and y for which $0 = 7$ is true, so the system has no solution. The solution set is $\varnothing$. The system of equations is inconsistent.

The graphs of the equations are parallel lines. In fact, we see this when we write the equations in simplified "$Y = \cdots$" form to enter them on a grapher. The slopes are the same, $\left(\frac{1}{3}\right)$, but the y-intercepts are different, $\left(0, \frac{5}{6}\right)$ and $\left(0, -\frac{1}{3}\right)$.

$$y_1 = \tfrac{1}{3}x - \tfrac{1}{3}, \ \ y_2 = \tfrac{1}{3}x + \tfrac{5}{6}$$

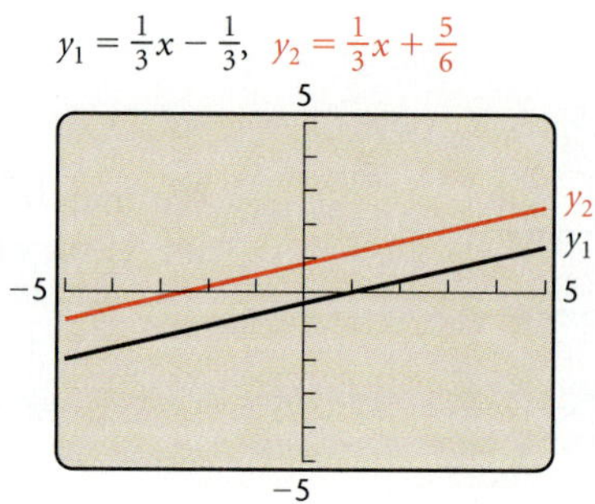

b) We multiply equation (1) by -2 and add:

$$-4x - 6y = -12 \qquad \text{Multiplying equation (1) by } -2$$
$$\underline{4x + 6y = 12} \qquad (2)$$
$$0 = 0. \qquad \text{Adding}$$

We obtain the equation $0 = 0$, which is true for all values of x and y. This tells us that the system of equations is dependent, so there are infinitely many solutions. That is, any solution of one equation of the system is also a solution of the other. The graphs of the equations are identical. (In fact, when we write the equations in simplified "$Y = \cdots$" form to enter them on a grapher, we get identical equations.)

Using either equation, we can write $y = -\frac{2}{3}x + 2$, so we can write the solutions of the system as ordered pairs of the form $\left(x, -\frac{2}{3}x + 2\right)$. Any real value that we choose for x then gives us a value for y and thus an ordered pair in the solution set. Some of the solutions are $(-3, 4)$, $(0, 2)$, and $(6, -2)$. Similarly, we can write $x = -\frac{3}{2}y + 3$, so the solutions can also be expressed as $\left(-\frac{3}{2}y + 3, y\right)$.

$$y_1 = -\tfrac{2}{3}x + 2, \ \ y_2 = -\tfrac{2}{3}x + 2$$

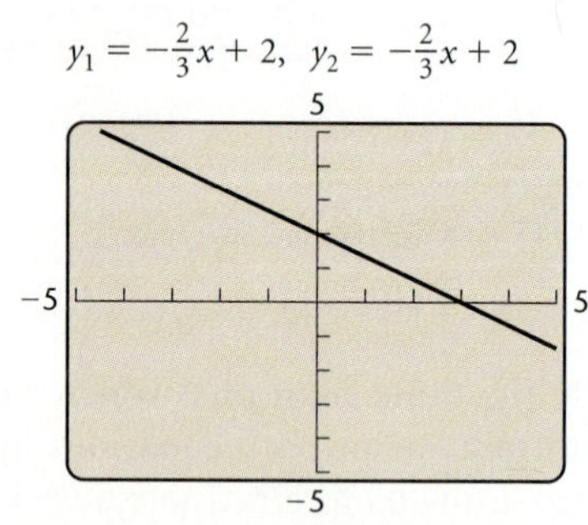

Applications

Frequently the most challenging and time-consuming step in the problem-solving process is translating a situation to mathematical language. However, in many cases, this task is facilitated if we translate to more than one equation in more than one variable.

Example 4 *Motion.* A Boeing 747 flies the 3000-mi distance from Los Angeles to New York, with a tail wind, in 5 hr. The return trip, against the wind, takes 6 hr. Find the speed of the plane and the speed of the wind.

SOLUTION   We will use the five-step problem-solving process. (See Section R.9.)

1. **Familiarize.** We first make a drawing. Let p = the speed of the plane in still air and w = the speed of the wind, both in miles per hour. The distances are the same. When the plane flies east with the wind, its speed is $p + w$. When it flies west against the wind, its speed is $p - w$. We organize this information in a table, the columns of the table coming from the motion formula $d = rt$.

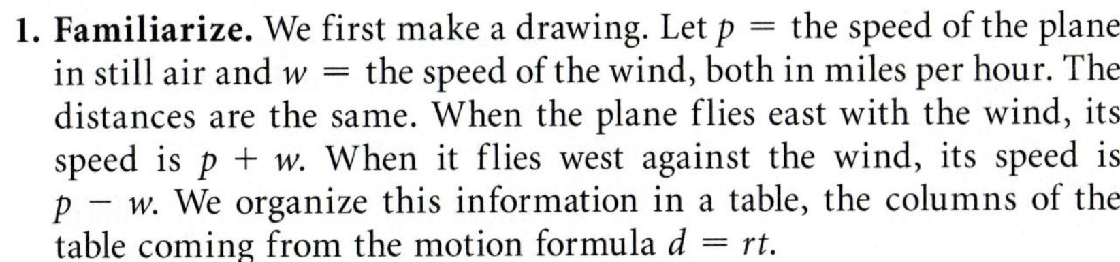
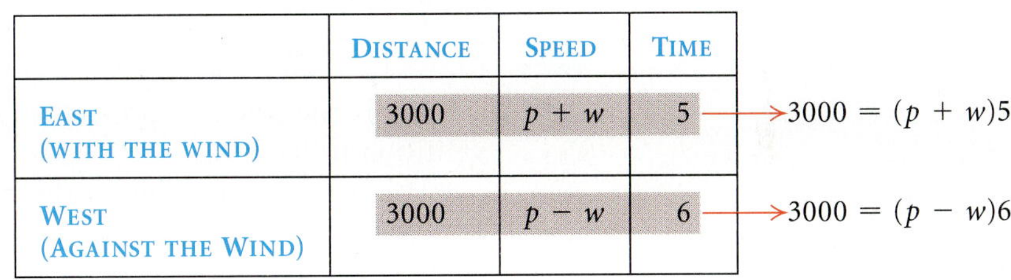
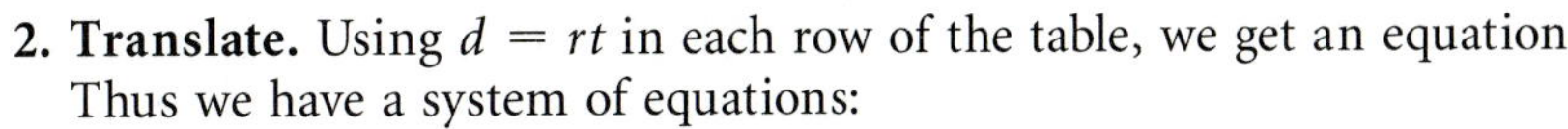

	DISTANCE	SPEED	TIME	
EAST (WITH THE WIND)	3000	$p + w$	5	$\rightarrow 3000 = (p + w)5$
WEST (AGAINST THE WIND)	3000	$p - w$	6	$\rightarrow 3000 = (p - w)6$

2. **Translate.** Using $d = rt$ in each row of the table, we get an equation. Thus we have a system of equations:

$$3000 = (p + w)5 = 5p + 5w,$$
$$3000 = (p - w)6 = 6p - 6w.$$

3. **Carry out.** We solve the system

$$5p + 5w = 3000, \qquad (1)$$
$$6p - 6w = 3000. \qquad (2)$$

ALGEBRAIC SOLUTION

Note that we can simplify these equations by multiplying equation (1) by $\frac{1}{5}$ and equation (2) by $\frac{1}{6}$. Then we have

$$p + w = 600 \qquad (1a)$$
$$p - w = 500. \qquad (2a)$$

We add these equations to eliminate w:

$$
\begin{array}{r}
p + w = 600 \\
\underline{p - w = 500} \\
2p \quad\;\; = 1100 \\
p = 550.
\end{array}
$$

Next, we back-substitute to find w:

$$550 + w = 600 \qquad \text{Substituting in equation (1a)}$$

$$w = 50.$$

GRAPHICAL SOLUTION

We replace p with x and w with y, graph $y_1 = 600 - x$ and $y_2 = x - 500$, and find the point of intersection of the graphs.

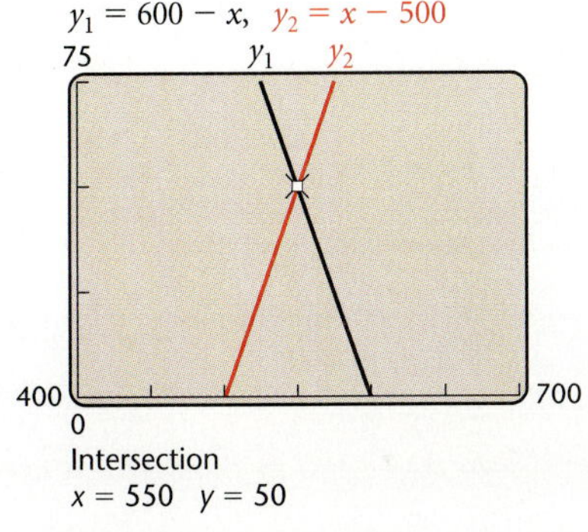

4. Check. When the plane is traveling east, its speed is $550 + 50$, or 600 mph, and the trip takes 3000/600, or 5 hr. When the plane travels west, its speed is $550 - 50$, or 500 mph, and the trip takes 3000/500, or 6 hr. Our solution checks.

5. State. The solution of the system of equations is $(550, 50)$. The speed of the plane is 550 mph, and the speed of the wind is 50 mph.

Example 5 *Supply and Demand.* Suppose that the price and the supply of the Great Tunes portable CD player are related by the equation

$$y = 200 - 25x,$$

where y is the price, in dollars, at which the seller is willing to supply x million units. Also suppose that the price and the demand for the same model of CD player are related by the equation

$$y = 90 + 30x,$$

where y is the price, in dollars, at which the consumer is willing to buy x million units.

The *equilibrium point* for this product is the pair (x, y) that is a solution of both equations. The *equilibrium price* is the price at which the amount of the product that the seller is willing to supply is the same as the amount demanded by the consumer. Find the equilibrium point for this product.

SOLUTION

1., 2. Familiarize and **Translate.** We are given a system of equations in the statement of the problem, so no further translation is necessary.

$$y = 200 - 25x, \qquad (1)$$
$$y = 90 + 30x. \qquad (2)$$

We substitute some values for x in each equation to get an idea of the corresponding prices. When $x = 1$,

$$y = 200 - 25 \cdot 1 = 175, \qquad \text{Substituting in equation (1)}$$
$$y = 90 + 30 \cdot 1 = 120. \qquad \text{Substituting in equation (2)}$$

This indicates that the price when 1 million units are supplied is higher than the price when 1 million units are demanded.

When $x = 4$,

$$y = 200 - 25 \cdot 4 = 100, \qquad \text{Substituting in equation (1)}$$
$$y = 90 + 30 \cdot 4 = 210. \qquad \text{Substituting in equation (2)}$$

In this case, the price related to supply is lower than the price related to demand. It would appear that the x-value we are looking for is between 1 and 4.

3. Carry out. We use substitution:

$$200 - 25x = 90 + 30x \qquad \text{Substituting } 200 - 25x \text{ for } y \text{ in equation (2)}$$
$$110 = 55 \qquad \text{Adding } 25x \text{ and subtracting } 90 \text{ on both sides}$$
$$2 = x. \qquad \text{Dividing by } 55 \text{ on both sides}$$

We now back-substitute 2 for x in either equation and find y:

$$y = 200 - 25 \cdot 2 \qquad \text{Substituting in equation (1)}$$
$$y = 150.$$

4. **Check.** We can check algebraically by substituting 2 for x and 150 for y in both equations. Since the equations were given in the statement of the problem, we can also check graphically, as shown below.

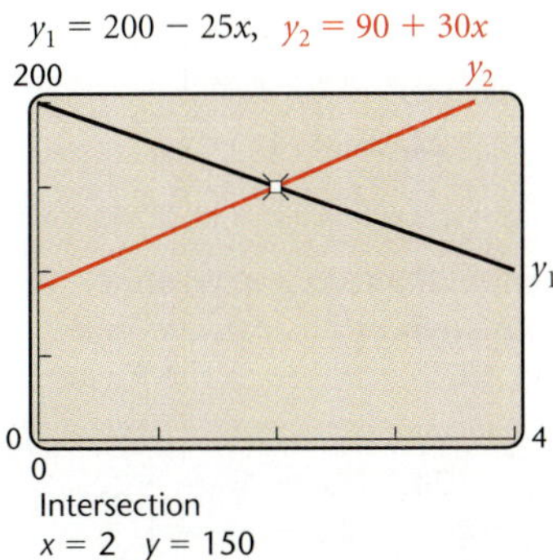

5. **State.** The equilibrium point is (2, \$150). That is, the equilibrium supply is 2 million units and the equilibrium price is \$150.

There is another algebraic method for solving systems of two linear equations in two variables known as *Cramer's rule*. It is presented in Appendix A. Systems of equations can be solved on some graphers using the SIMULT feature.

4.1 Exercise Set

In Exercises 1–6, match each system of equations with one of the graphs (a)–(f), which follow.

a)
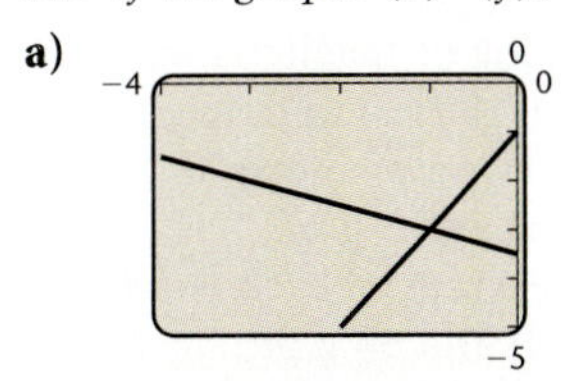

b)
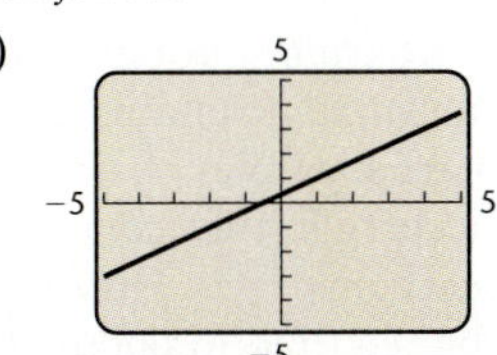

c)
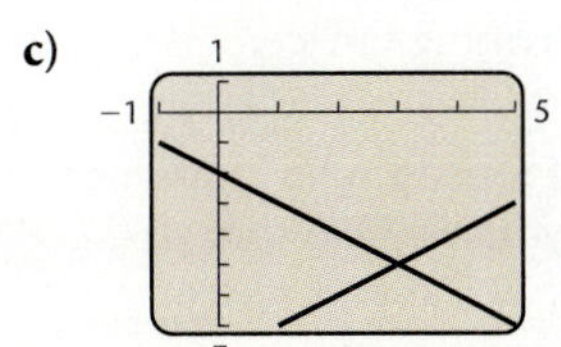

d)
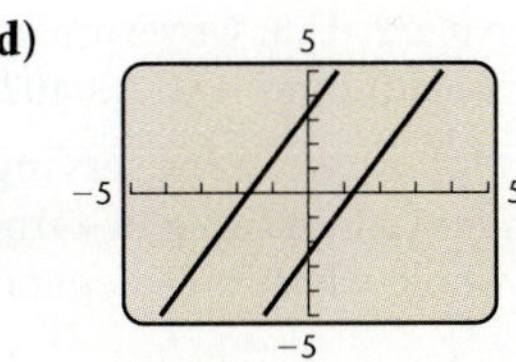

e)
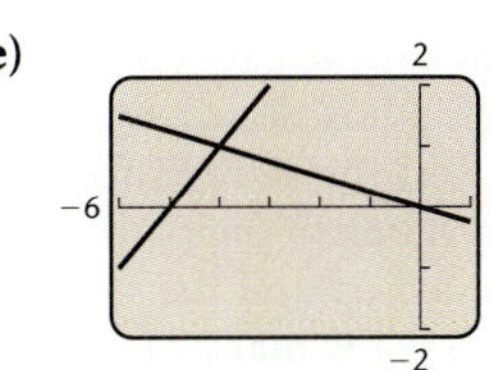

f)
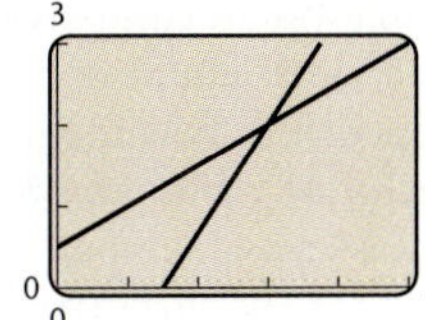

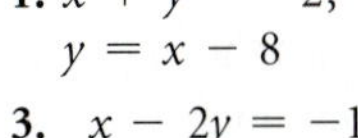
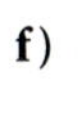

1. $x + y = -2$,
   $y = x - 8$

2. $x - y = -5$,
   $x = -4y$

3. $x - 2y = -1$,
   $4x - 3y = 6$

4. $2x - y = 1$,
   $x + 2y = -7$

5. $2x - 3y = -1$,
   $-4x + 6y = 2$

6. $4x - 2y = 5$,
   $6x - 3y = -10$

Solve graphically.

7. $x + y = 2$,
   $3x + y = 0$

8. $x + y = 1$,
   $3x + y = 7$

9. $\frac{1}{2}x + y = \frac{3}{2}$,
   $2x + 8y = 13$

10. $x + 2y = -\frac{2}{3}$,
    $\frac{1}{4}x - \frac{1}{2}y = 1$

11. $y + 1 = 2x$,
    $y - 1 = 2x$

12. $2x - y = 1$,
    $3y = 6x - 3$

Solve using the substitution method. Use a grapher to check your answer.

13. $x + y = 9,$
$\quad 2x - 3y = -2$

14. $3x - y = 5,$
$\quad x + y = \frac{1}{2}$

15. $x - 2y = 7,$
$\quad x = y + 4$

16. $x + 4y = 6,$
$\quad x = -3y + 3$

17. $y = 2x - 6,$
$\quad 5x - 3y = 16$

18. $3x + 5y = 2,$
$\quad 2x - y = -3$

19. $x - 5y = 4,$
$\quad y = 7 - 2x$

20. $5x + 3y = -1,$
$\quad x + y = 1$

21. $2x - 3y = 5,$
$\quad 5x + 4y = 1$

22. $3x + 4y = 6,$
$\quad 2x + 3y = 5$

Solve using the elimination method. Also determine whether each system is consistent or inconsistent and dependent or independent. Use a grapher to check your answer.

23. $x + 2y = 7,$
$\quad x - 2y = -5$

24. $3x + 4y = -2,$
$\quad -3x - 5y = 1$

25. $x - 3y = 2,$
$\quad 6x + 5y = -34$

26. $x + 3y = 0,$
$\quad 20x - 15y = 75$

27. $0.3x - 0.2y = -0.9,$
$\quad 0.2x - 0.3y = -0.6$

28. $0.2x - 0.3y = 0.3,$
$\quad 0.4x + 0.6y = -0.2$

29. $3x - 12y = 6,$
$\quad 2x - 8y = 4$

30. $2x + 6y = 7,$
$\quad 3x + 9y = 10$

31. $\frac{1}{5}x + \frac{1}{2}y = 6,$
$\quad \frac{3}{5}x - \frac{1}{2}y = 2$

32. $\frac{2}{3}x + \frac{3}{5}y = -17,$
$\quad \frac{1}{2}x - \frac{1}{3}y = -1$

33. $2x = 5 - 3y,$
$\quad 4x = 11 - 7y$

34. $7(x - y) = 14,$
$\quad 2x = y + 5$

35. *Sales Promotions.* During a one-month promotional campaign, Video Village gave either a free video rental or a 12-serving box of microwave popcorn to new members. It cost the store $1 for each free rental and $2 for each box of popcorn. In all, 48 new members were signed up and the store's cost for the incentives was $86. How many of each incentive were given away?

36. *Concert Ticket Prices.* One evening 1500 concert tickets were sold for the Fairmont Summer Jazz Festival. Tickets cost $25 for covered pavilion seats and $15 for lawn seats. Total receipts were $28,500. How many of each type of ticket were sold?

37. *Museum Admission Prices.* Admission to the Indianapolis Children's Museum costs twice as much for adults as for children. Admission to the museum for 4 adults and 16 children from the Tiny Tots Day Care Center cost $72. Find the cost of each adult's admission and each child's admission.

38. *Mail-Order Business.* A mail-order lacrosse equipment business shipped 120 packages one day. Customers are charged $3.50 for each standard-delivery package and $7.50 for each express-delivery package. Total shipping charges for the day were $596. How many of each kind of package were shipped?

39. *Supply and Demand.* The supply and demand for a particular model of electronic organizer are related to price by the equations

$$y = 175 - 5x,$$
$$y = 70 + 2x,$$

respectively, where y is the price, in dollars, and x is the number of units, in thousands. Find the equilibrium point for this product.

40. *Supply and Demand.* The supply and demand for a particular model of treadmill are related to price by the equations

$$y = 500 - 25x,$$
$$y = 240 + 40x,$$

respectively, where y is the price, in dollars, and x is the number of units, in thousands. Find the equilibrium point for this product.

The point at which a company's costs equal its revenues is the break-even point. In Exercises 41–44, C represents the production cost, in dollars, of x units of a product and R represents the revenue, in dollars, from the sale of x units. Find the number of units that must be produced and sold in order to break even. That is, find the value of x for which C = R.

41. $C = 14x + 350,$
$\quad R = 16.5x$

42. $C = 8.5x + 75,$
$\quad R = 10x$

43. $C = 15x + 12,000,$
$\quad R = 18x - 6000$

44. $C = 3x + 400,$
$\quad R = 7x - 600$

45. *Nutrition.* A one-cup serving of spaghetti with meatballs contains 260 Cal (calories) and 32 g of carbohydrates. A one-cup serving of chopped iceberg lettuce contains 5 Cal and 1 g of carbohydrates. How many servings of each would be required to obtain 400 Cal and 50 g of carbohydrates? (*Source: Home and Garden Bulletin No. 72*, U.S. Government Printing Office, Washington, D.C. 20402)

46. *Nutrition.* One serving of tomato soup contains 100 Cal and 18 g of carbohydrates. One slice of whole wheat bread contains 70 Cal and 13 g of carbohydrates. How many servings of each would be required to obtain 230 Cal and 42 g of carbohydrates? (*Source: Home and Garden Bulletin No. 72,* U.S. Government Printing Office, Washington, D.C. 20402)

47. *Motion.*   A Leisure Time Cruises riverboat travels 46 km downstream in 2 hr. It travels 51 km upstream in 3 hr. Find the speed of the boat and the speed of the stream.

48. *Motion.*   A DC10 travels 3000 km with a tail wind in 3 hr. It travels 3000 km with a head wind in 4 hr. Find the speed of the plane and the speed of the wind.

49. *Investment.*   Bernadette inherited $15,000 and invested it in two municipal bonds, which pay 7% and 9% simple interest. The annual interest is $1230. Find the amount invested at each rate.

50. *Tee Shirt Sales.*   Mack's Tee Shirt Shack sold 36 shirts one day. All short-sleeved tee shirts cost $12 and all long-sleeved tee shirts cost $18. Total receipts for the day were $522. How many of each kind of shirt were sold?

51. *Coffee Mixtures.*   The owner of The Daily Grind coffee shop mixes French roast coffee worth $9.00 per pound with Kenyan coffee worth $7.50 per pound in order to get 10 lb of a mixture worth $8.40 per pound. How much of each type of coffee was used?

52. *Snack Mixtures.*   At Sunda's Snacks, caramel corn worth $2.50 per pound is mixed with honey roasted peanuts worth $7.50 per pound in order to get 20 lb of a mixture worth $3 per pound. How much of each snack was used?

53. *Commissions.*   Jackson Manufacturing offers its sales representatives a choice between being paid a commission of 8% of sales or being paid a monthly salary of $1500 plus a commission of 1% of sales. For what monthly sales do the two plans pay the same amount?

54. *Motion.*   Two private airplanes travel toward each other from cities that are 780 km apart at speeds of 190 km/h and 200 km/h. They left at the same time. In how many hours will they meet?

Skill Maintenance

55. Evaluate $2x - 3y$ for $x = -2$ and $y = 5$.

56. Evaluate $-x + 5y - 4z$ for $x = -1$, $y = 5$, and $z = 2$.

57. Graph: $y = 2x^2 - 3x + 4$.

58. Graph: $y = 0.4x^2 + 0.01x - 0.05$.

Synthesis

59. ◈ Write a problem for a classmate to solve that can be translated to a system of equations. Devise the problem so that the solution is "The concert promoter sold 200 reserved seat tickets and 950 general admission tickets."

60. ◈ Explain in your own words when the elimination method for solving a system of equations is preferable to the substitution method.

61. *Motion.*   Nancy jogs and walks to campus each day. She averages 4 km/h walking and 8 km/h jogging. The distance from home to the campus is 6 km and she makes the trip in 1 hr. How far does she jog on each trip?

62. *Mail-Order Business.*   A mail-order catalog advertises a limited-time sale, offering 1 turtleneck for $15 and 2 turtlenecks for $25. A total of 1250 turtlenecks are sold and $16,750 is taken in. How many customers ordered 2 turtlenecks?

63. *Motion.*   A train leaves Union Station for Central Station, 216 km away, at 9 A.M. One hour later, a train leaves Central Station for Union Station. They meet at noon. If the second train had started at 9 A.M. and the first train at 10:30 A.M., they would still have met at noon. Find the speed of each train.

64. *Antifreeze Mixtures.*   An automobile radiator contains 16 L of antifreeze and water. This mixture is 30% antifreeze. How much of this mixture should be drained and replaced with pure antifreeze so that the final mixture will be 50% antifreeze?

65. *Value of a Horse.*   A stablehand agreed to work for 1 yr. At the end of that year, he was to receive $240 and one horse. After 7 months, the stablehand quit the job but still received the horse and $100. What is the value of the horse?

66. *Ticket Line.*   You are in line at a ticket window. There are 2 more people ahead of you in line than there are behind you. In the entire line, there are three times as many people as there are behind you. How many people are ahead of you?

67. Two solutions of the equation $y = mx + b$ are $(-2, 3)$ and $(4, -5)$. Find m and b.

68. Two solutions of the equation $Ax + By = 1$ are $(3, -1)$ and $(-4, -2)$. Find A and B.

69. *Gas Mileage.*   A 1994 Dodge Stealth gets 18 miles per gallon (mpg) in city driving and 24 mpg in highway driving. The car is driven 465 mi on a full tank of 23 gal of gasoline. How many miles were driven in the city and how many were driven on the highway?

70. *Motion.*   Heather is out hiking and is standing on a railroad bridge, as shown in the figure at right. A train is approaching from the direction shown by the arrow. If Heather runs at a speed of 10 mph toward the train, she will reach point P on the bridge at the same moment that the train does. If she runs to point Q at the other end of the bridge at a speed of 10 mph, she will reach point Q also at the same moment that the train does. How fast, in miles per hour, is the train traveling?

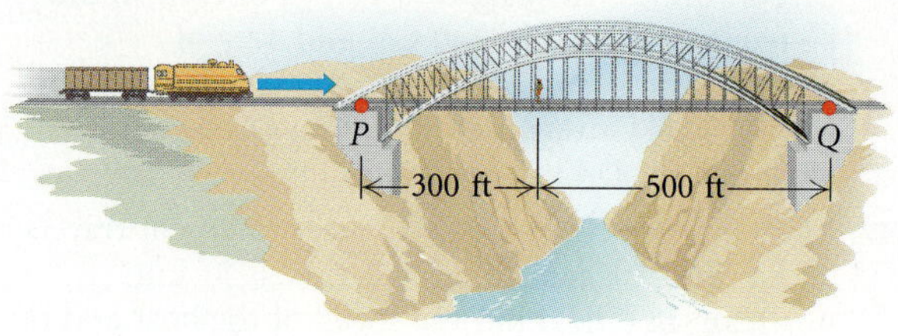

4.2

Systems of Equations in Three or More Variables

- *Solve systems of linear equations in three or more variables.*
- *Use systems of three equations to solve applied problems.*
- *Model a situation using a quadratic function.*

A **linear equation in three variables** is an equation equivalent to one of the form $Ax + By + Cz = D$, where A, B, C, and D are real numbers and A, B, and C are not all 0. A **solution of a system of three equations in three variables** is an ordered triple that makes all three equations true. For example, the triple $(2, -1, 0)$ is a solution of the system of equations

$$4x + 2y + 5z = 6,$$
$$2x - y + z = 5,$$
$$3x + 2y - z = 4.$$

We can verify this by substituting 2 for x, -1 for y, and 0 for z in each equation.

Solving Systems of Equations in Three or More Variables

We will solve systems of equations in three or more variables using an algebraic method called **Gaussian elimination**, named for the German mathematician Karl Friedrich Gauss (1777–1855). Our goal is to transform the original system to an equivalent one of the form

$$Ax + By + Cz = D,$$
$$Ey + Fz = G,$$
$$Hz = K.$$

Then we solve the third equation for z and back-substitute to find y and then x.

Each of the following operations can be used to transform the original system to an equivalent system in the desired form.

> **1.** Interchange any two equations.
>
> **2.** Multiply both sides of one of the equations by a nonzero constant.
>
> **3.** Add a nonzero multiple of one equation to another equation.

Example 1 Solve the following system:

$$
\begin{aligned}
x - 2y + 3z &= 11, &&(1) \\
4x + 2y - 3z &= 4, &&(2) \\
3x + 3y - z &= 4. &&(3)
\end{aligned}
$$

SOLUTION   We multiply equation (1) by -4 and add it to equation (2). We also multiply equation (1) by -3 and add it to equation (3). These operations produce an equivalent system of equations in which the x-terms have been eliminated from both the second and the third equations.

$$
\begin{array}{l}
-4x + 8y - 12z = -44 \longleftarrow \qquad x - 2y + 3z = 11, \longrightarrow -3x + 6y - 9z = -33 \\
\underline{4x + 2y - 3z = 4}\ \text{Eq. (2)} \qquad \longrightarrow 10y - 15z = -40, \qquad \underline{3x + 3y - z = 4}\ \text{Eq. (3)} \\
10y - 15z = -40 \qquad\qquad\quad 9y - 10z = -29 \longleftarrow \qquad 9y - 10z = -29
\end{array}
$$

Now we multiply the last equation by 10 to make the y-coefficient a multiple of the y-coefficient in the second equation:

$$
\begin{aligned}
x - 2y + 3z &= 11, &&(1) \\
10y - 15z &= -40, &&(4) \\
90y - 100z &= -290. &&(5)
\end{aligned}
$$

Next we multiply equation (4) by -9 and add it to equation (5).

$$
\begin{array}{l}
-90y + 135z = 360 \longleftarrow \qquad x - 2y + 3z = 11, \qquad\qquad (1) \\
\underline{90y - 100z = -290}\ \text{Eq. (5)} \qquad\quad 10y - 15z = -40, \qquad\qquad (4) \\
35z = 70 \longrightarrow 35z = 70. \qquad\qquad\qquad\qquad (6)
\end{array}
$$

Now we solve equation (6) for z:

$$
\begin{aligned}
35z &= 70 \\
z &= 2.
\end{aligned}
$$

Then we back-substitute 2 for z in equation (4) and solve for y:

$$
\begin{aligned}
10y - 15 \cdot 2 &= -40 \\
10y - 30 &= -40 \\
10y &= -10 \\
y &= -1.
\end{aligned}
$$

Finally, we back-substitute -1 for y and 2 for z in equation (1) and solve for x:

$$
\begin{aligned}
x - 2(-1) + 3 \cdot 2 &= 11 \\
x + 2 + 6 &= 11 \\
x &= 3.
\end{aligned}
$$

We can check the triple $(3, -1, 2)$ in each of the three original equations. Since it makes all three equations true, the solution is $(3, -1, 2)$. ▬

Example 2 Solve the following system:

$$\begin{aligned}
x + y + z &= 7, &\quad (1)\\
3x - 2y + z &= 3, &\quad (2)\\
x + 6y + 3z &= 25. &\quad (3)
\end{aligned}$$

SOLUTION We multiply equation (1) by -3 and add it to equation (2). We also multiply equation (1) by -1 and add it to equation (3).

$$\begin{aligned}
x + y + z &= 7, &\quad (1)\\
-5y - 2z &= -18, &\quad (4)\\
5y + 2z &= 18 &\quad (5)
\end{aligned}$$

Now we add equation (4) to equation (5):

$$\begin{aligned}
x + y + z &= 7,\\
-5y - 2z &= -18,\\
0 &= 0.
\end{aligned}$$

The equation $0 = 0$ tells us that equation (3) of the original system is dependent on the first two equations. Thus the original system is equivalent to

$$\begin{aligned}
x + y + z &= 7,\\
3x - 2y + z &= 3
\end{aligned}$$

and has infinitely many solutions. To find an expression for these solutions, we first solve equation (4) for either y or z. In this case, we choose to solve for y:

$$\begin{aligned}
-5y - 2z &= -18\\
-5y &= 2z - 18\\
y &= -\tfrac{2}{5}z + \tfrac{18}{5}.
\end{aligned}$$

Then we back-substitute in equation (1) to find an expression for x in terms of z:

$$\begin{aligned}
x - \tfrac{2}{5}z + \tfrac{18}{5} + z &= 7\\
x + \tfrac{3}{5}z + \tfrac{18}{5} &= 7\\
x + \tfrac{3}{5}z &= \tfrac{17}{5}\\
x &= -\tfrac{3}{5}z + \tfrac{17}{5}.
\end{aligned}$$

The solutions of the system of equations are ordered triples of the form $\left(-\tfrac{3}{5}z + \tfrac{17}{5}, -\tfrac{2}{5}z + \tfrac{18}{5}, z\right)$, where z can be any real number. Any real number that we use for z then gives us values for x and y and thus an ordered triple in the solution set. ▬

If we get a false equation, such as $0 = -5$, at some stage of the elimination process, we conclude that the original system is inconsistent. That is, it has no solutions.

Just as with systems of equations in two variables, systems of three linear equations in three variables can also be solved using Cramer's rule. See Appendix A.

Although systems of three linear equations in three variables do not lend themselves well to graphical solutions, it is of interest to picture some possible solutions. The graph of a linear equation in three variables is a plane. Thus the solution set of such a system is the intersection of three planes. Some possibilities are shown below.

Program

SYSLINEQ: This program solves 2×2, 3×3, and 4×4 systems of linear equations.

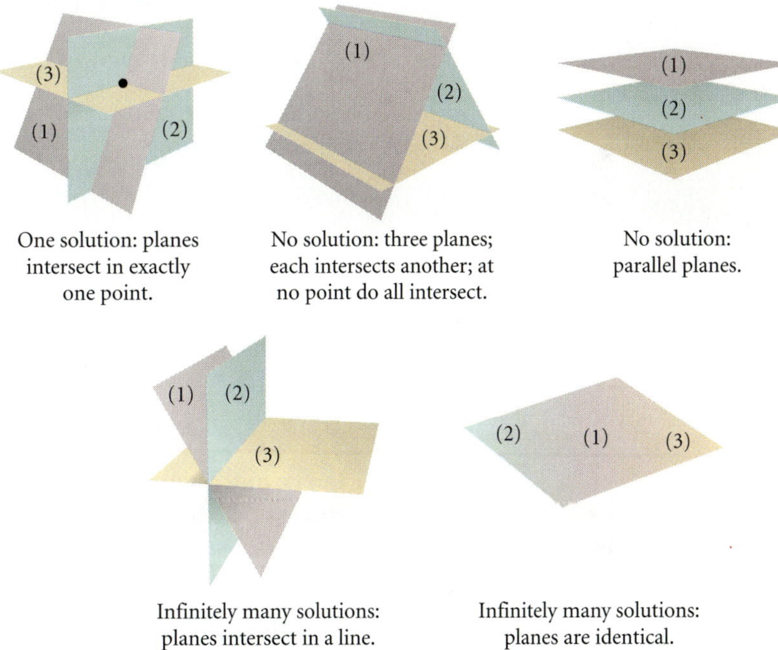

One solution: planes intersect in exactly one point.

No solution: three planes; each intersects another; at no point do all intersect.

No solution: parallel planes.

Infinitely many solutions: planes intersect in a line.

Infinitely many solutions: planes are identical.

If you have access to a grapher that graphs equations in three variables, use it to graph the systems in Examples 1 and 2.

Applications

Systems of equations in three or more variables allow us to solve many problems in fields such as business, the social and natural sciences, and engineering.

Example 3   *Investment.* Moira inherited $15,000 and invested part of it in a money market account, part in municipal bonds, and part in a mutual fund. After 1 yr, she received a total of $730 in simple interest from the three investments. The money market account paid 4% annually, the bonds paid 5% annually, and the mutual fund paid 6% annually. There was $2000 more invested in the mutual fund than in bonds. Find the amount that Moira invested in each category.

SOLUTION

1. **Familiarize.** We let x, y, and z represent the amounts invested in the money market account, the bonds, and the mutual fund, respectively. Then the amounts of income produced annually by each investment are given by 4%x, 5%y, and 6%z, or 0.04x, 0.05y, and 0.06z.

2. **Translate.** The fact that a total of $15,000 is invested gives us one equation:

$$x + y + z = 15{,}000.$$

Since the total interest is $730, we have a second equation:

$$0.04x + 0.05y + 0.06z = 730.$$

Another statement in the problem gives us a third equation.

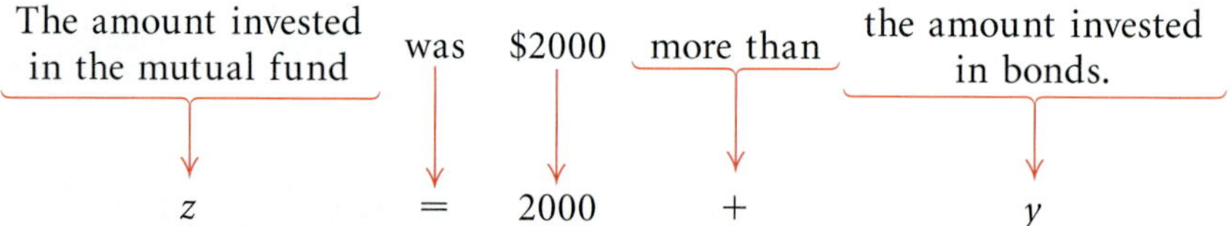

We now have a system of three equations:

$$x + y + z = 15{,}000, \qquad\qquad x + y + z = 15{,}000,$$
$$0.04x + 0.05y + 0.06z = 730, \quad \text{or} \quad 4x + 5y + 6z = 73{,}000,$$
$$z = 2000 + y; \qquad\qquad\qquad\quad -y + z = 2000.$$

3. **Carry out.** Solving the system of equations, we get (7000, 3000, 5000).

4. **Check.** The sum of the numbers is 15,000. The income produced is

$$0.04(7000) + 0.05(3000) + 0.06(5000) = 280 + 150 + 300, \quad \text{or} \quad 730.$$

Also the amount invested in the mutual fund, $5000, is $2000 more than the amount invested in bonds, $3000. Our solution checks in the original problem.

5. **State.** Moira invested $7000 in a money market account, $3000 in municipal bonds, and $5000 in a mutual fund.

Mathematical Models and Applications

In a situation in which a quadratic function will serve as a mathematical model, we may wish to find an equation, or formula, for the function. For a linear model, we can find an equation if we know two data points. For a quadratic function, we need three data points.

Example 4   *The Cost of Operating an Automobile at Various Speeds.* Under certain conditions, it is found that the cost of operating an automobile as a function of speed is approximated by a quadratic function. Use the data shown below to find an equation of the function. Then use the equation to determine the cost of operating the automobile at 60 mph and at 80 mph.

SPEED (IN MILES PER HOUR)	OPERATING COST PER MILE (IN CENTS)
10	22
20	20
50	20

SOLUTION   Letting x = the speed and $f(x)$ = the cost, we use the three data points $(10, 22)$, $(20, 20)$, and $(50, 20)$ to find a, b, and c in the equation $f(x) = ax^2 + bx + c$. First we substitute:

For $(10, 22)$: $22 = a \cdot 10^2 + b \cdot 10 + c$;

For $(20, 20)$: $20 = a \cdot 20^2 + b \cdot 20 + c$;

For $(50, 20)$: $20 = a \cdot 50^2 + b \cdot 50 + c$.

We now have a system of equations in the variables a, b, and c:

$$100a + 10b + c = 22,$$
$$400a + 20b + c = 20,$$
$$2500a + 50b + c = 20.$$

Solving this system of equations, we obtain $(0.005, -0.35, 25)$. Thus,

$$f(x) = 0.005x^2 - 0.35x + 25.$$

To determine the cost of operating an automobile at 60 mph, we find $f(60)$:

$$f(60) = 0.005(60)^2 - 0.35(60) + 25 = 22¢.$$

To find the cost of operating an automobile at 80 mph, we find $f(80)$:

$$f(80) = 0.005(80)^2 - 0.35(80) + 25 = 29¢.$$

The graph of the cost function is shown at left.

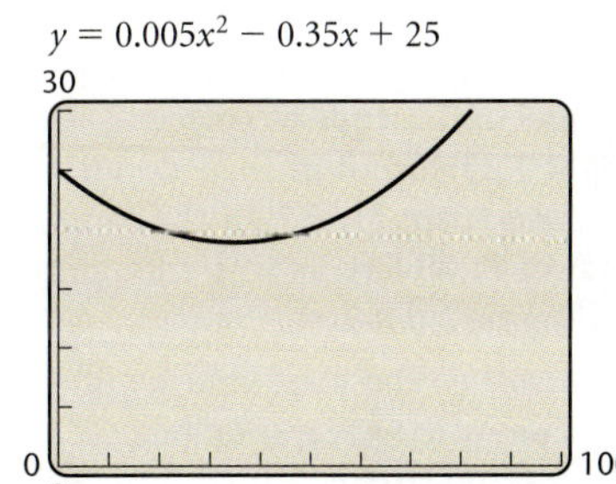

The function in Example 4 can also be found using the QUADRATIC REGRESSION feature in the STAT package on a grapher as we did in Section 2.3. Note that the method of Example 4 works when we have exactly three data points, whereas the QUADRATIC REGRESSION feature on a grapher can be used for three or more points.

4.2 Exercise Set

Solve each of the following systems.

1. $x + y + z = 2,$
$6x - 4y + 5z = 31,$
$5x + 2y + 2z = 13$

2. $x + 6y + 3z = 4,$
$2x + y + 2z = 3,$
$3x - 2y + z = 0$

3. $x - y + 2z = -3,$
$x + 2y + 3z = 4,$
$2x + y + z = -3$

4. $x + y + z = 6,$
$2x - y - z = -3,$
$x - 2y + 3z = 6$

5. $x + 2y - z = 5,$
$2x - 4y + z = 0,$
$3x + 2y + 2z = 3$

6. $2x + 3y - z = 1,$
$x + 2y + 5z = 4,$
$3x - y - 8z = -7$

7. $x + 2y - z = -8,$
$2x - y + z = 4,$
$8x + y + z = 2$

8. $x + 2y - z = 4,$
$4x - 3y + z = 8,$
$5x - y = 12$

9. $2x + y - 3z = 1,$
$x - 4y + z = 6,$
$4x - 7y - z = 13$

10. $x + 3y + 4z = 1,$
$3x + 4y + 5z = 3,$
$x + 8y + 11z = 2$

11. $4a + 9b = 8,$
$8a + 6c = -1,$
$6b + 6c = -1$

12. $3p + 2r = 11,$
$q - 7r = 4,$
$p - 6q = 1$

13. $w + x + y + z = 2,$
$w + 2x + 2y + 4z = 1,$
$-w + x - y - z = -6,$
$-w + 3x + y - z = -2$

14. $w + x - y + z = 0,$
$-w + 2x + 2y + z = 5,$
$-w + 3x + y - z = -4,$
$-2w + x + y - 3z = -7$

15. *Mail-Order Business.* Computer Warehouse charges $3 for shipping orders up to 10 lb, $5 for orders from 10 lb up to 15 lb, and $7.50 for orders of 15 lb or more. One day shipping charges for 150 orders totaled $680. The number of orders under 10 lb was three times the number of orders weighing 15 lb or more. Find the number of packages shipped at each rate.

16. *Mail-Order Business.* Natural Fibers Clothing charges $4 for shipping orders of $30 or less, $6 for orders from $30 to $70, and $7 for orders over $70. One week shipping charges for 600 orders totaled $3340. Eighty more orders for $30 or less were shipped than orders for more than $70. Find the number of orders shipped at each rate.

17. *Nutrition.* A hospital dietician must plan a lunch menu that provides 485 Cal, 41.5 g of carbohydrates, and 35 mg of calcium. A 3-oz serving of broiled ground beef contains 245 Cal, 0 g of carbohydrates, and 9 mg of calcium. One baked potato contains 145 Cal, 34 g of carbohydrates, and 8 mg of calcium. A one-cup serving of strawberries contains 45 Cal, 10 g of carbohydrates, and 21 mg of calcium. How many servings of each are required to provide the desired nutritional values? (*Source: Home and Garden Bulletin No. 72*, U.S. Government Printing Office, Washington, D.C. 20402)

18. *Nutrition.* A diabetic patient wishes to prepare a meal consisting of roasted chicken breast, mashed potatoes, and peas. A 3-oz serving of roasted, skinless chicken breast contains 140 Cal, 27 g of protein, and 64 mg of sodium. A one-cup serving of mashed potatoes contains 160 Cal, 4 g of protein, and 636 mg of sodium, and a one-cup serving of peas contains 125 Cal, 8 g of protein, and 139 mg of sodium. How many servings of each should be used if the meal is to contain 415 Cal, 50.5 g of protein, and 553 mg of sodium? (*Source: Home and Garden Bulletin No. 72*, U.S. Government Printing Office, Washington, D.C. 20402)

19. *Investment.* Jamal earns a year-end bonus of $5000 and puts it in 3 one-year investments that pay $302 in simple interest. Part is invested at 4%, part at 6%, and part at 7%. There is $1500 more invested at 7% than at 4%. Find the amount invested at each rate.

20. *Investment.* Casey receives $336 per year in simple interest from three investments. Part is invested at 8%, part at 9%, and part at 10%. There is $500 more invested at 9% than at 8%. The amount invested at 10% is three times the amount invested at 9%. Find the amount invested at each rate.

21. *Price Increases.* Orange juice, a raisin bagel, and a cup of coffee from Kelly's Koffee Kart cost a total of $3. Kelly posts a notice announcing that, effective next week, the price of orange juice will increase 50% and the price of bagels will increase 20%. After the increase, the same purchase will cost a total of $3.75, and orange juice will cost twice as much as coffee. Find the price of each item before the increase.

22. *Cost of Snack Food.* Martin and Eva pool their loose change to buy snacks on their coffee break. One day, they spent $1.85 on 1 carton of milk, 2 donuts, and 1 cup of coffee. The next day, they spent $2.30 on 3 donuts and 2 cups of coffee. The third day, they bought 1 carton of milk, 1 donut, and 2 cups of coffee and spent $1.75. On the fourth day, they have a total of $1.80 left. Is this enough to buy 2 cartons of milk and 2 donuts?

23. *Passenger Transportation.* The total volume of passenger traffic on domestic airways, by bus (excluding school buses and urban transit buses), and by railroads in the United States in a recent year was 405 billion passenger-miles. (One passenger-mile is the transportation of one passenger the distance of one mile.) The volume of bus traffic was 10 billion passenger-miles more than the volume of railroad traffic. The total volume of railroad traffic was 329 billion passenger-miles less than the volume of traffic on domestic airways. What was the volume of each type of passenger traffic? (*Source: Transportation in America*, Eno Transportation Foundation, Inc., Landsdowne, VA)

24. *Cheese Consumption.* The total per capita consumption (the average amount consumed per person) of cheddar, mozzarella, and Swiss cheese in the United States in a recent year was 18.1 lb. The total consumption of mozzarella and Swiss cheese was 0.3 lb less than that of cheddar cheese. The consumption of mozzarella cheese was 6.5 lb more than that of Swiss cheese. What was the per capita consumption of each type of cheese? (*Source:* U.S. Department of Agriculture, Economic Research Service, *Food Consumption, Prices, and Expenditures*)

25. *Golf.* On an 18-hole golf course, there are par-3 holes, par-4 holes, and par-5 holes. A golfer who shoots par on every hole has a score of 72. The sum of the number of par-3 holes and the number of par-5 holes is 8. How many of each type of hole are there on the golf course?

26. *Golf.* On an 18-hole golf course, there are par-3 holes, par-4 holes, and par-5 holes. A golfer who shoots par on every hole has a score of 70. There are twice as many par-4 holes as there are par-5 holes. How many of each type of hole are there on the golf course?

27. *Cost of a New Car.* The following table shows the average new car transaction price in the United States as a percentage of disposable income for three recent years, represented as years since 1960.

YEAR		PERCENT OF DISPOSABLE INCOME
(1960)	0	43
(1973)	13	31
(1992)	32	43

Source: Kemper Reports, Fall/Winter, 1995, p. 11.

a) Find a quadratic function $f(x) = ax^2 + bx + c$ that fits the data.
b) Use the function to predict the price of a new car as a percentage of disposable income in the year 2000.

28. *Number of Marriages.* The following table shows the number of marriages, in thousands, in the mountain states in three recent years, presented as years since 1980.

YEAR		NUMBER OF MARRIAGES (IN THOUSANDS)
(1980)	0	242
(1985)	5	234
(1993)	13	255

Source: U.S. National Center for Health Statistics, *Vital Statistics of the United States,* annual.

a) Find a quadratic function $f(x) = ax^2 + bx + c$ that fits the data.
b) Use the function to predict the number of marriages in the mountain states in 2000.

29. *Milk Consumption.* The following table shows the per capita milk consumption, in pounds, in the United States in three recent years, represented as years since 1975.

YEAR		MILK CONSUMPTION (IN POUNDS)
(1975)	0	539
(1985)	10	594
(1992)	17	565

Source: U.S. Department of Agriculture, Economic Research Service, *Food Consumption, Prices, and Expenditures.*

a) Find a quadratic function $f(x) = ax^2 + bx + c$ that fits the data.
b) Use the function to predict the per capita consumption of milk in 2005.

30. *Crude Steel Production.* The following table shows the crude steel production, in millions of metric tons, in the United States in three recent years, represented as years since 1980.

YEAR		CRUDE STEEL PRODUCTION (IN MILLIONS OF METRIC TONS)
(1980)	0	80
(1988)	8	92
(1993)	13	89

Source: Statistical Office of the United Nations, New York, NY, *Statistical Yearbook* and *Industrial Commodity Statistics Yearbook,* annuals.

a) Find a quadratic function $f(x) = ax^2 + bx + c$ that fits the data.
b) Use the function to predict the crude steel production in the United States in 2000.

Skill Maintenance

For $A = 2x - 3y$ and $B = -x + 5y$, *find each of the following.*

31. $2A$

32. $-A$

33. $A + B$

34. $-3A + B$

Synthesis

35. ◆ Given two linear equations in three variables, $Ax + By + Cz = D$ and $Ex + Fy + Gz = H$, explain how you would find a third equation such that the system composed of the three equations is dependent.

36. ◆ Write a problem for a classmate to solve that can be translated to a system of three equations in three variables.

In Exercises 37 and 38, let u represent 1/x, v represent 1/y, and w represent 1/z. Solve first for u, v, and w. Then solve the system.

37.
$$\frac{2}{x} - \frac{1}{y} - \frac{3}{z} = -1,$$
$$\frac{2}{x} - \frac{1}{y} + \frac{1}{z} = -9,$$
$$\frac{1}{x} + \frac{2}{y} - \frac{4}{z} = 17$$

38.
$$\frac{2}{x} + \frac{2}{y} - \frac{3}{z} = 3,$$
$$\frac{1}{x} - \frac{2}{y} - \frac{3}{z} = 9,$$
$$\frac{7}{x} - \frac{2}{y} + \frac{9}{z} = -39$$

39. Find the sum of the angle measures at the tips of the star.

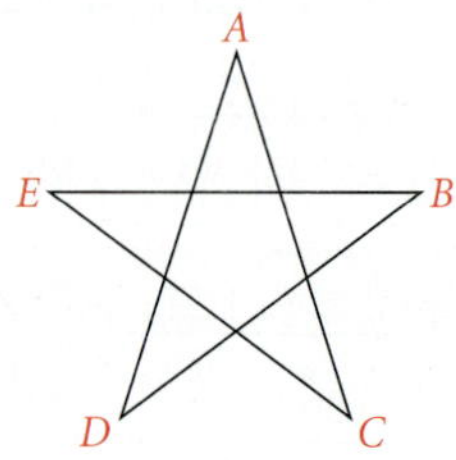

40. *Transcontinental Railroad.* Use the following facts to find the year in which the first U.S. transcontinental railroad was completed. The sum of the digits in the year is 24. The units digit is 1 more than the hundreds digit. Both the tens and the units digits are multiples of three.

In Exercises 41 and 42, three solutions of an equation are given. Use a system of three equations in three variables to find the constants and write the equation.

41. $Ax + By + Cz = 12;$
$\left(1, \frac{3}{4}, 3\right), \left(\frac{4}{3}, 1, 2\right),$ and $(2, 1, 1)$

42. $y = B - Mx - Nz;$
$(1, 1, 2), (3, 2, -6),$ and $\left(\frac{3}{2}, 1, 1\right)$

In Exercises 43 and 44, four solutions of the equation $y = ax^3 + bx^2 + cx + d$ are given. Use a system of four equations in four variables to find the constants and write the equation.

43. $(-2, 59), (-1, 13), (1, -1),$ and $(2, -17)$

44. $(-2, -39), (-1, -12), (1, -6),$ and $(3, 16)$

45. *Theater Attendance.* A performance at the Bingham Performing Arts Center was attended by 100 people. The audience consisted of adults, students, and children. The ticket prices were $10 for adults, $3 for students, and 50 cents for children. The total amount of money taken in was $100. How many adults, students, and children were in attendance? Does there seem to be some information missing? Do some careful reasoning.

4.3

Matrices and Systems of Equations

• *Solve systems of equations using matrices.*

Matrices and Row-Equivalent Operations

In this section, we consider additional techniques for solving systems of equations. You have probably observed that when we solve a system of equations, we perform computations with the coefficients and the constants and continually rewrite the variables. We can streamline the solution process by omitting the variables until a solution is found. For example, the system

$$2x - 3y = 7,$$
$$x + 4y = -2$$

can be written more simply as

$$\begin{bmatrix} 2 & -3 & \bigm| & 7 \\ 1 & 4 & \bigm| & -2 \end{bmatrix}.$$

The vertical line replaces the equals signs.

The rectangular array of numbers above is called a **matrix** (pl., **matrices**). In fact, it is called an **augmented matrix** for the given system of equations, because it contains not only the coefficients but also the constant terms. The matrix

$$\begin{bmatrix} 2 & -3 \\ 1 & 4 \end{bmatrix}$$

is called the **coefficient matrix** of the system.

The **rows** of a matrix are horizontal, and the **columns** are vertical. The augmented matrix above has 2 rows and 3 columns, and the coefficient matrix has 2 rows and 2 columns. A matrix with m rows and n columns is said to be of **order** $m \times n$. Thus the order of the augmented matrix above is 2×3, and the order of the coefficient matrix is 2×2. When $m = n$, a matrix is said to be **square**. The coefficient matrix above is a square matrix. The numbers 2 and 4 lie on the **main diagonal** of the coefficient matrix. The numbers in a matrix are called **entries**.

Gaussian Elimination with Matrices

In Section 4.2, we described a series of operations that can be used to transform a system of equations to an equivalent system. Each of these operations corresponds to one that can be used to produce *row-equivalent matrices*.

Row-Equivalent Operations

1. Interchange any two rows.
2. Multiply each entry in a row by the same nonzero constant.
3. Add a nonzero multiple of one row to another row.

We can use these operations on the augmented matrix of a system of equations to solve the system.

Example 1 Solve the following system:

$$
\begin{aligned}
2x - y + 4z &= -3, \\
x - 2y - 10z &= -6, \\
3x \phantom{{}- 2y} + 4z &= 7.
\end{aligned}
$$

SOLUTION   First we write the augmented matrix, writing 0 for the missing y-term in the last equation:

$$
\left[\begin{array}{ccc|c}
2 & -1 & 4 & -3 \\
1 & -2 & -10 & -6 \\
3 & 0 & 4 & 7
\end{array}\right].
$$

Our goal is to find a row-equivalent matrix of the form

$$
\left[\begin{array}{ccc|c}
1 & a & b & c \\
0 & 1 & d & e \\
0 & 0 & 1 & f
\end{array}\right].
$$

The variables can then be reinserted to form equations from which we can complete the solution by working from the bottom equation to the top and using back-substitution.

The first step is to multiply and/or interchange rows so that each number in the first column below the first number is a multiple of that number. In this case, we interchange the first and second rows to obtain a 1 in the upper left-hand corner.

$$
\left[\begin{array}{ccc|c}
1 & -2 & -10 & -6 \\
2 & -1 & 4 & -3 \\
3 & 0 & 4 & 7
\end{array}\right]
\qquad
\begin{array}{l}
\text{New row 1 = row 2} \\[4pt]
\text{New row 2 = row 1}
\end{array}
$$

Next we multiply the first row by -2 and add it to the second row. We also multiply the first row by -3 and add it to the third row.

$$
\left[\begin{array}{ccc|c}
1 & -2 & -10 & -6 \\
0 & 3 & 24 & 9 \\
0 & 6 & 34 & 25
\end{array}\right]
\qquad
\begin{array}{l}
\text{New row 2 = } -2(\text{row 1}) + \text{row 2} \\[4pt]
\text{New row 3 = } -3(\text{row 1}) + \text{row 3}
\end{array}
$$

Now we multiply the second row by $\tfrac{1}{3}$ to get a 1 in the second row, second column.

$$
\left[\begin{array}{ccc|c}
1 & -2 & -10 & -6 \\
0 & 1 & 8 & 3 \\
0 & 6 & 34 & 25
\end{array}\right]
\qquad
\text{New row 2 = } \tfrac{1}{3}(\text{row 2})
$$

Then we multiply the second row by -6 and add it to the third row.

$$
\left[\begin{array}{ccc|c}
1 & -2 & -10 & -6 \\
0 & 1 & 8 & 3 \\
0 & 0 & -14 & 7
\end{array}\right]
\qquad
\text{New row 3 = } -6(\text{row 2}) + \text{row 3}
$$

Finally we multiply the third row by $-\frac{1}{14}$ to get a 1 in the third row, third column.

$$\left[\begin{array}{ccc|c} 1 & -2 & -10 & -6 \\ 0 & 1 & 8 & 3 \\ 0 & 0 & 1 & -\frac{1}{2} \end{array}\right] \qquad \text{New row 3} = -\tfrac{1}{14}(\text{row 3})$$

Now we can write the system of equations that corresponds to the last matrix above:

$$\begin{aligned} x - 2y - 10z &= -6, & (1) \\ y + 8z &= 3, & (2) \\ z &= -\tfrac{1}{2}. & (3) \end{aligned}$$

We back-substitute $-\frac{1}{2}$ for z in equation (2) and solve for y:

$$\begin{aligned} y + 8\left(-\tfrac{1}{2}\right) &= 3 \\ y - 4 &= 3 \\ y &= 7. \end{aligned}$$

Next we back-substitute 7 for y and $-\frac{1}{2}$ for z in equation (1) and solve for x:

$$\begin{aligned} x - 2 \cdot 7 - 10\left(-\tfrac{1}{2}\right) &= -6 \\ x - 14 + 5 &= -6 \\ x - 9 &= -6 \\ x &= 3. \end{aligned}$$

The triple $\left(3, 7, -\tfrac{1}{2}\right)$ checks in the original system of equations, so it is the solution.

The procedure followed in Example 1 is called **Gaussian elimination** for matrices. The last matrix above is in **row-echelon form**. To be in this form, a matrix must have the following properties.

Row-Echelon Form

1. If a row does not consist entirely of 0's, then the first nonzero element in the row is a 1 (called a **leading 1**).

2. For any two successive nonzero rows, the leading 1 in the lower row is farther to the right than the leading 1 in the higher row.

3. All the rows consisting entirely of 0's are at the bottom of the matrix.

If a fourth property is also satisfied, a matrix is said to be in **reduced row-echelon form**:

4. Each column that contains a leading 1 has 0's everywhere else.

Example 2   Which of the following matrices are in row-echelon form? Which, if any, are in reduced row-echelon form?

a) $\begin{bmatrix} 1 & -3 & 5 & -2 \\ 0 & 1 & -4 & 3 \\ 0 & 0 & 1 & 10 \end{bmatrix}$ **b)** $\begin{bmatrix} 0 & -1 & 2 \\ 0 & 1 & 5 \end{bmatrix}$ **c)** $\begin{bmatrix} 1 & -2 & -6 & 4 & 7 \\ 0 & 3 & 5 & -8 & -1 \\ 0 & 0 & 1 & 9 & 2 \end{bmatrix}$

d) $\begin{bmatrix} 1 & 0 & 0 & -2.4 \\ 0 & 1 & 0 & 0.8 \\ 0 & 0 & 1 & 5.6 \end{bmatrix}$ **e)** $\begin{bmatrix} 1 & 0 & 0 & 0 & \frac{2}{3} \\ 0 & 1 & 0 & 0 & -\frac{1}{4} \\ 0 & 0 & 1 & 0 & \frac{6}{7} \\ 0 & 0 & 0 & 0 & 0 \end{bmatrix}$ **f)** $\begin{bmatrix} 1 & -4 & 2 & 5 \\ 0 & 0 & 0 & 0 \\ 0 & 1 & -3 & -8 \end{bmatrix}$

SOLUTION   The matrices in (a), (d), and (e) satisfy the row-echelon criteria and, thus, are in row-echelon form. In (b) and (c), the first nonzero elements of the first and second rows, respectively, are not 1. In (f), the row consisting entirely of 0's is not at the bottom of the matrix. Thus the matrices in (b), (c), and (f) are not in row-echelon form. In (d) and (e), not only are the row-echelon criteria met but each column that contains a leading 1 also has 0's elsewhere, so these matrices are in reduced row-echelon form.

Gauss–Jordan Elimination

We have seen that with Gaussian elimination we perform row-equivalent operations on a matrix to obtain a row-equivalent matrix in row-echelon form. When we continue to apply these operations until we have a matrix in *reduced* row-echelon form, we are using **Gauss–Jordan elimination**. This method is named for Karl Friedrich Gauss and Wilhelm Jordan (1842–1899).

Example 3   Use Gauss–Jordan elimination to solve the system of equations in Example 1.

SOLUTION   Using Gaussian elimination in Example 1, we obtained the matrix

$$\left[\begin{array}{ccc|c} 1 & -2 & -10 & -6 \\ 0 & 1 & 8 & 3 \\ 0 & 0 & 1 & -\frac{1}{2} \end{array}\right].$$

We continue to perform row-equivalent operations until we have a matrix in reduced row-echelon form. First we multiply the third row by -8 and add it to the second row. We also multiply the third row by 10 and add it to the first row.

$$\left[\begin{array}{ccc|c} 1 & -2 & 0 & -11 \\ 0 & 1 & 0 & 7 \\ 0 & 0 & 1 & -\frac{1}{2} \end{array}\right] \quad \begin{array}{l} \text{New row 1} = 10(\text{row 3}) + \text{row 1} \\ \text{New row 2} = -8(\text{row 3}) + \text{row 2} \end{array}$$

Then we multiply the second row by 2 and add it to the first row.

$$\left[\begin{array}{ccc|c} 1 & 0 & 0 & 3 \\ 0 & 1 & 0 & 7 \\ 0 & 0 & 1 & -\frac{1}{2} \end{array}\right] \qquad \text{New row 1} = 2(\text{row 2}) + \text{row 1}$$

Writing the system of equations that corresponds to this matrix, we have

$$\begin{aligned} x &= 3, \\ y &= 7, \\ z &= -\tfrac{1}{2}. \end{aligned}$$

We can actually read the solution, $\left(3, 7, -\frac{1}{2}\right)$, directly from the reduced row-echelon matrix.

Example 4 Solve the following system:

$$\begin{aligned} 3x - 4y - z &= 6, \\ 2x - y + z &= -1, \\ 4x - 7y - 3z &= 13. \end{aligned}$$

SOLUTION We write the augmented matrix and use Gauss–Jordan elimination.

$$\left[\begin{array}{ccc|c} 3 & -4 & -1 & 6 \\ 2 & -1 & 1 & -1 \\ 4 & -7 & -3 & 13 \end{array}\right]$$

We begin by multiplying the second and third rows by 3 so that each number in the first column below the first number, 3, is a multiple of that number.

$$\left[\begin{array}{ccc|c} 3 & -4 & -1 & 6 \\ 6 & -3 & 3 & -3 \\ 12 & -21 & -9 & 39 \end{array}\right] \qquad \begin{array}{l} \text{New row 2} = 3(\text{row 2}) \\ \text{New row 3} = 3(\text{row 3}) \end{array}$$

Next we multiply the first row by -2 and add it to the second row. We also multiply the first row by -4 and add it to the third row.

$$\left[\begin{array}{ccc|c} 3 & -4 & -1 & 6 \\ 0 & 5 & 5 & -15 \\ 0 & -5 & -5 & 15 \end{array}\right] \qquad \begin{array}{l} \text{New row 2} = -2(\text{row 1}) + \text{row 2} \\ \text{New row 3} = -4(\text{row 1}) + \text{row 3} \end{array}$$

Now we add the second row to the third row.

$$\left[\begin{array}{ccc|c} 3 & -4 & -1 & 6 \\ 0 & 5 & 5 & -15 \\ 0 & 0 & 0 & 0 \end{array}\right] \qquad \text{New row 3} = \text{row 2} + \text{row 3}$$

We can stop at this stage because we have a row consisting entirely of 0's. The last row of the matrix corresponds to the equation $0 = 0$, which is true for all values of x, y, and z. Consequently, we have a dependent system of equations that is equivalent to

$$3x - 4y - z = 6,$$
$$5y + 5z = -15.$$

Solving the second equation for y gives us

$$y = -z - 3.$$

Substituting for y in the first equation, we get

$$3x - 4(-z - 3) - z = 6$$
$$x = -z - 2.$$

Then the solutions of this system are of the form

$$(-z - 2, -z - 3, z),$$

where z can be any real number.

Similarly, if we obtain a row whose only nonzero entry occurs in the last column, we have an inconsistent system of equations. For example, in the matrix

$$\begin{bmatrix} 1 & 0 & 3 & -2 \\ 0 & 1 & 5 & 4 \\ 0 & 0 & 0 & 6 \end{bmatrix},$$

the last row corresponds to the false equation $0 = 6$, so we know the original system of equations has no solution.

Row-equivalent operations on matrices can be performed using the MATRIX feature on most graphers. Refer to your user's manual for instructions. Then use a grapher to perform the row-equivalent operations in Examples 1, 3, and 4.

4.3 Exercise Set

Determine the order of each matrix.

1. $\begin{bmatrix} 1 & -6 \\ -3 & 2 \\ 0 & 5 \end{bmatrix}$

2. $\begin{bmatrix} 7 \\ -5 \\ -1 \\ 3 \end{bmatrix}$

3. $[2 \ -4 \ 0 \ 9]$

4. $[-8]$

5. $\begin{bmatrix} 1 & -5 & -8 \\ 6 & 4 & -2 \\ -3 & 0 & 7 \end{bmatrix}$

6. $\begin{bmatrix} 13 & 2 & -6 & 4 \\ -1 & 18 & 5 & -12 \end{bmatrix}$

Write the augmented matrix for each system of equations.

7. $2x - y = 7,$
$\quad\ x + 4y = -5$

8. $3x + 2y = 8,$
$\quad\ 2x - 3y = 15$

9. $\ \ x - 2y + 3z = 12,$
$\quad 2x \quad\ - 4z = 8,$
$\quad\quad\ 3y + z = 7$

10. $\quad x + y - z = 7,$
$\quad\quad\quad 3y + 2z = 1,$
$\quad -2x - 5y \quad\ = 6$

Write the system of equations that corresponds to each augmented matrix.

11. $\begin{bmatrix} 3 & -5 & 1 \\ 1 & 4 & -2 \end{bmatrix}$

12. $\begin{bmatrix} 1 & 2 & -6 \\ 4 & 1 & -3 \end{bmatrix}$

13. $\begin{bmatrix} 2 & 1 & -4 & | & 12 \\ 3 & 0 & 5 & | & -1 \\ 1 & -1 & 1 & | & 2 \end{bmatrix}$

14. $\begin{bmatrix} -1 & -2 & 3 & | & 6 \\ 0 & 4 & 1 & | & 2 \\ 2 & -1 & 0 & | & 9 \end{bmatrix}$

Solve each system of equations using Gaussian elimination or Gauss–Jordan elimination.

15. $4x + 2y = 11,$
$3x - y = 2$

16. $2x + y = 1,$
$3x + 2y = -2$

17. $5x - 2y = -3,$
$2x + 5y = -24$

18. $2x + y = 1,$
$3x - 6y = 4$

19. $3x + 4y = 7,$
$-5x + 2y = 10$

20. $5x - 3y = -2,$
$4x + 2y = 5$

21. $3x + 2y = 6,$
$2x - 3y = -9$

22. $x - 4y = 9,$
$2x + 5y = 5$

23. $x - 3y = 8,$
$-2x + 6y = 3$

24. $4x - 8y = 12,$
$-x + 2y = -3$

25. $-2x + 6y = 4,$
$3x - 9y = -6$

26. $6x + 2y = -10,$
$-3x - y = 6$

27. $x + 2y - 3z = 9,$
$2x - y + 2z = -8,$
$3x - y - 4z = 3$

28. $x - y + 2z = 0,$
$x - 2y + 3z = -1,$
$2x - 2y + z = -3$

29. $4x - y - 3z = 1,$
$8x + y - z = 5,$
$2x + y + 2z = 5$

30. $3x + 2y + 2z = 3,$
$x + 2y - z = 5,$
$2x - 4y + z = 0$

31. $x - 2y + 3z = -4,$
$3x + y - z = 0,$
$2x + 3y - 5z = 1$

32. $2x - 3y + 2z = 2,$
$x + 4y - z = 9,$
$-3x + y - 5z = 5$

33. $2x - 4y - 3z = 3,$
$x + 3y + z = -1,$
$5x + y - 2z = 2$

34. $x + y - 3z = 4,$
$4x + 5y + z = 1,$
$2x + 3y + 7z = -7$

35. $p + q + r = 1,$
$p + 2q + 3r = 4,$
$4p + 5q + 6r = 7$

36. $m + n + t = 9,$
$m - n - t = -15,$
$3m + n + t = 2$

37. $a + b - c = 7,$
$a - b + c = 5,$
$3a + b - c = -1$

38. $a - b + c = 3,$
$2a + b - 3c = 5,$
$4a + b - c = 11$

39. $-2w + 2x + 2y - 2z = -10,$
$w + x + y + z = -5,$
$3w + x - y + 4z = -2,$
$w + 3x - 2y + 2z = -6$

40. $-w + 2x - 3y + z = -8,$
$-w + x + y - z = -4,$
$w + x + y + z = 22,$
$-w + x - y - z = -14$

Use Gaussian elimination or Gauss–Jordan elimination in Exercises 41–44.

41. *Time of Return.* The Houlihans pay their babysitter $5 per hour before 11 P.M. and $7.50 per hour after 11 P.M. One evening they went out for 5 hr and paid the sitter $30. What time did they come home?

42. *Advertising Expense.* Cabletron Systems, Inc., spent a total of $17.5 million on advertising in fiscal years 1992, 1993, and 1994. The amount spent in 1994 was three fourths the total amount spent in 1992 and 1993. The amount spent in 1993 was $2.4 million less than the amount spent in 1994. How much was spent on advertising each year? (*Source:* Cabletron Systems Annual Report, 1994)

43. *Borrowing.* Gonzalez Manufacturing borrowed $30,000 to buy a new piece of equipment. Part of the money was borrowed at 8%, part at 10%, and part at 12%. The annual interest was $3040, and the total amount borrowed at 8% and 10% was twice the amount borrowed at 12%. How much was borrowed at each rate?

44. *Stamp Purchase.* Ricardo spent $18.30 on 32¢ and 23¢ stamps. He bought a total of 60 stamps. How many of each type did he buy?

Skill Maintenance

45. Find the opposite of -8.

46. Solve: $y + 3 = -2$.

47. Simplify: $3 - 12$.

48. Simplify: $5(-1) + 4 \cdot 3 - 8 \cdot 2$.

Synthesis

49. ◈ Solve the following system of equations using matrices and Gaussian elimination. Then solve it again using Gauss–Jordan elimination. Do you prefer one method over the other? Why or why not?

$$3x + 4y + 2z = 0,$$
$$x - y - z = 10,$$
$$2x + 3y + 3z = -10$$

50. ◈ Explain in your own words why the augmented matrix below represents a dependent system of equations.

$$\begin{bmatrix} 1 & -3 & 2 & | & -5 \\ 0 & 1 & -4 & | & 8 \\ 0 & 0 & 0 & | & 0 \end{bmatrix}$$

In Exercises 51 and 52, three solutions of the equation $y = ax^2 + bx + c$ are given. Use a system of three equations in three variables and Gaussian elimination or Gauss–Jordan elimination to find the constants and write the equation.

51. $(-3, 12)$, $(-1, -7)$, and $(1, -2)$

52. $(-1, 0)$, $(1, -3)$, and $(3, -22)$

53. Find two different row-echelon forms of

$$\begin{bmatrix} 1 & 5 \\ 3 & 2 \end{bmatrix}.$$

54. Consider the system of equations

$$\begin{aligned} x - y + 3z &= -8, \\ 2x + 3y - z &= 5, \\ 3x + 2y + 2kz &= -3k. \end{aligned}$$

For what value(s) of k, if any, will the system have

a) no solution?
b) exactly one solution?
c) infinitely many solutions?

Solve using matrices.

55. $y = x + z$,
$3y + 5z = 4$,
$x + 4 = y + 3z$

56. $x + y = 2z$,
$2x - 5z = 4$,
$x - z = y + 8$

57. $x - 4y + 2z = 7$,
$3x + y + 3z = -5$

58. $x - y - 3z = 3$,
$-x + 3y + z = -7$

59. $4x + 5y = 3$,
$-2x + y = 9$,
$3x - 2y = -15$

60. $2x - 3y = -1$,
$-x + 2y = -2$,
$3x - 5y = 1$

4.4

Matrix Operations

- *Add, subtract, and multiply matrices when possible.*
- *Write a matrix equation equivalent to a system of equations.*

In Section 4.3, we used matrices to solve systems of equations. Matrices are useful in many other types of applications as well. In this section, we study matrices and some of their properties.

A capital letter is generally used to name a matrix, and lower-case letters with double subscripts generally denote its entries. For example, a_{47}, read "a sub four seven," indicates the entry in the fourth row and the seventh column. A general term is represented by a_{ij}. The notation a_{ij} indicates the entry in row i and column j. In general, we can write a matrix as

$$\mathbf{A} = [a_{ij}] = \begin{bmatrix} a_{11} & a_{12} & a_{13} & \cdots & a_{1n} \\ a_{21} & a_{22} & a_{23} & \cdots & a_{2n} \\ a_{31} & a_{32} & a_{33} & \cdots & a_{3n} \\ \vdots & \vdots & \vdots & & \vdots \\ a_{m1} & a_{m2} & a_{m3} & \cdots & a_{mn} \end{bmatrix}.$$

The matrix above has m rows and n columns. That is, its order is $m \times n$.

Two matrices are **equal** if they have the same order and corresponding entries are equal.

Enter the following matrices on a grapher.

$$\mathbf{A} = \begin{bmatrix} 2 & -3 \\ -1 & 4 \end{bmatrix}, \qquad \mathbf{B} = \begin{bmatrix} -3 & 7 \\ 1 & -5 \end{bmatrix},$$

$$\mathbf{C} = \begin{bmatrix} 4 & 0 & -2 \\ 10 & -4 & -6 \\ -1 & 5 & 3 \end{bmatrix}, \qquad \mathbf{D} = \begin{bmatrix} 1 & -1 & 8 \\ 0 & 3 & -4 \\ 2 & -5 & 7 \end{bmatrix}$$

Use the MATRIX feature of the grapher to find $\mathbf{A} + \mathbf{B}$, $\mathbf{A} - \mathbf{B}$, $\mathbf{C} + \mathbf{D}$, and $\mathbf{C} - \mathbf{D}$. What relationship do you observe between corresponding entries of each pair of matrices and the entries of the sum? the difference?

What happens when you try to find $\mathbf{A} + \mathbf{C}$?

Matrix Addition and Subtraction

To add or subtract matrices, we add or subtract their corresponding entries. The matrices must have the same order for this to be possible.

Addition and Subtraction of Matrices

Given two $m \times n$ matrices $\mathbf{A} = [a_{ij}]$ and $\mathbf{B} = [b_{ij}]$, their sum is

$$\mathbf{A} + \mathbf{B} = [a_{ij} + b_{ij}]$$

and their difference is

$$\mathbf{A} - \mathbf{B} = [a_{ij} - b_{ij}].$$

Addition of matrices is both commutative and associative.

Example 1 Find $\mathbf{A} + \mathbf{B}$ for each of the following.

a) $\mathbf{A} = \begin{bmatrix} -5 & 0 \\ 4 & \frac{1}{2} \end{bmatrix}, \quad \mathbf{B} = \begin{bmatrix} 6 & -3 \\ 2 & 3 \end{bmatrix}$

b) $\mathbf{A} = \begin{bmatrix} 1 & 3 \\ -1 & 5 \\ 6 & 0 \end{bmatrix}, \quad \mathbf{B} = \begin{bmatrix} -1 & -2 \\ 1 & -2 \\ -3 & 1 \end{bmatrix}$

Solution Each pair of matrices has the same order. We add the corresponding entries.

a) $\mathbf{A} + \mathbf{B} = \begin{bmatrix} -5 & 0 \\ 4 & \frac{1}{2} \end{bmatrix} + \begin{bmatrix} 6 & -3 \\ 2 & 3 \end{bmatrix}$

$$= \begin{bmatrix} -5 + 6 & 0 + (-3) \\ 4 + 2 & \frac{1}{2} + 3 \end{bmatrix} = \begin{bmatrix} 1 & -3 \\ 6 & 3\frac{1}{2} \end{bmatrix}$$

b) $\mathbf{A} + \mathbf{B} = \begin{bmatrix} 1 & 3 \\ -1 & 5 \\ 6 & 0 \end{bmatrix} + \begin{bmatrix} -1 & -2 \\ 1 & -2 \\ -3 & 1 \end{bmatrix}$

$= \begin{bmatrix} 1 + (-1) & 3 + (-2) \\ -1 + 1 & 5 + (-2) \\ 6 + (-3) & 0 + 1 \end{bmatrix} = \begin{bmatrix} 0 & 1 \\ 0 & 3 \\ 3 & 1 \end{bmatrix}$

Use a grapher to verify these results.

Example 2 Find $\mathbf{C} - \mathbf{D}$ for each of the following.

a) $\mathbf{C} = \begin{bmatrix} 1 & 2 \\ -2 & 0 \\ -3 & -1 \end{bmatrix}$, $\mathbf{D} = \begin{bmatrix} 1 & -1 \\ 1 & 3 \\ 2 & 3 \end{bmatrix}$ b) $\mathbf{C} = \begin{bmatrix} 5 & -6 \\ -3 & 4 \end{bmatrix}$, $\mathbf{D} = \begin{bmatrix} -4 \\ 1 \end{bmatrix}$

SOLUTION

a) We subtract corresponding entries:

$\mathbf{C} - \mathbf{D} = \begin{bmatrix} 1 & 2 \\ -2 & 0 \\ -3 & -1 \end{bmatrix} - \begin{bmatrix} 1 & -1 \\ 1 & 3 \\ 2 & 3 \end{bmatrix}$

$= \begin{bmatrix} 1 - 1 & 2 - (-1) \\ -2 - 1 & 0 - 3 \\ -3 - 2 & -1 - 3 \end{bmatrix} = \begin{bmatrix} 0 & 3 \\ -3 & -3 \\ -5 & -4 \end{bmatrix}$

b) The matrices do not have the same order, so we cannot subtract.

Use a grapher to verify these results.

Interactive Discovery

Enter each of the following matrices on a grapher.

$\mathbf{A} = \begin{bmatrix} -2 & 4 \\ 1 & -5 \end{bmatrix}$, $\mathbf{B} = \begin{bmatrix} 2 & -4 \\ -1 & 5 \end{bmatrix}$,

$\mathbf{C} = \begin{bmatrix} 0 & -3 & -2 \\ 8 & 4 & -4 \\ -1 & -5 & 7 \end{bmatrix}$, $\mathbf{D} = \begin{bmatrix} 0 & 3 & 2 \\ -8 & -4 & 4 \\ 1 & 5 & -7 \end{bmatrix}$

What relationship do you observe between the corresponding entries of $\mathbf{A}$ and $\mathbf{B}$? of $\mathbf{C}$ and $\mathbf{D}$? Use the MATRIX feature of the grapher to find $\mathbf{A} + \mathbf{B}$ and $\mathbf{C} + \mathbf{D}$. What relationship do you observe between each pair of matrices and its sum?

The **opposite**, or **additive inverse**, of a matrix is obtained by replacing each entry by its opposite.

Example 3 Find $-\mathbf{A}$ and $\mathbf{A} + (-\mathbf{A})$ for

$\mathbf{A} = \begin{bmatrix} 1 & 0 & 2 \\ 3 & -1 & 5 \end{bmatrix}.$

SOLUTION   To find $-\mathbf{A}$, we replace each entry of A by its opposite.

$$-\mathbf{A} = \begin{bmatrix} -1 & 0 & -2 \\ -3 & 1 & -5 \end{bmatrix},$$

$$\mathbf{A} + (-\mathbf{A}) = \begin{bmatrix} 1 & 0 & 2 \\ 3 & -1 & 5 \end{bmatrix} + \begin{bmatrix} -1 & 0 & -2 \\ -3 & 1 & -5 \end{bmatrix}$$

$$= \begin{bmatrix} 0 & 0 & 0 \\ 0 & 0 & 0 \end{bmatrix}$$

Use a grapher to verify these results.

Interactive Discovery

Enter each of the following matrices on a grapher.

$$\mathbf{A} = \begin{bmatrix} 1 & -3 \\ -1 & 10 \end{bmatrix}, \qquad \mathbf{B} = \begin{bmatrix} 0 & 0 \\ 0 & 0 \end{bmatrix},$$

$$\mathbf{C} = \begin{bmatrix} 3 & -3 & 4 \\ -8 & -1 & -7 \end{bmatrix}, \qquad \mathbf{D} = \begin{bmatrix} 0 & 0 & 0 \\ 0 & 0 & 0 \end{bmatrix}$$

Use the MATRIX feature of the grapher to find $\mathbf{A} + \mathbf{B}$ and $\mathbf{C} + \mathbf{D}$. What is the outcome when a matrix consisting entirely of 0's is added to another matrix of the same order?

A matrix having 0's for all its entries is called a **zero matrix**. When a zero matrix is added to a second matrix of the same order, the second matrix is unchanged. Thus a zero matrix is an **additive identity**. For example,

$$\begin{bmatrix} 0 & 0 & 0 \\ 0 & 0 & 0 \end{bmatrix}$$

is the additive identity for any 2×3 matrix.

Scalar Multiplication

When we find the product of a number and a matrix, we obtain a *scalar product*.

Scalar Product

The *scalar product* of a number k and a matrix $\mathbf{A}$ is the matrix denoted $k\mathbf{A}$, obtained by multiplying each entry of $\mathbf{A}$ by the number k. The number k is called a *scalar*.

Example 4   Find $3\mathbf{A}$ and $(-1)\mathbf{A}$ for

$$\mathbf{A} = \begin{bmatrix} -3 & 0 \\ 4 & 5 \end{bmatrix}.$$

SOLUTION We have

$$3\mathbf{A} = 3\begin{bmatrix} -3 & 0 \\ 4 & 5 \end{bmatrix} = \begin{bmatrix} 3(-3) & 3 \cdot 0 \\ 3 \cdot 4 & 3 \cdot 5 \end{bmatrix} = \begin{bmatrix} -9 & 0 \\ 12 & 15 \end{bmatrix},$$

$$(-1)\mathbf{A} = -1\begin{bmatrix} -3 & 0 \\ 4 & 5 \end{bmatrix} = \begin{bmatrix} -1(-3) & -1 \cdot 0 \\ -1 \cdot 4 & -1 \cdot 5 \end{bmatrix} = \begin{bmatrix} 3 & 0 \\ -4 & -5 \end{bmatrix}.$$

Use a grapher to verify these results.

Example 5 *Production.* Mitchell Fabricators manufactures three styles of bicycle frames in its two plants. The following table shows the number of each style produced at each plant in April.

	MOUNTAIN BIKE	RACING BIKE	TOURING BIKE
NORTH PLANT	150	120	100
SOUTH PLANT	180	90	130

a) Write a 2 × 3 matrix **A** that represents the information in the table.

b) The manufacturer increased production by 20% in May. Find a matrix **M** that represents the increased production figures.

c) Find the matrix **A** + **M** and tell what it represents.

SOLUTION

a) Write the entries in the table in a 2 × 3 matrix **A**.

$$\mathbf{A} = \begin{bmatrix} 150 & 120 & 100 \\ 180 & 90 & 130 \end{bmatrix}$$

b) The production in May will be represented by **A** + 20%**A**, or **A** + 0.2**A**, or 1.2**A**. Thus,

$$\mathbf{M} = (1.2)\begin{bmatrix} 150 & 120 & 100 \\ 180 & 90 & 130 \end{bmatrix} = \begin{bmatrix} 180 & 144 & 120 \\ 216 & 108 & 156 \end{bmatrix}.$$

c) $\mathbf{A} + \mathbf{M} = \begin{bmatrix} 150 & 120 & 100 \\ 180 & 90 & 130 \end{bmatrix} + \begin{bmatrix} 180 & 144 & 120 \\ 216 & 108 & 156 \end{bmatrix}$

$$= \begin{bmatrix} 330 & 264 & 220 \\ 396 & 198 & 286 \end{bmatrix}$$

The matrix **A** + **M** represents the total production of each of the three types of frames at each plant in April and May.

The properties of matrix addition and scalar multiplication are similar to the properties of addition and multiplication of real numbers.

Properties of Matrix Addition and Scalar Multiplication

For any $m \times n$ matrices **A**, **B**, and **C** and any scalars k and l:

$$\mathbf{A} + \mathbf{B} = \mathbf{B} + \mathbf{A}. \qquad \textit{Commutative Property}$$

$$\mathbf{A} + (\mathbf{B} + \mathbf{C}) = (\mathbf{A} + \mathbf{B}) + \mathbf{C}. \qquad \textit{Associative Property of Addition}$$

$$(kl)\mathbf{A} = k(l\mathbf{A}). \qquad \textit{Associative Property of Scalar Multiplication}$$

There exists a unique matrix **0** such that:

$$\mathbf{A} + \mathbf{0} = \mathbf{0} + \mathbf{A} = \mathbf{A}. \qquad \textit{Additive Identity Property}$$

There exists a unique matrix $-\mathbf{A}$ such that:

$$\mathbf{A} + (-\mathbf{A}) = -\mathbf{A} + \mathbf{A} = \mathbf{0}. \qquad \textit{Additive Inverse Property}$$

$$k(\mathbf{A} + \mathbf{B}) = k\mathbf{A} + k\mathbf{B}. \qquad \textit{Distributive Property}$$

$$(k + l)\mathbf{A} = k\mathbf{A} + l\mathbf{A}. \qquad \textit{Distributive Property}$$

Products of Matrices

Matrix multiplication is defined in such a way that it can be used in solving systems of equations and in many applications.

Matrix Multiplication

For an $m \times n$ matrix $\mathbf{A} = [a_{ij}]$ and an $n \times p$ matrix $\mathbf{B} = [b_{ij}]$, the *product* $\mathbf{AB} = [c_{ij}]$ is an $m \times p$ matrix, where

$$c_{ij} = a_{i1} \cdot b_{1j} + a_{i2} \cdot b_{2j} + a_{i3} \cdot b_{3j} + \cdots + a_{in} \cdot b_{nj}.$$

In other words, the entry c_{ij} in **AB** is obtained by multiplying the entries in row i of **A** by the corresponding entries in column j of **B** and adding the results. Note that we can multiply two matrices only when the number of columns in the first matrix is equal to the number of rows in the second matrix.

Example 6   For

$$\mathbf{A} = \begin{bmatrix} 3 & 1 & -1 \\ 2 & 0 & 3 \end{bmatrix}, \qquad \mathbf{B} = \begin{bmatrix} 1 & 6 \\ 3 & -5 \\ -2 & 4 \end{bmatrix}, \quad \text{and} \quad \mathbf{C} = \begin{bmatrix} 4 & -6 \\ 1 & 2 \end{bmatrix},$$

find each of the following.

a) **AB**

b) **BA**

c) **BC**

d) **AC**

SOLUTION

a) **A** is a 2×3 matrix and **B** is a 3×2 matrix, so **AB** will be a 2×2 matrix.

$$\mathbf{AB} = \begin{bmatrix} 3 & 1 & -1 \\ 2 & 0 & 3 \end{bmatrix} \begin{bmatrix} 1 & 6 \\ 3 & -5 \\ -2 & 4 \end{bmatrix}$$

$$= \begin{bmatrix} 3 \cdot 1 + 1 \cdot 3 + (-1)(-2) & 3 \cdot 6 + 1(-5) + (-1)(4) \\ 2 \cdot 1 + 0 \cdot 3 + 3(-2) & 2 \cdot 6 + 0(-5) + 3 \cdot 4 \end{bmatrix}$$

$$= \begin{bmatrix} 8 & 9 \\ -4 & 24 \end{bmatrix}$$

b) **B** is a 3×2 matrix and **A** is a 2×3 matrix, so **BA** will be a 3×3 matrix.

$$\mathbf{BA} = \begin{bmatrix} 1 & 6 \\ 3 & -5 \\ -2 & 4 \end{bmatrix} \begin{bmatrix} 3 & 1 & -1 \\ 2 & 0 & 3 \end{bmatrix}$$

$$= \begin{bmatrix} 1 \cdot 3 + 6 \cdot 2 & 1 \cdot 1 + 6 \cdot 0 & 1(-1) + 6 \cdot 3 \\ 3 \cdot 3 + (-5)(2) & 3 \cdot 1 + (-5)(0) & 3(-1) + (-5)(3) \\ -2 \cdot 3 + 4 \cdot 2 & -2 \cdot 1 + 4 \cdot 0 & -2(-1) + 4 \cdot 3 \end{bmatrix}$$

$$= \begin{bmatrix} 15 & 1 & 17 \\ -1 & 3 & -18 \\ 2 & -2 & 14 \end{bmatrix}$$

c) **B** is a 3×2 matrix and **C** is a 2×2 matrix, so **BC** will be a 3×2 matrix.

$$\mathbf{BC} = \begin{bmatrix} 1 & 6 \\ 3 & -5 \\ -2 & 4 \end{bmatrix} \begin{bmatrix} 4 & -6 \\ 1 & 2 \end{bmatrix}$$

$$= \begin{bmatrix} 1 \cdot 4 + 6 \cdot 1 & 1(-6) + 6 \cdot 2 \\ 3 \cdot 4 + (-5)(1) & 3(-6) + (-5)(2) \\ -2 \cdot 4 + 4 \cdot 1 & -2(-6) + 4 \cdot 2 \end{bmatrix}$$

$$= \begin{bmatrix} 10 & 6 \\ 7 & -28 \\ -4 & 20 \end{bmatrix}$$

d) The product **AC** is not defined because the number of columns of **A** (3) is not equal to the number of rows of **C** (2).

Use a grapher to verify these results. What message appeared on the grapher in part (d)?

Note in Example 6 that $\mathbf{AB} \neq \mathbf{BA}$. Multiplication of matrices is generally not commutative.

Example 7 *Dairy Profit.* Dalton's Dairy produces no-fat ice cream and frozen yogurt. The table below shows the number of gallons of each product that are shipped to the dairy's three retail outlets one week.

	STORE 1	STORE 2	STORE 3
NO-FAT ICE CREAM	100	80	120
FROZEN YOGURT	160	120	100

On each gallon of no-fat ice cream, the dairy's profit is $4, and on each gallon of frozen yogurt, it is $3. Use matrices to find the total profit on these items at each store for the given week.

SOLUTION   We can write the table showing the distribution of the products as a 2×3 matrix:

$$\mathbf{D} = \begin{bmatrix} 100 & 80 & 120 \\ 160 & 120 & 100 \end{bmatrix}.$$

The profit per gallon for each product can also be written as a matrix:

$$\mathbf{P} = [4 \quad 3].$$

Then the total profit at each store is given by the matrix product $\mathbf{PD}$:

$$\mathbf{PD} = [4 \quad 3]\begin{bmatrix} 100 & 80 & 120 \\ 160 & 120 & 100 \end{bmatrix}$$

$$= [4 \cdot 100 + 3 \cdot 160 \quad 4 \cdot 80 + 3 \cdot 120 \quad 4 \cdot 120 + 3 \cdot 100]$$

$$= [880 \quad 680 \quad 780].$$

Thus the total profit on no-fat ice cream and frozen yogurt for the given week was $880 at store 1, $680 at store 2, and $780 at store 3. You can verify these results on a grapher.

A matrix that consists of a single row, like $\mathbf{P}$ in Example 7, is called a **row matrix**. Similarly, a matrix that consists of a single column, like

$$\begin{bmatrix} 8 \\ -3 \\ 5 \end{bmatrix},$$

is called a **column matrix**.

We have already seen that matrix multiplication is generally not commutative. Nevertheless, matrix multiplication does have some properties that are similar to those for multiplication of real numbers.

> **Properties of Matrix Multiplication**
>
> For matrices **A**, **B**, and **C**, assuming that the indicated operation is possible:
>
> $$\mathbf{A(BC) = (AB)C}.\qquad \text{Associative Property of Multiplication}$$
>
> $$\mathbf{A(B + C) = AB + AC}.\qquad \text{Distributive Property}$$
>
> $$\mathbf{(B + C)A = BA + CA}.\qquad \text{Distributive Property}$$

Matrix Equations

We can write a matrix equation equivalent to a system of equations.

Example 8   Write a matrix equation equivalent to the following system of equations:

$$\begin{aligned}
4x + 2y - \;\;z &= 3, \\
9x \qquad\;\; + \;\;z &= 5, \\
4x + 5y - 2z &= 1.
\end{aligned}$$

SOLUTION   We write the coefficients on the left in a matrix. We then write the product of that matrix and the column matrix containing the variables, and set the result equal to the column matrix containing the constants on the right:

$$\begin{bmatrix} 4 & 2 & -1 \\ 9 & 0 & 1 \\ 4 & 5 & -2 \end{bmatrix} \begin{bmatrix} x \\ y \\ z \end{bmatrix} = \begin{bmatrix} 3 \\ 5 \\ 1 \end{bmatrix}.$$

If we let

$$\mathbf{A} = \begin{bmatrix} 4 & 2 & -1 \\ 9 & 0 & 1 \\ 4 & 5 & -2 \end{bmatrix}, \qquad \mathbf{X} = \begin{bmatrix} x \\ y \\ z \end{bmatrix}, \quad \text{and} \quad \mathbf{B} = \begin{bmatrix} 3 \\ 5 \\ 1 \end{bmatrix},$$

we can write this matrix equation as $\mathbf{AX = B}$.

In the next section, we will solve systems of equations using a matrix equation like the one in Example 8.

4.4 Exercise Set

Find x, y, and z.

1. $[5 \quad x] = [y \quad -3]$

2. $\begin{bmatrix} 6x \\ 25 \end{bmatrix} = \begin{bmatrix} -9 \\ 5y \end{bmatrix}$

3. $\begin{bmatrix} 3 & 2x \\ y & -8 \end{bmatrix} = \begin{bmatrix} 3 & -2 \\ 1 & -8 \end{bmatrix}$

4. $\begin{bmatrix} x - 1 & 4 \\ y + 3 & -7 \end{bmatrix} = \begin{bmatrix} 0 & 4 \\ -2 & -7 \end{bmatrix}$

For Exercises 5–20, let

$$A = \begin{bmatrix} 1 & 2 \\ 4 & 3 \end{bmatrix}, \qquad B = \begin{bmatrix} -3 & 5 \\ 2 & -1 \end{bmatrix},$$

$$C = \begin{bmatrix} 1 & -1 \\ -1 & 1 \end{bmatrix}, \qquad D = \begin{bmatrix} 1 & 1 \\ 1 & 1 \end{bmatrix},$$

$$E = \begin{bmatrix} 1 & 3 \\ 2 & 6 \end{bmatrix}, \qquad F = \begin{bmatrix} 3 & 3 \\ -1 & -1 \end{bmatrix},$$

$$O = \begin{bmatrix} 0 & 0 \\ 0 & 0 \end{bmatrix}, \qquad I = \begin{bmatrix} 1 & 0 \\ 0 & 1 \end{bmatrix}.$$

Find each of the following. Use a grapher to check your answers.

5. $A + B$

6. $B + A$

7. $E + O$

8. $2A$

9. $3F$

10. $(-1)D$

11. $3F + 2A$

12. $A - B$

13. $B - A$

14. AB

15. BA

16. OF

17. CD

18. EF

19. AI

20. IA

21. *Budget.* For the month of June, Nelia budgets $150 for food, $80 for clothes, and $40 for entertainment.

 a) Write a 1×3 matrix B that represents these items.
 b) After receiving a raise, Nelia increases the amount budgeted for each item in July by 5%. Find a matrix R that represents the new amounts.
 c) Find $B + R$ and tell what the entries represent.

22. *Produce.* The produce manager at Dugan's Market orders 40 lb of tomatoes, 20 lb of zucchini, and 30 lb of onions from a local farmer one week.

 a) Write a 1×3 matrix A that represents these items.
 b) The following week the produce manager increases her order by 10%. Find a matrix B that represents this order.
 c) Find $A + B$ and tell what the entries represent.

23. *Nutrition.* A 3-oz serving of roasted, skinless chicken breast contains 140 Cal, 27 g of protein, 3 g of fat, 13 mg of calcium, and 64 mg of sodium. One-half cup of potato salad contains 180 Cal, 4 g of protein, 11 g of fat, 24 mg of calcium, and 662 mg of sodium. One broccoli spear contains 50 Cal, 5 g of protein, 1 g of fat, 82 mg of calcium, and 20 mg of sodium. (*Source: Home and Garden Bulletin No. 72, U.S. Government Printing Office, Washington, D.C. 20402*)

 a) Write 1×5 matrices C, P, and B that represent the nutritional values of each food.

b) Find $C + 2P + 3B$ and tell what the entries represent.

24. *Nutrition.* One slice of cheese pizza contains 290 Cal, 15 g of protein, 9 g of fat, and 39 g of carbohydrates. One-half cup of gelatin dessert contains 70 Cal, 2 g of protein, 0 g of fat, and 17 g of carbohydrates. One cup of whole milk contains 150 Cal, 8 g of protein, 8 g of fat, and 11 g of carbohydrates. (*Source: Home and Garden Bulletin No. 72, U.S. Government Printing Office, Washington, D.C. 20402*)

 a) Write 1×4 matrices P, G, and M that represent the nutritional values of each food.
 b) Find $3P + 2G + 2M$ and tell what the entries represent.

Use a grapher to find each product, if possible.

25. $\begin{bmatrix} -1 & 0 & 7 \\ 3 & -5 & 2 \end{bmatrix} \begin{bmatrix} 6 \\ -4 \\ 1 \end{bmatrix}$

26. $[6 \quad -1 \quad 2] \begin{bmatrix} 1 & 4 \\ -2 & 0 \\ 5 & -3 \end{bmatrix}$

27. $\begin{bmatrix} -2 & 4 \\ 5 & 1 \\ -1 & -3 \end{bmatrix} \begin{bmatrix} 3 & -6 \\ -1 & 4 \end{bmatrix}$

28. $\begin{bmatrix} 2 & -1 & 0 \\ 0 & 5 & 4 \end{bmatrix} \begin{bmatrix} -3 & 1 & 0 \\ 0 & 2 & -1 \\ 5 & 0 & 4 \end{bmatrix}$

29. $\begin{bmatrix} 1 \\ -5 \\ 3 \end{bmatrix} \begin{bmatrix} -6 & 5 & 8 \\ 0 & 4 & -1 \end{bmatrix}$

30. $\begin{bmatrix} 2 & 0 & 0 \\ 0 & -1 & 0 \\ 0 & 0 & 3 \end{bmatrix} \begin{bmatrix} 0 & -4 & 3 \\ 2 & 1 & 0 \\ -1 & 0 & 6 \end{bmatrix}$

31. $\begin{bmatrix} 1 & -4 & 3 \\ 0 & 8 & 0 \\ -2 & -1 & 5 \end{bmatrix} \begin{bmatrix} 3 & 0 & 0 \\ 0 & -4 & 0 \\ 0 & 0 & 1 \end{bmatrix}$

32. $\begin{bmatrix} 4 \\ -5 \end{bmatrix} \begin{bmatrix} 2 & 0 \\ 6 & -7 \\ 0 & -3 \end{bmatrix}$

33. *Food Service Management.* The food service manager at a large hospital is concerned about maintaining reasonable food costs. The table below shows the cost per serving, in cents, for items on four menus.

MENU	MEAT	POTATO	VEGETABLE	SALAD	DESSERT
1	45.29	6.63	10.94	7.42	8.01
2	53.78	4.95	9.83	6.16	12.56
3	47.13	8.47	12.66	8.29	9.43
4	51.64	7.12	11.57	9.35	10.72

On a particular day a dietician orders 65 meals from menu 1, 48 from menu 2, 93 from menu 3, and 57 from menu 4.

a) Write the information in the table as a 4 × 5 matrix **M**.

b) Write a row matrix **N** that represents the number of each menu ordered.

c) Find the product **NM**.

d) State what the entries of **NM** represent.

34. *Food Service Management.* A college food service manager uses a table like the one below to show the number of units of ingredients, by weight, required for various menu items.

	WHITE CAKE	BREAD	COFFEE CAKE	SUGAR COOKIES
FLOUR	1	2.5	0.75	0.5
MILK	0	0.5	0.25	0
EGGS	0.75	0.25	0.5	0.5
BUTTER	0.5	0	0.5	1

The cost per unit of each ingredient is 15 cents for flour, 28 cents for milk, 54 cents for eggs, and 83 cents for butter.

a) Write the information in the table as a 4 × 4 matrix **M**.

b) Write a row matrix **C** that represents the cost per unit of each ingredient.

c) Find the product **CM**.

d) State what the entries of **CM** represent.

35. *Production Cost.* Karin supplies two small campus coffee shops with homemade chocolate chip cookies, oatmeal cookies, and peanut butter cookies. The table below shows the number of each type of cookie, in dozens, that Karin sold in one week.

	MUGSY'S COFFEE SHOP	THE COFFEE CLUB
CHOCOLATE CHIP	8	15
OATMEAL	6	10
PEANUT BUTTER	4	3

Karin spends $3 for the ingredients for one dozen chocolate chip cookies, $1.50 for the ingredients for one dozen oatmeal cookies, and $2 for the ingredients for one dozen peanut butter cookies.

a) Write the information in the table as a 3 × 2 matrix **S**.

b) Write a row matrix **C** that represents the cost, per dozen, of the ingredients for each type of cookie.

c) Find the product **CS**.

d) State what the entries of **CS** represent.

36. *Profit.* A manufacturer produces exterior plywood, interior plywood, and fiberboard, which are shipped to two distributors. The table below shows the number of units of each type of product that are shipped to each warehouse.

	DISTRIBUTOR 1	DISTRIBUTOR 2
EXTERIOR PLYWOOD	900	500
INTERIOR PLYWOOD	450	1000
FIBERBOARD	600	700

The profits from each unit of exterior plywood, interior plywood, and fiberboard are $5, $8, and $4, respectively.

a) Write the information in the table as a 3 × 2 matrix **M**.

b) Write a row matrix **P** that represents the profit, per unit, of each type of product.

c) Find the product **PM**.

d) State what the entries of **PM** represent.

37. *Profit.* In Exercise 35, suppose that Karin's profits on one dozen chocolate chip, oatmeal, and peanut butter cookies are $6, $4.50, and $5.20, respectively.

a) Write a row matrix **P** that represents this information.

b) Use the matrices **S** and **P** to find Karin's total profit from each snack bar.

38. *Production Cost.* In Exercise 36, suppose that the manufacturer's production costs for each unit of exterior plywood, interior plywood, and fiberboard are $20, $25, and $15, respectively.

    a) Write a row matrix **C** that represents this information.

    b) Use the matrices **M** and **C** to find the total production cost for the products shipped to each distributor.

Write a matrix equation equivalent to the system of equations.

39. $2x - 3y = 7,$
$\quad x + 5y = -6$

40. $-x + y = 3,$
$\quad 5x - 4y = 16$

41. $x + y - 2z = 6,$
$\quad 3x - y + z = 7,$
$\quad 2x + 5y - 3z = 8$

42. $3x - y + z = 1,$
$\quad x + 2y - z = 3,$
$\quad 4x + 3y - 2z = 11$

43. $3x - 2y + 4z = 17,$
$\quad 2x + y - 5z = 13$

44. $3x + 2y + 5z = 9,$
$\quad 4x - 3y + 2z = 10$

45. $-4w + x - y + 2z = 12,$
$\quad w + 2x - y - z = 0,$
$\quad -w + x + 4y - 3z = 1,$
$\quad 2w + 3x + 5y - 7z = 9$

46. $12w + 2x + 4y - 5z = 2,$
$\quad -w + 4x - y + 12z = 5,$
$\quad 2w - x + 4y = 13,$
$\quad 2x + 10y + z = 5$

Skill Maintenance

Solve.

47. $5x = 45$ **48.** $-13x = 52$

49. $4x = -28$ **50.** $ax = b,$ for x

Synthesis

51. ◈ Is it true that if $\mathbf{AB} = \mathbf{0}$, for matrices **A** and **B**, then $\mathbf{A} = \mathbf{0}$ or $\mathbf{B} = \mathbf{0}$? Why or why not?

52. ◈ Explain how Karin could use the matrix products found in Exercises 35 and 37 in making business decisions.

For Exercises 53–56, let

$$\mathbf{A} = \begin{bmatrix} -1 & 0 \\ 2 & 1 \end{bmatrix} \quad and \quad \mathbf{B} = \begin{bmatrix} 1 & -1 \\ 0 & 2 \end{bmatrix}.$$

53. Show that
$$(\mathbf{A} + \mathbf{B})(\mathbf{A} - \mathbf{B}) \neq \mathbf{A}^2 - \mathbf{B}^2,$$
where
$$\mathbf{A}^2 = \mathbf{AA} \quad and \quad \mathbf{B}^2 = \mathbf{BB}.$$

54. Show that
$$(\mathbf{A} + \mathbf{B})(\mathbf{A} + \mathbf{B}) \neq \mathbf{A}^2 + 2\mathbf{AB} + \mathbf{B}^2.$$

55. Show that
$$(\mathbf{A} + \mathbf{B})(\mathbf{A} - \mathbf{B}) = \mathbf{A}^2 + \mathbf{BA} - \mathbf{AB} - \mathbf{B}^2.$$

56. Show that
$$(\mathbf{A} + \mathbf{B})(\mathbf{A} + \mathbf{B}) = \mathbf{A}^2 + \mathbf{BA} + \mathbf{AB} + \mathbf{B}^2.$$

In Exercises 57–61, let

$$\mathbf{A} = \begin{bmatrix} a_{11} & a_{12} & a_{13} & \cdots & a_{1n} \\ a_{21} & a_{22} & a_{23} & \cdots & a_{2n} \\ a_{31} & a_{32} & a_{33} & \cdots & a_{3n} \\ \vdots & \vdots & \vdots & & \vdots \\ a_{m1} & a_{m2} & a_{m3} & \cdots & a_{mn} \end{bmatrix},$$

$$\mathbf{B} = \begin{bmatrix} b_{11} & b_{12} & b_{13} & \cdots & b_{1n} \\ b_{21} & b_{22} & b_{23} & \cdots & b_{2n} \\ b_{31} & b_{32} & b_{33} & \cdots & b_{3n} \\ \vdots & \vdots & \vdots & & \vdots \\ b_{m1} & b_{m2} & b_{m3} & \cdots & b_{mn} \end{bmatrix},$$

$$and\ \mathbf{C} = \begin{bmatrix} c_{11} & c_{12} & c_{13} & \cdots & c_{1n} \\ c_{21} & c_{22} & c_{23} & \cdots & c_{2n} \\ c_{31} & c_{32} & c_{33} & \cdots & c_{3n} \\ \vdots & \vdots & \vdots & & \vdots \\ c_{m1} & c_{m2} & c_{m3} & \cdots & c_{mn} \end{bmatrix},$$

and let k and l be any scalars.

57. Prove that $\mathbf{A} + \mathbf{B} = \mathbf{B} + \mathbf{A}.$

58. Prove that $\mathbf{A} + (\mathbf{B} + \mathbf{C}) = (\mathbf{A} + \mathbf{B}) + \mathbf{C}.$

59. Prove that $(kl)\mathbf{A} = k(l\mathbf{A}).$

60. Prove that $k(\mathbf{A} + \mathbf{B}) = k\mathbf{A} + k\mathbf{B}.$

61. Prove that $(k + l)\mathbf{A} = k\mathbf{A} + l\mathbf{A}.$

4.5

Inverses of Matrices

- *Find the inverse of a square matrix, if it exists.*
- *Use inverses of matrices to solve systems of equations.*

In this section, we continue our study of matrix algebra, finding the **multiplicative inverse**, or simply **inverse**, of a square matrix, if it exists. Then we use such inverses to solve systems of equations.

Interactive Discovery

Enter the following matrices on a grapher.

$$\mathbf{A} = \begin{bmatrix} 2 & -3 \\ -1 & 4 \end{bmatrix}, \qquad \mathbf{B} = \begin{bmatrix} 1 & 0 \\ 0 & 1 \end{bmatrix},$$

$$\mathbf{C} = \begin{bmatrix} 4 & 0 & -2 \\ 10 & -4 & -6 \\ -1 & 5 & 3 \end{bmatrix}, \qquad \mathbf{D} = \begin{bmatrix} 1 & 0 & 0 \\ 0 & 1 & 0 \\ 0 & 0 & 1 \end{bmatrix}$$

Use the MATRIX feature of the grapher to find **AB**, **BA**, **CD**, and **DC**. What is the effect of multiplying by **B**? by **D**?

The Identity Matrix

Recall that, for real numbers, $a \cdot 1 = 1 \cdot a = 1$; 1 is the multiplicative identity. A multiplicative identity matrix is very similar to the number 1.

Identity Matrix

For any positive integer n, the $n \times n$ *identity matrix* is an $n \times n$ matrix with 1's on the main diagonal and 0's elsewhere and is denoted by

$$\mathbf{I} = \begin{bmatrix} 1 & 0 & 0 & \cdots & 0 \\ 0 & 1 & 0 & \cdots & 0 \\ 0 & 0 & 1 & \cdots & 0 \\ \vdots & \vdots & \vdots & & \vdots \\ 0 & 0 & 0 & \cdots & 1 \end{bmatrix}.$$

Then $\mathbf{AI} = \mathbf{IA} = \mathbf{A}$, for any $n \times n$ matrix **A**.

Example 1   For

$$\mathbf{A} = \begin{bmatrix} 4 & -7 \\ -3 & 2 \end{bmatrix}$$

and $\mathbf{I} = \begin{bmatrix} 1 & 0 \\ 0 & 1 \end{bmatrix},$

find each of the following.

a) **AI**   b) **IA**

SOLUTION

a) $\mathbf{AI} = \begin{bmatrix} 4 & -7 \\ -3 & 2 \end{bmatrix}\begin{bmatrix} 1 & 0 \\ 0 & 1 \end{bmatrix}$

$= \begin{bmatrix} 4 \cdot 1 - 7 \cdot 0 & 4 \cdot 0 - 7 \cdot 1 \\ -3 \cdot 1 + 2 \cdot 0 & -3 \cdot 0 + 2 \cdot 1 \end{bmatrix}$

$= \begin{bmatrix} 4 & -7 \\ -3 & 2 \end{bmatrix} = \mathbf{A}$

b) $\mathbf{IA} = \begin{bmatrix} 1 & 0 \\ 0 & 1 \end{bmatrix}\begin{bmatrix} 4 & -7 \\ -3 & 2 \end{bmatrix}$

$= \begin{bmatrix} 1 \cdot 4 + 0(-3) & 1(-7) + 0 \cdot 2 \\ 0 \cdot 4 + 1(-3) & 0(-7) + 1 \cdot 2 \end{bmatrix} = \begin{bmatrix} 4 & -7 \\ -3 & 2 \end{bmatrix} = \mathbf{A}$

The Inverse of a Matrix

Recall that for every nonzero real number a, there is a multiplicative inverse $1/a$ such that $a(1/a) = (1/a)a = 1$. The multiplicative inverse of a matrix behaves in a similar manner.

Inverse of a Matrix

For an $n \times n$ matrix $\mathbf{A}$, if there is a matrix $\mathbf{A}^{-1}$ for which $\mathbf{A}^{-1} \cdot \mathbf{A} = \mathbf{I} = \mathbf{A} \cdot \mathbf{A}^{-1}$, then $\mathbf{A}^{-1}$ is the *inverse* of $\mathbf{A}$.

Note that $\mathbf{A}^{-1}$ is read "$\mathbf{A}$ inverse."

Example 2 Verify that

$$\mathbf{B} = \begin{bmatrix} 4 & -3 \\ 3 & -2 \end{bmatrix} \quad \text{is the inverse of} \quad \mathbf{A} = \begin{bmatrix} -2 & 3 \\ -3 & 4 \end{bmatrix}.$$

SOLUTION We show that $\mathbf{BA} = \mathbf{I} = \mathbf{AB}$.

$$\mathbf{BA} = \begin{bmatrix} 4 & -3 \\ 3 & -2 \end{bmatrix}\begin{bmatrix} -2 & 3 \\ -3 & 4 \end{bmatrix} = \begin{bmatrix} 1 & 0 \\ 0 & 1 \end{bmatrix}$$

$$\mathbf{AB} = \begin{bmatrix} -2 & 3 \\ -3 & 4 \end{bmatrix}\begin{bmatrix} 4 & -3 \\ 3 & -2 \end{bmatrix} = \begin{bmatrix} 1 & 0 \\ 0 & 1 \end{bmatrix}$$

We can find the inverse of a square matrix, if it exists, by using row-equivalent operations as in the Gauss–Jordan elimination method. For example, consider the matrix

$$\mathbf{A} = \begin{bmatrix} -2 & 3 \\ -3 & 4 \end{bmatrix}.$$

To find its inverse, we first form an **augmented matrix** consisting of **A** on the left side and the 2×2 identity matrix on the right side:

$$\left[\begin{array}{cc|cc} -2 & 3 & 1 & 0 \\ -3 & 4 & 0 & 1 \end{array}\right].$$

The 2×2 The 2×2
matrix **A** identity matrix

Then we attempt to transform the augmented matrix to one of the form

$$\left[\begin{array}{cc|cc} 1 & 0 & a & b \\ 0 & 1 & c & d \end{array}\right].$$

The 2×2 The matrix $\mathbf{A}^{-1}$
identity matrix

The matrix on the right, $\begin{bmatrix} a & b \\ c & d \end{bmatrix}$, is $\mathbf{A}^{-1}$.

Example 3 Find $\mathbf{A}^{-1}$, where

$$\mathbf{A} = \begin{bmatrix} -2 & 3 \\ -3 & 4 \end{bmatrix}.$$

SOLUTION First we write the augmented matrix. Then we transform it to the desired form.

$$\left[\begin{array}{cc|cc} -2 & 3 & 1 & 0 \\ -3 & 4 & 0 & 1 \end{array}\right]$$

$$\left[\begin{array}{cc|cc} 1 & -\frac{3}{2} & -\frac{1}{2} & 0 \\ -3 & 4 & 0 & 1 \end{array}\right] \qquad \text{New row } 1 = -\tfrac{1}{2}(\text{row } 1)$$

$$\left[\begin{array}{cc|cc} 1 & -\frac{3}{2} & -\frac{1}{2} & 0 \\ 0 & -\frac{1}{2} & -\frac{3}{2} & 1 \end{array}\right] \qquad \text{New row } 2 = 3(\text{row } 1) + \text{row } 2$$

$$\left[\begin{array}{cc|cc} 1 & -\frac{3}{2} & -\frac{1}{2} & 0 \\ 0 & 1 & 3 & -2 \end{array}\right] \qquad \text{New row } 2 = -2(\text{row } 2)$$

$$\left[\begin{array}{cc|cc} 1 & 0 & 4 & -3 \\ 0 & 1 & 3 & -2 \end{array}\right] \qquad \text{New row } 1 = \tfrac{3}{2}(\text{row } 2) + \text{row } 1$$

Thus,

$$\mathbf{A}^{-1} = \begin{bmatrix} 4 & -3 \\ 3 & -2 \end{bmatrix},$$

which we verified in Example 2.

A grapher can be used to find the inverse of a matrix quickly. Check the user's manual for the procedure. Then use a grapher to verify the result of Example 3.

If a matrix has an inverse, we say that it is **invertible**, or **nonsingular**. When we cannot obtain the identity matrix on the left using the Gauss–Jordan method, then no inverse exists. This occurs when we obtain a row consisting entirely of 0's in either of the two matrices in the augmented matrix. In this case, we say that **A** is a **singular matrix**. A grapher will produce an error message similar to "ERR: SINGULAR MATRIX" in this situation.

Solving Systems of Equations

We can write a system of n linear equations in n variables as a matrix equation $\mathbf{AX} = \mathbf{B}$. (See Section 4.4.) If **A** has an inverse, then the system of equations has a unique solution that can be found by solving for **X**, as follows:

$$\mathbf{AX} = \mathbf{B}$$

$\mathbf{A}^{-1}(\mathbf{AX}) = \mathbf{A}^{-1}\mathbf{B}$ Multiplying by $\mathbf{A}^{-1}$ on the left on both sides

$(\mathbf{A}^{-1}\mathbf{A})\mathbf{X} = \mathbf{A}^{-1}\mathbf{B}$ Using the associative property of matrix multiplication

$\mathbf{IX} = \mathbf{A}^{-1}\mathbf{B}$ $\mathbf{A}^{-1}\mathbf{A} = \mathbf{I}$

$\mathbf{X} = \mathbf{A}^{-1}\mathbf{B}.$ $\mathbf{IX} = \mathbf{X}$

Matrix Solutions of Systems of Equations

For a system of n linear equations in n variables, $\mathbf{AX} = \mathbf{B}$, if **A** is an invertible matrix, then the unique solution of the system is given by

$$\mathbf{X} = \mathbf{A}^{-1}\mathbf{B}.$$

Since matrix multiplication is not commutative in general, care must be taken to multiply *on the left* by $\mathbf{A}^{-1}$.

Example 4   Use an inverse matrix to solve the following system of equations:

$$\begin{aligned} x + 2y - z &= -2, \\ 3x + 5y + 3z &= 3, \\ 2x + 4y + 3z &= 1. \end{aligned}$$

SOLUTION   We write an equivalent matrix equation, $\mathbf{AX} = \mathbf{B}$:

$$\begin{bmatrix} 1 & 2 & -1 \\ 3 & 5 & 3 \\ 2 & 4 & 3 \end{bmatrix} \begin{bmatrix} x \\ y \\ z \end{bmatrix} = \begin{bmatrix} -2 \\ 3 \\ 1 \end{bmatrix}.$$

Then we use a grapher to find $\mathbf{A}^{-1}\mathbf{B}$. We get

$$\mathbf{X} = \mathbf{A}^{-1}\mathbf{B} = \begin{bmatrix} -0.6 & 2 & -2.2 \\ 0.6 & -1 & 1.2 \\ -0.4 & 0 & 0.2 \end{bmatrix} \begin{bmatrix} -2 \\ 3 \\ 1 \end{bmatrix} = \begin{bmatrix} 5 \\ -3 \\ 1 \end{bmatrix},$$

or

$$\begin{bmatrix} x \\ y \\ z \end{bmatrix} = \begin{bmatrix} 5 \\ -3 \\ 1 \end{bmatrix},$$

so the solution is $x = 5$, $y = -3$, and $z = 1$, or $(5, -3, 1)$.

When a grapher is used, it is not actually necessary to enter the matrix $\mathbf{A}^{-1}$. After the matrices $\mathbf{A}$ and $\mathbf{B}$ are entered on a grapher, the computation $[\mathbf{A}]^{-1} * [\mathbf{B}]$ is entered and only the result

$$\begin{bmatrix} 5 \\ -3 \\ 1 \end{bmatrix}$$

is displayed.

4.5 Exercise Set

Determine whether $\mathbf{B}$ *is the inverse of* $\mathbf{A}$.

1. $\mathbf{A} = \begin{bmatrix} 1 & -3 \\ -2 & 7 \end{bmatrix}$, $\mathbf{B} = \begin{bmatrix} 7 & 3 \\ 2 & 1 \end{bmatrix}$

2. $\mathbf{A} = \begin{bmatrix} 3 & 2 \\ 4 & 3 \end{bmatrix}$, $\mathbf{B} = \begin{bmatrix} 3 & -2 \\ -4 & 3 \end{bmatrix}$

3. $\mathbf{A} = \begin{bmatrix} -1 & -1 & 6 \\ 1 & 0 & -2 \\ 1 & 0 & -3 \end{bmatrix}$, $\mathbf{B} = \begin{bmatrix} 2 & 3 & 2 \\ 3 & 3 & 4 \\ 1 & 1 & 1 \end{bmatrix}$

4. $\mathbf{A} = \begin{bmatrix} -2 & 0 & -3 \\ 5 & 1 & 7 \\ -3 & 0 & 4 \end{bmatrix}$, $\mathbf{B} = \begin{bmatrix} 4 & 0 & -3 \\ 1 & 1 & 1 \\ -3 & 0 & 2 \end{bmatrix}$

Use the Gauss–Jordan method to find $\mathbf{A}^{-1}$, *if it exists. Check your answers on a grapher by finding* $\mathbf{A}^{-1}\mathbf{A}$ *and* $\mathbf{A}\mathbf{A}^{-1}$.

5. $\mathbf{A} = \begin{bmatrix} 3 & 2 \\ 5 & 3 \end{bmatrix}$

6. $\mathbf{A} = \begin{bmatrix} 3 & 5 \\ 1 & 2 \end{bmatrix}$

7. $\mathbf{A} = \begin{bmatrix} 11 & 3 \\ 7 & 2 \end{bmatrix}$

8. $\mathbf{A} = \begin{bmatrix} 8 & 5 \\ 5 & 3 \end{bmatrix}$

9. $\mathbf{A} = \begin{bmatrix} 6 & 9 \\ 4 & 6 \end{bmatrix}$

10. $\mathbf{A} = \begin{bmatrix} -4 & -6 \\ 2 & 3 \end{bmatrix}$

11. $\mathbf{A} = \begin{bmatrix} 3 & 1 & 0 \\ 1 & 1 & 1 \\ 1 & -1 & 2 \end{bmatrix}$

12. $\mathbf{A} = \begin{bmatrix} 1 & 0 & 1 \\ 2 & 1 & 0 \\ 1 & -1 & 1 \end{bmatrix}$

13. $\mathbf{A} = \begin{bmatrix} 1 & -1 & 2 \\ 0 & 1 & 3 \\ 2 & 1 & -2 \end{bmatrix}$

14. $\mathbf{A} = \begin{bmatrix} 1 & -1 & 2 \\ 0 & 1 & 2 \\ 1 & -3 & -4 \end{bmatrix}$

15. $\mathbf{A} = \begin{bmatrix} 1 & -4 & 8 \\ 1 & -3 & 2 \\ 2 & -7 & 10 \end{bmatrix}$

16. $\mathbf{A} = \begin{bmatrix} -2 & 5 & 3 \\ 4 & -1 & 3 \\ 7 & -2 & 5 \end{bmatrix}$

Use a grapher to find $\mathbf{A}^{-1}$, *if it exists.*

17. $\mathbf{A} = \begin{bmatrix} 4 & -3 \\ 1 & -2 \end{bmatrix}$

18. $\mathbf{A} = \begin{bmatrix} 0 & -1 \\ 1 & 0 \end{bmatrix}$

19. $\mathbf{A} = \begin{bmatrix} 2 & 3 & 2 \\ 3 & 3 & 4 \\ -1 & -1 & -1 \end{bmatrix}$

20. $\mathbf{A} = \begin{bmatrix} 1 & 2 & 3 \\ 2 & -1 & -2 \\ -1 & 3 & 3 \end{bmatrix}$

21. $A = \begin{bmatrix} 1 & 2 & -1 \\ -2 & 0 & 1 \\ 1 & -1 & 0 \end{bmatrix}$

22. $A = \begin{bmatrix} 7 & -1 & -9 \\ 2 & 0 & -4 \\ -4 & 0 & 6 \end{bmatrix}$

23. $A = \begin{bmatrix} 1 & 1 & 0 \\ 1 & 0 & -1 \\ 0 & -1 & 1 \end{bmatrix}$

24. $A = \begin{bmatrix} 1 & 0 & 0 \\ -2 & 1 & 0 \\ 1 & -2 & 1 \end{bmatrix}$

25. $A = \begin{bmatrix} 1 & 3 & -1 \\ 0 & 2 & -1 \\ 1 & 1 & 0 \end{bmatrix}$

26. $A = \begin{bmatrix} -1 & 0 & -1 \\ -1 & 1 & 0 \\ 0 & 1 & 1 \end{bmatrix}$

27. $A = \begin{bmatrix} 1 & 2 & 3 & 4 \\ 0 & 1 & 3 & -5 \\ 0 & 0 & 1 & -2 \\ 0 & 0 & 0 & -1 \end{bmatrix}$

28. $A = \begin{bmatrix} -2 & -3 & 4 & 1 \\ 0 & 1 & 1 & 0 \\ 0 & 4 & -6 & 1 \\ -2 & -2 & 5 & 1 \end{bmatrix}$

29. $A = \begin{bmatrix} 1 & -14 & 7 & 38 \\ -1 & 2 & 1 & -2 \\ 1 & 2 & -1 & -6 \\ 1 & -2 & 3 & 6 \end{bmatrix}$

30. $A = \begin{bmatrix} 10 & 20 & -30 & 15 \\ 3 & -7 & 14 & -8 \\ -7 & -2 & -1 & 2 \\ 4 & 4 & -3 & 1 \end{bmatrix}$

Solve each system of equations using the inverse of the coefficient matrix of the equivalent matrix equation.

31. $\begin{aligned} 4x + 3y &= 2, \\ x - 2y &= 6 \end{aligned}$

32. $\begin{aligned} 2x - 3y &= 7, \\ 4x + y &= -7 \end{aligned}$

33. $\begin{aligned} 5x + y &= 2, \\ 3x - 2y &= -4 \end{aligned}$

34. $\begin{aligned} x - 6y &= 5, \\ -x + 4y &= -5 \end{aligned}$

35. $\begin{aligned} x + y &= 7, \\ 2x - y &= 2 \end{aligned}$

36. $\begin{aligned} 2x + 5y &= -3, \\ 3x + 7y &= -5 \end{aligned}$

37. $\begin{aligned} x \quad\ + z &= 1, \\ 2x + y \quad &= 3, \\ x - y + z &= 4 \end{aligned}$

38. $\begin{aligned} x + 2y + 3z &= -1, \\ 2x - 3y + 4z &= 2, \\ -3x + 5y - 6z &= 4 \end{aligned}$

39. $\begin{aligned} 2x + 3y + 4z &= 2, \\ x - 4y + 3z &= 2, \\ 5x + y + z &= -4 \end{aligned}$

40. $\begin{aligned} x + y \quad\ &= 2, \\ 3x \quad\ + 2z &= 5, \\ 2x + 3y - 3z &= 9 \end{aligned}$

41. $\begin{aligned} x + y + z &= 6, \\ 2x - y - z &= 9, \\ 3x - 2y + z &= -5 \end{aligned}$

42. $\begin{aligned} x + y + z &= 1, \\ 2x - y + 3z &= -2, \\ 3x + 2y - 2z &= 15 \end{aligned}$

43. $\begin{aligned} 2w - 3x + 4y - 5z &= 0, \\ 3w - 2x + 7y - 3z &= 2, \\ w + x - y + z &= 1, \\ -w - 3x - 6y + 4z &= 6 \end{aligned}$

44. $\begin{aligned} 5w - 4x + 3y - 2z &= -6, \\ w + 4x - 2y + 3z &= -5, \\ 2w - 3x + 6y - 9z &= 14, \\ 3w - 5x + 2y - 4z &= -3 \end{aligned}$

45. $\begin{aligned} w + x + y + z &= 0, \\ 2w - 3x + 2y + 5z &= 3, \\ w - 2x - y + 2z &= 7, \\ 3w - x - 2y - 7z &= 5 \end{aligned}$

46. $\begin{aligned} w + x - y + 2z &= -7, \\ 2w - x + y - z &= 7, \\ -w - x + 2y - 4z &= 9, \\ 3w + 2x + y + z &= 0 \end{aligned}$

47. *Sales.* Stefan sold a total of 145 Italian sausages and hot dogs from his curbside pushcart and collected $242.05. He sold 45 more hot dogs than sausages. How many of each did he sell?

48. *Price of School Supplies.* Miranda bought 4 lab record books and 3 highlighters for $13.93. Victor bought 3 lab record books and 2 highlighters for $10.25. Find the price of each item.

49. *Cost.* Evergreen Landscaping bought 4 tons of topsoil, 3 tons of mulch, and 6 tons of pea gravel for $2825. The next week the firm bought 5 tons of topsoil, 2 tons of mulch, and 5 tons of pea gravel for $2663. Pea gravel costs $17 less per ton than topsoil. Find the price per ton for each item.

50. *Investment.* Selena receives $537.75 per year in simple interest from three investments totaling $8500. Part is invested at 5.15%, part at 6.05%, and the rest at 7.2%. There is $1500 more invested at 7.2% than at 5.15%. Find the amount invested at each rate.

Skill Maintenance

Graph.

51. $y = x - 3$

52. $2x + 3y = 6$

53. $y = -5$

54. $x - y = 7,$
$2x - 3y = 15$

Synthesis

55. ◈ For square matrices **A** and **B**, is it true, in general, that $(\mathbf{A} + \mathbf{B})^{-1} = \mathbf{A}^{-1} + \mathbf{B}^{-1}$? Explain.

56. ◈ For square matrices **A** and **B**, is it true, in general, that $(\mathbf{AB})^{-1} = \mathbf{A}^{-1}\mathbf{B}^{-1}$? Explain.

State the conditions under which $\mathbf{A}^{-1}$ exists. Then find a formula for $\mathbf{A}^{-1}$.

57. $\mathbf{A} = [x]$

58. $\mathbf{A} = \begin{bmatrix} x & 0 \\ 0 & y \end{bmatrix}$

59. $\mathbf{A} = \begin{bmatrix} 0 & 0 & x \\ 0 & y & 0 \\ z & 0 & 0 \end{bmatrix}$

60. $\mathbf{A} = \begin{bmatrix} x & 1 & 1 & 1 \\ 0 & y & 0 & 0 \\ 0 & 0 & z & 0 \\ 0 & 0 & 0 & w \end{bmatrix}$

4.6
Systems of Inequalities and Linear Programming

- *Graph linear inequalities.*
- *Graph systems of linear inequalities.*
- *Solve linear programming problems.*

A graph of an inequality is a drawing that represents its solutions. We have already seen that an inequality in one variable can be graphed on a number line. An inequality in two variables can be graphed on a coordinate plane.

Graphs of Linear Inequalities

A statement like $5x - 4y < 20$ is a linear inequality in two variables.

> **Linear Inequality in Two Variables**
>
> A *linear inequality in two variables* is an inequality that can be written in the form
>
> $$Ax + By < C,$$
>
> where A, B, and C are real numbers and A and B are not both zero. The symbol $<$ may be replaced by $\leq$, $>$, or $\geq$.

A solution of a linear inequality in two variables is an ordered pair (x, y) for which the inequality is true. For example, $(1, 3)$ is a solution of $5x - 4y < 20$ because $5 \cdot 1 - 4 \cdot 3 < 20$, or $-7 < 20$, is true. On the other hand, $(2, -6)$ is not a solution of $5x - 4y < 20$ because $5 \cdot 2 - 4 \cdot (-6) \not< 20$.

The **solution set** of an inequality is the set of all the ordered pairs that make it true. A **graph of an inequality** represents its solution set.

<table>
<tr><td>

Interactive Discovery

</td><td>

Graph $y_1 = x - 4$. Select three or four points (x, y) in the region below the graph and compare the values of y_1 and $x - 4$ for each point. Do the same for three or four points in the region above the graph. Which points satisfy the inequality $y_1 < x - 4$? Shade that region. (Most graphers are capable of shading a region above or below a graph or between two graphs. Consult your user's manual for the keystrokes to use.) What inequality is satisfied by the points in the region that is not shaded?

</td></tr>
</table>

Example 1 Graph: $y < x + 3$.

SOLUTION We begin by graphing the related equation $y = x + 3$. We use a dashed line because the inequality symbol is $<$. This indicates that the line itself is not in the solution set of the inequality.

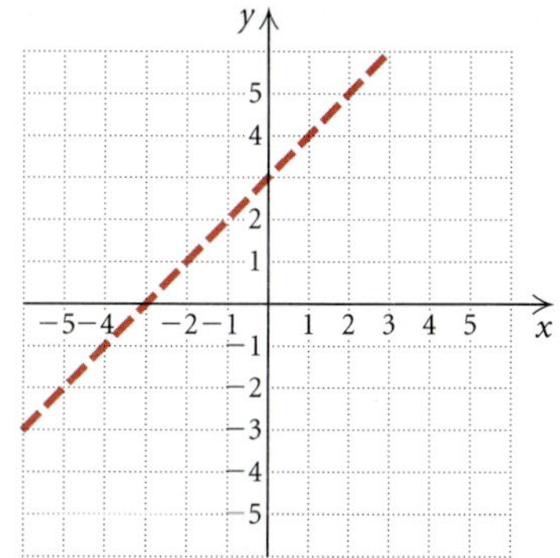

Note that the line divides the coordinate plane into two regions called **half-planes**, one of which satisfies the inequality. Either *all* points in a half-plane are in the solution set of the inequality or *none* are.

To determine which half-plane satisfies the inequality, we try a test point in either region. The point $(0, 0)$ is usually a convenient choice so long as it does not lie on the line.

$$y < x + 3$$
$$0 \;?\; 0 + 3$$
$$0 \;|\; 3 \qquad \text{TRUE}$$

Since $(0, 0)$ satisfies the inequality, so do all points in the half-plane that contains $(0, 0)$. We shade this region to show the solution set of the inequality.

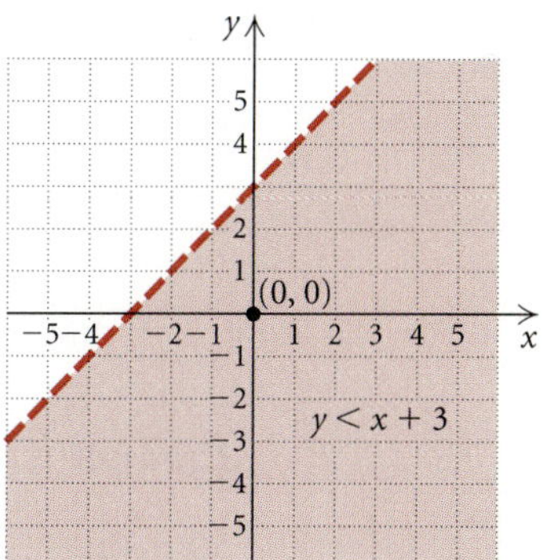

In general, we use the following procedure to graph linear inequalities in two variables.

To graph a linear inequality in two variables:

1. Replace the inequality symbol with an equals sign and graph this related equation. If the inequality symbol is $<$ or $>$, draw the line dashed. If the inequality symbol is $\leq$ or $\geq$, draw the line solid.

2. The graph consists of a half-plane on one side of the line and, if the line is solid, the line as well. To determine which half-plane to shade, test a point not on the line in the original inequality. If that point is a solution, shade the half-plane containing that point. If not, shade the opposite half-plane.

Example 2   Graph: $3x + 4y \geq 12$.

SOLUTION

1. First we graph the related equation $3x + 4y = 12$. We use a solid line because the inequality symbol is $\geq$. This indicates that the line is included in the solution set.

2. To determine which half-plane to shade, we test a point in either region. We choose $(0, 0)$.

$$3x + 4y \geq 12$$
$$3 \cdot 0 + 4 \cdot 0 \ ? \ 12$$
$$0 \ | \ 12 \quad \text{FALSE}$$

Because $(0, 0)$ is *not* a solution, all the points in the half-plane that does *not* contain $(0, 0)$ are solutions. We shade that region.

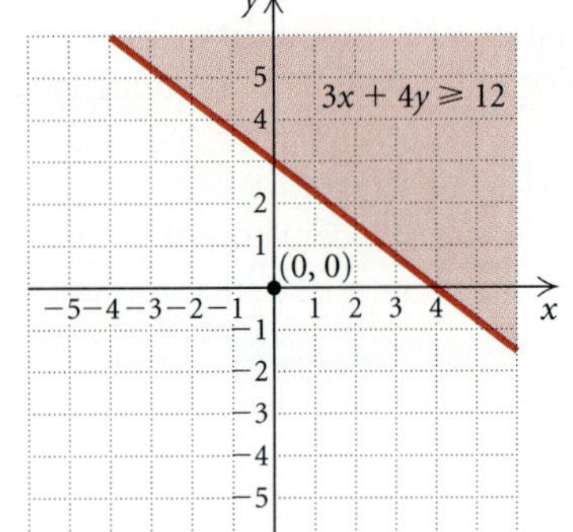

This inequality can also be graphed using a grapher with a SHADE feature.

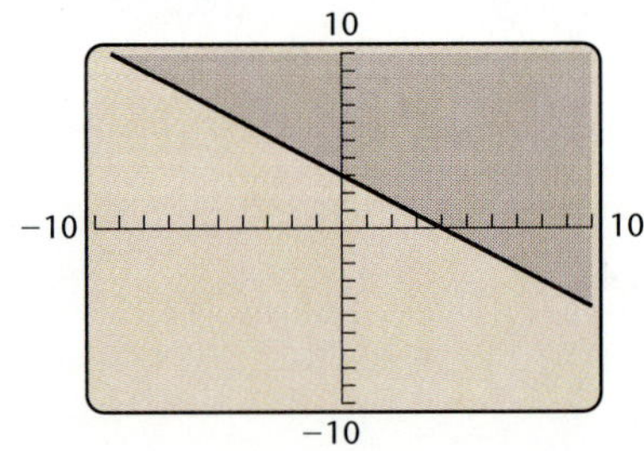

Example 3   Graph $x > -3$ on a plane.

SOLUTION

1. First we graph the related equation $x = -3$. We use a dashed line because the inequality symbol is $>$. This indicates that the line is not included in the solution set.

2. The inequality tells us that all points (x, y) for which $x > -3$ are so-lutions. These are the points to the right of the line. We can also use a test point to determine the solutions. We choose $(5, 1)$.

$$\frac{x > -3}{}$$

5 ? −3   TRUE

Because $(5, 1)$ is a solution, we shade the region containing that point—that is, the region to the right of the dashed line.

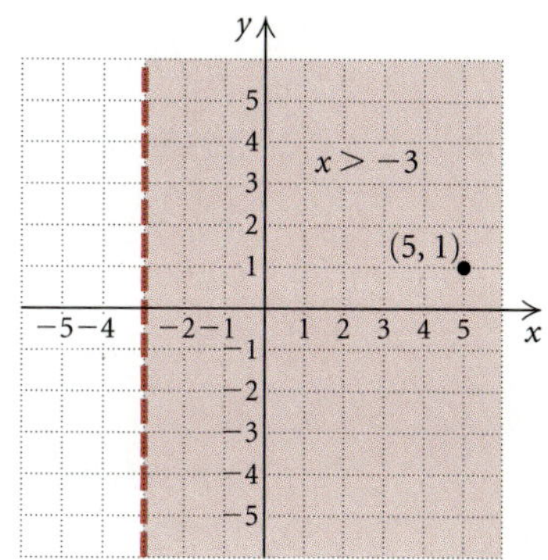

Example 4   Graph $y \leq 4$ on a plane.

SOLUTION

1. First we graph the related equation $y = 4$. We use a solid line because the inequality symbol is $\leq$.

2. The inequality tells us that all points (x, y) for which $y \leq 4$ are solu-tions of the inequality. These are the points below the line. We can also use a test point to determine the solutions. We choose $(-2, 5)$.

$$\frac{y \leq 4}{}$$

5 ? 4   FALSE

Because $(-2, 5)$ is not a solution, we shade the half-plane that does not contain that point.

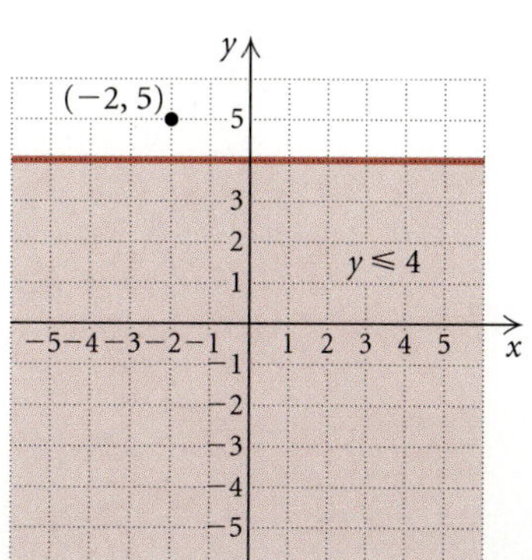

We can also use a grapher with the SHADE feature to graph this inequality.

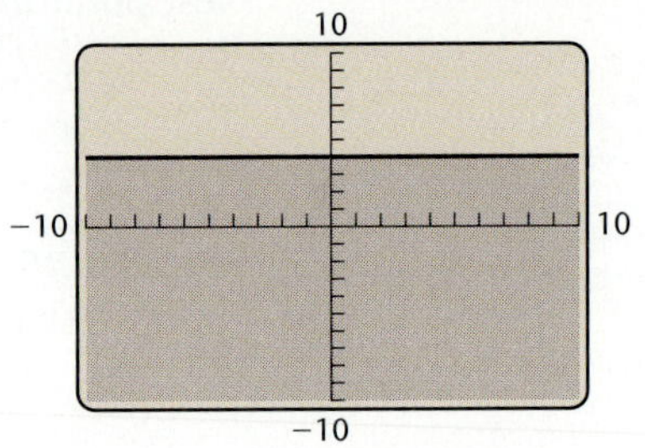

Systems of Linear Inequalities

A system of inequalities in two variables consists of two or more inequalities in two variables considered simultaneously. For example,

$$x + y \le 4,$$
$$x - y \ge 2$$

is a system of two *linear* inequalities in two variables.

A solution of a system of inequalities is an ordered pair that is a solution of each inequality in the system. To graph a system of linear inequalities, we graph each inequality and determine the region that is common to all the solution sets.

Example 5 Graph the solution set of the system

$$x + y \le 4,$$
$$x - y \ge 2.$$

SOLUTION We graph $x + y \le 4$ by first graphing the equation $x + y = 4$ using a solid line. Next, we choose $(0, 0)$ as a test point and find that it is a solution of $x + y \le 4$, so we shade the half-plane containing $(0, 0)$ using red. Next we graph $x - y = 2$ using a solid line. We find that $(0, 0)$ is not a solution of $x - y \ge 2$, so we shade the half-plane that does not contain $(0, 0)$ using blue. The arrows at the ends of each line help to indicate the half-plane that contains each solution set.

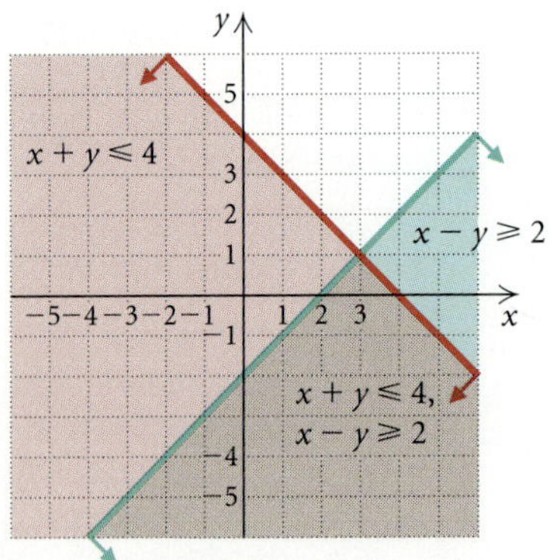

The solution set of the system of equations is the region shaded both blue and red, or purple, including parts of the lines $x + y = 4$ and $x - y = 2$.

A system of inequalities may have a graph that consists of a polygon and its interior. As we will see later in this section, it is important in many applications to be able to find the vertices of such a polygon.

Example 6   Graph the following system of inequalities and find the coordinates of any vertices formed:

$$3x - y \leq 6, \quad (1)$$
$$y - 3 \leq 0, \quad (2)$$
$$x + y \geq 0. \quad (3)$$

SOLUTION   We graph the related equations $3x - y = 6$, $y - 3 = 0$, and $x + y = 0$ using solid lines. The half-plane containing the solution set for each inequality is indicated by the arrows at the ends of each line. We shade the region common to all three solution sets.

To find the vertices, we solve three systems of equations. The system of equations from inequalities (1) and (2) is

$$3x - y = 6,$$
$$y - 3 = 0.$$

Solving, we obtain the vertex $(3, 3)$.

The system of equations from inequalities (1) and (3) is

$$3x - y = 6,$$
$$x + y = 0.$$

Solving, we obtain the vertex $\left(\frac{3}{2}, -\frac{3}{2}\right)$.

The system of equations from inequalities (2) and (3) is

$$y - 3 = 0,$$
$$x + y = 0.$$

Solving, we obtain the vertex $(-3, 3)$.

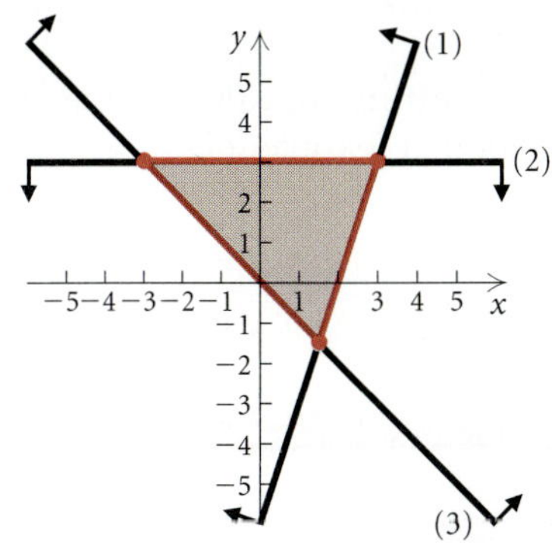

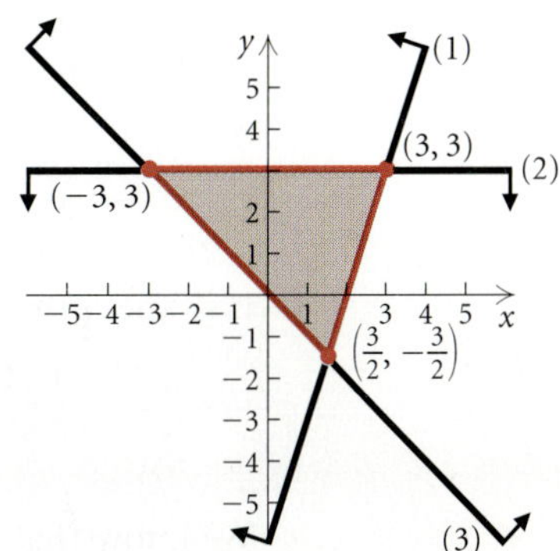

We could also graph the system of related equations on a grapher and use the INTERSECT feature or the TRACE and ZOOM features to find the vertices. However, most graphers are not capable of shading a region determined by three or more graphs like the one above.

Applications: Linear Programming

In many applications, we want to find a maximum or minimum value. In business, for example, we might want to maximize profit and minimize cost. **Linear programming** can tell us how to do this.

In our study of linear programming, we will consider linear functions of two variables that are to be maximized or minimized subject to several conditions, or **constraints**. These constraints are expressed as inequalities. The solution set of the system of inequalities made up of the constraints contains all the **feasible solutions** of a linear programming problem. The function that we want to maximize or minimize is called the **objective function**.

It can be shown that the maximum and minimum values of the objective function occur at a vertex of the region of feasible solutions. Thus we have the following procedure.

Linear Programming Procedure

To find the maximum or minimum value of a linear objective function subject to a set of constraints:

1. Graph the region of feasible solutions.

2. Determine the coordinates of the vertices of the region.

3. Evaluate the objective function at each vertex. The largest and smallest of those values are the maximum and minimum values of the function, respectively.

Example 7 *Maximizing Profit.* Dovetail Carpentry Shop makes bookcases and desks. Each bookcase requires 5 hr of woodworking and 4 hr of finishing. Each desk requires 10 hr of woodworking and 3 hr of finishing. Each month the shop has 600 hr of labor available for woodworking and 240 hr for finishing. The profit on each bookcase is $40 and on each desk is $75. How many of each product should be made each month in order to maximize profit?

SOLUTION We let $x =$ the number of bookcases to be produced and $y =$ the number of desks. Then the profit P is given by the function

$$P = 40x + 75y.$$

To emphasize that P is a function of two variables, we sometimes write $P(x, y) = 40x + 75y.$

We know that x bookcases require $5x$ hr of woodworking and y desks require $10y$ hr of woodworking. Since there is no more than 600 hr of labor available for woodworking, we have one constraint:

$$5x + 10y \le 600.$$

Similarly, the bookcases and desks require $4x$ hr and $3y$ hr of finishing, respectively. There is no more than 240 hr of labor available for finishing, so we have a second constraint:

$$4x + 3y \le 240.$$

We also know that $x \ge 0$ and $y \ge 0$ because the carpentry shop cannot make a negative number of either product.

Thus we want to maximize the objective function

$$P = 40x + 75y$$

subject to the constraints

$$5x + 10y \le 600,$$
$$4x + 3y \le 240,$$
$$x \ge 0,$$
$$y \ge 0.$$

We graph the system of inequalities and determine the vertices.

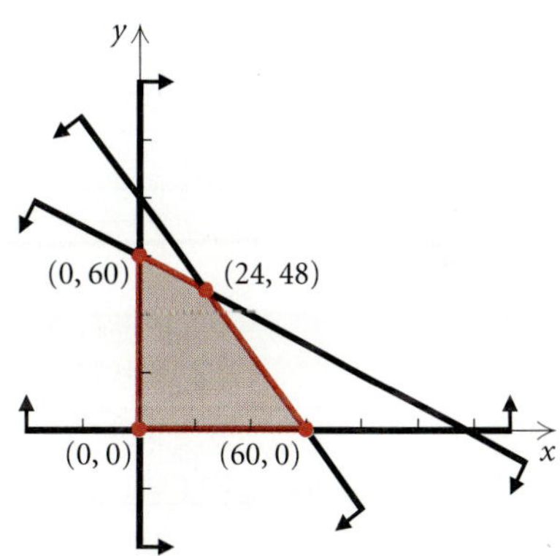

Now we evaluate the objective function P at each vertex.

Vertices (x, y)	Profit $P = 40x + 75y$	
$(0, 0)$	$P = 40 \cdot 0 + 75 \cdot 0 = 0$	
$(60, 0)$	$P = 40 \cdot 60 + 75 \cdot 0 = 2400$	
$(24, 48)$	$P = 40 \cdot 24 + 75 \cdot 48 = 4560$	⟵ Maximum
$(0, 60)$	$P = 40 \cdot 0 + 75 \cdot 60 = 4500$	

The carpentry shop will make a maximum profit of \$4560 when 24 bookcases and 48 desks are produced.

4.6 Exercise Set

In Exercises 1–8, match each inequality with one of the graphs (a)–(h), which follow.

a)
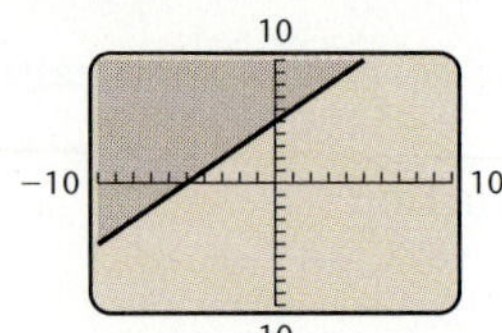

b)
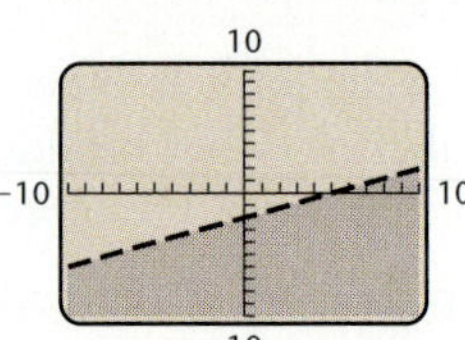

c)
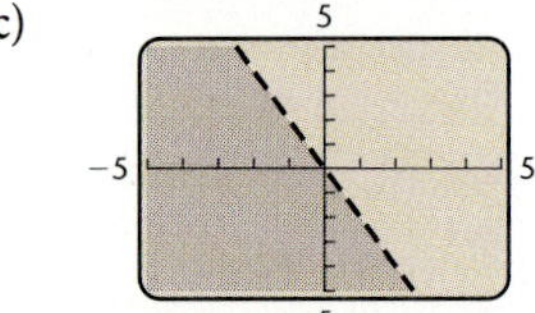

d)
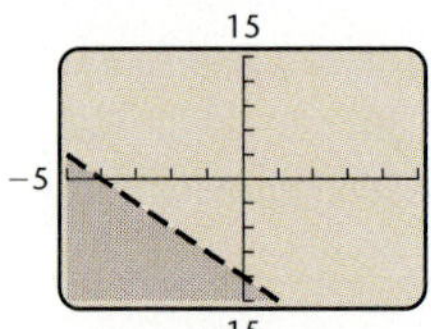

e)
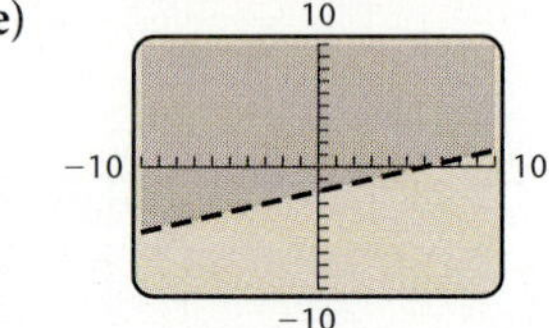

f)
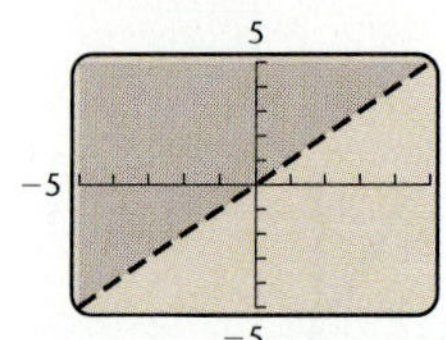

g)
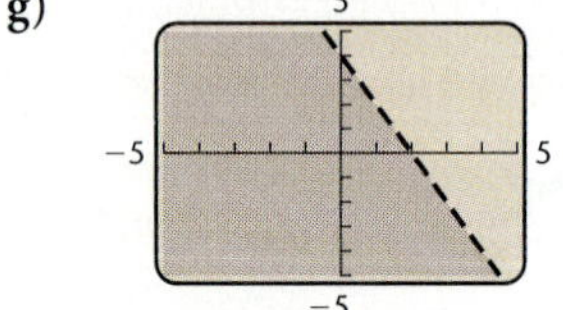

h)
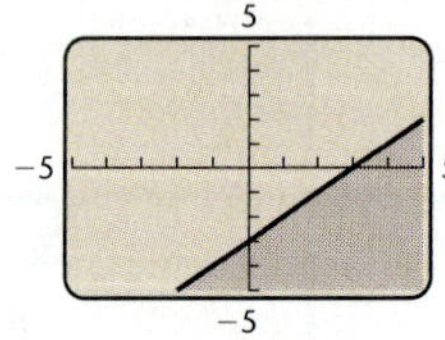

1. $y > x$

2. $y < -2x$

3. $y \leq x - 3$

4. $y \geq x + 5$

5. $2x + y < 4$

6. $3x + y < -12$

7. $2x - 5y > 10$

8. $3x - 9y < 18$

Graph.

9. $y > 2x$

10. $2y < x$

11. $y + x \geq 0$

12. $y - x < 0$

13. $y > x - 3$

14. $y \leq x + 4$

15. $x + y < 4$

16. $x - y \geq 5$

17. $3x - 2y \leq 6$

18. $2x - 5y < 10$

19. $3y + 2x \geq 6$

20. $2y + x \leq 4$

21. $3x - 2 \leq 5x + y$

22. $2x - 6y \geq 8 + 2y$

23. $x < -4$

24. $y \geq 5$

25. $y > -3$

26. $x \leq 5$

27. $-4 < y < -1$
(*Hint*: Think of this as $-4 < y$ and $y < -1$.)

28. $-3 < x < 3$
(*Hint*: Think of this as $-3 < x$ and $x < 3$.)

29. $y \geq |x|$

30. $y \leq |x + 2|$

In Exercises 31–36, match each system of inequalities with one of the graphs (a)–(f), which follow.

a)
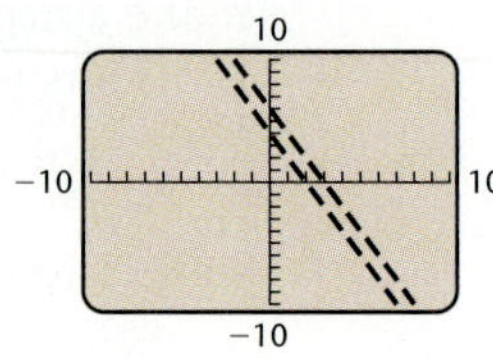

b)
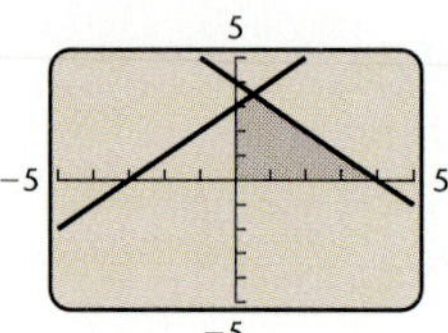

c)
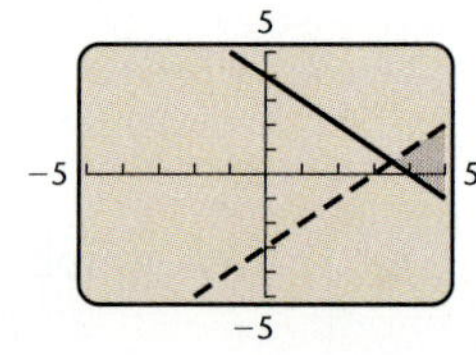

d)
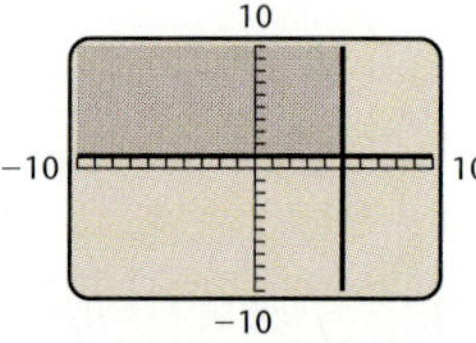

e)
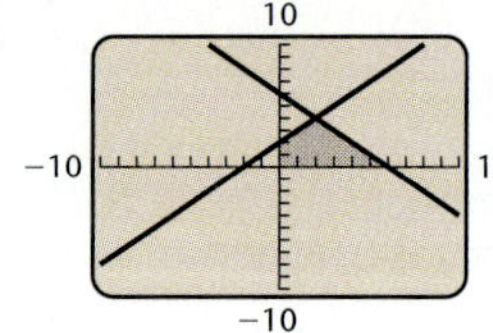

f)
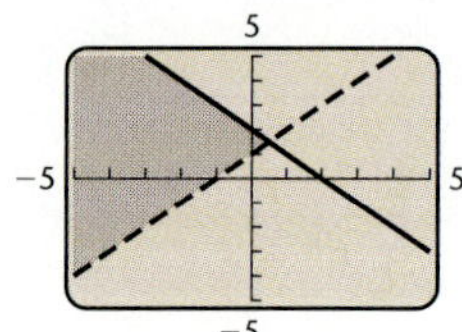

31. $y > x + 1,$
$y \leq 2 - x$

32. $y < x - 3,$
$y \geq 4 - x$

33. $2x + y < 4,$
$4x + 2y > 12$

34. $x \leq 5,$
$y \geq 1$

35. $x + y \leq 4,$
$x - y \geq -3,$
$x \geq 0,$
$y \geq 0$

36. $x - y \geq -2,$
$x + y \leq 6,$
$x \geq 0,$
$y \geq 0$

Graph the system of inequalities. Then find the coordinates of the vertices.

37. $y \leq x,$
$y \geq 3 - x$

38. $y \geq x,$
$y \leq x - 5$

39. $y \geq x,$
$y \leq x - 4$

40. $y \geq x,$
$y \leq 2 - x$

41. $y \geq -3,$
$x \geq 1$

42. $y \leq -2,$
$x \geq 2$

43. $x \leq 3,$
$y \geq 2 - 3x$

44. $x \geq -2,$
$y \leq 3 - 2x$

45. $x + y \leq 1,$
$x - y \leq 2$

46. $y + 3x \geq 0,$
$y + 3x \leq 2$

47. $2y - x \leq 2,$
$\quad y + 3x \geq -1$

48. $\quad\quad y \leq 2x + 1,$
$\quad\quad y \geq -2x + 1,$
$\quad\quad x - 2 \leq 0$

49. $x - y \leq 2,$
$\quad x + 2y \geq 8,$
$\quad y - 4 \leq 0$

50. $x + 2y \leq 12,$
$\quad 2x + y \leq 12,$
$\quad\quad x \geq 0,$
$\quad\quad y \geq 0$

51. $4y - 3x \geq -12,$
$\quad 4y + 3x \geq -36,$
$\quad\quad\quad y \leq 0,$
$\quad\quad\quad x \leq 0$

52. $8x + 5y \leq 40,$
$\quad x + 2y \leq 8,$
$\quad\quad x \geq 0,$
$\quad\quad y \geq 0$

53. $3x + 4y \geq 12,$
$\quad 5x + 6y \leq 30,$
$\quad\quad 1 \leq x \leq 3$

54. $y - x \geq 1,$
$\quad y - x \leq 3,$
$\quad 2 \leq x \leq 5$

Find the maximum and the minimum values of the function and the values of x and y for which they occur.

55. $P = 17x - 3y + 60,$ subject to
$$6x + 8y \leq 48,$$
$$0 \leq y \leq 4,$$
$$0 \leq x \leq 7.$$

56. $Q = 28x - 4y + 72,$ subject to
$$5x + 4y \geq 20,$$
$$0 \leq y \leq 4,$$
$$0 \leq x \leq 3.$$

57. $F = 5x + 36y,$ subject to
$$5x + 3y \leq 34,$$
$$3x + 5y \leq 30,$$
$$x \geq 0,$$
$$y \geq 0.$$

58. $G = 16x + 14y,$ subject to
$$3x + 2y \leq 12,$$
$$7x + 5y \leq 29,$$
$$x \geq 0,$$
$$y \geq 0.$$

59. *Maximizing Income.* Golden Harvest Foods makes jumbo biscuits and regular biscuits. The oven can cook at most 200 biscuits per day. Each jumbo biscuit requires 2 oz of flour, each regular biscuit requires 1 oz of flour, and there is 300 oz of flour available. The income from each jumbo biscuit is $0.10 and from each regular biscuit is $0.08. How many of each size biscuit should be made in order to maximize income? What is the maximum income?

60. *Maximizing Mileage.* Omar owns a car and a moped. He can afford 12 gal of gasoline to be split between the car and the moped. Omar's car gets 20 mpg and holds at most 10 gal of gas. His moped gets 100 mpg and holds at most 3 gal of gas. How many gallons of gasoline should each vehicle use if Omar wants to travel as far as possible? What is the maximum number of miles that he can travel?

61. *Maximizing Profit.* Norris Mill can convert logs into lumber and plywood. In a given week, the mill can turn out 400 units of production, of which 100 units of lumber and 150 units of plywood are required by regular customers. The profit is $20 per unit of lumber and $30 per unit of plywood. How many units of each should the mill produce in order to maximize the profit?

62. *Maximizing Profit.* Sunnydale Farm includes 240 acres of cropland. The farm owner wishes to plant this acreage in corn and oats. The profit per acre in corn production is $40 and in oats is $30. A total of 320 hr of labor is available. Each acre of corn requires 2 hr of labor, whereas each acre of oats requires 1 hr of labor. How should the land be divided between corn and oats in order to yield the maximum profit? What is the maximum profit?

63. *Minimizing Cost.* An animal feed to be mixed from soybean meal and oats must contain at least 120 lb of protein, 24 lb of fat, and 10 lb of mineral ash. Each 100-lb sack of soybean meal costs $15 and contains 50 lb of protein, 8 lb of fat, and 5 lb of mineral ash. Each 100-lb sack of oats costs $5 and contains 15 lb of protein, 5 lb of fat, and 1 lb of mineral ash. How many sacks of each should be used to satisfy the minimum requirements at minimum cost?

64. *Minimizing Cost.* Suppose that in the preceding problem the oats were replaced by alfalfa, which costs $8 per 100 lb and contains 20 lb of protein, 6 lb of fat, and 8 lb of mineral ash. How much of each is now required in order to minimize the cost?

65. *Maximizing Income.* Clayton is planning to invest up to $40,000 in corporate and municipal bonds. The least he is allowed to invest in corporate bonds is $6000, and he does not want to invest more than $22,000 in corporate bonds. He also does not want to invest more than $30,000 in municipal bonds. The interest is 8% on corporate bonds and $7\frac{1}{2}$% on municipal bonds. This is simple interest for one year. How much should he invest in each type of bond in order to maximize his income? What is the maximum income?

66. *Maximizing Income.* Margaret is planning to invest up to $22,000 in certificates of deposit at City Bank and People's Bank. She wants to invest at least $2000 but no more than $14,000 at City Bank. People's Bank does not insure more than a $15,000 investment, so she will invest no more than that in People's Bank. The interest is 6% at City Bank and $6\frac{1}{2}$% at People's Bank. This is simple interest for one year. How much should she invest in each bank in order to maximize her income? What is the maximum income?

67. *Minimizing Transportation Cost.* An airline with two types of airplanes, P_1 and P_2, has contracted with a tour group to provide transportation for a minimum of 2000 first-class, 1500 tourist-class, and 2400 economy-class passengers. For a certain trip, airplane P_1 costs $12 thousand to operate and can accommodate 40 first-class, 40 tourist-class, and 120 economy-class passengers, whereas airplane P_2 costs $10 thousand to operate and can accommodate 80 first-class, 30 tourist-class, and 40 economy-class passengers. How many of each type of airplane should be used in order to minimize the operating cost?

68. *Minimizing Transportation Cost.* Suppose that in the preceding problem a new airplane P_3 becomes available, having an operating cost for the same trip of $15 thousand and accommodating 40 first-class, 40 tourist-class, and 80 economy-class passengers. If airplane P_1 were replaced by airplane P_3, how many of P_2 and P_3 should be used in order to minimize the operating cost?

69. *Maximizing Profit.* It takes Just Sew 2 hr of cutting and 4 hr of sewing to make a knit suit. It takes 4 hr of cutting and 2 hr of sewing to make a worsted suit. At most 20 hr per day is available for cutting and at most 16 hr per day is available for sewing. The profit is $34 on a knit suit and $31 on a worsted suit. How many of each kind of suit should be made each day in order to maximize profit? What is the maximum profit?

70. *Maximizing Profit.* Cambridge Metal Works manufactures two sizes of gears. The smaller gear requires 4 hr of machining and 1 hr of polishing and yields a profit of $25. The larger gear requires 1 hr of machining and 1 hr of polishing and yields a profit of $10. The firm has available at most 24 hr per day for machining and 9 hr per day for polishing. How many of each type of gear should be produced each day in order to maximize profit? What is the maximum profit?

71. *Minimizing Nutrition Cost.* Suppose it takes 12 units of carbohydrates and 6 units of protein to satisfy Jacob's minimum weekly requirements. A particular type of meat contains 2 units of carbohydrates and 2 units of protein per pound. A particular cheese contains 3 units of carbohydrates and 1 unit of protein per pound. The meat costs $3.50 per pound and the cheese costs $4.60 per pound. How many pounds of each are needed in order to minimize the cost and still meet the minimum requirements?

72. *Minimizing Salary Cost.* The Spring Hill school board is analyzing education costs for Hill Top School. It wants to hire teachers and teacher's aides to make up a faculty that satisfies its needs at minimum cost. The average annual salary for a teacher is $35,000 and for a teacher's aide is $18,000. The school building can accommodate a faculty of no more than 50 but needs at least 20 faculty members to function properly. The school must have at least 12 aides, but the number of teachers must be at least twice the number of aides in order to accommodate the expectations of the community. How many teachers and teacher's aides should be hired in order to minimize salary costs?

73. *Maximizing Animal Support in a Forest.* A certain area of forest is populated by two species of animals, which scientists refer to as A and B for simplicity. The forest supplies two kinds of food, referred to as F_1 and F_2. For one year, species A requires 1 unit of F_1 and 0.5 unit of F_2. Species B requires 0.2 unit of F_1 and 1 unit of F_2. The forest can normally supply at most 600 units of F_1 and 525 units of F_2 per year. What is the maximum total number of these animals that the forest can support?

74. *Maximizing Animal Support in a Forest.* In reference to Exercise 73, if there is a wet spring, then supplies of food increase to 1080 units of F_1 and 810 units of F_2. In this case, what is the maximum total number of these animals that the forest can support?

Skill Maintenance

Factor.

75. $x^2 - 2x - 15$

76. $2x^2 + x - 3$

Add.

77. $\dfrac{3}{x + 1} + \dfrac{2}{x - 4}$

78. $\dfrac{4}{x - 2} + \dfrac{1}{x^2 + 3}$

Synthesis

79. ◆ Write an applied linear programming problem for a classmate to solve. Devise your problem so that the answer is "The bakery will make a maximum profit when 5 dozen pies and 12 dozen cookies are baked."

80. ◆ Describe how the graph of a linear inequality differs from the graph of a linear equation.

Graph the system of inequalities.

81. $y \geq x^2 - 2$, $y \leq 2 - x^2$

82. $y < x + 1$, $y \geq x^2$

Graph the inequality.

83. $|x + y| \leq 1$

84. $|x| + |y| \leq 1$

85. $|x| > |y|$

86. $|x - y| > 0$

87. *Allocation of Resources.* Significant Sounds manufactures two types of speaker assemblies. The less expensive assembly, which sells for \$350, consists of one midrange speaker and one tweeter. The more expensive speaker assembly, which sells for \$600, consists of one woofer, one midrange speaker, and two tweeters. The manufacturer has in stock 44 woofers, 60 midrange speakers, and 90 tweeters. How many of each type of speaker assembly should be made in order to maximize income? What is the maximum income?

88. *Allocation of Resources.* Sitting Pretty Furniture produces chairs and sofas. The chairs require 20 ft of wood, 1 lb of foam rubber, and 2 yd^2 of fabric. The sofas require 100 ft of wood, 50 lb of foam rubber, and 20 yd^2 of fabric. The manufacturer has in stock 1900 ft of wood, 500 lb of foam rubber, and 240 yd^2 of fabric. The chairs can be sold for \$80 each and the sofas for \$300 each. How many of each should be produced in order to maximize income? What is the maximum income?

4.7
Partial Fractions

• *Decompose rational expressions into partial fractions.*

There are situations in calculus in which it is useful to write a rational expression as a sum of two or more simpler rational expressions. For example, in the equation

$$\frac{4x - 13}{2x^2 + x - 6} = \frac{3}{x + 2} + \frac{-2}{2x - 3},$$

each fraction on the right side is called a **partial fraction**. The expression on the right side is the **partial fraction decomposition** of the rational expression on the left side. In this section, we learn how such decompositions are created.

Interactive Discovery

Look at the table of values for

$$y_1 = \frac{13x + 5}{3x^2 - 7x - 6} \quad \text{and} \quad y_2 = \frac{1}{3x + 2} + \frac{4}{x - 3}.$$

Then do the same for the following pairs of equations.

$$y_1 = \frac{3x^2 - 3x - 2}{(x + 1)(x - 1)^2}, \qquad y_2 = \frac{1}{x + 1} + \frac{2}{x - 1} + \frac{-1}{(x - 1)^2};$$

$$y_1 = \frac{2x^2 + 4x + 5}{(x^2 + 1)(x + 2)}, \qquad y_2 = \frac{x + 2}{x^2 + 1} + \frac{1}{x + 2}$$

What do you observe about the relationship between each pair of equations?

Partial Fraction Decompositions

The procedure for finding the partial fraction decomposition of a rational expression involves factoring its denominator into linear and quadratic factors.

Procedure for Decomposing a Rational Expression into Partial Fractions

Consider any rational expression $P(x)/Q(x)$ such that $P(x)$ and $Q(x)$ have no common factor other than 1 or -1.

1. If the degree of $P(x)$ is greater than or equal to the degree of $Q(x)$, divide to express $P(x)/Q(x)$ as a quotient $+$ remainder/$Q(x)$ and follow steps (2)–(5) to decompose the resulting rational expression.

2. If the degree of $P(x)$ is less than the degree of $Q(x)$, factor $Q(x)$ into linear factors of the form $(px + q)^n$ and/or quadratic factors of the form $(ax^2 + bx + c)^m$. Any quadratic factor $ax^2 + bx + c$ must be *irreducible*, meaning that it cannot be factored into linear factors with real coefficients.

3. Assign to each linear factor $(px + q)^n$ the sum of n partial fractions:

$$\frac{A_1}{px + q} + \frac{A_2}{(px + q)^2} + \cdots + \frac{A_n}{(px + q)^n}.$$

4. Assign to each quadratic factor $(ax^2 + bx + c)^m$ the sum of m partial fractions:

$$\frac{B_1 x + C_1}{ax^2 + bx + c} + \frac{B_2 x + C_2}{(ax^2 + bx + c)^2} + \cdots + \frac{B_m x + C_m}{(ax^2 + bx + c)^m}.$$

5. Apply algebraic methods, as illustrated in the following examples, to find the constants in the numerators of the partial fractions.

Example 1   Decompose into partial fractions:

$$\frac{4x - 13}{2x^2 + x - 6}.$$

SOLUTION   The degree of the numerator is less than the degree of the denominator. We begin by factoring the denominator: $(x + 2)(2x - 3)$. We know that there are constants A and B such that

$$\frac{4x - 13}{(x + 2)(2x - 3)} = \frac{A}{x + 2} + \frac{B}{2x - 3}.$$

To determine A and B, we add the expressions on the right:

$$\frac{4x - 13}{(x + 2)(2x - 3)} = \frac{A(2x - 3) + B(x + 2)}{(x + 2)(2x - 3)}.$$

Next, we equate the numerators:

$$4x - 13 = A(2x - 3) + B(x + 2).$$

Since the last equation containing A and B is true for all x, we can substitute any value of x and still have a true equation. In order to have $2x - 3 = 0$, we choose $x = \frac{3}{2}$. This gives us

$$4\left(\tfrac{3}{2}\right) - 13 = A\left(2 \cdot \tfrac{3}{2} - 3\right) + B\left(\tfrac{3}{2} + 2\right)$$
$$-7 = 0 + \tfrac{7}{2}B.$$

Solving, we obtain $B = -2$.

In order to have $x + 2 = 0$, we choose $x = -2$, which gives us

$$4(-2) - 13 = A[2(-2) - 3] + B(-2 + 2).$$

Solving, we obtain $A = 3$.

The decomposition is as follows:

$$\frac{4x - 13}{2x^2 + x - 6} = \frac{3}{x + 2} + \frac{-2}{2x - 3}, \quad \text{or} \quad \frac{3}{x + 2} - \frac{2}{2x - 3}.$$

To check, we can add to see if we get the expression on the left. We can also perform a partial check by graphing

$$y_1 = \frac{4x - 13}{2x^2 + x - 6} \quad \text{and} \quad y_2 = \frac{3}{x + 2} - \frac{2}{2x - 3}.$$

If the graphs are identical, the decomposition is probably correct. As we see in the following figure, the graphs appear to be the same.

We can also use the TABLE feature, comparing values of

$$y_1 = \frac{4x - 13}{2x^2 + x - 6} \quad \text{and} \quad y_2 = \frac{3}{x + 2} - \frac{2}{2x - 3}$$

for the same values of x. Yet another check using a grapher is accomplished by graphing

$$y = \frac{4x - 13}{2x^2 + x - 6} - \left(\frac{3}{x + 2} - \frac{2}{2x - 3}\right).$$

If the partial fraction decomposition is correct, the right-hand side of the equation will be 0 and the graph will be $y = 0$, or the x-axis. ▬

Example 2 Decompose into partial fractions:

$$\frac{7x^2 - 29x + 24}{(2x - 1)(x - 2)^2}.$$

SOLUTION   The degree of the numerator is less than the degree of the denominator. The decomposition has the following form:

$$\frac{7x^2 - 29x + 24}{(2x - 1)(x - 2)^2} = \frac{A}{2x - 1} + \frac{B}{x - 2} + \frac{C}{(x - 2)^2}.$$

As in Example 1, we add and equate the numerators. This gives us

$$7x^2 - 29x + 24 = A(x - 2)^2 + B(2x - 1)(x - 2) + C(2x - 1).$$

Since the equation containing A, B, and C is true for all x, we can substitute any value of x and still have a true equation. In order to have $2x - 1 = 0$, we let $x = \frac{1}{2}$. This gives us

$$7\left(\tfrac{1}{2}\right)^2 - 29 \cdot \tfrac{1}{2} + 24 = A\left(\tfrac{1}{2} - 2\right)^2 + 0.$$

Solving we obtain $A = 5$.

In order to have $x - 2 = 0$, we let $x = 2$. Substituting gives us

$$7(2)^2 - 29(2) + 24 = 0 + C(2 \cdot 2 - 1).$$

Solving, we obtain $C = -2$.

To find B, we choose any value for x except $\frac{1}{2}$ or 2 and replace A with 5 and C with -2. We let $x = 1$:

$$\begin{aligned}
7 \cdot 1^2 - 29 \cdot 1 + 24 &= 5(1 - 2)^2 + B(2 \cdot 1 - 1)(1 - 2) \\
&\quad + (-2)(2 \cdot 1 - 1) \\
2 &= 5 - B - 2 \\
B &= 1.
\end{aligned}$$

The decomposition is as follows:

$$\frac{7x^2 - 29x + 24}{(2x - 1)(x - 2)^2} = \frac{5}{2x - 1} + \frac{1}{x - 2} - \frac{2}{(x - 2)^2}.$$

We can check the result on a grapher using one of the methods discussed in Example 1. Here we let

$$y_1 = \frac{7x^2 - 29x + 24}{(2x - 1)(x - 2)^2} \quad \text{and} \quad y_2 = \frac{5}{2x - 1} + \frac{1}{x - 2} - \frac{2}{(x - 2)^2}$$

and check a table of values.

X	Y₁	Y₂
−5	−.6382	−.6382
−4	−.7778	−.7778
−3	−.9943	−.9943
−2	−1.375	−1.375
−1	−2.222	−2.222
0	−6	−6
1	2	2

X = −5

Since $y_1 = y_2$ for given values of x as we scroll through the table, the decomposition appears to be correct.

Example 3 Decompose into partial fractions:

$$\frac{6x^3 + 5x^2 - 7}{3x^2 - 2x - 1}.$$

SOLUTION The degree of the numerator is greater than that of the denominator. Thus we divide and find an equivalent expression:

$$
\begin{array}{r}
2x + 3 \\
3x^2 - 2x - 1{\overline{\smash{\big)}\,6x^3 + 5x^2 \phantom{{}- 2x} - 7}} \\
\underline{6x^3 - 4x^2 - 2x \phantom{{}- 7}} \\
9x^2 + 2x - 7 \\
\underline{9x^2 - 6x - 3} \\
8x - 4
\end{array}
$$

The original expression is thus equivalent to

$$2x + 3 + \frac{8x - 4}{3x^2 - 2x - 1}.$$

We decompose the fraction to get

$$\frac{8x - 4}{(3x + 1)(x - 1)} = \frac{5}{3x + 1} + \frac{1}{x - 1}.$$

The final result is

$$2x + 3 + \frac{5}{3x + 1} + \frac{1}{x - 1}.$$

We can check the result on a grapher using one of the methods discussed in Example 1. Here we graph

$$y = \frac{6x^3 + 5x^2 - 7}{3x^2 - 2x - 1} - \left(2x + 3 + \frac{5}{3x + 1} + \frac{1}{x - 1}\right).$$

The graph appears to be $y = 0$, or the x-axis, so the decomposition is probably correct.

Systems of equations can be used to decompose rational expressions. Let's reconsider Example 2.

Example 4 Decompose into partial fractions:

$$\frac{7x^2 - 29x + 24}{(2x - 1)(x - 2)^2}.$$

SOLUTION The decomposition has the following form:

$$\frac{A}{2x - 1} + \frac{B}{x - 2} + \frac{C}{(x - 2)^2}.$$

We first add:

$$\frac{7x^2 - 29x + 24}{(2x - 1)(x - 2)^2} = \frac{A}{2x - 1} + \frac{B}{x - 2} + \frac{C}{(x - 2)^2}$$

$$= \frac{A(x - 2)^2}{(2x - 1)(x - 2)^2} + \frac{B(2x - 1)(x - 2)}{(2x - 1)(x - 2)^2}$$

$$+ \frac{C(2x - 1)}{(2x - 1)(x - 2)^2}.$$

Then we equate numerators:

$$7x^2 - 29x + 24$$
$$= A(x - 2)^2 + B(2x - 1)(x - 2) + C(2x - 1)$$
$$= A(x^2 - 4x + 4) + B(2x^2 - 5x + 2) + C(2x - 1)$$
$$= Ax^2 - 4Ax + 4A + 2Bx^2 - 5Bx + 2B + 2Cx - C,$$

or

$$7x^2 - 29x + 24$$
$$= (A + 2B)x^2 + (-4A - 5B + 2C)x + (4A + 2B - C).$$

Next, we equate corresponding coefficients:

$$7 = A + 2B,$$ **The coefficients of the x^2-terms must be the same.**

$$-29 = -4A - 5B + 2C,$$ **The coefficients of the x-terms must be the same.**

$$24 = 4A + 2B - C.$$ **The constant terms must be the same.**

We now have a system of three equations. You should confirm that the solution of the system is

$$A = 5, \quad B = 1, \quad \text{and} \quad C = -2.$$

The decomposition is as follows:

$$\frac{7x^2 - 29x + 24}{(2x - 1)(x - 2)^2} = \frac{5}{2x - 1} + \frac{1}{x - 2} - \frac{2}{(x - 2)^2}.$$

Example 5 Decompose into partial fractions:

$$\frac{11x^2 - 8x - 7}{(2x^2 - 1)(x - 3)}.$$

SOLUTION   The decomposition has the following form:

$$\frac{11x^2 - 8x - 7}{(2x^2 - 1)(x - 3)} = \frac{Ax + B}{2x^2 - 1} + \frac{C}{x - 3}.$$

Adding and equating numerators, we get

$$11x^2 - 8x - 7 = (Ax + B)(x - 3) + C(2x^2 - 1)$$
$$= Ax^2 - 3Ax + Bx - 3B + 2Cx^2 - C,$$

or $11x^2 - 8x - 7 = (A + 2C)x^2 + (-3A + B)x + (-3B - C).$

We then equate corresponding coefficients:

$$11 = A + 2C,\qquad \text{The coefficients of the } x^2\text{-terms}$$
$$-8 = -3A + B,\qquad \text{The coefficients of the } x\text{-terms}$$
$$-7 = -3B - C.\qquad \text{The constant terms}$$

We solve this system of three equations and obtain

$$A = 3,\qquad B = 1,\quad \text{and}\quad C = 4.$$

The decomposition is as follows:

$$\frac{11x^2 - 8x - 7}{(2x^2 - 1)(x - 3)} = \frac{3x + 1}{2x^2 - 1} + \frac{4}{x - 3}.$$

We can check the result on a grapher.

$$y_1 = \frac{11x^2 - 8x - 7}{(2x^2 - 1)(x - 3)}, \quad y_2 = \frac{3x + 1}{2x^2 - 1} + \frac{4}{x - 3}$$

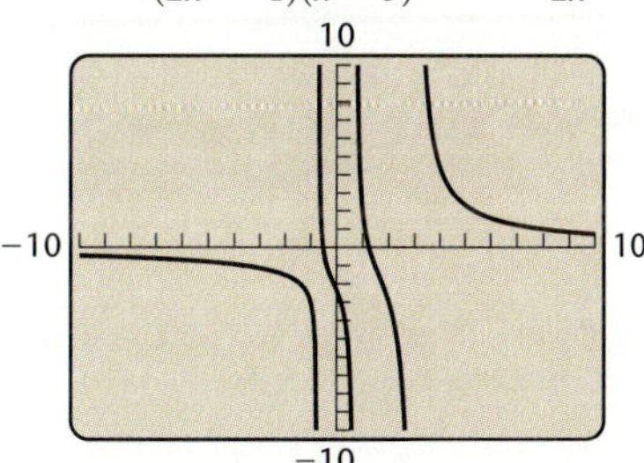

The graphs appear to be the same.

<table>
<tr><td>

4.7 | *Exercise Set*

</td><td>

3. $\dfrac{7x - 1}{6x^2 - 5x + 1}$

5. $\dfrac{3x^2 - 11x - 26}{(x^2 - 4)(x + 1)}$

7. $\dfrac{9}{(x + 2)^2(x - 1)}$

</td><td>

4. $\dfrac{13x + 46}{12x^2 - 11x - 15}$

6. $\dfrac{5x^2 + 9x - 56}{(x - 4)(x - 2)(x + 1)}$

8. $\dfrac{x^2 - x - 4}{(x - 2)^3}$

</td></tr>
</table>

Decompose into partial fractions. Check your answers on a grapher.

1. $\dfrac{x + 7}{(x - 3)(x + 2)}$ **2.** $\dfrac{2x}{(x + 1)(x - 1)}$ **9.** $\dfrac{2x^2 + 3x + 1}{(x^2 - 1)(2x - 1)}$

10. $\dfrac{x^2 - 10x + 13}{(x^2 - 5x + 6)(x - 1)}$

11. $\dfrac{x^4 - 3x^3 - 3x^2 + 10}{(x + 1)^2(x - 3)}$

12. $\dfrac{10x^3 - 15x^2 - 35x}{x^2 - x - 6}$

13. $\dfrac{-x^2 + 2x - 13}{(x^2 + 2)(x - 1)}$

14. $\dfrac{26x^2 + 208x}{(x^2 + 1)(x + 5)}$

15. $\dfrac{6 + 26x - x^2}{(2x - 1)(x + 2)^2}$

16. $\dfrac{5x^3 + 6x^2 + 5x}{(x^2 - 1)(x + 1)^3}$

17. $\dfrac{6x^3 + 5x^2 + 6x - 2}{2x^2 + x - 1}$

18. $\dfrac{2x^3 + 3x^2 - 11x - 10}{x^2 + 2x - 3}$

19. $\dfrac{2x^2 - 11x + 5}{(x - 3)(x^2 + 2x - 5)}$

20. $\dfrac{3x^2 - 3x - 8}{(x - 5)(x^2 + x - 4)}$

21. $\dfrac{-4x^2 - 2x + 10}{(3x + 5)(x + 1)^2}$

22. $\dfrac{26x^2 - 36x + 22}{(x - 4)(2x - 1)^2}$

23. $\dfrac{36x + 1}{12x^2 - 7x - 10}$

24. $\dfrac{-17x + 61}{6x^2 + 39x - 21}$

25. $\dfrac{-4x^2 - 9x + 8}{(3x^2 + 1)(x - 2)}$

26. $\dfrac{11x^2 - 39x + 16}{(x^2 + 4)(x - 8)}$

Skill Maintenance

Factor.

27. $4x^2 - 25y^2$

28. $3x^2 + 10x - 8$

Complete the square and write the resulting expression in the form $(x + a)^2$.

29. $x^2 + 8x$

30. $x^2 - x$

Synthesis

31. ◆ Consider the three methods of checking the partial fraction decomposition of a rational expression in Example 1. Which do you prefer? Why?

32. ◆ What would you say to a classmate who tells you that the partial fraction decomposition of

$$\frac{3x^2 - 8x + 9}{(x + 3)(x^2 - 5x + 6)}$$

is

$$\frac{2}{x + 3} + \frac{x - 1}{x^2 - 5x + 6}?$$

Explain.

Decompose into partial fractions.

33. $\dfrac{x}{x^4 - a^4}$

34. $\dfrac{9x^3 - 24x^2 + 48x}{(x - 2)^4(x + 1)}$

[*Hint:* Let the expression equal

$$\frac{A}{x + 1} + \frac{P(x)}{(x - 2)^4}$$

and find $P(x)$.]

35. $\dfrac{1 + \ln x^2}{(\ln x + 2)(\ln x - 3)^2}$

36. $\dfrac{1}{e^{-x} + 3 + 2e^x}$

37. ◆ Explain the error in the following process.

$$\frac{x^2 + 4}{(x + 2)(x + 1)} = \frac{A}{x + 2} + \frac{B}{x + 1}$$
$$= \frac{A(x + 1) + B(x + 2)}{(x + 2)(x + 1)}$$

Then

$$x^2 + 4 = A(x + 1) + B(x + 2).$$

When $x = -1$:

$$(-1)^2 + 4 = A(-1 + 1) + B(-1 + 2)$$
$$5 = B.$$

When $x = -2$:

$$(-2)^2 + 4 = A(-2 + 1) + B(-2 + 2)$$
$$8 = -A$$
$$-8 = A.$$

Thus,

$$\frac{x^2 + 4}{(x + 2)(x + 1)} = \frac{-8}{x + 2} + \frac{5}{x + 1}.$$

4 Summary and Review

Important Properties and Formulas

Row-Equivalent Operations

1. Interchange any two rows.
2. Multiply each entry in a row by the same nonzero constant.
3. Add a nonzero multiple of one row to another row.

Row-Echelon Form

1. If a row does not consist entirely of 0's, then the first nonzero element in the row is a 1 (called a leading 1).
2. For any two successive nonzero rows, the leading 1 in the lower row is farther to the right than the leading 1 in the higher row.
3. All the rows consisting entirely of 0's are at the bottom of the matrix.

If a fourth property is also satisfied, a matrix is said to be in reduced row-echelon form:

4. Each column that contains a leading 1 has 0's everywhere else.

Properties of Matrix Addition and Scalar Multiplication

For any $m \times n$ matrices $\mathbf{A}$, $\mathbf{B}$, and $\mathbf{C}$ and any scalars k and l:

Commutative Property:
$$\mathbf{A} + \mathbf{B} = \mathbf{B} + \mathbf{A}.$$
Associative Property of Addition:
$$\mathbf{A} + (\mathbf{B} + \mathbf{C}) = (\mathbf{A} + \mathbf{B}) + \mathbf{C}.$$
Associative Property of Scalar Multiplication:
$$(kl)\mathbf{A} = k(l\mathbf{A}).$$
Additive Identity Property:
There exists a unique matrix $\mathbf{0}$ such that
$$\mathbf{A} + \mathbf{0} = \mathbf{0} + \mathbf{A} = \mathbf{A}.$$

Additive Inverse Property:
There exists a unique matrix $-\mathbf{A}$ such that
$$\mathbf{A} + (-\mathbf{A}) = -\mathbf{A} + \mathbf{A} = \mathbf{0}.$$
Distributive Properties:
$$k(\mathbf{A} + \mathbf{B}) = k\mathbf{A} + k\mathbf{B},$$
$$(k + l)\mathbf{A} = k\mathbf{A} + l\mathbf{A}.$$

Properties of Matrix Multiplication

For matrices $\mathbf{A}$, $\mathbf{B}$, and $\mathbf{C}$, assuming that the indicated operation is possible:

Associative Property of Multiplication:
$$\mathbf{A}(\mathbf{BC}) = (\mathbf{AB})\mathbf{C}$$
Distributive Properties:
$$\mathbf{A}(\mathbf{B} + \mathbf{C}) = \mathbf{AB} + \mathbf{AC},$$
$$(\mathbf{B} + \mathbf{C})\mathbf{A} = \mathbf{BA} + \mathbf{CA}$$

Matrix Solutions of Systems of Equations

For a system of n linear equations in n variables, $\mathbf{AX} = \mathbf{B}$, if $\mathbf{A}$ is an invertible matrix, then the unique solution of the system is given by $\mathbf{X} = \mathbf{A}^{-1}\mathbf{B}$.

To Graph a Linear Inequality in Two Variables:

1. Replace the inequality symbol with an equals sign and graph this related equation. If the inequality symbol is $<$ or $>$, draw the line dashed. If the inequality symbol is $\leq$ or $\geq$, draw the line solid.
2. The graph consists of a half-plane on one side of the line and, if the line is solid, the line as well. To determine which half-plane to shade, test a point not on the line in the original inequality. If that point is a solution, shade the half-plane containing that point. If not, shade the opposite half-plane.

(continued)

Linear Programming Procedure

To find the maximum or minimum value of a linear objective function subject to a set of constraints:

1. Graph the region of feasible solutions.

2. Determine the coordinates of the vertices of the region.

3. Evaluate the objective function at each vertex. The largest and smallest of those values are the maximum and minimum values of the function, respectively.

Procedure for Decomposing a Rational Expression into Partial Fractions

Consider any rational expression $P(x)/Q(x)$ such that $P(x)$ and $Q(x)$ have no common factor other than 1 or -1.

1. If the degree of $P(x)$ is greater than or equal to the degree of $Q(x)$, divide to express $P(x)/Q(x)$ as a quotient $+$ remainder/$Q(x)$ and follow steps (2)–(5) to decompose the resulting rational expression.

2. If the degree of $P(x)$ is less than the degree of $Q(x)$, factor $Q(x)$ into linear factors of the form $(px + q)^n$ and/or quadratic factors of the form $(ax^2 + bx + c)^m$. Any quadratic factor $ax^2 + bx + c$ must be irreducible, meaning that it cannot be factored into linear factors with real coefficients.

3. Assign to each linear factor $(px + q)^n$ the sum of n partial fractions:

$$\frac{A_1}{px + q} + \frac{A_2}{(px + q)^2} + \cdots + \frac{A_n}{(px + q)^n}.$$

4. Assign to each quadratic factor $(ax^2 + bx + c)^m$ the sum of m partial fractions:

$$\frac{B_1 x + C_1}{ax^2 + bx + c} + \frac{B_2 x + C_2}{(ax^2 + bx + c)^2} + \cdots + \frac{B_m x + C_m}{(ax^2 + bx + c)^m}.$$

5. Apply algebraic methods to find the constants in the numerators of the partial fractions.

REVIEW EXERCISES

Match each of the following with one of the graphs (a)–(h), which follow.

a) 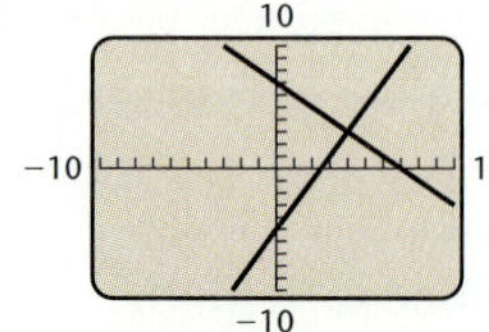

b) 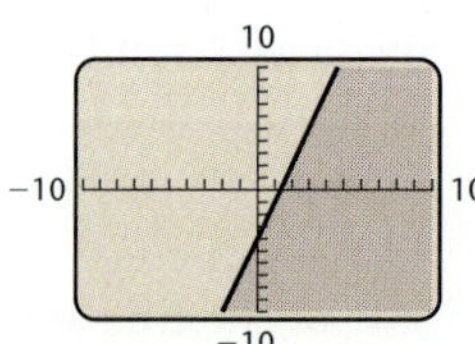

c) 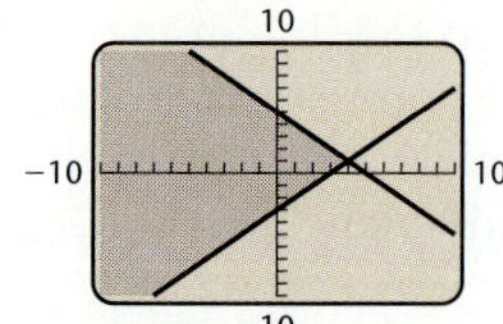

d) 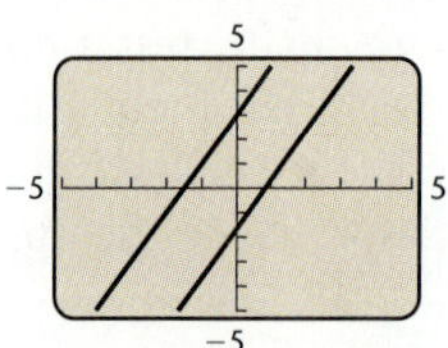

e) 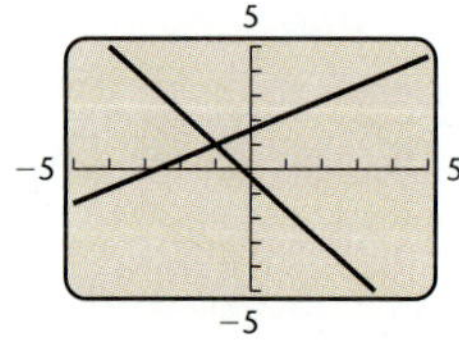

f) 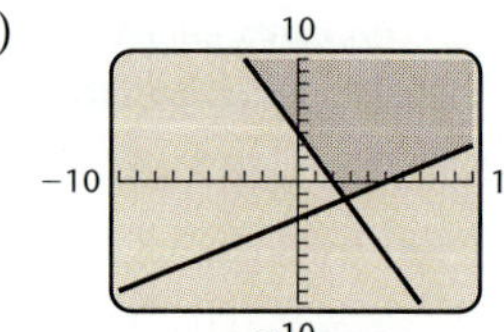

g) 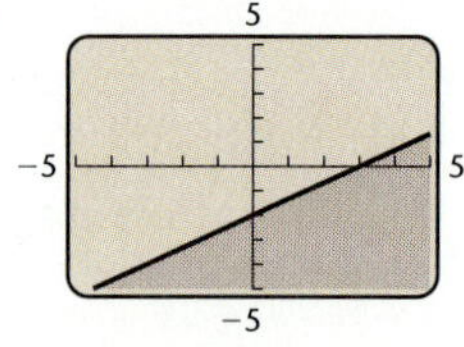

h) 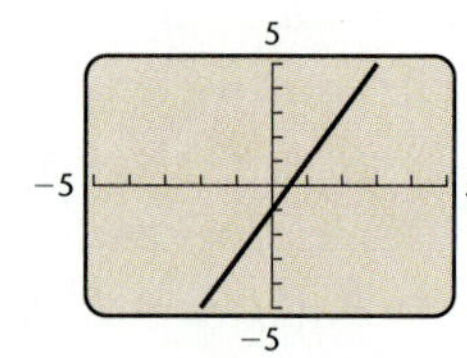

1. $x + y = 7,$
   $2x - y = 5$

2. $3x - 5y = -8,$
   $4x + 3y = -1$

3. $y = 2x - 1,$
   $4x - 2y = 2$

4. $6x - 3y = 5,$
   $y = 2x + 3$

5. $y \leq 3x - 4$

6. $2x - 3y \geq 6$

7. $x - y \leq 3,$
$x + y \leq 5$

8. $2x + y \geq 4,$
$3x - 5y \leq 15$

Solve.

9. $5x - 3y = -4,$
$3x - y = -4$

10. $2x + 3y = 2,$
$5x - y = -29$

11. $x + 5y = 12,$
$5x + 25y = 12$

12. $x - y = -2,$
$-3x + 3y = 6$

13. $2x - 4y + 3z = -3,$
$-5x + 2y - z = 7,$
$3x + 2y - 2z = 4$

14. $x + 5y + 3z = 0,$
$3x - 2y + 4z = 0,$
$2x + 3y - z = 0$

15. $x - y = 5,$
$y - z = 6,$
$z - w = 7,$
$x + w = 8$

16. Classify each of the systems in Exercises 9–15 as consistent or inconsistent.

17. Classify each of the systems in Exercises 9–15 as dependent or independent.

Solve each system of equations using Gaussian elimination or Gauss–Jordan elimination.

18. $x + 2y = 5,$
$2x - 5y = -8$

19. $3x + 4y + 2z = 3,$
$5x - 2y - 13z = 3,$
$4x + 3y - 3z = 6$

20. $3x + 5y + z = 0,$
$2x - 4y - 3z = 0,$
$x + 3y + z = 0$

21. $w + x + y + z = -2,$
$-3w - 2x + 3y + 2z = 10,$
$2w + 3x + 2y - z = -12,$
$2w + 4x - y + z = 1$

22. *Coins.* The value of 75 coins, consisting of nickels and dimes, is \$5.95. How many of each kind are there?

23. *Investment.* The Mendez family invested \$5000, part at 10% and the remainder at 10.5%. The annual income from both investments is \$517. What is the amount invested at each rate?

24. *Nutrition.* A dietician must plan a breakfast menu that provides 460 Cal, 9 g of fat, and 55 mg of calcium. One plain bagel contains 200 Cal, 2 g of fat, and 29 mg of calcium. A one-tablespoon serving of cream cheese contains 100 Cal, 10 g of fat, and 24 mg of calcium. One banana contains 105 Cal, 1 g of fat, and 7 mg of calcium. How many servings of each are required to provide the desired nutritional values? (*Source: Home and Garden Bulletin No. 72,* U.S. Government Printing Office, Washington, D.C. 20402)

25. *Test Scores.* A student has a total of 225 on three tests. The sum of the scores on the first and second tests exceeds the score on the third test by 61. The first score exceeds the second by 6. Find the three scores.

26. *Ice Milk Consumption.* The table below shows the per capita ice milk consumption, in pounds, in the United States in three recent years, represented as years since 1985.

YEAR	ICE MILK CONSUMPTION (IN POUNDS)
(1985) 0	6.9
(1990) 5	7.7
(1992) 7	7.1

Source: U.S. Department of Agriculture, Economic Research Service, *Food Consumption, Prices, and Expenditures,* annual.

a) Find a quadratic function $f(x) = ax^2 + bx + c$ that fits the data.

b) Use the function to predict the per capita ice milk consumption in 1998.

For Exercises 26–33, let

$$A = \begin{bmatrix} 1 & -1 & 0 \\ 2 & 3 & -2 \\ -2 & 0 & 1 \end{bmatrix},$$

$$B = \begin{bmatrix} -1 & 0 & 6 \\ 1 & -2 & 0 \\ 0 & 1 & -3 \end{bmatrix},$$

and

$$C = \begin{bmatrix} -2 & 0 \\ 1 & 3 \end{bmatrix}.$$

Find each of the following, if possible.

27. $A + B$

28. $-3A$

29. $-A$

30. AB

31. $B + C$

32. $A - B$

33. $2A - B$

34. $A + 3B$

35. *Food Service Management.* The table below shows the cost per serving, in cents, for items on four menus that are served at an elder-care facility.

MENU	MEAT	POTATO	VEGETABLE	SALAD	DESSERT
1	46.1	5.9	10.1	8.5	11.4
2	54.6	4.6	9.6	7.6	10.6
3	48.9	5.5	12.7	9.4	9.3
4	51.3	4.8	11.3	6.9	12.7

On a particular day, a dietician orders 32 meals from menu 1, 19 from menu 2, 43 from menu 3, and 38 from menu 4.

a) Write the information in the table as a 4 × 5 matrix $\mathbf{M}$.

b) Write a row matrix $\mathbf{N}$ that represents the number of each menu ordered.

c) Find the product $\mathbf{NM}$.

d) State what the entries of $\mathbf{NM}$ represent.

Find $\mathbf{A}^{-1}$, if it exists.

36. $\mathbf{A} = \begin{bmatrix} -2 & 0 \\ 1 & 3 \end{bmatrix}$

37. $\mathbf{A} = \begin{bmatrix} 0 & 0 & 3 \\ 0 & -2 & 0 \\ 4 & 0 & 0 \end{bmatrix}$

38. $\mathbf{A} = \begin{bmatrix} 1 & 0 & 0 & 0 \\ 0 & 4 & -5 & 0 \\ 0 & 2 & 2 & 0 \\ 0 & 0 & 0 & 1 \end{bmatrix}$

39. Write a matrix equation equivalent to this system of equations:

$$3x - 2y + 4z = 13,$$
$$x + 5y - 3z = 7,$$
$$2x - 3y + 7z = -8.$$

Solve each system of equations using the inverse of the coefficient matrix of the equivalent matrix equation.

40. $2x + 3y = 5,$
$3x + 5y = 11$

41. $5x - y + 2z = 17,$
$3x + 2y - 3z = -16,$
$4x - 3y - z = 5$

42. $w - x - y + z = -1,$
$2w + 3x - 2y - z = 2,$
$-w + 5x + 4y - 2z = 3,$
$3w - 2x + 5y + 3z = 4$

Graph.

43. $y \le 3x + 6$

44. $4x - 3y \ge 12$

45. Graph this system of inequalities and find the coordinates of any vertices formed.

$$2x + y \ge 9,$$
$$4x + 3y \ge 23,$$
$$x + 3y \ge 8,$$
$$x \ge 0,$$
$$y \ge 0$$

46. Find the maximum and minimum values of $T = 6x + 10y$ subject to

$$x + y \le 10,$$
$$5x + 10y \ge 50,$$
$$x \ge 2,$$
$$y \ge 0.$$

47. *Maximizing a Test Score.* Marita is taking a test that contains questions in group A worth 7 points each and questions in group B worth 12 points each. The total number of questions answered must be at least 8. If Marita knows that group A questions take 8 min each and group B questions take 10 min each and the maximum time for the test is 80 min, how many questions from each group must she answer correctly in order to maximize her score? What is the maximum score?

Decompose into partial fractions.

48. $\dfrac{5}{(x + 2)^2(x + 1)}$

49. $\dfrac{-8x + 23}{2x^2 + 5x - 12}$

Synthesis

50. ◆ Write a problem for a classmate to solve that can be translated to a system of equations. Devise the problem so that the solution is "The caterer sold 20 cheese trays and 35 seafood trays."

51. ◆ For square matrices $\mathbf{A}$ and $\mathbf{B}$, is it true, in general, that $(\mathbf{AB})^2 = \mathbf{A}^2\mathbf{B}^2$? Explain.

52. One year, Don invested a total of $40,000, part at 12%, part at 13%, and the rest at $14\frac{1}{2}\%$. The total amount of interest received on the investments was $5370. The interest received on the $14\frac{1}{2}\%$ investment was $1050 more than the interest received on the 13% investment. How much was invested at each rate?

Solve.

53. $\dfrac{2}{3x} + \dfrac{4}{5y} = 8,$

$\dfrac{5}{4x} - \dfrac{3}{2y} = -6$

54. $\dfrac{3}{x} - \dfrac{4}{y} + \dfrac{1}{z} = -2,$

$\dfrac{5}{x} + \dfrac{1}{y} - \dfrac{2}{z} = 1,$

$\dfrac{7}{x} + \dfrac{3}{y} + \dfrac{2}{z} = 19$

Graph.

55. $|x| - |y| \le 1$

56. $|xy| > 1$

Conic Sections 5

APPLICATION

For a student recreation building at Southport Community College, an architect wants to lay out a rectangular piece of ground that has a perimeter of 204 m and an area of 2565 m^2. The dimensions of the piece of ground are solutions of the nonlinear system of equations

$$2x + 2y = 204,$$
$$xy = 2565.$$

I n this chapter, we study *conic sections.* These curves are formed by the intersection of a cone and a plane. Conic sections and their properties were first studied by the Greeks. Today they have many applications, as we will see.

X	Y₁	Y₂
42	60	61.071
43	59	59.651
44	58	58.295
45	57	57
46	56	55.761
47	55	54.574
48	54	53.438

X = 45

$y_1 = (204 - 2x)/2, \ y_2 = 2565/x$

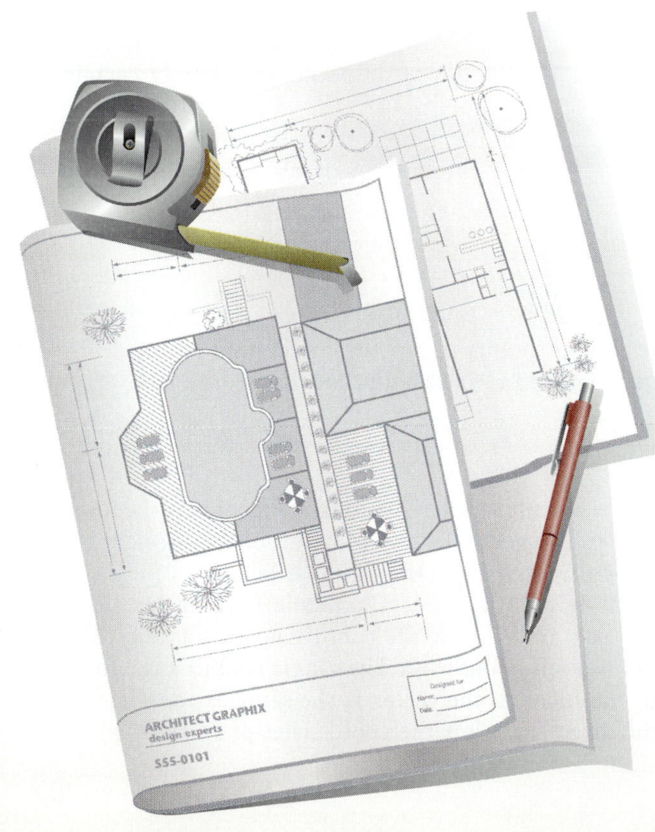

5.1 The Parabola

5.2 The Circle and the Ellipse

5.3 The Hyperbola

5.4 Nonlinear Systems of Equations

SUMMARY AND REVIEW

5.1
The Parabola

• *Given an equation of a parabola, complete the square, if necessary, and then find the vertex, the focus, and the directrix and graph the parabola.*

A **conic section** is formed when a right circular cone with two parts, called *nappes*, is intersected by a plane. One of four types of curves can be formed: a parabola, a circle, an ellipse, or a hyperbola.

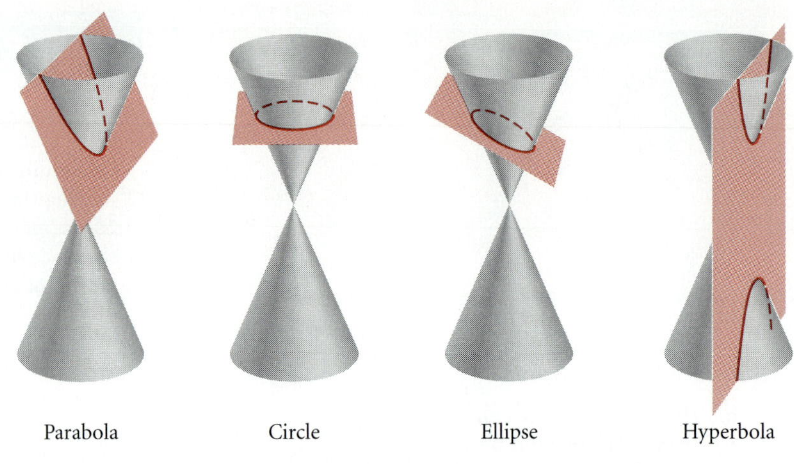

Conic Sections

Parabolas

Conic sections can be defined algebraically using second-degree equations of the form $Ax^2 + Bxy + Cy^2 + Dx + Ey + F = 0$. In addition, they can be defined geometrically as a set of points that satisfy certain conditions.

In Section 2.2, we saw that the graph of the quadratic function $f(x) = ax^2 + bx + c$, $a \neq 0$, is a parabola. A parabola can be defined geometrically.

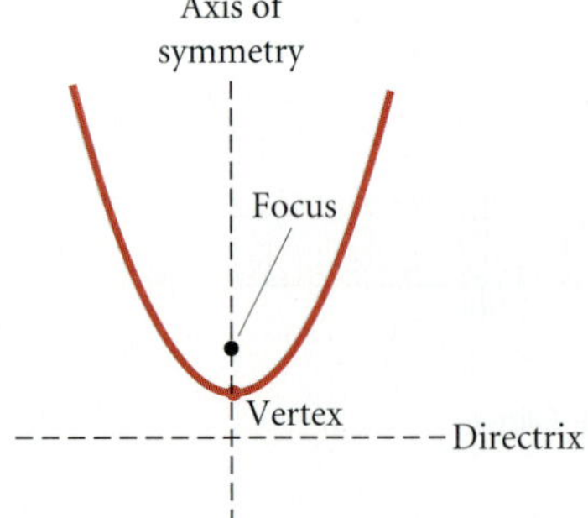

> **Parabola**
>
> A *parabola* is the set of all points in a plane equidistant from a fixed line (the *directrix*) and a fixed point not on the line (the *focus*).

The line that is perpendicular to the directrix and contains the focus is the **axis of symmetry**. The **vertex** is the midpoint of the segment between the focus and the directrix. (See the figure at left.)

Let's derive the standard equation of a parabola with vertex $(0, 0)$ and directrix $y = -p$, $p > 0$. We place the coordinate axes as shown in the figure at the top of the following page. The y-axis contains the focus F. The distance from the focus to the vertex is the same as the distance from the vertex to the directrix. Thus the coordinates of F are $(0, p)$.

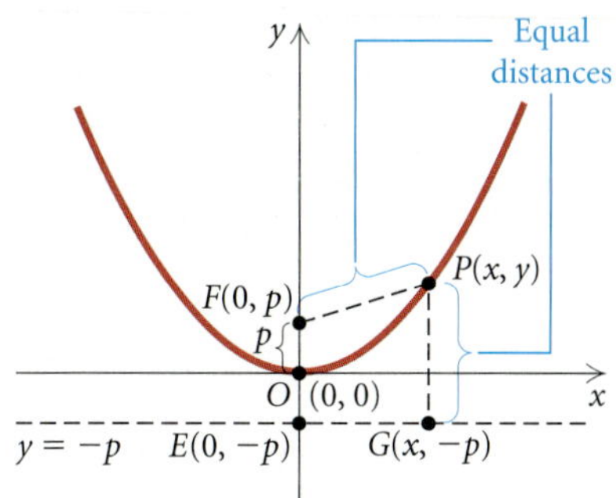

Let $P(x, y)$ be any point of the parabola and consider $\overline{PG}$ perpendicular to the line $y = -p$. The coordinates of G are $(x, -p)$. By the definition of a parabola,

$$PF = PG.$$ The distance from P to the focus is the same as the distance from P to the directrix.

Then using the distance formula, we have

$$\sqrt{(x - 0)^2 + (y - p)^2} = \sqrt{(x - x)^2 + (y + p)^2}$$
$$x^2 + y^2 - 2py + p^2 = y^2 + 2py + p^2 \qquad \text{Squaring both sides}$$
$$x^2 = 4py.$$

We have shown that if $P(x, y)$ is on the parabola shown above, then its coordinates satisfy this equation. The converse is also true, but we will not prove it here.

Note that if $p > 0$, as above, the graph opens up. If $p < 0$, the graph opens down.

The equation of a parabola with vertex $(0, 0)$ and directrix $x = -p$ is derived similarly. Such a parabola opens either right $(p > 0)$ or left $(p < 0)$.

Standard Equation of a Parabola with Vertex at the Origin

The standard equation of a parabola with vertex $(0, 0)$ and directrix $y = -p$ is

$$x^2 = 4py.$$

The focus is $(0, p)$ and the y-axis is the axis of symmetry.

The standard equation of a parabola with vertex $(0, 0)$ and directrix $x = -p$ is

$$y^2 = 4px.$$

The focus is $(p, 0)$ and the x-axis is the axis of symmetry.

Example 1 Find the focus and the directrix of the parabola $y = -\frac{1}{12}x^2$. Then graph the parabola.

SOLUTION   We write $y = -\frac{1}{12}x^2$ in the form $x^2 = 4py$:

$$-\frac{1}{12}x^2 = y \qquad \text{Given equation}$$
$$x^2 = -12y \qquad \text{Multiplying both sides by } -12$$
$$x^2 = 4(-3)y. \qquad \text{Standard form}$$

Thus, $p = -3$, so the focus is $(0, p)$, or $(0, -3)$. The directrix is $y = -p = -(-3) = 3$.

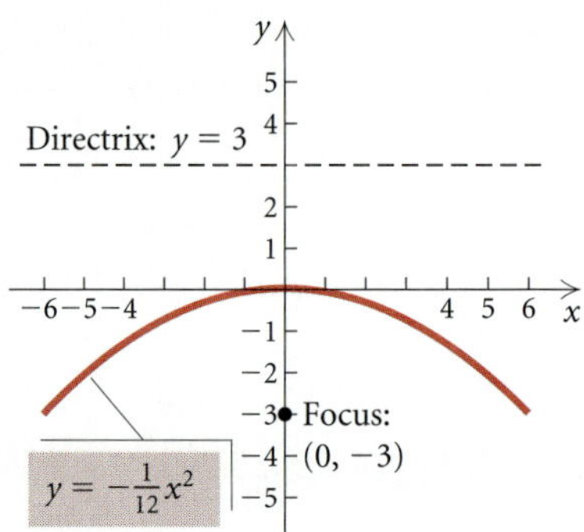

Example 2   Find an equation of the parabola with vertex $(0, 0)$ and focus $(5, 0)$. Then graph the parabola.

SOLUTION   The focus is on the x-axis so the line of symmetry is the x-axis. Thus the equation is of the type

$$y^2 = 4px.$$

Since the focus is 5 units to the right of the vertex, $p = 5$ and the equation is

$$y^2 = 4(5)x, \quad \text{or}$$
$$y^2 = 20x.$$

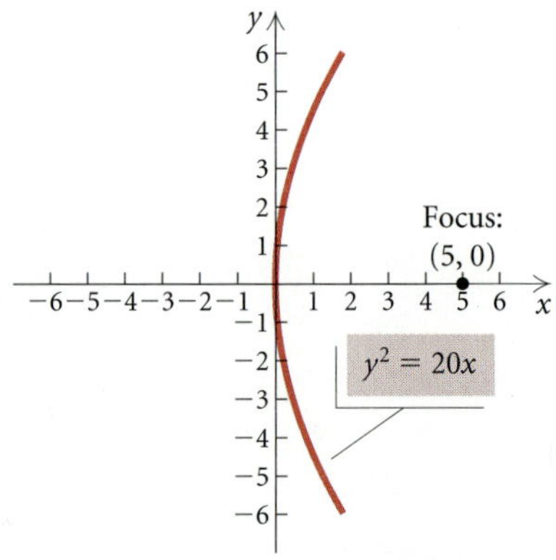

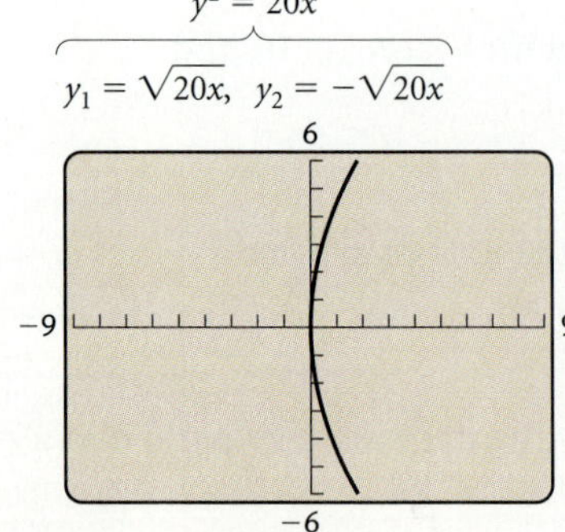

We can check the graph on a grapher using a squared viewing window. But first it might be necessary to solve for y:

$$y^2 = 20x$$
$$y = \pm\sqrt{20x}. \qquad \text{Using the principle of square roots}$$

We now graph $y_1 = \sqrt{20x}$ and $y_2 = -\sqrt{20x}$. On some graphers, it is possible to graph $y_1 = \sqrt{20x}$ and $y_2 = -y_1$ by using the Y-VARS menu.

Finding Standard Form by Completing the Square

If a parabola with vertex at the origin is translated $|h|$ units horizontally and $|k|$ units vertically, it has an equation as follows.

Standard Equation of a Parabola with Vertex (h, k) and Vertical Axis of Symmetry

The standard equation of a parabola with vertex (h, k) and vertical axis of symmetry is

$$(x - h)^2 = 4p(y - k),$$

where the vertex is (h, k), the focus is $(h, k + p)$, and the directrix is $y = k - p$.

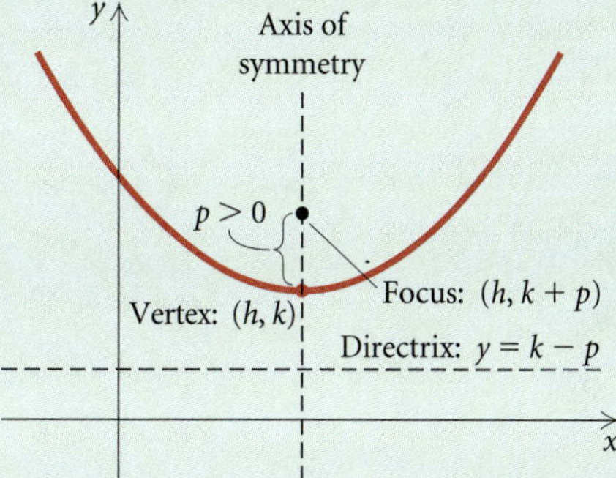

(When $p < 0$, the parabola opens down.)

Standard Equation of a Parabola with Vertex (h, k) and Horizontal Axis of Symmetry

The standard equation of a parabola with vertex (h, k) and horizontal axis of symmetry is

$$(y - k)^2 = 4p(x - h),$$

where the vertex is (h, k), the focus is $(h + p, k)$, and the directrix is $x = h - p$.

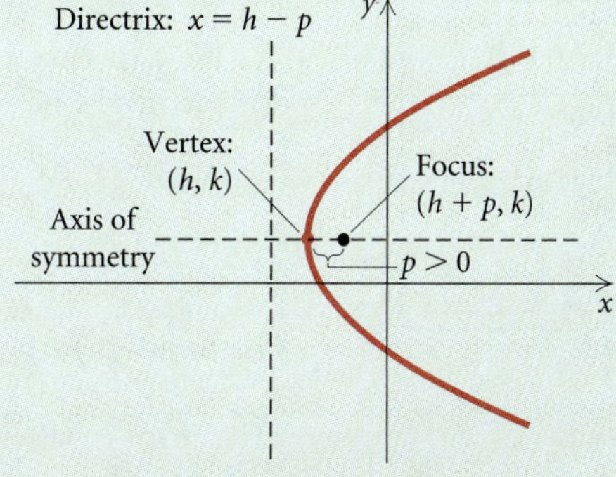

(When $p < 0$, the parabola opens left.)

We can complete the square on equations of the form

$$y = ax^2 + bx + c \quad \text{or} \quad x = ay^2 + by + c$$

in order to write them in standard form.

Example 3   For the parabola

$$x^2 + 6x + 4y + 5 = 0,$$

find the vertex, the focus, and the directrix. Then draw the graph.

SOLUTION   We first complete the square:

$$x^2 + 6x + 4y + 5 = 0$$

$$x^2 + 6x = -4y - 5 \qquad \text{Subtracting } 4y \text{ and } 5 \text{ on both sides}$$

$$x^2 + 6x + 9 = -4y - 5 + 9 \qquad \text{Adding 9 on both sides to complete the square on the left side}$$

$$x^2 + 6x + 9 = -4y + 4$$

$$(x + 3)^2 = -4(y - 1) \qquad \text{Factoring}$$

$$[(x - (-3)]^2 = 4(-1)(y - 1). \qquad \text{Writing standard form: } (x - h)^2 = 4p(y - k)$$

We now have the following:

Vertex (h, k): $(-3, 1)$;

Focus $(h, k + p)$: $(-3, 1 + (-1))$, or $(-3, 0)$;

Directrix $y = k - p$: $y = 1 - (-1)$, or $y = 2$.

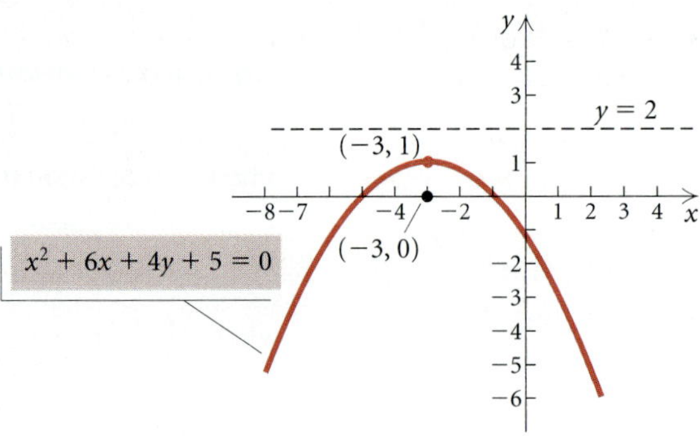

We can check the graph on a grapher using a squared viewing window. It might be necessary to solve for y first:

$$x^2 + 6x + 4y + 5 = 0$$

$$4y = -x^2 - 6x - 5$$

$$y = \tfrac{1}{4}(-x^2 - 6x - 5).$$

The hand-drawn graph appears to be correct.

$y = \tfrac{1}{4}(-x^2 - 6x - 5)$

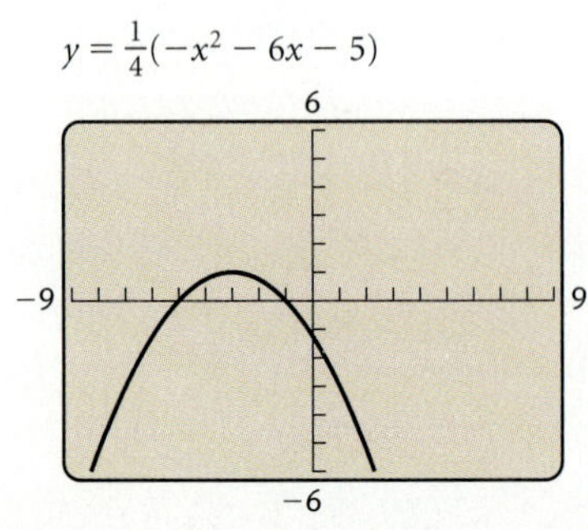

Example 4   For the parabola

$$y^2 - 2y - 8x - 31 = 0,$$

find the vertex, the focus, and the directrix. Then draw the graph.

SOLUTION   We first complete the square:

$$y^2 - 2y - 8x - 31 = 0$$
$$y^2 - 2y = 8x + 31 \qquad \text{Adding } 8x \text{ and } 31 \text{ on both sides}$$
$$y^2 - 2y + 1 = 8x + 31 + 1 \qquad \text{Adding } 1 \text{ on both sides to complete the square on the left side}$$
$$y^2 - 2y + 1 = 8x + 32$$
$$(y - 1)^2 = 8(x + 4)$$
$$(y - 1)^2 = 4(2)[x - (-4)]. \qquad \text{Writing standard form: } (y - k)^2 = 4p(x - h)$$

We now have the following:

Vertex (h, k): $(-4, 1)$;

Focus $(h + p, k)$: $(-4 + 2, 1)$, or $(-2, 1)$;

Directrix $x = h - p$: $x = -4 - 2$, or $x = -6$.

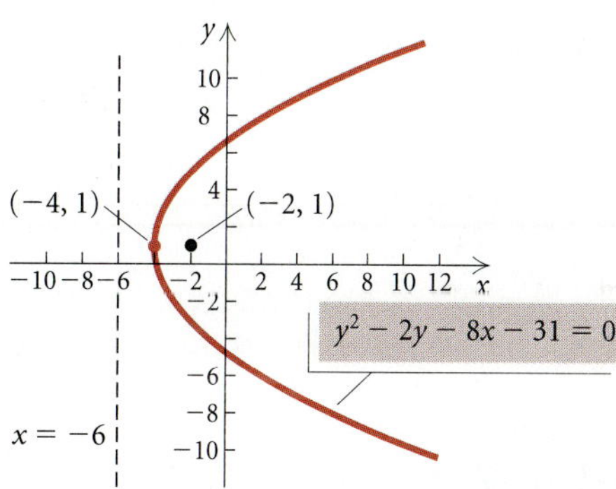

We can check the graph on a grapher using a squared window. We solve the original equation for y using the quadratic formula:

$$y^2 - 2y - 8x - 31 = 0$$
$$y^2 - 2y + (-8x - 31) = 0$$
$$a = 1, \quad b = -2, \quad c = -8x - 31$$
$$y = \frac{-(-2) \pm \sqrt{(-2)^2 - 4 \cdot 1(-8x - 31)}}{2 \cdot 1}$$
$$y = \frac{2 \pm \sqrt{32x + 128}}{2}.$$

We now graph

$$y_1 = \frac{2 + \sqrt{32x + 128}}{2} \quad \text{and} \quad y_2 = \frac{2 - \sqrt{32x + 128}}{2}.$$

The hand-drawn graph appears to be correct.

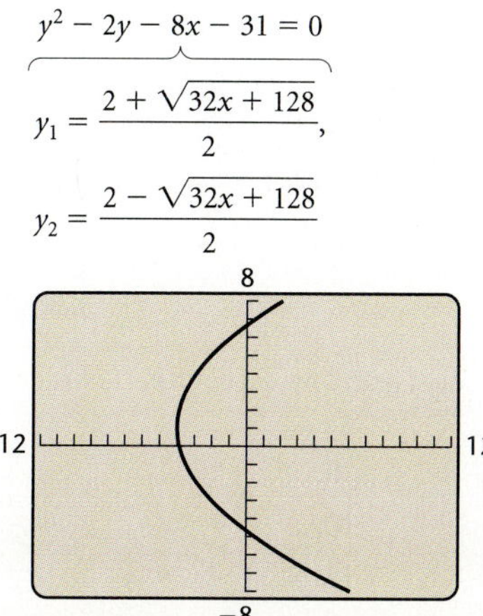

Applications

Parabolas have many applications. For example, cross sections of car headlights, flashlights, and searchlights are parabolas. The bulb is located at the focus and light from that point is reflected outward parallel to the axis of symmetry. Satellite dishes and field microphones used at sporting events often have parabolic cross sections. Incoming radio waves or sound waves, parallel to the axis, are reflected into the focus. Cables hung between structures in suspension bridges, such as the Golden Gate Bridge, form parabolas. When a cable supports only its own weight, however, it forms a curve called a *catenary* rather than a parabola.

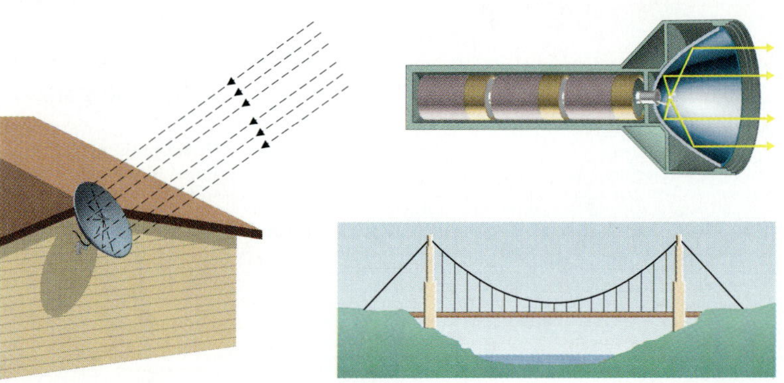

5.1 | *Exercise Set*

In Exercises 1–6, match each equation with one of the graphs (a)–(f), which follow.

a)
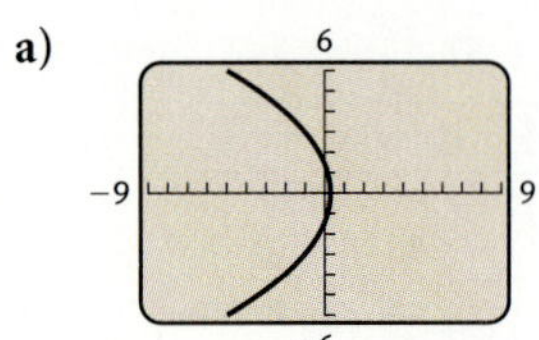

b)
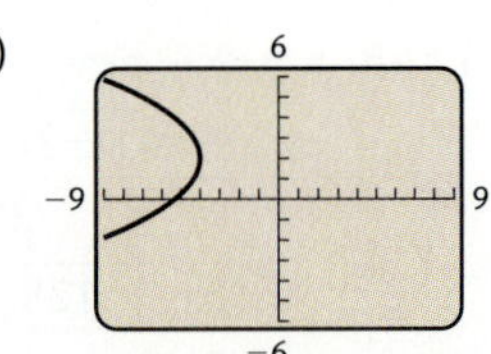

c)
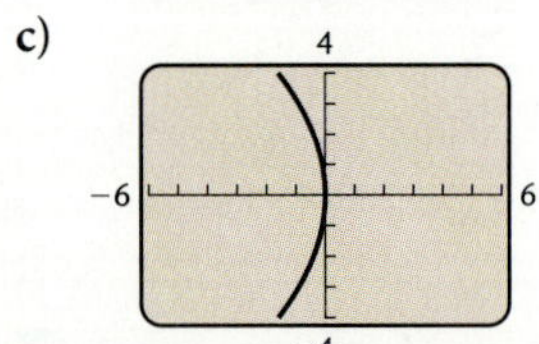

d)
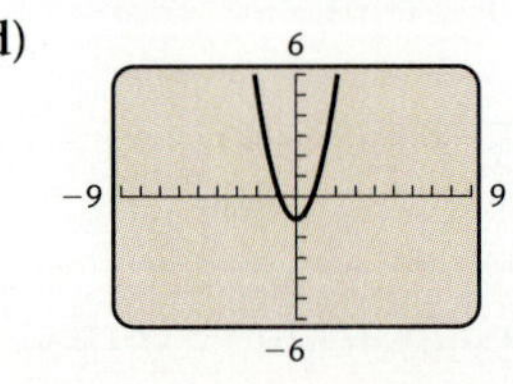

e)
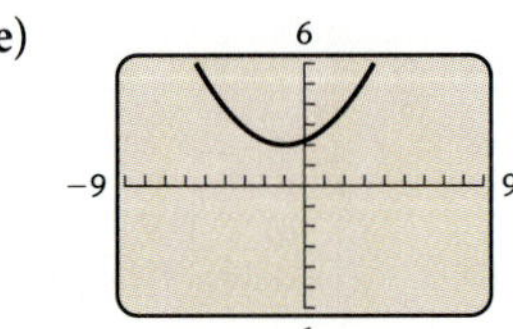

f)
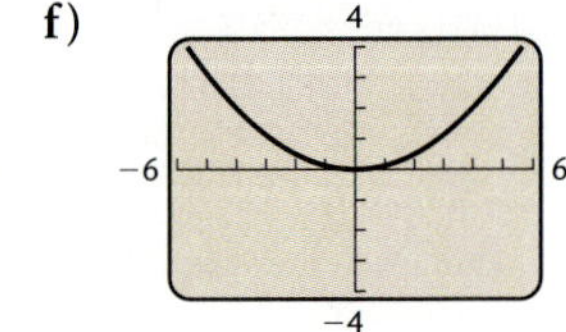

1. $x^2 = 8y$

2. $y^2 = -10x$

3. $(y - 2)^2 = -3(x + 4)$

4. $(x + 1)^2 = 5(y - 2)$

5. $13x^2 - 8y - 9 = 0$

6. $41x + 6y^2 = 12$

Find the vertex, the focus, and the directrix. Then draw the graph.

7. $x^2 = 20y$ 8. $x^2 = 16y$

9. $y^2 = -6x$ 10. $y^2 = -2x$

11. $x^2 - 4y = 0$ 12. $y^2 + 4x = 0$

13. $x = 2y^2$ 14. $y = \frac{1}{2}x^2$

Find an equation of a parabola satisfying the given conditions.

15. Focus $(4, 0)$, directrix $x = -4$

16. Focus $\left(0, \frac{1}{4}\right)$, directrix $y = -\frac{1}{4}$

17. Focus $(0, -\pi)$, directrix $y = \pi$

18. Focus $(-\sqrt{2}, 0)$, directrix $x = \sqrt{2}$

19. Focus $(3, 2)$, directrix $x = -4$

20. Focus $(-2, 3)$, directrix $y = -3$

Find the vertex, the focus, and the directrix. Then draw the graph.

21. $(x + 2)^2 = -6(y - 1)$

22. $(y - 3)^2 = -20(x + 2)$

23. $x^2 + 2x + 2y + 7 = 0$

24. $y^2 + 6y - x + 16 = 0$

25. $x^2 - y - 2 = 0$

26. $x^2 - 4x - 2y = 0$

27. $y = x^2 + 4x + 3$

28. $y = x^2 + 6x + 10$

29. $y^2 - y - x + 6 = 0$

30. $y^2 + y - x - 4 = 0$

31. *Satellite Dish.*   An engineer designs a satellite dish with a parabolic cross section. The dish is 15 ft wide at the opening and the focus is placed 4 ft from the vertex.

a) Position a coordinate system with the origin at the vertex and the x-axis on the parabola's axis of symmetry and find an equation of the parabola.

b) Find the depth of the satellite dish at the vertex.

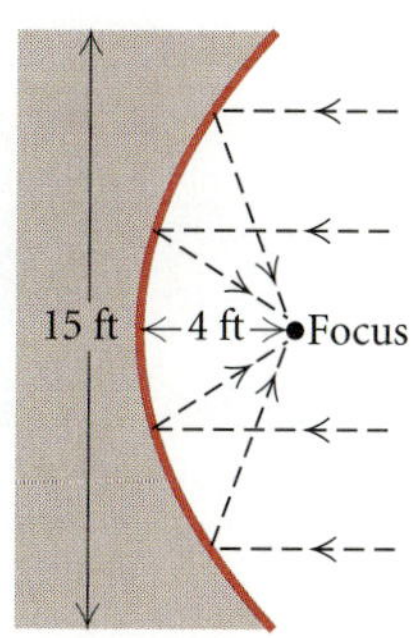

32. *Headlight Mirror.*   A car headlight mirror has a parabolic cross section with diameter 6 in. and depth 1 in.

a) Position a coordinate system with the origin at the vertex and the x-axis on the parabola's axis of symmetry and find an equation of the parabola.

b) How far from the vertex should the bulb be positioned if it is to be placed at the focus?

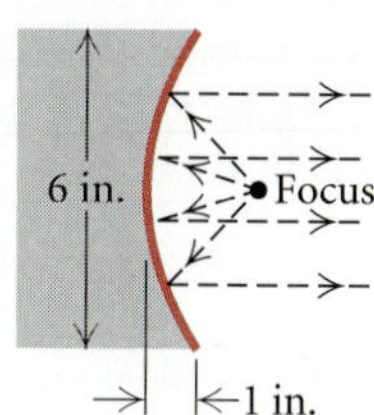

33. *Spotlight.*   A spotlight has a parabolic cross section that is 4 ft wide at the opening and 1.5 ft deep at the vertex. How far from the vertex is the focus?

34. *Field Microphone.*   A field microphone used at a football game has a parabolic cross section and is 18 in. deep. The focus is 4 in. from the vertex. Find the width of the microphone at the opening.

Skill Maintenance

Complete the square and write the result as the square of a binomial.

35. $x^2 + 10x$

36. $y^2 - 9y$

Find the center and the radius of each circle.

37. $(x - 1)^2 + (y + 2)^2 = 9$

38. $(x + 3)^2 + (y - 5)^2 = 36$

Synthesis

39. ◈ Is a parabola always the graph of a function? Why or why not?

40. ◈ Explain how the distance formula is used to find the standard equation of a parabola.

41. Find an equation of the parabola with a vertical axis of symmetry and vertex $(-1, 2)$ and containing the point $(-3, 1)$.

42. Find an equation of a parabola with a horizontal axis of symmetry and vertex $(-2, 1)$ and containing the point $(-3, 5)$.

Use a grapher to find the vertex, the focus, and the directrix of each of the following.

43. $4.5x^2 - 7.8x + 9.7y = 0$

44. $134.1y^2 + 43.4x - 316.6y - 122.4 = 0$

45. *Suspension Bridge.*   The cables of a suspension bridge are 50 ft above the roadbed at the ends of the bridge and 10 ft above it in the center of the bridge. The roadbed is 200 ft long. Vertical cables are to be spaced every 20 ft along the bridge. Calculate the lengths of these vertical cables.

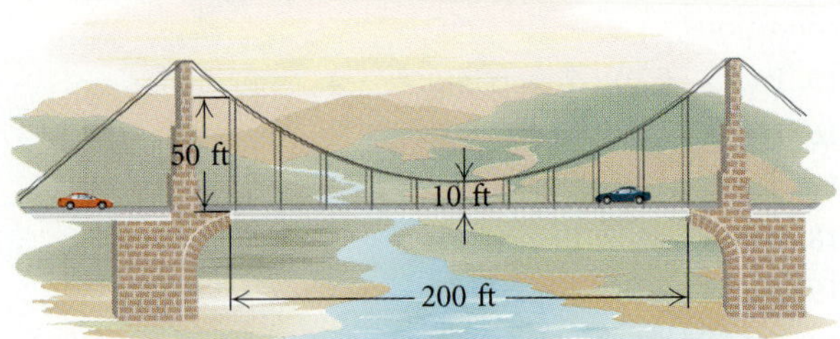

5.2
The Circle and the Ellipse

- *Given an equation of a circle, complete the square, if necessary, and then find the center and the radius and graph the circle.*
- *Given an equation of an ellipse, complete the square, if necessary, and then find the center, the vertices, and the foci and graph the ellipse.*

Circles

We can define a circle geometrically.

Circle

A *circle* is the set of all points in a plane that are at a fixed distance from a fixed point (the *center*) in the plane.

Circles were introduced in Section 1.5. Recall the standard equation of a circle with center (h, k) and radius r.

Standard Equation of a Circle

The standard equation of a circle with center (h, k) and radius r is

$$(x - h)^2 + (y - k)^2 = r^2.$$

Example 1   For the circle

$$x^2 + y^2 - 16x + 14y + 32 = 0,$$

find the center and the radius. Then graph the circle.

SOLUTION   First we complete the square twice:

$$x^2 + y^2 - 16x + 14y + 32 = 0$$
$$x^2 - 16x \qquad + y^2 + 14y \qquad = -32$$
$$x^2 - 16x + 64 + y^2 + 14y + 49 = -32 + 64 + 49$$

Adding 64 and 49 on both sides to complete the square twice on the left side

$$(x - 8)^2 + (y + 7)^2 = 81$$
$$(x - 8)^2 + [y - (-7)]^2 = 9^2. \qquad \text{Writing standard form}$$

The center is $(8, -7)$ and the radius is 9.

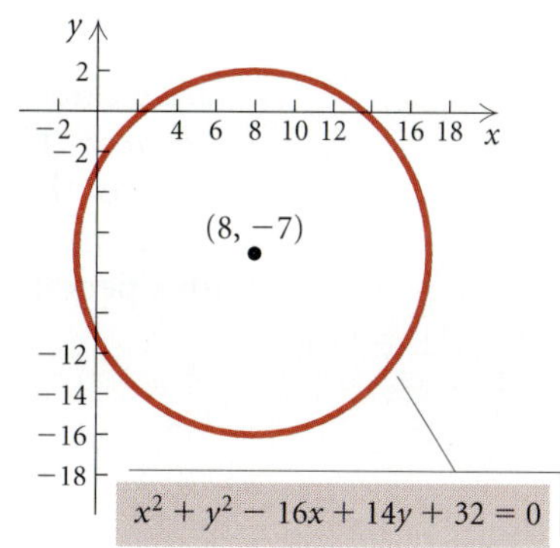

To use a grapher to graph the circle, it might be necessary to solve for y first. The original equation can be solved using the quadratic formula, or the standard form of the equation can be solved using the principle of square roots. The second alternative is illustrated here:

$$(x - 8)^2 + (y + 7)^2 = 81$$
$$(y + 7)^2 = 81 - (x - 8)^2$$
$$y + 7 = \pm\sqrt{81 - (x - 8)^2} \qquad \text{Using the principle of square roots}$$
$$y = -7 \pm \sqrt{81 - (x - 8)^2}.$$

Then we graph

$$y_1 = -7 + \sqrt{81 - (x - 8)^2}$$

and

$$y_2 = -7 - \sqrt{81 - (x - 8)^2}$$

in a squared viewing window.

The hand-drawn graph appears to be correct.

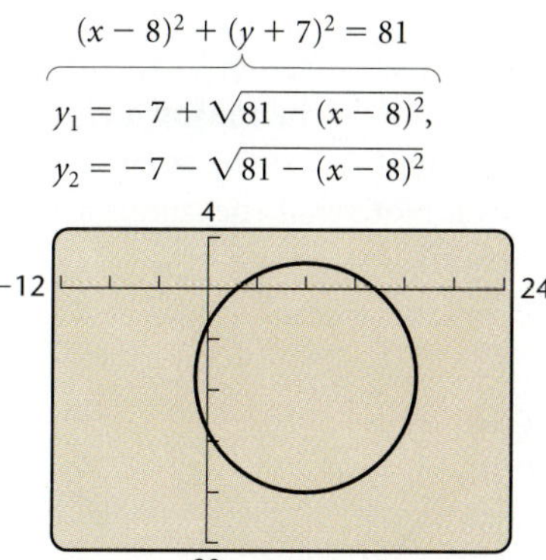

Some graphers have a DRAW feature that provides a quick way to graph a circle when the center and the radius are known.

Ellipses

We have studied two conic sections, the parabola and the circle. Now we turn our attention to a third, the *ellipse*.

> ### Ellipse
>
> An *ellipse* is the set of all points in a plane, the sum of whose distances from two fixed points (the *foci*) is constant. The *center* of an ellipse is the midpoint of the segment between the foci.

We can draw an ellipse by first placing two thumbtacks in a piece of cardboard. These are the foci (singular, *focus*). We then attach a piece of string to the tacks. Its length is the constant sum of the distances $d_1 + d_2$ from the foci to any point on the ellipse. Next, we trace a curve with a pencil held tight against the string. The figure traced is an ellipse.

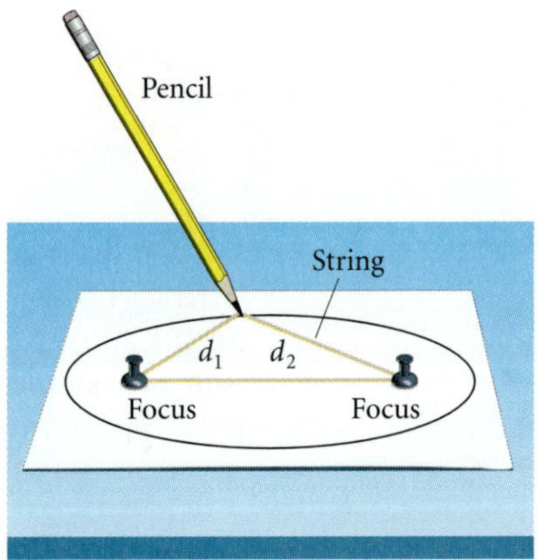

Let's first consider the ellipse shown below with center at the origin. The points F_1 and F_2 are the foci. The segment $\overline{A'A}$ is the **major axis**, and the points A' and A are the **vertices**. The segment $\overline{B'B}$ is the **minor axis**, and the points B and B' are the **y-intercepts**. Note that the major axis of an ellipse is always longer than the minor axis.

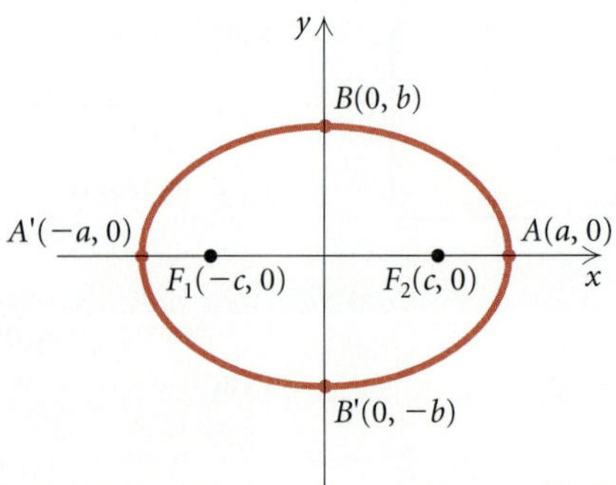

Standard Equation of an Ellipse with Center at the Origin

Major Axis Horizontal

$$\frac{x^2}{a^2} + \frac{y^2}{b^2} = 1, \ a > b > 0$$

Vertices: $(-a, 0), (a, 0)$

y-intercepts: $(0, -b), (0, b)$

Foci: $(-c, 0), (c, 0)$, where $c^2 = a^2 - b^2$

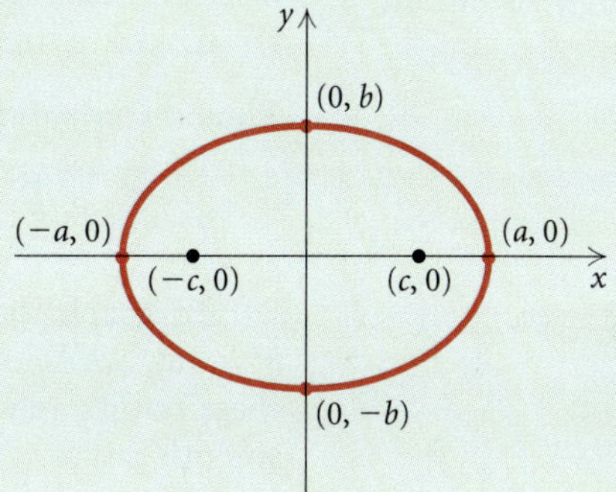

Major Axis Vertical

$$\frac{x^2}{b^2} + \frac{y^2}{a^2} = 1, \ a > b > 0$$

Vertices: $(0, -a), (0, a)$

x-intercepts: $(-b, 0), (b, 0)$

Foci: $(0, -c), (0, c)$, where $c^2 = a^2 - b^2$

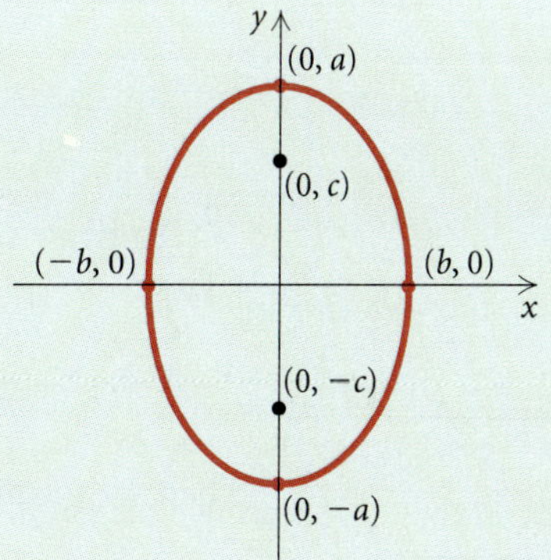

Example 2 Find the standard equation of the ellipse with vertices $(-5, 0)$ and $(5, 0)$ and foci $(-3, 0)$ and $(3, 0)$. Then graph the ellipse.

SOLUTION   Since the foci are on the x-axis and the origin is the midpoint of the segment between them, the major axis is horizontal and $(0, 0)$ is the center of the ellipse. Thus the equation is of the form

$$\frac{x^2}{a^2} + \frac{y^2}{b^2} = 1.$$

Since the vertices are $(-5, 0)$ and $(5, 0)$ and the foci are $(-3, 0)$ and $(3, 0)$, we know that $a = 5$ and $c = 3$. These values can be used to find b^2:

$$c^2 = a^2 - b^2$$
$$3^2 = 5^2 - b^2$$
$$9 = 25 - b^2$$
$$b^2 = 16.$$

Thus the equation of the ellipse is

$$\frac{x^2}{25} + \frac{y^2}{16} = 1.$$

To graph the ellipse, we plot the vertices $(-5, 0)$ and $(5, 0)$. Since $b^2 = 16$, we know that the y-intercepts are $(0, -4)$ and $(0, 4)$. We plot these points as well and connect the four points we have plotted with a smooth curve.

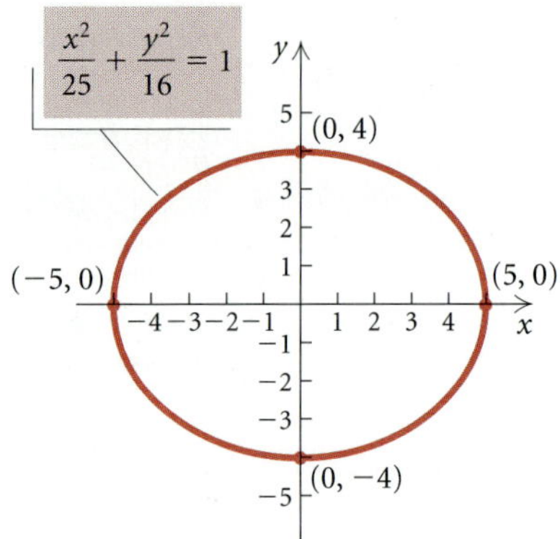

To draw the graph using a grapher, it might be necessary to solve for y first:

$$y = \pm\sqrt{\frac{400 - 16x^2}{25}}.$$

Then we graph

$$y_1 = -\sqrt{\frac{400 - 16x^2}{25}} \quad \text{and} \quad y_2 = \sqrt{\frac{400 - 16x^2}{25}}$$

or

$$y_1 = -\sqrt{\frac{400 - 16x^2}{25}} \quad \text{and} \quad y_2 = -y_1$$

in a squared viewing window.

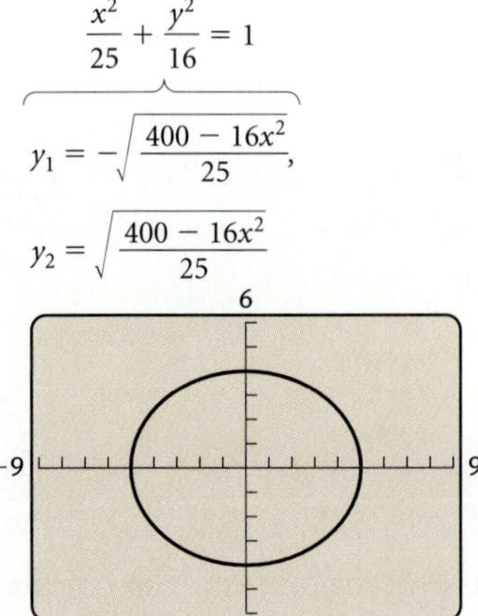

Example 3 For the ellipse

$$9x^2 + 4y^2 = 36,$$

find the vertices and the foci. Then draw the graph.

SOLUTION We first find standard form:

$$9x^2 + 4y^2 = 36$$

$$\frac{9x^2}{36} + \frac{4y^2}{36} = \frac{36}{36} \qquad \text{Dividing by 36 on both sides to get 1 on the right side}$$

$$\frac{x^2}{4} + \frac{y^2}{9} = 1$$

$$\frac{x^2}{2^2} + \frac{y^2}{3^2} = 1. \qquad \text{Writing standard form}$$

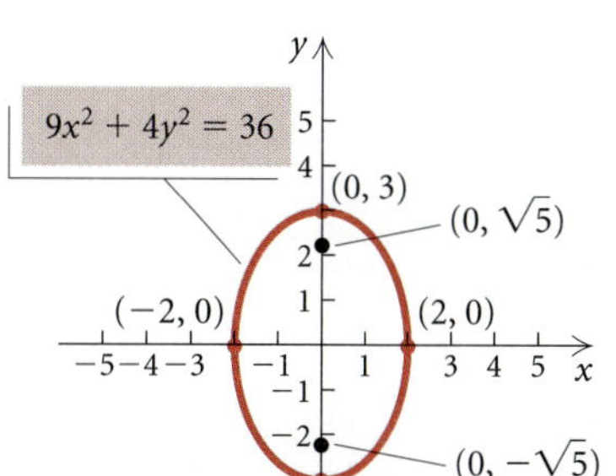

Thus, $a = 3$ and $b = 2$. The major axis is vertical, so the vertices are $(0, -3)$ and $(0, 3)$. Since we know that $c^2 = a^2 - b^2$, we have $c^2 = 9 - 4$, so $c = \sqrt{5}$ and the foci are $(0, -\sqrt{5})$ and $(0, \sqrt{5})$.

To graph the ellipse, we plot the vertices. Note also that since $b = 2$, the x-intercepts are $(-2, 0)$ and $(2, 0)$. We plot these points as well and connect the four points we have plotted with a smooth curve.

If the center of an ellipse is not at the origin but at some point (h, k), then we can think of an ellipse with center at the origin being translated $|h|$ units left or right and $|k|$ units up or down.

Standard Equation of an Ellipse with Center at (h, k)
Major Axis Horizontal

$$\frac{(x - h)^2}{a^2} + \frac{(y - k)^2}{b^2} = 1, \ a > b > 0$$

Vertices: $(h - a, k), (h + a, k)$
Length of minor axis: $2b$
Foci: $(h - c, k), (h + c, k)$, where $c^2 = a^2 - b^2$

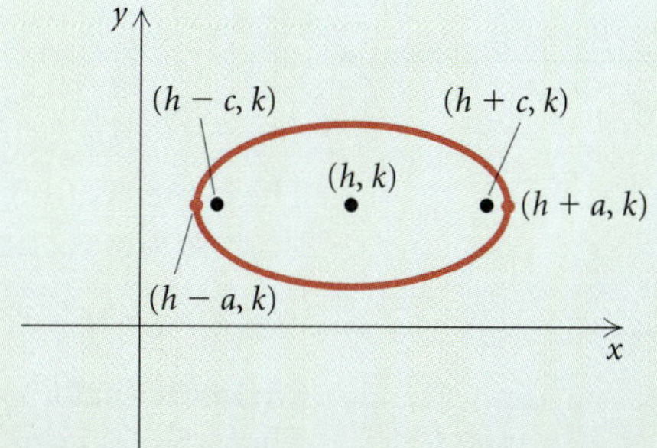

(continued)

Major Axis Vertical

$$\frac{(x - h)^2}{b^2} + \frac{(y - k)^2}{a^2} = 1, \; a > b > 0$$

Vertices: $(h, k - a)$, $(h, k + a)$

Length of minor axis: $2b$

Foci: $(h, k - c)$, $(h, k + c)$, where $c^2 = a^2 - b^2$

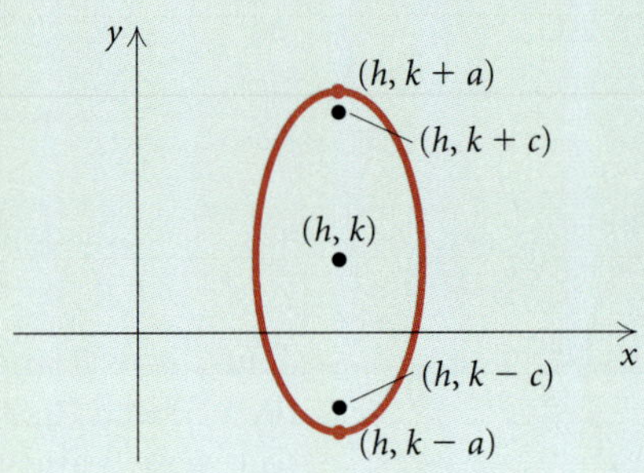

Example 4   For the ellipse

$$4x^2 + y^2 + 24x - 2y + 21 = 0,$$

find the center, the vertices, and the foci. Then draw the graph.

SOLUTION   First we complete the square to get standard form:

$$4x^2 + y^2 + 24x - 2y + 21 = 0$$
$$4(x^2 + 6x \quad\;\;) + (y^2 - 2y \quad\;\;) = -21$$
$$4(x^2 + 6x + 9) + (y^2 - 2y + 1) = -21 + 4 \cdot 9 + 1$$

> **Completing the square twice by adding $4 \cdot 9$ and 1 on both sides**

$$4(x + 3)^2 + (y - 1)^2 = 16$$
$$\frac{1}{16}[4(x + 3)^2 + (y - 1)^2] = \frac{1}{16} \cdot 16$$
$$\frac{(x + 3)^2}{4} + \frac{(y - 1)^2}{16} = 1$$
$$\frac{[x - (-3)]^2}{2^2} + \frac{(y - 1)^2}{4^2} = 1.$$

> **Writing standard form:**
> $$\frac{(x - h)^2}{b^2} + \frac{(y - k)^2}{a^2} = 1$$

The center is $(-3, 1)$. Note that $a = 4$ and $b = 2$. The major axis is vertical, so the vertices are 4 units above and below the center:

$$(-3, 1 + 4) \;\text{ and }\; (-3, 1 - 4), \quad \text{or} \quad (-3, 5) \;\text{ and }\; (-3, -3).$$

We know that $c^2 = a^2 - b^2$, so $c^2 = 16 - 4 = 12$ and $c = \sqrt{12} = 2\sqrt{3}$. Then the foci are $2\sqrt{3}$ units above and below the center:

$$(-3, 1 + 2\sqrt{3}) \;\text{ and }\; (-3, 1 - 2\sqrt{3}).$$

To graph the ellipse, we plot the vertices. Note also that since $b = 2$, two other points on the graph are the endpoints of the minor axis, 2 units right and left of the center:

$$(-3 + 2, 1) \text{ and } (-3 - 2, 1), \quad \text{or} \quad (-1, 1) \text{ and } (-5, 1).$$

We plot these points as well and connect the four points with a smooth curve.

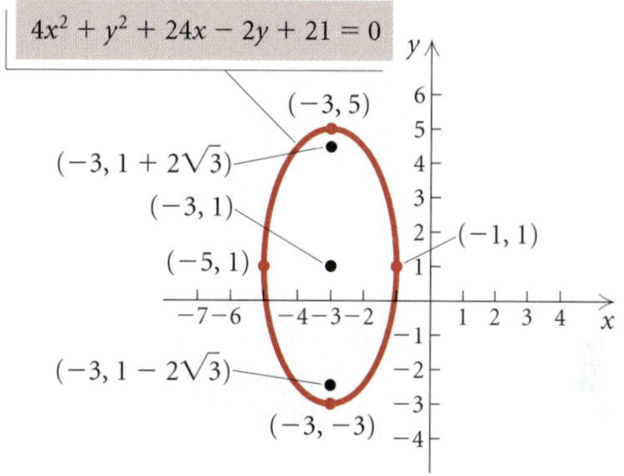

Applications

One of the most exciting recent applications of an ellipse is a medical device called a *lithotripter*. This machine uses underwater shock waves to crush kidney stones. The waves originate at one focus of an ellipse and are reflected to the kidney stone, which is positioned at the other focus. Recovery time following the use of this technique is much shorter than with conventional surgery and the mortality rate is far lower.

A room with an ellipsoidal ceiling is known as a *whispering gallery*. In such a room, a word whispered at one focus can be clearly heard at the other. Whispering galleries are found in the rotunda of the Capitol Building in Washington, D.C., and in the Mormon Tabernacle in Salt Lake City.

Ellipses have many other applications. Planets travel around the sun in elliptical orbits with the sun at one focus, for example, and satellites travel around the earth in elliptical orbits as well.

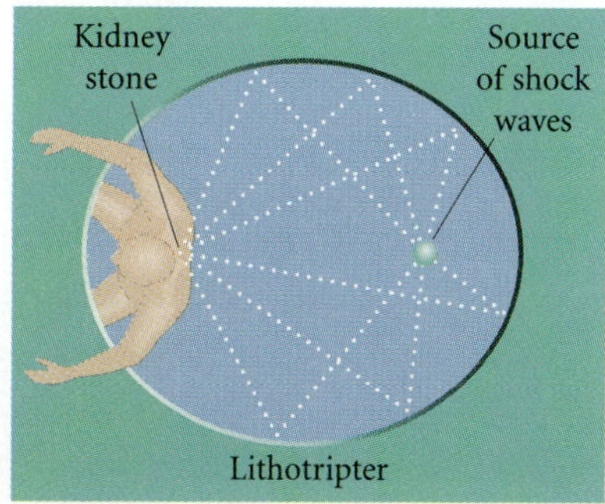

Lithotripter

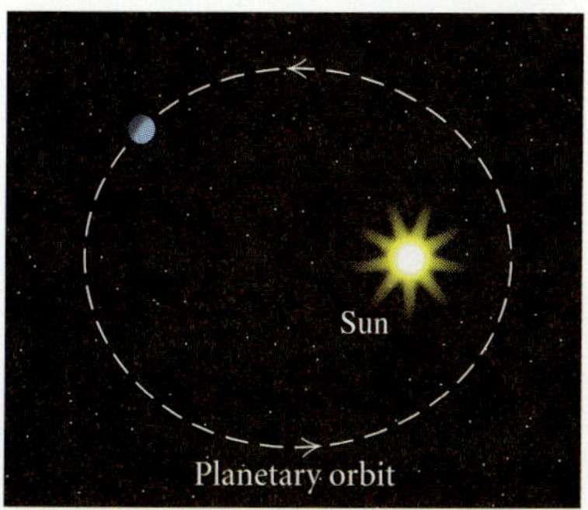

Planetary orbit

5.2 Exercise Set

In Exercises 1–6, match each equation with one of the graphs (a)–(f), which follow.

a)
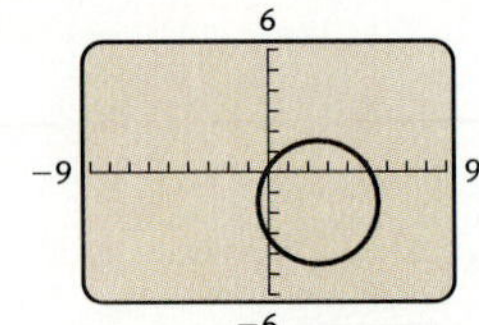

b)
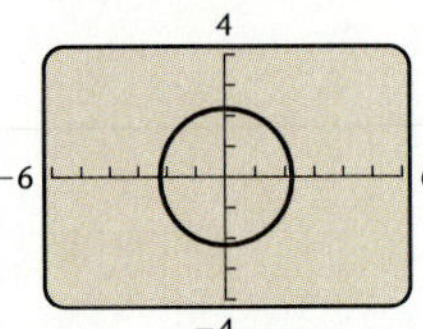

c)
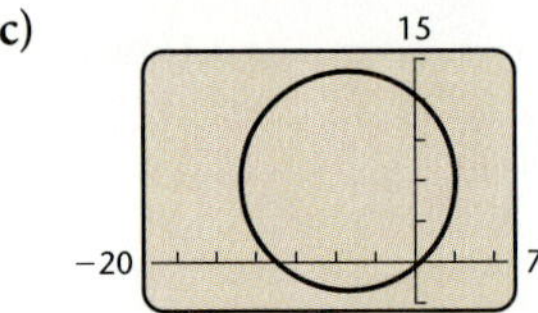

d)
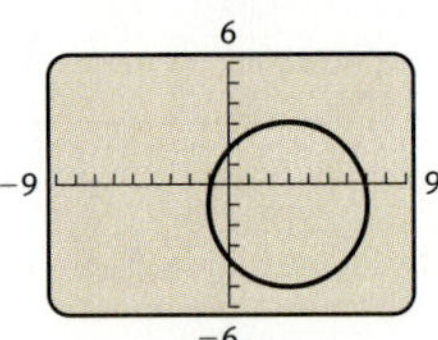

e)
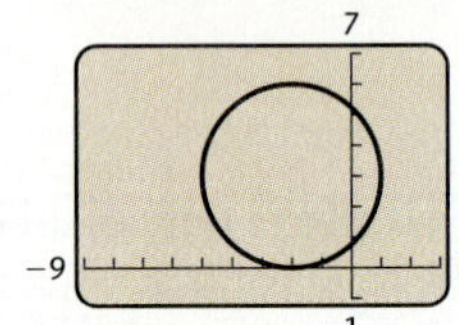

f)
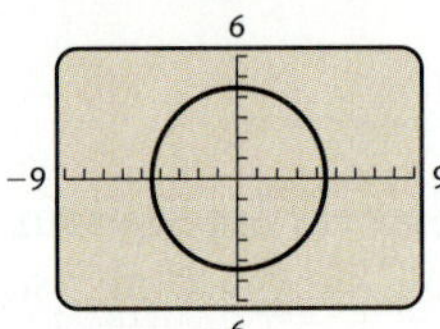

1. $x^2 + y^2 = 5$

2. $y^2 = 20 - x^2$

3. $x^2 + y^2 - 6x + 2y = 6$

4. $x^2 + y^2 + 10x - 12y = 3$

5. $x^2 + y^2 - 5x + 3y = 0$

6. $x^2 + 4x - 2 = 6y - y^2 - 6$

Find the center and the radius of the circle with the given equation. Then draw the graph.

7. $x^2 + y^2 - 14x + 4y = 11$

8. $x^2 + y^2 + 2x - 6y = -6$

9. $x^2 + y^2 + 4x - 6y - 12 = 0$

10. $x^2 + y^2 - 8x - 2y - 19 = 0$

11. $x^2 + y^2 + 6x - 10y = 0$

12. $x^2 + y^2 - 7x - 2y = 0$

13. $x^2 + y^2 - 9x = 7 - 4y$

14. $y^2 - 6y - 1 = 8x - x^2 + 3$

In Exercises 15–18, match each equation with one of the graphs (a)–(d), which follow.

a)
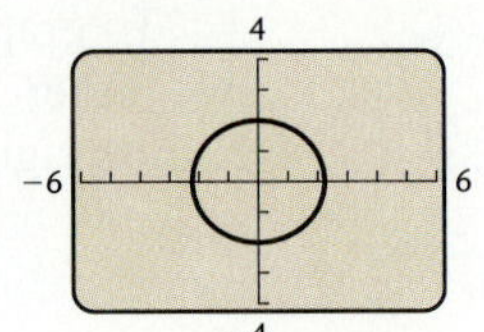

b)
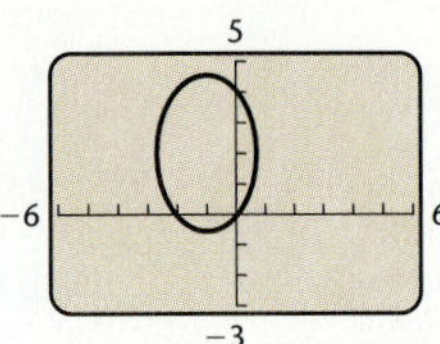

c)
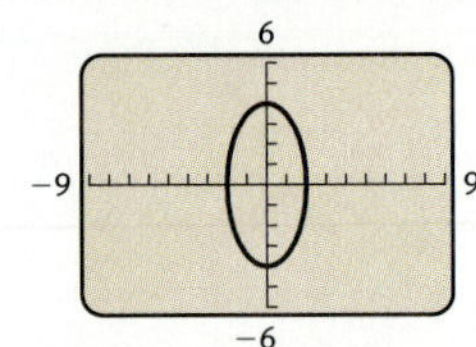

d)
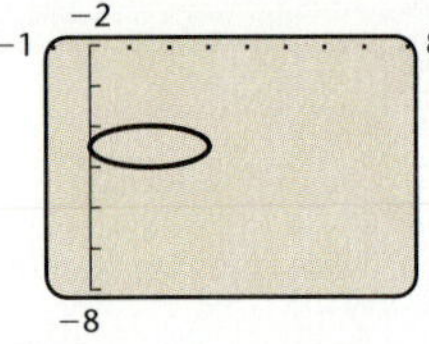

15. $16x^2 + 4y^2 = 64$

16. $4x^2 + 5y^2 = 20$

17. $x^2 + 9y^2 - 6x + 90y = -225$

18. $9x^2 + 4y^2 + 18x - 16y = 11$

Find the vertices and the foci of the ellipse with the given equation. Then draw the graph.

19. $\dfrac{x^2}{4} + \dfrac{y^2}{1} = 1$

20. $\dfrac{x^2}{25} + \dfrac{y^2}{36} = 1$

21. $16x^2 + 9y^2 = 144$

22. $9x^2 + 4y^2 = 36$

23. $2x^2 + 3y^2 = 6$

24. $5x^2 + 7y^2 = 35$

25. $4x^2 + 9y^2 = 1$

26. $25x^2 + 16y^2 = 1$

Find an equation of an ellipse satisfying the given conditions.

27. Vertices: $(-7, 0)$ and $(7, 0)$; foci: $(-3, 0)$ and $(3, 0)$

28. Vertices: $(0, -6)$ and $(0, 6)$; foci: $(0, -4)$ and $(0, 4)$

29. Vertices: $(0, -8)$ and $(0, 8)$; length of minor axis: 10

30. Vertices: $(-5, 0)$ and $(5, 0)$; length of minor axis: 6

31. Foci: $(-2, 0)$ and $(2, 0)$; length of major axis: 6

32. Foci: $(0, -3)$ and $(0, 3)$; length of major axis: 10

Find the center, the vertices, and the foci of each ellipse. Then draw the graph.

33. $\dfrac{(x - 1)^2}{9} + \dfrac{(y - 2)^2}{4} = 1$

34. $\dfrac{(x - 1)^2}{1} + \dfrac{(y - 2)^2}{4} = 1$

35. $\dfrac{(x + 3)^2}{25} + \dfrac{(y - 5)^2}{36} = 1$

36. $\dfrac{(x-2)^2}{16} + \dfrac{(y+3)^2}{25} = 1$

37. $3(x+2)^2 + 4(y-1)^2 = 192$

38. $4(x-5)^2 + 3(y-4)^2 = 48$

39. $4x^2 + 9y^2 - 16x + 18y - 11 = 0$

40. $x^2 + 2y^2 - 10x + 8y + 29 = 0$

41. $4x^2 + y^2 - 8x - 2y + 1 = 0$

42. $9x^2 + 4y^2 + 54x - 8y + 49 = 0$

*The **eccentricity** of an ellipse is defined as $e = c/a$. For an ellipse, $0 < c < a$, so $0 < e < 1$. When e is close to 0, an ellipse appears to be nearly circular. When e is close to 1, an ellipse is very flat.*

43. Observe the shapes of the ellipses in Examples 2 and 4. Which ellipse has the smaller eccentricity? Confirm your answer by computing the eccentricity of each ellipse.

44. Which ellipse below has the smaller eccentricity? (Assume that the coordinate systems have the same scale.)

a)　　　　　　　　　**b)**

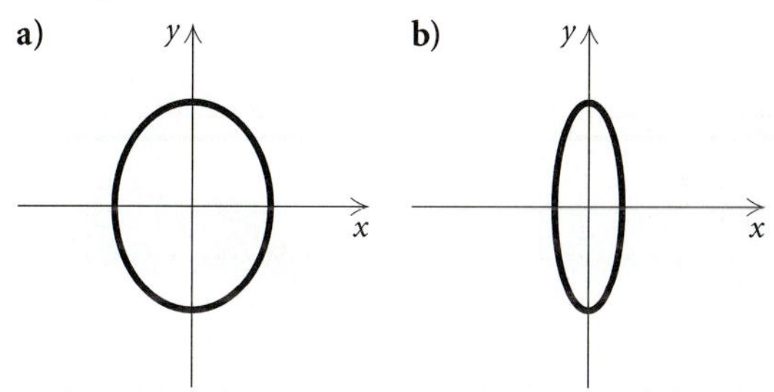

45. Find an equation of an ellipse with vertices $(0, -4)$ and $(0, 4)$ and $e = \frac{1}{4}$.

46. Find an equation of an ellipse with vertices $(-3, 0)$ and $(3, 0)$ and $e = \frac{7}{10}$.

47. *Bridge Supports.* The bridge support shown in the figure below is the top half of an ellipse. Assuming that a coordinate system is superimposed on the drawing in such a way that the center of the ellipse is at point Q, find an equation of the ellipse.

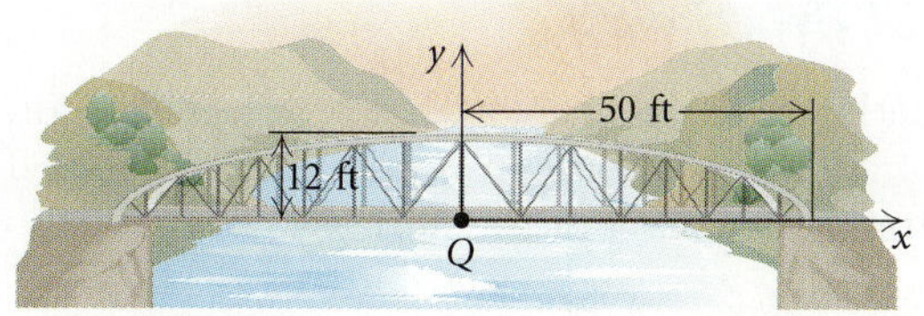

48. *The Ellipse.* In Washington, D.C., there is a large grassy area south of the White House known as the Ellipse. It is actually an ellipse with major axis of length 1048 ft and minor axis of length 898 ft. Assuming that a coordinate system is superimposed on the area in such a way that the center is at the origin and the major and minor axes are on the x- and y-axes of the coordinate system, respectively, find an equation of the ellipse.

49. *The Earth's Orbit.* The maximum distance of the earth from the sun is 9.3×10^7 miles. The minimum distance is 9.1×10^7 miles. The sun is at one focus of the elliptical orbit. Find the distance from the sun to the other focus.

50. *Carpentry.* A carpenter is cutting a 3-ft by 4-ft elliptical sign from a 3-ft by 4-ft piece of plywood. The ellipse will be drawn using a string attached to the board at the foci of the ellipse.

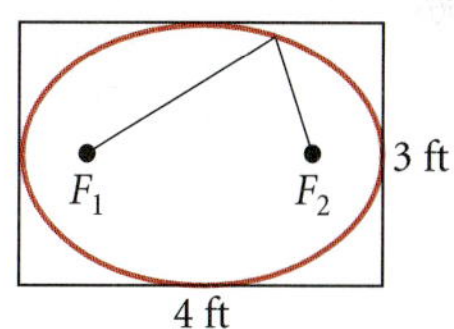

a) How far from the ends of the board should the string be attached?

b) How long should the string be?

Skill Maintenance

Graph.

51. $y = \frac{5}{2}x$

52. $y = -\frac{3}{4}x$

53. $y - 1 = \frac{1}{3}(x + 3)$

54. $y + 2 = -\frac{5}{4}(x - 5)$

Synthesis

55. ◆ Explain why function notation is not used in this section.

56. ◆ Would you prefer to graph the ellipse

$$\dfrac{(x-3)^2}{4} + \dfrac{(y+5)^2}{9} = 1$$

by hand or using a grapher? Explain why you answered as you did.

Find an equation of an ellipse satisfying the given conditions.

57. Vertices: $(3, -4)$, $(3, 6)$; endpoints of minor axis: $(1, 1)$, $(5, 1)$

58. Vertices: $(-1, -1)$, $(-1, 5)$;
endpoints of minor axis: $(-3, 2)$, $(1, 2)$

59. Vertices: $(-3, 0)$ and $(3, 0)$; passing through $\left(2, \frac{22}{3}\right)$

60. Center: $(-2, 3)$; major axis vertical;
length of major axis: 4;
length of minor axis: 1

Use a grapher to find the center and the vertices of each of the following.

61. $4x^2 + 9y^2 - 16.025x + 18.0927y - 11.346 = 0$

62. $9x^2 + 4y^2 + 54.063x - 8.016y + 49.872 = 0$

63. *Bridge Arch.*   A bridge with a semielliptical arch spans a river as shown below. What is the clearance 6 ft from the riverbank?

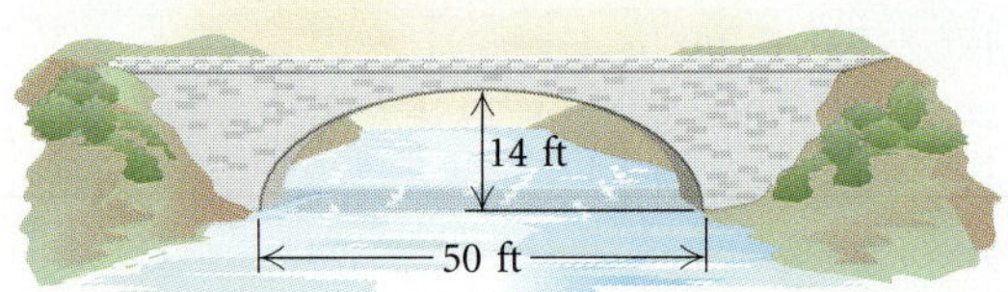

5.3
The Hyperbola

• *Given an equation of a hyperbola, complete the square, if necessary, and then find the center, the vertices, and the foci and graph the hyperbola.*

The last type of conic section that we will study is the *hyperbola*.

> ### Hyperbola
>
> A *hyperbola* is the set of all points in a plane for which the absolute value of the difference of the distances from two fixed points (the *foci*) is constant. The midpoint of the segment between the foci is the *center* of the hyperbola.

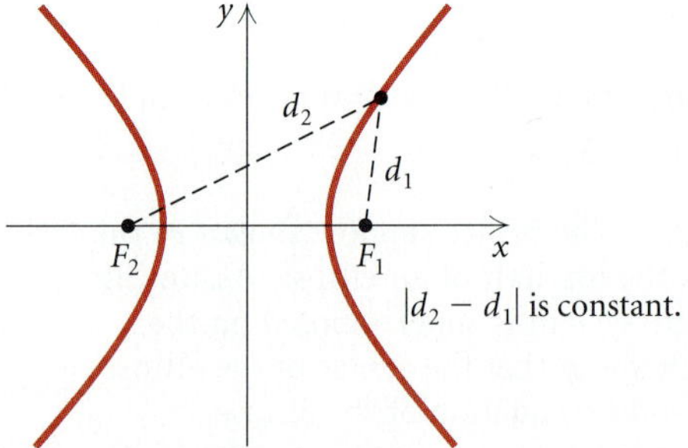

Standard Equations of Hyperbolas

We first consider the equation of a hyperbola with center at the origin. In the figure at the top of the following page, F_1 and F_2 are the foci. The segment $\overline{V_2 V_1}$ is the **transverse axis** and the points V_2 and V_1 are the **vertices**.

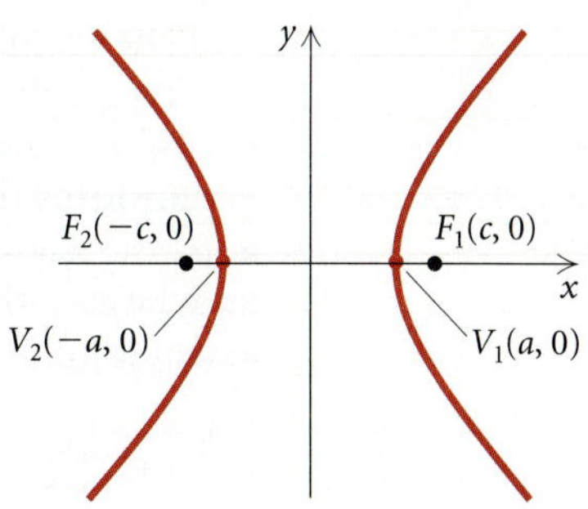

Standard Equation of a Hyperbola with Center at the Origin

Transverse Axis Horizontal

$$\frac{x^2}{a^2} - \frac{y^2}{b^2} = 1$$

Vertices: $(-a, 0), (a, 0)$

Foci: $(-c, 0), (c, 0)$, where $c^2 = a^2 + b^2$

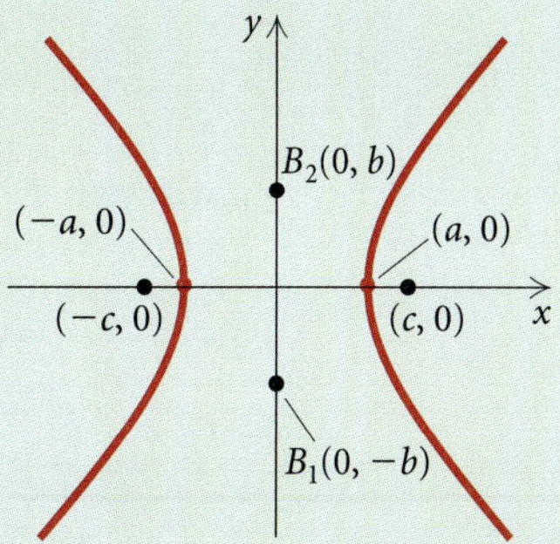

Transverse Axis Vertical

$$\frac{y^2}{a^2} - \frac{x^2}{b^2} = 1$$

Vertices: $(0, -a), (0, a)$

Foci: $(0, -c), (0, c)$, where $c^2 = a^2 + b^2$

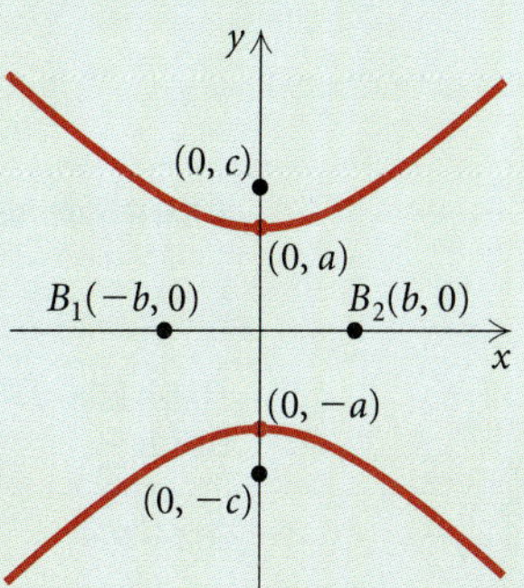

The segment $\overline{B_1B_2}$ is the **conjugate axis** of the hyperbola.

To graph a hyperbola with a horizontal transverse axis, it is helpful to begin by graphing the lines $y = -(b/a)x$ and $y = (b/a)x$. These are the **asymptotes** of the hyperbola. For a hyperbola with a vertical transverse axis, the asymptotes are $y = -(a/b)x$ and $y = (a/b)x$. As $|x|$ gets larger and larger, the graph of the hyperbola gets closer and closer to the asymptotes.

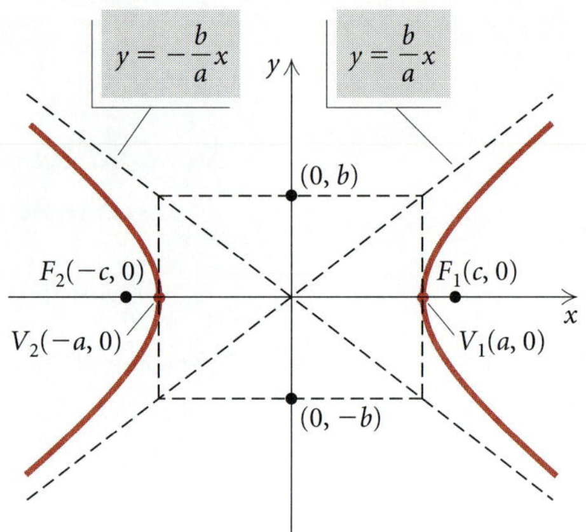

Example 1 Find an equation of the hyperbola with vertices $(0, -4)$ and $(0, 4)$ and foci $(0, -6)$ and $(0, 6)$.

SOLUTION   We know that $a = 4$ and $c = 6$. We find b^2.

$$c^2 = a^2 + b^2$$
$$6^2 = 4^2 + b^2$$
$$36 = 16 + b^2$$
$$20 = b^2$$

Since the vertices and the foci are on the y-axis, we know that the transverse axis is vertical. We can now write the equation of the hyperbola:

$$\frac{y^2}{a^2} - \frac{x^2}{b^2} = 1$$
$$\frac{y^2}{16} - \frac{x^2}{20} = 1.$$

Example 2 For the hyperbola given by

$$9x^2 - 16y^2 = 144,$$

find the vertices, the foci, and the asymptotes. Then graph the hyperbola.

SOLUTION   First, we find standard form:

$$9x^2 - 16y^2 = 144$$

$$\frac{1}{144}(9x^2 - 16y^2) = \frac{1}{144} \cdot 144$$

Multiplying by $\frac{1}{144}$ to get 1 on the right side

$$\frac{x^2}{16} - \frac{y^2}{9} = 1$$

$$\frac{x^2}{4^2} - \frac{y^2}{3^2} = 1.$$

Writing standard form

The hyperbola has a horizontal transverse axis, so the vertices are $(-a, 0)$ and $(a, 0)$, or $(-4, 0)$ and $(4, 0)$. From the standard form of the equation, we know that $a^2 = 4^2$, or 16, and $b^2 = 3^2$, or 9. We find the foci:

$$c^2 = a^2 + b^2$$
$$c^2 = 16 + 9$$
$$c^2 = 25$$
$$c = 5.$$

Thus the foci are $(-5, 0)$ and $(5, 0)$.

Next we find the asymptotes:

$$y = \frac{b}{a}x = \frac{3}{4}x$$

and

$$y = -\frac{b}{a}x = -\frac{3}{4}x.$$

To draw the graph, we sketch the asymptotes first. This is easily done by drawing the rectangle with horizontal sides passing through $(0, 3)$ and $(0, -3)$ and vertical sides through $(4, 0)$ and $(-4, 0)$. Then we draw and extend the diagonals of this rectangle. The two extended diagonals are the asymptotes of the hyperbola. Next, we plot the vertices and draw the branches of the hyperbola outward from the vertices toward the asymptotes.

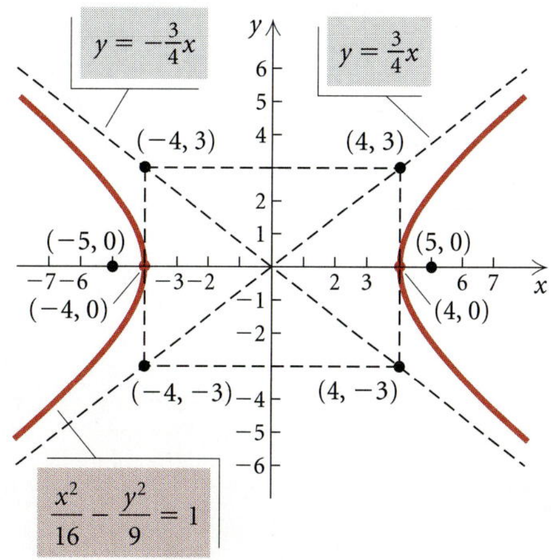

We can use a grapher as a check. It might be necessary to solve for y first and then graph the top and bottom halves of the hyperbola in the same squared viewing window.

$$\frac{x^2}{16} - \frac{y^2}{9} = 1$$

$$y_1 = \sqrt{\frac{9x^2 - 144}{16}}, \quad y_2 = -\sqrt{\frac{9x^2 - 144}{16}}$$

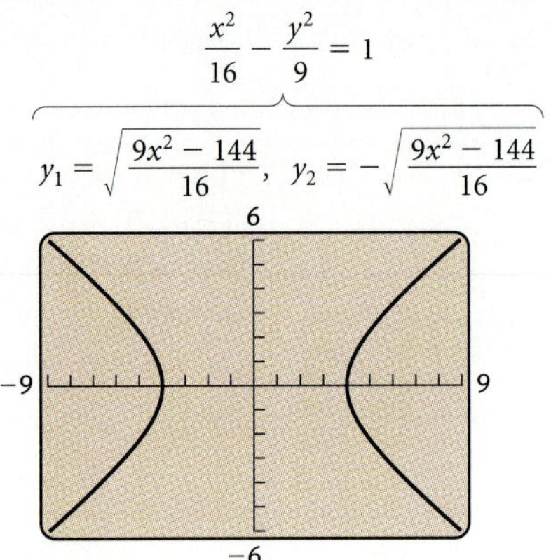

If a hyperbola with center at the origin is translated $|h|$ units left or right and $|k|$ units up or down, the center is at the point (h, k).

Standard Equation of a Hyperbola with Center (h, k)
Transverse Axis Horizontal

$$\frac{(x - h)^2}{a^2} - \frac{(y - k)^2}{b^2} = 1$$

Vertices: $(h - a, k), (h + a, k)$

Asymptotes: $y - k = \dfrac{b}{a}(x - h), \; y - k = -\dfrac{b}{a}(x - h)$

Foci: $(h - c, k), (h + c, k)$, where $c^2 = a^2 + b^2$

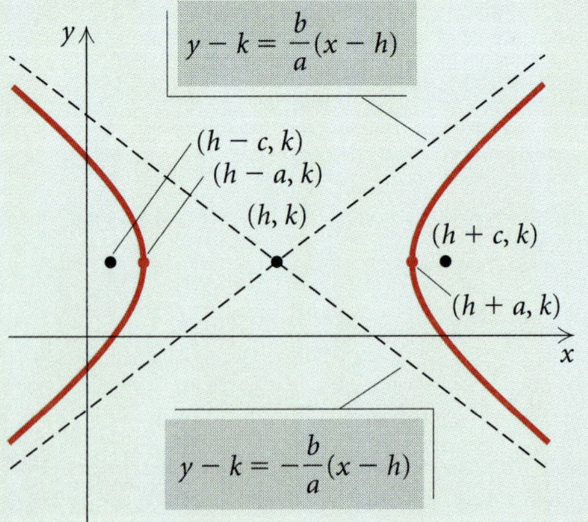

(continued)

Transverse Axis Vertical

$$\frac{(y - k)^2}{a^2} - \frac{(x - h)^2}{b^2} = 1$$

Vertices: $(h, k - a)$, $(h, k + a)$

Asymptotes: $y - k = \dfrac{a}{b}(x - h)$, $y - k = -\dfrac{a}{b}(x - h)$

Foci: $(h, k - c)$, $(h, k + c)$, where $c^2 = a^2 + b^2$

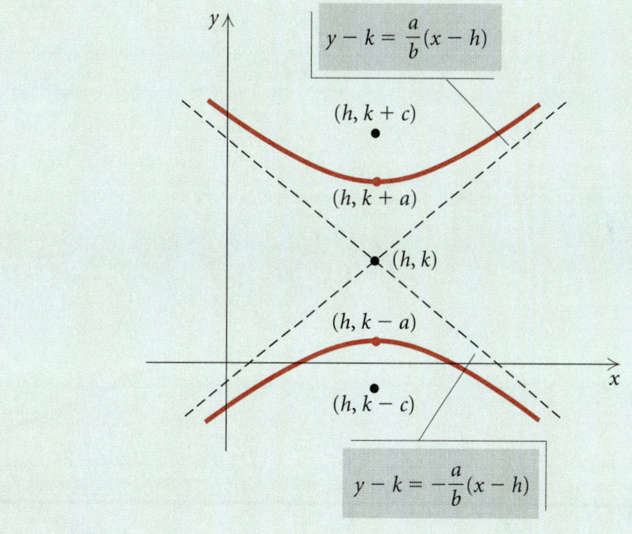

Example 3 For the hyperbola given by

$$4y^2 - x^2 + 24y + 4x + 28 = 0,$$

find the center, the vertices, and the foci. Then draw the graph.

SOLUTION First, we complete the square to get standard form:

$$4y^2 - x^2 + 24y + 4x + 28 = 0$$

$$4(y^2 + 6y \qquad) - (x^2 - 4x \qquad) = -28$$

$$4(y^2 + 6y + 9 - 9) - (x^2 - 4x + 4 - 4) = -28$$

$$4(y^2 + 6y + 9) - (x^2 - 4x + 4) = -28 + 36 - 4$$

$$4(y + 3)^2 - (x - 2)^2 = 4$$

$$\frac{(y + 3)^2}{1} - \frac{(x - 2)^2}{4} = 1 \qquad \textbf{Dividing by 4}$$

$$\frac{[y - (-3)]^2}{1^2} - \frac{(x - 2)^2}{2^2} = 1. \qquad \textbf{Standard form}$$

The center is $(2, -3)$. Note that $a = 1$ and $b = 2$. The transverse axis is vertical, so the vertices are 1 unit below and above the center:

$$(2, -3 - 1) \text{ and } (2, -3 + 1), \quad \text{or} \quad (2, -4) \text{ and } (2, -2).$$

We know that $c^2 = a^2 + b^2$, so $c^2 = 1 + 4 = 5$ and $c = \sqrt{5}$. Thus the foci are $\sqrt{5}$ units below and above the center:

$$(2, -3 - \sqrt{5}) \quad \text{and} \quad (2, -3 + \sqrt{5}).$$

The asymptotes are

$$y - (-3) = \frac{1}{2}(x - 2) \quad \text{and} \quad y - (-3) = -\frac{1}{2}(x - 2),$$

or

$$y + 3 = \frac{1}{2}(x - 2) \quad \text{and} \quad y + 3 = -\frac{1}{2}(x - 2).$$

We sketch the asymptotes, plot the vertices, and draw the graph.

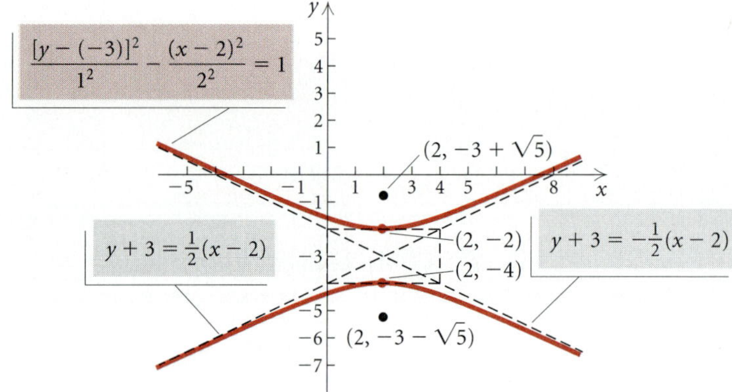

Applications

Some comets travel in hyperbolic paths with the sun at one focus. Such comets pass by the sun only one time unlike those with elliptical orbits, which reappear at intervals. A cross section of an amphitheater might be one branch of a hyperbola. A cross section of a nuclear cooling tower might also be a hyperbola.

One other important application of hyperbolas is in the long range navigation system LORAN. This system uses transmitting stations in three locations to send out simultaneous signals to a ship or aircraft. The difference in the arrival times of the signals from one pair of transmitters is recorded on the ship or aircraft. This difference is also recorded for signals from another pair of transmitters. For each pair, a computation is performed to determine the difference in the distances from each member of the pair to the ship or aircraft. If each pair of differences is kept constant, two hyperbolas can be drawn. Each has one of the pairs of transmitters as foci and the ship or aircraft lies on the intersection of two of their branches.

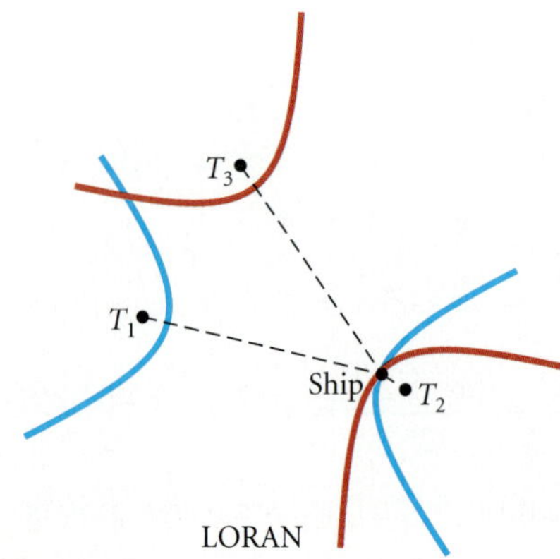

5.3 | *Exercise Set*

In Exercises 1–6, match each equation with one of the graphs (a)–(f), which follow.

a) 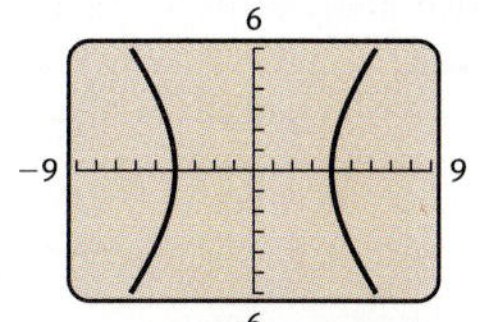

b) 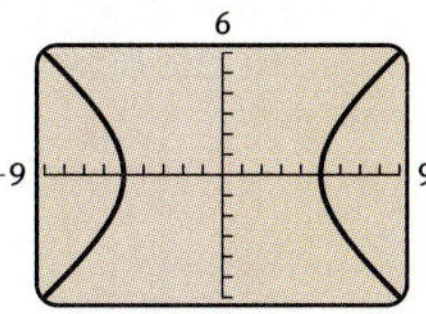

c) 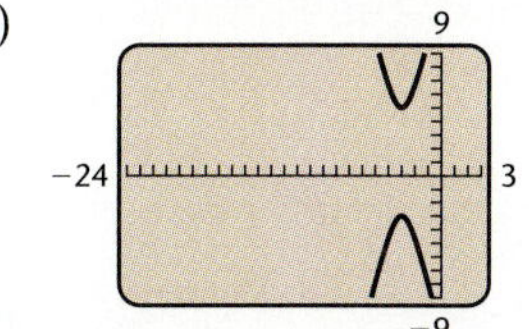

d) 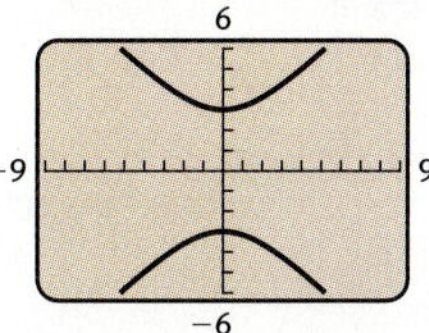

e) 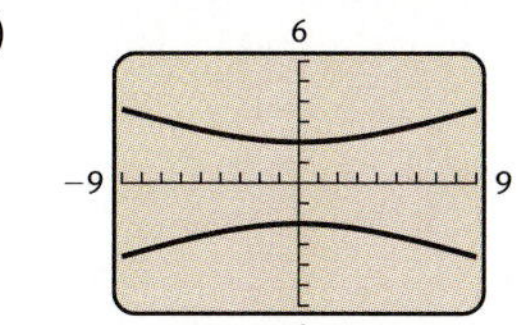

f) 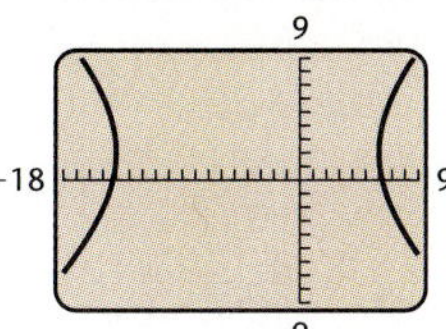

1. $\dfrac{x^2}{25} - \dfrac{y^2}{9} = 1$

2. $\dfrac{y^2}{4} - \dfrac{x^2}{36} = 1$

3. $\dfrac{(y - 1)^2}{16} - \dfrac{(x + 3)^2}{1} = 1$

4. $\dfrac{(x + 4)^2}{100} - \dfrac{(y - 2)^2}{81} = 1$

5. $25x^2 - 16y^2 = 400$

6. $y^2 - x^2 = 9$

Find an equation of a hyperbola satisfying the given conditions.

7. Vertices at $(0, 3)$ and $(0, -3)$; foci at $(0, 5)$ and $(0, -5)$

8. Vertices at $(1, 0)$ and $(-1, 0)$; foci at $(2, 0)$ and $(-2, 0)$

9. Asymptotes $y = \frac{3}{2}x$, $y = -\frac{3}{2}x$; one vertex $(2, 0)$

10. Asymptotes $y = \frac{5}{4}x$, $y = -\frac{5}{4}x$; one vertex $(0, 3)$

Find the center, the vertices, the foci, and the asymptotes. Then draw the graph.

11. $\dfrac{x^2}{4} - \dfrac{y^2}{4} = 1$

12. $\dfrac{x^2}{1} - \dfrac{y^2}{9} = 1$

13. $\dfrac{(x - 2)^2}{9} - \dfrac{(y + 5)^2}{1} = 1$

14. $\dfrac{(x - 5)^2}{16} - \dfrac{(y + 2)^2}{9} = 1$

15. $\dfrac{(y + 3)^2}{4} - \dfrac{(x + 1)^2}{16} = 1$

16. $\dfrac{(y + 4)^2}{25} - \dfrac{(x + 2)^2}{16} = 1$

17. $x^2 - 4y^2 = 4$

18. $4x^2 - y^2 = 16$

19. $9y^2 - x^2 = 81$

20. $y^2 - 4x^2 = 4$

21. $x^2 - y^2 = 2$

22. $x^2 - y^2 = 3$

23. $y^2 - x^2 = \frac{1}{4}$

24. $y^2 - x^2 = \frac{1}{9}$

Find the center, the vertices, the foci, and the asymptotes of each hyperbola. Then draw the graph.

25. $x^2 - y^2 - 2x - 4y - 4 = 0$

26. $4x^2 - y^2 + 8x - 4y - 4 = 0$

27. $36x^2 - y^2 - 24x + 6y - 41 = 0$

28. $9x^2 - 4y^2 + 54x + 8y + 41 = 0$

29. $9y^2 - 4x^2 - 18y + 24x - 63 = 0$

30. $x^2 - 25y^2 + 6x - 50y = 41$

31. $x^2 - y^2 - 2x - 4y = 4$

32. $9y^2 - 4x^2 - 54y - 8x + 41 = 0$

33. $y^2 - x^2 - 6x - 8y - 29 = 0$

34. $x^2 - y^2 = 8x - 2y - 13$

*The **eccentricity** of a hyperbola is defined as $e = c/a$. For a hyperbola, $c > a > 0$, so $e > 1$. When e is close to 1, a hyperbola appears to be very narrow. As the eccentricity increases, the hyperbola becomes "wider."*

35. Observe the shapes of the hyperbolas in Examples 2 and 3. Which hyperbola has the larger eccentricity? Confirm your answer by computing the eccentricity of each hyperbola.

36. Which hyperbola below has the larger eccentricity? (Assume that the coordinate systems have the same scale.)

a)

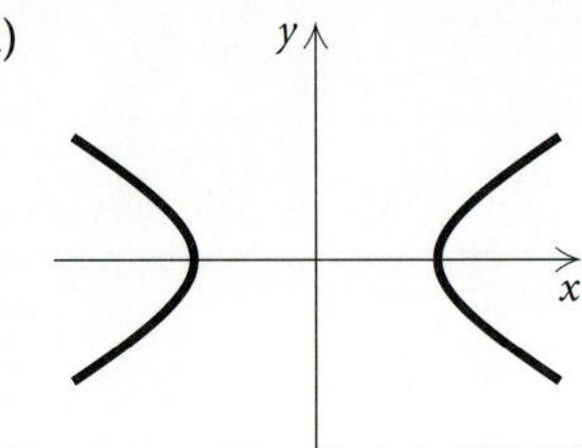

b)

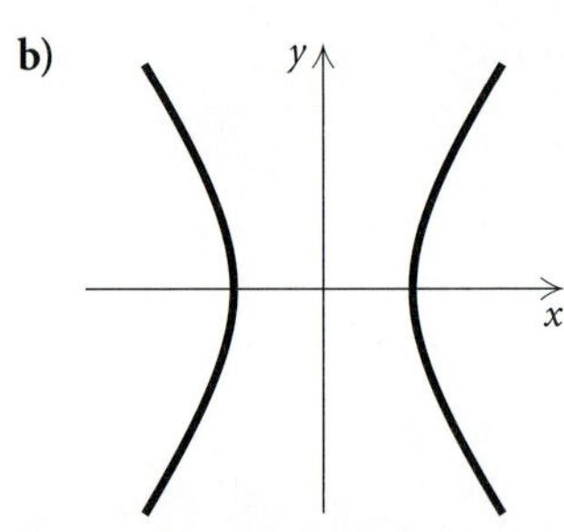

37. Find an equation of a hyperbola with vertices $(3, 7)$ and $(-3, 7)$ and $e = \frac{5}{3}$.

38. Find an equation of a hyperbola with vertices $(-1, 3)$ and $(-1, 7)$ and $e = 4$.

39. *Hyperbolic Mirror.* Certain telescopes contain both a parabolic and a hyperbolic mirror. In the telescope shown in the figure below, the parabola and the hyperbola share focus F_1, which is 14 m above the vertex of the parabola. The hyperbola's second focus F_2 is 2 m above the parabola's vertex. The vertex of the hyperbolic mirror is 1 m below F_1. Position a coordinate system with the origin at the center of the hyperbola and with the foci on the y-axis. Then find the equation of the hyperbola.

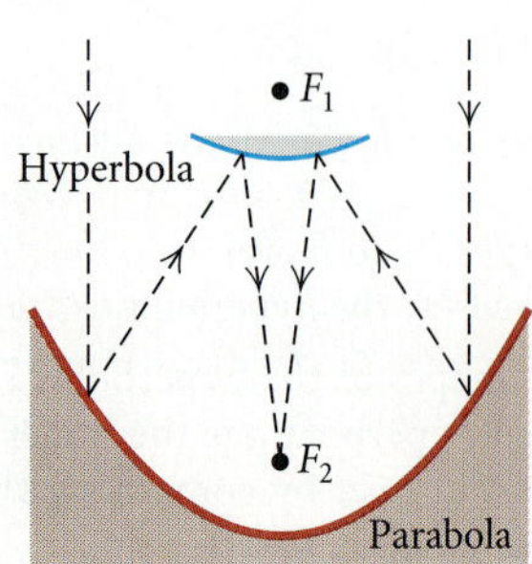

40. *Nuclear Cooling Tower.* A cross section of a nuclear cooling tower is a hyperbola with equation

$$\frac{x^2}{90^2} - \frac{y^2}{130^2} = 1.$$

The tower is 450 ft tall and the distance from the top of the tower to the center of the hyperbola is half the distance from the base of the tower to the center of the hyperbola. Find the diameter of the top and the base of the tower.

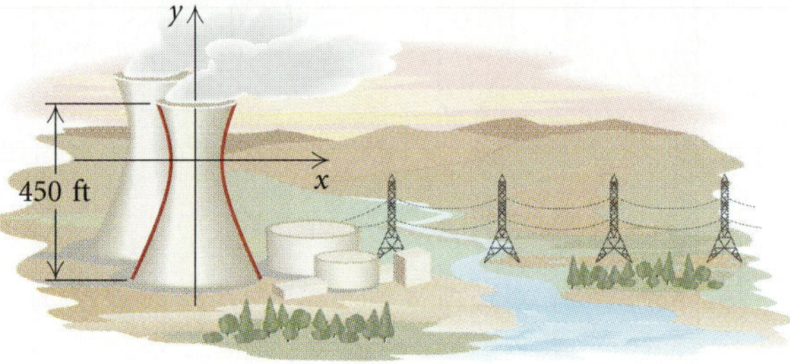

Skill Maintenance

Solve.

41. $x + y = 5$,
  $x - y = 7$

42. $3x - 2y = 5$,
  $5x + 2y = 3$

43. $2x - 3y = 7$,
  $3x + 5y = 1$

44. $3x + 2y = -1$,
  $2x + 3y = 6$

Synthesis

45. ◈ How does the graph of a parabola differ from the graph of one branch of a hyperbola?

46. ◈ Would you prefer to graph the hyperbola

$$\frac{(y + 5)^2}{16} + \frac{(x - 4)^2}{36} = 1$$

by hand or using a grapher? Explain why you answered as you did.

Find an equation of a hyperbola satisfying the given conditions.

47. Vertices at $(3, -8)$ and $(3, -2)$; asymptotes $y = 3x - 14$, $y = -3x + 4$

48. Vertices at $(-9, 4)$ and $(-5, 4)$; asymptotes $y = 3x + 25$, $y = -3x - 17$

Use a grapher to find the center, the vertices, and the asymptotes.

49. $5x^2 - 3.5y^2 + 14.6x - 6.7y + 3.4 = 0$

50. $x^2 - y^2 - 2.046x - 4.088y - 4.228 = 0$

51. *Navigation.*   Two radio transmitters positioned
300 mi apart send simultaneous signals to a ship
that is 200 mi offshore, sailing parallel to the
shoreline. The signal from transmitter S reaches the
ship 200 microseconds later than the signal from
transmitter T. The signals travel at a speed of
186,000 miles per second, or 0.186 mile per
microsecond. Find the equation of the hyperbola
with foci S and T on which the ship is located.
(*Hint:* For any point on the hyperbola, the absolute
value of the difference of its distances from the foci
is $2a$.)

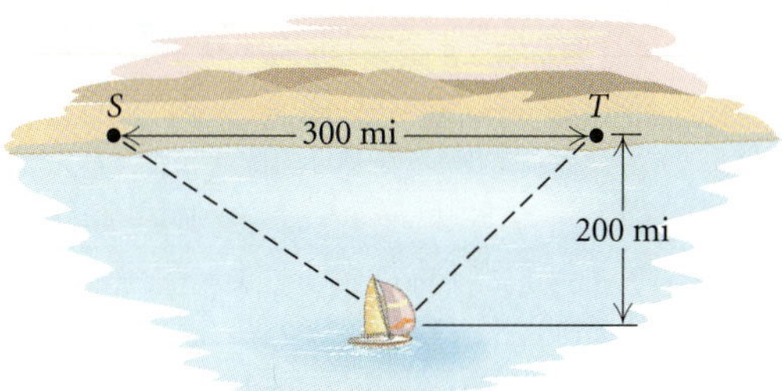

5.4

Nonlinear Systems of Equations

- *Solve a nonlinear system of equations.*
- *Use nonlinear systems of equations to solve applied problems.*

The systems of equations that we have studied so far have been composed
of linear equations. Now we consider systems of two equations in two
variables in which at least one equation is not linear.

Interactive Discovery

Graph the circle $x^2 + y^2 = 5$ and the line $y = x - 4$ in the window
$[-6, 6, -4, 4]$. How many points of intersection are there? What conclusion can you draw about the number of real-number solutions of the system of equations composed of $x^2 + y^2 = 5$ and $y = x - 4$?

Now graph $x^2 + y^2 = 5$ and $y = -0.5x + 2.5$ in the same window
and determine the number of points of intersection. What does each
point of intersection represent?

Finally, graph $x^2 + y^2 = 5$ and $y = x$ in the same window. How
many points of intersection are there? What does each represent?

Nonlinear Systems of Equations

The graph of a nonlinear system of equations can have no point of intersection or one or more points of intersection. The coordinates of each
point of intersection are a solution of the system of equations. When no
point of intersection exists, the system of equations has no real-number
solution.

Real-number solutions of nonlinear systems of equations can be
found algebraically, using the substitution or elimination method, or
graphically. However, since a graph shows *only* real-number solutions,
solutions involving imaginary numbers must always be found algebraically.

Example 1 Solve the following system of equations:

$$x^2 + y^2 = 25, \qquad (1) \qquad \text{(The graph is a circle.)}$$
$$3x - 4y = 0. \qquad (2) \qquad \text{(The graph is a line.)}$$

ALGEBRAIC SOLUTION

We use the substitution method. First, we solve equation (2) for x:

$$x = \tfrac{4}{3}y. \qquad (3) \qquad \textbf{We could have solved for } y \textbf{ instead.}$$

Next, we substitute $\tfrac{4}{3}y$ for x in equation (1) and solve for y:

$$\left(\tfrac{4}{3}y\right)^2 + y^2 = 25$$
$$\tfrac{16}{9}y^2 + y^2 = 25$$
$$\tfrac{25}{9}y^2 = 25$$
$$y^2 = 9 \qquad \textbf{Multiplying by } \tfrac{9}{25}$$
$$y = \pm 3.$$

Now we substitute these numbers for y in equation (3) and solve for x:

$$x = \tfrac{4}{3}(3) = 4, \qquad \textbf{The pair (4, 3) appears to be a solution.}$$
$$x = \tfrac{4}{3}(-3) = -4. \qquad \textbf{The pair } (-4, -3) \textbf{ appears to be a solution.}$$

CHECK: For (4, 3):

$$x^2 + y^2 = 25 \qquad\qquad 3x - 4y = 0$$
$$4^2 + 3^2 \;\overset{?}{=}\; 25 \qquad\qquad 3(4) - 4(3) \;\overset{?}{=}\; 0$$
$$16 + 9 \qquad\qquad\qquad\quad 12 - 12$$
$$25 \;\big|\; 25 \;\; \text{TRUE} \qquad\qquad\quad 0 \;\big|\; 0 \;\; \text{TRUE}$$

For $(-4, -3)$:

$$x^2 + y^2 = 25 \qquad\qquad 3x - 4y = 0$$
$$(-4)^2 + (-3)^2 \;\overset{?}{=}\; 25 \qquad\qquad 3(-4) - 4(-3) \;\overset{?}{=}\; 0$$
$$16 + 9 \qquad\qquad\qquad\qquad -12 + 12$$
$$25 \;\big|\; 25 \;\; \text{TRUE} \qquad\qquad\quad 0 \;\big|\; 0 \;\; \text{TRUE}$$

The pairs (4, 3) and $(-4, -3)$ check, so they are the solutions.

GRAPHICAL SOLUTION

We graph both equations in the same viewing window and note that there are two points of intersection. We can find their coordinates using the INTERSECT feature or the TRACE and ZOOM features.

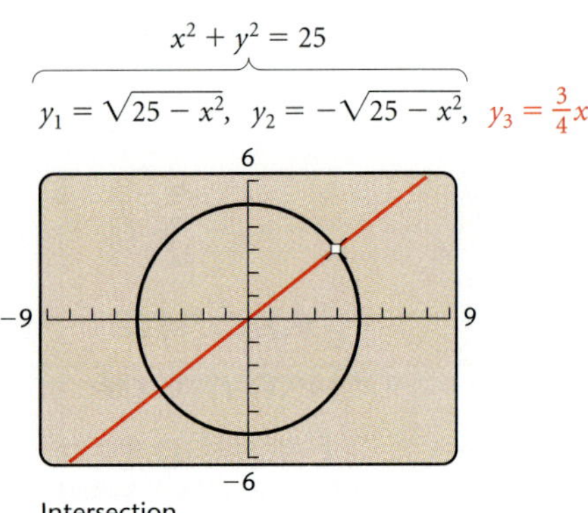

The graph can also be used to check the solutions that we found algebraically.

In the algebraic solution above, suppose that to find x, we had substituted 3 and -3 in equation (1) rather than equation (3). If $y = 3$, $y^2 = 9$, and if $y = -3$, $y^2 = 9$, so both substitutions can be performed at the same time:

$$x^2 + y^2 = 25 \qquad (1)$$
$$x^2 + (\pm 3)^2 = 25$$
$$x^2 + 9 = 25$$
$$x^2 = 16$$
$$x = \pm 4.$$

Thus, if $y = 3$, $x = 4$ or $x = -4$, and if $y = -3$, $x = 4$ or $x = -4$. The possible solutions are $(4, 3)$, $(-4, 3)$, $(4, -3)$, and $(-4, -3)$. A check reveals that $(4, -3)$ and $(-4, 3)$ are not solutions of equation (2). Had we graphed the system of equations before solving algebraically, it would have been clear that there are only two real-number solutions.

Example 2   Solve the following system of equations:

$$x + y = 5, \qquad (1) \qquad \text{(The graph is a line.)}$$
$$y = 3 - x^2. \qquad (2) \qquad \text{(The graph is a parabola.)}$$

We use the substitution method, substituting $3 - x^2$ for y in the first equation:

$$x + 3 - x^2 = 5$$
$$-x^2 + x - 2 = 0$$
$$x^2 - x + 2 = 0.$$

Next, we use the quadratic formula:

$$x = \frac{-b \pm \sqrt{b^2 - 4ac}}{2a} = \frac{-(-1) \pm \sqrt{(-1)^2 - 4(1)(2)}}{2(1)}$$
$$= \frac{1 \pm \sqrt{1 - 8}}{2} = \frac{1 \pm \sqrt{-7}}{2} = \frac{1 \pm i\sqrt{7}}{2} = \frac{1}{2} \pm \frac{\sqrt{7}}{2}i.$$

Now, we substitute these values for x in equation (1) and solve for y:

$$\frac{1}{2} + \frac{\sqrt{7}}{2}i + y = 5$$
$$y = 5 - \frac{1}{2} - \frac{\sqrt{7}}{2}i = \frac{9}{2} - \frac{\sqrt{7}}{2}i$$

and

$$\frac{1}{2} - \frac{\sqrt{7}}{2}i + y = 5$$
$$y = 5 - \frac{1}{2} + \frac{\sqrt{7}}{2}i = \frac{9}{2} + \frac{\sqrt{7}}{2}i.$$

The solutions are

$$\left(\frac{1}{2} + \frac{\sqrt{7}}{2}i, \ \frac{9}{2} - \frac{\sqrt{7}}{2}i\right) \quad \text{and} \quad \left(\frac{1}{2} - \frac{\sqrt{7}}{2}i, \ \frac{9}{2} + \frac{\sqrt{7}}{2}i\right).$$

There are no real-number solutions.

We graph both equations in the same viewing window.

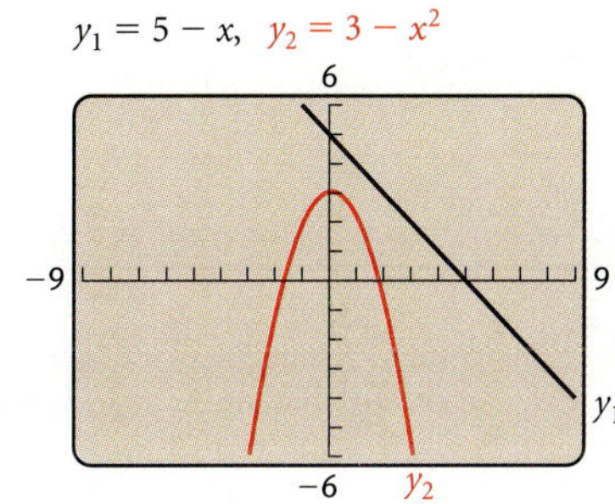

Note that there are no points of intersection. This indicates that there are no real-number solutions. Algebra must be used, as at left, to find the complex-number solutions.

Example 3 Solve the following system of equations:

$$2x^2 + 5y^2 = 39, \qquad (1) \qquad \text{(The graph is an ellipse.)}$$
$$3x^2 - y^2 = -1. \qquad (2) \qquad \text{(The graph is a hyperbola.)}$$

ALGEBRAIC SOLUTION

We use the elimination method. First we multiply equation (2) by 5 and add to eliminate the y^2-term:

$$2x^2 + 5y^2 = 39 \qquad (1)$$
$$\underline{15x^2 - 5y^2 = -5} \qquad \text{Multiplying (2) by 5}$$
$$17x^2 \qquad\quad = 34 \qquad \text{Adding}$$
$$x^2 = 2$$
$$x = \pm\sqrt{2}.$$

If $x = \sqrt{2}$, $x^2 = 2$, and if $x = -\sqrt{2}$, $x^2 = 2$. Thus substituting $\sqrt{2}$ or $-\sqrt{2}$ for x in equation (2) gives us

$$3(\pm\sqrt{2})^2 - y^2 = -1$$
$$3 \cdot 2 - y^2 = -1$$
$$6 - y^2 = -1$$
$$-y^2 = -7$$
$$y^2 = 7$$
$$y = \pm\sqrt{7}.$$

Thus, for $x = \sqrt{2}$, we have $y = \sqrt{7}$ or $y = -\sqrt{7}$, and for $x = -\sqrt{2}$, we have $y = \sqrt{7}$ or $y = -\sqrt{7}$. The possible solutions are $(\sqrt{2}, \sqrt{7})$, $(\sqrt{2}, -\sqrt{7})$, $(-\sqrt{2}, \sqrt{7})$, and $(-\sqrt{2}, -\sqrt{7})$. All four pairs check, so they are the solutions.

GRAPHICAL SOLUTION

We graph both equations in the same viewing window and note that there are four points of intersection. We can use the INTERSECT feature or the TRACE and ZOOM features to find their coordinates.

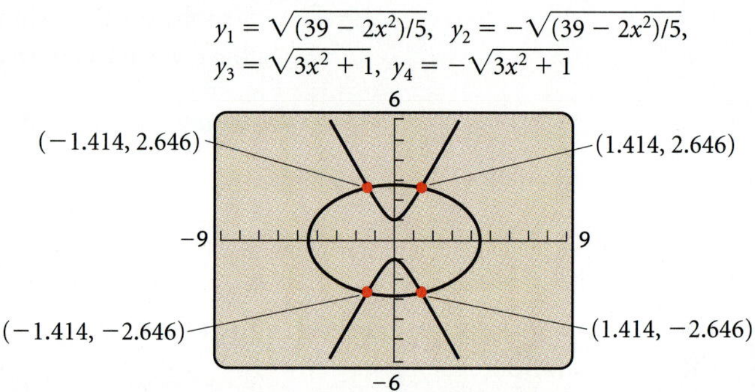

Note that the algebraic method yields exact solutions whereas the graphical method yields decimal approximations of the solutions.

Example 4   Solve the following system of equations:

$$x^2 - 3y^2 = 6, \qquad (1)$$
$$xy = 3. \qquad (2)$$

ALGEBRAIC SOLUTION

We use the substitution method. First, we solve equation (2) for y:

$$xy = 3$$
$$y = \frac{3}{x}. \qquad (3)$$

Then we substitute $3/x$ for y in equation (1) and solve for x:

$$x^2 - 3\left(\frac{3}{x}\right)^2 = 6$$
$$x^2 - 3 \cdot \frac{9}{x^2} = 6$$
$$x^2 - \frac{27}{x^2} = 6$$
$$x^4 - 27 = 6x^2 \qquad \text{Multiplying by } x^2$$
$$x^4 - 6x^2 - 27 = 0$$
$$u^2 - 6u - 27 = 0 \qquad \text{Letting } u = x^2$$
$$(u - 9)(u + 3) = 0 \qquad \text{Factoring}$$
$$u = 9 \quad or \quad u = -3 \qquad \text{Principle of zero products}$$
$$x^2 = 9 \quad or \quad x^2 = -3$$
$$x = \pm 3 \quad or \quad x = \pm i\sqrt{3}.$$

Since $y = 3/x$,

when $x = 3$, $\qquad y = \dfrac{3}{3} = 1;$

when $x = -3$, $\qquad y = \dfrac{3}{-3} = -1;$

when $x = i\sqrt{3}$, $\qquad y = \dfrac{3}{i\sqrt{3}} = \dfrac{3}{i\sqrt{3}} \cdot \dfrac{-i\sqrt{3}}{-i\sqrt{3}} = -i\sqrt{3};$

when $x = -i\sqrt{3}$, $\quad y = \dfrac{3}{-i\sqrt{3}} = \dfrac{3}{-i\sqrt{3}} \cdot \dfrac{i\sqrt{3}}{i\sqrt{3}} = i\sqrt{3}.$

The pairs $(3, 1)$, $(-3, -1)$, $(i\sqrt{3}, -i\sqrt{3})$, and $(-i\sqrt{3}, i\sqrt{3})$ check, so they are the solutions.

GRAPHICAL SOLUTION

We graph both equations in the same viewing window and find the coordinates of their points of intersection.

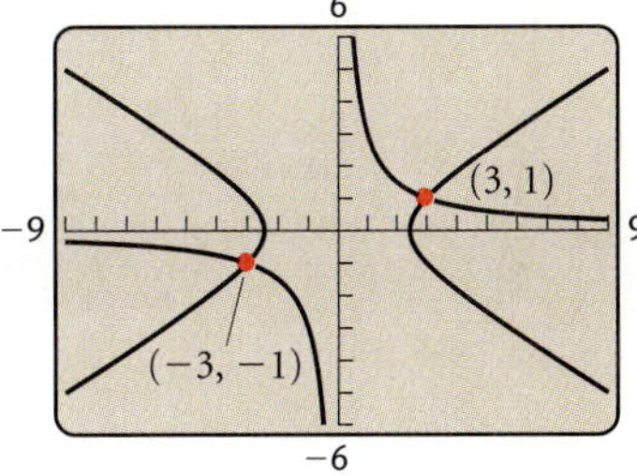

Again, note that the graphical method yields only the real-number solutions of the system of equations. The algebraic method must be used to find *all* the solutions.

Modeling and Problem Solving

Example 5   *Dimensions of a Piece of Land.* For a student recreation building at Southport Community College, an architect wants to lay out a rectangular piece of land that has a perimeter of 204 m and an area of 2565 m^2. Find the dimensions of the piece of land.

SOLUTION

1. Familiarize. We make a drawing and label it, letting l = the length, in meters, and w = the width, in meters.

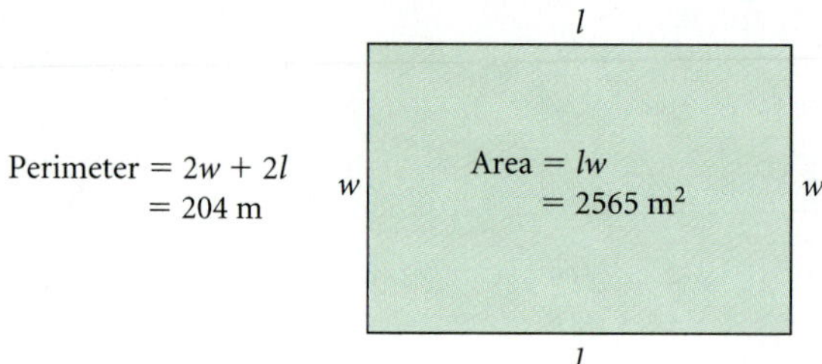

2. Translate. We now have the following:

Perimeter: $2w + 2l = 204$,   (1)

Area: $lw = 2565.$ (2)

3. Carry out. We solve the system of equations both algebraically and graphically.

ALGEBRAIC SOLUTION

We solve the system of equations

$$2w + 2l = 204,$$
$$lw = 2565.$$

Solving the second equation for l gives us $l = 2565/w$. We then substitute $2565/w$ for l in the first equation and solve for w:

$$2w + 2\left(\frac{2565}{w}\right) = 204$$

$$2w^2 + 2(2565) = 204w \qquad \text{Multiplying by } w$$

$$2w^2 - 204w + 2(2565) = 0$$

$$w^2 - 102w + 2565 = 0 \qquad \text{Multiplying by } \tfrac{1}{2}$$

$$(w - 57)(w - 45) = 0$$

$$w = 57 \quad or \quad w = 45.$$

$\qquad\qquad$ **Principle of zero products**

If $w = 57$, then $l = 2565/w = 2565/57 = 45$. If $w = 45$, then $l = 2565/w = 2565/45 = 57$. Since length is generally considered to be longer than width, we have the solution $l = 57$ and $w = 45$, or $(57, 45)$.

GRAPHICAL SOLUTION

We replace l with x and w with y, graph $y_1 = (204 - 2x)/2$ and $y_2 = 2565/x$, and find the point(s) of intersection of the graphs.

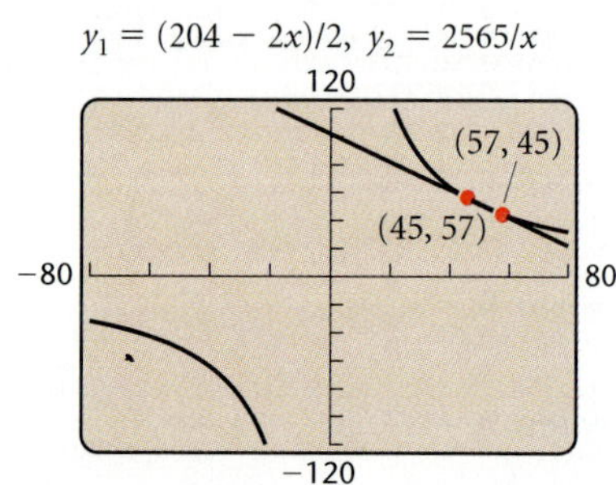

As in the algebraic solution, we have two possible solutions: $(45, 57)$ and $(57, 45)$. Since length (x) is generally considered to be longer than width (y), we have the solution $(57, 45)$.

4. Check. If $l = 57$ and $w = 45$, the perimeter is $2 \cdot 57 + 2 \cdot 45$, or 204. The area is $57 \cdot 45$, or 2565. The numbers check.

5. State. The length of the piece of land is 57 m and the width is 45 m.

5.4 | *Exercise Set*

In Exercises 1–6, match each system of equations with one of the graphs (a)–(f), which follow.

a)
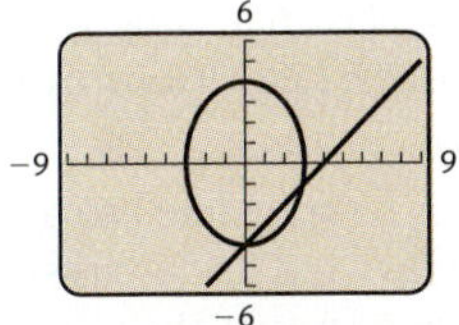

b)
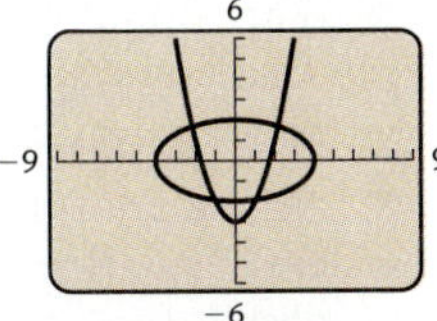

c)
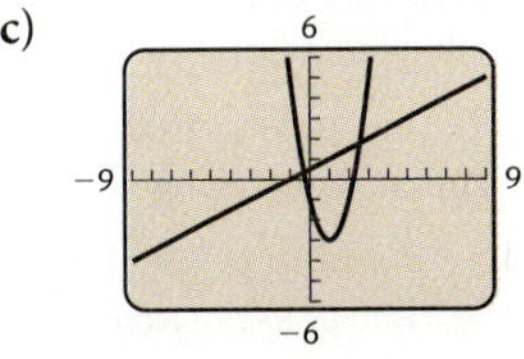

d)
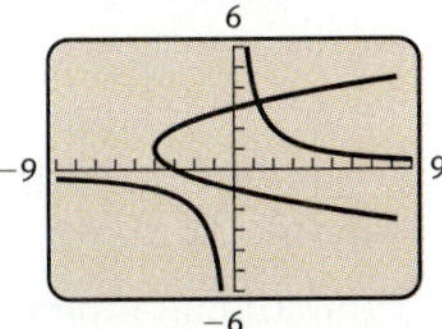

e)
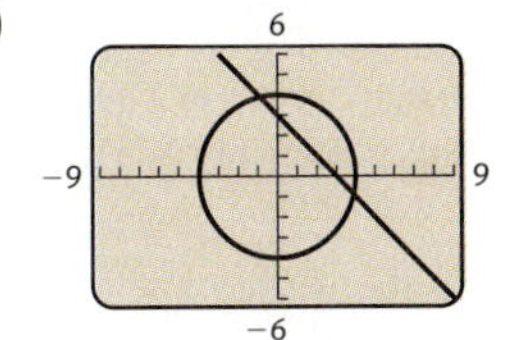

f)
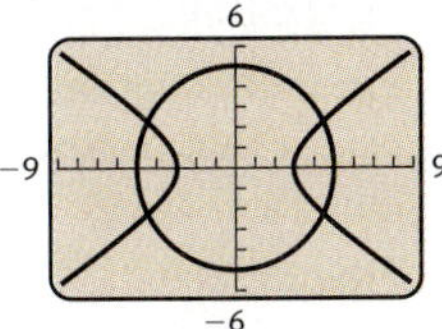

1. $x^2 + y^2 = 16,$
$x + y = 3$

2. $16x^2 + 9y^2 = 144,$
$x - y = 4$

3. $y = x^2 - 4x - 2,$
$2y - x = 1$

4. $4x^2 - 9y^2 = 36,$
$x^2 + y^2 = 25$

5. $y = x^2 - 3,$
$x^2 + 4y^2 = 16$

6. $y^2 - 2y = x + 3,$
$xy = 4$

Solve.

7. $x^2 + y^2 = 25,$
$y - x = 1$

8. $x^2 + y^2 = 100,$
$y - x = 2$

9. $4x^2 + 9y^2 = 36,$
$3y + 2x = 6$

10. $9x^2 + 4y^2 = 36,$
$3x + 2y = 6$

11. $x^2 + y^2 = 25,$
$y^2 = x + 5$

12. $y = x^2,$
$x = y^2$

13. $x^2 + y^2 = 9,$
$x^2 - y^2 = 9$

14. $y^2 - 4x^2 = 4,$
$4x^2 + y^2 = 4$

15. $y^2 - x^2 = 9,$
$2x - 3 = y$

16. $x + y = -6,$
$xy = -7$

17. $y^2 = x + 3,$
$2y = x + 4$

18. $y = x^2,$
$3x = y + 2$

19. $x^2 + y^2 = 25,$
$xy = 12$

20. $x^2 - y^2 = 16,$
$x + y^2 = 4$

21. $x^2 + y^2 = 4,$
$16x^2 + 9y^2 = 144$

22. $x^2 + y^2 = 25,$
$25x^2 + 16y^2 = 400$

23. $x^2 + 4y^2 = 25,$
$x + 2y = 7$

24. $y^2 - x^2 = 16,$
$2x - y = 1$

25. $x^2 - xy + 3y^2 = 27,$
$x - y = 2$

26. $2y^2 + xy + x^2 = 7,$
$x - 2y = 5$

27. $x^2 + y^2 = 16,$
$y^2 - 2x^2 = 10$

28. $x^2 + y^2 = 14,$
$x^2 - y^2 = 4$

29. $x^2 + y^2 = 5,$
$xy = 2$

30. $x^2 + y^2 = 20,$
$xy = 8$

31. $3x + y = 7,$
$4x^2 + 5y = 56$

32. $2y^2 + xy = 5,$
$4y + x = 7$

33. $a + b = 7,$
$ab = 4$

34. $p + q = -4,$
$pq = -5$

35. $x^2 + y^2 = 13,$
$xy = 6$

36. $x^2 + 4y^2 = 20,$
$xy = 4$

37. $x^2 + y^2 + 6y + 5 = 0,$
$x^2 + y^2 - 2x - 8 = 0$

38. $2xy + 3y^2 = 7,$
$3xy - 2y^2 = 4$

39. $2a + b = 1,$
$b = 4 - a^2$

40. $4x^2 + 9y^2 = 36,$
$x + 3y = 3$

41. $a^2 + b^2 = 89,$
$a - b = 3$

42. $xy = 4,$
$x + y = 5$

43. $xy - y^2 = 2,$
$2xy - 3y^2 = 0$

44. $4a^2 - 25b^2 = 0,$
$2a^2 - 10b^2 = 3b + 4$

45. $m^2 - 3mn + n^2 + 1 = 0,$
$3m^2 - mn + 3n^2 = 13$

46. $ab - b^2 = -4,$
$ab - 2b^2 = -6$

47. $x^2 + y^2 = 5,$
$x - y = 8$

48. $4x^2 + 9y^2 = 36,$
$y - x = 8$

49. $a^2 + b^2 = 14,$
$ab = 3\sqrt{5}$

50. $x^2 + xy = 5,$
$2x^2 + xy = 2$

51. $x^2 + y^2 = 25,$
$9x^2 + 4y^2 = 36$

52. $x^2 + y^2 = 1,$
$9x^2 - 16y^2 = 144$

53. $5y^2 - x^2 = 1,$
$xy = 2$

54. $x^2 - 7y^2 = 6,$
$xy = 1$

55. *Picture Frame Dimensions.* Frank's Frame Shop is building a frame for a rectangular oil painting with a perimeter of 28 cm and a diagonal of 10 cm. Find the dimensions of the painting.

56. *Landscaping.*   Green Leaf Landscaping is planting a rectangular wildflower garden with a perimeter of 6 m and a diagonal of $\sqrt{5}$ m. Find the dimensions of the garden.

57. *Pamphlet Design.*   A graphic artist is designing a rectangular advertising pamphlet that is to have an area of 20 in^2 and a perimeter of 18 in. Find the dimensions of the pamphlet.

58. *Sign Dimensions.*   Peden's Advertising is building a rectangular sign with an area of 2 yd^2 and a perimeter of 6 yd. Find the dimensions of the sign.

59. *Banner Design.*   A rectangular banner with an area of $\sqrt{3}$ m^2 is being designed to advertise an exhibit at the Madison Art League. The length of a diagonal is 2 m. Find the dimensions of the banner.

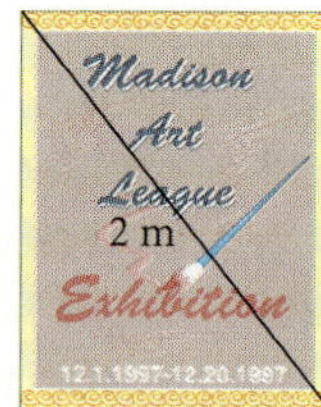

Area = $\sqrt{3}$ m^2

60. *Carpentry.*   A carpenter wants to make a rectangular tabletop with an area of $\sqrt{2}$ m^2 and a diagonal of $\sqrt{3}$ m. Find the dimensions of the tabletop.

61. *Fencing.*   It will take 210 yd of fencing to enclose a rectangular dog run. The area of the run is 2250 yd^2. What are the dimensions of the run?

62. *Office Dimensions.*   The diagonal of the floor of a rectangular office cubicle is 1 ft longer than the length of the cubicle and 3 ft longer than twice the width. Find the dimensions of the cubicle.

63. *Seed Test Plots.*   The Burton Seed Company has two square test plots. The sum of their areas is 832 ft^2 and the difference of their areas is 320 ft^2. Find the length of a side of each plot.

64. *Investment.*   Jenna made an investment for 1 yr that earned $7.50 simple interest. If the principal had been $25 more and the interest rate 1% less, the interest would have been the same. Find the principal and the rate.

Skill Maintenance

Evaluate each expression for the given value of n.

65. $n - \dfrac{1}{n}$, for $n = 5$

66. $(3n + 1)^2$, for $n = 3$

67. $\dfrac{n + 3}{n + 5}$, for $n = 20$

68. $(-1)^n n^2$, for $n = 7$

Synthesis

69. ◈ What would you say to a classmate who tells you that any system of nonlinear equations can be solved graphically?

70. ◈ Write a problem that can be translated to a system of nonlinear equations, and ask a classmate to solve it. Devise the problem so that the solution is "The dimensions of the rectangle are 6 ft by 8 ft."

71. Find an equation of the circle that passes through the points $(2, 4)$ and $(3, 3)$ and whose center is on the line $3x - y = 3$.

72. Find an equation of the circle that passes through the points $(-2, 3)$ and $(-4, 1)$ and whose center is on the line $5x + 8y = -2$.

73. Find an equation of an ellipse centered at the origin that passes through the points $(1, \sqrt{3}/2)$ and $(\sqrt{3}, 1/2)$.

74. Find an equation of a hyperbola of the type
$$\frac{x^2}{b^2} - \frac{y^2}{a^2} = 1$$
that passes through the points $(-3, -3\sqrt{5}/2)$ and $(-3/2, 0)$.

75. Find an equation of the circle that passes through the points $(4, 6)$, $(-6, 2)$, and $(1, -3)$.

76. Find an equation of the circle that passes through the points $(2, 3)$, $(4, 5)$, and $(0, -3)$.

77. Show that a hyperbola does not intersect its asymptotes. That is, solve the system of equations
$$\frac{x^2}{a^2} - \frac{y^2}{b^2} = 1,$$
$$y = \frac{b}{a}x \quad \left(\text{or } y = -\frac{b}{a}x\right).$$

78. *Numerical Relationship.*   Find two numbers whose product is 2 and the sum of whose reciprocals is $\dfrac{33}{8}$.

79. *Numerical Relationship.*   The square of a number exceeds twice the square of another number by $\dfrac{1}{8}$. The sum of their squares is $\dfrac{5}{16}$. Find the numbers.

80. *Box Dimensions.*   Four squares with sides 5 in. long are cut from the corners of a rectangular metal sheet that has an area of 340 in^2. The edges are bent up to form an open box with a volume of 350 in^3. Find the dimensions of the box.

81. *Numerical Relationship.*   The sum of two numbers is 1, and their product is 1. Find the sum of their cubes. There is a method to solve this problem that is easier than solving a nonlinear system of equations. Can you discover it?

82. Solve for x and y:
$$x^2 - y^2 = a^2 - b^2,$$
$$x - y = a - b.$$

Solve.

83. $x^3 + y^3 = 72,$
$x + y = 6$

84. $a + b = \dfrac{5}{6},$
$\dfrac{a}{b} + \dfrac{b}{a} = \dfrac{13}{6}$

85. $p^2 + q^2 = 13,$
$\dfrac{1}{pq} = -\dfrac{1}{6}$

86. $x^2 + y^2 = 4,$
$(x - 1)^2 + y^2 = 4$

87. $5^{x+y} = 100,$
$3^{2x-y} = 1000$

88. $e^x - e^{x+y} = 0,$
$e^y - e^{x-y} = 0$

Solve using a grapher.

89. $y - \ln x = 2,$
$y = x^2$

90. $y = \ln (x + 4),$
$x^2 + y^2 = 6$

91. $e^x - y = 1,$
$3x + y = 4$

92. $y - e^{-x} = 1,$
$y = 2x + 5$

93. $y = e^x,$
$x - y = -2$

94. $y = e^{-x},$
$x + y = 3$

95. $x^2 + y^2 = 19{,}380{,}510.36,$
$27{,}942.25x - 6.125y = 0$

96. $2x + 2y = 1660,$
$xy = 35{,}325$

97. $14.5x^2 - 13.5y^2 - 64.5 = 0,$
$5.5x - 6.3y - 12.3 = 0$

98. $13.5xy + 15.6 = 0,$
$5.6x - 6.7y - 42.3 = 0$

99. $0.319x^2 + 2688.7y^2 = 56{,}548,$
$0.306x^2 - 2688.7y^2 = 43{,}452$

100. $18.465x^2 + 788.723y^2 = 6408,$
$106.535x^2 - 788.723y^2 = 2692$

CHAPTER

5 Summary and Review

Important Properties and Formulas

Standard Equation of a Parabola with Vertex at the Origin

The standard equation of a parabola with vertex $(0, 0)$ and directrix $y = -p$ is

$$x^2 = 4py.$$

The focus is $(0, p)$ and the y-axis is the axis of symmetry.

The standard equation of a parabola with vertex $(0, 0)$ and directrix $x = -p$ is

$$y^2 = 4px.$$

The focus is $(p, 0)$ and the x-axis is the axis of symmetry.

Standard Equation of a Parabola with Vertex (h, k) and Vertical Axis of Symmetry

The standard equation of a parabola with vertex (h, k) and vertical axis of symmetry is

$$(x - h)^2 = 4p(y - k),$$

where the vertex is (h, k), the focus is $(h, k + p)$, and the directrix is $y = k - p.$

Standard Equation of a Parabola with Vertex (h, k) and Horizontal Axis of Symmetry

The standard equation of a parabola with vertex (h, k) and horizontal axis of symmetry is

$$(y - k)^2 = 4p(x - h),$$

where the vertex is (h, k), the focus is $(h + p, k)$, and the directrix is $x = h - p.$

(continued)

Standard Equation of a Circle

The standard equation of a circle with center (h, k) and radius r is

$$(x - h)^2 + (y - k)^2 = r^2.$$

Standard Equation of an Ellipse with Center at the Origin

Major axis horizontal

$$\frac{x^2}{a^2} + \frac{y^2}{b^2} = 1, \ a > b > 0$$

Vertices: $(-a, 0), (a, 0)$

y-intercepts: $(0, -b), (0, b)$

Foci: $(-c, 0), (c, 0)$, where $c^2 = a^2 - b^2$

Major axis vertical

$$\frac{x^2}{b^2} + \frac{y^2}{a^2} = 1, \ a > b > 0$$

Vertices: $(0, -a), (0, a)$

x-intercepts: $(-b, 0), (b, 0)$

Foci: $(0, -c), (0, c)$, where $c^2 = a^2 - b^2$

Standard Equation of an Ellipse with Center at (h, k)

Major axis horizontal

$$\frac{(x - h)^2}{a^2} + \frac{(y - k)^2}{b^2} = 1, \ a > b > 0$$

Vertices: $(h + a, k), (h - a, k)$

Length of minor axis: $2b$

Foci: $(h - c, k), (h + c, k)$, where $c^2 = a^2 - b^2$

Major axis vertical

$$\frac{(x - h)^2}{b^2} + \frac{(y - k)^2}{a^2} = 1, \ a > b > 0$$

Vertices: $(h, k - a), (h, k + a)$

Length of minor axis: $2b$

Foci: $(h, k - c), (h, k + c)$, where $c^2 = a^2 - b^2$

Standard Equation of a Hyperbola with Center at the Origin

Transverse axis horizontal

$$\frac{x^2}{a^2} - \frac{y^2}{b^2} = 1$$

Vertices: $(-a, 0), (a, 0)$

Foci: $(-c, 0), (c, 0)$, where $c^2 = a^2 + b^2$

Transverse axis vertical

$$\frac{y^2}{a^2} - \frac{x^2}{b^2} = 1$$

Vertices: $(0, -a), (0, a)$

Foci: $(0, -c), (0, c)$, where $c^2 = a^2 + b^2$

Standard Equation of a Hyperbola with Center at (h, k)

Transverse axis horizontal

$$\frac{(x - h)^2}{a^2} - \frac{(y - k)^2}{b^2} = 1$$

Vertices: $(h - a, k), (h + a, k)$

Asymptotes: $y - k = \dfrac{b}{a}(x - h),$

$$y - k = -\frac{b}{a}(x - h)$$

Foci: $(h - c, k), (h + c, k)$, where $c^2 = a^2 + b^2$

Transverse axis vertical

$$\frac{(y - k)^2}{a^2} - \frac{(x - h)^2}{b^2} = 1$$

Vertices: $(h, k - a), (h, k + a)$

Asymptotes: $y - k = \dfrac{a}{b}(x - h),$

$$y - k = -\frac{a}{b}(x - h)$$

Foci: $(h, k - c), (h, k + c)$, where $c^2 = a^2 + b^2$

REVIEW EXERCISES

In Exercises 1–8, match each equation with one of the graphs (a)–(h), which follow.

a)
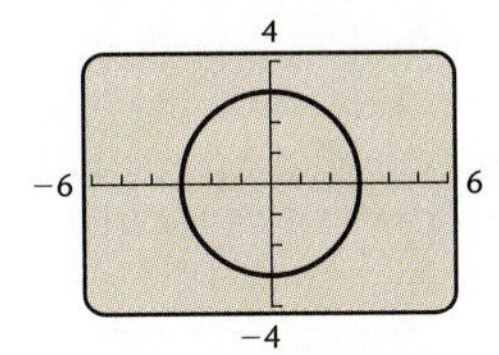

b)
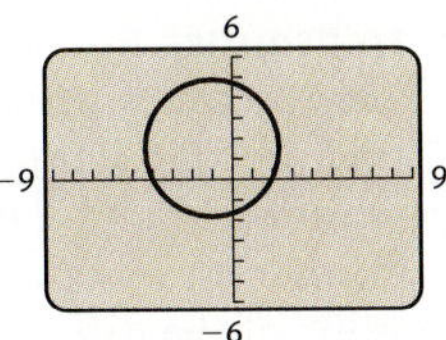

c)
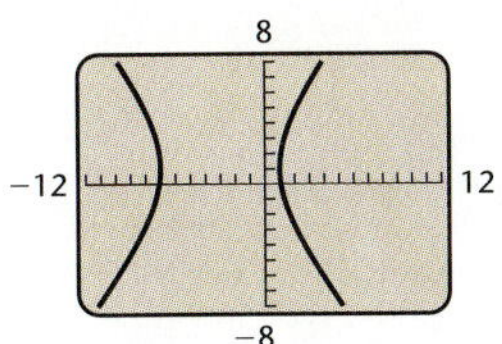

d)
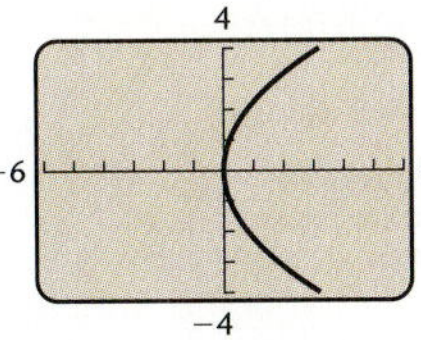

e)
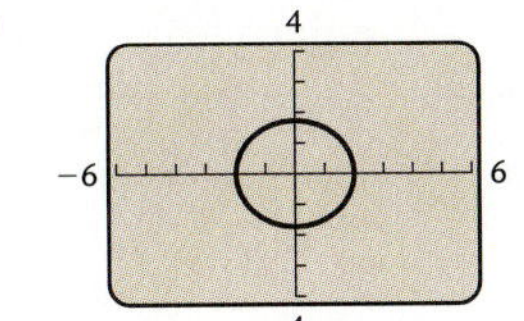

f)
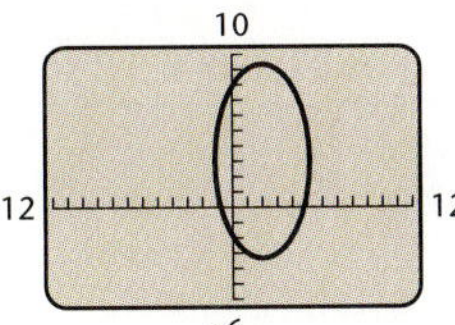

g)
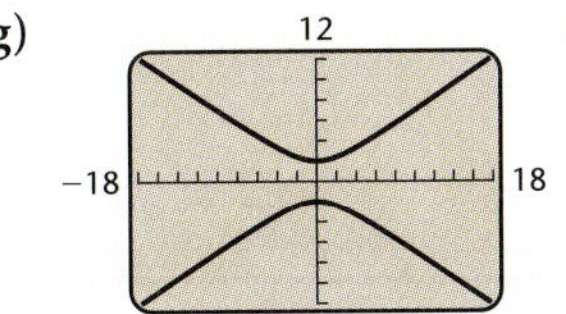

h)
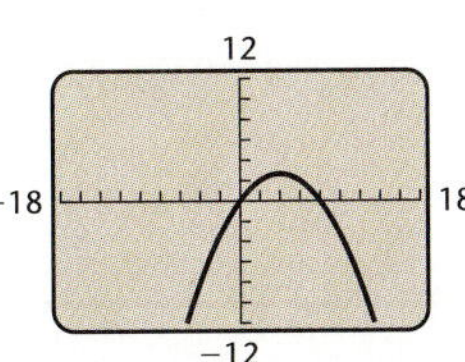

1. $y^2 = 5x$

2. $y^2 = 9 - x^2$

3. $3x^2 + 4y^2 = 12$

4. $9y^2 - 4x^2 = 36$

5. $x^2 + y^2 + 2x - 3y = 8$

6. $4x^2 + y^2 - 16x - 6y = 15$

7. $x^2 - 8x + 6y = 0$

8. $\dfrac{(x + 3)^2}{16} - \dfrac{(y - 1)^2}{25} = 1$

9. Find an equation of the parabola with directrix $y = \frac{3}{2}$ and focus $\left(0, -\frac{3}{2}\right)$.

10. Find the focus, the vertex, and the directrix of the parabola given by
$$y^2 = -12x.$$

11. Find the vertex, the focus, and the directrix of the parabola given by
$$x^2 + 10x + 2y + 9 = 0.$$

12. Find the center, the vertices, and the foci of the ellipse given by
$$16x^2 + 25y^2 - 64x + 50y - 311 = 0.$$
Then draw the graph.

13. Find an equation of the ellipse having vertices $(0, -4)$ and $(0, 4)$ with minor axis of length 6.

14. Find the center, the vertices, the foci, and the asymptotes of the hyperbola given by
$$x^2 - 2y^2 + 4x + y - \tfrac{1}{8} = 0.$$

15. *Spotlight.* A spotlight has a parabolic cross section that is 2 ft wide at the opening and 1.5 ft deep at the vertex. How far from the vertex is the focus?

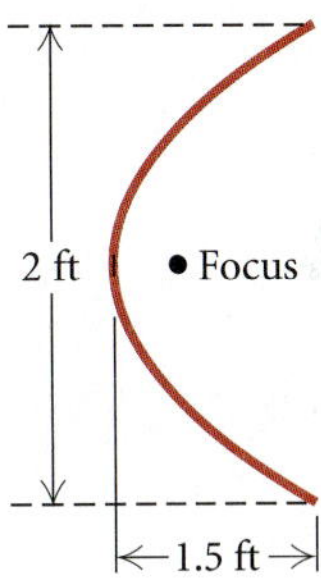

Solve.

16. $x^2 - 16y = 0,$
$\quad x^2 - y^2 = 64$

17. $4x^2 + 4y^2 = 65,$
$\quad 6x^2 - 4y^2 = 25$

18. $x^2 - y^2 = 33,$
$\quad x + y = 11$

19. $x^2 - 2x + 2y^2 = 8,$
$\quad 2x + y = 6$

20. $x^2 - y = 3,$
$\quad 2x - y = 3$

21. $x^2 + y^2 = 25,$
$\quad x^2 - y^2 = 7$

22. $x^2 - y^2 = 3,$
$\quad y = x^2 - 3$

23. $x^2 + y^2 = 18,$
$\quad 2x + y = 3$

24. $x^2 + y^2 = 100,$
$\quad 2x^2 - 3y^2 = -120$

25. $x^2 + 2y^2 = 12,$
$\quad xy = 4$

26. *Numerical Relationship.* The sum of two numbers is 11 and the sum of their squares is 65. Find the numbers.

27. *Dimensions of a Rectangle.* A rectangle has a perimeter of 38 m and an area of 84 m². What are the dimensions of the rectangle?

28. *Numerical Relationship.* Find two positive integers whose sum is 12 and the sum of whose reciprocals is $\frac{3}{8}$.

29. *Perimeter.* The perimeter of a square is 12 cm more than the perimeter of another square. The area of the first square exceeds the area of the other by 39 cm². Find the perimeter of each square.

30. *Radius of a Circle.*   The sum of the areas of two circles is 130π ft². The difference of the areas is 112π ft². Find the radius of each circle.

Synthesis

31. ◈ What would you say to a classmate who tells you that an algebraic solution of a nonlinear system of equations is always preferable to a graphical solution?

32. ◈ Is a circle a special type of ellipse? Why or why not?

33. Find an equation of the ellipse containing the point $(-1/2, 3\sqrt{3}/2)$ and with vertices $(0, -3)$ and $(0, 3)$.

34. Find two numbers whose product is 4 and the sum of whose reciprocals is $\frac{65}{56}$.

35. Find an equation of the circle that passes through the points $(10, 7)$, $(-6, 7)$, and $(-8, 1)$.

36. *Navigation.*   Two radio transmitters positioned 400 mi apart send simultaneous signals to a ship that is 250 mi offshore, sailing parallel to the shoreline. The signal from transmitter A reaches the ship 300 microseconds before the signal from transmitter B. The signals travel at a speed of 186,000 miles per second, or 0.186 mile per microsecond. Find the equation of the hyperbola with foci A and B on which the ship is located. (*Hint:* For any point on the hyperbola, the absolute value of the difference of its distances from the foci is $2a$.)

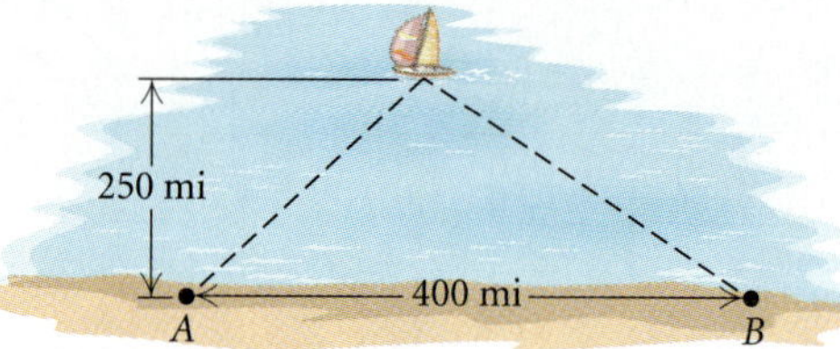

Sequences, Series, and Combinatorics 6

The following sequence involves world bicycle production in years starting with 1988. We consider year 1 to be 1988. The sequence is

$$105, \ 95, \ 90, \ 96, \ 103, \ 108.$$

We can visualize the sequence in a table and a scatterplot, or xy-graph.

A desire to calculate odds in games of chance gave rise to the *theory of probability,* which today has many applications to business, medicine, sociology, and science.

The first part of this chapter is devoted to *sequence* and *series.* For example, when we list world bicycle production for various years, a *sequence* is being formed. When the members of a sequence are numbers, we can think of adding them. Such a sum is called a *series.* We also study a method of proof known as *mathematical induction,* which enables us to prove many important mathematical results.

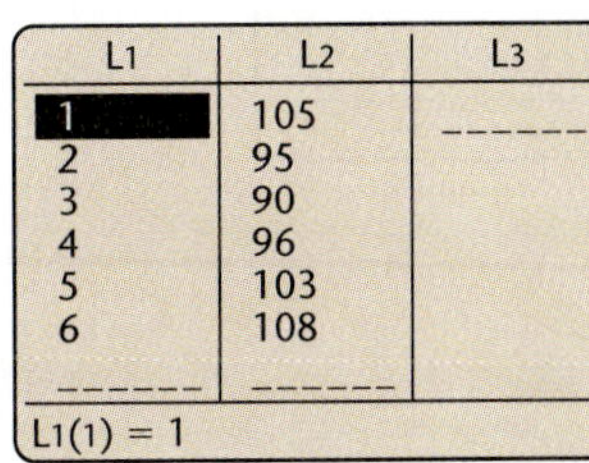

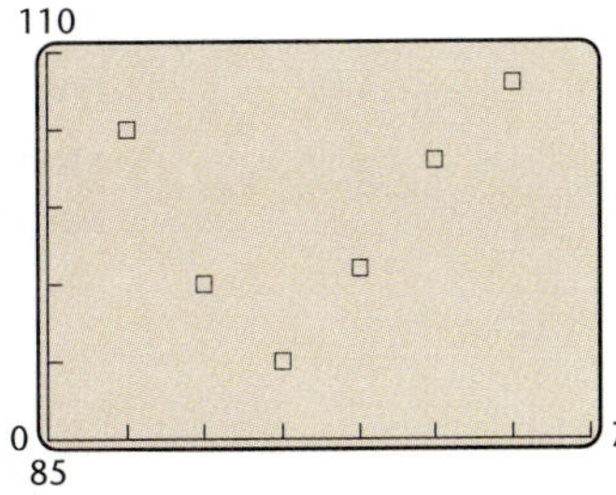

6.1 Sequences and Series

6.2 Arithmetic Sequences and Series

6.3 Geometric Sequences and Series

6.4 Mathematical Induction

6.5 Combinatorics: Permutations

6.6 Combinatorics: Combinations

6.7 The Binomial Theorem

6.8 Probability

SUMMARY AND REVIEW

6.1

Sequences and Series

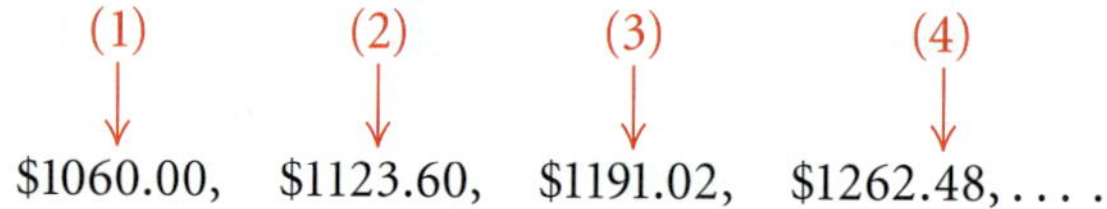

- *Find terms of sequences given the nth term.*
- *Look for a pattern in a sequence and try to determine a general term.*
- *Convert between sigma notation and other notation for a series.*
- *Given a recursively defined sequence, construct terms.*

In this section, we discuss sets of numbers, considered in order, and their sums.

Sequences

Suppose that \$1000 is invested at 6%, compounded annually. The amounts to which the account will grow after 1 yr, 2 yr, 3 yr, 4 yr, and so on, are as follows:

$$
\begin{array}{cccc}
(1) & (2) & (3) & (4) \\
\downarrow & \downarrow & \downarrow & \downarrow \\
\$1060.00, & \$1123.60, & \$1191.02, & \$1262.48, \ldots .
\end{array}
$$

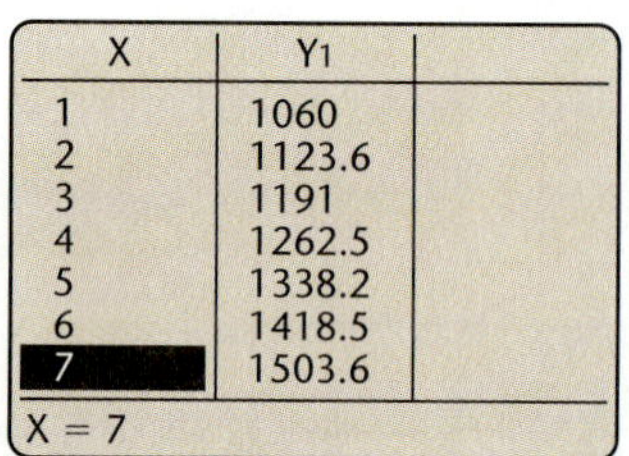

(We found these numbers using the compound interest formula, $A = P(1 + i)^n$, expressed as $y_1 = 1000(1.06)^x$, in order to create the table on a grapher. Note the rounding error in the table at left.) Note that we can think of this ordered list of numbers as a function such that the input **1** corresponds to the number \$1060.00, **2** to the number \$1123.60, **3** to the number \$1191.02, **4** to the number \$1262.48, and so on. A **sequence** is thus a *function*, where the domain is a set of consecutive positive integers beginning with 1. (Sometimes sequences start with 0, but we will seldom consider such cases.)

If we continue to compute the amounts of money in the account forever, we obtain an **infinite sequence**:

$1060.00, \$1123.60, \$1191.02, \$1262.48, \$1338.23, \$1418.52,$
$1503.63, \$1593.85, \ldots .$

The three dots at the end "..." indicate that the sequence goes on without stopping. If we stop after a certain number of years, we obtain a **finite sequence**:

$1060.00, \$1123.60, \$1191.02, \$1262.48.$

Sequences

An *infinite sequence* is a function having for its domain the set of positive integers:

$$\{1, 2, 3, 4, 5, \ldots\}.$$

A *finite sequence* is a function having for its domain a set of positive integers:

$$\{1, 2, 3, 4, 5, \ldots, n\}$$

for some positive integer n.

As another example, consider the sequence given by the formula

$$a(n) = 2^n, \quad \text{or} \quad a_n = 2^n.$$

The notation a_n means the same as $a(n)$ but is more commonly used with sequences. Some of the function values (also known as the **terms** of the sequence) are as follows:

$$a_1 = 2^1 = 2,$$
$$a_2 = 2^2 = 4,$$
$$a_3 = 2^3 = 8,$$
$$a_4 = 2^4 = 16,$$
$$a_5 = 2^5 = 32.$$

X	Y₁	
1	2	
2	4	
3	8	
4	16	
5	32	
6	64	
7	128	
X = 1		

The first term of the sequence is a_1, the fifth term is a_5, and the nth term, or **general term**, is a_n. This sequence can also be denoted in the following ways:

a) $2, 4, 8, \ldots$;

b) $a_1, a_2, a_3, a_4, \ldots, a_n, \ldots$;

c) $2, 4, 8, \ldots, 2^n, \ldots$.

Example 1 Find the first 4 terms and the 37th term of the sequence whose general term is given by

$$a_n = \frac{(-1)^n}{(n+1)}.$$

SOLUTION We have the following:

$$a_1 = \frac{(-1)^1}{1+1} = -\frac{1}{2},$$
$$a_2 = \frac{(-1)^2}{2+1} = \frac{1}{3},$$
$$a_3 = \frac{(-1)^3}{3+1} = -\frac{1}{4},$$
$$a_4 = \frac{(-1)^4}{4+1} = \frac{1}{5},$$
$$a_{37} = \frac{(-1)^{37}}{37+1} = -\frac{1}{38}.$$

Thus the first 4 terms are $-\frac{1}{2}, \frac{1}{3}, -\frac{1}{4},$ and $\frac{1}{5}$, and the 37th term is $-\frac{1}{38}$.

Note in Example 1 that the power $(-1)^n$ causes the signs of the terms to alternate between positive and negative, depending on whether n is even or odd.

Finding the General Term

When only the first few terms of a sequence are known, we do not know for sure what the general term is, but we can make a prediction by looking for a pattern.

Example 2   For each of the following sequences, predict the general term.

a) 1, 4, 9, 16, 25, ...
b) 1, $\sqrt{2}$, $\sqrt{3}$, 2, ...
c) -1, 3, -9, 27, -81, ...
d) 2, 4, 8, ...

SOLUTION

RESULT	REASON
a) n^2	These are squares of consecutive positive integers, so the general term may be n^2.
b) $\sqrt{n}$	These are square roots of consecutive integers, so the general term may be $\sqrt{n}$.
c) $(-1)^n[3^{n-1}]$	These are powers of 3 with alternating signs, so the general term may be $(-1)^n[3^{n-1}]$.
d) Uncertain	If we see the pattern of powers of 2, we will see 16 as the next term and guess 2^n for the general term. Then the sequence could be written with more terms as

$$2, 4, 8, 16, 32, 64, 128, \ldots .$$

If we see that we can get the second term by adding 2, the third term by adding 4, and the next term by adding 6, and so on, we will see 14 as the next term. A general term for the sequence is then $n^2 - n + 2$, and the sequence can be written with more terms as

$$2, 4, 8, 14, 22, 32, 44, 58, \ldots .$$

Example 2(d) illustrates that, in fact, you can never be certain about the general term. The fewer the given terms, the greater the uncertainty.

Visualizing Sequences

There are many ways in which we can see and use sequences. The following sequence involves world bicycle production in years starting with 1988. We consider year 1 to be 1988. The sequence is

$$105, \quad 95, \quad 90, \quad 96, \quad 103, \quad 108.$$

We can visualize the sequence in a table, a bar graph, and a scatterplot, or xy-graph, as follows.

	TABLE	
YEAR	PRODUCTION (IN MILLIONS)	
1	105	
2	95	
3	90	
4	96	
5	103	
6	108	

Source: UN Interbike Directory.

BAR GRAPH

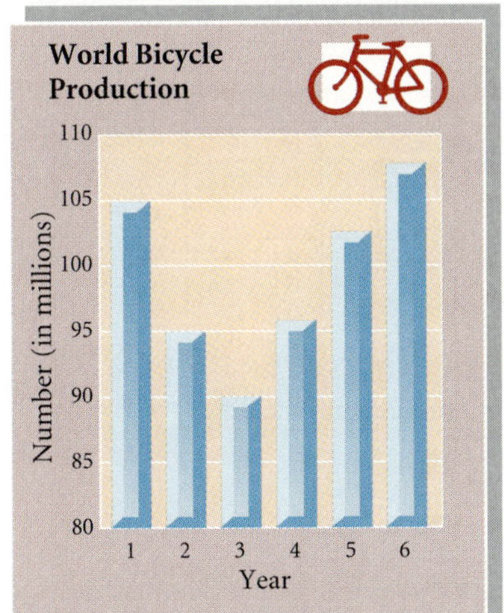

xy-GRAPH (SCATTERPLOT)

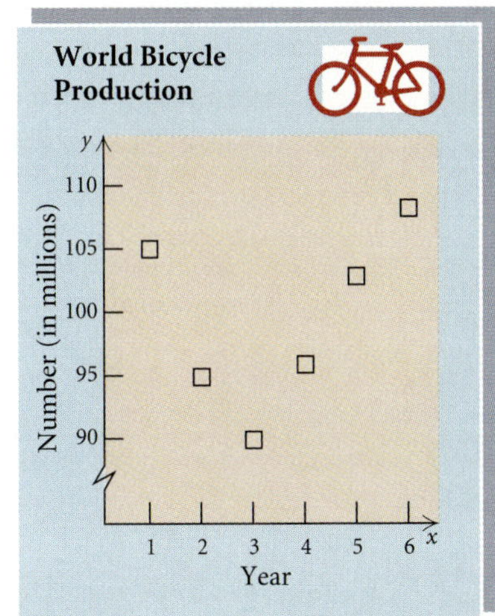

On a grapher, we can create lists and tables and draw graphs. Some graphers have special procedures in order to deal with sequences. For example, we might use a SEQUENCE mode, which allows us to give start (first) and stop (last) values of n. Generally, we start with a value of 1, but there might be situations in which we would start with 0. Then a STEP would be specified, which establishes the distance between values of n. Generally, we use Step = 1, giving 1, 2, 3, 4, ..., but there may be situations in which we might want to count by two's: 0, 2, 4, 6, ..., and so on. The window in a SEQUENCE mode will generally allow us to make these adjustments in sequences and in size of viewing window. Consult the grapher manual for more details. If available, use DOT mode rather than CONNECTED mode.

Another way to work with sequences on graphers is to use a TABLE feature or a software package that creates spreadsheets, such as Lotus 1-2-3 or Excel. A sequence is specified by a formula, a first value and a last value of n, and a step size. We can handle sequences in much the same way as we do other functions. The only difference is often that instead of entering an expression such as "$x/(x + 1)$" at an "$f(x) =$" or a "$y =$" prompt, we must enter a formula such as "$n/(n + 1)$" as a "$U_n =$" prompt. We can write the general term of a sequence by considering a function whose domain is, say, the entire set of real numbers, such as

$$f(x) = x^2 + 1,$$

and restricting its domain to the natural numbers to consider it as

$$a_n = f(n) = n^2 + 1.$$

Program

LOAN: This program can be used to compute various parts of mortgage payments. Payments made to repay a loan form a sequence.

Example 3 Construct a table of values and a graph for the first 10 terms of the sequence whose general term is given by

$$a_n = \frac{n}{n + 1}.$$

Use both a grapher and a spreadsheet program, if available.

SOLUTION   We use {1, 2, 3, 4, 5, 6, 7, 8, 9, 10} as the domain.

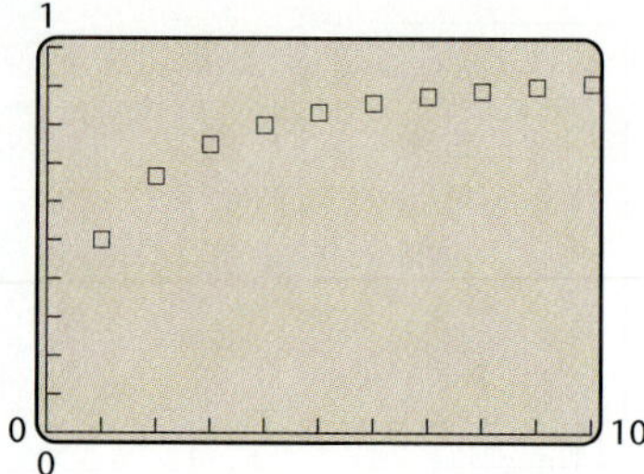

n	Un
1	0.5
2	0.66667
3	0.75
4	0.8
5	0.83333
6	0.85714
7	0.875
8	0.88889
9	0.9
10	0.90909

Sums and Series

Series

Given the infinite sequence

$$a_1, a_2, a_3, a_4, \ldots, a_n, \ldots,$$

the sum of the terms

$$a_1 + a_2 + a_3 + \cdots + a_n + \cdots$$

is called an *infinite series*. A *partial sum* is the sum of the first n terms:

$$a_1 + a_2 + a_3 + \cdots + a_n.$$

A partial sum is also called a *finite series*, or *nth partial sum*, and is denoted S_n.

Consider the sequence

$$3, 5, 7, 9, \ldots, 2n + 1, \ldots.$$

We construct some partial sums:

$$S_1 = 3,$$ **This is the first term of the given sequence.**

$$S_2 = 3 + 5 = 8,$$ **The sum of the first two terms (second partial sum)**

$$S_3 = 3 + 5 + 7 = 15,$$ **The sum of the first three terms (third partial sum)**

$$S_4 = 3 + 5 + 7 + 9 = 24.$$ **The sum of the first four terms (fourth partial sum)**

Note that if we write these partial sums in order, we create a new sequence:

$$3, 8, 15, 24, \ldots.$$

Example 4 For the sequence $-2, 4, -6, 8, -10, 12, -14, \ldots,$ find each of the following.

a) S_3 **b)** S_5

SOLUTION

a) $S_3 = -2 + 4 + (-6) = -4$

b) $S_5 = -2 + 4 + (-6) + 8 + (-10) = -6$

Sigma Notation: Σ

The Greek letter Σ (sigma) can be used to simplify notation when the general term of a series is a formula. For example, the sum of the first four terms of the sequence $3, 5, 7, 9, \ldots, 2k + 1, \ldots$ can be named as follows, using what is called **sigma notation**, or **summation notation**:

$$\sum_{k=1}^{4} (2k + 1).$$

This is read "the sum as k goes from 1 to 4 of $(2k + 1)$." The letter k is called the **index of summation**. Sometimes the index of summation starts at a number other than 1, and sometimes other letters are used rather than k.

Example 5 Find and evaluate each of the following sums.

a) $\displaystyle\sum_{k=1}^{5} k^3$

b) $\displaystyle\sum_{k=0}^{4} (1)^k 5^k$

c) $\displaystyle\sum_{i=8}^{11} \left(2 + \frac{1}{i}\right)$

SOLUTION

a) We replace k by 1, 2, 3, 4, and 5. Then we add the results.

$$\sum_{k=1}^{5} k^3 = 1^3 + 2^3 + 3^3 + 4^3 + 5^3 = 1 + 8 + 27 + 64 + 125 = 225$$

b) $\displaystyle\sum_{k=0}^{4} (-1)^k 5^k = (-1)^0 5^0 + (-1)^1 5^1 + (-1)^2 5^2 + (-1)^3 5^3 + (-1)^4 5^4$

$$= 1 - 5 + 25 - 125 + 625 = 521$$

c) $\displaystyle\sum_{i=8}^{11} \left(2 + \frac{1}{i}\right) = \left(2 + \frac{1}{8}\right) + \left(2 + \frac{1}{9}\right) + \left(2 + \frac{1}{10}\right) + \left(2 + \frac{1}{11}\right)$

$$= 8\frac{1691}{3960}$$

Some graphers have a SUM SEQ feature that will generate partial sums of sequences automatically.

Example 6   Write sigma notation for each of the following.

a) $-2 + 4 - 6 + 8 - 10$

b) $1 + 4 + 8 + 16 + 32 + 64 + \cdots$

c) $x + \dfrac{x^2}{2} + \dfrac{x^3}{3} + \dfrac{x^4}{4}$

SOLUTION

RESULT	REASON
a) $-2 + 4 - 6 + 8 - 10 = \displaystyle\sum_{k=1}^{5} (-1)^k(2k)$	These are even integers with alternating signs. Therefore, the general term is $(-1)^k(2k)$, beginning with $k = 1$.
b) $1 + 4 + 8 + 16 + 32 + 64 + \cdots = \displaystyle\sum_{k=0}^{\infty} 2^k$	This is a sum of powers of 2, and it is also an infinite series. We use the symbol ∞ to represent infinity.
c) $x + \dfrac{x^2}{2} + \dfrac{x^3}{3} + \dfrac{x^4}{4} = \displaystyle\sum_{k=1}^{4} \dfrac{x^k}{k}$	This is a sum of expressions x^k/k.

Recursive Definitions

A sequence may be defined by **recursive definition**. Such a definition lists the first term, or the first few terms, and then tells how to determine the remaining terms from the given terms.

Example 7   Find the first 5 terms of the sequence defined by

$$a_1 = 5, \qquad a_{k+1} = 2a_k - 3, \quad \text{for } k \geq 1.$$

SOLUTION

$$a_1 = 5,$$

$$a_2 = 2a_1 - 3 = 2 \cdot 5 - 3 = 7,$$

$$a_3 = 2a_2 - 3 = 2 \cdot 7 - 3 = 11,$$

$$a_4 = 2a_3 - 3 = 2 \cdot 11 - 3 = 19,$$

$$a_5 = 2a_4 - 3 = 2 \cdot 19 - 3 = 35.$$

The recursive definition in Example 7 can also be expressed as follows:

$$a_0 = 5, \qquad a_k = 2a_{k-1} - 3, \quad \text{for } k \geq 1,$$

where a_1 is found by substituting 1 for k.

Many graphers have the capability of working with recursively defined sequences. For Example 7, the recursive function might be entered as $U_n = 2 * U_{n-1} - 3$ with $U_n\text{Start} = 5$.

Example 8 *The Fibonacci Sequence.* One of the most famous recursively defined sequences is the *Fibonacci sequence*. In 1202, Leonardo Fibonacci (also called Leonardo da Pisa), an Italian mathematician, proposed a model for rabbit population growth as follows. We start with one pair of rabbits, one female and one male. These rabbits mature to an age of reproductivity and produce a new pair, again one female and one male. The former pair endures until each pair can reproduce, and then each produces a new pair. This pattern continues. The population of rabbits can be modeled by the following recursively defined sequence:

$$a_1 = 1, \qquad a_2 = 1, \qquad a_{k+1} = a_k + a_{k-1}, \quad \text{for } k \geq 2,$$

where $a_k =$ the total number of pairs of rabbits after $k - 2$ reproductions for $k \geq 2$. Find the first 7 terms of the Fibonacci sequence.

SOLUTION We have

$$a_1 = 1,$$
$$a_2 = 1,$$
$$a_3 = a_2 + a_1 = 1 + 1 = 2,$$
$$a_4 = a_3 + a_2 = 2 + 1 = 3,$$
$$a_5 = a_4 + a_3 = 3 + 2 = 5,$$
$$a_6 = a_5 + a_4 = 5 + 3 = 8,$$
$$a_7 = a_6 + a_5 = 8 + 5 = 13.$$

The following figure illustrates the Fibonacci sequence.

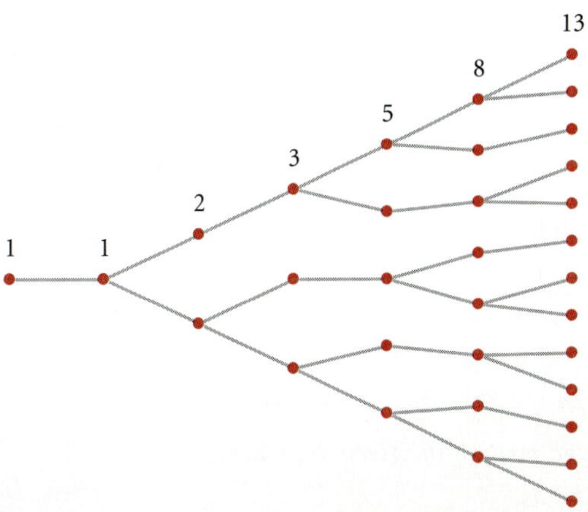

DNA Defined

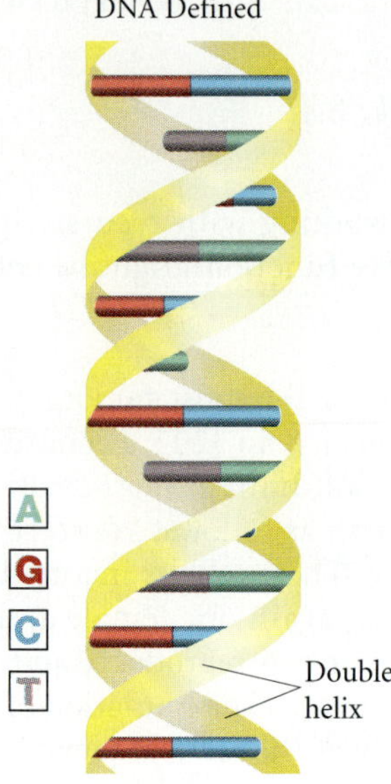

Other Applications of Sequences

DIGITAL TECHNOLOGY Digital systems use sequences of *bits*, or binary digits (0 and 1), which are used in the base-two number system. Since computer components are typically two-state mechanisms (ON–OFF), the bit is the fundamental unit of information. The compact disc (CD), fax machine, modem, and CD-ROM are other devices that use digital technology.

DNA In 1953, James Watson and Francis Crick discovered that DNA contains the "blueprint" for the construction of each individual human. The genetic code in DNA is the chemical sequence by which hereditary data are translated from genes into proteins, which serve as structural material and chemical regulators for the human body. The genetic code is composed of a sequence of four molecules: adenine (A), cytosine (C), guanine (G), and thymine (T). (These letters could just as well be represented by numbers.) The complete collection of human DNA is called the human *genome*.

6.1 Exercise Set

In each of the following, the nth term of a sequence is given. Find the first 4 terms, a_{10}, and a_{15}.

1. $a_n = 4n - 1$

2. $a_n = (n - 1)(n - 2)(n - 3)$

3. $a_n = \dfrac{n}{n - 1}$, $n \geq 2$

4. $a_n = n^2 - 1$, $n \geq 3$

5. $a_n = \dfrac{n^2 - 1}{n^2 + 1}$

6. $a_n = \left(-\dfrac{1}{2}\right)^{n-1}$

7. $a_n = (-1)^n n^2$

8. $a_n = (-1)^{n-1}(3n - 5)$

9. $a_n = 5 + \dfrac{(-2)^{n+1}}{2^n}$

10. $a_n = \dfrac{2n - 1}{n^2 + 2n}$

Find the indicated term of the given sequence.

11. $a_n = 5n - 6$; a_8

12. $a_n = (3n - 4)(2n + 5)$; a_7

13. $a_n = (2n + 3)^2$; a_6

14. $a_n = (-1)^{n-1}(4.6n - 18.3)$; a_{12}

15. $a_n = 5n^2(4n - 100)$; a_{11}

16. $a_n = \left(1 + \dfrac{1}{n}\right)^2$; a_{80}

17. $a_n = \ln e^n$; a_{67}

18. $a_n = 2 - \dfrac{1000}{n}$; a_{100}

Predict the general, or nth term, a_n, of each sequence. Answers may vary.

19. 2, 4, 6, 8, 10, ...

20. 3, 9, 27, 81, 243, ...

21. −2, 6, −18, 54, ...

22. −2, 3, 8, 13, 18, ...

23. $\dfrac{2}{3}, \dfrac{3}{4}, \dfrac{4}{5}, \dfrac{5}{6}, \dfrac{6}{7}, \ldots$

24. $\sqrt{2}, 2, \sqrt{6}, 2\sqrt{2}, \sqrt{10}, \ldots$

25. $1 \cdot 2, 2 \cdot 3, 3 \cdot 4, 4 \cdot 5, \ldots$

26. −1, −4, −7, −10, −13, ...

27. 0, log 10, log 100, log 1000, ...

28. $\ln e^2$, $\ln e^3$, $\ln e^4$, $\ln e^5$, ...

Find the indicated partial sums for each sequence.

29. 1, 2, 3, 4, 5, 6, 7, ...; S_3 and S_7

30. 1, −3, 5, −7, 9, −11, ...; S_2 and S_5

31. 2, 4, 6, 8, ...; S_4 and S_5

32. $1, \frac{1}{4}, \frac{1}{9}, \frac{1}{16}, \frac{1}{25}, \ldots;$ S_1 and S_5

Find and evaluate each sum.

33. $\displaystyle\sum_{k=1}^{5} \frac{1}{2k}$

34. $\displaystyle\sum_{i=1}^{6} \frac{1}{2i + 1}$

35. $\displaystyle\sum_{i=0}^{6} 2^i$

36. $\displaystyle\sum_{k=4}^{7} \sqrt{2k - 1}$

37. $\displaystyle\sum_{k=7}^{10} \ln k$

38. $\displaystyle\sum_{k=1}^{4} \pi k$

39. $\displaystyle\sum_{k=1}^{8} \frac{k}{k + 1}$

40. $\displaystyle\sum_{i=1}^{5} \frac{i - 1}{i + 3}$

41. $\displaystyle\sum_{i=1}^{5} (-1)^i$

42. $\displaystyle\sum_{k=0}^{5} (-1)^{k+1}$

43. $\displaystyle\sum_{k=1}^{8} (-1)^{k+1}3k$

44. $\displaystyle\sum_{k=0}^{7} (-1)^k 4^{k+1}$

45. $\displaystyle\sum_{k=0}^{6} \frac{2}{k^2 + 1}$

46. $\displaystyle\sum_{i=1}^{10} i(i + 1)$

47. $\displaystyle\sum_{k=0}^{5} (k^2 - 2k + 3)$

48. $\displaystyle\sum_{k=1}^{10} \frac{1}{k(k + 1)}$

49. $\displaystyle\sum_{i=0}^{10} \frac{2^i}{2^i + 1}$

50. $\displaystyle\sum_{k=0}^{3} (-2)^{2k}$

Write sigma notation.

51. $5 + 10 + 15 + 20 + 25 + \cdots$

52. $7 + 14 + 21 + 28 + 35 + \cdots$

53. $2 - 4 + 8 - 16 + 32 - 64$

54. $3 + 6 + 9 + 12 + 15$

55. $-\dfrac{1}{2} + \dfrac{2}{3} - \dfrac{3}{4} + \dfrac{4}{5} - \dfrac{5}{6} + \dfrac{6}{7}$

56. $\dfrac{1}{1^2} + \dfrac{1}{2^2} + \dfrac{1}{3^2} + \dfrac{1}{4^2} + \dfrac{1}{5^2}$

57. $4 - 9 + 16 - 25 + \cdots + (-1)^n n^2$

58. $9 - 16 + 25 + \cdots + (-1)^{n+1} n^2$

59. $\dfrac{1}{1 \cdot 2} + \dfrac{1}{2 \cdot 3} + \dfrac{1}{3 \cdot 4} + \dfrac{1}{4 \cdot 5} + \cdots$

60. $\dfrac{1}{1 \cdot 2^2} + \dfrac{1}{2 \cdot 3^2} + \dfrac{1}{3 \cdot 4^2} + \dfrac{1}{4 \cdot 5^2} + \cdots$

Find the first 4 terms of each recursively defined sequence.

61. $a_1 = 4, \ a_{k+1} = 1 + \dfrac{1}{a_k}$

62. $a_1 = 256, \ a_{k+1} = \sqrt{a_k}$

63. $a_1 = 6561, \ a_{k+1} = (-1)^k \sqrt{a_k}$

64. $a_1 = e^Q, \ a_{k+1} = \ln a_k$

65. $a_1 = 2, \ a_2 = 3, \ a_{k+1} = a_k + a_{k-1}$

66. $a_1 = -10, \ a_2 = 8, \ a_{k+1} = a_k - a_{k-1}$

Construct a table of values and a graph for the first 10 terms of each sequence. Use both a grapher and a spreadsheet program, if available.

67. $a_n = \left(1 + \dfrac{1}{n}\right)^n$

68. $a_n = \sqrt{n + 1} - \sqrt{n}$

69. $a_1 = 2, \ a_{k+1} = \sqrt{1 + \sqrt{a_k}}$

70. $a_1 = 2, \ a_{k+1} = \dfrac{1}{2}\left(a_k + \dfrac{2}{a_k}\right)$

71. *The Gap.* The Gap, Inc., is a national clothing firm that has experienced tremendous growth (*Source:* Gap, Inc., annual report). Total sales, in millions of dollars, can be estimated by the sequence model

$$a_n = 436.27n + 1577.84, \quad n = 1, 2, 3, \ldots.$$

The year 1990 corresponds to $n = 1$.

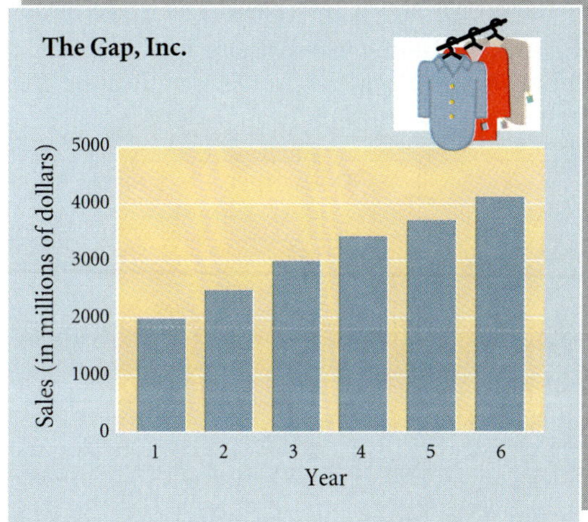

a) Find the first 10 terms of this sequence.

b) Estimate the total sales from 1990 to 1999 by evaluating the sum

$$\sum_{n=1}^{10} (436.27n + 1577.84).$$

72. *Driver Fatalities by Age.* The number of licensed drivers per 100,000 who died in motor vehicle accidents at age n can be estimated by the sequence model

$$a_n = 0.018n^2 - 1.7n + 48.7, \quad n = 15, 16, 17, \ldots, 90.$$

a) Make a table or list of the first 10 terms of the sequence.

b) Construct a graph of the sequence.

c) Find the number of fatalities per 100,000 of those drivers whose age is 16 yr, 23 yr, 50 yr, 75 yr, and 85 yr.

73. *Compound Interest.*   Suppose that $1000 is invested at 6.2%, compounded annually. The value of the investment after n years is given by the sequence model

$$a_n = \$1000(1.062)^n, \quad n = 1, 2, 3, \ldots.$$

a) Find the first 10 terms of the sequence.
b) Find the amount of the investment after 20 yr.

74. *Salvage Value.*   The value of an office machine is $5200. Its salvage value each year is 75% of its value the year before. Give a sequence that lists the salvage value of the machine for each year of a 10-yr period.

75. *Bacteria Growth.*   A single cell of bacteria divides into two every 15 min. Suppose that the same rate of division is maintained for 4 hr. Give a sequence that lists the number of cells after successive 15-min periods.

76. *Salary Sequence.*   Torrey is paid $6.30 per hour for working at Red Freight Limited. Each year he receives a $0.40 hourly raise. Give a sequence that lists Torrey's hourly salary over a 10-yr period.

77. *Household Discretionary Income by Age.*   The following table contains factual data relating household discretionary income to age.

AGE, x	HOUSEHOLD DISCRETIONARY INCOME, y
25	$10,223
34	15,607
54	21,173
64	18,257
70	13,747

Source: U.S. Bureau of the Census.

a) Use the REGRESSION feature of a grapher to fit a quadratic sequence function

$$a_n = an^2 + bn + c$$

to the data.
b) Find the household discretionary income of people whose age is 16 yr, 20 yr, 40 yr, 60 yr, and 75 yr.

Skill Maintenance

78. Solve for S_n: $2S_n = n(a_1 + a_n)$.

79. Collect like terms:

$a_1 + (a_1 + d) + (a_2 + 2d) + (a_n - 2d) + (a_n - d) + a_n$.

Simplify.

80. $\log_a a$

81. $\log_a 1$

Synthesis

82. ◆ The Fibonacci sequence has intrigued mathematicians for centuries. In fact, a journal called the *Fibonacci Quarterly* is devoted to publishing the results pertaining to such sequences. Do some library research on the connection of the Fibonacci sequence to the idea of the "Golden Section."

83. ◆

a) Find the first few terms of the sequence $a_n = n^2 - n + 41$ and describe the pattern you observe.
b) Does the pattern you found in part (a) hold for all choices of n? Why or why not?

Find the first 5 terms of the sequence, and then find S_5.

84. $a_n = \dfrac{1}{2^n} \log 1000^n$

85. $a_n = i^n, \ i = \sqrt{-1}$

86. $a_n = \ln (1 \cdot 2 \cdot 3 \cdots n)$

For each sequence, find a formula for S_n.

87. $a_n = \ln n$

88. $a_n = \dfrac{1}{n} - \dfrac{1}{n + 1}$

6.2
Arithmetic Sequences and Series

- *For any arithmetic sequence, find the nth term when n is given and n when the nth term is given, and given two terms, find the common difference and construct the sequence.*
- *Find the sum of the first n terms of an arithmetic sequence.*
- *Insert arithmetic means between two numbers.*

If we begin with a particular first term and then add the same number successively, we obtain an **arithmetic sequence**. In this section, we study arithmetic sequences and series.

Arithmetic Sequences

Consider this sequence:

$$2, 5, 8, 11, 14, 17, \ldots .$$

Note that adding 3 to any term produces the following term. In other words, the difference between any term and the preceding one is 3. This is an *arithmetic* (pronounced: ăr′ĭth-mĕt′-ĭk) *sequence*.

Arithmetic Sequence

A sequence is *arithmetic* if there exists a number d, called the *common difference*, such that

$$a_{n+1} = a_n + d, \quad \text{or} \quad a_{n+1} - a_n = d, \quad \text{for } n \geq 1.$$

Arithmetic sequences are also called **arithmetic progressions**.

Example 1   For each of the following arithmetic sequences, identify the first term, a_1, and the common difference, d.

a) 4, 9, 14, 19, 24, ...   b) 34, 27, 20, 13, 6, -1, -8, ...
c) $2, 2\frac{1}{2}, 3, 3\frac{1}{2}, 4, 4\frac{1}{2}, \ldots$

SOLUTION   To find the first term, a_1, we determine the first term listed. To find the common difference, d, we choose any term beyond the first and subtract the preceding term from it.

SEQUENCE	FIRST TERM, a_1	COMMON DIFFERENCE, d
a) 4, 9, 14, 19, 24, ...	4	5 $(9 - 4 = 5)$
b) 34, 27, 20, 13, 6, -1, -8, ...	34	-7 $(20 - 27 = -7)$
c) $2, 2\frac{1}{2}, 3, 3\frac{1}{2}, 4, 4\frac{1}{2}, \ldots$	2	$\frac{1}{2}$ $\left(2\frac{1}{2} - 2 = \frac{1}{2}\right)$

We obtain the common difference by subtracting a_1 from a_2. Had we subtracted a_2 from a_3 or a_3 from a_4, we would have obtained the same values for d. Thus we can check by adding d to each term in a sequence to see if we progress correctly to the next term.

CHECK:

a) $4 + 5 = 9$, $9 + 5 = 14$, $14 + 5 = 19$, $19 + 5 = 24$

b) $34 + (-7) = 27$, $27 + (-7) = 20$, $20 + (-7) = 13$,
$13 + (-7) = 6$, $6 + (-7) = -1$, $-1 + (-7) = -8$

c) $2 + \frac{1}{2} = 2\frac{1}{2}$, $2\frac{1}{2} + \frac{1}{2} = 3$, $3 + \frac{1}{2} = 3\frac{1}{2}$

To find a formula for the general, or *n*th, term of any arithmetic sequence, we denote the common difference by *d*, write out the first few terms, and look for a pattern:

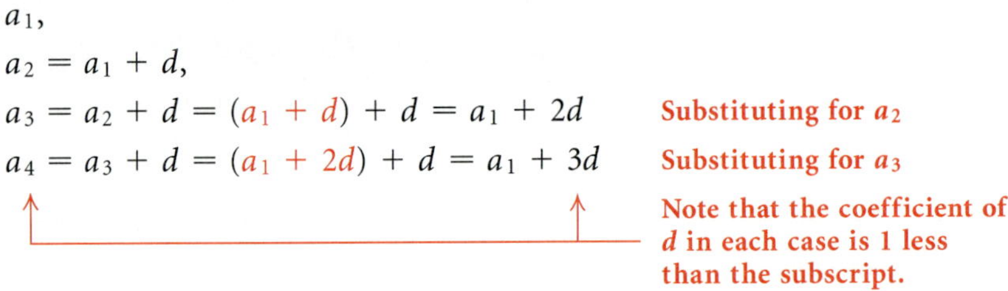

$a_1,$

$a_2 = a_1 + d,$

$a_3 = a_2 + d = (a_1 + d) + d = a_1 + 2d$ Substituting for a_2

$a_4 = a_3 + d = (a_1 + 2d) + d = a_1 + 3d$ Substituting for a_3

Note that the coefficient of *d* in each case is 1 less than the subscript.

Generalizing, we obtain the following formula.

---

nth Term of an Arithmetic Sequence

The *n*th *term* of an arithmetic sequence is given by

$$a_n = a_1 + (n - 1)d, \quad \text{for any } n \geq 1.$$

Program

AP: This program finds the *n*th term of an arithmetic sequence.

Example 2 Find the 14th term of the arithmetic sequence 4, 7, 10, 13, ...

SOLUTION We first note that $a_1 = 4$, $d = 3$, and $n = 14$. Then using the formula for the *n*th term, we obtain

$$a_n = a_1 + (n - 1)d$$
$$a_{14} = 4 + (14 - 1) \cdot 3 = 4 + 13 \cdot 3 = 4 + 39 = 43.$$

The 14th term is 43.

Example 3 In the sequence of Example 2, which term is 301? That is, find *n* if $a_n = 301$.

SOLUTION We substitute $a_n = 301$, $a_1 = 4$, and $d = 3$ into the formula for the *n*th term and solve for *n*:

$$a_n = a_1 + (n - 1)d$$
$$301 = 4 + (n - 1) \cdot 3 \qquad \text{Substituting}$$
$$301 = 4 + 3n - 3$$
$$301 = 3n + 1$$
$$300 = 3n$$
$$100 = n.$$

Solving for *n*

The term 301 is the 100th term of the sequence.

Given two terms and their places in an arithmetic sequence, we can construct the sequence.

Example 4   The 3rd term of an arithmetic sequence is 8, and the 16th term is 47. Find a_1 and d and construct the sequence.

SOLUTION   We know that $a_3 = 8$ and $a_{16} = 47$. Thus we would have to add d 13 times to get 47 from 8. That is,

$$8 + 13d = 47. \qquad \textit{a_3 and a_{16} are 13 terms apart.}$$

Solving $8 + 13d = 47$, we obtain

$$13d = 39$$
$$d = 3.$$

Since $a_3 = 8$, we subtract d twice to get a_1. Thus,

$$a_1 = 8 - 2 \cdot 3 = 2.$$

The sequence is 2, 5, 8, 11, ... Note that we could also subtract d 15 times from a_{16} in order to find a_1.

In general, d should be subtracted $n - 1$ times from a_n in order to find a_1.

Interactive Discovery

Graph the first 10 terms of each sequence whose general term is given. Which are arithmetic? Why? Can you determine whether a sequence is arithmetic by looking at its general term? its graph?

$$a_n = 2n + 3, \qquad a_n = n^2 - 100,$$
$$a_n = -0.5n + 20, \qquad a_n = \ln n$$

The formula $a_n = a_1 + (n - 1)d$ and the preceding Interactive Discovery lead us to the following result:

> A sequence is arithmetic$\longleftrightarrow$the general term is a linear function.

Sum of the First n Terms of an Arithmetic Sequence

Consider the arithmetic sequence

$$3, 5, 7, 9, \dots .$$

When we add the first 4 terms of the sequence, we get S_4, which is

$$3 + 5 + 7 + 9, \quad \text{or} \quad 24.$$

This sum is called an **arithmetic series**. To find a formula for the sum of the first n terms, S_n, of an arithmetic series, we first denote an arithmetic sequence, as follows:

> This term is two terms back from the last. If you add d to this term, the result is the next-to-last term, $a_n - d$.

$$a_1, \quad (a_1 + d), \quad (a_1 + 2d), \quad \ldots, \quad (a_n - 2d), \quad (a_n - d), \quad a_n.$$

> This is the next-to-last term. If you add d to this term, the result is a_n.

Then S_n is given by

$$S_n = a_1 + (a_1 + d) + (a_1 + 2d) + \cdots + (a_n - 2d)$$
$$+ (a_n - d) + a_n. \tag{1}$$

Reversing the order of the addition gives us

$$S_n = a_n + (a_n - d) + (a_n - 2d) + \cdots + (a_1 + 2d)$$
$$+ (a_1 + d) + a_1. \tag{2}$$

If we add corresponding terms of each side of equations (1) and (2), we get

$$2S_n = [a_1 + a_n] + [(a_1 + d) + (a_n - d)] + [(a_1 + 2d) + (a_n - 2d)]$$
$$+ \cdots + [(a_n - 2d) + (a_1 + 2d)]$$
$$+ [(a_n - d) + (a_1 + d)] + [a_n + a_1].$$

There are n pairs of square brackets.

This simplifies to

$$2S_n = [a_1 + a_n] + [a_1 + a_n] + [a_1 + a_n] + \cdots + [a_n + a_1]$$
$$+ [a_n + a_1] + [a_n + a_1].$$

Since $a_1 + a_n$ is being added n times, it follows that

$$2S_n = n(a_1 + a_n),$$

from which we get the following formula.

Sum of the First n Terms

The sum of the first n terms of an arithmetic sequence is given by

$$S_n = \frac{n}{2}(a_1 + a_n).$$

Example 5 Find the sum of the first 100 natural numbers.

SOLUTION The sum is

$$1 + 2 + 3 + \cdots + 99 + 100.$$

This is the sum of the first 100 terms of the arithmetic sequence for which

$$a_1 = 1, \qquad a_n = 100, \quad \text{and} \quad n = 100.$$

Thus substituting into the formula

$$S_n = \frac{n}{2}(a_1 + a_n),$$

we get

$$S_{100} = \frac{100}{2}(1 + 100) = 50(101) = 5050.$$

The sum of the first 100 natural numbers is 5050.

Example 6 Find the sum of the first 15 terms of the arithmetic sequence 4, 7, 10, 13, ...

SOLUTION Note that $a_1 = 4$, $d = 3$, and $n = 15$. Before using the formula

$$S_n = \frac{n}{2}(a_1 + a_n),$$

we find the last term, a_{15}:

$$a_{15} = 4 + (15 - 1)3 \qquad \text{Substituting into the formula } a_n = a_1 + (n - 1)d$$
$$= 4 + 14 \cdot 3 = 46.$$

Thus,

$$S_{15} = \frac{15}{2}(4 + 46) \qquad \text{Substituting into } S_n = \frac{n}{2}(a_1 + a_n)$$

$$= \frac{15}{2}(50) = 375.$$

The sum of the first 15 terms is 375.

Example 7 Find the sum: $\displaystyle\sum_{k=1}^{130}(4k + 5)$.

SOLUTION It is helpful to first write out a few terms:

$$9 + 13 + 17 + \cdots .$$

It appears that this is an arithmetic series coming from an arithmetic sequence with $a_1 = 9$, $d = 4$, and $n = 130$. Before using the formula

$$S_n = \frac{n}{2}(a_1 + a_n),$$

we find the last term, a_{130}:

$$a_{130} = 9 + (130 - 1)4 \qquad \text{Substituting into the formula } a_n = a_1 + (n - 1)d$$
$$= 9 + 129 \cdot 4 = 525.$$

Thus,

$$S_{130} = \frac{130}{2}(9 + 525) \qquad \text{Substituting into } S_n = \frac{n}{2}(a_1 + a_n)$$

$$= 34{,}710.$$

Applications

The translation of some applications and problem-solving situations may involve arithmetic sequences or series. Let's look at some examples.

Example 8 *Hourly Wages.* Gloria accepts a job, starting with an hourly wage of $14.25, and is promised a raise of 15¢ per hour every 2 months for 5 yr. At the end of 5 yr, what will Gloria's hourly wage be?

SOLUTION

1. **Familiarize.** It helps to first write down the hourly wage for several 2-month time periods:

Beginning:	$14.25,
After two months:	14.40,
After four months:	14.55,

   and so on.

   What appears is a sequence of numbers:

   $$14.25, \ 14.40, \ 14.55, \ \dots .$$

   Is it arithmetic? Yes, because adding $0.15 each time gives us the next term.

   We ask ourselves what we know about arithmetic sequences. The pertinent formulas are

   $$a_n = a_1 + (n - 1)d \quad \text{and} \quad S_n = \frac{n}{2}(a_1 + a_n).$$

   In this case, we are not looking for a sum, so it is probably the first formula that will lead to our answer. We want to know the last term of an arithmetic sequence, so we need to know a_1, n, and d. From our list above, we see that

   $$a_1 = 14.25 \quad \text{and} \quad d = 0.15.$$

   What is n? That is, how many terms are in the sequence? Each year there are 6 raises, since Gloria gets a raise every 2 months. There are 5 yr, so the total number of raises will be $5 \cdot 6$, or 30. There will be 31 terms: the original wage and 30 increased rates.

2. **Translate.** We want to find $a_n = a_1 + (n - 1)d$ for the arithmetic sequence in which $a_1 = 14.25$, $d = 0.15$, and $n = 31$.

3. **Carry out.** Substituting gives us

   $$a_{31} = 14.25 + (31 - 1) \cdot 0.15 = \$18.75.$$

4. Check. We can check the calculations. We can also calculate in a slightly different way for another check. For example, at the end of 1 yr, there will be 6 raises, for a total raise of $0.90. At the end of 5 yr, the total raise will be 5 × $0.90, or $4.50. If we add that to the original wage of $14.25, we obtain $18.75. The answer checks.

5. State. At the end of 5 yr, Gloria's hourly wage will be $18.75.

Example 8 is one in which the calculations or the translation could be done in a number of ways. There is often a variety of ways in which a problem can be solved. In this chapter, we will concentrate on the use of sequences and series and their related formulas.

Example 9   *Total in a Stack.* A stack of telephone poles has 30 poles in the bottom row. There are 29 poles in the second row, 28 in the next row, and so on. How many poles are in the stack if there are 5 poles in the top row?

SOLUTION

1. Familiarize. A picture will help in this case. The following figure shows the ends of the poles and the way in which they stack. There are 30 poles on the bottom, and we see that there are fewer in each succeeding row. How many poles are there altogether?

Since the number of poles goes from 30 in a row up to 5 in the top row, there must be 26 rows. We want the sum

$$30 + 29 + 28 + \cdots + 5.$$

Thus we have an arithmetic series. We recall, or look up if necessary, the formula

$$S_n = \frac{n}{2}(a_1 + a_n).$$

2. Translate. We want to find the sum of the first 26 terms of an arithmetic sequence in which $a_1 = 30$ and $a_{26} = 5$.

3. Carry out. Substituting, we get

$$S_{26} = \frac{26}{2}(30 + 5) = 455.$$

4. Check. In this case, we can check by repeating the calculations. A longer, harder way would be to do the entire addition: $30 + 29 + 28 + \cdots + 5$.

5. State. There are 455 poles in the stack.

Arithmetic Means

If p, m, and q form an arithmetic sequence, it can be shown (see Exercise 66) that $m = (p + q)/2$. The number m is the **arithmetic mean**, or **average**, of p and q. Given two numbers p and q, if we find k other numbers $m_1, m_2, \ldots, m_k$ such that

$$p, m_1, m_2, \ldots, m_k, q$$

forms an arithmetic sequence, we say that we have "inserted k arithmetic means between p and q."

Example 10   Insert three arithmetic means between 4 and 13.

SOLUTION   We look for numbers m_1, m_2, and m_3 such that 4, m_1, m_2, m_3, 13 is an arithmetic sequence. In this case, $a_1 = 4$, $n = 5$, and $a_5 = 13$. We then use the formula $a_n = a_1 + (n - 1)d$ and solve for d:

$$a_n = a_1 + (n - 1)d;$$

$$d = \frac{a_n - a_1}{n - 1}$$

$$= \frac{13 - 4}{5 - 1} = 2\frac{1}{4}.$$

Then using $d = 2\frac{1}{4}$, we compute the values of m_1, m_2, and m_3:

$$m_1 = a_1 + d = 4 + 2\frac{1}{4} = 6\frac{1}{4},$$
$$m_2 = m_1 + d = 6\frac{1}{4} + 2\frac{1}{4} = 8\frac{1}{2},$$
$$m_3 = m_2 + d = 8\frac{1}{2} + 2\frac{1}{4} = 10\frac{3}{4}.$$

6.2 | Exercise Set

Find the first term and the common difference.

1. 3, 8, 13, 18, ...

2. \$1.08, \$1.16, \$1.24, \$1.32, ...

3. 9, 5, 1, -3, ...

4. $-8, -5, -2, 1, 4, \ldots$

5. $\frac{3}{2}, \frac{9}{4}, 3, \frac{15}{4}, \ldots$

6. $\frac{3}{5}, \frac{1}{10}, -\frac{2}{5}, \ldots$

7. \$316, \$313, \$310, \$307, ...

8. Find the 11th term of the arithmetic sequence 0.07, 0.12, 0.17, ...

9. Find the 12th term of the arithmetic sequence 2, 6, 10, ...

10. Find the 17th term of the arithmetic sequence 7, 4, 1, ...

11. Find the 14th term of the arithmetic sequence $3, \frac{7}{3}, \frac{5}{3}, \ldots$.

12. Find the 13th term of the arithmetic sequence $1200, $964.32, $728.64, \ldots$.

13. Find the 10th term of the arithmetic sequence $2345.78, $2967.54, $3589.30, \ldots$.

14. In the sequence of Exercise 9, what term is the number 106?

15. In the sequence of Exercise 8, what term is the number 1.67?

16. In the sequence of Exercise 10, what term is -296?

17. In the sequence of Exercise 11, what term is -27?

18. Find a_{20} when $a_1 = 14$ and $d = -3$.

19. Find a_1 when $d = 4$ and $a_8 = 33$.

20. Find d when $a_1 = 8$ and $a_{11} = 26$.

21. Find n when $a_1 = 25$, $d = -14$, and $a_n = -507$.

22. In an arithmetic sequence, $a_{17} = -40$ and $a_{28} = -73$. Find a_1 and d. Write the first 5 terms of the sequence.

23. In an arithmetic sequence, $a_{17} = \frac{25}{3}$ and $a_{28} = \frac{95}{6}$. Find a_1 and d. Write the first 5 terms of the sequence.

24. Find the sum of the first 14 terms of the series $11 + 7 + 3 + \cdots$.

25. Find the sum of the first 20 terms of the series $5 + 8 + 11 + 14 + \cdots$.

26. Find the sum of the first 300 natural numbers.

27. Find the sum of the first 400 even natural numbers.

28. Find the sum of the odd numbers 1 to 199, inclusive.

29. Find the sum of the multiples of 7 from 7 to 98, inclusive.

30. Find the sum of all multiples of 4 that are between 14 and 523.

31. If an arithmetic series has $a_1 = 2$, $d = 5$, and $n = 20$, what is S_n?

32. If an arithmetic series has $a_1 = 7$, $d = -3$, and $n = 32$, what is S_n?

Find the sum.

33. $\sum\limits_{k=1}^{40} (2k + 3)$

34. $\sum\limits_{k=5}^{20} 8k$

35. $\sum\limits_{k=0}^{19} \dfrac{k - 3}{4}$

36. $\sum\limits_{k=2}^{50} (2000 - 3k)$

37. $\sum\limits_{k=12}^{57} \dfrac{7 - 4k}{13}$

38. $\sum\limits_{k=101}^{200} (1.14k - 2.8) - \sum\limits_{k=1}^{5} \left(\dfrac{k + 4}{10} \right)$

39. *Pole Stacking.* How many poles will be in a stack of telephone poles if there are 50 in the first layer, 49 in the second, and so on, with 6 in the top layer?

40. *Investment Return.* Max is an investment counselor. He sets up an investment situation for a client that will return $5000 the first year, $6125 the second year, $7250 the third year, and so on, for 25 yr. How much is received from the investment altogether?

41. *Garden Plantings.* A gardener is making a planting in the shape of a trapezoid. It will have 35 plants in the front row, 31 in the second row, 27 in the third row, and so on. If the pattern is consistent, how many plants will there be in the last row? How many plants are there altogether?

42. *Band Formation.* A formation of a marching band has 14 marchers in the front row, 16 in the second row, 18 in the third row, and so on, for 25 rows. How many marchers are in the last row? How many marchers are there altogether?

43. *Total Savings.* If 10¢ is saved on October 1, 20¢ is saved on October 2, 30¢ on October 3, and so on, how much is saved during the 31 days of October?

44. *Parachutist Free Fall.* It has been found that when a parachutist jumps from an airplane, the distances, in feet, which the parachutist falls in each successive second before pulling the ripcord to release the parachute are as follows:

$$16, 48, 80, 112, 144, \ldots .$$

Is this sequence arithmetic? What is the common difference? What is the total distance fallen after 10 sec?

45. *Theater Seating.*   Theaters are often built with more seats per row as the rows move toward the back. Suppose that the first balcony of a theater has 28 seats in the first row, 32 in the second, 36 in the third, and so on, for 20 rows. How many seats are in the first balcony altogether?

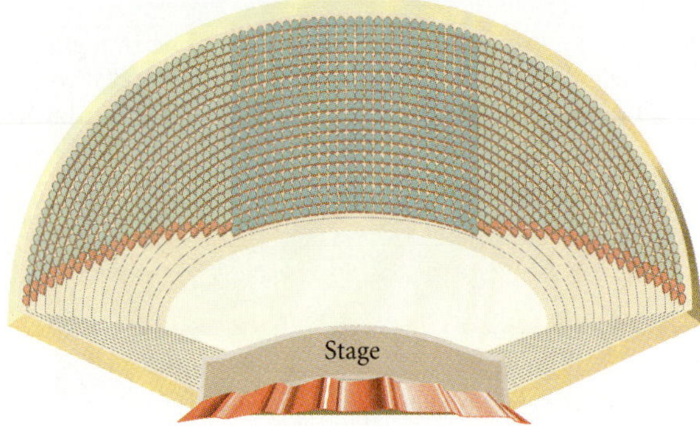

46. Insert four arithmetic means between 4 and 13.

47. Insert three arithmetic means between -3 and 5.

48. Insert ten arithmetic means between 27 and 300.

49. *Raw Material Production.*   In an industrial situation, it took 3 units of raw materials to produce 1 unit of a product. The raw material needs thus formed the sequence

$$3, 6, 9, \ldots, 3n, \ldots .$$

Is this sequence arithmetic? What is the common difference?

50. *Small Group Interaction.*   In a social science study, Stephan found the following data regarding an interaction measurement r_n for groups of size n.*

n	r_n
3	0.5908
4	0.6080
5	0.6252
6	0.6424
7	0.6596
8	0.6768
9	0.6940
10	0.7112

Is this sequence arithmetic? What is the common difference?

*Stephan, Frederick, "The Relative Rate of Communication Between Members of Small Groups," *American Sociological Review,* **17** (1952): 482–486.

Skill Maintenance

Simplify.

51. $64\left(-\dfrac{1}{2}\right)^9$

52. $4 \cdot 5^6$

53. $\dfrac{0.01(2^{30} - 1)}{2 - 1}$

54. Solve for S_n: $S_n - rS_n = a_1 - a_1 r^n$.

Synthesis

55. ◈ The sum of the first n terms of an arithmetic sequence is given by

$$S_n = \frac{n}{2}[2a_1 + (n - 1)d].$$

Compare this formula to

$$S_n = \frac{n}{2}(a_1 + a_n).$$

Discuss the reasons for the use of one formula over the other.

56. ◈ It is said that as a young child, the mathematician Karl F. Gauss (1777–1855) was able to compute the sum $1 + 2 + 3 + \cdots + 100$ very quickly in his head to the amazement of a teacher. Explain how Gauss might have done this had he possessed some knowledge of arithmetic sequences and series. Then give a formula for the sum of the first n natural numbers.

57. Find a formula for the sum of the first n odd natural numbers:

$$1 + 3 + 5 + \cdots + (2n - 1).$$

58. Find three numbers in an arithmetic sequence such that the sum of the first and third is 10 and the product of the first and second is 15.

59. Find the first term and the common difference for the arithmetic sequence for which

$$a_2 = 40 - 3q \quad \text{and} \quad a_4 = 10p + q.$$

60. Find the first 10 terms of the arithmetic sequence for which

$$a_1 = \$8760 \quad \text{and} \quad d = -\$798.23.$$

Then find the sum of the first 10 terms.

61. The zeros of this polynomial function form an arithmetic sequence. Find them.

$$f(x) = x^4 + 4x^3 - 84x^2 - 176x + 640$$

62. Insert enough arithmetic means between 1 and 50 so that the sum of the resulting series will be 459.

63. Suppose that the lengths of the sides of a right triangle form an arithmetic sequence. Prove that the triangle is similar to a right triangle whose sides have lengths 3, 4, and 5.

64. *Straight-Line Depreciation.*   A company buys an office machine for $5200 on January 1 of a given year. The machine is expected to last for 8 yr, at the end of which time its **trade-in value**, or **salvage value** will be $1100. If the company's accountant figures the decline in value to be the same each year, then its **book values**, or **salvage values**, after t years, $0 \leq t \leq 8$, form an arithmetic sequence given by

$$a_t = C - t\left(\frac{C - S}{N}\right),$$

where $C =$ the original cost of the item ($5200), $N =$ the number of years of expected life (8), and $S =$ the salvage value ($1100).

a) Find the formula for a_t for the straight-line depreciation of the office machine.

b) Find the salvage value after 0 yr, 1 yr, 2 yr, 3 yr, 4 yr, 7 yr, and 8 yr.

65. Prove that the general term of an arithmetic sequence is a linear function.

66. Prove that if p, m, and q form an arithmetic sequence, then

$$m = \frac{p + q}{2}.$$

6.3
Geometric Sequences and Series

- *Identify the common ratio of a geometric sequence, and find a given term and the sum of the first n terms.*
- *Find the sum of an infinite geometric series, if it exists.*

If we begin with a particular first term and then multiply by the same number successively, we obtain a **geometric sequence**. In this section, we study geometric sequences and series.

Geometric Sequences

Consider this sequence:

$$2, 6, 18, 54, 162, \ldots.$$

Note that multiplying each term by 3 produces the next term. Sequences in which each term can be multiplied by a certain number to get the next term are called **geometric**. We call this number the **common ratio** because it can be found by dividing any term by the preceding term.

Geometric Sequence

A sequence is *geometric* if there is a number r, called the *common ratio*, such that

$$\frac{a_{n+1}}{a_n} = r, \quad \text{or} \quad a_{n+1} = a_n r, \quad \text{for any integer } n \geq 1.$$

A geometric sequence is also called a **geometric progression**.

Example 1 For each of the following geometric sequences, identify the common ratio.

a) 3, 6, 12, 24, 48, ...

b) $1, -\dfrac{1}{2}, \dfrac{1}{4}, -\dfrac{1}{8}, \ldots$

c) $5200, $3900, $2925, $2193.75, ...

d) $1000, $1060, $1123.60, ...

SOLUTION

SEQUENCE	COMMON RATIO
a) 3, 6, 12, 24, 48, ...	$2 \quad \left(\frac{6}{3} = 2, \frac{12}{6} = 2, \text{ and so on}\right)$
b) $1, -\dfrac{1}{2}, \dfrac{1}{4}, -\dfrac{1}{8}, \ldots$	$-\dfrac{1}{2} \quad \left(\dfrac{-\frac{1}{2}}{1} = -\dfrac{1}{2}, \dfrac{\frac{1}{4}}{-\frac{1}{2}} = -\dfrac{1}{2}, \text{ and so on}\right)$
c) $5200, $3900, $2925, $2193.75, ...	$0.75 \quad \left(\dfrac{\$3900}{\$5200} = 0.75, \dfrac{\$2925}{\$3900} = 0.75, \text{ and so on}\right)$
d) $1000, $1060, $1123.60, ...	$1.06 \quad \left(\dfrac{\$1060}{\$1000} = 1.06, \dfrac{\$1123.60}{\$1060} = 1.06, \text{ and so on}\right)$

We now find a formula for the general, or *n*th, term of a geometric sequence. Let a_1 be the first term and r the common ratio. The first few terms are as follows:

$$a_1,$$
$$a_2 = a_1 r,$$
$$a_3 = a_2 r = (a_1 r)r = a_1 r^2, \qquad \text{Substituting } a_1 r \text{ for } a_2$$
$$a_4 = a_3 r = (a_1 r^2)r = a_1 r^3. \qquad \text{Substituting } a_1 r^2 \text{ for } a_3$$

Note that the exponent is 1 less than the subscript.

Generalizing, we obtain the following.

---

nth Term of a Geometric Sequence

The *n*th term of a geometric sequence is given by

$$a_n = a_1 r^{n-1}, \quad \text{for any integer } n \geq 1.$$

Program

GP: This program finds the *n*th term of a geometric sequence.

Example 2 Find the 7th term of the geometric sequence 4, 20, 100, ...

SOLUTION We first note that

$$a_1 = 4 \quad \text{and} \quad n = 7.$$

To find the common ratio, we can divide any term (other than the first) by the preceding term. Since the second term is 20 and the first is 4, we get

$$r = \frac{20}{4}, \quad \text{or} \quad 5.$$

Then using the formula $a_n = a_1 r^{n-1}$, we have

$$a_7 = 4 \cdot 5^{7-1} = 4 \cdot 5^6 = 4 \cdot 15{,}625 = 62{,}500.$$

Thus the 7th term is 62,500.

Example 3   Find the 10th term of the geometric sequence 64, -32, 16, $-8, \ldots$.

SOLUTION   We first note that

$$a_1 = 64, \qquad n = 10, \quad \text{and} \quad r = \frac{-32}{64}, \text{ or } -\frac{1}{2}.$$

Then using the formula $a_n = a_1 r^{n-1}$, we have

$$a_{10} = 64 \cdot \left(-\frac{1}{2}\right)^{10-1} = 64 \cdot \left(-\frac{1}{2}\right)^9 = 2^6 \cdot \left(-\frac{1}{2^9}\right) = -\frac{1}{8}.$$

Thus the 10th term is $-\frac{1}{8}$.

Sum of the First n Terms of a Geometric Sequence

Next we develop a formula for the sum S_n of the first n terms of a geometric sequence:

$$a_1, \ a_1 r, \ a_1 r^2, \ a_1 r^3, \ \ldots, \ a_1 r^{n-1}, \ \ldots .$$

The associated geometric series is given by

$$S_n = a_1 + a_1 r + a_1 r^2 + a_1 r^3 + \cdots + a_1 r^{n-1}. \tag{1}$$

We want to find a formula that allows us to find this sum without a great amount of adding. If we multiply on both sides of equation (1) by r, we have

$$r S_n = a_1 r + a_1 r^2 + a_1 r^3 + a_1 r^4 + \cdots + a_1 r^n. \tag{2}$$

Subtracting corresponding terms of equation (2) from equation (1), we see that the red terms add to 0, leaving

$$S_n - r S_n = a_1 - a_1 r^n,$$

or

$$S_n(1 - r) = a_1(1 - r^n). \qquad \text{Factoring}$$

Dividing on both sides by $1 - r$ gives us the following formula.

Sum of the First n Terms

The sum of the first n terms of a geometric sequence is given by

$$S_n = \frac{a_1(1 - r^n)}{1 - r}, \quad \text{for any } r \neq 1.$$

Example 4 Find the sum of the first 7 terms of the geometric sequence 3, 15, 75, 375, ...

SOLUTION We first note that

$$a_1 = 3, \quad n = 7, \quad \text{and} \quad r = \tfrac{15}{3}, \text{ or } 5.$$

Then using the formula

$$S_n = \frac{a_1(1 - r^n)}{1 - r},$$

we have

$$S_7 = \frac{3(1 - 5^7)}{1 - 5} = \frac{3(1 - 78{,}125)}{-4}$$
$$= 58{,}593.$$

Thus the sum of the first 7 terms is 58,593.

Example 5 Find the sum: $\displaystyle\sum_{k=1}^{11} (0.3)^k$.

SOLUTION This is a geometric series with $a_1 = 0.3$, $r = 0.3$, and $n = 11$. Thus,

$$S_{11} = \frac{0.3(1 - 0.3^{11})}{1 - 0.3}$$
$$\approx 0.42857.$$

Infinite Geometric Series

Suppose we consider the sum of the terms of an infinite geometric sequence, such as 2, 4, 8, 16, 32, ... We get what is called an **infinite geometric series**:

$$2 + 4 + 8 + 16 + 32 + \cdots.$$

We want to examine what happens to the sums S_n as n gets larger and larger. We can use a grapher to complete a table and a graph of the sequence S_n.

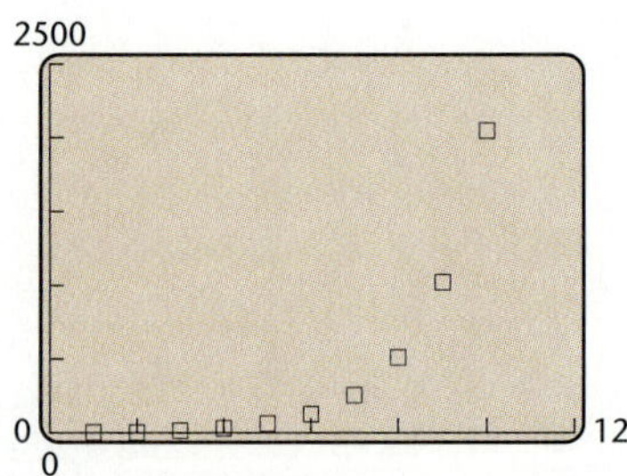

n	S_n
1	2
2	6
3	14
4	30
5	62
6	126
7	254
8	510
9	1022
10	2046

We could extend the table and the graph out for much larger values of n—for example, $S_{100} \approx 2.535 \times 10^{30}$. Note that as n gets larger and larger, the sum of the first n terms, S_n, becomes larger and larger without bound. That is, $S_n \to \infty$.

Many infinite sums get closer and closer to some specific number. Here is an example:

$$\frac{1}{2} + \frac{1}{4} + \frac{1}{8} + \frac{1}{16} + \cdots + \frac{1}{2^n} + \cdots .$$

To examine what happens to the sums S_n as n gets larger and larger, we again use a grapher and complete the following table and graph of the sequence S_n.

n	S_n
1	0.5
2	0.75
3	0.875
4	0.9375
5	0.96875
6	0.984375
7	0.992188
8	0.996094
9	0.998047
10	0.999023

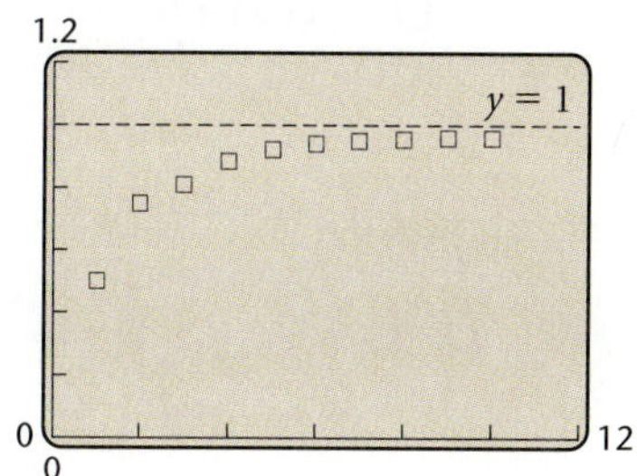

We could extend the table and the graph out for much larger values of n—for example, $S_{25} \approx 0.9999999702$. Note that the value of S_n is less than 1 for any value of n, but as n gets larger and larger, the values of S_n get closer and closer to 1. That is, $S_n \to 1$ and 1 is the **sum of the infinite geometric series**. The sum of an infinite geometric series, if it exists, is denoted S_∞. In this case, $S_\infty = 1$.

Interactive Discovery

Use a grapher that creates tables and graphs sequences. For each of the following infinite series, make a table and a graph and try to find S_∞, if it exists. Look for a pattern connecting the value of the common ratio r with the existence of a value for S_∞.

SERIES	COMMON RATIO, r	S_∞
$1 + 7 + 49 + 343 + \cdots$		
$1 + (-1) + 1 + (-1) + \cdots$		
$\dfrac{1}{2} + \dfrac{1}{4} + \dfrac{1}{8} + \dfrac{1}{16} + \dfrac{1}{32} + \cdots$		
$625 + 250 + 100 + 40 + \cdots$		

It can be shown (but we will not do it here) that the sum of the terms of an infinite geometric series exists if and only if $|r| < 1$ (that is, the absolute value of the common ratio is less than 1).

To find a formula for the sum of an infinite geometric series, we first consider the sum of the first n terms:

$$S_n = \frac{a_1(1 - r^n)}{1 - r} = \frac{a_1 - a_1 r^n}{1 - r}.$$

Using the distributive law

Let's look at values of r^n for a value of r for which $-1 < r < 1$ and see what happens when n gets larger and larger.

Interactive Discovery

Use a grapher that creates tables and graphs sequences. Let $r = 0.65$. Complete the following table and graph of the sequence whose general term is $(0.65)^n$.

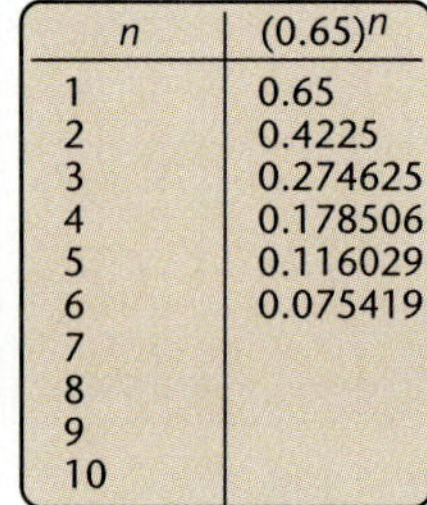

n	$(0.65)^n$
1	0.65
2	0.4225
3	0.274625
4	0.178506
5	0.116029
6	0.075419
7	
8	
9	
10	

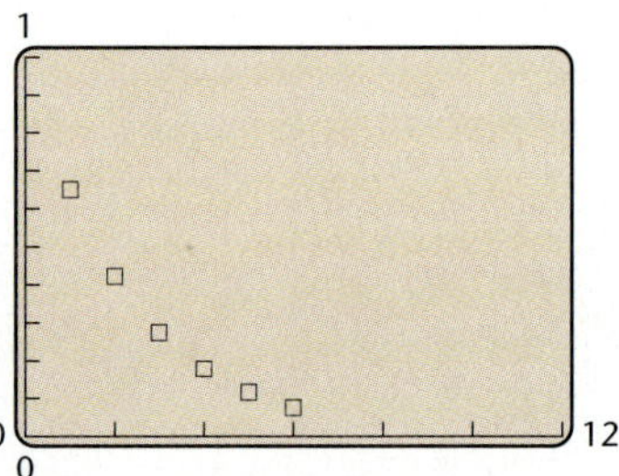

Describe what happens to $(0.65)^n$ as values of n get larger and larger.

We see that values of r^n get closer and closer to 0 as n gets larger. As r^n gets closer and closer to 0, so does $a_1 r^n$. Thus, S_n gets closer and closer to $a_1/(1 - r)$.

Limit or Sum of an Infinite Geometric Series

When $|r| < 1$, the limit or sum of an infinite geometric series is given by

$$S_\infty = \frac{a_1}{1 - r}.$$

Example 6 Determine whether each of the following infinite geometric series has a limit. If a limit exists, find it.

a) $1 + 3 + 9 + 27 + \cdots$

b) $-2 + 1 - \frac{1}{2} + \frac{1}{4} - \frac{1}{8} + \cdots$

SOLUTION

a) Here $r = 3$, so $|r| = |3| = 3$, and since $|r| > 1$, the series *does not* have a limit.

b) Here $r = -\frac{1}{2}$, so $|r| = \left|-\frac{1}{2}\right| = \frac{1}{2}$, and since $|r| < 1$, the series *does* have a limit. We find the limit by

$$S_\infty = \frac{a_1}{1 - r} = \frac{-2}{1 - \left(-\frac{1}{2}\right)} = \frac{-2}{\frac{3}{2}} = -\frac{4}{3}.$$

If you wish, you can also check these results on a grapher using the TABLE and SEQUENCE GRAPH features.

Example 7 Find fractional notation for $0.78787878\ldots$, or $0.\overline{78}$.

SOLUTION We can express this as

$$0.78 + 0.0078 + 0.000078 + \cdots.$$

Then we see that this is an infinite geometric series, where $a_1 = 0.78$ and $r = 0.01$. Since $|r| < 1$, this series has a limit:

$$S_\infty = \frac{a_1}{1 - r} = \frac{0.78}{1 - 0.01} = \frac{0.78}{0.99} = \frac{78}{99}, \quad \text{or} \quad \frac{26}{33}.$$

Thus fractional notation for $0.78787878\ldots$ is $\frac{26}{33}$. You can check this on your calculator.

Applications

The translation of some application and problem-solving situations may involve geometric sequences or series. Examples 9 and 10 in particular show applications in business and economics.

Example 8 *A Daily Doubling Salary.* Suppose someone offered you a job for the month of September (30 days) under the following conditions. You will be paid $0.01 for the first day, $0.02 for the second, $0.04 for the third, and so on, doubling your previous day's salary each day. How much would you earn? (Would you take the job? Make a conjecture before reading further.)

SOLUTION

1. **Familiarize.** You earn $0.01 the first day, $0.01(2) the second day, $0.01(2)(2) the third day, and so on.

2. **Translate.** The amount earned is the geometric series

$$\$0.01 + \$0.01(2) + \$0.01(2^2) + \$0.01(2^3) + \cdots + \$0.01(2^{29}),$$

where

$$a_1 = \$0.01, \qquad r = 2, \quad \text{and} \quad n = 30.$$

3. **Carry out.** Using the formula

$$S_n = \frac{a_1(1 - r^n)}{1 - r},$$

we have

$$S_{30} = \frac{\$0.01(1 - 2^{30})}{1 - 2} \approx \$10{,}737{,}418.23.$$

4. Check. The calculations can be repeated.

5. State. The pay exceeds $10.7 million for the month. Most people would probably take the job!

Example 9   *The Amount of an Annuity.* An **annuity** is a sequence of equal payments made at equal time intervals. The payments do earn interest. For example, fixed deposits in a savings account are an example of an annuity. Suppose someone makes a sequence of yearly deposits of $1000 each in a savings account on which interest is compounded annually at 8%. The total amount in the account at the end of 5 yr is called the **amount of the annuity**. Find that amount.

SOLUTION

1. Familiarize. The following time diagram can help visualize the problem. Note that no deposit is made until the end of the first year.

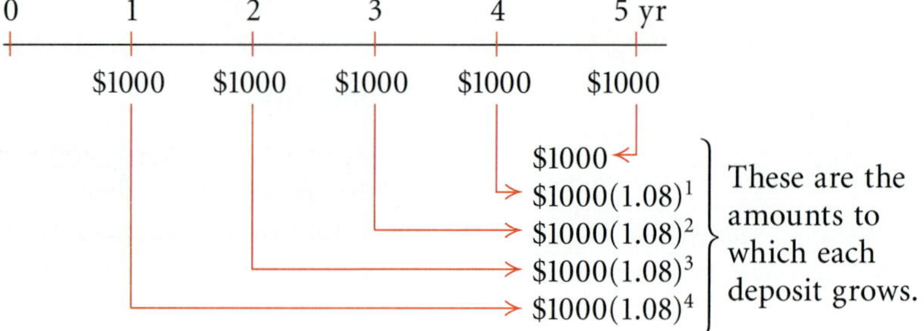

2. Translate. The amount of the annuity is the geometric series

$$\$1000 + \$1000(1.08)^1 + \$1000(1.08)^2 + \$1000(1.08)^3 + \$1000(1.08)^4,$$

where

$$a_1 = \$1000, \quad n = 5, \quad \text{and} \quad r = 1.08.$$

3. Carry out. We could simply carry this out by adding the terms, but we want to make use of the formulas developed in this section. (If we had 40 terms in the sequence, it would be more difficult to add the terms.) Using the formula

$$S_n = \frac{a_1(1 - r^n)}{1 - r},$$

we have

$$S_5 = \frac{\$1000(1 - 1.08^5)}{1 - 1.08}$$
$$\approx \$5866.60.$$

4. Check. The calculations can be repeated.

5. State. The amount of the annuity is $5866.60.

Example 10 *The Economic Multiplier.* The NCAA finals have a tremendous economic effect on the economy of the host city. Suppose 30,000 people visit the city and spend $500 each while there. Then assume that 80% of that money is spent again in the city, and then 80% of that money is spent again, and so on. This is known as the **economic multiplier effect**. Find the total effect on the economy.

SOLUTION

1. **Familiarize.** According to certain economic theory, the money that this effectively puts into the economy can be calculated as the sum of an infinite geometric series as follows. The *initial effect* is 30,000 · $500, or $15,000,000.

2. **Translate.** The *total effect* is given by the series

$$\$15{,}000{,}000 + \$15{,}000{,}000(0.80) + \$15{,}000{,}000(0.80^2) + \cdots .$$

3. **Carry out.** Using the formula for the sum of an infinite geometric series, we have

$$S_\infty = \frac{a_1}{1 - r} = \frac{\$15{,}000{,}000}{1 - 0.80} = \$75{,}000{,}000.$$

4. **Check.** The calculations can be repeated.

5. **State.** The total effect on the economy of the spending from the tournament is $75,000,000.

Do you see why cities work so hard to be chosen to host the NCAA finals?

6.3 Exercise Set

Find the common ratio.

1. 2, 4, 8, 16, ...

2. 18, -6, 2, $-\frac{2}{3}$, ...

3. -1, 1, -1, 1, ...

4. -8, -0.8, -0.08, -0.008, ...

5. $\frac{2}{3}$, $-\frac{4}{3}$, $\frac{8}{3}$, $-\frac{16}{3}$, ...

6. 75, 15, 3, $\frac{3}{5}$, ...

7. 6.275, 0.6275, 0.06275, ...

8. $\dfrac{1}{x}$, $\dfrac{1}{x^2}$, $\dfrac{1}{x^3}$, ...

9. 5, $\dfrac{5a}{2}$, $\dfrac{5a^2}{4}$, $\dfrac{5a^3}{8}$, ...

10. $780, $858, $943.80, $1038.18, ...

Find the indicated term.

11. 2, 4, 8, 16, ...; the 7th term

12. 2, -10, 50, -250, ...; the 9th term

13. 2, $2\sqrt{3}$, 6, ...; the 9th term

14. 1, -1, 1, -1, ...; the 57th term

15. $\dfrac{7}{625}$, $-\dfrac{7}{25}$, ...; the 23rd term

16. $1000, $1060, $1123.60, ...; the 5th term

Find the nth, or general, term.

17. 1, 3, 9, ...

18. 25, 5, 1, ...

19. 1, -1, 1, -1, ...

20. -2, 4, -8, ...

21. $\dfrac{1}{x}$, $\dfrac{1}{x^2}$, $\dfrac{1}{x^3}$, ...

22. 5, $\dfrac{5a}{2}$, $\dfrac{5a^2}{4}$, $\dfrac{5a^3}{8}$, ...

23. Find the sum of the first 7 terms of the geometric series

$$6 + 12 + 24 + \cdots .$$

24. Find the sum of the first 10 terms of the geometric series

$$16 - 8 + 4 - \cdots .$$

25. Find the sum of the first 9 terms of the geometric series

$$\tfrac{1}{18} - \tfrac{1}{6} + \tfrac{1}{2} - \cdots .$$

26. Find the sum of the geometric series

$$-8 + 4 + (-2) + \cdots + \left(-\tfrac{1}{32}\right).$$

Find the sum, if it exists. If you wish, you can also check the results on a grapher using the TABLE *and* SEQUENCE GRAPH *features.*

27. $4 + 2 + 1 + \cdots$

28. $7 + 3 + \tfrac{9}{7} + \cdots$

29. $25 + 20 + 16 + \cdots$

30. $100 - 10 + 1 - \tfrac{1}{10} + \cdots$

31. $8 + 40 + 200 + \cdots$

32. $-6 + 3 - \tfrac{3}{2} + \tfrac{3}{4} - \cdots$

33. $0.6 + 0.06 + 0.006 + \cdots$

34. $\displaystyle\sum_{k=0}^{10} 3^k$

35. $\displaystyle\sum_{k=1}^{11} 15\left(\tfrac{2}{3}\right)^k$

36. $\displaystyle\sum_{k=0}^{50} 200(1.08)^k$

37. $\displaystyle\sum_{k=1}^{\infty} \left(\tfrac{1}{2}\right)^{k-1}$

38. $\displaystyle\sum_{k=1}^{\infty} 2^k$

39. $\displaystyle\sum_{k=1}^{\infty} 12.5^k$

40. $\displaystyle\sum_{k=1}^{\infty} 400(1.0625)^k$

41. $\displaystyle\sum_{k=1}^{\infty} \$500(1.11)^{-k}$

42. $\displaystyle\sum_{k=1}^{\infty} \$1000(1.06)^{-k}$

43. $\displaystyle\sum_{k=1}^{\infty} 16(0.1)^{k-1}$

44. $\displaystyle\sum_{k=1}^{\infty} \tfrac{8}{3}\left(\tfrac{1}{2}\right)^{k-1}$

Find fractional notation.

45. $0.131313\ldots$, or $0.\overline{13}$

46. $0.2222\ldots$, or $0.\overline{2}$

47. $8.999\overline{9}$

48. $6.16\overline{16}$

49. $3.4125\overline{125}$

50. $12.7809\overline{809}$

51. *Bouncing Ping-Pong Ball.*   A ping-pong ball is dropped from a height of 16 ft and always rebounds $\tfrac{1}{4}$ of the distance fallen.

   a) How high does it rebound the 6th time?

   b) Find the total sum of the rebound heights of the ball.

52. *Daily Doubling Salary.*   Suppose someone offered you a job for the month of February (28 days) under the following conditions. You will be paid $0.01 the 1st day, $0.02 the 2nd, $0.04 the 3rd, and so on, doubling your previous day's salary each day. How much would you earn altogether?

53. *Bungee Jumping.*   A bungee jumper rebounds 60% of the height jumped. A bungee jump is made using a cord that stretches to 200 ft.

   a) After jumping and then rebounding 9 times, how far has a bungee jumper traveled upward (the total rebound distance)?

   b) About how far will a jumper have traveled upward (bounced) before coming to rest?

54. *Population Growth.*   Gaintown has a present population of 100,000, and the population is increasing by 3% each year.

   a) What will the population be in 15 yr?

   b) How long will it take for the population to double?

55. *Amount of an Annuity.* To create a college fund, a parent makes a sequence of 18 yearly deposits of $1000 each in a savings account on which interest is compounded annually at 6.2%. Find the amount of the annuity.

56. *Amount of an Annuity.* A sequence of yearly payments of P dollars is invested at the end of each of N years at interest rate i, compounded annually. The total amount in the account, or the amount of the annuity, is V.

a) Show that

$$V = \frac{P[(1 + i)^N - 1]}{i}.$$

b) Suppose that interest is compounded n times per year and deposits are made every compounding period. Show that the formula for V is then given by

$$V = \frac{P\left[\left(1 + \dfrac{i}{n}\right)^{nN} - 1\right]}{i/n}.$$

57. *Loan Repayment.* A family borrows $120,000. The loan is to be repaid in 13 yr at 12% interest, compounded annually. How much will be repaid at the end of 13 yr?

58. *Doubling Paper Folds.* A piece of paper is 0.01 in. thick. It is cut and stacked repeatedly in such a way that its thickness is doubled each time for 20 times. How thick is the result?

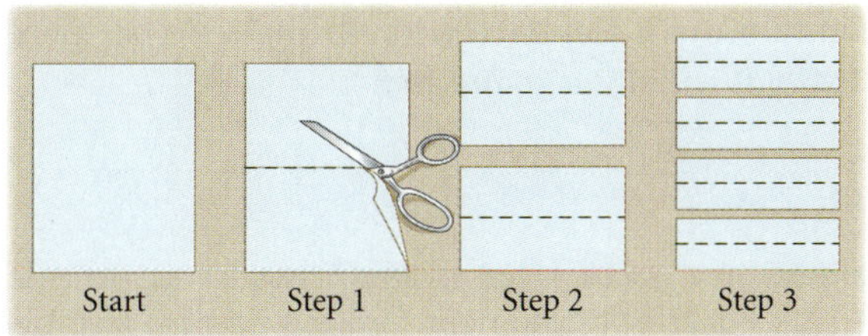

59. *The Economic Multiplier.* The government is making a $13,000,000,000 expenditure for educational improvement. If 85% of this is spent again, and so on, what is the total effect on the economy?

60. *Advertising Effect.* The cereal company Box-o-Vim is about to market a new low-fat great-tasting cereal in a city of 5,000,000 people. They plan an advertising campaign that they think will induce 30% of the people to buy the product. They estimate that if those people like the product, they will induce 30% · 30% · 5,000,000 more to buy the

product, and those will induce 30% · 30% · 30% · 5,000,000, and so on. In all, how many people will buy the product as a result of the advertising campaign? What percentage of the population is this?

Skill Maintenance

Simplify.

61. $k^2 + [2(k + 1) - 1]$

62. $\dfrac{2^k - 1}{2^k} + \dfrac{1}{2^{k+1}}$

63. $2k(k + 1) + 4(k + 1)$

64. $\dfrac{3k(k + 1)}{2} + 3(k + 1)$

65. $\dfrac{k}{k + 1} + \dfrac{1}{(k + 1)(k + 2)}$

66. $\dfrac{k^2(k + 1)^2}{4} + (k + 1)^3$

Synthesis

67. ◈ Write a problem for a classmate to solve. Devise the problem so that a geometric series is involved and the solution is: "The total amount in the bank is $900(1.08)^{40}$, or about $19,552."

68. ◈ The infinite series

$$S_\infty = 2 + \frac{1}{2} + \frac{1}{2 \cdot 3} + \frac{1}{2 \cdot 3 \cdot 4} + \frac{1}{2 \cdot 3 \cdot 4 \cdot 5} + \cdots$$

is not geometric, but it does have a sum. Consider $S_1, S_2, S_3, S_4, S_5,$ and S_6. Construct a table and a graph of the sequence. Expand the sequence of sums, if needed. Make a conjecture about the value of S_∞ and explain your reasoning.

69. Prove that

$$\sqrt{3} - \sqrt{2}, \quad 4 - \sqrt{6}, \quad \text{and} \quad 6\sqrt{3} - 2\sqrt{2}$$

form a geometric sequence.

70. Consider the sequence

$$4, \ 20.4, \ 104.04, \ 530.604, \ \ldots \ .$$

What is the error in using $a_{277} = 4(5.1)^{276}$ to find the 277th term?

71. Consider the sequence

$$x + 3, \ x + 7, \ 4x - 2, \ \ldots \ .$$

a) If the sequence is arithmetic, find x and then determine each of the 3 terms and the 4th term.

b) If the sequence is geometric, find x and then determine each of the 3 terms and the 4th term.

72. Find the sum of the first n terms of
$$1 + x + x^2 + \cdots .$$

73. Find the sum of the first n terms of
$$x^2 - x^3 + x^4 - x^5 + \cdots .$$

In Exercises 74 and 75, assume that $a_1, a_2, a_3, \ldots$ is a geometric sequence.

74. Prove that $a_1^2, a_2^2, a_3^2, \ldots,$ is a geometric sequence.

75. Prove that $\ln a_1, \ln a_2, \ln a_3, \ldots,$ is an arithmetic sequence.

76. Prove that $5^{a_1}, 5^{a_2}, 5^{a_3}, \ldots,$ is a geometric sequence, if $a_1, a_2, a_3, \ldots,$ is an arithmetic sequence.

77. The sides of a square are 16 cm long. A second square is inscribed by joining the midpoints of the sides, successively. In the second square, we repeat the process, inscribing a third square. If this process is continued indefinitely, what is the sum of all the areas of all the squares? (*Hint:* Use an infinite geometric series.)

<table><tr><td>

6.4

Mathematical Induction

</td><td>

- *List the statements of an infinite sequence that is defined by a formula.*
- *Do proofs by mathematical induction.*

</td></tr></table>

In this section, we learn to prove a sequence of mathematical statements using a procedure called *mathematical induction*.

Sequences of Statements

Infinite sequences of statements occur often in mathematics. In an infinite sequence of statements, there is a statement for each natural number. For example, consider the sequence of statements represented by the following:

"For each x between 0 and 1, $0 < x^n < 1$."

Let's think of this as $S(n)$, or S_n. Substituting natural numbers for n gives a sequence of statements. We list a few of them.

Statement 1, S_1:   For x between 0 and 1, $0 < x^1 < 1$.

Statement 2, S_2:   For x between 0 and 1, $0 < x^2 < 1$.

Statement 3, S_3:   For x between 0 and 1, $0 < x^3 < 1$.

Statement 4, S_4:   For x between 0 and 1, $0 < x^4 < 1$.

In this context, the symbols S_1, S_2, S_3, and so on, do not represent sums.

Example 1 List the first four statements in the sequence obtainable from each of the following.

a) $\log n < n$

b) $1 + 3 + 5 + \cdots + (2n - 1) = n^2$

SOLUTION

a) This time, S_n is "$\log n < n$."

$$S_1: \quad \log 1 < 1$$
$$S_2: \quad \log 2 < 2$$
$$S_3: \quad \log 3 < 3$$
$$S_4: \quad \log 4 < 4$$

b) This time, S_n is "$1 + 3 + 5 + \cdots + (2n - 1) = n^2$."

$$S_1: \quad 1 = 1^2$$
$$S_2: \quad 1 + 3 = 2^2$$
$$S_3: \quad 1 + 3 + 5 = 3^2$$
$$S_4: \quad 1 + 3 + 5 + 7 = 4^2$$

Proving Infinite Sequences of Statements

We now develop a method of proof, called **mathematical induction**, which we can use to try to prove that all statements in an infinite sequence of statements are true. The statements usually have the form:

"For all natural numbers n, S_n",

where S_n is some mathematical sentence such as those of the preceding examples. Of course, we cannot prove each statement of an infinite sequence individually. Instead, we try to show that "whenever S_k holds, then S_{k+1} must hold." We abbreviate this as $S_k \rightarrow S_{k+1}$. (This is also read "*If* S_k, *then* S_{k+1}," or "S_k *implies* S_{k+1}.") Suppose that we could somehow establish that this holds for all natural numbers k. Then we would have the following:

$$S_1 \longrightarrow S_2 \qquad \text{meaning "if } S_1 \text{ is true, then } S_2 \text{ is true";}$$
$$S_2 \longrightarrow S_3 \qquad \text{meaning "if } S_2 \text{ is true, then } S_3 \text{ is true";}$$
$$S_3 \longrightarrow S_4 \qquad \text{meaning "if } S_3 \text{ is true, then } S_4 \text{ is true";}$$

and so on, indefinitely.

Even knowing that $S_k \rightarrow S_{k+1}$, we would still not be certain whether there is *any* k for which S_k is true. All we would know is that "if S_k is true, then S_{k+1} is true." Suppose now that S_k is true for some k, say,

$k = 1$. We then must have the following.

S_1 is true.	We have verified, or proved, this.
$S_1 \longrightarrow S_2$	This means that whenever S_1 is true, S_2 is true.
Therefore, S_2 is true.	
$S_2 \longrightarrow S_3$	This means that whenever S_2 is true, S_3 is true.
Therefore, S_3 is true.	
and so on.	

We conclude that S_n is true for all natural numbers n.

This leads us to the principle of mathematical induction, which we use to prove the types of statements considered here.

> ### The Principle of Mathematical Induction
>
> We can prove an infinite sequence of statements S_n by showing the following.
>
> **(1)** *Basis step.* S_1 is true.
>
> **(2)** *Induction step.* For all natural numbers k, $S_k \rightarrow S_{k+1}$.

Mathematical induction is analogous to lining up a sequence of dominoes. The induction step tells us that if any one domino is knocked over, then the one next to it will be hit and knocked over. The basis step tells us that the first domino can indeed be knocked over. Note that in order for all dominoes to fall, *both* conditions must be satisfied.

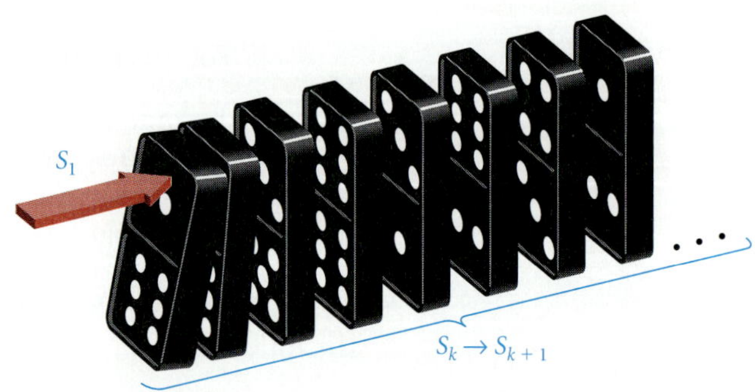

> When you are learning to do proofs by mathematical induction, it is helpful to first write out S_n, S_1, S_k, and S_{k+1}. This helps to identify what is to be assumed and what is to be deduced.

Example 2   Prove: For every natural number n,

$$1 + 3 + 5 + \cdots + (2n - 1) = n^2.$$

PROOF   We first list S_n, S_1, S_k, and S_{k+1}.

S_n: $1 + 3 + 5 + \cdots + (2n - 1) = n^2$

S_1: $1 = 1^2$

S_k: $1 + 3 + 5 + \cdots + (2k - 1) = k^2$

S_{k+1}: $1 + 3 + 5 + \cdots + (2k - 1) + [2(k + 1) - 1] = (k + 1)^2$

(1) *Basis step.* S_1, as listed, is true.

(2) *Induction step.* We let k be any natural number. We assume S_k to be true and try to show that it implies that S_{k+1} is true. Now S_k is

$$1 + 3 + 5 + \cdots + (2k - 1) = k^2.$$

Starting with the left side of S_{k+1} and substituting k^2 for $1 + 3 + 5 + \cdots + (2k - 1)$, we have

$$\underbrace{1 + 3 + \cdots + (2k - 1)} + [2(k + 1) - 1]$$

$$= k^2 + [2(k + 1) - 1] = k^2 + 2k + 1 = (k + 1)^2.$$

We have derived S_{k+1} from S_k. Thus we have shown that for all natural numbers k, $S_k \rightarrow S_{k+1}$. This completes the induction step. It and the basis step tell us that the proof is complete.   ▬

Example 3   Prove: For every natural number n,

$$\frac{1}{2} + \frac{1}{4} + \frac{1}{8} + \cdots + \frac{1}{2^n} = \frac{2^n - 1}{2^n}.$$

PROOF   We first list S_n, S_1, S_k, and S_{k+1}.

S_n: $\dfrac{1}{2} + \dfrac{1}{4} + \dfrac{1}{8} + \cdots + \dfrac{1}{2^n} = \dfrac{2^n - 1}{2^n}$

S_1: $\dfrac{1}{2} = \dfrac{2^1 - 1}{2^1}$

S_k: $\dfrac{1}{2} + \dfrac{1}{4} + \dfrac{1}{8} + \cdots + \dfrac{1}{2^k} = \dfrac{2^k - 1}{2^k}$

S_{k+1}: $\dfrac{1}{2} + \dfrac{1}{4} + \dfrac{1}{8} + \cdots + \dfrac{1}{2^k} + \dfrac{1}{2^{k+1}} = \dfrac{2^{k+1} - 1}{2^{k+1}}$

(1) *Basis step.* We show S_1 to be true as follows:

$$\frac{2^1 - 1}{2^1} = \frac{2 - 1}{2} = \frac{1}{2}.$$

(2) *Induction step.* We let k be any natural number. We assume S_k to be true and try to show that it implies that S_{k+1} is true. Now S_k is

$$\frac{1}{2} + \frac{1}{4} + \frac{1}{8} + \cdots + \frac{1}{2^k} = \frac{2^k - 1}{2^k}.$$

Starting with the left side of S_{k+1} and substituting

$$\frac{2^k - 1}{2^k} \quad \text{for} \quad \frac{1}{2} + \frac{1}{4} + \cdots + \frac{1}{2^k},$$

we have

$$\underbrace{\frac{1}{2} + \frac{1}{4} + \frac{1}{8} + \cdots + \frac{1}{2^k}} + \frac{1}{2^{k+1}}$$

$$= \frac{2^k - 1}{2^k} + \frac{1}{2^{k+1}} = \frac{2^k - 1}{2^k} \cdot \frac{2}{2} + \frac{1}{2^{k+1}} = \frac{(2^k - 1) \cdot 2 + 1}{2^{k+1}}$$

$$= \frac{2^{k+1} - 1}{2^{k+1}}.$$

We have derived S_{k+1} from S_k. Thus we have shown that for all natural numbers k, $S_k \rightarrow S_{k+1}$. This completes the induction step. It and the basis step tell us that the proof is complete.

Example 4 Prove: For every natural number n, $n < 2^n$.

PROOF We first list S_n, S_1, S_k, and S_{k+1}.

S_n: $\quad n < 2^n$

S_1: $\quad 1 < 2^1$

S_k: $\quad k < 2^k$

S_{k+1}: $\quad k + 1 < 2^{k+1}$

(1) *Basis step.* S_1, as listed, is true since $2^1 = 2$ and $1 < 2$.

(2) *Induction step.* We let k be any natural number. We assume S_k to be true and try to show that it implies that S_{k+1} is true. Now

$$k < 2^k \qquad \text{This is } S_k.$$
$$2k < 2 \cdot 2^k \qquad \text{Multiplying on both sides by 2}$$
$$2k < 2^{k+1} \qquad \text{Adding exponents on the right}$$
$$k + k < 2^{k+1}. \qquad \text{Rewriting } 2k \text{ as } k + k$$

Since k is any natural number, we know that $1 \leq k$. Thus,

$$k + 1 \leq k + k. \qquad \text{Adding } k \text{ on both sides}$$

Putting the results $k + 1 \leq k + k$ and $k + k < 2^{k+1}$ together gives us

$$k + 1 < 2^{k+1}. \qquad \text{This is } S_{k+1}.$$

We have derived S_{k+1} from S_k. Thus we have shown that for all natural numbers k, $S_k \rightarrow S_{k+1}$. This completes the induction step. It and the basis step tell us that the proof is complete.

Interactive Discovery

Use a TABLE feature or spreadsheet to make a table for a sequence given by $y = 2^n$ for values for n from 1 to 100. Does the inequality of Example 4 hold? Then check the inequality using a grapher.

Mathematicians often spend considerable time searching for a rule, or sequence of statements, that can be proved using mathematical induction. Although we have not discussed how such rules are found, an interesting geometric interpretation may have been used to discover the formula in Example 2:

$$1 = \bullet = 1^2$$

$$1 + 3 = \quad = 2^2$$

$$1 + 3 + 5 = \quad = 3^2$$

$$1 + 3 + 5 + 7 = \quad = 4^2, \text{ and so on.}$$

<table><tr><td>

6.4 | *Exercise Set*

List the first five statements in the sequence obtainable from each of the following. Determine whether each of the statements is true or false.

1. $n^2 < n^3$

2. $n^2 - n + 41$ is prime. Find a value for n for which the statement is false.

3. A polygon of n sides has $[n(n - 3)]/2$ diagonals.

4. The sum of the angles of a polygon of n sides is $(n - 2) \cdot 180°$.

Use mathematical induction to prove each of the following.

5. $2 + 4 + 6 + \cdots + 2n = n(n + 1)$

6. $4 + 8 + 12 + \cdots + 4n = 2n(n + 1)$

7. $1 + 5 + 9 + \cdots + (4n - 3) = n(2n - 1)$

8. $3 + 6 + 9 + \cdots + 3n = \dfrac{3n(n + 1)}{2}$

</td><td>

9. $2 + 4 + 8 + \cdots + 2^n = 2(2^n - 1)$

10. $2 \leq 2^n$

11. $n < n + 1$

12. $3^n < 3^{n+1}$

13. $2n \leq 2^n$

14. $\dfrac{1}{1 \cdot 2} + \dfrac{1}{2 \cdot 3} + \cdots + \dfrac{1}{n(n + 1)} = \dfrac{n}{n + 1}$

15. $\dfrac{1}{1 \cdot 2 \cdot 3} + \dfrac{1}{2 \cdot 3 \cdot 4} + \dfrac{1}{3 \cdot 4 \cdot 5} + \cdots$

$$+ \dfrac{1}{n(n + 1)(n + 2)} = \dfrac{n(n + 3)}{4(n + 1)(n + 2)}$$

16. If x is any real number greater than 1, then for any natural number n, $x \leq x^n$.

The following formulas can be used to find sums of powers of natural numbers. Use mathematical induction to prove each of the following.

17. $1 + 2 + 3 + \cdots + n = \dfrac{n(n + 1)}{2}$

18. $1^2 + 2^2 + 3^2 + \cdots + n^2 = \dfrac{n(n + 1)(2n + 1)}{6}$

</td></tr></table>

19. $1^3 + 2^3 + 3^3 + \cdots + n^3 = \dfrac{n^2(n+1)^2}{4}$

20. $1^4 + 2^4 + 3^4 + \cdots + n^4$
$$= \dfrac{n(n+1)(2n+1)(3n^2+3n-1)}{30}$$

21. $1^5 + 2^5 + 3^5 + \cdots + n^5$
$$= \dfrac{n^2(n+1)^2(2n^2+2n-1)}{12}$$

Use mathematical induction to prove each of the following.

22. $\displaystyle\sum_{i=1}^{n} (3i - 1) = \dfrac{n(3n+1)}{2}$

23. $\displaystyle\sum_{i=1}^{n} i(i+1) = \dfrac{n(n+1)(n+2)}{3}$

24. $\left(1 + \dfrac{1}{1}\right)\left(1 + \dfrac{1}{2}\right)\left(1 + \dfrac{1}{3}\right) \cdots \left(1 + \dfrac{1}{n}\right) = n + 1$

25. The sum of n terms of an arithmetic sequence:
$$a_1 + (a_1 + d) + (a_1 + 2d) + \cdots + [a_1 + (n-1)d]$$
$$= \dfrac{n}{2}[2a_1 + (n-1)d]$$

Skill Maintenance

Simplify.

26. $\dfrac{8 \cdot 7 \cdot 6 \cdot 5 \cdot 4 \cdot 3 \cdot 2 \cdot 1}{4 \cdot 3 \cdot 2 \cdot 1}$

27. $\dfrac{8 \cdot 7 \cdot 6 \cdot 5 \cdot 4 \cdot 3 \cdot 2 \cdot 1}{5 \cdot 4 \cdot 3 \cdot 2 \cdot 1}$

28. $\dfrac{10 \cdot 9 \cdot 8 \cdot 7 \cdot 6 \cdot 5 \cdot 4 \cdot 3 \cdot 2 \cdot 1}{7 \cdot 6 \cdot 5 \cdot 4 \cdot 3 \cdot 2 \cdot 1}$

29. $\dfrac{10 \cdot 9 \cdot 8 \cdot 7 \cdot 6 \cdot 5 \cdot 4 \cdot 3 \cdot 2 \cdot 1}{8 \cdot 7 \cdot 6 \cdot 5 \cdot 4 \cdot 3 \cdot 2 \cdot 1}$

Synthesis

Use mathematical induction to prove each of the following.

30. The sum of n terms of a geometric sequence:
$$a_1 + a_1 r + a_1 r^2 + \cdots + a_1 r^{n-1} = \dfrac{a_1 - a_1 r^n}{1 - r}$$

31. $x + y$ is a factor of $x^{2n} - y^{2n}$

32. $x + y$ is a factor of $x^{2n-1} + y^{2n-1}$

33. $\dfrac{x^n - y^n}{x - y} = x^{n-1} + yx^{n-2} + \cdots + y^{n-2}x + y^{n-1}$

34. ◆ Write an explanation of the idea behind mathematical induction for a fellow student.

35. ◈ Find two statements not considered in this section that are not true for all natural numbers. Then try to find where a proof by mathematical induction fails.

Prove each of the following using mathematical induction. Do the basis step for $n = 2$.

36. For every natural number $n \geq 2$,
$$2n + 1 < 3^n.$$

37. For every natural number $n \geq 2$,
$$\log_a (b_1 b_2 \cdots b_n)$$
$$= \log_a b_1 + \log_a b_2 + \cdots + \log_a b_n.$$

38. For every natural number $n \geq 2$,
$$\left(1 - \dfrac{1}{2^2}\right)\left(1 - \dfrac{1}{3^2}\right) \cdots \left(1 - \dfrac{1}{n^2}\right) = \dfrac{n+1}{2n}.$$

39. For every natural number $n \geq 2$,
$$\dfrac{1}{\sqrt{1}} + \dfrac{1}{\sqrt{2}} + \dfrac{1}{\sqrt{3}} + \cdots + \dfrac{1}{\sqrt{n}} > \sqrt{n}.$$

Prove each of the following for any complex numbers $z_1, z_2, \ldots, z_n$, where $i^2 = -1$ and $\bar{z}$ is the conjugate of z (see Section 2.2).

40. $\overline{z^n} = \bar{z}^n$

41. $\overline{z_1 + z_2 + \cdots + z_n} = \overline{z_1} + \overline{z_2} + \cdots + \overline{z_n}$

42. $\overline{z_1 z_2 \cdots z_n} = \overline{z_1} \cdot \overline{z_2} \cdots \overline{z_n}$

43. i^n is either 1, -1, i, or $-i$.

For any integers a and b, b is a factor of a if there exists an integer c such that $a = bc$. Prove each of the following for any natural number n.

44. 2 is a factor of $n^2 + n$.

45. 3 is a factor of $n^3 + 2n$.

46. 5 is a factor of $n^5 - n$.

47. 3 is a factor of $n(n+1)(n+2)$

48. *The Tower of Hanoi Problem.* There are three pegs on a board. On one peg are n disks, each smaller than the one on which it rests. The problem is to move this pile of disks to another peg. The final order must be the same, but you can move only one disk at a time and can never place a larger disk on a smaller one.

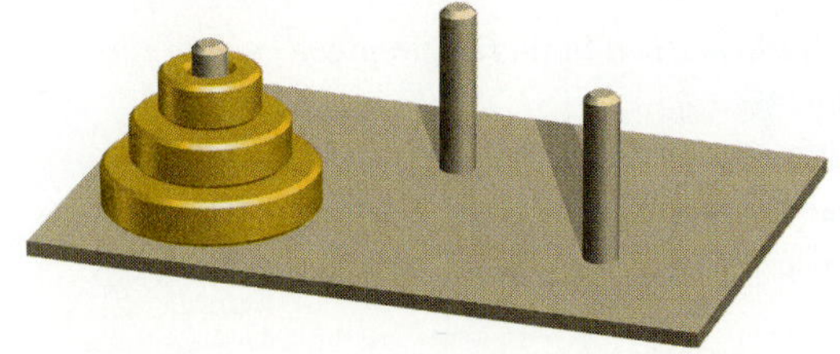

a) What is the *least* number of moves needed to move 3 disks? 4 disks? 2 disks? 1 disk?

b) Conjecture a formula for the *least* number of moves needed to move n disks. Prove it by mathematical induction.

49. Consider this statement: For every natural number n, $n = n + 1$.

a) Can you prove the basis step? If so, do it.

b) Can you prove the induction step? If so, do it.

c) Is the statement true? This illustrates the need for the basis step in an induction proof.

6.5
Combinatorics: Permutations

- *Evaluate factorial and permutation notation and solve related applications.*

In order to study probability, it is first necessary to learn about the theory of counting, called **combinatorics**. Such a study concerns itself with determining the number of ways in which a set can be arranged or combined, certain objects can be chosen, or a succession of events can occur.

Permutations

In this section, we will consider the part of combinatorics called *permutations*.

> The study of permutations involves *order* and *arrangements*.

Example 1   How many 3-letter code symbols can be formed with the letters A, B, C *without* repetition (that is, without a letter repeated)?

SOLUTION   Examples of such symbols are ABC, CBA, ACB, and so on. Consider placing the letters in the frames shown at left. We can select any of the 3 letters for the first letter in the symbol. Once this letter has been selected, the second must be selected from the 2 remaining letters. The third letter is already determined, since only 1 possibility is left. The possibilities can be arrived at using a **tree diagram**, as shown below.

TREE DIAGRAM	OUTCOMES
A → B → C	ABC
A → C → B	ACB
B → A → C	BAC
B → C → A	BCA
C → A → B	CAB
C → B → A	CBA

Each outcome represents one permutation of the letters A, B, C.

There are $3 \cdot 2 \cdot 1$, or 6, possibilities. The set of all of them are as follows:

$$\{ABC, ACB, BAC, BCA, CAB, CBA\}.$$

Suppose that we perform an experiment such as selecting letters (as in the preceding example), flipping a coin, or drawing a card. The results are called **outcomes**. An **event** is a set of outcomes. The following concerns the counting of events that occur together, or are combined.

The Fundamental Counting Principle

Given a combined action, or *event*, in which the first action can be performed in n_1 ways, the second action can be performed in n_2 ways, and so on, the total number of ways in which the combined action can be performed is the product

$$n_1 \cdot n_2 \cdot n_3 \cdot \, \cdots \, \cdot n_k.$$

Example 2   How many 3-letter code symbols can be formed with the letters A, B, C, D, and E *with* repetition (that is, allowing letters to be repeated)?

SOLUTION   Since repetition is allowed, there are 5 choices for the first letter, 5 choices for the second, and 5 for the third. Thus, by the fundamental counting principle, there are $5 \cdot 5 \cdot 5$, or 125 code symbols.

Permutation

A *permutation* of a set of n objects is an ordered arrangement of all n objects.

Consider, for example, a set of 4 objects

$$\{A, B, C, D\}.$$

To find the number of ordered arrangements of the set, we select a first letter: There are 4 choices. Then we select a second letter: There are 3 choices. Then we select a third letter: There are 2 choices. Finally, there is 1 choice for the last selection. Thus, by the fundamental counting principle, there are $4 \cdot 3 \cdot 2 \cdot 1$, or 24, permutations of a set of 4 objects.

We can find a formula for the total number of permutations of all objects in a set of n objects. We have n choices for the first selection, $n - 1$ choices for the second, $n - 2$ for the third, and so on. For the nth selection, there is only 1 choice.

> **A Formula for the Total Number of Permutations of n Objects**
>
> The total number of permutations of n objects, denoted $_nP_n$, is given by
>
> $$_nP_n = n(n-1)(n-2) \cdots 3 \cdot 2 \cdot 1.$$

Example 3 Find each of the following.

a) $_4P_4$ 　　　　　　　　　　　　　　　　　　**b)** $_7P_7$

SOLUTION

Start with 4.

a) $_4P_4 = 4 \cdot 3 \cdot 2 \cdot 1 = 24$

　　　　　　　　　　　　　　4 factors

b) $_7P_7 = 7 \cdot 6 \cdot 5 \cdot 4 \cdot 3 \cdot 2 \cdot 1 = 5040$

Example 4 In how many different ways can 9 packages be placed in 9 mailboxes, one package in a box?

SOLUTION We have

$$_9P_9 = 9 \cdot 8 \cdot 7 \cdot 6 \cdot 5 \cdot 4 \cdot 3 \cdot 2 \cdot 1 = 362{,}880.$$

Factorial Notation

We will use products such as

$$7 \cdot 6 \cdot 5 \cdot 4 \cdot 3 \cdot 2 \cdot 1$$

so often that it is convenient to adopt a notation for them. For the product

$$7 \cdot 6 \cdot 5 \cdot 4 \cdot 3 \cdot 2 \cdot 1, \quad \text{we write} \quad 7!, \quad \text{read} \quad \text{“7 factorial.”}$$

We now define factorial notation for natural numbers and for 0.

> **Factorial Notation**
>
> For any natural number n, we define
>
> $$n! = n(n-1)(n-2) \cdots 3 \cdot 2 \cdot 1.$$
>
> For the number 0, we define
>
> $$0! = 1.$$

We define 0! as 1 so that certain formulas can be stated concisely and with a consistent pattern.

Here are some examples.

$$7! = 7 \cdot 6 \cdot 5 \cdot 4 \cdot 3 \cdot 2 \cdot 1 = 5040$$
$$6! = \quad\; 6 \cdot 5 \cdot 4 \cdot 3 \cdot 2 \cdot 1 = \;\;720$$
$$5! = \qquad\; 5 \cdot 4 \cdot 3 \cdot 2 \cdot 1 = \;\;120$$
$$4! = \qquad\qquad 4 \cdot 3 \cdot 2 \cdot 1 = \;\;\;\;24$$
$$3! = \qquad\qquad\qquad 3 \cdot 2 \cdot 1 = \;\;\;\;\;6$$
$$2! = \qquad\qquad\qquad\qquad 2 \cdot 1 = \;\;\;\;\;2$$
$$1! = \qquad\qquad\qquad\qquad\qquad 1 = \;\;\;\;\;1$$
$$0! = \qquad\qquad\qquad\qquad\qquad 1 = \;\;\;\;\;1$$

Interactive Discovery

Most graphers contain keys for $_nP_n$ and $n!$. Consult your manual. Complete the following table and look for a pattern.

n	$n!$	$_nP_n$
1		
2		
8		
10		
11		
32		

You have probably discovered the following.

$$_nP_n = n!$$

We will often need to manipulate factorial notation. For example, note that

$$8! = 8 \cdot 7 \cdot 6 \cdot 5 \cdot 4 \cdot 3 \cdot 2 \cdot 1$$
$$= 8 \cdot (7 \cdot 6 \cdot 5 \cdot 4 \cdot 3 \cdot 2 \cdot 1)$$
$$= 8 \cdot 7!.$$

Generalizing, we get the following.

For any natural number n, $n! = n(n - 1)!$.

By using this result repeatedly, we can further manipulate factorial notation.

Example 5   Rewrite 7! with a factor of 5!.

SOLUTION   We have

$$7! = 7 \cdot 6 \cdot 5!.$$

In general, we have the following.

> For any natural numbers k and n, with $k < n$,
> $$n! = \underbrace{n(n-1)(n-2)\cdots[n-(k-1)]}_{k \text{ factors}} \cdot \underbrace{(n-k)!}_{n-k \text{ factors}}$$

Permutations of n Objects Taken k at a Time

Consider a set of 5 objects

$$\{A, B, C, D, E\}.$$

How many ordered arrangements are there having 3 members without repetition? Examples of such an arrangement are EBA, CAB, and BCD. There are 5 choices for the first object, 4 choices for the second, and 3 choices for the third. By the fundamental counting principle, there are

$$5 \cdot 4 \cdot 3,$$

or

60 *permutations* of a set of 5 objects taken 3 at a time.

Note that

$$5 \cdot 4 \cdot 3 = \frac{5 \cdot 4 \cdot 3 \cdot 2 \cdot 1}{2 \cdot 1}, \quad \text{or} \quad \frac{5!}{2!}.$$

> ### Permutation of n Objects Taken k at a Time
>
> A *permutation* of a set of n objects taken k at a time is an ordered arrangement of k objects taken from the set.

Consider a set of n objects and the selection of an ordered arrangement of k of them. There would be n choices for the first object. Then there would remain $n-1$ choices for the second, $n-2$ choices for the third, and so on. We make k choices in all, so there are k factors in the product. By the fundamental counting principle, the total number of permutations is

$$\underbrace{n(n-1)(n-2)\cdots[n-(k-1)]}_{k \text{ factors}}.$$

We can express this in another way by multiplying by 1, as follows:

$$n(n-1)(n-2) \cdots [n-(k-1)] \cdot \frac{(n-k)!}{(n-k)!}$$

$$= \frac{n(n-1)(n-2) \cdots [n-(k-1)](n-k)!}{(n-k)!}$$

$$= \frac{n!}{(n-k)!}.$$

This gives us the following.

Formulas for the Number of Permutations of n Objects Taken k at a Time

The number of permutations of a set of n objects taken k at a time, denoted $_nP_k$, is given by

$$_nP_k = \underbrace{n(n-1)(n-2) \cdots [n-(k-1)]}_{k \text{ factors}} \qquad (1)$$

$$= \frac{n!}{(n-k)!}. \qquad (2)$$

Formula (1) is most useful in applications, but formula (2) will be useful in Section 6.6.

Example 6 Compute $_8P_4$ using both formulas.

SOLUTION Using formula (1), we have

The 8 tells where to start.

$$_8P_4 = 8 \cdot 7 \cdot 6 \cdot 5 = 1680.$$

The 4 tells how many factors.

Using formula (2), we have

$$_8P_4 = \frac{8!}{(8-4)!}$$

$$= \frac{8!}{4!}$$

$$= \frac{8 \cdot 7 \cdot 6 \cdot 5 \cdot 4 \cdot 3 \cdot 2 \cdot 1}{4 \cdot 3 \cdot 2 \cdot 1}$$

$$= 8 \cdot 7 \cdot 6 \cdot 5 = 1680.$$

Example 7 *Flags of Nations.* The flags of many nations consist of three vertical stripes similar to the one shown here. For example, the flag of Ireland, shown below, has its first stripe green, second white, and third gold.

Suppose the following 9 colors are available:

{black, yellow, red, blue, white, gold, orange, pink, purple}.

How many different flags of 3 colors can be made without repetition of colors? This assumes that the order in which the stripes appear is considered.

SOLUTION   We are determining the number of permutations of 9 objects taken 3 at a time. There is no repetition of colors. Using formula (1), we get

$$_9P_3 = 9 \cdot 8 \cdot 7 = 504.$$

Example 8   *Batting Orders.* A baseball manager arranges the batting order as follows: The 4 infielders will bat first. Then the 3 outfielders, the catcher, and the pitcher will follow, not necessarily in that order. How many different batting orders are possible?

SOLUTION   The infielders can bat in 4! different ways, the rest in 5! different ways. Then by the fundamental counting principle, we have

$$_4P_4 \cdot {}_5P_5 = 4! \cdot 5!, \quad \text{or} \quad 2880 \text{ possible batting orders.}$$

You probably also have an $\boxed{{}_nP_k}$ key on your calculator or grapher. The notations $P(n, k)$ and $P_{n,k}$ are also used. Consult your manual and then check the results of Examples 6–8.

If we allow repetition, a situation like the following can occur.

Example 9   How many 5-letter code symbols can be formed with the letters A, B, C, and D if we allow a letter to occur more than once?

SOLUTION   We have 5 spaces:

We can select the first letter in 4 ways, the second in 4 ways, and so on. Thus there are 4^5, or 1024 arrangements.

> The number of distinct arrangements of n objects taken k at a time, allowing repetition, is n^k.

Permutations of Sets with Nondistinguishable Objects

Consider a set of 7 marbles, 4 of which are blue and 3 of which are red. Although the marbles are all different, when they are lined up, one red marble will look just like any other red marble. In this sense, we say that the red marbles are nondistinguishable and the blue marbles are nondistinguishable.

We know that there are 7! permutations of this set. Many of them will look alike, however. We develop a formula for finding the number of distinguishable permutations.

Consider a set of n objects in which n_1 are of one kind, n_2 are of a second kind, ..., n_k are of a kth kind. The total number of permutations of the set is $n!$, but this includes many that are indistinguishable. Let N be the total number of distinguishable permutations. For each of these N permutations, there are $n_1!$ actual permutations, obtained by permuting the objects of the first kind. For each of these $N \cdot n_1!$ permutations, there are $n_2!$ nondistinguishable permutations, obtained by permuting the objects of the second kind, and so on. By the fundamental counting principle, the total number of permutations, including those that are nondistinguishable, is

$$N \cdot n_1! \cdot n_2! \cdot \cdots \cdot n_k!.$$

Then we have $N \cdot n_1! \cdot n_2! \cdot \cdots \cdot n_k! = n!$. Solving for N, we obtain

$$N = \frac{n!}{n_1! \cdot n_2! \cdot \cdots \cdot n_k!}.$$

Now, to finish our problem with the marbles, we have

$$N = \frac{7!}{4! \, 3!} = \frac{7 \cdot 6 \cdot 5 \cdot 4 \cdot 3 \cdot 2 \cdot 1}{4 \cdot 3 \cdot 2 \cdot 1 \cdot 3 \cdot 2 \cdot 1} = \frac{7 \cdot 5}{1}, \quad \text{or} \quad 35$$

distinguishable permutations of the marbles.

In general:

> For a set of n objects in which n_1 are of one kind, n_2 are of another kind, ..., n_k are of a kth kind, the number of distinguishable permutations is
>
> $$\frac{n!}{n_1! \cdot n_2! \cdot \cdots \cdot n_k!}.$$

Example 10   In how many distinguishable ways can the letters of the word CINCINNATI be arranged?

SOLUTION   There are 2 C's, 3 I's, 3 N's, 1 A, and 1 T for a total of 10 letters. Thus,

$$N = \frac{10!}{2! \cdot 3! \cdot 3! \cdot 1! \cdot 1!}, \quad \text{or} \quad 50{,}400.$$

The letters of the word CINCINNATI can be arranged in 50,400 distinguishable ways.

6.5 Exercise Set

Evaluate. Do your work by hand and then check it on a grapher.

1. $_6P_6$

2. $_4P_3$

3. $_{10}P_7$

4. $_{10}P_3$

5. $5!$

6. $7!$

7. $0!$

8. $1!$

9. $\dfrac{9!}{5!}$

10. $\dfrac{9!}{4!}$

11. $(8-3)!$

12. $(8-5)!$

13. $\dfrac{10!}{7!\,3!}$

14. $\dfrac{7!}{(7-2)!}$

15. $_8P_0$

16. $_{13}P_1$

Evaluate. Use a grapher.

17. $_{52}P_4$

18. $_{52}P_5$

Evaluate.

19. $_nP_3$

20. $_nP_2$

21. $_nP_1$

22. $_nP_0$

In each of the following exercises, give your answer using permutation notation, factorial notation, or other operations. Then evaluate.

How many permutations are there of the letters in each of the following words, if all the letters are used without repetition?

23. MARVIN

24. JUDY

25. UNDERMOST

26. COMBINES

27. How many permutations are there of the letters of the word UNDERMOST if the letters are taken 4 at a time?

28. How many permutations are there of the letters of the word COMBINES if the letters are taken 5 at a time?

29. How many 5-digit numbers can be formed using the digits 2, 4, 6, 8, and 9 without repetition? with repetition?

30. In how many ways can 7 athletes be arranged in a straight line?

31. How many distinguishable code symbols can be formed from the letters of the word BUSINESS? BIOLOGY? MATHEMATICS?

32. A professor is going to grade her 24 students on a curve. She will give 3 A's, 5 B's, 9 C's, 4 D's, and 3 F's. In how many ways can she do this?

33. *Phone Numbers.* How many 7-digit phone numbers can be formed with the digits 0, 1, 2, 3, 4, 5, 6, 7, 8, and 9, assuming that the first number cannot be 0 or 1? Accordingly, how many telephone numbers can there be within a given area code, before the area needs to be split with a new area code?

34. *Program Planning.* A program is planned to have 5 rock numbers and 4 speeches. In how many ways can this be done if a rock number and a speech are to alternate and a rock number is to come first?

35. Suppose the expression $a^2b^3c^4$ is rewritten without exponents. In how many ways can this be done?

36. *Coin Arrangements.* A penny, a nickel, a dime, and a quarter are arranged in a straight line.

  a) Considering just the coins, in how many ways can they be lined up?

  b) Considering the coins and heads and tails, in how many ways can they be lined up?

37. How many code symbols can be formed using 5 out of 6 letters of A, B, C, D, E, F if the letters:

  a) are not repeated?

  b) can be repeated?

  c) are not repeated but must begin with D?

  d) are not repeated but must begin with DE?

38. *License Plates.* A state forms its license plates by first listing a number that corresponds to the county in which the owner of the car resides (the names of the counties are alphabetized and the number is its location in that order). Then the plate lists a letter of the alphabet, and this is followed by a number from 1 to 9999. How many such plates are possible if there are 80 counties?

39. *Zip Codes.* A zip code in Dallas, Texas, is 75247. A zip code in Cambridge, Massachusetts, is 02142.

  a) How many zip codes are possible if any of the digits 0 to 9 can be used?

  b) If each post office has its own zip code, how many possible post offices can there be?

40. *Zip-Post Codes.*   Zip-post codes use a 9-digit number like 75247-5456, in which the zip code is augmented with a post office box number.

a) How many 9-digit zip-post codes are possible?
b) There are about 266 million people in the United States. Are there enough zip-post codes?

41. *Social Security Numbers.*   A social security number is a 9-digit number like 243-47-0825.

a) How many different social security numbers can there be?
b) There are about 266 million people in the United States. Can each person have a unique social security number?

Skill Maintenance

Simplify.

42. $\dfrac{8 \cdot 7 \cdot 6 \cdot 5 \cdot 4 \cdot 3 \cdot 2 \cdot 1}{3 \cdot 2 \cdot 1 \cdot 5 \cdot 4 \cdot 3 \cdot 2 \cdot 1}$

43. $\dfrac{9 \cdot 8 \cdot 7 \cdot 6 \cdot 5 \cdot 4 \cdot 3 \cdot 2 \cdot 1}{4 \cdot 3 \cdot 2 \cdot 1 \cdot 5 \cdot 4 \cdot 3 \cdot 2 \cdot 1}$

44. $\dfrac{10 \cdot 9 \cdot 8 \cdot 7 \cdot 6 \cdot 5 \cdot 4 \cdot 3 \cdot 2 \cdot 1}{4 \cdot 3 \cdot 2 \cdot 1 \cdot 6 \cdot 5 \cdot 4 \cdot 3 \cdot 2 \cdot 1}$

45. $\dfrac{44 \cdot 43 \cdot 42 \cdot 41 \cdot 40 \cdot 39}{6 \cdot 5 \cdot 4 \cdot 3 \cdot 2 \cdot 1}$

Synthesis

46. ◈ How "long" is 15!? You own 15 different books and decide to actually make up all the possible arrangements of the books on a shelf. About how long, in years, would it take you if you make one arrangement per second? Write out the reasoning you used for this problem in the form of a paragraph.

47. ◈ *Circular Arrangements.*   In how many ways can the numbers on a clock face be arranged? See if you can derive a formula for the number of distinct circular arrangements of n objects. Explain your reasoning.

Solve for n.

48. $_nP_5 = 7 \cdot {}_nP_4$

49. $_nP_4 = 8 \cdot {}_{n-1}P_3$

50. $_nP_5 = 9 \cdot {}_{n-1}P_4$

51. $_nP_4 = 8 \cdot {}_nP_3$

52. Show that $n! = n(n - 1)(n - 2)(n - 3)!$.

53. *Single-Elimination Tournaments.*   In a single-elimination sports tournament consisting of n teams, a team is eliminated when it loses one game. How many games are required to complete the tournament?

54. *Double-Elimination Tournaments.*   In a double-elimination softball tournament consisting of n teams, a team is eliminated when it loses two games. At most, how many games are required to complete the tournament?

6.6
Combinatorics: Combinations

• *Evaluate combination notation and solve related applications.*

We now consider counting techniques in which order is not considered.

Combinations

If you play cards, you know that in most situations the *order* in which you hold cards is *not important!* That is,

The hand

is "equivalent" to these hands.

We may sometimes make selections from a set *without regard to order.* Such selections are called **combinations**.

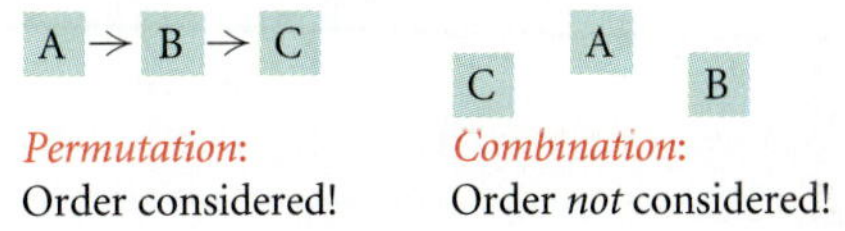

Permutation: *Combination:*
Order considered!   Order *not* considered!

Example 1 Find all the combinations of 3 elements taken from the set of 5 elements {A, B, C, D, E}.

SOLUTION   The combinations are

$$\{A, B, C\}, \quad \{A, B, D\},$$
$$\{A, B, E\}, \quad \{A, C, D\},$$
$$\{A, C, E\}, \quad \{A, D, E\},$$
$$\{B, C, D\}, \quad \{B, C, E\},$$
$$\{B, D, E\}, \quad \{C, D, E\}.$$

There are 10 combinations of 5 objects taken 3 at a time.

When we find all the combinations of 5 objects taken 3 at a time, we are finding all the 3-element subsets. When a set is named, the order of the listing is *not* considered. Thus,

$$\{A, C, B\} \quad \text{names the same set as} \quad \{A, B, C\}.$$

Subset

The set A is a subset of B, denoted $A \subseteq B$, if every element of A is an element of B.

$A \subseteq B$

A is a subset of B.

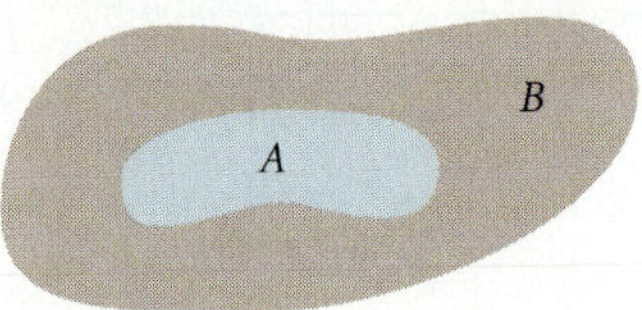

Combination

A *combination* containing k objects is a subset containing k objects.

The elements of a subset are not ordered. When thinking of *combinations,* do *not* think about order!

Example 2   Find all the subsets of {A, B, C}. Identify these as combinations. How many subsets are there in all?

SOLUTION   The empty set has 0 elements in it. It is denoted $\varnothing$. The empty set is a subset of every set. In this case, it is the combination of 3 objects taken 0 at a time. There is 1 such combination.

The following are all the 1-element subsets of {A, B, C}:

$$\{A\}, \quad \{B\}, \quad \{C\}.$$

These are the combinations of 3 objects taken 1 at a time. There are 3 such combinations.

The following are all the 2-element subsets of {A, B, C}:

$$\{A, B\}, \quad \{A, C\}, \quad \{B, C\}.$$

These are the combinations of 3 objects taken 2 at a time. There are 3 such combinations.

The following are all the 3-element subsets of {A, B, C}:

$$\{A, B, C\}.$$

These are the combinations of 3 objects taken 3 at a time. There is only 1 such combination. A set is always a subset of itself.

The total number of subsets is $1 + 3 + 3 + 1$, or 8.

We want to develop a formula for computing the number of combinations of n objects taken k at a time without actually listing the combinations or subsets.

Combination Notation

The number of combinations of n objects taken k at a time is denoted $_nC_k$.

We call $_nC_k$ **combination notation**. We want to derive a general formula for $_nC_k$ for any $k \leq n$. First, it is true that $_nC_n = 1$, because a set with n objects has only 1 subset with n objects, the set itself. Second, $_nC_1 = n$, because a set with n objects has n subsets with 1 element each. Finally, $_nC_0 = 1$, because a set with n objects has only one subset with 0 elements, namely the empty set $\varnothing$. To consider other possibilities, let's return to Example 1 and compare the number of combinations with the number of permutations.

	COMBINATIONS		PERMUTATIONS					
	$\{A, B, C\} \longrightarrow$	ABC	BCA	CAB	CBA	BAC	ACB	
	$\{A, B, D\} \longrightarrow$	ABD	BDA	DAB	DBA	BAD	ADB	
	$\{A, B, E\} \longrightarrow$	ABE	BEA	EAB	EBA	BAE	AEB	
	$\{A, C, D\} \longrightarrow$	ACD	CDA	DAC	DCA	CAD	ADC	
$_5C_3$ of these	$\{A, C, E\} \longrightarrow$	ACE	CEA	EAC	ECA	CAE	AEC	$3! \cdot {}_5C_3$ of these
	$\{A, D, E\} \longrightarrow$	ADE	DEA	EAD	EDA	DAE	AED	
	$\{B, C, D\} \longrightarrow$	BCD	CDB	DBC	DCB	CBD	BDC	
	$\{B, C, E\} \longrightarrow$	BCE	CEB	EBC	ECB	CBE	BEC	
	$\{B, D, E\} \longrightarrow$	BDE	DEB	EBD	EDB	DBE	BED	
	$\{C, D, E\} \longrightarrow$	CDE	DEC	ECD	EDC	DCE	CED	

Note that each combination of 3 objects, say $\{A, C, E\}$, yields $3!$, or 6, permutations, as shown above. It follows that

$$3! \cdot {}_5C_3 = 60 = {}_5P_3 = 5 \cdot 4 \cdot 3,$$

so

$$_5C_3 = \frac{_5P_3}{3!} = \frac{5 \cdot 4 \cdot 3}{3 \cdot 2 \cdot 1} = 10.$$

In general, the number of combinations of n objects taken k at a time, $_nC_k$, times the number of permutations of these objects, $k!$, must equal the number of permutations of n objects taken k at a time:

$$k! \cdot {}_nC_k = {}_nP_k$$

$$_nC_k = \frac{_nP_k}{k!}$$

$$= \frac{1}{k!} \cdot {}_nP_k = \frac{1}{k!} \cdot \frac{n!}{(n-k)!} = \frac{n!}{k!(n-k)!}.$$

Combinations of *n* Objects Taken *k* at a Time

The total number of combinations of n objects taken k at a time, denoted $_nC_k$, is given by

$$_nC_k = \frac{n!}{k!(n-k)!},\tag{1}$$

or

$$_nC_k = \frac{_nP_k}{k!} = \frac{n(n-1)(n-2)\cdots[n-(k-1)]}{k!}.\tag{2}$$

Another kind of notation for $_nC_k$ is **binomial coefficient notation**. The reason for such terminology will be seen later.

Binomial Coefficient Notation

$$\binom{n}{k} = {}_nC_k$$

You should be able to use either notation and either formula.

Example 3 Evaluate $\binom{7}{5}$, using formulas (1) and (2).

SOLUTION

a) By formula (1),

$$\binom{7}{5} = \frac{7!}{5!\,2!} = \frac{7\cdot6\cdot5\cdot4\cdot3\cdot2\cdot1}{5\cdot4\cdot3\cdot2\cdot1\cdot2\cdot1} = \frac{7\cdot6}{2\cdot1} = 21.$$

b) By formula (2),

$$\binom{7}{5} = \frac{7\cdot6\cdot5\cdot4\cdot3}{5\cdot4\cdot3\cdot2\cdot1} = \frac{7\cdot6}{2\cdot1} = 21.$$

The 7 tells where to start.

The 5 tells how many factors there are in both the numerator and the denominator and where to start the denominator.

The method in Example 3(b), using formula (2), is easier to carry out, but in some situations formula (1) becomes useful. Be sure to keep in mind that $\binom{n}{k}$ does not mean $n \div k$, or n/k.

Example 4   Evaluate $\binom{n}{0}$ and $\binom{n}{2}$.

SOLUTION   We use formula (1) for the first expression and formula (2) for the second. Then

$$\binom{n}{0} = \frac{n!}{0!(n-0)!} = \frac{n!}{1 \cdot n!} = 1,$$

using formula (1), and

$$\binom{n}{2} = \frac{n(n-1)}{2!} = \frac{n(n-1)}{2}, \quad \text{or} \quad \frac{n^2-n}{2},$$

using formula (2).

Most calculators and graphers also have a $\boxed{{}_nC_k}$ key. Let's use it to make a discovery.

Interactive Discovery

Use a grapher to complete the table at right. Look for a pattern.

n	k	${}_nC_k$	${}_nC_{n-k}$	$\binom{n}{k}$	$\binom{n}{n-k}$
8	3				
10	7				
9	1				
7	0				
13	10				

Note that

$$\binom{7}{2} = \frac{7 \cdot 6}{2 \cdot 1} = 21,$$

so that using the result of Example 3 gives us

$$\binom{7}{5} = \binom{7}{2}.$$

This says that the number of 5-element subsets of a set of 7 objects is the same as the number of 2-element subsets of a set of 7 objects. When 5 elements are chosen from a set, one also chooses *not* to include 2 elements. To see this, consider such a set:

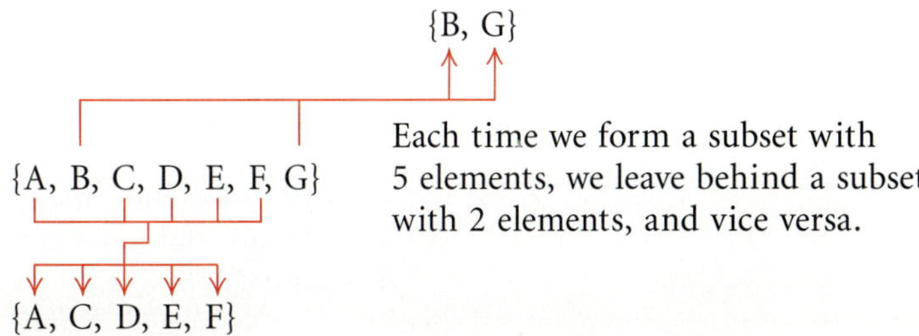

In general, we have the following.

Subsets of Size k and of Size n − k

$$\binom{n}{k} = \binom{n}{n-k} \quad \text{and} \quad {}_nC_k = {}_nC_{n-k}$$

The number of subsets of size k of a set with n objects is the same as the number of subsets of size $n - k$. The number of combinations of n objects taken k at a time is the same as the number of combinations of n objects taken $n - k$ at a time.

This result provides an alternative way to compute. For example, it is much easier to compute ${}_{52}C_4$ than to compute ${}_{52}C_{48}$, though if you were using a grapher, the difference would not be worth considering.

We now solve problems involving combinations.

Example 5 *Michigan Lotto.* The state of Michigan runs a 6-out-of-49-number lotto twice a week that pays at least $2 million. You purchase a card for $1 and pick any 6 numbers from 1 to 49. If your numbers match those that the state draws, you win.

a) How many 6-number combinations are there for drawing?

b) Suppose it takes 10 min to pick your numbers and buy a ticket. How many tickets can you buy in 4 days?

c) How many people would you have to hire to buy all the tickets and ensure that you win?

SOLUTION

a) No order is implied here. You pick any 6 numbers from 1 to 49. Thus the number of combinations is

$$\begin{aligned}
{}_{49}C_6 &= \binom{49}{6} \\
&= \frac{49 \cdot 48 \cdot 47 \cdot 46 \cdot 45 \cdot 44}{6 \cdot 5 \cdot 4 \cdot 3 \cdot 2 \cdot 1} \\
&= 13{,}983{,}816.
\end{aligned}$$

b) In 4 days, there are $4 \cdot 24 \cdot 60$, or 5760 min, so you could buy 5760/10, or 576 tickets in that entire time period.

c) You would need to hire 13,983,816/576, or about 24,278 people to buy all the tickets and ensure a win. (This presumes lottery tickets can be bought 24 hours a day, which is questionable.) ▬

Example 6 How many committees can be formed from a group of 5 governors and 7 senators if each committee consists of 3 governors and 4 senators?

SOLUTION   The 3 governors can be selected in $_5C_3$ ways and the 4 senators can be selected in $_7C_4$ ways. If we use the fundamental counting principle, it follows that the number of possible committees is

$$_5C_3 \cdot {_7C_4} = 10 \cdot 35 = 350.$$

6.6 Exercise Set

Evaluate. Do your work by hand and then check it on a grapher.

1. $_{13}C_2$

2. $_9C_6$

3. $\binom{13}{11}$

4. $\binom{9}{3}$

5. $\binom{7}{1}$

6. $\binom{8}{8}$

7. $\dfrac{_5P_3}{3!}$

8. $\dfrac{_{10}P_5}{5!}$

9. $\binom{6}{0}$

10. $\binom{6}{1}$

11. $\binom{6}{2}$

12. $\binom{6}{3}$

13. $\binom{7}{0} + \binom{7}{1} + \binom{7}{2} + \binom{7}{3} + \binom{7}{4} + \binom{7}{5} + \binom{7}{6} + \binom{7}{7}$

14. $\binom{6}{0} + \binom{6}{1} + \binom{6}{2} + \binom{6}{3} + \binom{6}{4} + \binom{6}{5} + \binom{6}{6}$

Evaluate. Use a grapher.

15. $_{52}C_4$

16. $_{52}C_5$

17. $\binom{27}{11}$

18. $\binom{37}{8}$

Evaluate.

19. $\binom{n}{1}$

20. $\binom{n}{3}$

21. $\binom{m}{m}$

22. $\binom{t}{4}$

In each of the following exercises, give an expression for the answer using permutation notation, combination notation, factorial notation, or other operations. Then evaluate.

23. *Fraternity Officers.* There are 23 students in a fraternity. How many sets of 4 officers can be selected?

24. *League Games.* How many games can be played in a 9-team sports league if each team plays all other teams once? twice?

25. *Test Options.* On a test, a student is to select 10 out of 13 questions. In how many ways can this be done?

26. *Test Options.* Of the first 10 questions on a test, a student must answer 7. Of the second 5 questions, the student must answer 3. In how many ways can this be done?

27. *Lines and Triangles from Points.* How many lines are determined by 8 points, no 3 of which are collinear? How many triangles are determined by the same points?

28. *Senate Committees.* Suppose the Senate of the United States consists of 58 Republicans and 42 Democrats. How many committees can be formed consisting of 6 Republicans and 4 Democrats?

29. *Poker Hands.* How many 5-card poker hands are possible with a 52-card deck?

30. *Bridge Hands.* How many 13-card bridge hands are possible with a 52-card deck?

31. *Baskin-Robbins Ice Cream.* Baskin-Robbins, a national firm, sells ice cream in 31 flavors.

a) How many 2-dip cones are possible if order of flavors is to be considered and no flavor is repeated?

b) How many 2-dip cones are possible if order is to be considered and a flavor can be repeated?

c) How many 2-dip cones are possible if order is not considered and no flavor is repeated?

Skill Maintenance

Simplify or expand.

32. $(a + b)^0$

33. $(a + b)^1$

34. $(a + b)^2$

35. $(a + b)^3$

36. $(a - b)^3$

37. $(a + b)^4$

Synthesis

38. ◆ Explain why a "combination" lock should really be called a "permutation" lock?

39. ◆ Give an explanation that you might use with a fellow student to explain that

$$\binom{n}{k} = \binom{n}{n - k}.$$

40. *Full House.*   A full house in poker consists of a pair (two of a kind) and three of a kind. How many full houses are there that consist of 3 aces and 2 queens? (See Section 9.8 for a description of a 52-card deck.)

41. *Flush.*   A flush in poker consists of a 5-card hand with all cards of the same suit. How many 5-card hands (flushes) are there that consist of all diamonds?

42. There are n points on a circle. How many quadrilaterals can be inscribed with these points as vertices?

43. *League Games.*   How many games are played in a league with n teams if each team plays each other team once? twice?

Solve for n.

44. $\binom{n + 1}{3} = 2 \cdot \binom{n}{2}$ **45.** $\binom{n}{n - 2} = 6$

46. $\binom{n}{3} = 2 \cdot \binom{n - 1}{2}$ **47.** $\binom{n + 2}{4} = 6 \cdot \binom{n}{2}$

48. How many line segments are determined by the n vertices of an n-agon? Of these, how many are diagonals? Use mathematical induction to prove the result for the diagonals.

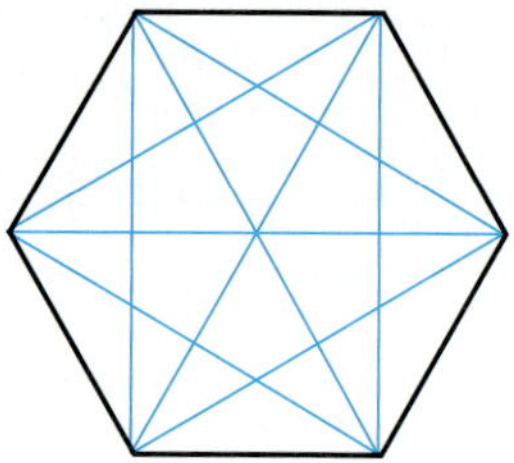

49. Prove that

$$\binom{n}{k - 1} + \binom{n}{k} = \binom{n + 1}{k}$$

for any natural numbers n and k, $k \leq n$.

6.7

The Binomial Theorem

- *Expand a power of a binomial using Pascal's triangle or factorial notation.*
- *Find a specific term of a binomial expansion.*
- *Find the total number of subsets of a set of n objects.*

In this section, we consider ways of expanding a binomial $(a + b)^n$.

Binomial Expansions Using Pascal's Triangle

Consider the following expanded powers of $(a + b)^n$, where $a + b$ is any binomial and n is a whole number. Look for patterns.

$$(a + b)^0 = 1$$
$$(a + b)^1 = a + b$$
$$(a + b)^2 = a^2 + 2ab + b^2$$
$$(a + b)^3 = a^3 + 3a^2b + 3ab^2 + b^3$$
$$(a + b)^4 = a^4 + 4a^3b + 6a^2b^2 + 4ab^3 + b^4$$
$$(a + b)^5 = a^5 + 5a^4b + 10a^3b^2 + 10a^2b^3 + 5ab^4 + b^5$$

Each expansion is a polynomial. There are some patterns to be noted.

1. There is one more term than the power of the exponent, n. That is, there are $n + 1$ terms in the expansion of $(a + b)^n$.

2. In each term, the sum of the exponents is n, the power to which the binomial is raised.

3. The exponents of a start with n, the power of the binomial, and decrease to 0. The last term has no factor of a. The first term has no factor of b, so powers of b start with 0 and increase to n.

4. The coefficients start at 1 and increase through certain values about "half"-way and then decrease through these same values back to 1. Let's explore the coefficients further.

Suppose we want to find an expansion of $(a + b)^8$. The patterns we noted above indicate that there are 9 terms in the expansion:

$$a^8 + c_1a^7b + c_2a^6b^2 + c_3a^5b^3 + c_4a^4b^4$$
$$+ c_5a^3b^5 + c_6a^2b^6 + c_7ab^7 + b^8.$$

How can we determine the value of each coefficient, c_i? We can answer this question in two different ways. The first method seems to be the easiest, but is not always. It involves writing down the coefficients in a triangular array, as follows. We form what is known as **Pascal's triangle**:

$$
\begin{array}{cccccccccccc}
(a + b)^0\text{:} & & & & & & 1 & & & & & \\
(a + b)^1\text{:} & & & & & 1 & & 1 & & & & \\
(a + b)^2\text{:} & & & & 1 & & 2 & & 1 & & & \\
(a + b)^3\text{:} & & & 1 & & 3 & & 3 & & 1 & & \\
(a + b)^4\text{:} & & 1 & & 4 & & 6 & & 4 & & 1 & \\
(a + b)^5\text{:} & 1 & & 5 & & 10 & & 10 & & 5 & & 1
\end{array}
$$

There are many patterns in the triangle. Find as many as you can.

Perhaps you discovered a way to write the next row of numbers, given the numbers in the row above it. There are always 1's on the outside. Each remaining number is the sum of the two numbers above. Let's try to find an expansion for $(a + b)^6$ by adding another row using the patterns we have discovered:

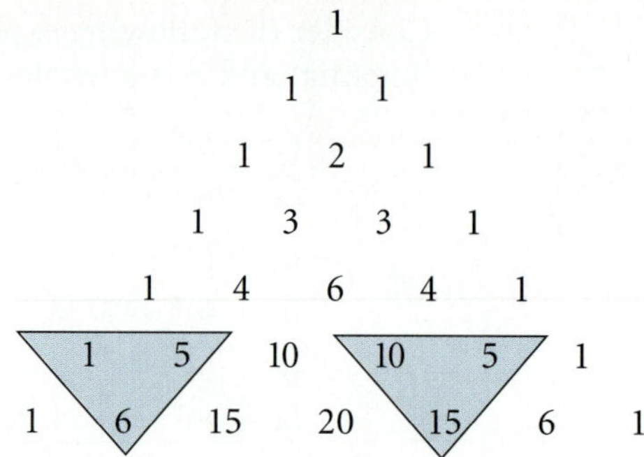

We see that in the last row

> the 1st and last numbers are **1**;
> the 2nd number is $1 + 5$, or **6**;
> the 3rd number is $5 + 10$, or **15**;
> the 4th number is $10 + 10$, or **20**;
> the 5th number is $10 + 5$, or **15**; and
> the 6th number is $5 + 1$, or **6**.

Thus the expansion for $(a + b)^6$ is

$$(a + b)^6 = 1a^6 + 6a^5b + 15a^4b^2 + 20a^3b^3 + 15a^2b^4 + 6ab^5 + 1b^6.$$

To find an expansion for $(a + b)^8$, we complete two more rows of Pascal's triangle:

$$
\begin{array}{ccccccccccccccc}
&&&&&&& 1 &&&&&&& \\
&&&&&& 1 && 1 &&&&&& \\
&&&&& 1 && 2 && 1 &&&&& \\
&&&& 1 && 3 && 3 && 1 &&&& \\
&&& 1 && 4 && 6 && 4 && 1 &&& \\
&& 1 && 5 && 10 && 10 && 5 && 1 && \\
& 1 && 6 && 15 && 20 && 15 && 6 && 1 & \\
1 && 7 && 21 && 35 && 35 && 21 && 7 && 1 \\
\end{array}
$$
$$1 \quad 8 \quad 28 \quad 56 \quad 70 \quad 56 \quad 28 \quad 8 \quad 1$$

Thus the expansion of $(a + b)^8$ is

$$(a + b)^8 = a^8 + 8a^7b^1 + 28a^6b^2 + 56a^5b^3 + 70a^4b^4 + 56a^3b^5$$
$$+ 28a^2b^6 + 8ab^7 + b^8.$$

We can generalize our results as follows.

> **The Binomial Theorem Using Pascal's Triangle**
>
> For any binomial $a + b$ and any natural number n,
> $$(a + b)^n = c_0 a^n b^0 + c_1 a^{n-1} b^1 + c_2 a^{n-2} b^2 + \cdots + c_{n-1} a^1 b^{n-1}$$
> $$+ c_n a^0 b^n,$$
>
> where the numbers $c_0, c_1, c_2, \ldots, c_{n-1}, c_n$ are from the $(n + 1)$st row of Pascal's triangle.

Example 1   Expand: $(u - v)^5$.

Solution   We note that $a = u$, $b = -v$, and $n = 5$. We use the 6th row of Pascal's triangle:

$$1 \qquad 5 \qquad 10 \qquad 10 \qquad 5 \qquad 1$$

Then we have

$$(u - v)^5 = [u + (-v)]^5$$
$$= 1(u)^5 + 5(u)^4(-v)^1 + 10(u)^3(-v)^2 + 10(u)^2(-v)^3$$
$$+ 5(u)(-v)^4 + 1(-v)^5$$
$$= u^5 - 5u^4 v + 10u^3 v^2 - 10u^2 v^3 + 5uv^4 - v^5.$$

Note that the signs of the terms alternate between $+$ and $-$. When the power of $-v$ is odd, the sign is $-$.

Interactive Discovery

Graph each of the following functions. See if you can discover an identity from your graphs. Confirm your results using the binomial theorem.

$$y_1 = (x - 2)^4,$$
$$y_2 = x^4 + 8x^3 + 24x^2 + 32x + 16,$$
$$y_3 = x^4 - 8x^3 + 24x^2 - 32x - 16,$$
$$y_4 = x^4 + 8x^3 - 24x^2 + 32x - 16,$$
$$y_5 = x^4 - 8x^3 + 24x^2 - 32x + 16,$$
$$y_6 = x^4 - 8x^3 + 24x^2 - 31x + 16$$

Example 2   Expand: $\left(2t + \dfrac{3}{t} \right)^6$.

Solution   We note that $a = 2t$, $b = 3/t$, and $n = 6$. We use the 7th row of Pascal's triangle:

$$1 \qquad 6 \qquad 15 \qquad 20 \qquad 15 \qquad 6 \qquad 1$$

Then we have

$$\left(2t + \frac{3}{t}\right)^6 = (2t)^6 + 6(2t)^5\left(\frac{3}{t}\right)^1 + 15(2t)^4\left(\frac{3}{t}\right)^2$$

$$+ 20(2t)^3\left(\frac{3}{t}\right)^3 + 15(2t)^2\left(\frac{3}{t}\right)^4 + 6(2t)^1\left(\frac{3}{t}\right)^5 + \left(\frac{3}{t}\right)^6$$

$$= 64t^6 + 6(32t^5)\left(\frac{3}{t}\right) + 15(16t^4)\left(\frac{9}{t^2}\right) + 20(8t^3)\left(\frac{27}{t^3}\right)$$

$$+ 15(4t^2)\left(\frac{81}{t^4}\right) + 6(2t)\left(\frac{243}{t^5}\right) + \frac{729}{t^6}$$

$$= 64t^6 + 576t^4 + 2160t^2 + 4320 + 4860t^{-2}$$

$$+ 2916t^{-4} + 729t^{-6}.$$

Binomial Expansion Using Factorial Notation

Suppose we want to find the expansion of $(a + b)^{11}$. The disadvantage in using Pascal's triangle is that we must compute all the preceding rows of the table to obtain the row needed for the expansion. The following method avoids this difficulty. It will also enable us to find a specific term—say, the 8th term—without computing all the other terms of the expansion. This method is useful in such courses as finite mathematics, calculus, and statistics, and uses the *binomial coefficient notation* $\binom{n}{k}$ developed in Section 6.6.

We can restate the binomial theorem as follows.

The Binomial Theorem Using Factorial Notation

For any binomial $a + b$ and any natural number n,

$$(a + b)^n = \binom{n}{0}a^n b^0 + \binom{n}{1}a^{n-1}b^1 + \binom{n}{2}a^{n-2}b^2 + \cdots$$

$$+ \binom{n}{n-1}a^1 b^{n-1} + \binom{n}{n}a^0 b^n$$

$$= \sum_{k=0}^{n} \binom{n}{k}a^{n-k}b^k.$$

The binomial theorem can be proved by mathematical induction, but we will not do so here. This form shows why $\binom{n}{k}$ is called a *binomial coefficient.*

Example 3   Expand: $(x^2 - 2y)^5$.

SOLUTION Note that $a = x^2$, $b = -2y$, and $n = 5$. Then using the binomial theorem, we have

$$(x^2 - 2y)^5 = \binom{5}{0}(x^2)^5 + \binom{5}{1}(x^2)^4(-2y) + \binom{5}{2}(x^2)^3(-2y)^2$$

$$+ \binom{5}{3}(x^2)^2(-2y)^3 + \binom{5}{4}x^2(-2y)^4 + \binom{5}{5}(-2y)^5$$

$$= \frac{5!}{0!\,5!}x^{10} + \frac{5!}{1!\,4!}x^8(-2y) + \frac{5!}{2!\,3!}x^6(-2y)^2 + \frac{5!}{3!\,2!}x^4(-2y)^3$$

$$+ \frac{5!}{4!\,1!}x^2(-2y)^4 + \frac{5!}{5!\,0!}(-2y)^5$$

$$= x^{10} - 10x^8y + 40x^6y^2 - 80x^4y^3 + 80x^2y^4 - 32y^5.$$

Example 4 Expand: $\left(\dfrac{2}{x} + 3\sqrt{x}\right)^4$.

SOLUTION Note that $a = 2/x$, $b = 3\sqrt{x}$, and $n = 4$. Then using the binomial theorem, we have

$$\left(\frac{2}{x} + 3\sqrt{x}\right)^4 = \binom{4}{0}\left(\frac{2}{x}\right)^4 + \binom{4}{1}\left(\frac{2}{x}\right)^3(3\sqrt{x}) + \binom{4}{2}\left(\frac{2}{x}\right)^2(3\sqrt{x})^2$$

$$+ \binom{4}{3}\left(\frac{2}{x}\right)(3\sqrt{x})^3 + \binom{4}{4}(3\sqrt{x})^4$$

$$= \frac{4!}{0!\,4!}\left(\frac{2}{x}\right)^4 + \frac{4!}{1!\,3!}\left(\frac{2}{x}\right)^3(3\sqrt{x}) + \frac{4!}{2!\,2!}\left(\frac{2}{x}\right)^2(3\sqrt{x})^2$$

$$+ \frac{4!}{3!\,1!}\left(\frac{2}{x}\right)(3\sqrt{x})^3 + \frac{4!}{4!\,0!}(3\sqrt{x})^4$$

$$= \frac{16}{x^4} + \frac{96}{x^{5/2}} + \frac{216}{x} + 216\sqrt{x} + 81x^2.$$

Interactive Discovery

Describe how you might use a grapher to check the result of Example 4.

Finding a Specific Term

Suppose we want to determine only a particular term of an expansion. The method we have developed will allow us to find such a term without computing all the rows of Pascal's triangle or all the preceding coefficients.

Note that in the binomial theorem, $\binom{n}{0}a^nb^0$ gives us the 1st term, $\binom{n}{1}a^{n-1}b^1$ gives us the 2nd term, $\binom{n}{2}a^{n-2}b^2$ gives us the 3rd term, and so on. This can be generalized as follows.

> **Finding the $(k + 1)$st Term**
>
> The $(k + 1)$st term of $(a + b)^n$ is $\binom{n}{k}a^{n-k}b^k$.

Example 5   Find the 5th term in the expansion of $(2x - 5y)^6$.

SOLUTION   First, we note that $5 = 4 + 1$. Thus, $k = 4$, $a = 2x$, $b = -5y$, and $n = 6$. Then the 5th term of the expansion is

$$\binom{6}{4}(2x)^{6-4}(-5y)^4, \quad \text{or} \quad \frac{6!}{4!\,2!}(2x)^2(-5y)^4, \quad \text{or} \quad 37{,}500x^2y^4.$$

Example 6   Find the 8th term in the expansion of $(3x - 2)^{10}$.

SOLUTION   First, we note that $8 = 7 + 1$. Thus, $k = 7$, $a = 3x$, $b = -2$, and $n = 10$. Then the 8th term of the expansion is

$$\binom{10}{7}(3x)^{10-7}(-2)^7, \quad \text{or} \quad \frac{10!}{7!\,3!}(3x)^3(-128), \quad \text{or} \quad -414{,}720x^3.$$

Total Number of Subsets

Suppose a set has n objects. The number of subsets containing k elements is $\binom{n}{k}$ by a result of Section 6.6. The total number of subsets of a set is the number of subsets with 0 elements, plus the number of subsets with 1 element, plus the number of subsets with 2 elements, and so on. The total number of subsets of a set with n elements is

$$\binom{n}{0} + \binom{n}{1} + \binom{n}{2} + \cdots + \binom{n}{n}.$$

Let's find the expansion of $(1 + 1)^n$:

$$(1 + 1)^n = \binom{n}{0} \cdot 1^n + \binom{n}{1} \cdot 1^{n-1} \cdot 1^1 + \binom{n}{2} \cdot 1^{n-2} \cdot 1^2$$
$$+ \cdots + \binom{n}{n} \cdot 1^n$$
$$= \binom{n}{0} + \binom{n}{1} + \binom{n}{2} + \cdots + \binom{n}{n}.$$

Thus the total number of subsets is $(1 + 1)^n$, or 2^n. We have proved the following.

> **Total Number of Subsets**
>
> The total number of subsets of a set with n elements is 2^n.

Example 7   The set {A, B, C, D, E} has how many subsets?

SOLUTION   The set has 5 elements, so the number of subsets is 2^5, or 32.

Example 8   Wendy's, a national restaurant firm, offers the following condiments for its hamburgers:

> {*catsup, mustard, mayonnaise, tomato,*
>
> *lettuce, onions, pickle, relish, cheese*}.

How many different kinds of hamburgers can Wendy's serve, excluding size of hamburger or number of patties?

SOLUTION   The condiments on each hamburger is a subset of the set of all possible condiments, the empty set being a plain hamburger. The total number of possible hamburgers is

$$\binom{9}{0} + \binom{9}{1} + \binom{9}{2} + \cdots + \binom{9}{9} = 2^9 = 512.$$

Thus Wendy's serves hamburgers in 512 different ways.

We end this section with a discussion of why 0! is defined to be 1. In the binomial expansion, we want $\binom{n}{0}$ to equal 1 and we also want the definition

$$\binom{n}{k} = \frac{n!}{k!(n-k)!}$$

to hold for all whole numbers n and k, where $k \le n$. Thus we must have

$$\binom{n}{0} = \frac{n!}{0!(n-0)!} = \frac{n!}{0!\,n!} = 1.$$

This will be satisfied if 0! is defined to be 1.

6.7 Exercise Set

Expand.

1. $(x + 5)^4$

2. $(x - 1)^4$

3. $(x - 3)^5$

4. $(x + 2)^9$

5. $(x - y)^5$

6. $(x + y)^8$

7. $(5x + 4y)^6$

8. $(2x - 3y)^5$

9. $\left(2t + \dfrac{1}{t}\right)^7$

10. $\left(3y - \dfrac{1}{y}\right)^4$

11. $(x^2 - 1)^5$

12. $(1 + 2q^3)^8$

13. $(\sqrt{5} + t)^6$

14. $(x - \sqrt{2})^6$

15. $\left(a - \dfrac{2}{a}\right)^9$

16. $(1 + 3)^n$

17. $(\sqrt{2} + 1)^6 - (\sqrt{2} - 1)^6$

18. $(1 - \sqrt{2})^4 + (1 + \sqrt{2})^4$

19. $(x^{-2} + x^2)^4$

20. $\left(\dfrac{1}{\sqrt{x}} - \sqrt{x}\right)^6$

Find the indicated term of the binomial expansion.

21. 3rd; $(a + b)^7$

22. 6th; $(x + y)^8$

23. 6th; $(x - y)^{10}$

24. 5th; $(p - 2q)^9$

25. 12th; $(a - 2)^{14}$

26. 11th; $(x - 3)^{12}$

27. 5th; $(2x^3 - \sqrt{y})^8$

28. 4th; $\left(\dfrac{1}{b^2} + \dfrac{b}{3}\right)^7$

29. Middle; $(2u - 3v^2)^{10}$

30. Middle two; $(\sqrt{x} + \sqrt{3})^5$

Determine the number of subsets of each of the following.

31. A set of 7 elements

32. A set of 6 members

33. The set of letters of the Greek alphabet, which contains 24 letters

34. The set of letters of the English alphabet, which contains 26 letters

35. What is the degree of $(x^5 + 3)^4$?

36. What is the degree of $(2 - 5x^3)^7$?

Expand each of the following, where $i^2 = -1$.

37. $(3 + i)^5$ **38.** $(1 + i)^6$

39. $(\sqrt{2} - i)^4$ **40.** $\left(\dfrac{\sqrt{3}}{2} - \dfrac{1}{2}i\right)^{11}$

41. Find a formula for $(a - b)^n$. Use sigma notation.

42. Expand and simplify:
$$\frac{(x + h)^{13} - x^{13}}{h}.$$

43. Expand and simplify:
$$\frac{(x + h)^n - x^n}{h}.$$

Use sigma notation.

Skill Maintenance

Simplify.

44. $\dfrac{\dfrac{13 \cdot 12}{2 \cdot 1}}{\dfrac{52 \cdot 51}{2 \cdot 1}}$ **45.** $\dfrac{\dfrac{6 \cdot 5}{3 \cdot 2} \cdot \dfrac{4 \cdot 3}{2 \cdot 1}}{\dfrac{10 \cdot 9 \cdot 8}{3 \cdot 2 \cdot 1}}$

Synthesis

46. ◈ Blaise Pascal (1623–1662) was a French scientist and philosopher who founded the modern theory of probability. Do some research on Pascal and see if you can find out how he discovered his famous "triangle of numbers."

47. ◈ Discuss the pros and cons of each method of finding a binomial expansion. Give examples of when you might use one method rather than the other.

Solve for x.

48. $\displaystyle\sum_{k=0}^{8} \binom{8}{k} x^{8-k}3^k = 0$

49. $\displaystyle\sum_{k=0}^{4} \binom{4}{k} 5^{4-k}x^k = 64$

50. $\displaystyle\sum_{k=0}^{5} \binom{5}{k} (-1)^k x^{5-k}3^k = 32$

51. $\displaystyle\sum_{k=0}^{4} \binom{4}{k} (-1)^k x^{4-k}6^k = 81$

52. *Hitting Probability.* At one point in a recent season, Barry Bonds of the San Francisco Giants had a batting average of 0.313. Suppose he came to bat 5 times in a game. The probability (likelihood) of his getting exactly 3 hits is the 3rd term of the binomial expansion of $(0.313 + 0.687)^5$.

a) Find that term to determine the probability.

b) The probability that Bonds gets at most 3 hits in 5 at-bats is found by adding the last 4 terms of the binomial expansion of $(0.313 + 0.687)^5$. Find that probability.

53. *Probability of Being Widowed or Divorced.* The probability that a woman will be either widowed or divorced is 85%. Suppose 8 women are interviewed. The probability that exactly 5 of them will be either widowed or divorced is the 6th term of the binomial expansion of $(0.15 + 0.85)^8$.

a) Find that probability.

b) The probability that at least 6 of the women will be either widowed or divorced is found by adding the last 3 terms of the binomial expansion of $(0.15 + 0.85)^8$. Find that probability.

54. Find the middle term of $(8u + 3v^2)^{10}$.

55. Find the two middle terms of $(\sqrt{x} - \sqrt{y})^5$.

56. Find the term of
$$\left(\frac{3x^2}{2} - \frac{1}{3x}\right)^{12}$$
that does not contain x.

57. Find the middle term of $(x^2 - 6y^{3/2})^6$.

58. Find the ratio of the 4th term of
$$\left(p^2 - \frac{1}{2}p\sqrt[3]{q}\right)^5$$
to the 3rd term.

59. Find the term of
$$\left(\sqrt[3]{x} - \frac{1}{\sqrt{x}}\right)^7$$
containing $1/x^{1/6}$.

60. *Money Combinations.*   A money clip contains one each of the following bills: \$1, \$2, \$5, \$10, \$20, \$50, and \$100. How many different sums of money can be formed using the bills?

Find the sum.

61. $_{100}C_0 + {}_{100}C_1 + \cdots + {}_{100}C_{100}$

62. $_nC_0 + {}_nC_1 + \cdots + {}_nC_n$

Simplify.

63. $\displaystyle\sum_{k=0}^{23} \binom{23}{k}(\log_a x)^{23-k}(\log_a t)^k$

64. $\displaystyle\sum_{k=0}^{15} \binom{15}{k}i^{30-2k}$

65. Use mathematical induction and the property

$$\binom{n}{r-1} + \binom{n}{r} = \binom{n+1}{r}$$

to prove the binomial theorem.

6.8
Probability

• *Compute the probability of a simple event.*

When a coin is tossed, we can reason that the chances, or likelihood, that it will fall heads are 1 out of 2, or the **probability** that it will fall heads is $\frac{1}{2}$. Of course, this does not mean that if a coin is tossed 10 times it will necessarily fall heads 5 times. If the coin is a "fair coin" and it is tossed a great many times, however, it will fall heads very nearly half of the time. Here we give an introduction to two kinds of probability, **experimental** and **theoretical**.

Experimental and Theoretical Probability

If we toss a coin a great number of times—say, 1000—and count the number of times it falls heads, we can determine the probability of it falling heads. If it falls heads 503 times, we would calculate the probability of it falling heads to be

$$\frac{503}{1000}, \quad \text{or} \quad 0.503.$$

This is an **experimental** determination of probability. Such a determination of probability is discovered by the observation and study of data and is quite common and very useful. Here, for example, are some probabilities that have been determined *experimentally:*

1. The probability that a woman will get breast cancer in her lifetime is $\frac{1}{11}$.

2. If you kiss someone who has a cold, the probability of your catching a cold is 0.07.

3. A person who has just been released from prison has an 80% probability of returning.

If we consider a coin and reason that it is just as likely to fall heads as tails, we would calculate the probability to be $\frac{1}{2}$. This is a **theoretical** determination of probability. Here, for example, are some probabilities that have been determined *theoretically,* using mathematics:

1. If there are 30 people in a room, the probability that two of them have the same birthday (excluding year) is 0.706.

2. While on a trip, you meet someone, and after a period of conversation, discover that you have a common acquaintance. The typical reaction, "It's a small world!", is actually not appropriate, because the probability of such an occurrence is quite high—just over 22%.

In summary, experimental and theoretical probabilities are determined by making observations and gathering data. Theoretical probabilities are determined by reasoning mathematically. Examples of experimental and theoretical probability like those above, especially those we do not expect, lead us to see the value of a study of probability. You might ask, "What is the *true* probability?" In fact, there is none. Experimentally, we can determine probabilities within certain limits. These may or may not agree with the probabilities that we obtain theoretically. There are situations in which it is much easier to determine one of these types of probabilities than the other. For example, it would be quite difficult to arrive at the probability of catching a cold using theoretical probability.

Computing Experimental Probabilities

We first consider experimental determination of probability. The basic principle we use in computing such probabilities is as follows.

Principle P (Experimental)

An experiment is performed in which n observations are made. If a situation E, or event, occurs m times out of n observations, then we say that the *experimental probability* of the event, $P(E)$, is given by

$$P(E) = \frac{m}{n}.$$

Example 1 *Sociological Survey.* The authors of this text conducted an experiment to determine the number of people who are left-handed, right-handed, or both. The results are shown in the following graph.

a) Determine the probability that a person is right-handed.

b) Determine the probability that a person is ambidextrous (uses both hands with equal ability).

c) Determine the probability that a person is left-handed.

d) For most tournaments held by the Professional Bowlers Association, there are 160 bowlers. On the basis of the data in this experiment, how many of the bowlers would you expect to be left-handed?

SOLUTION

a) The number of people who are right-handed is 82, the number who are left-handed is 17, and the number who are ambidextrous is 1. The total number of observations is $82 + 17 + 1$, or 100. Thus the probability that a person is right-handed is

$$P = \frac{82}{100}, \quad \text{or} \quad 0.82, \quad \text{or} \quad 82\%.$$

b) The probability that a person is ambidextrous is P, where

$$P = \frac{1}{100}, \quad \text{or} \quad 0.01, \quad \text{or} \quad 1\%.$$

c) The probability that a person is left-handed is P, where

$$P = \frac{17}{100}, \quad \text{or} \quad 0.17, \quad \text{or} \quad 17\%.$$

d) There are 160 bowlers, and from part (c) we can expect 17% to be left-handed. Since

$$17\% \text{ of } 160 = 0.17 \cdot 160 = 27.2,$$

we can expect that about 27 of the bowlers will be left-handed.

Example 2 *Quality Control.* It is very important for a manufacturer to maintain the quality of its products. In fact, companies hire quality control inspectors to ensure this process. The goal is to produce as few defective products as possible. But since a company is producing thousands of products every day, it cannot afford to check every product to see if it is defective. To find out what percentage of its products are defective, the company checks a smaller sample.

The U.S. Department of Agriculture requires that 80% of the seeds that a company produces must sprout. To find out about the quality of the seeds it produces, a company takes 500 seeds from those it has produced and plants them. It finds that 417 of the seeds sprout.

a) What is the probability that a seed will sprout?

b) Did the seeds pass government standards?

SOLUTION

a) We know that 500 seeds were planted and 417 sprouted. The probability of a seed sprouting is P, where

$$P = \frac{417}{500} = 0.834, \quad \text{or} \quad 83.4\%.$$

b) Since the percentage of seeds exceeded the 80% requirement, the company determines that it is producing quality seeds.

Example 3   *Television Ratings.* Television networks are always concerned about the percentage of homes that have TVs and are watching their programs. A sample of the homes are contacted by attaching an electronic device to the TVs of about 1400 homes across the country. Viewing information is then fed into a computer. The following are the results of a recent survey.

NETWORK	CBS	ABC	NBC	Fox	Other, or not watching
NUMBER OF HOMES WATCHING	196	224	237	126	617

What is the probability that a home was tuned to NBC during the time period? to Fox?

SOLUTION   The probability that a home was tuned to NBC is P, where

$$P = \frac{237}{1400} \approx 0.169 = 16.9\%.$$

The probability that a home was tuned to Fox is P, where

$$P = \frac{126}{1400} = 0.09 = 9.0\%.$$

The percentages found in Example 3 are called *ratings*.

Theoretical Probability

Suppose we perform an experiment such as flipping a coin, throwing a dart, drawing a card from a deck, or checking an item off an assembly line for quality. The results of such an experiment are called **outcomes**. The set of all possible outcomes is called the **sample space**. An **event** is a set of outcomes, that is, a subset of the sample space.

Example 4   *Dart Throwing.* Consider this dartboard. Assume that the experiment is "throwing a dart" and that the dart hits the board. Find each of the following.

a) The outcomes

b) The sample space

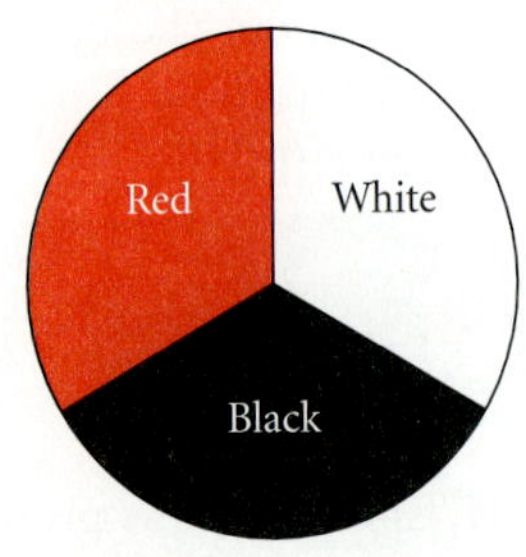

SOLUTION

a) The outcomes are *hitting black* (B), *hitting red* (R), and *hitting white* (W).

b) The sample space is {*hitting black, hitting red, hitting white*}, which can be simply stated as {B, R, W}.

Example 5   *Die Rolling.* A die (pl., dice) is a cube, with six faces, each containing a number of dots from 1 to 6 on each side.

A die is rolled. Find each of the following.

a) The outcomes

b) The sample space

SOLUTION

a) The outcomes are 1, 2, 3, 4, 5, 6.

b) The sample space is {1, 2, 3, 4, 5, 6}.

We denote the probability that an event E occurs as $P(E)$. For example, "a coin falling heads" may be denoted H. Then $P(H)$ represents the probability of the coin falling heads. When all the outcomes of an experiment have the same probability of occurring, we say that they are *equally likely*. To see the distinction between events that are equally likely and those that are not, consider the dartboards shown below.

For board A, the events *hitting black, red,* and *white* are equally likely, but for board B, they are not. A sample space that can be expressed as a union of equally likely events can allow us to calculate probabilities of other events.

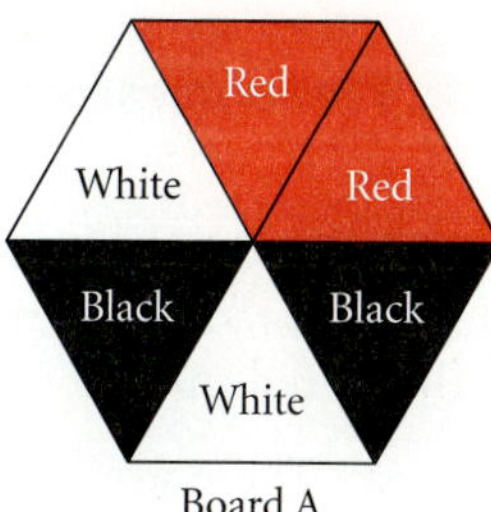

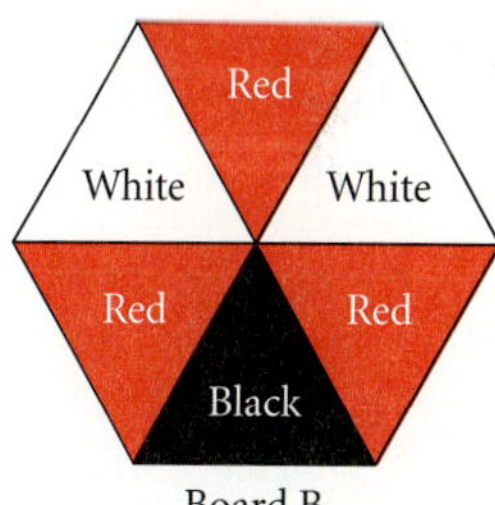

Principle P (Theoretical)

If an event E can occur m ways out of n possible equally likely outcomes of a sample space S, then the *theoretical probability* of the event, $P(E)$, is given by

$$P(E) = \frac{m}{n}.$$

Example 6   What is the probability of rolling a 3 on a die?

SOLUTION   On a fair die, there are 6 equally likely outcomes and there is 1 way to roll a 3. By Principle P, $P(3) = \frac{1}{6}$.

Example 7   What is the probability of rolling an even number on a die?

SOLUTION   The event is rolling an *even* number. It can occur 3 ways (getting 2, 4, or 6). The number of equally likely outcomes is 6. By Principle P, $P(\text{even}) = \frac{3}{6}$, or $\frac{1}{2}$.

We now use a number of examples related to a standard bridge deck of 52 cards. Such a deck is made up as shown in the following figure.

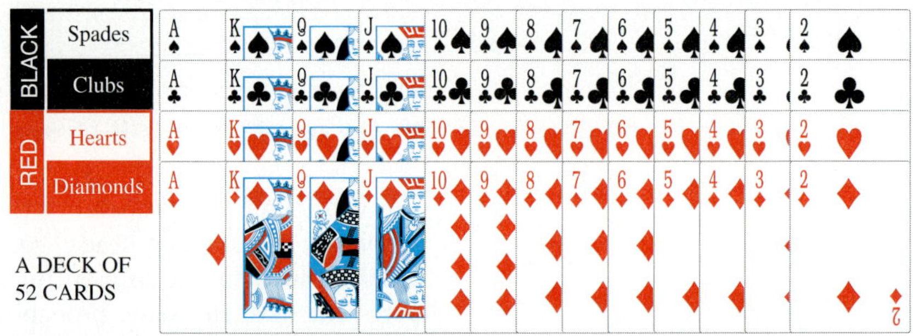

Example 8   What is the probability of drawing an ace from a well-shuffled deck of cards?

SOLUTION   Since there are 52 outcomes (the number of cards in the deck), they are equally likely (from a well-shuffled deck), and there are 4 ways to obtain an ace, by Principle P, we have

$$P(\text{drawing an ace}) = \frac{4}{52}, \quad \text{or} \quad \frac{1}{13}.$$

Example 9   Suppose we select, without looking, one marble from a bag containing 3 red marbles and 4 green marbles. What is the probability of selecting a red marble?

SOLUTION   There are 7 equally likely ways of selecting any marble, and since the number of ways of getting a red marble is 3, we have

$$P(\text{selecting a red marble}) = \frac{3}{7}.$$

The following are some results that follow from Principle P.

Probability Properties

a) If an event E cannot occur, then $P(E) = 0$.

b) If an event E is certain to occur (that is, every trial is a success), then $P(E) = 1$.

c) The probability that an event E will occur is a number from 0 to 1: $0 \le P(E) \le 1$.

For example, in coin tossing, the event that a coin will land on its edge has probability 0. The event that a coin falls either heads or tails has probability 1.

In the following examples, we use the combinatorics that we studied in Sections 6.5 and 6.6 to calculate theoretical probabilities.

Example 10 Suppose that 2 cards are drawn from a well-shuffled deck of 52 cards. What is the probability that both of them are spades?

SOLUTION The number of ways n of drawing 2 cards from a well-shuffled deck of 52 is $_{52}C_2$. Since 13 of the 52 cards are spades, the number of ways m of drawing 2 spades is $_{13}C_2$. Thus,

$$P(\text{getting 2 spades}) = \frac{m}{n} = \frac{_{13}C_2}{_{52}C_2} = \frac{78}{1326} = \frac{1}{17}.$$

Example 11 Suppose that 3 people are selected at random from a group that consists of 6 men and 4 women. What is the probability that 1 man and 2 women are selected?

SOLUTION The number of ways of selecting 3 people from a group of 10 is $_{10}C_3$. One man can be selected in $_6C_1$ ways, and 2 women can be selected in $_4C_2$ ways. By the fundamental counting principle, the number of ways of selecting 1 man and 2 women is $_6C_1 \cdot {_4C_2}$. Thus the probability that 1 man and 2 women are selected is

$$P = \frac{_6C_1 \cdot {_4C_2}}{_{10}C_3} = \frac{3}{10}.$$

Example 12 *Rolling Two Dice.* What is the probability of getting a total of 8 on a roll of a pair of dice?

SOLUTION On each die, there are 6 possible outcomes. The outcomes are paired so there are $6 \cdot 6$, or 36, possible ways in which the two can fall. (Assuming that the dice are different—say, one green and one purple—can help in visualizing this.)

	6	(1,6)	(2,6)	(3,6)	(4,6)	(5,6)	(6,6)
	5	(1,5)	(2,5)	(3,5)	(4,5)	(5,5)	(6,5)
	4	(1,4)	(2,4)	(3,4)	(4,4)	(5,4)	(6,4)
	3	(1,3)	(2,3)	(3,3)	(4,3)	(5,3)	(6,3)
	2	(1,2)	(2,2)	(3,2)	(4,2)	(5,2)	(6,2)
	1	(1,1)	(2,1)	(3,1)	(4,1)	(5,1)	(6,1)
		1	2	3	4	5	6

The pairs that total 8 are as shown in the figure above. Thus there are 5 possible ways of getting a total of 8, so the probability is $\frac{5}{36}$. ▬

The Origin and Uses of Probability

A desire to calculate odds in games of chance gave rise to the theory of probability. Today the theory of probability and its closely related field, mathematical statistics, have many applications, most of them not related to games of chance. Opinion polls, with such uses as predicting elections, are a familiar example. Quality control, in which a prediction about the percentage of faulty items manufactured is made without testing them all, is an important application, among many, in business. Still other applications are in the areas of the use of DNA blood testing for genetics and crime investigation, other areas of medicine, and the kinetic theory of gases.

6.8 Exercise Set

1. *Favorite Number.* In an actual survey conducted by the authors, 100 people were polled and asked to select a number from 1 to 5. The results are shown in the following table.

NUMBER OF CHOICES	1	2	3	4	5
NUMBER WHO CHOSE THAT NUMBER	18	24	23	23	12

a) What is the probability that the number chosen is 1? 2? 3? 4? 5?

b) What general conclusion might a psychologist make from the experiment?

2. *Mason Dots®.* Made by the Tootsie Industries of Chicago, Illinois, Mason Dots® is a gumdrop candy. A box was opened by the authors and was found to contain the following number of gumdrops:

Strawberry	7
Lemon	8
Orange	9
Cherry	4
Lime	5
Grape	6

If we take one gumdrop out of the box, what is the probability of getting a lemon? lime? orange? grape? strawberry? licorice?

3. *Junk Mail.* Have you ever wondered why you receive so much junk mail? In experimental studies, the U.S. Postal Service has found that the probability that a piece of advertising is opened and read is 78%. A business sends out 15,000 pieces of advertising. How many of these can the company expect to be opened and read?

4. *Linguistics.* An experiment was conducted by the authors to determine the relative occurrence of various letters of the English alphabet. The front page of a newspaper was considered. In all, there were 9136 letters. The number of occurrences of each letter of the alphabet is listed in the following table.

LETTER	NUMBER OF OCCURRENCES	PROBABILITY
A	853	$853/9136 \approx 9.3\%$
B	136	
C	273	
D	286	
E	1229	
F	173	
G	190	
H	399	
I	539	
J	21	
K	57	
L	417	
M	231	
N	597	
O	705	
P	238	
Q	4	
R	609	
S	745	
T	789	
U	240	
V	113	
W	127	
X	20	
Y	124	
Z	21	$21/9136 \approx 0.2\%$

a) Complete the table of probabilities with the percentage, to the nearest tenth of a percent, of the occurrence of each letter.

b) What is the probability of a vowel occurring?

c) What is the probability of a consonant occurring?

5. *Wheel of Fortune®.* The results of the experiment in Exercise 4 can be quite useful to a person playing the popular television game show *Wheel of Fortune.* Players guess letters in order to spell out a phrase, a person, or a thing.

a) What 5 consonants have the greatest probability of occurring?

b) What vowel has the greatest probability of occurring?

c) The winner of the main part of the show plays for a grand prize and at one time was allowed to guess 5 consonants and a vowel in order to

discover the secret wording. The 5 consonants R, S, T, L, N, and the vowel E seemed to be chosen most often. Do the results in parts (a) and (b) support such a choice?

6. *Card Drawing.* Suppose we draw a card from a well-shuffled deck of 52 cards.

a) How many equally likely outcomes are there?

What is the probability of drawing each of the following?

b) A queen   **c)** A heart
d) A 7   **e)** A red card
f) A 9 or a king   **g)** A black ace

7. *Marbles.* Suppose we select, without looking, one marble from a bag containing 4 red marbles and 10 green marbles. What is the probability of selecting each of the followng?

a) A red marble
b) A green marble
c) A purple marble
d) A red or a green marble

8. *Production Unit.* The sales force of a business consists of 10 men and 10 women. A production unit of 4 people is set up at random. What is the probability that 2 men and 2 women are chosen?

9. *Coin Drawing.* A sack contains 7 dimes, 5 nickels, and 10 quarters. Eight coins are drawn at random. What is the probability of getting 4 dimes, 3 nickels, and 1 quarter?

10. *Michigan Lotto.* Twice a week, the state of Michigan runs a 6-out-of-49-number lotto that pays at least $2 million. You purchase a card for $1 and pick any 6 numbers from 1 to 49. If your numbers match those that the state draws, you win.

a) How many 6-number combinations are there for drawing?

b) You buy 1 lottery ticket. What is your probability of winning?

Five-Card Poker Hands. *Suppose that 5 cards are drawn from a deck of 52 cards. What is the probability of drawing each of the following?*

11. 3 sevens and 2 kings

12. 5 aces

13. 5 spades

14. 4 aces and 1 five

15. *Random-Number Generator.* Many graphers have a **random-number generator**. This feature produces a random number in the interval [0, 1]. (Consult your manual.) We can use such a feature to simulate coin flipping. A number r such that

$0 \le r \le 0.5$ would indicate heads, H. A number r such that $0.5 < r \le 1.0$ would indicate tails, T. Use a random-number generator 100 times.

a) What is the experimental probability of getting heads?

b) What is the experimental probability of getting tails?

16. *Tossing Three Coins.*   Three coins are flipped. An outcome might be HTH.

a) Find the sample space.

What is the probability of getting each of the following?

b) Exactly one head
c) At most two tails
d) At least one head
e) Exactly two tails

Roulette.   *An American roulette wheel contains 38 slots numbered 00, 0, 1, 2, 3, . . . , 35, 36. Eighteen of the slots numbered 1–36 are colored red and 18 are colored black. The 00 and 0 slots are uncolored. The wheel is spun, and a ball is rolled around the rim until it falls into a slot. What is the probability that the ball falls in each of the following?*

17. A black slot
18. A red slot
19. A red or a black slot
20. The 00 slot
21. The 0 slot
22. Either the 00 or the 0 slot (in this case, the house always wins)

23. An odd-numbered slot

24. The number 24

25. *Dartboard.*   The figure below shows a dartboard. A dart is thrown and hits the board. Find the probabilities

$$P(\text{red}), \ P(\text{green}), \ P(\text{blue}), \ P(\text{yellow}).$$

Synthesis

26. ◆ Find at least one use of probability in today's newspaper. Make a report.

27. ◆ *Random Best-Selling Novels.*   Sir Arthur Stanley Eddington, an astronomer, once wrote in a satirical essay that if a monkey were left alone long enough with a typewriter and typed randomly, any great novel could be replicated.

What is the probability that the following passage could have been written by a monkey? Ignore capital letters and punctuation and consider only letters and spaces.

"It was the best of times, it was the worst of times, . . ." (Charles Dickens, 1859). Explain your answer.

Five-Card Poker Hands.   *Suppose that 5 cards are drawn from a deck of 52 cards. For the following exercises, give both a reasoned expression and an answer.*

28. *Royal Flush.*   A *royal flush* consists of a 5-card hand with A-K-Q-J-10 of the same suit.

a) How many royal flushes are there?
b) What is the probability of getting a royal flush?

29. *Straight Flush.*   A *straight flush* consists of 5 cards in sequence in the same suit, but excludes royal flushes. An ace can be used low, before a two, or high, following a king.

a) How many straight flushes are there?
b) What is the probability of getting a straight flush?

30. *Four of a Kind.* A *four-of-a-kind* is a 5-card hand in which 4 of the cards are of the same denomination, such as J-J-J-J-6, 7-7-7-7-A, or 2-2-2-2-5.

a) How many four-of-a-kind hands are there?
b) What is the probability of getting four of a kind?

31. *Full House.* A *full house* consists of a pair and 3 of a kind, such as Q-Q-Q-4-4.

a) How many full houses are there?
b) What is the probability of getting a full house?

32. *Three of a Kind.* A *three-of-a-kind* is a 5-card hand in which exactly 3 of the cards are of the same denomination and the other 2 are *not*, such as Q-Q-Q-10-7.

a) How many three-of-a-kind hands are there?
b) What is the probability of getting three of a kind?

33. *Flush.* An ordinary *flush* is a 5-card hand in which all the cards are of the same suit, but not all in sequence (not a straight flush or royal flush).

a) How many flushes are there?
b) What is the probability of getting a flush?

34. *Two Pairs.* A hand with *two pairs* is a hand like Q-Q-3-3-A.

a) How many are there?
b) What is the probability of getting two pairs?

35. *Straight.* An ordinary *straight* is any 5 cards in sequence, but not of the same suit—for example, 4 of spades, 5 of hearts, 6 of diamonds, 7 of hearts, and 8 of clubs.

a) How many straights are there?
b) What is the probability of getting a straight?

CHAPTER

6 *Summary and Review*

Important Properties and Formulas

Arithmetic Sequences and Series

General term: $a_n = a_{n-1} + d$

$$a_n = a_1 + (n-1)d$$

Common difference: d

Sum of the first n terms: $S_n = \dfrac{n}{2}(a_1 + a_n)$

Geometric Sequences and Series

General term: $a_{n+1} = a_n r$

$$a_n = a_1 r^{n-1}$$

Common ratio: r

Sum of the first n terms: $S_n = \dfrac{a_1(1 - r^n)}{1 - r}$

Sum of an infinite geometric series:

$$S_\infty = \dfrac{a_1}{1 - r}, \quad |r| < 1$$

The Principle of Mathematical Induction

(1) *Basis step:* Prove S_1 is true.

(2) *Induction step:* Prove for all numbers k, $S_k \rightarrow S_{k+1}$.

The Fundamental Counting Principle

$$n_1 \cdot n_2 \cdot n_3 \cdots n_k$$

Permutations of n Objects Taken n at a Time

$$_nP_n = n! = n(n-1)(n-2) \cdots 3 \cdot 2 \cdot 1$$

(continued)

Permutations of n Objects Taken k at a Time

$$_nP_k = \underbrace{n(n-1)(n-2)\cdots[n-(k-1)]}_{k \text{ factors}}$$

$$= \frac{n!}{(n-k)!}$$

Permutations of Sets With Some Nondistinguishable Objects

$$\frac{n!}{n_1! \cdot n_2! \cdot \cdots \cdot n_k!}$$

Combinations of n Objects Taken k at a Time

$$_nC_k = \binom{n}{k} = \frac{_nP_k}{k!} = \frac{n!}{k!(n-k)!}$$

$$= \frac{n(n-1)(n-2)\cdots[n-(k-1)]}{k!}$$

The Binomial Theorem

$$(a+b)^n = \sum_{k=0}^{n} \binom{n}{k} a^{n-k}b^k$$

The (k + 1)st Term of Binomial Expansion

The $(k+1)$st term of $(a+b)^n$ is $\binom{n}{k} a^{n-k}b^k$.

The total number of subsets of a set with n elements is 2^n.

Probability Principle P

$$P(E) = \frac{m}{n}$$

REVIEW EXERCISES

1. Find the first 4 terms, a_{11}, and a_{23}:

$$a_n = (-1)^n\left(\frac{n^2}{n^4+1}\right).$$

2. Predict the general, or nth, term. Answers may vary.
$$2, -5, 10, -17, 26, \ldots$$

3. Find and evaluate:

$$\sum_{k=1}^{4} \frac{(-1)^{k+1}3^k}{3^k-1}.$$

4. Construct a table of values and a graph for the first 10 terms of this sequence. Use both a grapher and a spreadsheet program, if available.

$$a_1 = 0.3, \quad a_{k+1} = 5a_k + 1$$

5. Write sigma notation:

$$0 + 3 + 8 + 15 + 24 + 35 + 48.$$

6. Find the 10th term of the arithmetic sequence
$$\tfrac{3}{4}, \tfrac{13}{12}, \tfrac{17}{12}, \ldots .$$

7. Find the 6th term of the arithmetic sequence
$$a - b, a, a + b, \ldots .$$

8. Find the sum of the first 18 terms of the arithmetic sequence
$$4, 7, 10, \ldots .$$

9. Find the sum of the first 200 natural numbers.

10. The 1st term in an arithmetic sequence is 5, and the 17th term is 53. Find the 3rd term.

11. The common difference in an arithmetic sequence is 3. The 10th term is 23. Find the first term.

12. For a geometric sequence, $a_1 = -2$, $r = 2$, and $a_n = -64$. Find n and S_n.

13. For a geometric sequence, $r = \tfrac{1}{2}$, $n = 5$, and $S_n = \tfrac{31}{2}$. Find a_1 and a_n.

Find the sum of each infinite geometric series, if it exists. If you wish, you can also check the results on a grapher using the TABLE *and* SEQUENCE GRAPH *features.*

14. $25 + 27.5 + 30.25 + 33.275 + \cdots$

15. $0.27 + 0.0027 + 0.000027 + \cdots$

16. $\tfrac{1}{2} - \tfrac{1}{6} + \tfrac{1}{18} - \cdots$

17. Find fractional notation for $2.\overline{43}$.

18. Insert four arithmetic means between 5 and 9.

19. *Bouncing Golfball.* A golfball is dropped from a height of 30 ft to the pavement. It always rebounds three fourths of the distance that it drops. How far (up and down) will the ball have traveled when it hits the pavement for the 6th time?

20. *The Amount of an Annuity.* To create a college fund, a parent makes a sequence of 18 yearly deposits of $2000 each in a savings account on which interest is compounded annually at 5.8%. Find the amount of the annuity.

21. *Total Gift.* You receive 10¢ on the first day of the year, 12¢ on the 2nd day, 14¢ on the 3rd day, and so on.

a) How much will you receive on the 365th day?
b) What is the sum of all these 365 gifts?

22. *The Economic Multiplier.* The government is making a $24,000,000,000 expenditure for travel to Mars. If 73% of this amount is spent again, and so on, what is the total effect on the economy?

Use mathematical induction to prove each of the following.

23. For every natural number n,

$$1 + 4 + 7 + \cdots + (3n - 2) = \frac{n(3n - 1)}{2}.$$

24. For every natural number n,

$$1 + 3 + 3^2 + \cdots + 3^{n-1} = \frac{3^n - 1}{2}.$$

25. For every natural number $n \geq 2$,

$$\left(1 - \frac{1}{2}\right)\left(1 - \frac{1}{3}\right) \cdots \left(1 - \frac{1}{n}\right) = \frac{1}{n}.$$

26. *Book Arrangements.* In how many ways can 6 books be arranged on a shelf?

27. *Flag Displays.* If 9 different signal flags are available, how many different displays are possible using 4 flags in a row?

28. *Prize Choices.* The winner of a contest can choose any 8 of 15 prizes. How many different sets of prizes can be chosen?

29. *Fraternity–Sorority Names.* The Greek alphabet contains 24 letters. How many fraternity or sorority names can be formed using 3 different letters?

30. *Letter Arrangements.* In how many distinguishable ways can the letters of the word TENNESSEE be arranged?

31. *Floor Plans.* A manufacturer of houses has 1 floor plan but achieves variety by having 3 different roofs, 4 different ways of attaching the garage, and 3 different types of entrance. Find the number of different houses that can be produced.

32. *Code Symbols.* How many code symbols can be formed using 5 out of 6 of the letters of G, H, I, J, K, L if the letters:

a) cannot be repeated?
b) can be repeated?
c) cannot be repeated but must begin with K?
d) cannot be repeated but must end with IGH?

33. Determine the number of subsets of a set containing 8 members.

Expand.

34. $(m + n)^7$

35. $(x - \sqrt{2})^5$

36. $(x^2 - 3y)^4$

37. $\left(a + \dfrac{1}{a}\right)^8$

38. $(1 + 5i)^6$, where $i^2 = -1$

39. Find the 4th term of $(a + x)^{12}$.

40. Find the 12th term of $(2a - b)^{18}$. Do not multiply out the factorials.

41. *Election Poll.* Before an election, a poll was conducted to see which candidate was favored. Three people were running for a particular office. During the polling, 86 favored candidate A, 97 favored B, and 23 favored C. Assuming that the poll is a valid indicator of the election, what is the probability that the election will be won by A? B? C?

42. *Rolling Dice.* What is the probability of getting a 10 on a roll of a pair of dice? on a roll of 1 die?

43. *Drawing a Card.* From a deck of 52 cards, 1 card is drawn at random. What is the probability that it is a club?

44. *Drawing Three Cards.* From a deck of 52 cards, 3 are drawn at random without replacement. What is the probability that 2 are aces and 1 is a king?

45. *Video Rentals.*   The following table contains factual data regarding annual video rental spending per VCR household.

YEAR, x	U.S. SPENDING ON VIDEO RENTALS, y (IN BILLIONS)
1. 1982	$141
3. 1984	139
5. 1986	127
7. 1988	124
9. 1990	120
10. 1991	113
11. 1992	114
12. 1993	119
13. 1994	122
14. 1995	129

Source: Pedonis, Suhler & Associates, Media Group Research.

a) Use the REGRESSION feature to fit a quadratic sequence function $a_n = an^2 + bn + c$ to the data.

b) Use the equation to predict the annual VCR video rental per household in 1998 and in 2002.

Synthesis

46. *Chain Business Deals.*   Chain letters have been outlawed by the government. Nevertheless, "chain" business deals still exist and they can be fraudulent. Suppose a salesperson is charged with the task of hiring 4 new salespersons. Each of them gives half of their profits to the person who hires them. Each of these people hires 4 new salespersons. Each of these gives half of his or her profits to the person who hired them. Half of these profits then go back to the original hiring person. Explain the lure of this business to someone who has managed several sequences of hirings. Explain the fallacy of such a business as well. Keep in mind that there are only 266 million people in this country.

47. ◈ Write an exercise for a classmate to solve. Design it so that the solution is $_9C_4$.

48. Explain why the following cannot be proved by mathematical induction: For every natural number n,

a) $3 + 5 + \cdots + (2n + 1) = (n + 1)^2$.
b) $1 + 3 + \cdots + (2n - 1) = n^2 + 3$.

49. Suppose that $a_1, a_2, \ldots, a_n$ and $b_1, b_2, \ldots, b_n$ are geometric sequences. Prove that $c_1, c_2, \ldots, c_n$ is a geometric sequence, where $c_n = a_n b_n$.

50. Suppose that $a_1, a_2, \ldots, a_n$ is an arithmetic sequence. Is $b_1, b_2, \ldots, b_n$ an arithmetic sequence if:

a) $b_n = |a_n|$?   **b)** $b_n = a_n + 8$?

c) $b_n = 7a_n$?   **d)** $b_n = \dfrac{1}{a_n}$?

e) $b_n = \log a_n$?   **f)** $b_n = a_n^3$?

51. The zeros of this polynomial function form an arithmetic sequence. Find them.

$$f(x) = x^4 - 4x^3 - 4x^2 + 16x$$

52. Write the first 3 terms of the infinite geometric series with $r = -\frac{1}{3}$ and $S_\infty = \frac{3}{8}$.

53. Simplify:

$$\sum_{k=0}^{10} (-1)^k \binom{10}{k}(\log x)^{10-k}(\log y)^k.$$

Solve for n.

54. $\dbinom{n}{6} = 3 \cdot \dbinom{n-1}{5}$ **55.** $\dbinom{n}{n-1} = 36$

56. Solve for a:

$$\sum_{k=0}^{5} \binom{5}{k} 9^{5-k} a^k = 0.$$

Appendixes

A

Determinants and Cramer's Rule

- *Evaluate determinants of square matrices.*
- *Use Cramer's rule to solve systems of equations.*

Determinants of Square Matrices

With every square matrix, we associate a number called its *determinant*.

Determinant of a 2 × 2 Matrix

The *determinant* of the matrix $\begin{bmatrix} a & c \\ b & d \end{bmatrix}$ is denoted $\begin{vmatrix} a & c \\ b & d \end{vmatrix}$ and is defined as

$$\begin{vmatrix} a & c \\ b & d \end{vmatrix} = ad - bc.$$

Example 1 Evaluate: $\begin{vmatrix} \sqrt{2} & -3 \\ -4 & -\sqrt{2} \end{vmatrix}$.

SOLUTION

$$\begin{vmatrix} \sqrt{2} & -3 \\ -4 & -\sqrt{2} \end{vmatrix}$$

The arrows indicate the products involved.

$$= \sqrt{2}(-\sqrt{2}) - (-4)(-3)$$
$$= -2 - 12 = -14$$

We now consider a way to evaluate determinants of square matrices of order 3 × 3 or higher.

Minor

For a square matrix $\mathbf{A} = [a_{ij}]$, the *minor* M_{ij} of an element a_{ij} is the determinant of the matrix formed by deleting the ith row and the jth column of $\mathbf{A}$.

Example 2 For the matrix $[a_{ij}] = \begin{bmatrix} -8 & 0 & 6 \\ 4 & -6 & 7 \\ -1 & -3 & 5 \end{bmatrix}$, find each of the following.

a) M_{11} **b)** M_{23}

SOLUTION

a) Delete the first row and the first column and find the determinant of the 2×2 matrix formed by the remaining elements.

$$\begin{bmatrix} -8 & 0 & 6 \\ 4 & -6 & 7 \\ -1 & -3 & 5 \end{bmatrix}$$

$$M_{11} = \begin{vmatrix} -6 & 7 \\ -3 & 5 \end{vmatrix}$$
$$= (-6) \cdot 5 - (-3) \cdot 7$$
$$= -30 - (-21)$$
$$= -30 + 21 = -9$$

b) Delete the second row and the third column and find the determinant of the 2×2 matrix formed by the remaining elements.

$$\begin{bmatrix} -8 & 0 & 6 \\ 4 & -6 & 7 \\ -1 & -3 & 5 \end{bmatrix}$$

$$M_{23} = \begin{vmatrix} -8 & 0 \\ -1 & -3 \end{vmatrix}$$
$$= -8(-3) - (-1)0 = 24$$

Cofactor

For a square matrix $\mathbf{A} = [a_{ij}]$, the *cofactor* A_{ij} of an element a_{ij} is given by

$$A_{ij} = (-1)^{i+j}M_{ij},$$

where M_{ij} is the minor of a_{ij}.

Example 3 For the matrix given in Example 2, find each of the following.

a) A_{11} **b)** A_{23}

SOLUTION

a) In Example 2, we found that $M_{11} = -9$. Then

$$A_{11} = (-1)^{1+1}(-9) = (1)(-9) = -9.$$

b) In Example 2, we found that $M_{23} = 24$. Then

$$A_{23} = (-1)^{2+3}(24) = (-1)(24) = -24.$$

Evaluating Determinants Using Cofactors

Consider the matrix $\mathbf{A}$ given by

$$\mathbf{A} = \begin{bmatrix} a_{11} & a_{12} & a_{13} \\ a_{21} & a_{22} & a_{23} \\ a_{31} & a_{32} & a_{33} \end{bmatrix}.$$

The determinant of the matrix, denoted $|\mathbf{A}|$, can be found by multiplying each element of the first column by its cofactor and adding:

$$|\mathbf{A}| = a_{11}A_{11} + a_{21}A_{21} + a_{31}A_{31}.$$

Because

$$A_{11} = (-1)^{1+1}M_{11} = M_{11},$$
$$A_{21} = (-1)^{2+1}M_{21} = -M_{21},$$

and $\quad A_{31} = (-1)^{3+1}M_{31} = M_{31},$

we can write

$$|\mathbf{A}| = a_{11} \cdot \begin{vmatrix} a_{22} & a_{23} \\ a_{32} & a_{33} \end{vmatrix} - a_{21} \cdot \begin{vmatrix} a_{12} & a_{13} \\ a_{32} & a_{33} \end{vmatrix} + a_{31} \cdot \begin{vmatrix} a_{12} & a_{13} \\ a_{22} & a_{23} \end{vmatrix}.$$

It can be shown that we can determine $|\mathbf{A}|$ by picking *any* row or column, multiplying each element in that row or column by its cofactor, and adding. This is called *expanding* across a row or down a column. We just expanded down the first column. We now define the determinant of a square matrix of any order.

Determinant of Any Square Matrix

For any square matrix $\mathbf{A}$ of order $n \times n$ $(n > 1)$, we define the *determinant* of $\mathbf{A}$, denoted $|\mathbf{A}|$, as follows. Choose any row or column. Multiply each element in that row or column by its cofactor and add the results. The determinant of a 1×1 matrix is simply the element of the matrix. The value of a determinant will be the same no matter how it is evaluated.

Example 4   Evaluate $|\mathbf{A}|$ by expanding across the third row.

$$\mathbf{A} = \begin{bmatrix} -8 & 0 & 6 \\ 4 & -6 & 7 \\ -1 & -3 & 5 \end{bmatrix}$$

SOLUTION   We have

$$|\mathbf{A}| = (-1)A_{31} + (-3)A_{32} + 5A_{33}$$

$$= (-1)(-1)^{3+1} \cdot \begin{vmatrix} 0 & 6 \\ -6 & 7 \end{vmatrix} + (-3)(-1)^{3+2} \cdot \begin{vmatrix} -8 & 6 \\ 4 & 7 \end{vmatrix}$$

$$+ 5(-1)^{3+3} \cdot \begin{vmatrix} -8 & 0 \\ 4 & -6 \end{vmatrix}$$

$$= (-1) \cdot 1 \cdot [0 \cdot 7 - (-6)6] + (-3)(-1)[-8 \cdot 7 - 4 \cdot 6]$$

$$+ 5 \cdot 1 \cdot [-8(-6) - 4 \cdot 0]$$

$$= -[36] + 3[-80] + 5[48]$$

$$= -36 - 240 + 240$$

$$= -36.$$

The value of this determinant is -36 no matter which row or column we expand upon.

Determinants can be evaluated on most graphers using the MATRIX package. After entering a matrix on the grapher, select the determinant operation from the MATRIX MATH menu and enter the name of the matrix. The grapher will return the value of the determinant of the matrix. For example, for

$$\mathbf{A} = \begin{bmatrix} 1 & 6 & -1 \\ -3 & -5 & 3 \\ 0 & 4 & 2 \end{bmatrix},$$

we have

```
det [A]
              26
```

Cramer's Rule

Determinants can be used to solve systems of linear equations. Consider a system of two linear equations:

$$a_1x + b_1y = c_1,$$
$$a_2x + b_2y = c_2.$$

Using the methods of Chapter 4, we obtain

$$x = \frac{c_1b_2 - c_2b_1}{a_1b_2 - a_2b_1} \quad \text{and} \quad y = \frac{a_1c_2 - a_2c_1}{a_1b_2 - a_2b_1}.$$

The numerators and denominators of these expressions can be written as determinants:

$$x = \frac{\begin{vmatrix} c_1 & b_1 \\ c_2 & b_2 \end{vmatrix}}{\begin{vmatrix} a_1 & b_1 \\ a_2 & b_2 \end{vmatrix}} \quad \text{and} \quad y = \frac{\begin{vmatrix} a_1 & c_1 \\ a_2 & c_2 \end{vmatrix}}{\begin{vmatrix} a_1 & b_1 \\ a_2 & b_2 \end{vmatrix}}.$$

If we let

$$D = \begin{vmatrix} a_1 & b_1 \\ a_2 & b_2 \end{vmatrix}, \qquad D_x = \begin{vmatrix} c_1 & b_1 \\ c_2 & b_2 \end{vmatrix}, \quad \text{and} \quad D_y = \begin{vmatrix} a_1 & c_1 \\ a_2 & c_2 \end{vmatrix},$$

we have

$$x = \frac{D_x}{D} \quad \text{and} \quad y = \frac{D_y}{D}.$$

This procedure for solving systems of equations is known as *Cramer's rule*.

Cramer's Rule for 2 × 2 Systems

The solution of the system of equations

$$a_1 x + b_1 y = c_1,$$
$$a_2 x + b_2 y = c_2$$

is given by

$$x = \frac{D_x}{D}, \qquad y = \frac{D_y}{D},$$

where

$$D = \begin{vmatrix} a_1 & b_1 \\ a_2 & b_2 \end{vmatrix}, \qquad D_x = \begin{vmatrix} c_1 & b_1 \\ c_2 & b_2 \end{vmatrix},$$

$$D_y = \begin{vmatrix} a_1 & c_1 \\ a_2 & c_2 \end{vmatrix}, \quad \text{and} \quad D \neq 0.$$

Note that the denominator D contains the coefficients of x and y, in the same position as in the original equations. For x, the numerator is obtained by replacing the x-coefficients in D (the a's) by the c's. For y, the numerator is obtained by replacing the y-coefficients in D (the b's) by the c's.

Example 5 Solve using Cramer's rule:

$$2x + 5y = 7,$$
$$5x - 2y = -3.$$

SOLUTION We have

$$x = \frac{\begin{vmatrix} 7 & 5 \\ -3 & -2 \end{vmatrix}}{\begin{vmatrix} 2 & 5 \\ 5 & -2 \end{vmatrix}} = \frac{7(-2) - (-3)5}{2(-2) - 5 \cdot 5} = -\frac{1}{29},$$

$$y = \frac{\begin{vmatrix} 2 & 7 \\ 5 & -3 \end{vmatrix}}{\begin{vmatrix} 2 & 5 \\ 5 & -2 \end{vmatrix}} = \frac{2(-3) - 5 \cdot 7}{-29} = \frac{41}{29}.$$

The solution is $\left(-\frac{1}{29}, \frac{41}{29}\right)$.

Cramer's rule works only when a system of equations has a unique solution. This occurs when $D \neq 0$. If $D = 0$, $D_x = 0$, and $D_y = 0$, then the system is dependent. If $D = 0$ and D_x and/or D_y is not 0, then the system is inconsistent.

Cramer's rule can be extended to a system of n linear equations in n variables. We consider a 3×3 system.

Cramer's Rule for 3 × 3 Systems

The solution of the system of equations

$$a_1 x + b_1 y + c_1 z = d_1,$$
$$a_2 x + b_2 y + c_2 z = d_2,$$
$$a_3 x + b_3 y + c_3 z = d_3$$

is given by

$$x = \frac{D_x}{D}, \qquad y = \frac{D_y}{D}, \qquad z = \frac{D_z}{D},$$

where

$$D = \begin{vmatrix} a_1 & b_1 & c_1 \\ a_2 & b_2 & c_2 \\ a_3 & b_3 & c_3 \end{vmatrix}, \qquad D_x = \begin{vmatrix} d_1 & b_1 & c_1 \\ d_2 & b_2 & c_2 \\ d_3 & b_3 & c_3 \end{vmatrix},$$

$$D_y = \begin{vmatrix} a_1 & d_1 & c_1 \\ a_2 & d_2 & c_2 \\ a_3 & d_3 & c_3 \end{vmatrix}, \qquad D_z = \begin{vmatrix} a_1 & b_1 & d_1 \\ a_2 & b_2 & d_2 \\ a_3 & b_3 & d_3 \end{vmatrix}, \quad \text{and } D \neq 0.$$

Note that the determinant D_x is obtained from D by replacing the x-coefficients by d_1, d_2, and d_3. A similar thing happens with D_y and D_z. When $D = 0$, Cramer's rule cannot be used. If $D = 0$ and D_x, D_y, and D_z are 0, the system is dependent. If $D = 0$ and one of D_x, D_y, or D_z is not 0, then the system is inconsistent.

Example 6   Solve using Cramer's rule:

$$x - 3y + 7z = 13,$$
$$x + y + z = 1,$$
$$x - 2y + 3z = 4.$$

SOLUTION   We have

$$D = \begin{vmatrix} 1 & -3 & 7 \\ 1 & 1 & 1 \\ 1 & -2 & 3 \end{vmatrix} = -10, \qquad D_x = \begin{vmatrix} 13 & -3 & 7 \\ 1 & 1 & 1 \\ 4 & -2 & 3 \end{vmatrix} = 20,$$

$$D_y = \begin{vmatrix} 1 & 13 & 7 \\ 1 & 1 & 1 \\ 1 & 4 & 3 \end{vmatrix} = -6, \qquad D_z = \begin{vmatrix} 1 & -3 & 13 \\ 1 & 1 & 1 \\ 1 & -2 & 4 \end{vmatrix} = -24.$$

Then

$$x = \frac{D_x}{D} = \frac{20}{-10} = -2,$$

$$y = \frac{D_y}{D} = \frac{-6}{-10} = \frac{3}{5},$$

$$z = \frac{D_z}{D} = \frac{-24}{-10} = \frac{12}{5}.$$

The solution is $\left(-2, \frac{3}{5}, \frac{12}{5}\right)$. In practice, it is not necessary to evaluate D_z. When we have found values for x and y, we can substitute them into one of the equations to find z.

A | Exercise Set

Evaluate each determinant algebraically.

1. $\begin{vmatrix} -2 & -\sqrt{5} \\ -\sqrt{5} & 3 \end{vmatrix}$

2. $\begin{vmatrix} \sqrt{5} & -3 \\ 4 & 2 \end{vmatrix}$

3. $\begin{vmatrix} x & 4 \\ x & x^2 \end{vmatrix}$

4. $\begin{vmatrix} y^2 & -2 \\ y & 3 \end{vmatrix}$

5. $\begin{vmatrix} 3 & 1 & 2 \\ -2 & 3 & 1 \\ 3 & 4 & -6 \end{vmatrix}$

6. $\begin{vmatrix} 3 & -2 & 1 \\ 2 & 4 & 3 \\ -1 & 5 & 1 \end{vmatrix}$

7. $\begin{vmatrix} x & 0 & -1 \\ 2 & x & x^2 \\ -3 & x & 1 \end{vmatrix}$

8. $\begin{vmatrix} x & 1 & -1 \\ x^2 & x & x \\ 0 & x & 1 \end{vmatrix}$

9. Use a grapher to check your answer to Exercise 1.

10. Use a grapher to check your answer to Exercise 2.

11. Use a grapher to check your answer to Exercise 5.

12. Use a grapher to check your answer to Exercise 6.

Use the following matrix for Exercises 13–20:

$$\mathbf{A} = \begin{bmatrix} 7 & -4 & -6 \\ 2 & 0 & -3 \\ 1 & 2 & -5 \end{bmatrix}.$$

13. Find M_{11}, M_{32}, and M_{22}.

14. Find M_{13}, M_{31}, and M_{23}.

15. Find A_{11}, A_{32}, and A_{22}.

16. Find A_{13}, A_{31}, and A_{23}.

17. Evaluate $|\mathbf{A}|$ by expanding across the second row.

18. Evaluate $|\mathbf{A}|$ by expanding down the second column.

19. Evaluate $|\mathbf{A}|$ by expanding down the third column.

20. Evaluate $|\mathbf{A}|$ on a grapher.

Use the following matrix for Exercises 21–27:

$$\mathbf{A} = \begin{bmatrix} 1 & 0 & 0 & -2 \\ 4 & 1 & 0 & 0 \\ 5 & 6 & 7 & 8 \\ -2 & -3 & -1 & 0 \end{bmatrix}$$

21. Find M_{41} and M_{33}.

22. Find M_{12} and M_{44}.

23. Find A_{24} and A_{43}.

24. Find A_{22} and A_{34}.

25. Evaluate $|\mathbf{A}|$ by expanding across the first row.

26. Evaluate $|\mathbf{A}|$ by expanding down the third column.

27. Evaluate $|\mathbf{A}|$ on a grapher.

28. Evaluate on a grapher:

$$\begin{vmatrix} 5 & -4 & 2 & -2 \\ 3 & -3 & -4 & 7 \\ -2 & 3 & 2 & 4 \\ -8 & 9 & 5 & -5 \end{vmatrix}.$$

Solve using Cramer's rule.

29. $-2x + 4y = 3,$
$\quad 3x - 7y = 1$

30. $5x - 4y = -3,$
$\quad 7x + 2y = 6$

31. $2x - y = 5,$
$\quad x - 2y = 1$

32. $3x + 4y = -2,$
$\quad 5x - 7y = 1$

33. $2x + 9y = -2,$
$\quad 4x - 3y = 3$

34. $2x + 3y = -1,$
$\quad 3x + 6y = -0.5$

35. $2x + 5y = 7,$
$\quad 3x - 2y = 1$

36. $3x + 2y = 7,$
$\quad 2x + 3y = -2$

37. $3x + 2y - z = 4,$
$\quad 3x - 2y + z = 5,$
$\quad 4x - 5y - z = -1$

38. $3x - y + 2z = 1,$
$\quad x - y + 2z = 3,$
$\quad -2x + 3y + z = 1$

39. $3x + 5y - z = -2,$
   $x - 4y + 2z = 13,$
   $2x + 4y + 3z = 1$

40. $3x + 2y + 2z = 1,$
   $5x - y - 6z = 3,$
   $2x + 3y + 3z = 4$

41. $x - 3y - 7z = 6,$
   $2x + 3y + z = 9,$
   $4x + y = 7$

42. $x - 2y - 3z = 4,$
   $3x - 2z = 8,$
   $2x + y + 4z = 13$

43. $6y + 6z = -1,$
   $8x + 6z = -1,$
   $4x + 9y = 8$

44. $3x + 5y = 2,$
   $2x - 3z = 7,$
   $4y + 2z = -1$

Synthesis

45. ◆ Explain why the system of equations

$$a_1 x + b_1 y = c_1,$$
$$a_2 x + b_2 y = c_2$$

is either dependent or inconsistent when

$$\begin{vmatrix} a_1 & b_1 \\ a_2 & b_2 \end{vmatrix} = 0.$$

46. ◆ If the lines $a_1 x + b_1 y = c_1$ and $a_2 x + b_2 y = c_2$ are parallel, what can you say about the values of

$$\begin{vmatrix} a_1 & b_1 \\ a_2 & b_2 \end{vmatrix}, \quad \begin{vmatrix} c_1 & b_1 \\ c_2 & b_2 \end{vmatrix}, \quad \text{and} \quad \begin{vmatrix} a_1 & c_1 \\ a_2 & c_2 \end{vmatrix}?$$

Solve.

47. $\begin{vmatrix} x & 5 \\ -4 & x \end{vmatrix} = 24$

48. $\begin{vmatrix} y & 2 \\ 3 & y \end{vmatrix} = y$

49. $\begin{vmatrix} x & -3 \\ -1 & x \end{vmatrix} \geq 0$

50. $\begin{vmatrix} y & -5 \\ -2 & y \end{vmatrix} < 0$

51. $\begin{vmatrix} x + 3 & 4 \\ x - 3 & 5 \end{vmatrix} = -7$

52. $\begin{vmatrix} m + 2 & -3 \\ m + 5 & -4 \end{vmatrix} = 3m - 5$

53. $\begin{vmatrix} 2 & x & 1 \\ 1 & 2 & -1 \\ 3 & 4 & -2 \end{vmatrix} = -6$

54. $\begin{vmatrix} x & 2 & x \\ 3 & -1 & 1 \\ 1 & -2 & 2 \end{vmatrix} = -10$

Rewrite each expression using a determinant. Answers may vary.

55. $2L + 2W$

56. $\pi r + \pi h$

57. $a^2 + b^2$

58. $\frac{1}{2} h(a + b)$

59. $2\pi r^2 + 2\pi rh$

60. $x^2 y^2 - Q^2$

61. Show that if a line contains the points (x_1, y_1) and (x_2, y_2), an equation of the line can be written as

$$\begin{vmatrix} x & y & 1 \\ x_1 & y_1 & 1 \\ x_2 & y_2 & 1 \end{vmatrix} = 0.$$

B
Parametric Equations

- *Graph parametric equations and determine an equivalent rectangular equation.*
- *Determine parametric equations for a rectangular equation.*

Graphing Parametric Equations

Much of our graphing of curves in this text has been with rectangular equations involving only two variables, x and y. We graphed sets of ordered pairs, (x, y), where y is a function of x. In this section, we introduce a third variable, t, such that x and y are each a function of t.

Consider a point P with coordinates (x, y) in the rectangular coordinate plane. As P moves in the plane, its location at time t is given by two functions

$$x = f(t) \quad \text{and} \quad y = g(t).$$

For example, let the equations be

$$x = \tfrac{1}{2}t \quad \text{and} \quad y = t^2 - 3, \quad t \geq 0.$$

We restrict t to nonnegative values since t represents time. When $t = 2$, we have

$$x = \tfrac{1}{2} \cdot 2 = 1 \quad \text{and} \quad y = 2^2 - 3 = 1.$$

Thus after 2 sec, the coordinates of P are $(1, 1)$. The table below lists other ordered pairs. We plot them and draw the curve for $t \geq 0$.

t	x	y	(x, y)
0	0	-3	$(0, -3)$
1	$\tfrac{1}{2}$	-2	$\left(\tfrac{1}{2}, -2\right)$
2	1	1	$(1, 1)$
3	$\tfrac{3}{2}$	6	$\left(\tfrac{3}{2}, 6\right)$

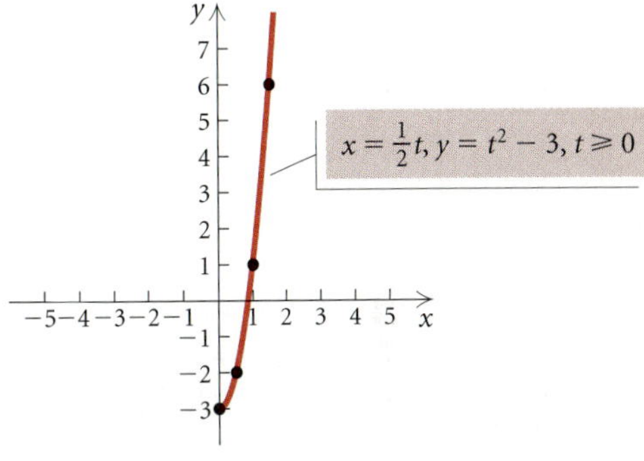

This curve appears to be part of a parabola. Let's verify this by finding the equivalent rectangular equation. Solving $x = \tfrac{1}{2}t$ for t, we get $t = 2x$. Substituting $2x$ for t in $y = t^2 - 3$, we have

$$y = (2x)^2 - 3 = 4x^2 - 3.$$

We know from Section 2.2 that this is a quadratic equation and that the graph is a parabola. Thus the curve is part of the parabola $y = 4x^2 - 3$.

The equations $x = \tfrac{1}{2}t$ and $y = t^2 - 3$ are called the **parametric equations** for the curve. The variable t is called the **parameter**. The equation $y = 4x^2 - 3$, with the restriction $x \geq 0$, is the equivalent rectangular equation for the curve. Since $t \geq 0$ and $x = \tfrac{1}{2}t$, the restriction $x \geq 0$ must be included. In this example, t represents time, but t, in general, can represent any real number in a specified interval. One advantage of parametric equations is that the restriction on the parameter t allows us to investigate a portion of the curve.

Parametric Equations

If f and g are continuous functions of t on an interval I, then the set of ordered pairs (x, y) such that $x = f(t)$ and $y = g(t)$ is a *plane curve*. The equations $x = f(t)$ and $y = f(t)$ are *parametric equations* for the curve. The variable t is the *parameter*.

Plane curves described with parametric equations can also be graphed on graphers. Consult your grapher's manual or the Graphing Calculator Manual that accompanies this text for specific instructions.

Example 1   Using a grapher, graph each of the following plane curves given their respective parametric equations and the restriction for the parameter. Then find the equivalent rectangular equation.

a) $x = t^2,\ y = t - 1;\ -1 \le t \le 4$

b) $x = \sqrt{t},\ y = 2t + 3;\ 0 \le t \le 3$

SOLUTION

a) When using a grapher set in PARAMETRIC mode, we must set minimum and maximum values for x, y, and t.

WINDOW
Tmin = -1
Tmax = 4
Tstep = $.1$
Xmin = -2
Xmax = 18
Xscl = 1
Ymin = -4
Ymax = 4
Yscl = 1

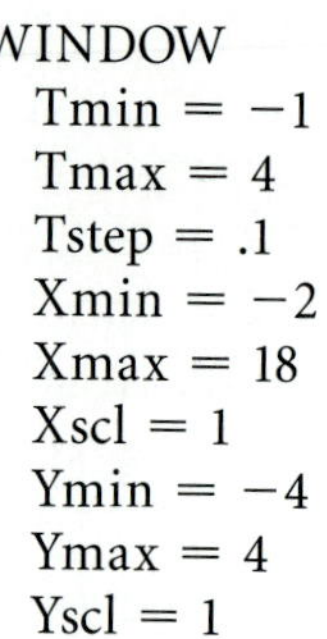
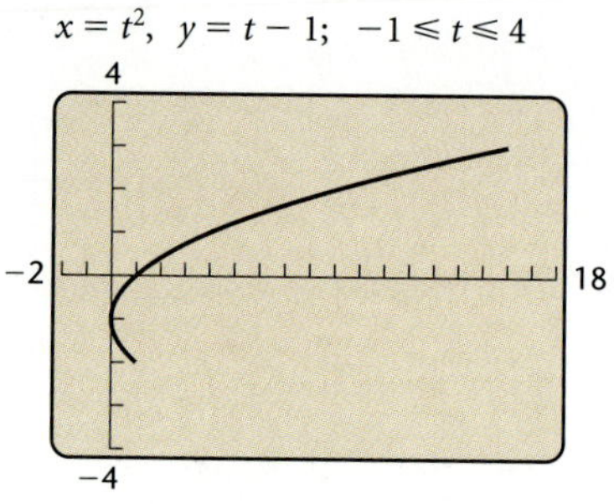

Using the TRACE feature, we can see the T, X, and Y values along the curve. To find an equivalent rectangular equation, we can solve either equation for t. Let's select the simpler equation $y = t - 1$ and solve for t:

$$y = t - 1$$
$$y + 1 = t.$$

We then substitute $y + 1$ for t in $x = t^2$:

$$x = t^2$$
$$x = (y + 1)^2. \qquad \textbf{Parabola}$$

This is an equation of a parabola that opens to the right. Given that $-1 \le t \le 4$, we have the corresponding restrictions on x and y: $0 \le x \le 16$ and $-2 \le y \le 3$. Thus the equivalent rectangular equation is

$$x = (y + 1)^2;\quad 0 \le x \le 16,\quad -2 \le y \le 3.$$

b) Using a grapher set in PARAMETRIC mode, we have

WINDOW
Tmin = 0
Tmax = 3
Tstep = $.1$
Xmin = -3
Xmax = 3
Xscl = 1
Ymin = -2
Ymax = 10
Yscl = 1

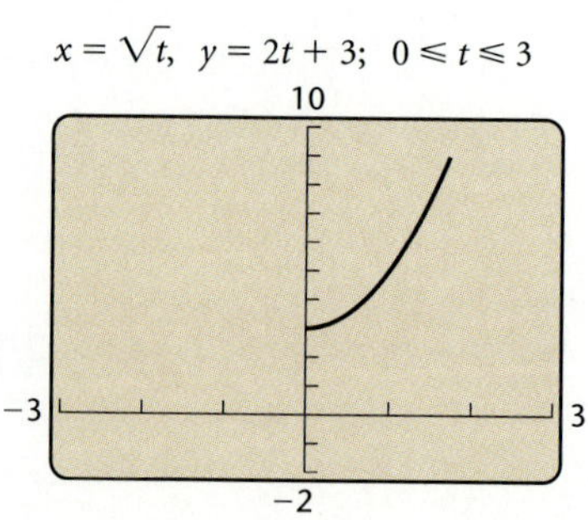

To find an equivalent rectangular equation, we first solve $x = \sqrt{t}$ for t:

$$x = \sqrt{t}$$
$$x^2 = t.$$

Then we substitute x^2 for t in $y = 2t + 3$:

$$y = 2t + 3$$
$$y = 2x^2 + 3. \qquad \text{Parabola}$$

When $0 \le t \le 3$, $0 \le x \le \sqrt{3}$ and $3 \le y \le 9$. The equivalent rectangular equation is

$$y = 2x^2 + 3; \quad 0 \le x \le \sqrt{3}, \quad 3 \le y \le 9.$$

Determining Parametric Equations for a Given Rectangular Equation

Many sets of parametric equations can represent the same plane curve. In fact, there are infinitely many such equations.

Example 2 Find four sets of parametric equations for the parabola

$$y = 4 - (x + 3)^2.$$

SOLUTION

If $x = t$, then $y = 4 - (t + 3)^2$.

If $x = t - 3$, then $y = 4 - (t - 3 + 3)^2$, or $4 - t^2$.

If $x = \dfrac{t}{3}$, then $y = 4 - \left(\dfrac{t}{3} + 3\right)^2$, or $-\dfrac{t^2}{9} - 2t - 5$.

In each of the above three sets, t is any real number. The following equations restrict t to any number greater than or equal to 0.

If $x = \sqrt{t} - 3$, then $y = 4 - (\sqrt{t} - 3 + 3)^2$, or $4 - t$; $t \ge 0$.

$$\text{B} \quad \textit{Exercise Set}$$

Using a grapher, graph the plane curve given by the set of parametric equations and the restriction for the parameter. Then find the equivalent rectangular equation.

1. $x = \frac{1}{2}t$, $y = 6t - 7$; $-1 \le t \le 6$

2. $x = t$, $y = 5 - t$; $-2 \le t \le 3$

3. $x = t^3$, $y = t - 4$; $-1 \le t \le 10$

4. $x = \sqrt{t}$, $y = 2t + 3$; $0 \le t \le 8$

5. $x = t^2$, $y = \sqrt{t}$; $0 \le t \le 4$

6. $x = t^3 + 1$, $y = t$; $-3 \le t \le 3$

7. $x = t + 3$, $y = \dfrac{1}{t + 3}$; $-2 \le t \le 2$

8. $x = 2t^3 + 1$, $y = 2t^3 - 1$; $-4 \le t \le 4$

9. $x = 2t - 1$, $y = t^2$; $-3 \le t \le 3$

10. $x = \frac{1}{3}t$, $y = t$; $-5 \le t \le 5$

11. $x = e^{-t}$, $y = e^t$; $-\infty < t < \infty$

12. $x = 2 \ln t$, $y = t^2$; $0 < t < \infty$

Find two sets of parametric equations for the rectangular equation.

13. $y = 4x - 3$

14. $y = x^2 - 1$

15. $y = (x - 2)^2 - 6x$

16. $y = x^3 + 3$

Synthesis

17. Determine a set of parametric equations for a line that passes through $(-2, 4)$ and $(-3, 8)$.

18. Determine a set of parametric equations for a circle with center $(-1, 2)$ and radius 5.

Answers

Introduction to Graphs and Graphers

1. 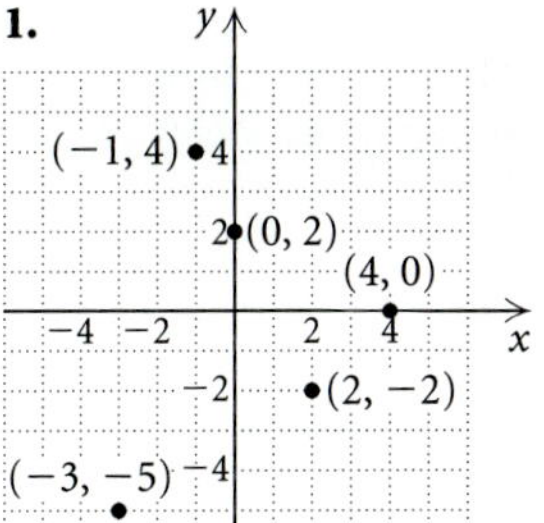

3. 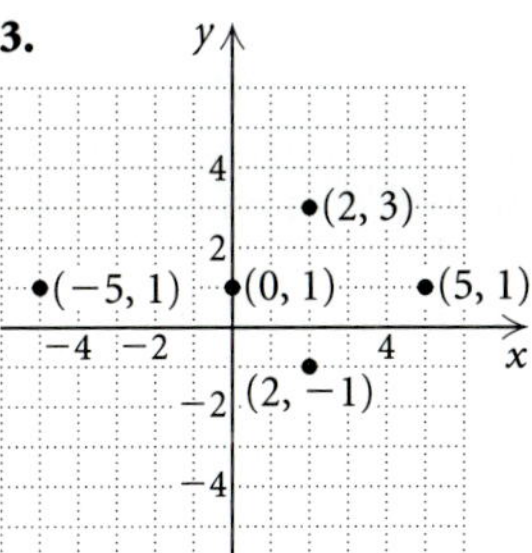

5. Yes; no   **7.** No; yes   **9.** No; yes

11. 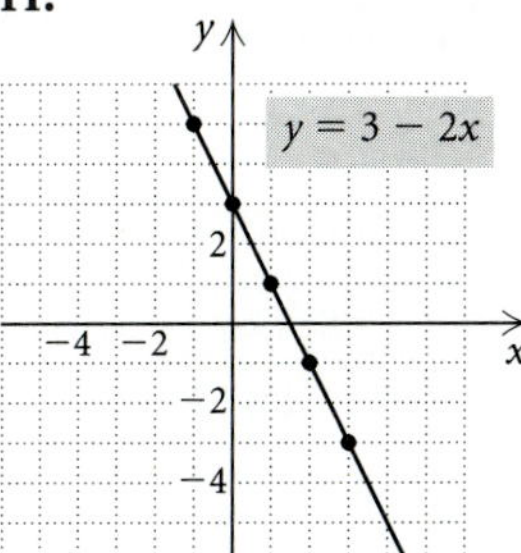

13. 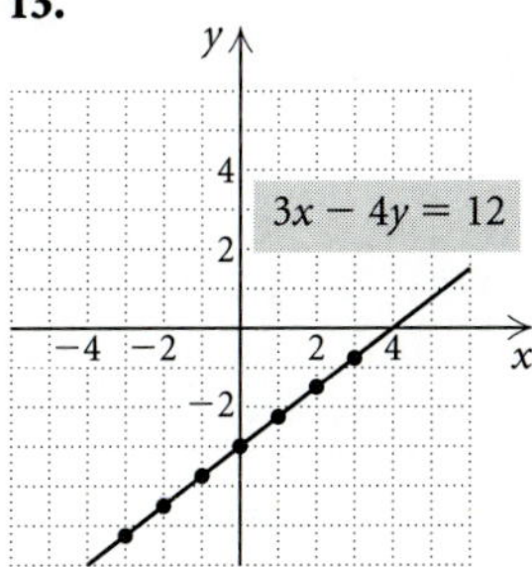

15. 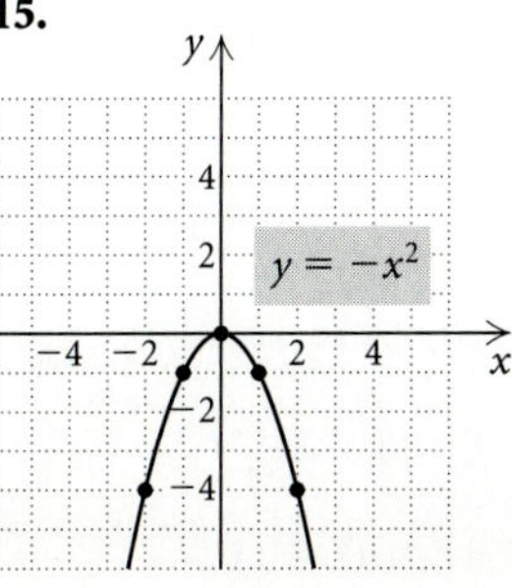

17. 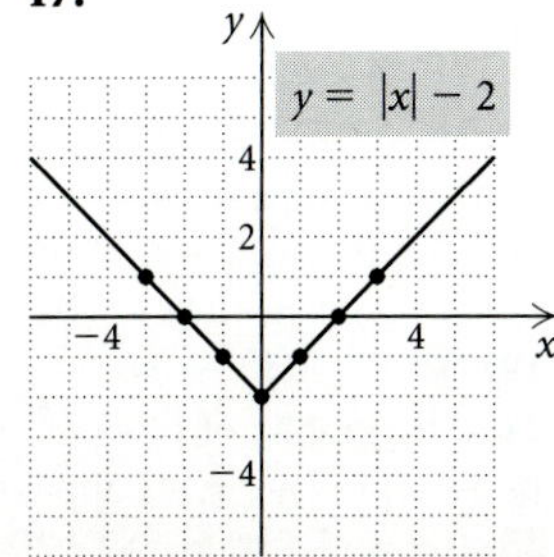

19. 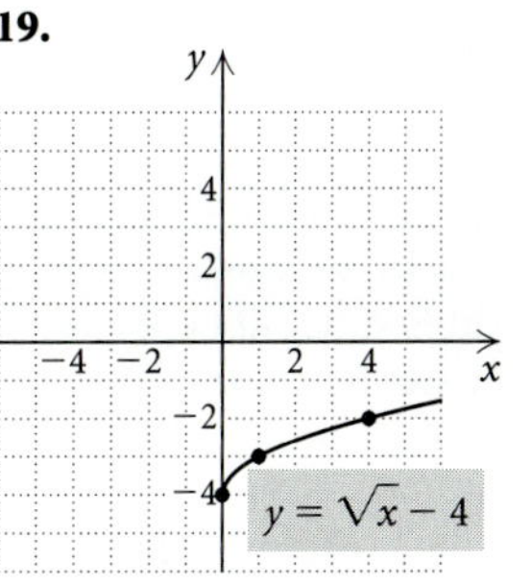

21. 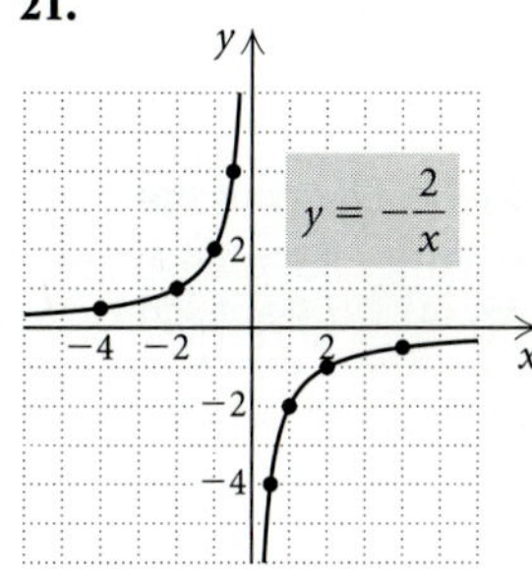

23. (f)   **25.** (c)   **27.** (b)   **29.** (d)

31. 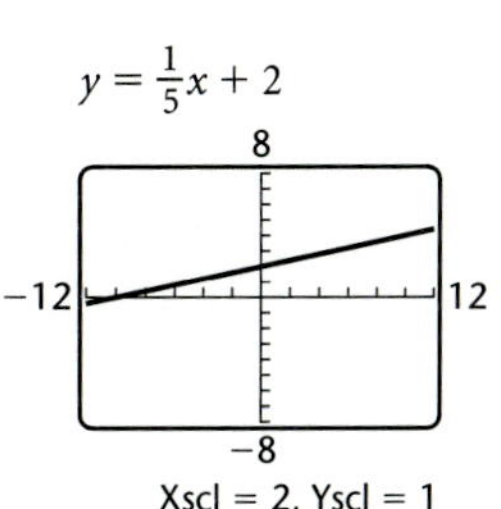

$y = \frac{1}{5}x + 2$

Xscl = 2, Yscl = 1

33. 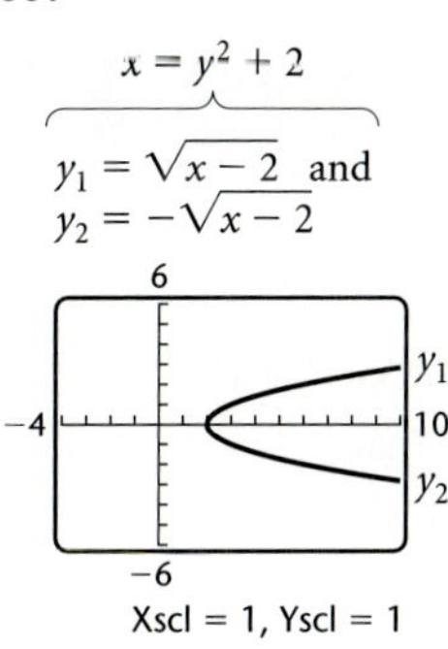

$x = y^2 + 2$

$y_1 = \sqrt{x - 2}$ and $y_2 = -\sqrt{x - 2}$

Xscl = 1, Yscl = 1

35. 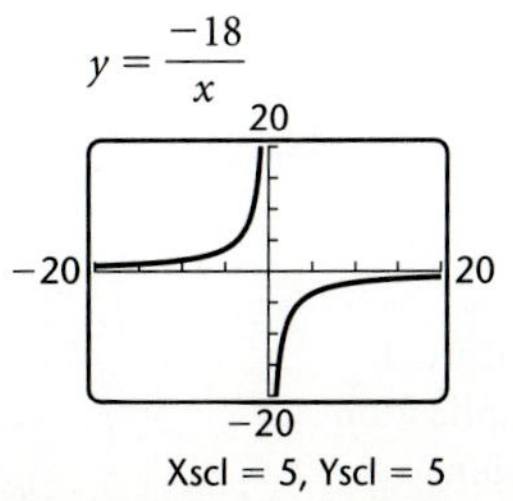

$y = \frac{-18}{x}$

Xscl = 5, Yscl = 5

37. 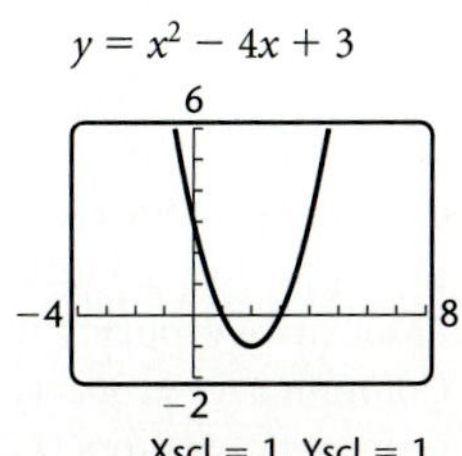

$y = x^2 - 4x + 3$

Xscl = 1, Yscl = 1

39.

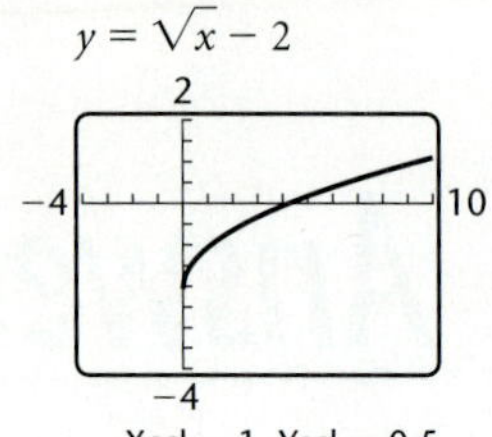

$y = \sqrt{x} - 2$

Xscl = 1, Yscl = 0.5

41.

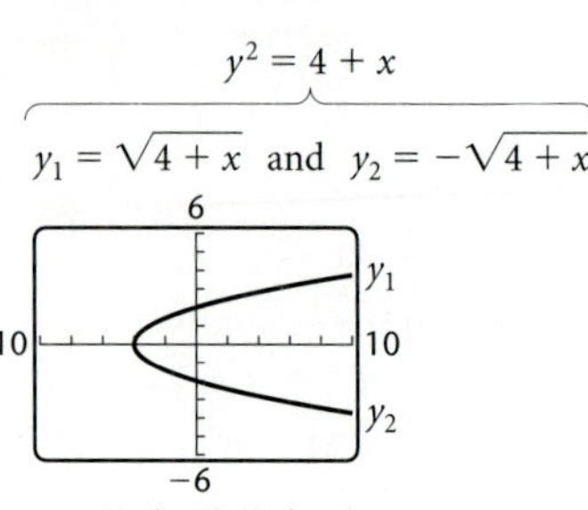

$y^2 = 4 + x$

$y_1 = \sqrt{4 + x}$ and $y_2 = -\sqrt{4 + x}$

Xscl = 2, Yscl = 1

43. (a) Does not show curvature; **(b)** best; **(c)** tick marks too close; **(d)** does not show curvature; ticks too close
45. (a) Does not show curvature; **(b)** best; **(c)** does not show curvature as well as (b); **(d)** cannot see the graph
47. Error, error, .6245, 1, 1.261, 1.4697, 1.6462; -32, -40, -53.33, -80, -160, error, 160   **49.** Yes   **51.** No
53. Yes   **55.** No   **57.** Yes   **59.** -8 **61.** -0.882
63. -5.8 **65.** 20.376   **67.** -0.104 **69.** ± 1.959
71. 1.489, 5.673   **73.** -2, 3
75. $(-4.378, 0)$, $(2.545, 0)$, $(-1.167, 0)$
77. $(-0.929, -3.388)$, $(1.080, -2.240)$, $(2.848, -1.229)$
79. $(-2.809, -2.528)$, $(0, 0)$, $(2.809, 2.528)$
81. Answer depends on the grapher and its features.
83. $y = x^2 + 1$; x: ± 9, ± 11; y: 26

Chapter R

EXERCISE SET R.1

1. $\sqrt[3]{8}$, 0, 9, $\sqrt{25}$
3. $\sqrt{7}$, 5.242242224..., $-\sqrt{14}$, $\sqrt[5]{5}$, $\sqrt[3]{4}$
5. -12, $5.\overline{3}$, $-\dfrac{7}{3}$, $\sqrt[3]{8}$, 0, -1.96, 9, $4\dfrac{2}{3}$, $\sqrt{25}$, $\dfrac{5}{7}$ **7.** $-\dfrac{3}{4}$
9. -5 **11.** True   **13.** False   **15.** True   **17.** False
19. False   **21.** True   **23.** True   **25.** True   **27.** False
29. 7.1   **31.** $\dfrac{5}{4}$ **33.** $8|b|$ **35.** $5|xz|$, or $5|x| \cdot |z|$
37. $0.02\left|\dfrac{x}{y}\right|$, or $\dfrac{0.02|x|}{|y|}$ **39.** 11   **41.** 6   **43.** 5.4
45. Commutative property of multiplication
47. Multiplicative identity property
49. Associative property of multiplication
51. Commutative property of multiplication
53. Commutative property of addition

55. Multiplicative inverse property   **57.** Identity
59. Not an identity   **61.** Identity   **63.** ◈
65. 0.124124412444...   **67.** -0.00999
69. The hypotenuse of a right triangle with legs of lengths 1 unit and 3 units has a length of $\sqrt{10}$ units.

EXERCISE SET R.2

1. 5^2, or 25   **3.** 1   **5.** 7^{-1}, or $\dfrac{1}{7}$ **7.** $6x^5$
9. $15a^{-1}b^5$, or $\dfrac{15b^5}{a}$ **11.** $72x^5$ **13.** b^3 **15.** x^3y^{-3}, or $\dfrac{x^3}{y^3}$
17. $8xy^{-5}$, or $\dfrac{8x}{y^5}$ **19.** $8a^3b^6$ **21.** $16x^{12}$ **23.** $\dfrac{c^2d^4}{25}$
25. $8^5a^{20}b^{-25}c^{10}$, or $\dfrac{8^5a^{20}c^{10}}{b^{25}}$ **27.** 4.05×10^5
29. 3.9×10^{-7} **31.** 1.6×10^{-5} **33.** 0.000083
35. 20,700,000   **37.** 4,690,000,000,000   **39.** 1.395×10^3
41. 8.0×10^{-14} **43.** 1.332×10^{14} **45.** About $\$1.2 \times 10^7$
47. 2   **49.** 2048   **51.** 5   **53.** $\$2883.67$ **55.** $\$8763.54$
57. ◈   **59.** x^{8t} **61.** t^{8x} **63.** $9x^{2a}y^{2b}$ **65.** $\$660.29$
67. For $x < -1$ or $x > 1$

EXERCISE SET R.3

1. $-5y^4$, $3y^3$, $7y^2$, $-y$, -4; 4   **3.** $3a^4b$, $-7a^3b^3$, $5ab$, -2; 6
5. $3x^2y - 5xy^2 + 7xy + 2$ **7.** $3x + 2y - 2z - 3$
9. $-2x^2 + 6x - 2$ **11.** $x^4 - 3x^3 - 4x^2 + 9x - 3$
13. $2a^4 - 2a^3b - a^2b + 4ab^2 - 3b^3$ **15.** $x^2 + 2x - 15$
17. $2a^2 + 13a + 15$
19. The graphs of $y_1 = (2x + 3)(x + 5)$ and $y_2 = 2x^2 + 13x + 15$ appear to coincide.
21. $4x^2 + 8xy + 3y^2$ **23.** $x^2 + 10x + 25$
25. $25y^2 - 30y + 9$
27. The graphs of $y_1 = (5x - 3)^2$ and $y_2 = 25x^2 - 30x + 9$ appear to coincide.   **29.** $4x^2 + 12xy + 9y^2$
31. $4x^4 - 12x^2y + 9y^2$ **33.** $a^2 - 9$
35. The graphs of $y_1 = (x + 3)(x - 3)$ and $y_2 = x^2 - 9$ appear to coincide.   **37.** $9x^2 - 4y^2$
39. $4x^2 + 12xy + 9y^2 - 16$ **41.** $x^4 - 1$ **43.** ◈
45. $a^{2n} - b^{2n}$ **47.** $a^{2n} + 2a^nb^n + b^{2n}$ **49.** $x^6 - 1$
51. $x^{a^2 - b^2}$ **53.** $a^2 + b^2 + c^2 + 2ab + 2ac + 2bc$
55. $a^4 + 4a^3b + 6a^2b^2 + 4ab^3 + b^4$

EXERCISE SET R.4

1. $2(x - 5)$ **3.** $3x^2(x^2 - 3)$ **5.** $4(a^2 - 3a + 4)$
7. $(b - 2)(a + c)$ **9.** $(x + 3)(x^2 + 6)$
11. $(y - 3)(y + 2)(y - 2)$
13. The graphs of $y_1 = x^3 + 3x^2 + 6x + 18$ and $y_2 = (x + 3)(x^2 + 6)$ appear to coincide.
15. $(p + 2)(p + 4)$ **17.** $(2n - 7)(n + 8)$
19. $(y^2 - 7)(y^2 + 3)$
21. The graphs of $y_1 = x^2 + 6x + 8$ and $y_2 = (x + 2)(x + 4)$ appear to coincide.
23. $(3x + 5)(3x - 5)$ **25.** $4x(y^2 + z)(y^2 - z)$

27. $(y - 3)^2$ **29.** $(1 - 4x)^2$
31. The graphs of $y_1 = 1 - 8x + 16x^2$ and $y_2 = (1 - 4x)^2$ appear to coincide.   **33.** $(x + 2)(x^2 - 2x + 4)$
35. $(m - 1)(m^2 + m + 1)$
37. The graphs of $y_1 = x^3 - 1$ and $y_2 = (x - 1)(x^2 + x + 1)$ appear to coincide.
39. $3ab(6a - 5b)$ **41.** $(x - 4)(x^2 + 5)$
43. $8(x + 2)(x - 2)$ **45.** Prime
47. The graphs of $y_1 = x^3 - 4x^2 + 5x - 20$ and $y_2 = (x - 4)(x^2 + 5)$ appear to coincide.
49. $(x + 4)(x + 5)$ **51.** $(y - 1)(y - 5)$
53. $(2a + 1)(a + 4)$ **55.** $(3x - 1)(2x + 3)$ **57.** $(y - 9)^2$
59. $(xy - 7)^2$ **61.** $4a(x + 7)(x - 2)$
63. $3(z - 2)(z^2 + 2z + 4)$
65. $2ab(2a^2 + 3b^2)(4a^4 - 6a^2b^2 + 9b^4)$
67. The graphs of $y_1 = 6x^2 + 7x - 3$ and $y_2 = (3x - 1)(2x + 3)$ appear to coincide.   **69.** ◈
71. $(y^2 + 12)(y^2 - 7)$ **73.** $\left(y + \frac{4}{7}\right)\left(y - \frac{2}{7}\right)$
75. $(t + 0.9)(t - 0.3)$ **77.** $h(3x^2 + 3xh + h^2)$
79. $(x - 4)(x + 8)$ **81.** $(3a - 3b - 2)(a - b + 4)$
83. $(x^n + 8)(x^n - 3)$ **85.** $(x + b)(x + a)$ **87.** $\left(\frac{1}{2}t - \frac{2}{5}\right)^2$
89. $(5y^m + x^n - 1)(5y^m - x^n + 1)$
91. $3(x^n - 2y^m)(x^{2n} + 2x^ny^m + 4y^{2m})$
93. $y(y - 1)^2(y - 2)$

EXERCISE SET R.5

1. $\{x \mid x$ is a real number$\}$
3. $\{x \mid x$ is a real number and $x \neq 0$ and $x \neq 1\}$
5. $\{x \mid x$ is a real number and $x \neq 2$ and $x \neq -2$ and $x \neq -5\}$
7.

X	Y₁	
−1	−3	
0	ERROR	
1	ERROR	
2	1.5	
3	1	
4	.75	
5	.6	

$X = 0$

9. $\dfrac{1}{x - y}$ **11.** $\dfrac{(x + 5)(2x + 3)}{7x}$ **13.** $\dfrac{a + 2}{a - 5}$ **15.** $m + n$
17. $\dfrac{3(x - 4)}{2(x + 4)}$ **19.** $\dfrac{1}{x + y}$ **21.** $\dfrac{x - y - z}{x + y + z}$
23. The graphs of $y_1 = \dfrac{x^2 - 2x - 35}{2x^3 - 3x^2} \cdot \dfrac{4x^3 - 9x}{7x - 49}$ and $y_2 = \dfrac{(x + 5)(2x + 3)}{7x}$ appear to coincide.   **25.** 1
27. $\dfrac{y - 2}{y - 1}$ **29.** $\dfrac{x + y}{2x - 3y}$ **31.** $\dfrac{3x - 4}{(x + 2)(x - 2)}$
33. $\dfrac{-y + 10}{(y - 5)(y + 4)}$ **35.** $\dfrac{4x - 8y}{(x + y)(x - y)}$
37. $\dfrac{3x - 4}{(x - 2)(x - 1)}$ **39.** $\dfrac{5a^2 + 10ab - 4b^2}{(a - b)(a + b)}$
41. $\dfrac{11x^2 - 18x + 8}{(2 + x)(2 - x)^2}$ **43.** 0

45. The graphs of $y_1 = \dfrac{3}{x + 2} + \dfrac{2}{x^2 - 4}$ and $y_2 = \dfrac{3x - 4}{(x + 2)(x - 2)}$ appear to coincide.   **47.** $\dfrac{x + y}{x}$
49. $\dfrac{a^2 - 1}{a^2 + 1}$ **51.** $\dfrac{c^2 - 2c + 4}{c}$ **53.** $\dfrac{xy}{x - y}$ **55.** $x - y$
57. $\dfrac{3(x - 1)^2(x + 2)}{(x + 3)(x - 3)(10 - x)}$ **59.** $\dfrac{1 + a}{1 - a}$ **61.** $\dfrac{b + a}{b - a}$
63. ◈   **65.** $2x + h$ **67.** $3x^2 + 3xh + h^2$ **69.** x^5
71. $\dfrac{(n + 1)(n + 2)(n + 3)}{2 \cdot 3}$ **73.** $\dfrac{x^3 + 2x^2 + 11x + 20}{(x + 1)(4 + 2x)}$

EXERCISE SET R.6

1. 11   **3.** $4|x|$ **5.** $|b + 1|$ **7.** $-3x$ **9.** $|x - 2|$ **11.** 2
13. $6\sqrt{5}$ **15.** $3\sqrt[3]{2}$ **17.** $8\sqrt{2}|c|d^2$
19. The values of $y_1 = \sqrt{x^2 - 4x + 4}$ and $y_2 = |x - 2|$ appear to be the same for any value of x.   **21.** $2x^2y\sqrt{6}$
23. $3x\sqrt[3]{4y}$ **25.** $2(x + 4)\sqrt[3]{(x + 4)^2}$ **27.** $\dfrac{m^2n^4}{2}$ **29.** 2
31. $\dfrac{1}{2x}$ **33.** $\dfrac{4a\sqrt[3]{a}}{3b}$ **35.** $\dfrac{x\sqrt{7x}}{6y^3}$ **37.** $51\sqrt{2}$
39. $-2x\sqrt{2} - 12\sqrt{5x}$ **41.** 1   **43.** $4 + 2\sqrt{3}$
45. The graphs of $y_1 = 8\sqrt{2x^2} - 6\sqrt{20x} - 5\sqrt{8x^2}$ and $y_2 = -2x\sqrt{2} - 12\sqrt{5x}$ appear to coincide.
47. About 13,709.5 ft   **49. (a)** $h = \dfrac{a}{2}\sqrt{3}$; **(b)** $A = \dfrac{a^2}{4}\sqrt{3}$
51. 8   **53.** $\dfrac{\sqrt{6}}{3}$ **55.** $\dfrac{\sqrt[3]{10}}{2}$ **57.** $\dfrac{2\sqrt[3]{18}}{3}$ **59.** $\dfrac{9 - 3\sqrt{5}}{2}$
61. $\dfrac{6\sqrt{m} + 6\sqrt{n}}{m - n}$ **63.** $\dfrac{6}{5\sqrt{3}}$ **65.** $\dfrac{7}{\sqrt[3]{98}}$ **67.** $\dfrac{11}{\sqrt{33}}$
69. $\dfrac{76}{27 - 9\sqrt{3} + 3\sqrt{5} - \sqrt{15}}$ **71.** $\dfrac{a - b}{3a(\sqrt{a} - \sqrt{b})}$
73. $\sqrt[4]{x^3}$ **75.** 8   **77.** $\dfrac{1}{5}$ **79.** $a\sqrt[4]{\dfrac{a}{b^3}}$
81. $13^{5/4}$, or $13\sqrt[4]{13}$ **83.** $20^{2/3}$ **85.** $11^{1/6}$ **87.** $5^{5/6}$
89. 4   **91.** $8a^2$ **93.** $\dfrac{x^3}{3b^{-2}}$, or $\dfrac{x^3b^2}{3}$ **95.** $x\sqrt[3]{y}$
97. The graphs of $y_1 = (2x^{3/2})(4x^{1/2})$ and $y_2 = 8x^2$ appear to coincide.   **99.** $\sqrt[6]{288}$ **101.** $\sqrt[12]{x^{11}y^7}$ **103.** $a\sqrt[6]{a^5}$
105. $(a + x)\sqrt[12]{(a + x)^{11}}$ **107.** $a^{-2}b^{-5}(b^{10} - a^5)$
109. $x^{-1/3}y^{-1/4}(y - x)$ **111.** $(2x - 3)^{-3}(x + 1)^{1/4}(3x - 2)$
113. ◈   **115.** $\dfrac{(2 + x^2)\sqrt{1 + x^2}}{1 + x^2}$ **117.** $a^{a/2}$
119. (a) $\{0, 1\}$; **(b)** $\{x \mid x > 1\}$; **(c)** $\{x \mid 0 < x < 1\}$

EXERCISE SET R.7

1. $\{4\}$ **3.** $\{8\}$ **5.** $\left\{\frac{4}{5}\right\}$ **7.** $\left\{-\frac{3}{2}\right\}$ **9.** $\left\{\frac{3}{2}, \frac{2}{3}\right\}$ **11.** $\left\{\frac{2}{3}, -1\right\}$
13. $\{0, 3\}$ **15.** $\left\{-\frac{1}{3}, 0, 2\right\}$ **17.** $\left\{-1, 1, -\frac{1}{7}\right\}$ **19.** $\left\{\frac{20}{9}\right\}$
21. $\{286\}$ **23.** $\{6\}$ **25.** $\varnothing$ **27.** $\varnothing$ **29.** $\varnothing$

31. $\{x \mid x \text{ is a real number and } x \neq 0 \text{ and } x \neq 6\}$ **33.** $\{2\}$
35. $\left\{\frac{5}{3}\right\}$ **37.** $\{\pm\sqrt{2}\}$ **39.** $\varnothing$ **41.** $\{-6\}$ **43.** $\{7\}$
45. $\varnothing$ **47.** $\{1\}$ **49.** $\{3, 7\}$ **51.** $\{-8\}$ **53.** $\{81\}$
55. $\{-7, 7\}$ **57.** $\varnothing$ **59.** $\{-3, 5\}$ **61.** $\left\{-\frac{1}{3}, \frac{1}{3}\right\}$ **63.** $\{0\}$
65. $\left\{-1, -\frac{1}{3}\right\}$ **67.** $\{-24, 44\}$ **69.** $\{-2, 4\}$ **71.** $b = \dfrac{2A}{h}$
73. $w = \dfrac{P - 2l}{2}$ **75.** $T_1 = \dfrac{P_1 V_1 T_2}{P_2 V_2}$ **77.** $R_2 = \dfrac{RR_1}{R_1 - R}$
79. $p = \dfrac{Fm}{m - F}$ **81.** ◆   **83.** ◆   **85.** $\{1\}$
87. $\{1, 5\}$ **89.** $\{-1\}$

EXERCISE SET R.8

1. $\{x \mid x > 3$, or $(3, \infty)$ **3.** $\left\{x \mid x \geq -\frac{5}{12}\right\}$, or $\left[-\frac{5}{12}, \infty\right)$
5. $\left\{y \mid y \geq \frac{22}{13}\right\}$, or $\left[\frac{22}{13}, \infty\right)$ **7.** $\left\{x \mid x \leq \frac{15}{34}\right\}$, or $\left(-\infty, \frac{15}{34}\right]$
9. $\{x \mid x < 1\}$, or $(-\infty, 1)$
11. The graph of $y_1 = 4x(x - 2)$ lies below the graph of
$y_2 = 2(2x - 1)(x - 3)$ for $x < 1$.   **13.** $[-3, 3]$
15. $[-14, -11)$ **17.** $(-\infty, -4]$ **19.** $(-\infty, 3.8)$
21. $(-\infty, 7) \cup (7, \infty)$ **23.** $(0, 5)$ **25.** $[-9, -4)$
27. $[x, x + h]$ **29.** (p, ∞) **31.** $[-3, 3)$ **33.** $[8, 10]$
35. $[-7, -1]$ **37.** $\left(-\frac{3}{2}, 2\right)$ **39.** $(1, 5]$ **41.** $\left(-\frac{11}{3}, \frac{13}{3}\right]$
43. $(-\infty, -2] \cup (1, \infty)$ **45.** $\left(-\infty, -\frac{7}{2}\right] \cup \left[\frac{1}{2}, \infty\right)$
47. $(-\infty, 9.6) \cup (10.4, \infty)$ **49.** $\left(-\infty, -\frac{57}{4}\right] \cup \left[-\frac{55}{4}, \infty\right)$
51. The graph of $y_1 = 6 - 2x$ lies both on or above the graph
of $y_2 = -4$ and below the graph of $y_3 = 4$ for $1 < x \leq 5$.
53. $(-7, 7)$;

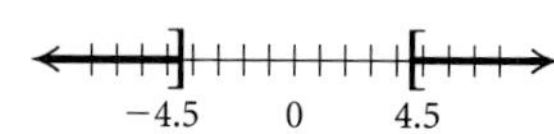

55. $(-\infty, -4.5] \cup [4.5, \infty)$;

57. $(-17, 1)$ **59.** $(-\infty, -17] \cup [1, \infty)$ **61.** $\left(-\frac{1}{4}, \frac{3}{4}\right)$
63. $\left(-\frac{1}{3}, \frac{1}{3}\right)$ **65.** $[-6, 3]$ **67.** $(-\infty, 4.9) \cup (5.1, \infty)$
69. $\left[-\frac{1}{2}, \frac{7}{2}\right]$ **71.** $\left[-\frac{7}{3}, 1\right]$ **73.** $(-\infty, -8) \cup (7, \infty)$ **75.** $\varnothing$
77. The graph of $y_1 = |x + 8|$ lies on or above the graph of
$y_2 = 9$ for $x \geq 1$ or $x \leq -17$.   **79.** ◆   **81.** ◆
83. $\left(-\frac{1}{4}, \frac{5}{9}\right]$ **85.** $\left(-\infty, \frac{1}{2}\right)$ **87.** $\{-\infty, \infty\}$ **89.** $\varnothing$

EXERCISE SET R.9

1. $0.832 billion on employee benefits; $3.239 billion on
wages and salaries   **3.** Length: 93 m; width: 68 m
5. About 11.5 sec   **7.** 2 cm   **9.** 26°, 130°, 24°
11. 280,526   **13.** Castulo: 8 km/h; Linette: 15 km/h
15. Amtrak: 80 mph; Central Railway: 66 mph
17. 98.3 mi   **19.** $650   **21.** 86 or higher
23. Mileages greater than $33\frac{1}{3}$ **25.** Calls longer than 2 hr
27. Gross sales greater than $18,000   **29.** ◆

31. 32 mph   **33.** $53\frac{6}{23}$ mph   **35.** $51\frac{3}{7}$ mph   **37.** $16

REVIEW EXERCISES, CHAPTER R

1. [R.1] $12, -3, -1, -19, 31, 0$ **2.** [R.1] $12, 31$
3. [R.1] $-43.89, 12, -3, -\frac{1}{5}, -1, -\frac{4}{3}, 7\frac{2}{3}, -19, 31, 0$
4. [R.1] $-43.89, 12, -3, -\frac{1}{5}, \sqrt{7}, \sqrt[3]{10}, -1, -\frac{4}{3}, 7\frac{2}{3}, -19,$
$31, 0$ **5.** [R.1] $\sqrt{7}, \sqrt[3]{10}$ **6.** [R.1] $12, 31, 0$ **7.** [R.1] 3.5
8. [R.1] $\frac{5}{6}a^2|b|$ **9.** [R.1] 10 **10.** [R.1] Not an identity
11. [R.2] 117 **12.** [R.2] -10 **13.** [R.2] $3,261,000$
14. [R.2] 0.00041 **15.** [R.2] 1.432×10^{-2}
16. [R.2] 4.321×10^4 **17.** [R.2] 7.8125×10^{-22}
18. [R.2] 5.46×10^{-32} **19.** [R.2] $-14a^{-2}b^7$, or $\dfrac{-14b^7}{a^2}$
20. [R.2] $6x^9y^{-6}z^6$, or $\dfrac{6x^9z^6}{y^6}$ **21.** [R.2] 3 **22.** [R.2] -2
23. [R.5] $\dfrac{b}{a}$ **24.** [R.5] $\dfrac{x + y}{xy}$ **25.** [R.6] -4
26. [R.6] $25x^4 - 10\sqrt{2}x^2 + 2$ **27.** [R.6] $13\sqrt{5}$
28. [R.3] $x^3 + t^3$
29. [R.3] $125a^3 + 300a^2b + 240ab^2 + 64b^3$
30. [R.3] $8xy^4 - 9xy^2 + 4x^2 + 2y - 7$
31. [R.4] $(x^2 - 3)(x + 2)$
32. [R.4] $3a(2a + 3b^2)(2a - 3b^2)$ **33.** [R.4] $(x + 12)^2$
34. [R.4] $x(9x - 1)(x + 4)$
35. [R.4] $(2x - 1)(4x^2 + 2x + 1)$
36. [R.4] $(3x^2 + 5y^2)(9x^4 - 15x^2y^2 + 25y^4)$
37. [R.4] $6(x + 2)(x^2 - 2x + 4)$
38. [R.4] $(x - 1)(2x + 3)(2x - 3)$
39. [R.4] $(3x - 5)^2$ **40.** [R.4] $3(6x^2 - x + 2)$
41. [R.4] $(3x + y + 4)(3x + y - 1)$ **42.** [R.5] 3
43. [R.5] $\dfrac{x - 5}{(x + 5)(x + 3)}$ **44.** [R.6] $y^3\sqrt[6]{y}$
45. [R.6] $\sqrt[3]{(a + b)^2}$ **46.** [R.6] $b\sqrt[5]{b^2}$ **47.** [R.6] $\dfrac{m^4n^2}{3}$
48. [R.6] $\dfrac{x - y}{x + 2\sqrt{xy} + y}$ **49.** [R.6] $x^{-1/2}y^{-3/4}(2x - 3y)$
50. [R.6] $(x - 2)^{-3/4}(3x + 5)^{3/2}(4x + 3)$
51. [R.6] About 18.8 ft   **52.** [R.7] $\{4\}$ **53.** [R.7] $\{-1\}$
54. [R.7] $\left\{-\frac{5}{2}, \frac{1}{3}\right\}$ **55.** [R.7] $\left\{-2, \frac{4}{3}\right\}$ **56.** [R.7] $\left\{\frac{27}{7}\right\}$
57. [R.7] $\left\{-\frac{1}{2}, \frac{9}{4}\right\}$ **58.** [R.7] $\{0, 3\}$ **59.** [R.7] $\{5\}$
60. [R.7] $\{1, 7\}$ **61.** [R.7] $\{-8, 1\}$ **62.** [R.8] $\left[-\frac{4}{3}, \frac{4}{3}\right]$
63. [R.8] $\left(\frac{2}{5}, 2\right]$ **64.** [R.8] $(-\infty, \infty)$
65. [R.8] $\left(-\infty, -\frac{5}{3}\right] \cup [1, \infty)$ **66.** [R.8] $\left(-\frac{2}{3}, 1\right)$
67. [R.8] $(-\infty, -6) \cup [-2, \infty)$ **68.** [R.7] $h = \dfrac{V^2}{2g}$
69. [R.7] $t = \dfrac{ab}{a + b}$ **70.** [R.9] 30 ft, 40 ft
71. [R.9] 94%   **72.** [R.9] 6 mph   **73.** [R.9] 80 km/h
74. [R.9] Mileages greater than 25

75. [R.9] Calls less than 2 hr

76. [R.9] ◆ Nisha is taking a geography class in which there are 4 tests. She has scores of 76, 82, and 78 on the first three tests. To get a B, she must average at least 80 on the 4 tests. What scores on the last test will give Nisha a B?

77. [R.5], [R.7] ◆ Two expressions are equivalent if they have the same value for all numbers in both domains. Two equations are equivalent if they have the same solution set.

78. [R.3] $x^{2n} + 6x^n - 40$ **79.** [R.3] $t^{2a} + 2 + t^{-2a}$

80. [R.3] $y^{2b} - z^{2c}$ **81.** [R.3] $a^{3n} - 3a^{2n}b^n + 3a^nb^{2n} - b^{3n}$

82. [R.4] $(y^n + 8)^2$ **83.** [R.4] $(x^t - 7)(x^t + 4)$

84. [R.4] $m^{3n}(m^n - 1)(m^{2n} + m^n + 1)$ **85.** [R.7] 256

Chapter 1

EXERCISE SET 1.1

1. Yes   **3.** Yes   **5.** No   **7.** Yes   **9.** Yes   **11.** Yes

13. No   **15.** Function; domain: $\{2, 3, 4\}$; range: $\{10, 15, 20\}$

17. Not a function; domain: $\{-7, -2, 0\}$; range: $\{3, 1, 4, 7\}$

19. Function; domain: $\{-2, 0, 2, 4, -3\}$; range: $\{1\}$

21. $f(-1) = 2$; $f(0) = 0$; $f(1) = -2$

23. (a) 1; **(b)** 6; **(c)** 22; **(d)** $3x^2 + 2x + 1$; **(e)** $3t^2 - 4t + 2$; **(f)** $6a + 3h - 2$

25. (a) 8; **(b)** -8; **(c)** $-x^3$; **(d)** $27y^3$; **(e)** $8 + 12h + 6h^2 + h^3$; **(f)** $3a^2 + 3ah + h^2$

27. (a) $\dfrac{1}{8}$; **(b)** 0; **(c)** does not exist; **(d)** $\dfrac{81}{53}$; **(e)** $\dfrac{x + h - 4}{x + h + 3}$;

(f) $\dfrac{7}{(x + h + 3)(x + 3)}$

29. 0; does not exist; does not exist as a real number; $\dfrac{1}{\sqrt{3}}$ or $\dfrac{\sqrt{3}}{3}$ **31.** All real numbers   **33.** $\{x \mid x \neq 0\}$

35. $\left\{x \mid x \geq -\frac{4}{7}\right\}$ **37.** $\{x \mid x \neq 5 \text{ and } x \neq -1\}$

39. $\{x \mid -3 \leq x \leq 3\}$

41. Domain: all real numbers; range: $[0, \infty)$

43. Domain: $[-3, 3]$; range: $[0, 3]$

45. Domain: all real numbers; range: all real numbers

47. Domain: $(-\infty, 7]$; range: $[0, \infty)$

49. Domain: $(-\infty, 0) \cup (0, \infty)$; range: $(-\infty, 0) \cup (0, \infty)$

51. No   **53.** Yes   **55.** Yes   **57.** No

59. Function; domain: $[0, 5]$; range: $[0, 3]$

61. Function; domain: $[-2\pi, 2\pi]$; range: $[-1, 1]$

63. (a) The domain of y_1 does not include the intervals $(-\infty, -1.2]$ and $[3.2, \infty)$. The range of y_1 appears to be $[0, 2]$. The domain of y_2 does not include 3.2. The range of y_2 appears to be all real numbers. **(b)** The largest output of y_1 seems to be 2, and the smallest 0. There does not appear to be a largest or smallest output of y_2.

65. 645 m; 0 m   **67.** $\left\{\frac{15}{2}\right\}$ **69.** $\left\{\frac{53}{2}\right\}$ **71.** ◆

73. $\dfrac{1}{\sqrt{x + h} + \sqrt{x}}$ **75.** $f(x) = x$, $g(x) = x + 1$

77. 35

EXERCISE SET 1.2

1. $-1.532, -0.347, 1.879$ **3.** $-1.414, 0, 1.414$ **5.** $-1, 0, 1$

7. $1.5, 9.5$ **9.** $-2, 3$

11. All numbers in the interval $[-1, 2]$

13. (a) $[-5, 1]$; **(b)** $[3, 5]$; **(c)** $[1, 3]$

15. (a) $[-3, -1]$, $[3, 5]$; **(b)** $[1, 3]$; **(c)** $[-5, -3)$

17. Increasing: $[0, \infty)$; decreasing: $(-\infty, 0]$; relative minimum: 0 at $x = 0$

19. Increasing: $(-\infty, 0]$; decreasing: $[0, \infty)$; relative maximum: 5 at $x = 0$

21. Increasing: $[3, \infty)$; decreasing: $(-\infty, 3]$; relative minimum: 1 at $x = 3$

23. Increasing: $[1, 3]$; decreasing: $(-\infty, 1]$, $[3, \infty)$; relative maximum: -4 at $x = 3$; relative minimum: -8 at $x = 1$

25. Increasing: $[-1.552, 0]$, $[1.552, \infty)$; decreasing: $(-\infty, -1.552]$, $[0, 1.552]$; relative maximum: 4.07 at $x = 0$; relative minima: -2.314 at $x = -1.552$, -2.314 at $x = 1.552$

27. (a)

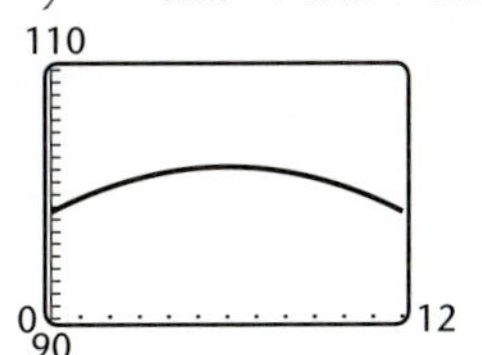

(b) 102.2; **(c)** 6 days after the illness began, the temperature is 102.2°F.

29. Increasing: $[-1, 1]$; decreasing: $(-\infty, -1]$, $[1, \infty)$; relative maximum: 4 at $x = 1$; relative minimum: -4 at $x = -1$

31. Increasing: $[-1.414, 1.414]$; decreasing: $[-2, -1.414]$, $[1.414, 2]$; relative maximum: 2 at $x = 1.414$; relative minimum: -2 at $x = -1.414$

33.

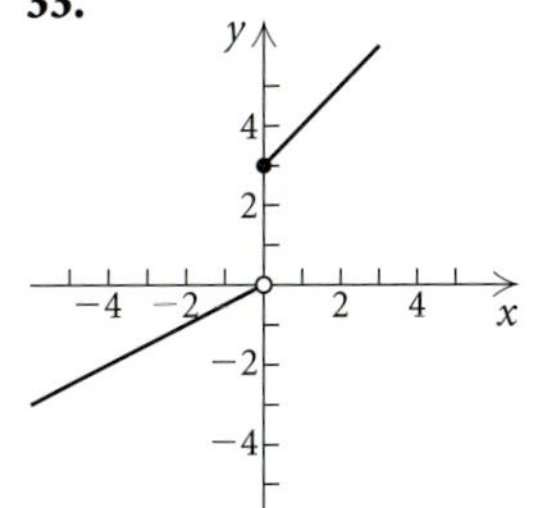

35.

37.

39.

41.

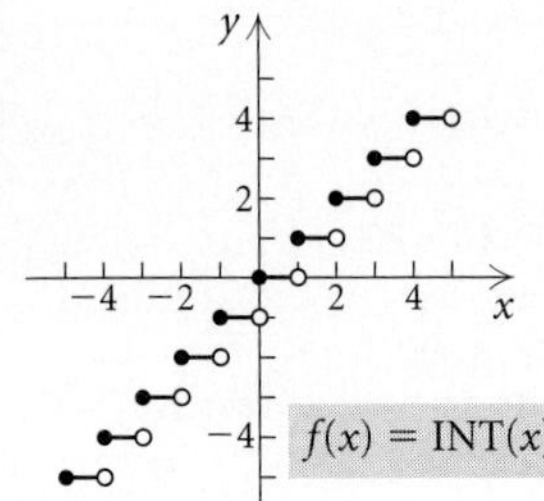

$f(x) = \text{INT}(x)$

43.

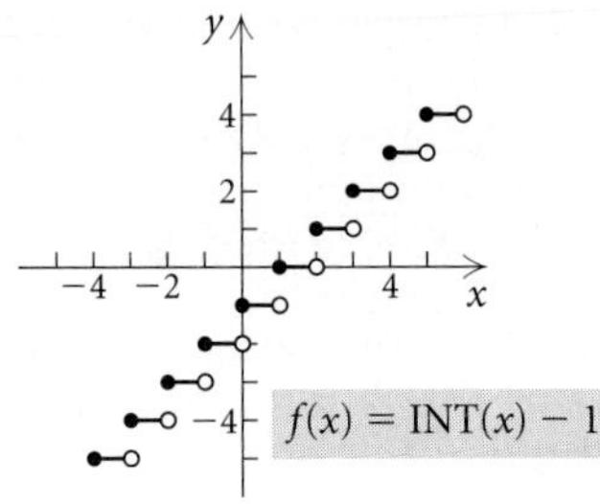

$f(x) = \text{INT}(x) - 1$

45.
$$y = \begin{cases} \sqrt[3]{x}, & \text{for } x \leq -1, \\ x^2 - 3x, & \text{for } -1 < x < 4, \\ \sqrt{x - 4}, & \text{for } x \geq 4 \end{cases}$$

47.
$$y = \begin{cases} \sqrt[3]{x + 27}, & \text{for } x < 1, \\ \left| 2 - \dfrac{x}{5} \right|, & \text{for } x \geq 1 \end{cases}$$

49. $d(t) = \sqrt{(120t)^2 + (400)^2}$ **51.** $A(x) = 24x - x^2$

53. $A(w) = 10w - \dfrac{w^2}{2}$ **55.** $d(s) = \dfrac{14}{s}$

57. (a) $V(x) = 4x^3 - 48x^2 + 144x$, or $4x(x - 6)^2$;
(b) $\{x \mid 0 < x < 6\}$;
(c) $y = 4x^3 - 48x^2 + 144x$ **(d)** 8 cm by 8 cm by 2 cm

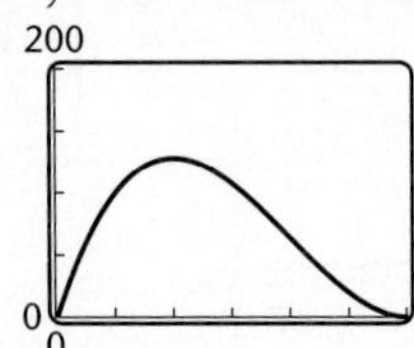

59. (a) $A(x) = x\sqrt{256 - x^2}$; **(b)** $\{x \mid 0 < x < 16\}$;
(c)

$y = x\sqrt{256 - x^2}$

(d) 11.314 ft by 11.314 ft   **61.** 7   **63.** ◈
65. Increasing: $[-1.933, 1.933]$; decreasing: $(-\infty, -1.933]$, $[1.933, \infty)$; relative maximum: 58.242 at $x = 1.933$; relative minimum: -58.242 at $x = -1.933$
67. Increasing: $[-5, -2]$, $[4, \infty)$; decreasing: $(-\infty, -5]$, $[-2, 4]$; relative maximum: 560 at $x = -2$; relative minima: 425 at $x = -5$, -304 at $x = 4$
69. (a)

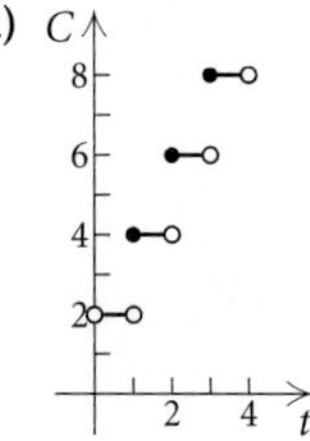

(b) $C(t) = 2(\text{INT}(t) + 1)$, $t > 0$
71. $\{x \mid -5 \leq x < -4 \text{ or } 5 \leq x < 6\}$
73. (a) $h(r) = \dfrac{30 - 5r}{3}$; **(b)** $V(r) = \pi r^2 \left(\dfrac{30 - 5r}{3} \right)$;
(c) $V(h) = \pi h \left(\dfrac{30 - 3h}{5} \right)^2$

EXERCISE SET 1.3

1. (a) Yes; **(b)** yes; **(c)** yes   **3. (a)** Yes; **(b)** no; **(c)** no
5. $-\frac{3}{5}$ **7.** $\frac{1}{5}$ **9.** $-\frac{5}{3}$ **11.** Not defined   **13.** $2a + h$
15. 6.7% grade; $y = 0.067x$
17. $\frac{3}{5}$; $(0, -7)$
19. $-\frac{1}{2}$; $(0, 5)$
21. $-\frac{3}{2}$; $(0, 5)$
23. $\frac{4}{3}$; $(0, -7)$
25. $y = \frac{2}{9}x + 4$ **27.** $y = -4x - 7$
29. $y = -4.2x + \frac{3}{4}$ **31.** $y = \frac{2}{9}x + \frac{19}{3}$
33. $y = 3x - 5$ **35.** $y = -\frac{3}{5}x - \frac{17}{5}$
37. $y = -3x + 2$ **39.** $y = -\frac{1}{2}x + \frac{7}{2}$
41. Parallel   **43.** Perpendicular
45. $y = \frac{2}{7}x + \frac{29}{7}$; $y = -\frac{7}{2}x + \frac{31}{2}$
47. $y = -0.3x - 2.1$; $y = \frac{10}{3}x + \frac{70}{3}$
49. $y = -\frac{3}{4}x + \frac{1}{4}$; $y = \frac{4}{3}x - 6$ **51.** $x = 3$; $y = -3$
53. (a) $W(h) = 3.5h - 110$; **(b)** 107 lb;

(c) $y = 3.5x - 110$

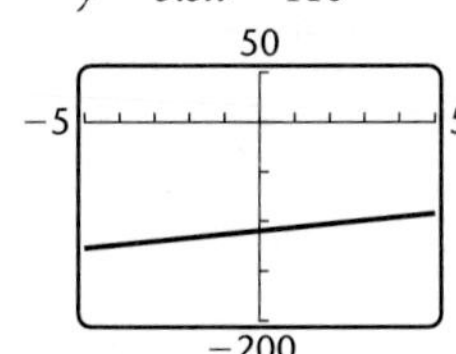

(d) $\{h \mid h > 31.43\}$, or $(31.43, \infty)$
55. (a) 115, 75, 135, 179; **(b)** $y = 2x + 115$

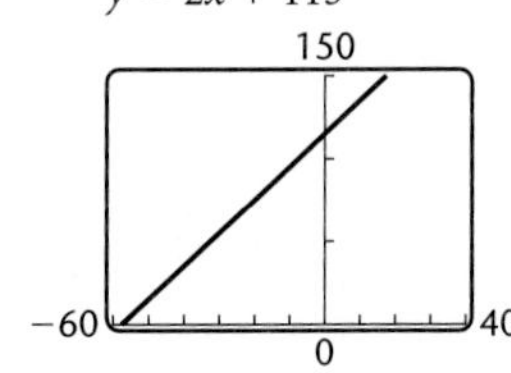

(c) Below $-57.5°$, stopping distance is negative; above $32°$, ice doesn't form.
57. (a) $\frac{11}{10}$. For each mile per hour faster that the car travels, it takes $\frac{11}{10}$ ft longer to stop; **(b)** 6, 11.5, 22.5, 55.5, 72;
(c)
$$y = \frac{11x + 5}{10}$$

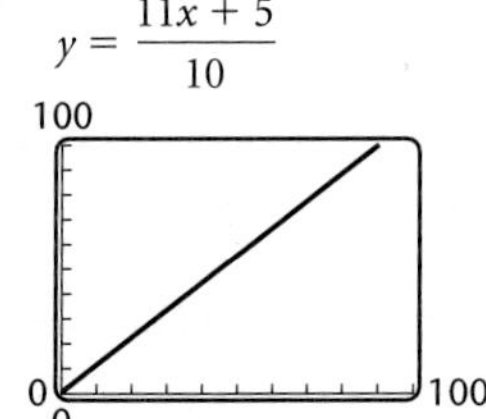

(d) $\{r \mid r > 0\}$, or $(0, \infty)$. If r is allowed to be 0, the function says that a stopped car travels $\frac{1}{2}$ ft before stopping.
59. $\dfrac{\text{Height of corn in feet}}{\text{Number of weeks after planting}}$
61. $C(t) = 60 + 40t$; $y = 60 + 40x$

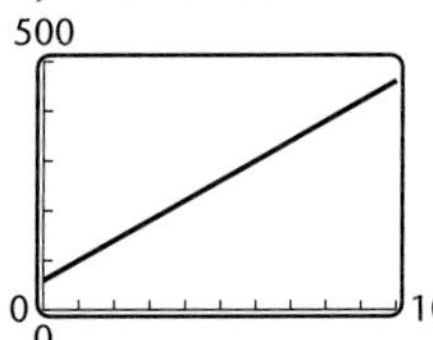

63. $C(x) = 800 + 3x$; $y = 800 + 3x$

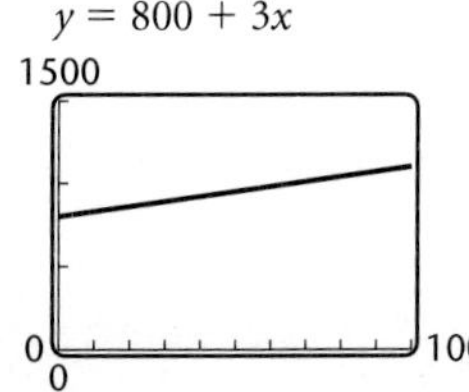

65. 16   **67.** 40   **69.** $a^2 + 3a$ **71.** ◈   **73.** -7.75
75. $C(F) = \frac{5}{9}(F - 32)$; $F(C) = \frac{9}{5}C + 32$ **77.** False
79. False   **81.** $f(x) = x + b$

1. Yes   **3.** No   **5.** No   **7.** Yes
9. (a) $y = 2.6x + 49.7$, 75.7; **(b)** $r = 0.9856$, a good fit
11. (a) $y = 645.7x + 9799.2$; \$14,319.10; \$15,610.50; \$18,193.30; **(b)** $r = 0.9994$, a good fit; **(c)** The president's assertion is about \$1000, or 6%, higher than the estimate from the model.
13. (a) $M = 0.2H + 156$; **(b)** 164, 169, 171, 173; **(c)** $r = 1$; the regression line fits the data perfectly and should be a good predictor.
15. (a) 22; **(b)** -22; **(c)** $4a^3 - 5a$; **(d)** $-4a^3 + 5a$ **17.** ◈
19. (a) $D = 0.000008d + 11.9$, $r = 0.0282$;
(b) The model does not fit the data and would not be a good predictor.

1. $\sqrt{10}$, 3.162   **3.** $\sqrt{45}$, 6.708   **5.** $\sqrt{14 + 6\sqrt{2}}$, 4.742
7. 6.5   **9.** Yes
11. The distances between the points are $\sqrt{10}$, 10, and $\sqrt{90}$. Since $(\sqrt{10})^2 + (\sqrt{90})^2 = 10^2$, the three points are vertices of a right triangle.   **13.** $\left(\frac{9}{2}, \frac{15}{2}\right)$ **15.** $\left(\frac{15}{2}, 2\right)$
17. $\left(\dfrac{\sqrt{3} - \sqrt{6}}{2}, -\dfrac{\sqrt{5}}{2}\right)$
19.

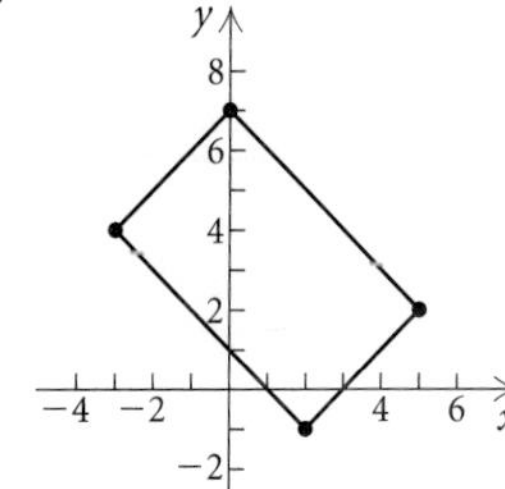

$\left(-\frac{1}{2}, \frac{3}{2}\right)$, $\left(\frac{7}{2}, \frac{1}{2}\right)$, $\left(\frac{5}{2}, \frac{9}{2}\right)$, $\left(-\frac{3}{2}, \frac{11}{2}\right)$; no
21. $\left(\dfrac{\sqrt{7} + \sqrt{2}}{2}, -\dfrac{1}{2}\right)$
23. Square the window; for example, $[-12, 9, -4, 10]$
25. $(x - 2)^2 + (y - 3)^2 = \frac{25}{9}$
27. $(x + 1)^2 + (y - 4)^2 = 25$
29. $(x - 2)^2 + (y - 1)^2 = 169$
31. $(x + 2)^2 + (y - 3)^2 = 4$
33. $(0, 0)$; 2; $x^2 + y^2 = 4$
$$y_1 = \sqrt{4 - x^2}, \quad y_2 = -\sqrt{4 - x^2}$$

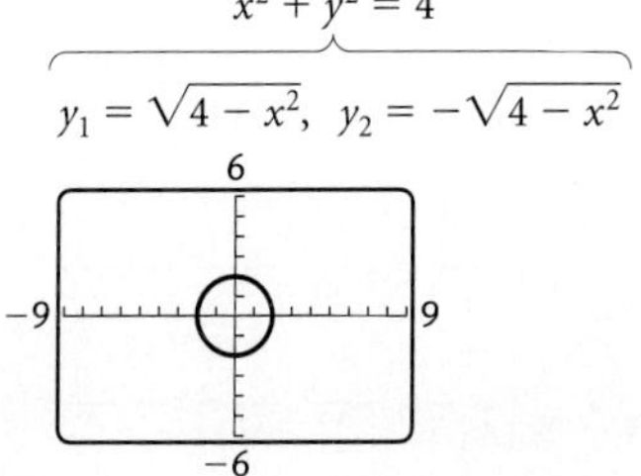

35. $(0, 3)$; 4;

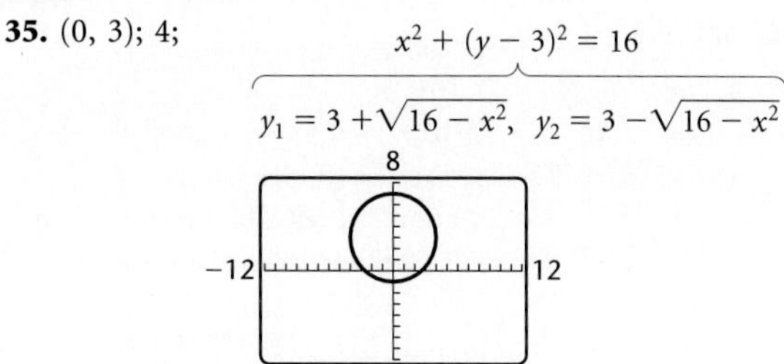

$$x^2 + (y - 3)^2 = 16$$
$$y_1 = 3 + \sqrt{16 - x^2}, \quad y_2 = 3 - \sqrt{16 - x^2}$$

37. $(1, 5)$; 6;

$$(x - 1)^2 + (y - 5)^2 = 36$$
$$y_1 = 5 + \sqrt{36 - (x - 1)^2}, \quad y_2 = 5 - \sqrt{36 - (x - 1)^2}$$

39. $(-4, -5)$; 3;

$$(x + 4)^2 + (y + 5)^2 = 9$$
$$y_1 = -5 + \sqrt{9 - (x + 4)^2}, \quad y_2 = -5 - \sqrt{9 - (x + 4)^2}$$

41. -35 **43.** $2ah + h^2$

45. $\sqrt{h^2 + h + 2a - 2\sqrt{a^2 + ah}}$,
$\left(\dfrac{2a + h}{2}, \dfrac{\sqrt{a} + \sqrt{a + h}}{2} \right)$

47. ◈   **49.** $(x - 2)^2 + (y + 7)^2 = 36$

51. $a_1 = 2.7$ ft, $a_2 = 37.3$ ft   **53.** Yes   **55.** Yes

57. $(0, 4)$ **59.** Let $P_1 = (x_1, y_1)$, $P_2 = (x_2, y_2)$, and
$M = \left(\dfrac{x_1 + x_2}{2}, \dfrac{y_1 + y_2}{2} \right)$. Let $d(AB)$ denote the distance
from point A to point B.

(a) $d(P_1 M) = \sqrt{ \left(\dfrac{x_1 + x_2}{2} - x_1 \right)^2 + \left(\dfrac{y_1 + y_2}{2} - y_1 \right)^2 }$

$= \dfrac{1}{2} \sqrt{(x_2 - x_1)^2 + (y_2 - y_1)^2}$;

$d(P_2 M) = \sqrt{ \left(\dfrac{x_1 + x_2}{2} - x_2 \right)^2 + \left(\dfrac{y_1 + y_2}{2} - y_2 \right)^2 }$

$= \dfrac{1}{2} \sqrt{(x_1 - x_2)^2 + (y_1 - y_2)^2}$

$= \dfrac{1}{2} \sqrt{(x_2 - x_1)^2 + (y_2 - y_1)^2} = d(P_1 M)$.

(b) $d(P_1 M) + d(P_2 M) = \dfrac{1}{2} \sqrt{(x_2 - x_1)^2 + (y_2 - y_1)^2}$

$+ \dfrac{1}{2} \sqrt{(x_2 - x_1)^2 + (y_2 - y_1)^2}$

$= \sqrt{(x_2 - x_1)^2 + (y_2 - y_1)^2}$

$= d(P_1 P_2)$.

Exercise Set 1.6

1. x-axis, no; y-axis, yes; origin, no
3. x-axis, yes; y-axis, no; origin, no
5. x-axis, no; y-axis, no; origin, yes
7. x-axis, no; y-axis, yes; origin, no
9. x-axis, no; y-axis, no; origin, no
11. x-axis, no; y-axis, yes; origin, no
13. x-axis, no; y-axis, no; origin, yes
15. x-axis, no; y-axis, no; origin, yes
17. x-axis, yes; y-axis, yes; origin, yes
19. x-axis, no; y-axis, yes; origin, no
21. x-axis, yes; y-axis, yes; origin, yes
23. x-axis, no; y-axis, no; origin, no
25. x-axis, no; y-axis, no; origin, yes   **27.** Even   **29.** Odd
31. Neither   **33.** Odd   **35.** Even   **37.** Odd   **39.** Even
41. Odd   **43.** Neither   **45.** Odd   **47.** Neither   **49.** Even
51. $3a^2 - 32a + 85$ **53.** $3a^2 + 22a + 40$ **55.** ◈
57. Odd   **59.** Neither   **61.** Odd
63. x-axis, yes; y-axis, yes; origin, yes
65. x-axis, yes; y-axis, no; origin, no
67.

$$y_1 = \tfrac{1}{2}x^4 - 5x^3 + 2,$$
$$y_2 = \tfrac{1}{2}x^4 + 5x^3 + 2$$

The graph of $f(x)$ is the reflection of the graph of $g(x)$ across
the y-axis.

Exercise Set 1.7

1. Start with the graph of $g(x) = x^2$. Shift it up 1 unit.

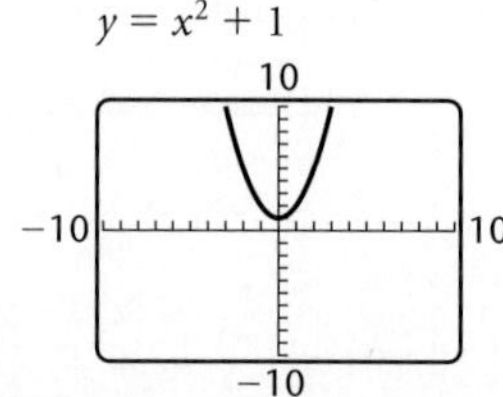

$$y = x^2 + 1$$

3. Start with the graph of $g(x) = 1/x$. Shift it up 3 units.

$$y = \frac{1}{x} + 3$$

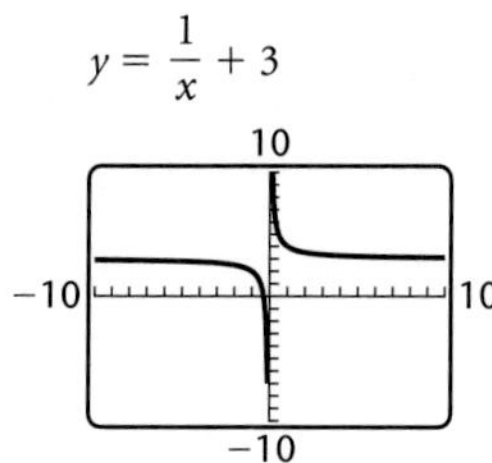

5. Start with the graph of $g(x) = \sqrt{x}$. Shift it down 5 units.

$$y = \sqrt{x} - 5$$

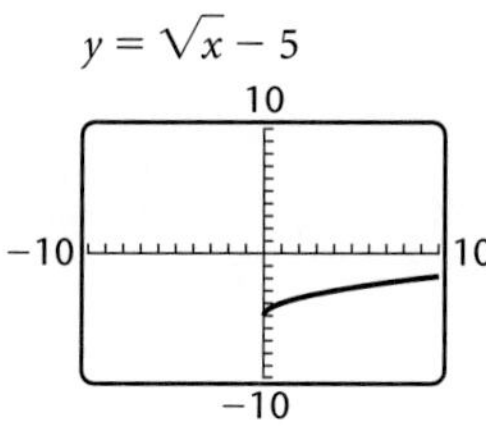

7. Start with the graph of $g(x) = x^2$. Reflect it across the x-axis.

$$y = -x^2$$

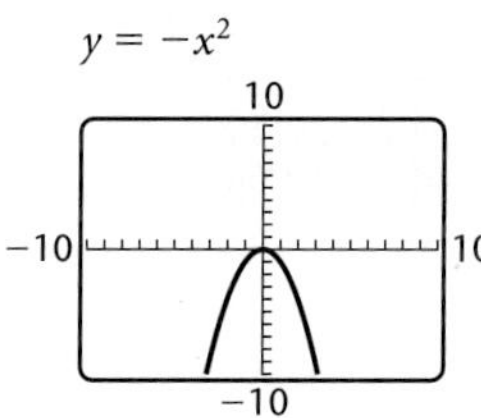

9. Start with the graph of $g(x) = |x|$. Shift it right 3 units.

$$y = |x - 3|$$

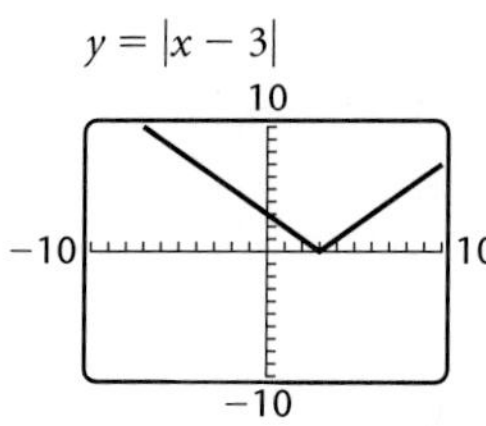

11. Start with the graph of $f(x) = x^3$. Shift it left 5 units.

$$y = (x + 5)^3$$

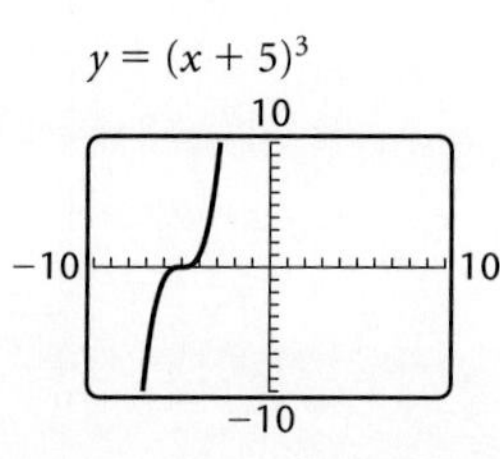

13. Start with the graph of $f(x) = x^2$. Shift it left 1 unit.

$$y = (x + 1)^2$$

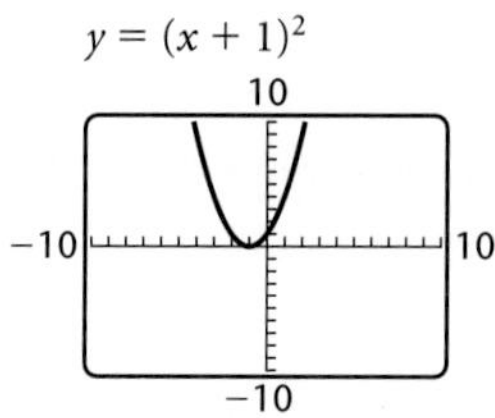

15. Start with the graph of $g(x) = \sqrt{x}$. Shrink it vertically by multiplying each y-coordinate by $\frac{1}{2}$.

$$y = \frac{1}{2}\sqrt{x}$$

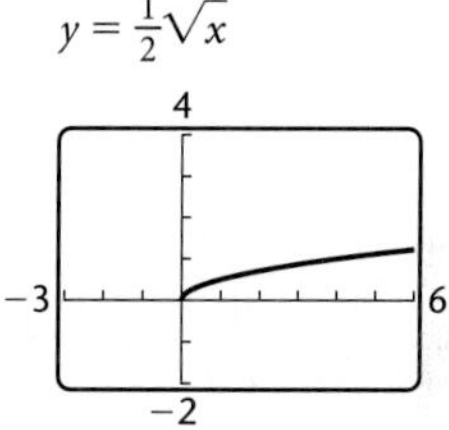

17. Start with the graph of $f(x) = 1/x$. Stretch it vertically by multiplying each y-coordinate by 2.

$$y = \frac{2}{x}$$

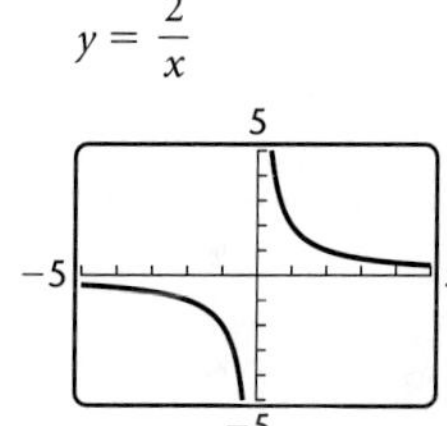

19. Start with the graph of $g(x) = \sqrt{x}$. Stretch it vertically by multiplying each y-coordinate by 3, then shift it down 5 units.

$$y = 3\sqrt{x} - 5$$

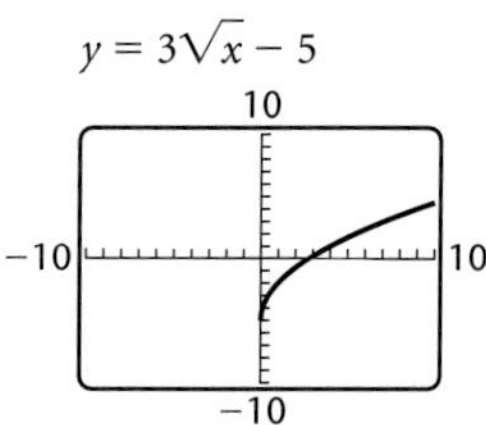

21. Start with the graph of $g(x) = x^2$. Shrink it vertically by multiplying each y-coordinate by $\frac{1}{2}$, then shift it right 3 units.

$$y = \frac{1}{2}(x - 3)^2$$

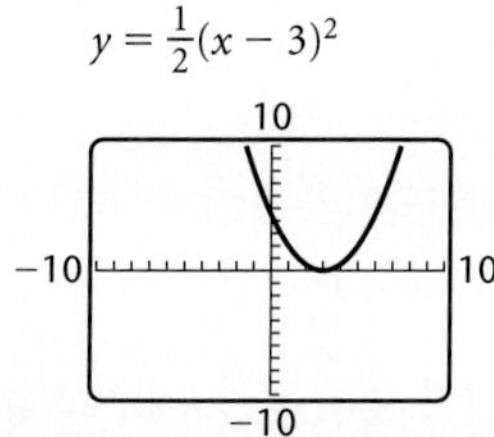

23. Start with the graph of $f(x) = |x|$. Stretch it horizontally by multiplying each x-coordinate by 3, then shift it down 4 units.

$$y = \left|\tfrac{1}{3}x\right| - 4$$

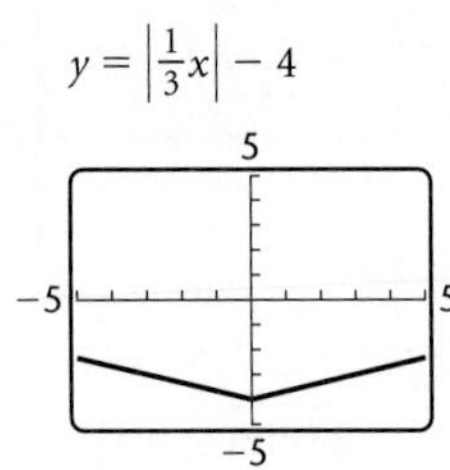

25. Start with the graph of $g(x) = x^2$. Shift it left 5 units and down 4 units.

$$y = (x + 5)^2 - 4$$

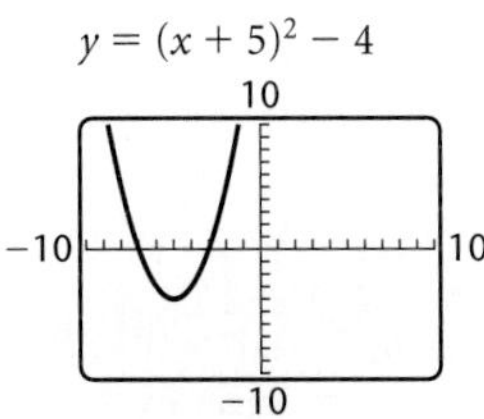

27. Start with the graph of $g(x) = x^2$. Shift it right 5 units, shrink it vertically by multiplying each y-coordinate by $\tfrac{1}{4}$, and reflect it across the x-axis.

$$y = -\tfrac{1}{4}(x - 5)^2$$

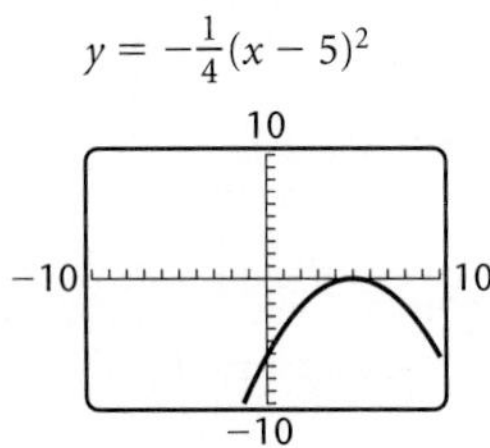

29. Start with the graph of $g(x) = 1/x$. Shift it left 3 units and up 2 units. Use DOT mode.

$$y = \frac{1}{x + 3} + 2$$

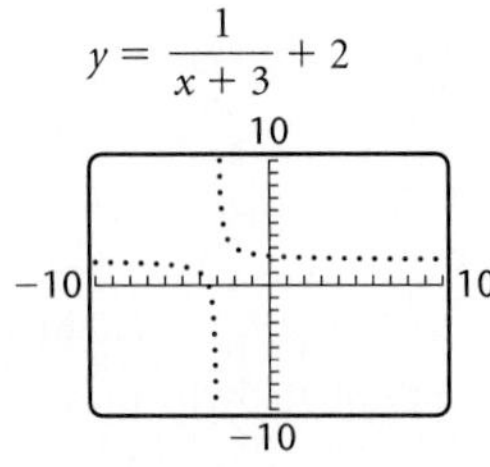

31. Start with the graph of $g(x) = x^3$. Shift it left 4 units and up 3 units.

$$y = (x + 4)^3 + 3$$

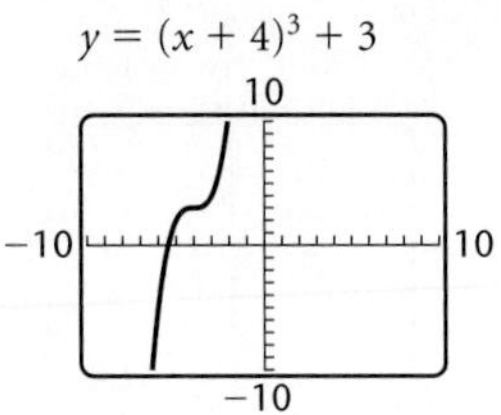

33. Start with the graph of $g(x) = \sqrt{x}$. Reflect it across the y-axis; then shift it up 5 units.

$$y = \sqrt{-x} + 5$$

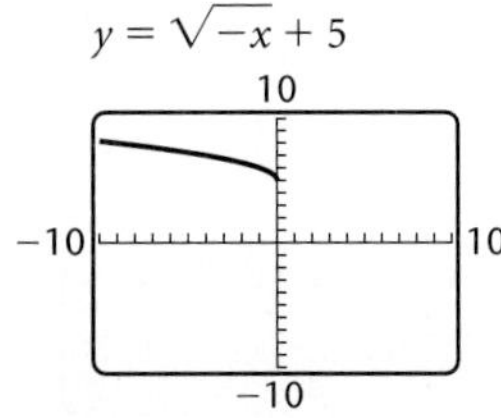

35. Start with the graph of $g(x) = x^2$. Shift it left 4 units, stretch it vertically by multiplying each y-coordinate by 3, and then shift it down 3 units.

$$y = 3(x + 4)^2 - 3$$

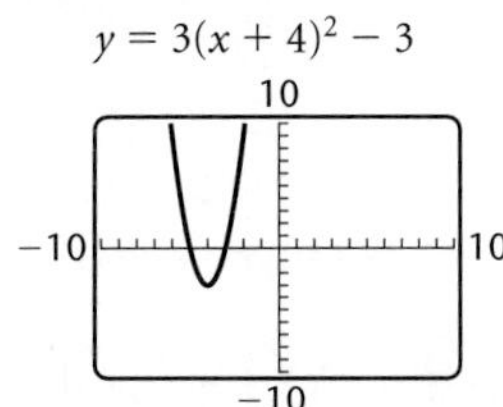

37. Start with the graph of $g(x) = x^3$. Shift it right 3 units, shrink it vertically by multiplying each y-coordinate by 0.43, and then shift it up 2.4 units.

$$y = 0.43(x - 3)^3 + 2.4$$

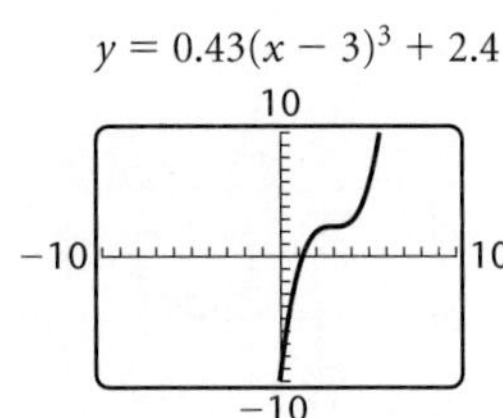

39. Start with the graph of $f(x) = x^2$. Shift it left 5.2 units, stretch it vertically by multiplying each y-coordinate by 2.8, and then shift it up 1.1 units.

$$y = 2.8(x + 5.2)^2 + 1.1$$

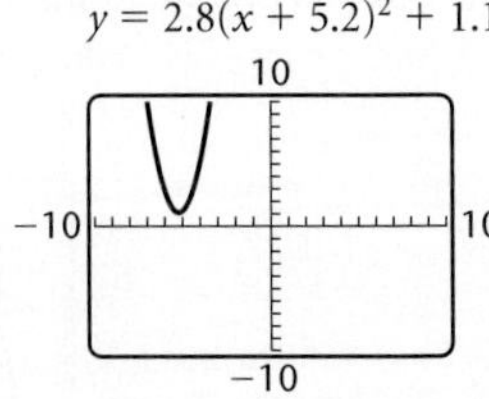

41. $f(x) = -(x - 8)^2$ **43.** $f(x) = |x + 7| + 2$

45. $f(x) = \dfrac{1}{2x} - 3$ **47.** $f(x) = -(x - 3)^2 + 4$

49. $f(x) = \sqrt{-(x + 2)} - 1$ **51.** $f(x) = 0.83(x + 4)^3$

53.

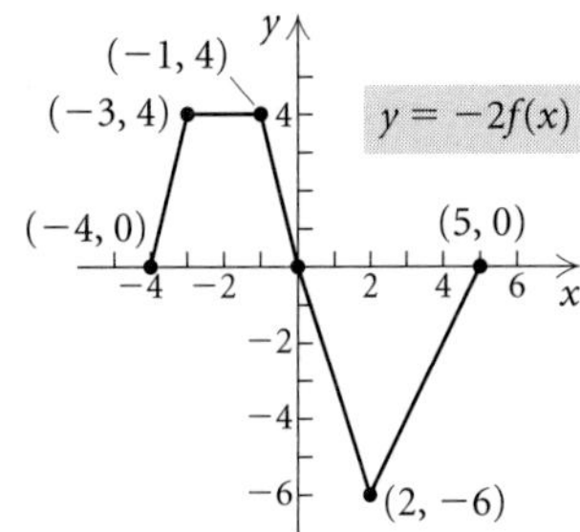

55.

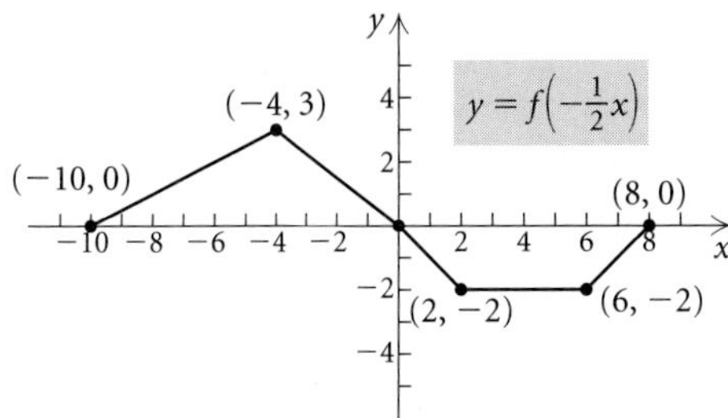

57.

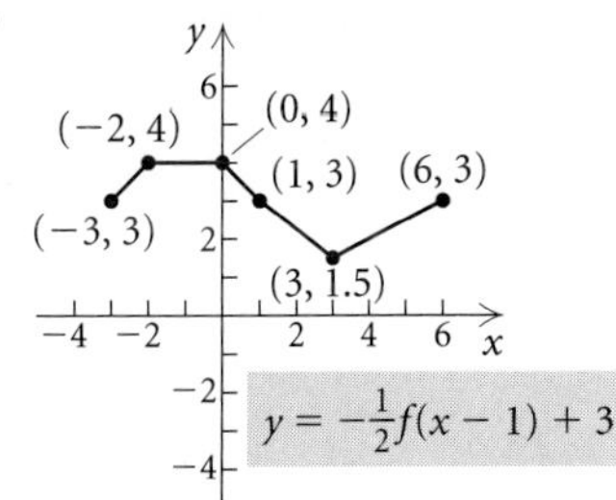

59. (f)   **61.** (f)   **63.** (d)   **65.** (c)   **67.** Use a grapher.
69. Use a grapher.
71. $f(-x) = 2(-x)^4 - 35(-x)^3 + 3(-x) - 5 = 2x^4 + 35x^3 - 3x - 5 = g(x)$ **73.** $g(x) = x^3 - 3x^2 + 2$
75. $k(x) = (x + 1)^3 - 3(x + 1)^2$ **77.** 6   **79.** 16
81. ◈   **83.** $g(x) = 7 - \sqrt{x - 4}$
85. Start with the graph of $g(x) = \text{INT}(x)$. Shift it right $\frac{1}{2}$ unit. Domain: all real numbers; range: all integers.
87. Start with the graph of $g(x) = \text{INT}(x)$. Shift it left 2 units, stretch it vertically by multiplying each y-coordinate by 3, and shift it up 1 unit. Domain: all real numbers; range: all integers that are 1 more than a multiple of 3.
89. Start with the graph of $f(x) = 1/x$. Reflect across the x-axis the portion of the graph for which the y-coordinates are negative. Domain: all real numbers except 0; range: $(0, \infty)$
91. Start with the graph of $g(x) = |x|$. Shift it down 5 units;

then reflect across the x-axis the portion of the graph for which the y-coordinates are negative. Domain: all real numbers; range: $[0, \infty)$ **93.** (3, 8); (3, 6); $\left(\frac{3}{2}, 4\right)$

EXERCISE SET 1.8

1. 33   **3.** 78   **5.** 5   **7.** 2   **9.** 0   **11.** 0
13. **(a)** Domain of f, all real numbers; domain of g, $[-4, \infty)$; domain of $f + g$, $f - g$, and fg, $[-4, \infty)$; domain of ff, all real numbers; domain of f/g, $(-4, \infty)$; domain of g/f, $[-4, 3) \cup (3, \infty)$; domain of $f \circ g$, $[-4, \infty)$; domain of $g \circ f$, $[-1, \infty)$; **(b)** $(f + g)(x) = x - 3 + \sqrt{x + 4}$; $(f - g)(x) = x - 3 - \sqrt{x + 4}$; $(fg)(x) = (x - 3)\sqrt{x + 4}$; $(ff)(x) = x^2 - 6x + 9$; $(f/g)(x) = (x - 3)/\sqrt{x + 4}$; $(g/f)(x) = \sqrt{x + 4}/(x - 3)$; $(f \circ g)(x) = \sqrt{x + 4} - 3$; $(g \circ f)(x) = \sqrt{x + 1}$
15. **(a)** Domain of f, g, $f + g$, $f - g$, fg, and ff, all real numbers; domain of f/g, all real numbers except -3 and $\frac{1}{2}$; domain of g/f, all real numbers except 0; domain of $f \circ g$ and $g \circ f$, all real numbers;
(b) $(f + g)(x) = x^3 + 2x^2 + 5x - 3$; $(f - g)(x) = x^3 - 2x^2 - 5x + 3$; $(fg)(x) = 2x^5 + 5x^4 - 3x^3$; $(ff)(x) = x^6$; $(f/g)(x) = x^3/(2x^2 + 5x - 3)$; $(g/f)(x) = (2x^2 + 5x - 3)/x^3$; $(f \circ g)(x) = (2x^2 + 5x - 3)^3$; $(g \circ f)(x) = 2x^6 + 5x^3 - 3$
17. $(f + g)(x) = 2x^2 - 2$; $(f - g)(x) = -6$; $(fg)(x) = x^4 - 2x^2 - 8$; $(f/g)(x) = (x^2 - 4)/(x^2 + 2)$; $(f \circ g)(x) = x^4 + 4x^2$
19. $(f + g)(x) = \sqrt{x - 7} + x^2 - 25$; $(f - g)(x) = \sqrt{x - 7} - x^2 + 25$; $(fg)(x) = \sqrt{x - 7}\,(x^2 - 25)$; $(f/g)(x) = \sqrt{x - 7}/(x^2 - 25)$; $(f \circ g)(x) = \sqrt{x^2 - 32}$
21. Domain of F: $[0, 9]$; domain of G: $[3, 10]$; domain of $F + G$: $[3, 9]$
23. $\{x \mid 3 \leq x \leq 9 \text{ and } x \neq 6 \text{ and } x \neq 8\}$, or $[3, 6) \cup (6, 8) \cup (8, 9]$
25.

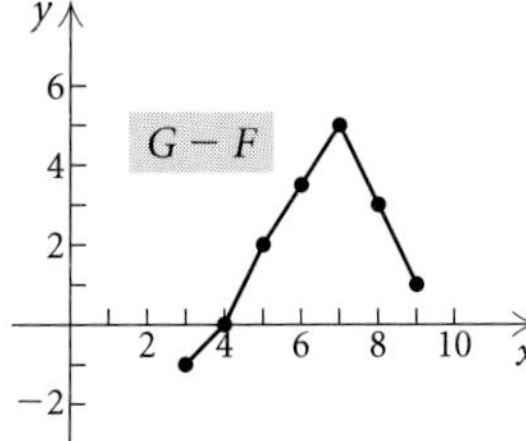

27. $(f \circ g)(x) = (g \circ f)(x) = x$
29. $(f \circ g)(x) = (g \circ f)(x) = x$
31. $(f \circ g)(x) = 20$; $(g \circ f)(x) = 0.05$
33. $(f \circ g)(x) = |x|$; $(g \circ f)(x) = x$
35. $(f \circ g)(x) = (g \circ f)(x) = x$
37. $(f \circ g)(x) = x^3 - 2x^2 - 4x + 6$; $(g \circ f)(x) = x^3 - 5x^2 + 3x + 8$

39. $f(x) = x^5$; $g(x) = 4 + 3x$

41. $f(x) = \dfrac{1}{x}$; $g(x) = (x - 2)^4$

43. $f(x) = \dfrac{x - 1}{x + 1}$; $g(x) = x^3$

45. $f(x) = x^6$; $g(x) = \dfrac{2 + x^3}{2 - x^3}$

47. $f(x) = \sqrt{x}$; $g(x) = \dfrac{x - 5}{x + 2}$

49. $f(x) = x^3 - 5x^2 + 3x - 1$; $g(x) = x + 2$
51. (a) $P(x) = -0.4x^2 + 57x - 13$; **(b)** $R(100) = 2000$;
$C(100) = 313$; $P(100) = 1687$; **(c)** Use a grapher.
53. (a) $r(t) = 3t$; **(b)** $A(r) = \pi r^2$; **(c)** $(A \circ r)(t) = 9\pi t^2$; the
area of the ripple in terms of time t **55.** $[-1, 2]$
57. -21 **59.** 
61. 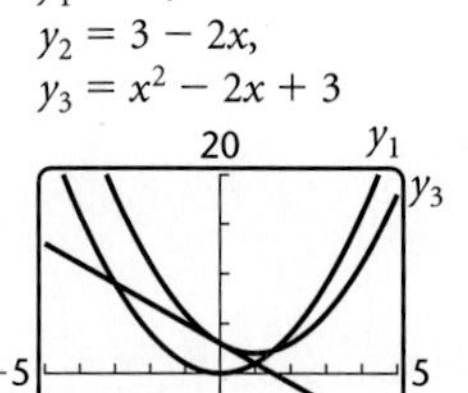   **63.** 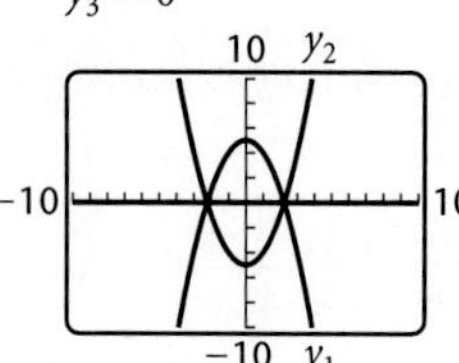

$y_1 = x^2$,
$y_2 = 3 - 2x$,
$y_3 = x^2 - 2x + 3$

$y_1 = 5 - x^2$,
$y_2 = x^2 - 5$,
$y_3 = 0$

65. Domain of $F \circ G$, $[3, 10]$; domain of $G \circ G$, $[5, 8]$;

domain of $F \circ F$, $[0, 6] \cup [8, 9]$ **67.** -2 **69.** $\dfrac{x - 1}{x}$; x

71. True   **73.** True

75. $E(-x) = \dfrac{f(-x) + f(-(-x))}{2} = \dfrac{f(-x) + f(x)}{2} = E(x)$

77. (a) $E(x) + O(x) = \dfrac{f(x) + f(-x)}{2} + \dfrac{f(x) - f(-x)}{2} =$

$\dfrac{2f(x)}{2} = f(x)$; **(b)** $f(x) = \dfrac{-22x^2 + \sqrt{x} + \sqrt{-x} - 20}{2} +$

$\dfrac{8x^3 + \sqrt{x} - \sqrt{-x}}{2}$

REVIEW EXERCISES, CHAPTER 1

1. [1.1] Not a function; domain: $\{3, 5, 7\}$; range: $\{1, 3, 5, 7\}$
2. [1.1] Function; domain: $\{-2, 0, 1, 2, 7\}$; range: $\{-7, -4, -2, 2, 7\}$ **3.** [1.1] No   **4.** [1.1] Yes   **5.** [1.1] No
6. [1.1] Yes   **7.** [1.1] All real numbers   **8.** [1.1] $\left(-\infty, \frac{7}{3}\right]$
9. [1.1] $\{x \mid x \neq 5 \text{ and } x \neq 1\}$
10. [1.1] $\{x \mid x \neq -4 \text{ and } x \neq 4\}$
11. [1.1] Domain: $[-5, 3]$; range: $[0, 4]$
12. [1.1] Domain: all real numbers; range: $[-5, \infty)$
13. [1.1] Domain: all real numbers; range: all real numbers
14. [1.1] Domain: all real numbers; range: $[-19, \infty)$
15. [1.1] **(a)** -3; **(b)** 9; **(c)** $a^2 - 3a - 1$; **(d)** $2x + h - 1$
16. [1.1] No   **17.** [1.1] Yes   **18.** [1.1] No   **19.** [1.1] Yes

20. [1.2] 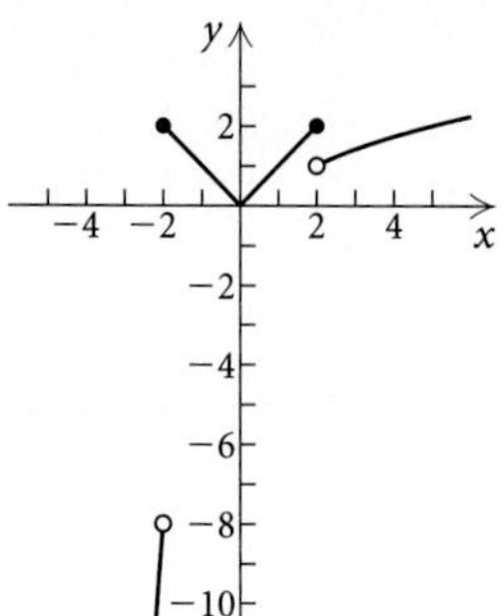

21. [1.2] 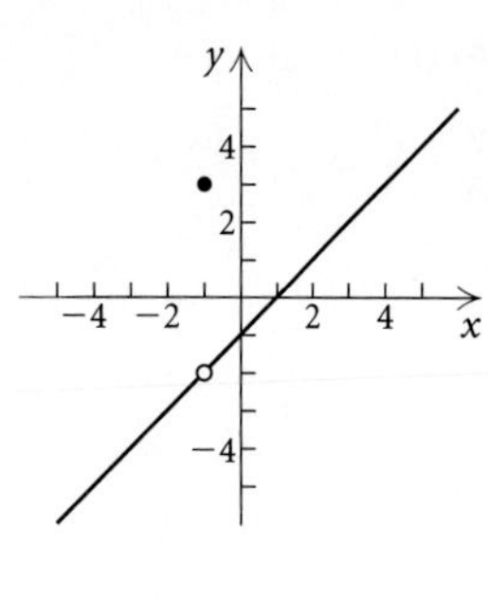

22. [1.2] 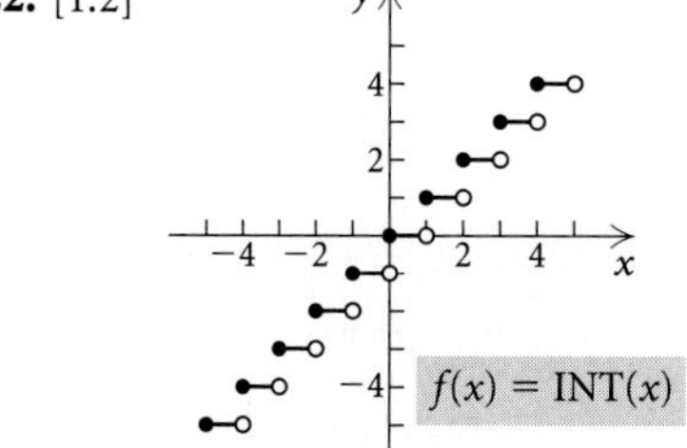

23. [1.2] 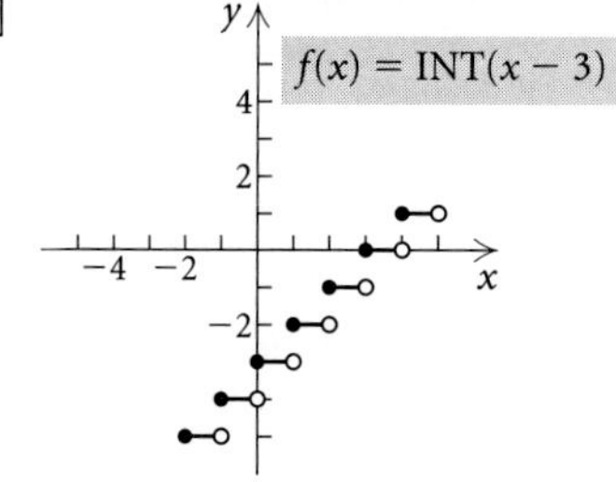

24. [1.1] **(a)** The domain of y_1 does not include the intervals
$(-\infty, -5.1]$ and $[3.1, \infty)$. The domain of y_2 appears to be all
real numbers. The range of y_1 appears to be $[0, 4]$. The
range of y_2 appears to be $[-5, 5]$. **(b)** The largest output of
y_1 appears to be 4, and the largest output of y_2 to be 5. The
smallest output of y_1 seems to be 0, and the smallest output
of y_2 to be -5.
25. [1.2] **(a)** -7.029; **(b)** increasing: $(-\infty, -0.860]$, $[6.339, \infty)$;
decreasing: $[-0.860, 6.339]$; **(c)** relative maximum:
$(-0.860, 710.094)$; relative minimum: $(6.339, 504.836)$
26. [1.2] **(a)** -5.732, -2.268; **(b)** increasing: $(-\infty, -4]$;
decreasing: $[-4, \infty)$; **(c)** relative maximum: $(-4, 15)$
27. [1.2] **(a)** -5, 0, 3; **(b)** increasing: $[-3.589, 2.089]$;
decreasing: $[-5, -3.589]$, $[2.089, 3]$; **(c)** relative maximum:
$(2.089, 4.247)$; relative minimum: $(-3.589, -8.755)$
28. [1.2] $A(x) = 2x\sqrt{4 - x^2}$

29. [1.2] **(a)** $A(x) = x^2 + \dfrac{432}{x}$; **(b)** $(0, \infty)$;

(c)
$$y = x^2 + \frac{432}{x}$$

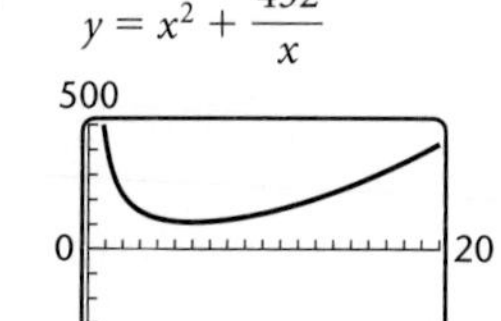

(d) $x = 6$ in.; $y = 3$ in.

30. [1.3] **(a)** Yes; **(b)** no; **(c)** no, strictly speaking, but data might be modeled by a linear regression function.
31. [1.3] **(a)** Yes; **(b)** yes; **(c)** yes   **32.** [1.3] -2; $(0, -7)$
33. [1.3] $y + 4 = -\frac{2}{3}(x - 0)$ **34.** [1.3] $y + 1 = 3(x + 2)$
35. [1.3] $y - 1 = \frac{1}{3}(x - 4)$, or $y + 1 = \frac{1}{3}(x + 2)$
36. [1.3] Parallel   **37.** [1.3] Neither
38. [1.3] Perpendicular   **39.** [1.3] $y = -\frac{2}{3}x - \frac{1}{3}$
40. [1.3] $y = \frac{3}{2}x - \frac{5}{2}$
41. [1.3] $C(t) = 25 + 20t$; $y = 25 + 20x$ $\$145$

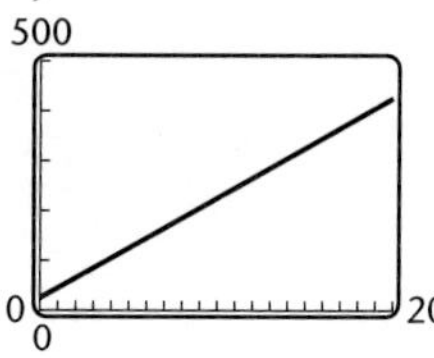

42. [1.3] **(a)** 70°C, 220°C, 10,020°C;
(b)
$$y = 10x + 20$$

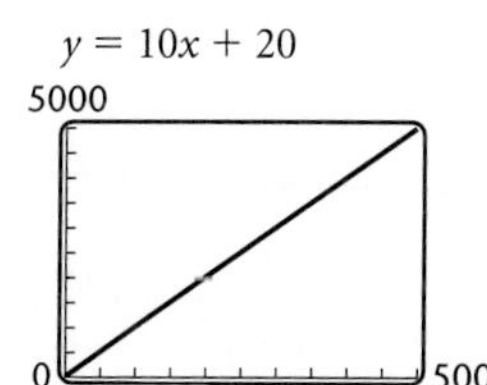

(c) [0, 5600]
43. [1.4] **(a)** $y = 2.4x + 17.5$, $\$36.7$ billion, $\$41.5$ billion, $\$65.5$ billion; **(b)** $r = 0.9938$, a good fit
44. [1.5] $\sqrt{34} \approx 5.831$ **45.** [1.5] $\left(\frac{1}{2}, \frac{11}{2}\right)$
46. [1.5] $(x + 2)^2 + (y - 6)^2 = 13$ **47.** [1.5] $(-1, 3)$; 4
48. [1.5] $(x - 2)^2 + (y - 4)^2 = 26$
49. [1.6] x-axis, yes; y-axis, yes; origin, yes
50. [1.6] x-axis, yes; y-axis, yes; origin, yes
51. [1.6] x-axis, no; y-axis, no; origin, no
52. [1.6] x-axis, no; y-axis, yes; origin, no
53. [1.6] x-axis, no; y-axis, no; origin, yes
54. [1.6] x-axis, no; y-axis, yes; origin, no   **55.** [1.6] Even
56. [1.6] Even   **57.** [1.6] Odd   **58.** [1.6] Even
59. [1.6] Even   **60.** [1.6] Neither   **61.** [1.6] Odd
62. [1.6] Even   **63.** [1.6] Even   **64.** [1.6] Odd
65. [1.7] $f(x) = (x + 3)^2$
66. [1.7] $f(x) = -\sqrt{x - 3} + 4$
67. [1.7] $f(x) = 2|x - 3|$

68. [1.7]

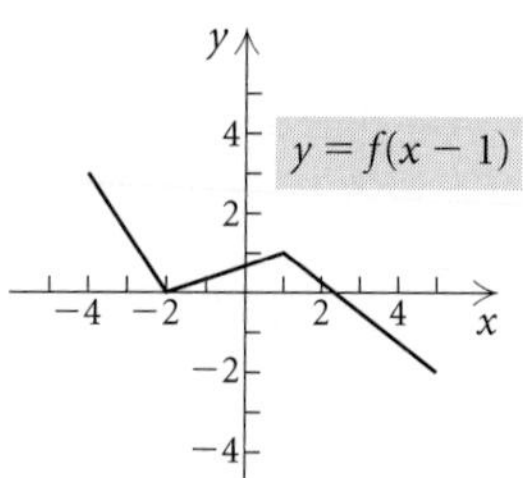

69. [1.7]

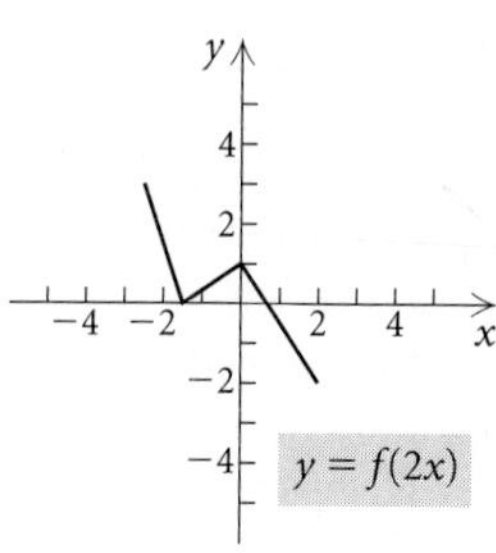

70. [1.7]

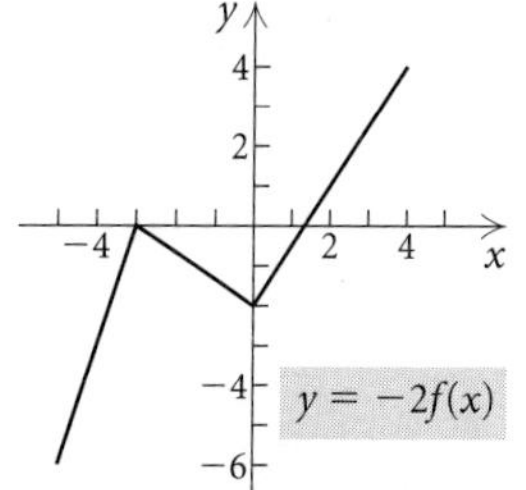

71. [1.7]

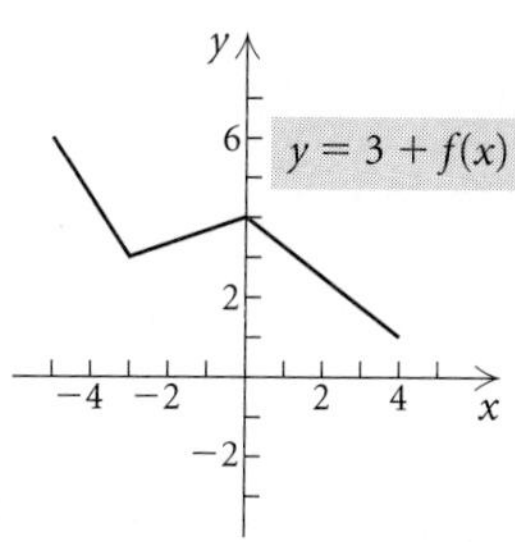

72. [1.8] **(a)** Domain of f, $\{x \mid x \neq 0\}$; domain of g, all real numbers; domain of $f + g$, $f - g$, and fg, $\{x \mid x \neq 0\}$;

domain of f/g, $\left\{x \mid x \neq 0 \text{ and } x \neq \frac{3}{2}\right\}$; domain of $f \circ g$,

$\left\{x \mid x \neq \frac{3}{2}\right\}$; domain of $g \circ f$, $\{x \mid x \neq 0\}$;

(b) $(f + g)(x) = \frac{4}{x^2} + 3 - 2x$; $(f - g)(x) = \frac{4}{x^2} - 3 + 2x$;

$fg(x) = \frac{12}{x^2} - \frac{8}{x}$; $(f/g)(x) = \frac{4}{x^2(3 - 2x)}$;

$(f \circ g)(x) = \frac{4}{(3 - 2x)^2}$; $(g \circ f)(x) = 3 - \frac{8}{x^2}$

73. [1.8] **(a)** Domain of f, g, $f + g$, $f - g$, and fg, all real numbers; domain of f/g, $\left\{x \mid x \neq \frac{1}{2}\right\}$; domain of $f \circ g$ and $g \circ f$, all real numbers; **(b)** $(f + g)(x) = 3x^2 + 6x - 1$; $(f - g)(x) = 3x^2 + 2x + 1$; $fg(x) = 6x^3 + 5x^2 - 4x$;

$(f/g)(x) = \frac{3x^2 + 4x}{2x - 1}$; $(f \circ g)(x) = 12x^2 - 4x - 1$;

$(g \circ f)(x) = 6x^2 + 8x - 1$
74. [1.8] $P(x) = -0.5x^2 + 105x - 6$
75. [1.8] $f(x) = \sqrt{x}$, $g(x) = 5x + 2$. Answers may vary.
76. [1.8] $f(x) = 4x^2 + 9$, $g(x) = 5x - 1$. Answers may vary.
77. [1.1], [1.7]  **(a)** $4x^3 - 2x + 9$;
(b) $4x^3 + 24x^2 + 46x + 35$; **(c)** $4x^3 - 2x + 42$. **(a)** Adds 2 to each function value; **(b)** adds 2 to each input before finding a function value; **(c)** adds the output for 2 to the output for x

78. [1.7] ◈ In the graph of $y = f(cx)$, the constant c stretches or shrinks the graph of $y = f(x)$ horizontally. The constant c in $y = cf(x)$ stretches or shrinks the graph of $y = f(x)$ vertically. For $y = f(cx)$, the x-coordinates of $y = f(x)$ are divided by c; for $y = cf(x)$, the y-coordinates of $y = f(x)$ are multiplied by c.

79. [1.7] ◈ **(a)** To draw the graph of y_2 from the graph of y_1, reflect across the x-axis the portions of the graph for which the y-coordinates are negative. **(b)** To draw the graph of y_2 from the graph of y_1, draw the portion of the graph of y_1 to the right of the y-axis; then draw its reflection across the y-axis.   **80.** [1.1] $\{x \mid x < 0\}$

81. [1.1] $\{x \mid x \neq -3 \text{ and } x \neq 0 \text{ and } x \neq 3\}$

82. [1.6] Let $f(x)$ and $g(x)$ be odd functions. Then by definition, $f(-x) = -f(x)$, or $f(x) = -f(-x)$, and $g(-x) = -g(x)$, or $g(x) = -g(-x)$. Thus, $(f + g)(x) = f(x) + g(x) = -f(-x) + [-g(-x)] = -[f(-x) + g(-x)] = -(f + g)(-x)$ and $f + g$ is odd.

83. [1.7] Reflect the graph of $y = f(x)$ across the x-axis and then across the y-axis.

Chapter 2

EXERCISE SET 2.1

1. (a) $-2, 3$; **(b)** none; **(c)** -6.25 at $x = 0.5$; **(d)** domain: all real numbers; range: $[-6.25, \infty)$

3. (a) $-5, -1, 1$; **(b)** 16.901 at $x = -3.431$; **(c)** -5.049 at $x = 0.097$; **(d)** domain: all real numbers; range: all real numbers

5. (a) $-1.831, -0.856, 3.188$; **(b)** 26.864 at $x = 1.703$; **(c)** -2.160 at $x = -1.370$; **(d)** domain: all real numbers; range: all real numbers

7. (a) $-4.796, 0, 4.796$; **(b)** 0.042 at $x = -2.769$; **(c)** -0.042 at $x = 2.769$; **(d)** domain: all real numbers; range: all real numbers

9.

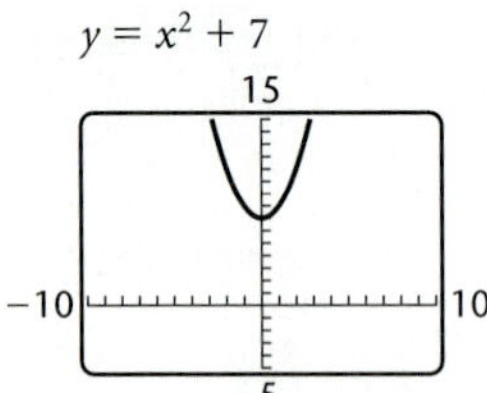

11.

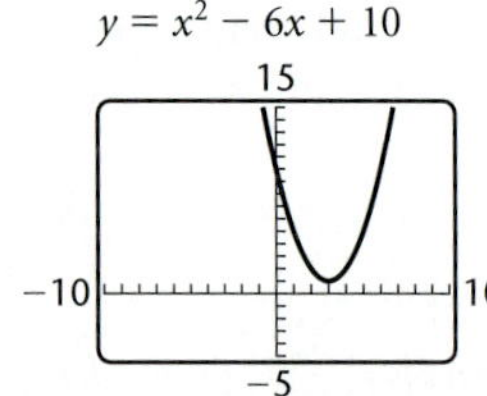

13.

15. $\sqrt{17}i$ **17.** $7i$ **19.** $-9i$ **21.** $6 - 2\sqrt{21}i$

23. $-2\sqrt{19} + 5\sqrt{5}i$ **25.** $-\sqrt{55}$ **27.** $\frac{5}{4}i$ **29.** $2 + 11i$

31. $5 - 12i$ **33.** $5 + 9i$ **35.** $5 + 4i$ **37.** $5 + 7i$

39. $13 - i$ **41.** $-11 + 16i$ **43.** $35 + 14i$ **45.** $-14 + 23i$

47. 41 **49.** $12 + 16i$ **51.** $-45 - 28i$

53. $\dfrac{-4\sqrt{3} + 10}{41} + \dfrac{5\sqrt{3} + 8}{41}i$ **55.** $-\dfrac{14}{13} + \dfrac{5}{13}i$

57. $\dfrac{15}{146} + \dfrac{33}{146}i$ **59.** $-\dfrac{1}{2} + \dfrac{1}{2}i$ **61.** $-\dfrac{1}{2} - \dfrac{13}{2}i$

63. $-i$ **65.** $-i$ **67.** 1 **69.** i **71.** 625

73. $(x + 4)^2$ **75.** $(x - 5)^2$ **77.** ◈   **79.** $\frac{12}{5} - \frac{1}{5}i$

81. $\frac{8}{29} + \frac{9}{29}i$ **83.** True   **85.** True   **87.** $a^2 + b^2$

EXERCISE SET 2.2

1. $\{2, 6\}$ **3.** $\left\{\frac{1}{7}, 6\right\}$ **5.** $\{-9, 9\}$ **7.** $\left\{-\frac{10}{3}, \frac{10}{3}\right\}$

9. $\{-\sqrt{10}i, \sqrt{10}i\}$ **11.** $\{-12, 6\}$ **13.** $\{2, 4\}$ **15.** $\{-7, 1\}$

17. $\{4 \pm \sqrt{7}\}$ **19.** $\{-4 \pm 3i\}$ **21.** $\left\{-2, \frac{1}{3}\right\}$

23. $\left\{\dfrac{3 \pm \sqrt{13}}{4}\right\}$ **25.** $\left\{\dfrac{5 \pm \sqrt{57}}{4}\right\}$ **27.** $\{-3, 5\}$

29. $\left\{-1, \dfrac{2}{5}\right\}$ **31.** $\left\{\dfrac{5 \pm \sqrt{7}}{3}\right\}$ **33.** $\left\{-\dfrac{1}{2} \pm \dfrac{\sqrt{7}}{2}i\right\}$

35. $\left\{\dfrac{4 \pm \sqrt{31}}{5}\right\}$ **37.** $\left\{\dfrac{5}{6} \pm \dfrac{\sqrt{23}}{6}i\right\}$ **39.** No   **41.** Yes

43. No   **45.** $\{-0.702, 5.702\}$ **47.** $\{-1.535, 0.869\}$

49. $\left\{\frac{5}{2}, 3\right\}$ **51.** $\{-8, 64\}$ **53.** $\{-2, -1, 1, 2\}$

55. $\left\{-\frac{3}{2}, -1, \frac{1}{2}, 1\right\}$ **57.** $\{16\}$

59. (a) $(4, -4)$; **(b)** $x = 4$; **(c)** minimum value: -4

61. (a) $\left(\frac{7}{2}, -\frac{1}{4}\right)$; **(b)** $x = \frac{7}{2}$; **(c)** minimum value: $-\frac{1}{4}$

63. (a) $\left(-\frac{3}{2}, \frac{7}{2}\right)$; **(b)** $x = -\frac{3}{2}$; **(c)** minimum value: $\frac{7}{2}$

65. (a) $\left(\frac{1}{2}, \frac{3}{2}\right)$; **(b)** $x = \frac{1}{2}$; **(c)** maximum value: $\frac{3}{2}$

67. (f)   **69.** (b)   **71.** (h)   **73.** (c)

75. (a) $(3, -4)$; **(b)** minimum value: -4; **(c)** $[-4, \infty)$; **(d)** increasing: $[3, \infty)$; decreasing: $(-\infty, 3]$

77. (a) $(-1, -18)$; **(b)** minimum value: -18; **(c)** $[-18, \infty)$; **(d)** increasing: $[-1, \infty)$; decreasing: $(-\infty, -1]$

79. (a) $\left(5, \frac{9}{2}\right)$; **(b)** maximum value: $\frac{9}{2}$; **(c)** $\left(-\infty, \frac{9}{2}\right]$; **(d)** increasing: $(-\infty, 5]$; decreasing: $[5, \infty)$

81. (a) $(-1, 2)$; **(b)** minimum value: 2; **(c)** $[2, \infty)$; **(d)** increasing: $[-1, \infty)$; decreasing: $(-\infty, -1]$

83. $\left\{-\frac{21}{4}, -1\right\}$ **85.** 6 mi   **87.** ◈

89. $\left\{\dfrac{-1 \pm \sqrt{4\sqrt{2} + 1}}{2}\right\}$ **91.** $\{3 \pm \sqrt{5}\}$

93. $\left\{\dfrac{-5 + \sqrt{61}}{18}\right\}$ **95.** $\{19\}$

97. $\left\{\dfrac{1}{2} - \dfrac{\sqrt{7}}{2}i, \dfrac{1}{2} + \dfrac{\sqrt{7}}{2}i, -2 - \sqrt{2}, -2 + \sqrt{2}\right\}$

99. (a) 2; (b) $\frac{11}{2}$ **101.** (a) 2; (b) $1 - i$ **103.** -236.25
105.

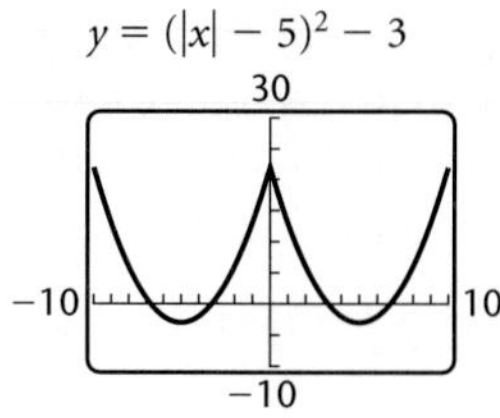

$$y = (|x| - 5)^2 - 3$$

EXERCISE SET 2.3

1. 0.866 sec   **3.** $3240   **5.** (a) 18.75%; (b) 10%
7. 3.5 in.   **9.** $b = h = 10$ cm   **11.** About 350.6 ft
13. $797 when $x = 40$ **15.** 4800 yd^2 **17.** (b)   **19.** (c)
21. (a)   **23.** (b)   **25.** (b)
27. (a) Yes; (b) no; (c) third differences are all -384;
(d) cubic; (e) $f(x) = -x^3 - x^2 + x + 5$; (f) $f(17) = -5180$;
$f(21) = -9676$
29. (a) Yes; (b) no; (c) fourth differences are all 1944;
(d) quartic; (e) $f(x) = x^4 - 10x^2 + x - 1$; $f(13) = 26{,}883$;
$x = -2.825, -1.500, 1.707,$ or 2.618 when $y = -19.94$
31. (a) $f(x) = -0.8630952381x^2 + 12.125x + 36.76785714$;
(b) 77.3 yr, 77.5 yr
33. (a) $f(x) = -2.775187775x^3 + 45.77200577x^2 -$
$190.2173752x + 2084.949495$;
(b) $y = -2.775187775x^3 + 45.77200577x^2 -$
     $190.2173752x + 2084.949495$

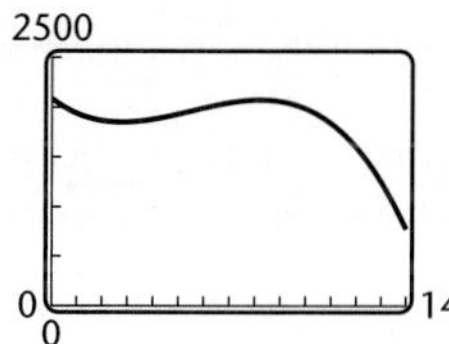

(c) $-$2694; (d) $f(x) = 1.586902681x^4 -$
$44.03465747x^3 + 400.5581051x^2 - 1316.011477x +$
3075.176768;
(e) $y = 1.586902681x^4 - 44.03465747x^3 +$
     $400.5581051x^2 - 1316.011477x + 3075.176768$

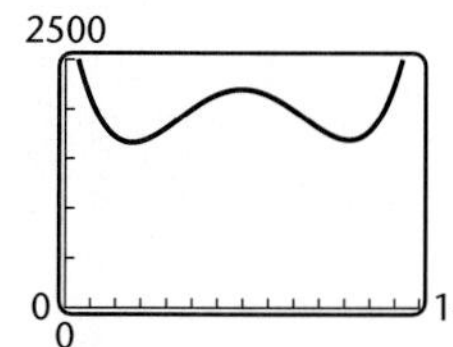

(f) $18,944
35. $x^4 + 5x^3 + 2x - 4$ **37.** $5x^4 - 17x^3 + 11x^2 - 16x + 3$
39. ◆   **41.** 7%
43. (a) $f(x) = 0.8586956522x^3 - 0.8452380952x^2 -$
$0.7546583851x + 31.10766046$; $f(-4) \approx -34.354$;
$f(4) \approx 69.522$; x is about -1.950 when y is 23;

(b) $g(x) = 1.558569182x^4 + 1.807389937x^3 -$
$15.7629717x^2 - 9.835017969x + 42.25965858$;
$g(-4) \approx 112.713$; $g(4) \approx 265.379$; x is about -3.301 or
-1.512 or 0.899 or 2.754 when $y = 23$
45. (a)

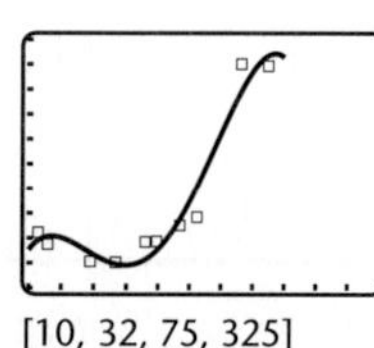

[10, 32, 75, 325]

(b) Linear seems inappropriate; quadratic, cubic, quartic
seem possible. (c) *Quadratic:* $y = 2.031904548x^2 -$
$59.04179598x + 527.2818092$; *cubic:* $y = -0.0273804413x^3 +$
$3.488800304x^2 - 83.76947384x + 660.2911579$; *quartic:*
$y = -0.0403607597x^4 + 2.842908639x^3 - 70.96865626x^2 +$
$749.2437238x - 2721.562456$
(d) Quadratic   Cubic

[10, 32, 75, 325]   [10, 32, 75, 325]

Quartic

[10, 32, 75, 325]

The quadratic and cubic equations seem to fit best. They
indicate that, for an advertising budget of about $16 million
or greater, box office revenue increases as the budget
increases. The quartic equation indicates that revenue
declines sharply as the advertising budget increases beyond
about $25.5 million.
(e) $584.74 million, $547.86 million, $-$49.72 million

EXERCISE SET 2.4

1. Yes; no; no   **3.** (a)
5. $P(x) = (x - 2)(x^2 + 8x + 15) + 0$
7. $P(x) = (x - 3)(x^2 + 9x + 26) + 48$
9. $P(x) = (x + 2)(x^2 - 2x + 4) - 16$
11. $P(x) = (x^2 + 4)(x^2 + 5) + 0$
13. $P(x) = (2x^2 - x + 1) \cdot$
$\left(\frac{5}{2}x^5 + \frac{5}{4}x^4 - \frac{5}{8}x^3 - \frac{39}{16}x^2 - \frac{29}{32}x + \frac{113}{64}\right) + \frac{171x - 305}{64}$
15. $Q(x) = 2x^3 + x^2 - 3x + 10$, $R(x) = -42$
17. $Q(x) = x^2 - 4x + 8$, $R(x) = -24$
19. $Q(x) = x^3 + x^2 + x + 1$, $R(x) = 0$
21. $Q(x) = 2x^3 + x^2 + \frac{7}{2}x + \frac{7}{4}$, $R(x) = -\frac{1}{8}$

23. $Q(x) = x^3 + x^2y + xy^2 + y^3$, $R(x) = 0$ **25.** 0; -60; 0
27. 5,935,988; -772 **29.** 0; 0; 65; $1 - 12\sqrt{2}$ **31.** Yes; no
33. No; no   **35.** $f(x) = (x - 1)(x + 2)(x + 3)$; 1, -2, -3
37. $f(x) = (x - 2)(x - 5)(x + 1)$; 2, 5, -1
39. $f(x) = (x - 2)(x - 3)(x + 4)$; 2, 3, -4
41. $f(x) = (x - 1)(x - 2)(x - 3)(x + 5)$; 1, 2, 3, -5
43. $y = \frac{3}{4}x + \frac{21}{4}$ **45.** $y = -\frac{2}{7}x + \frac{33}{7}$ **47.** ◈
49. (a) $x + 5$, $x + 3$, $x - 4$, $x - 6$, $x - 7$;
(b) $P(x) = -(x + 5)(x + 3)(x - 4)(x - 6)(x - 7)$;
(c) yes; two examples are $f(x) = c \cdot P(x)$; for any nonzero
constant c, and $g(x) = (x - a)P(x)$; **(d)** no
51. $\frac{14}{3}$ **53.** 0, 0.9636   **55.** $\{-1 \pm \sqrt{7}\}$ **57.** Yes
59. $x - 1 + 2i$, $R -9 + 4i$

EXERCISE SET 2.5

1. -3, multiplicity 2; 1, multiplicity 1
3. 0, multiplicity 3; 1, multiplicity 2; -4, multiplicity 1
5. $\pm\sqrt{3}$, ± 1, each has multiplicity 1
7. -3, -1, 1, each has multiplicity 1
9. $f(x) = x^3 - 6x^2 - x + 30$
11. $f(x) = x^3 + 3x^2 + 4x + 12$
13. $f(x) = x^3 - \sqrt{3}x^2 - 2x + 2\sqrt{3}$
15. $-3 - 4i$, $4 + \sqrt{5}$ **17.** $f(x) = x^3 - 4x^2 + 6x - 4$
19. $f(x) = x^3 - 5x^2 + 16x - 80$
21. $f(x) = x^4 + 4x^2 - 45$ **23.** i, 2, 3   **25.** $1 + 2i$, $1 - 2i$
27. ± 1 **29.** ± 1, ± 2, $\pm\frac{1}{3}$, $\pm\frac{1}{5}$, $\pm\frac{2}{3}$, $\pm\frac{2}{5}$, $\pm\frac{1}{15}$, $\pm\frac{2}{15}$
31. (a) Rational: -3, other: $\pm\sqrt{2}$; **(b)** -3, $-\sqrt{2}$, $\sqrt{2}$;
(c) $f(x) = (x + 3)(x + \sqrt{2})(x - \sqrt{2})$
33. (a) Rational: -2, 1; other: none; **(b)** -2, 1;
(c) $f(x) = (x + 2)(x - 1)^2$
35. (a) Rational: -1; other: $3 \pm 2\sqrt{2}i$; **(b)** -1, $3 \pm 2\sqrt{2}i$;
(c) $f(x) = (x + 1)(x - 3 - 2\sqrt{2}i)(x - 3 + 2\sqrt{2}i)$
37. (a) Rational: $-\frac{1}{5}$, 1; other: $\pm 2i$ **(b)** $-\frac{1}{5}$, 1, $\pm 2i$;
(c) $f(x) = (5x + 1)(x + 1)(x + 2i)(x - 2i)$
39. (a) Rational: -2, -1; other: $3 \pm \sqrt{13}$;
(b) -2, -1, $3 \pm \sqrt{13}$;
(c) $f(x) = (x + 2)(x + 1)(x - 3 - \sqrt{13})(x - 3 + \sqrt{13})$
41. (a) Rational: 2; other: $1 \pm \sqrt{3}$; **(b)** 2, $1 \pm \sqrt{3}$;
(c) $f(x) = (x - 2)(x - 1 - \sqrt{3})(x - 1 + \sqrt{3})$
43. (a) Rational: -2; other: $1 \pm \sqrt{3}i$; **(b)** -2, $1 \pm \sqrt{3}i$;
(c) $f(x) = (x + 2)(x - 1 - \sqrt{3}i)(x - 1 + \sqrt{3}i)$
45. (a) Rational: $\dfrac{1}{2}$; other: $\dfrac{1 \pm \sqrt{5}}{2}$; **(b)** $\dfrac{1}{2}$, $\dfrac{1 \pm \sqrt{5}}{2}$;
(c) $f(x) = \dfrac{1}{3}\left(x - \dfrac{1}{2}\right)\left(x - \dfrac{1 + \sqrt{5}}{2}\right)\left(x - \dfrac{1 - \sqrt{5}}{2}\right)$
47. None   **49.** None   **51.** None   **53.** -2, 1, 2
55. $f(-5) = -18$ and $f(-4) = 7$. By the intermediate value
theorem, since $f(-5)$ and $f(-4)$ have opposite signs, then
$f(x)$ has a zero between -5 and -4.
57. $\{-2\}$ **59.** $\frac{12}{13}$

61. ◈
63. (a) $\left\{-1, \frac{1}{2}, 3\right\}$; **(b)** $\left\{0, \frac{3}{2}, 4\right\}$; **(c)** $\left\{-3, -\frac{3}{2}, 1\right\}$; **(d)** $\left\{-\frac{1}{2}, \frac{1}{4}, \frac{3}{2}\right\}$
65. By the rational zeros theorem, only ± 1 and ± 5 can be
rational solutions of $x^2 - 5 = 0$. Since none of them is a
solution, the equation has no rational solutions. But $\sqrt{5}$ is a
solution, so $\sqrt{5}$ must be irrational.
67. -8, $-\frac{3}{2}$, 4, 7, 15

EXERCISE SET 2.6

1. (d); $x = 2$, $x = -2$, $y = 0$
3. (e); $x = 2$, $x = -2$, $y = 0$
5. (c); $x = 2$, $x = -2$, $y = 8x$
7.

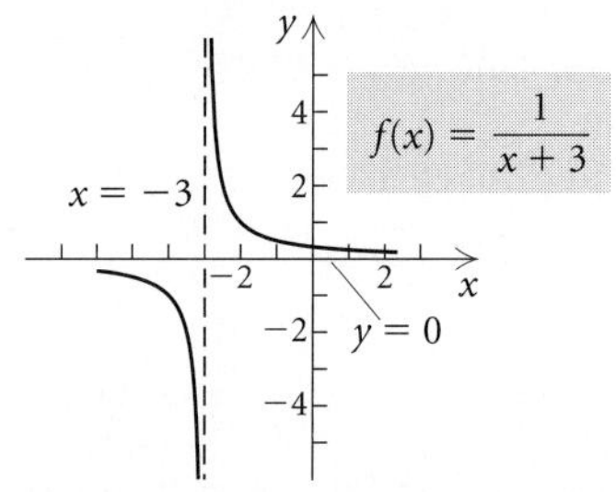

9.

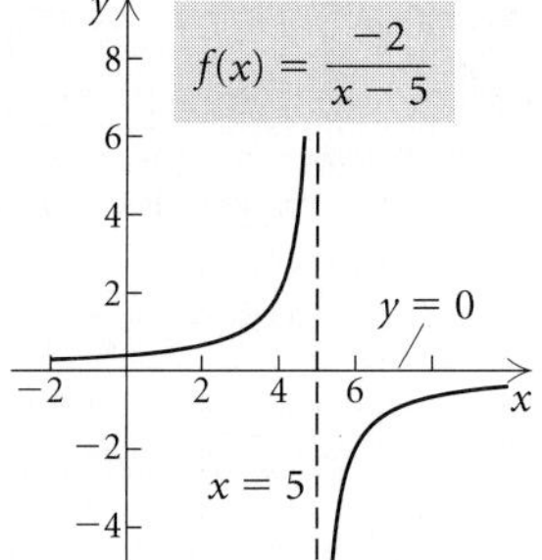

11.

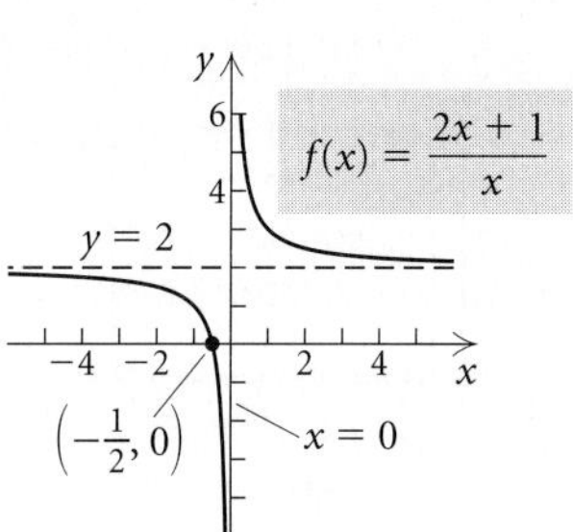

13.

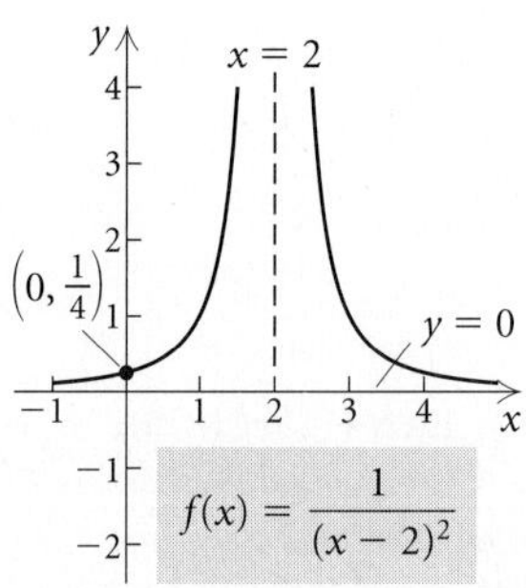

15.

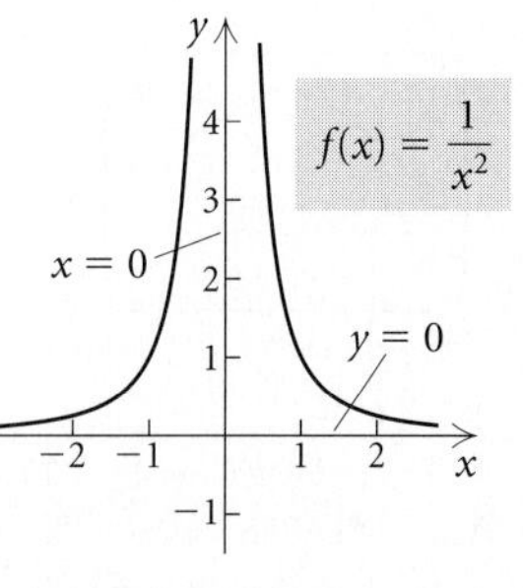

17.

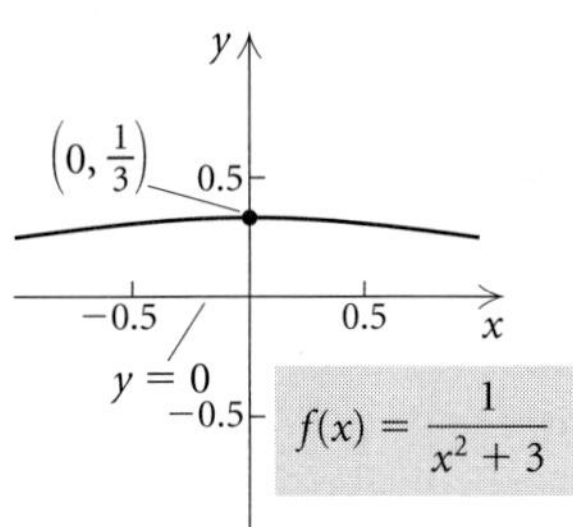

$$f(x) = \frac{1}{x^2 + 3}$$

19.

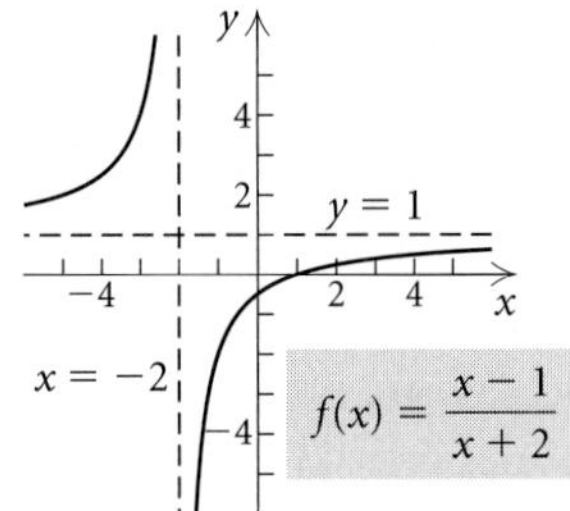

$$f(x) = \frac{x - 1}{x + 2}$$

31.

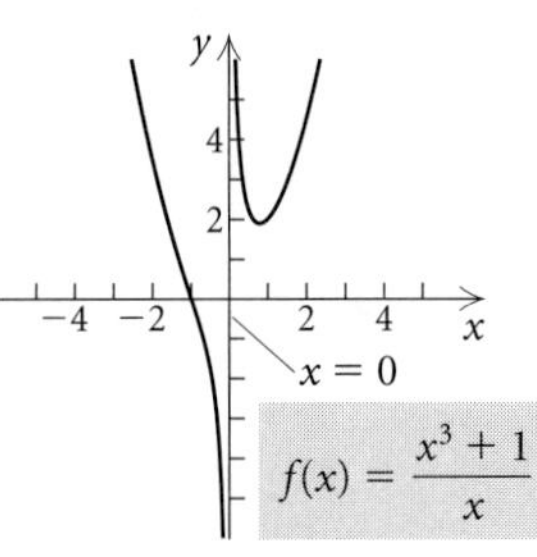

$$f(x) = \frac{x^3 + 1}{x}$$

33.

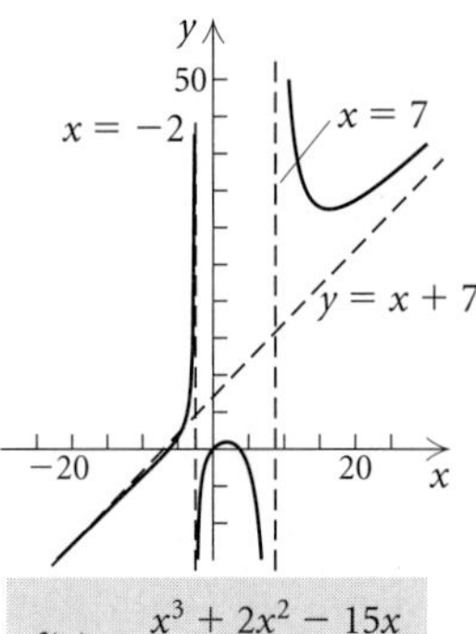

$$f(x) = \frac{x^3 + 2x^2 - 15x}{x^2 - 5x - 14}$$

21.

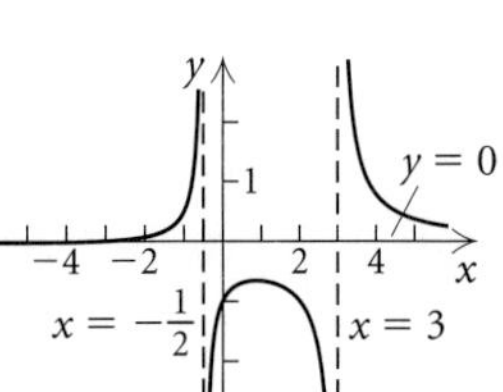

$$f(x) = \frac{x + 3}{2x^2 - 5x - 3}$$

23.

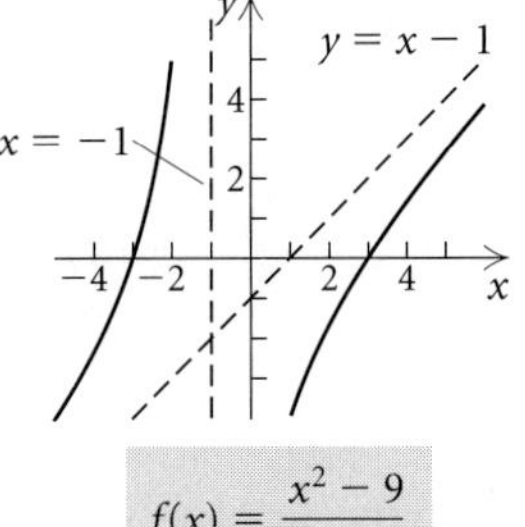

$$f(x) = \frac{x^2 - 9}{x + 1}$$

35.

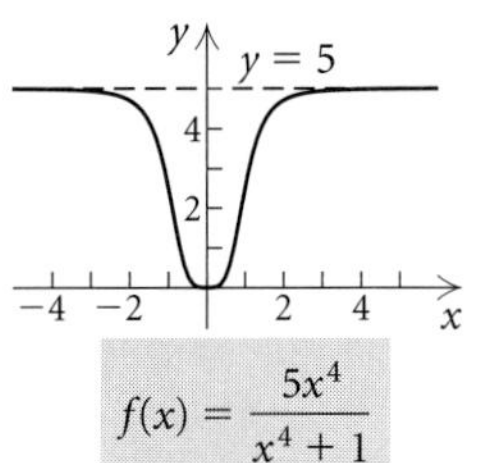

$$f(x) = \frac{5x^4}{x^4 + 1}$$

37.

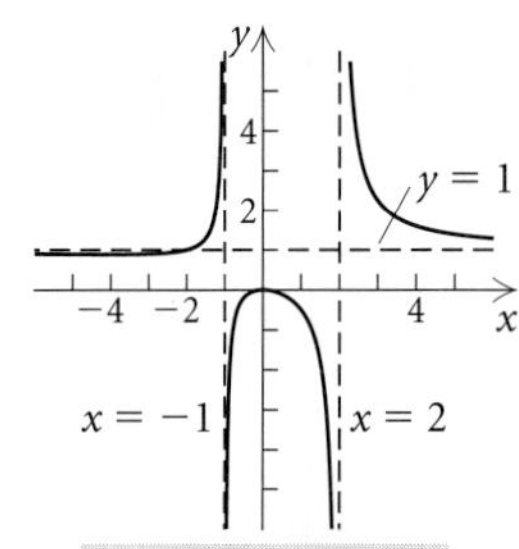

$$f(x) = \frac{x^2}{x^2 - x - 2}$$

25.

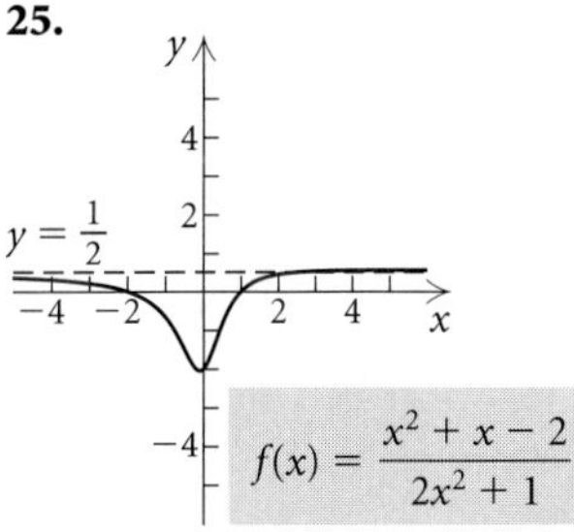

$$f(x) = \frac{x^2 + x - 2}{2x^2 + 1}$$

39. $f(x) = \dfrac{1}{x^2 - x - 20}$ **41.** $f(x) = \dfrac{3x^2 + 12x + 12}{2x^2 - 2x - 40}$

43. (a)

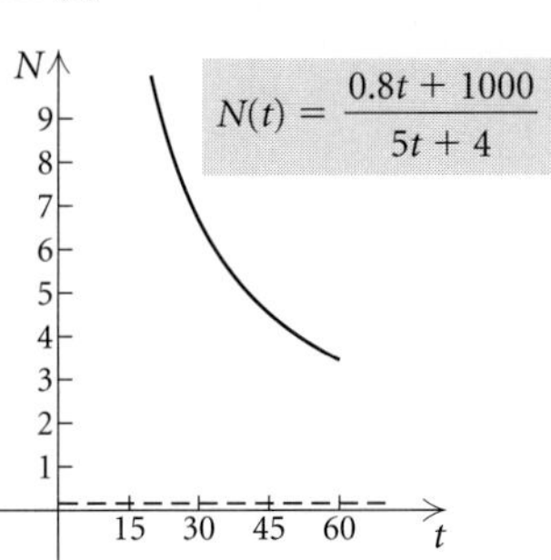

$$N(t) = \frac{0.8t + 1000}{5t + 4}$$

$N(t) \longrightarrow 0.16$ as $t \longrightarrow \infty$; **(b)** The medication never completely disappears from the body; a trace amount remains.

27.

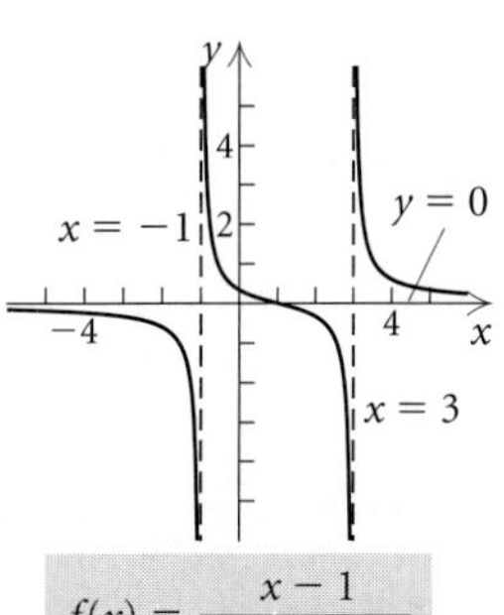

$$f(x) = \frac{x - 1}{x^2 - 2x - 3}$$

29.

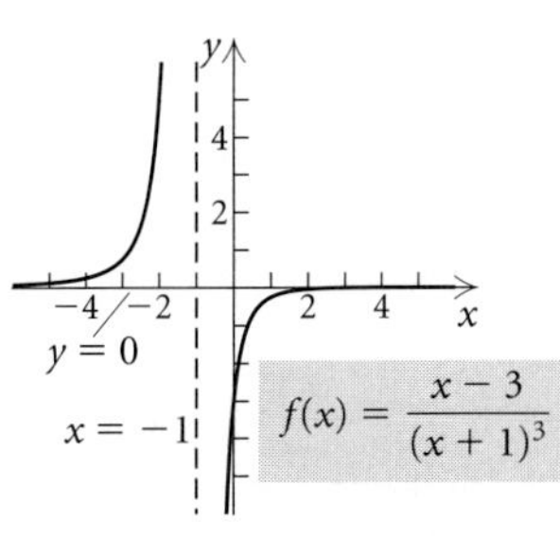

$$f(x) = \frac{x - 3}{(x + 1)^3}$$

45. (a)

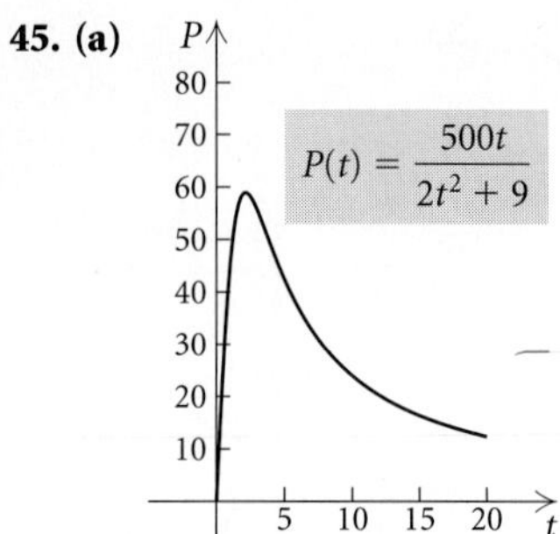

(b) $P(0) = 0$; $P(1) = 45{,}455$; $P(3) = 55{,}556$; $P(8) = 29{,}197$
(c) $P(t) \rightarrow 0$ as $t \rightarrow \infty$; **(d)** In time, no one lives in
Lordsburg. **(e)** 58,926   **47.** 22.5 W/m^2 **49.** $\{x \mid x > 3\}$
51. $\{3\}$ **53.** ◈   **55.** ◈   **57.** $y = x^3 + 4$
59.   **61.**

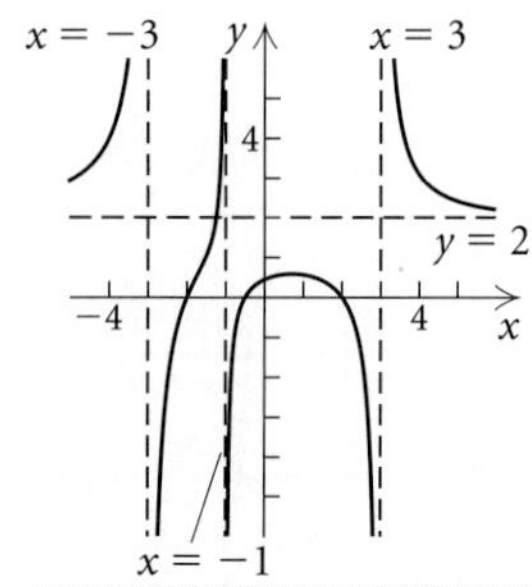

$$f(x) = \frac{2x^3 + x^2 - 8x - 4}{x^3 + x^2 - 9x - 9}$$

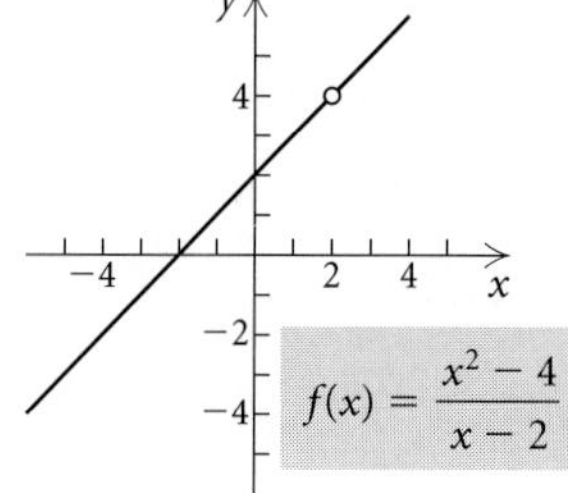

63.

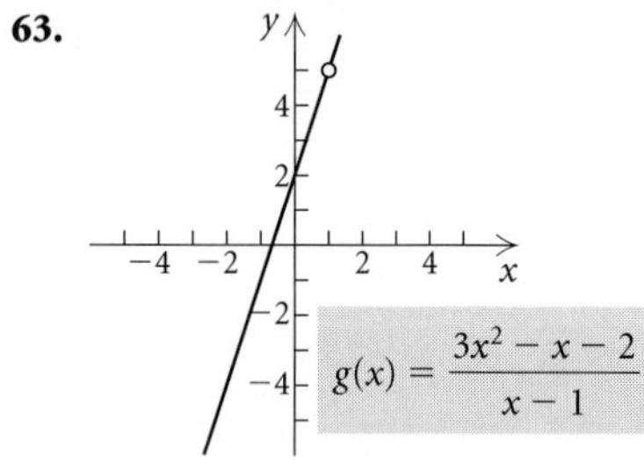

65. ◈   **67.** $(-\infty, -3) \cup (7, \infty)$

EXERCISE SET 2.7

1. $(-\infty, -5) \cup (-3, 2)$ **3.** $(-2, 0] \cup (2, \infty)$ **5.** $(-4, 1)$
7. $(-\infty, -2] \cup [4, \infty)$ **9.** $(-\infty, -2) \cup (1, \infty)$
11. $(-\infty, -5) \cup (5, \infty)$ **13.** $(-\infty, -2] \cup [2, \infty)$
15. $(-\infty, 3) \cup (3, \infty)$ **17.** $\varnothing$ **19.** $\left(-\infty, -\frac{5}{4}\right] \cup [0, 3]$
21. $[-3, -1] \cup [1, \infty)$ **23.** $(-\infty, -2) \cup (1, 3)$
25. $[-\sqrt{2}, -1] \cup [\sqrt{2}, \infty)$ **27.** $(-\infty, -1] \cup \left[\frac{3}{2}, 2\right]$
29. $(-\infty, 5]$ **31.** $(-\infty, -1.680) \cup (2.154, 5.526)$
33. $(-4, \infty)$ **35.** $\left(-\frac{5}{2}, \infty\right)$ **37.** $\left(-3, -\frac{1}{5}\right] \cup (1, \infty)$

39. $(-\infty, -3) \cup \left[\dfrac{5 - \sqrt{105}}{10}, -\dfrac{1}{3}\right) \cup \left[\dfrac{5 + \sqrt{105}}{10}, \infty\right)$
41. $\left(2, \frac{7}{2}\right]$ **43.** $(1 - \sqrt{2}, 0) \cup (1 + \sqrt{2}, \infty)$
45. $(-\infty, -3) \cup (1, 3) \cup \left[\frac{11}{3}, \infty\right)$ **47.** $(-\infty, \infty)$
49. $\left(-3, \dfrac{1 - \sqrt{61}}{6}\right) \cup \left(-\dfrac{1}{2}, 0\right) \cup \left(\dfrac{1 + \sqrt{61}}{6}, \infty\right)$
51. $(-1, 0) \cup \left(\frac{2}{7}, \frac{7}{2}\right)$
53. $[-6 - \sqrt{33}, -5) \cup [-6 + \sqrt{33}, 1) \cup (5, \infty)$
55. $(0.408, 2.449)$
57. (a) $\{x \mid 10 < x < 200\}$; **(b)** $\{x \mid 0 < x < 10 \ or \ x > 200\}$
59. $\{n \mid 9 \le n \le 23\}$ **61.** $b = \dfrac{a + 7}{3}$ **63.** 14   **65.** ◈
67. $\{3\}$ **69.** $(-\infty, \infty)$ **71.** $[-\sqrt{5}, \sqrt{5}]$ **73.** $\left[-\frac{3}{2}, \frac{3}{2}\right]$
75. $\left(-\infty, -\frac{1}{4}\right) \cup \left(\frac{1}{2}, \infty\right)$ **77.** $(-4, -2) \cup (-1, 1)$
79. $x^2 + x - 12 < 0$; answers may vary

REVIEW EXERCISES, CHAPTER 2

1. [2.1] **(a)** $-2.637, 1.137$; **(b)** $(-0.75, 7.125)$; **(c)** none;
(d) domain: all real numbers; range: $(-\infty, 7.125]$
2. [2.1] **(a)** $-3, -1.414, 1.414$; **(b)** $(-2.291, 2.303)$;
(c) $(0.291, -6.303)$; **(d)** domain: all real numbers; range: all
real numbers
3. [2.1] **(a)** $0, 1, 2$; **(b)** $(0.610, 0.202)$; **(c)** $(0, 0)$,
$(1.640, -0.620)$; **(d)** domain: all real numbers; range:
$[-0.620, \infty)$
4. [2.1] $-2\sqrt{10}i$ **5.** [2.1] $-4\sqrt{15}$ **6.** [2.1] $-\frac{7}{8}$
7. [2.1] $-18 - 26i$ **8.** [2.1] $\frac{11}{10} + \frac{3}{10}i$ **9.** [2.1] $1 - 4i$
10. [2.1] $2 - i$ **11.** [2.1] $-i$ **12.** [2.1] 3^{28}
13. [2.2] $x^2 - 3x + \frac{9}{4} = 18 + \frac{9}{4}$; $\left(x - \frac{3}{2}\right)^2 = \frac{81}{4}$; $x = \frac{3}{2} \pm \frac{9}{2}$;
$\{-3, 6\}$
14. [2.2] $x^2 - 4x = 2$; $x^2 - 4x + 4 = 2 + 4$; $(x - 2)^2 = 6$;
$x = 2 \pm \sqrt{6}$; $\{2 - \sqrt{6}, 2 + \sqrt{6}\}$ **15.** [2.2] $\left\{-2, \frac{4}{3}\right\}$
16. [2.2] $\{1 - 3i, 1 + 3i\}$ **17.** [2.2] $\{-3, 6\}$ **18.** [2.2] $\{1\}$
19. [2.2] $\left\{\dfrac{-1 - \sqrt{5}}{2}, \dfrac{-1 + \sqrt{5}}{2}, \dfrac{1 - \sqrt{5}}{2}, \dfrac{1 + \sqrt{5}}{2}\right\}$
20. [2.2] $\{-\sqrt{3}, 0, \sqrt{3}\}$ **21.** [2.5] $\left\{-2, -\frac{2}{3}, 3\right\}$
22. [2.5] $\{-5, -2, 2\}$ **23.** [2.2] **(a)** $\left(\frac{3}{8}, -\frac{7}{16}\right)$; **(b)** $x = \frac{3}{8}$;
(c) maximum: $-\frac{7}{16}$; **(d)** $\left(-\infty, -\frac{7}{16}\right]$
24. [2.2] **(a)** $(1, -2)$; **(b)** $x = 1$; **(c)** minimum: -2;
(d) $[-2, \infty)$ **25.** [2.2] **(d)**   **26.** [2.2] **(c)**   **27.** [2.2] **(b)**
28. [2.2] **(a)**   **29.** [2.3] **(a)** 4%; **(b)** 5%
30. [2.3] $35 - 5\sqrt{33}$ ft, or about 6.3 ft
31. [2.3] 6 ft by 6 ft
32. [2.3] $\dfrac{15 - \sqrt{115}}{2}$ cm, or about 2.1 cm
33. [2.3] 2.1 cm

34. [2.3] **(a)** *Linear:* $f(x) = 0.5408695652x - 30.30434783$; *quadratic:* $f(x) = 0.0030322581x^2 - 0.5764516129x + 57.53225806$; *cubic:* $f(x) = 0.0000247619x^3 - 0.0112857143x^2 + 2.002380952x - 82.14285714$; **(b)** the cubic function; **(c)** 298, 498

35. [2.4] $4x^2 - \frac{14}{5}x - \frac{17}{25}$, R 284/25

36. [2.4] $2x^3 + x + 1$, R $2x + 6$ **37.** [2.4] $\{-5, 1, 2\}$

38. [2.4] $\{-2 - \sqrt{11}, -5, -2 + \sqrt{11}, 4\}$

39. [2.4] $x^2 + 7x + 22$, R 120

40. [2.4] $x^3 + x^2 + x + 1$, R 0   **41.** [2.4] 120

42. [2.4] 0

43. [2.4] -3 is a root; $P(x) = (x + 5)(x + 3)(x - 1)$; other roots: $-5, 1$

44. [2.4] -2 is a root; $P(x) = (x + 2)(x + 1) \cdot [x - (-3 + \sqrt{10})][x - (-3 - \sqrt{10})]$
other roots: $-1, -3 \pm \sqrt{10}$

45. [2.5] $x^3 - x^2 - 10x - 8 = 0$; answers may vary

46. [2.5] $x^4 + 3x^3 - 7x^2 - 9x + 12 = 0$; answers may vary

47. [2.5] $x^6 - 13x^5 + 62x^4 - 132x^3 - 45x^2 + 745x - 858 = 0$; answers may vary

48. [2.5] $-\frac{5}{8}x^3 + \frac{5}{8}x^2 + \frac{25}{4}x + 5 = 0$

49. [2.5] $\frac{1}{2}x^4 + x^3 - \frac{11}{2}x^2 - 8x + 12 = 0$

50. [2.5] $-\frac{4}{5}x^3 - \frac{12}{5}x^2 + \frac{48}{5}x + 8 = 0$

51. [2.5] $\{-1, 2, 3\}$ **52.** [2.5] $\{-5, -3, 1\}$

53. [2.5] $\{-3, 2 - \sqrt{11}, 2 + \sqrt{11}, 3\}$

54. [2.5] **(a)** $\{-10, 1\}$; **(b)** $y = x^3 + 8x^2 - 19x + 10$

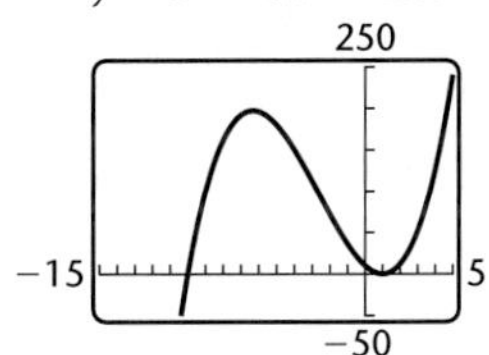

(c) $P(x) = (x - 1)^2(x + 10)$

55. [2.5] **(a)** $\{-4, 0, 3, 4\}$;

(b) $y = x^6 + x^5 - 28x^4 - 16x^3 + 192x^2$

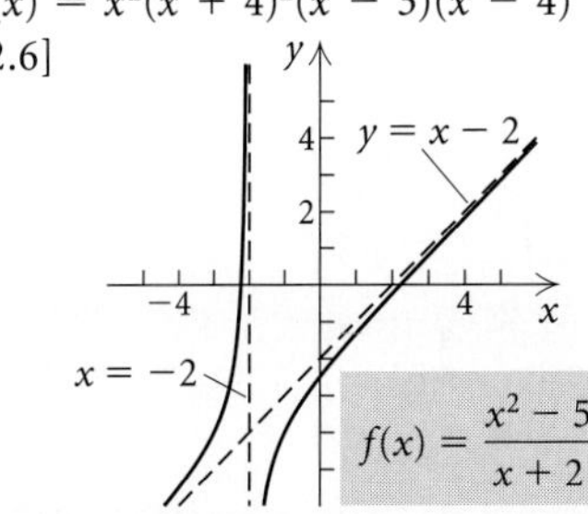

(c) $P(x) = x^2(x + 4)^2(x - 3)(x - 4)$

56. [2.6]

$y = x - 2$

$x = -2$

$f(x) = \dfrac{x^2 - 5}{x + 2}$

57. [2.6]

$x = 2$

$f(x) = \dfrac{5}{(x - 2)^2}$

$y = 0$

58. [2.6]

$y = 1$

$x = -4$ $x = 5$

$f(x) = \dfrac{x^2 + x - 6}{x^2 - x - 20}$

59. [2.6]

$y = 0$

$x = -3$ $x = 5$

$f(x) = \dfrac{x - 2}{x^2 - 2x - 15}$

60. [2.6] $f(x) = \dfrac{1}{x^2 - x - 6}$ **61.** [2.6] $f(x) = \dfrac{4x^2 + 12x}{x^2 - x - 6}$

62. [2.6] **(a)**

$N(t) = \dfrac{0.7t + 2000}{8t + 9}, \; t \geq 5$

$N(t) \rightarrow 0.0875$ as $t \rightarrow \infty$; **(b)** The medication never completely disappears from the body; a trace amount remains.   **63.** [2.7] $(-3, 3)$ **64.** [2.7] $\left(-\infty, -\frac{1}{2}\right) \cup (2, \infty)$

65. [2.7] $(-4, 1) \cup (2, \infty)$ **66.** [2.7] $\left(-\infty, -\frac{14}{3}\right) \cup (-3, \infty)$

67. [2.7] $(-5, -3.386) \cup (-2, 2) \cup (4, \infty)$

68. [2.3], [2.7] **(a)** 324 ft, 2.5 sec; **(b)** $t = 7$; **(c)** $(2, 3)$

69. [2.7] $\left[\dfrac{5 - \sqrt{15}}{2}, \dfrac{5 + \sqrt{15}}{2} \right]$

70. [2.1, [2.6] ◈ A polynomial function is a function that can be defined by a polynomial expression. A rational function is a function that can be defined as a quotient of two polynomials.

71. [2.6] ◈ Vertical asymptotes occur at any x-values that make the denominator zero. The graph of a rational function does not cross any vertical asymptotes. Horizontal asymptotes occur when the degree of the numerator is less than or equal to the degree of the denominator. Oblique asymptotes occur when the degree of the numerator is 1 greater than the degree of the denominator. Graphs of rational functions may cross horizontal or oblique asymptotes.

72. [2.5] Yes   **73.** [2.3] 9%

74. [2.7] $(-\infty, -1 - \sqrt{6}] \cup [-1 + \sqrt{6}, \infty)$

75. [2.7] $\left(-\infty, -\frac{1}{2}\right) \cup \left(\frac{1}{2}, \infty\right)$ **76.** [2.5] $\{1 + i, 1 - i, i, -i\}$

77. [2.7] $(-\infty, 2)$

78. [2.5] $(x - 1)\left(x + \dfrac{1}{2} - \dfrac{\sqrt{3}}{2}i\right)\left(x + \dfrac{1}{2} + \dfrac{\sqrt{3}}{2}i\right)$

79. [2.4] 7   **80.** [2.4] -4 **81.** [2.1] $(-\infty, -5] \cup [2, \infty)$

82. [2.1] $(-\infty, 1.1] \cup [2, \infty)$ **83.** [2.6] $\left(-1, \frac{3}{7}\right)$

Chapter 3

EXERCISE SET 3.1

1. $\{(8, 7), (8, -2), (-4, 3), (-8, 8)\}$

3. $\{(-1, -1), (4, -3)\}$ **5.** $x = 4y - 5$ **7.** $y^3x = -5$

9.

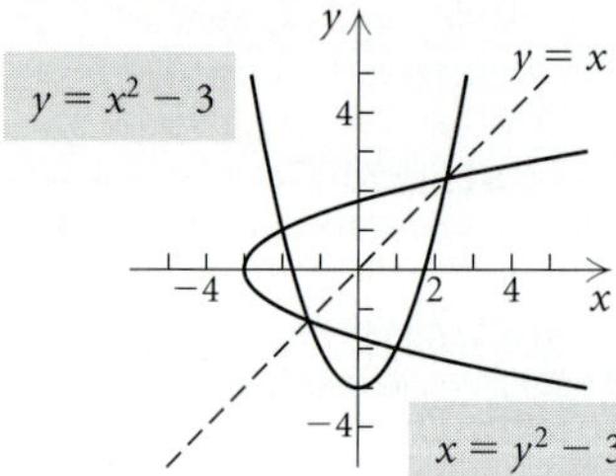

11.

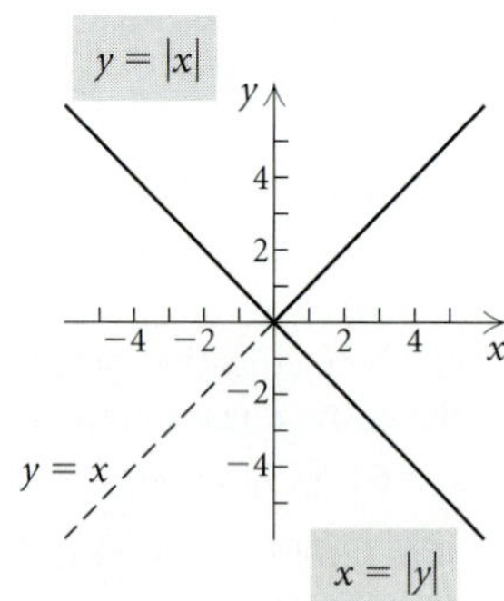

13. Yes   **15.** No   **17.** No   **19.** Yes   **21.** Yes   **23.** No

25. No   **27.** Yes

29.
$$y_1 = 0.8x + 1.7,$$
$$y_2 = \frac{x - 1.7}{0.8}$$

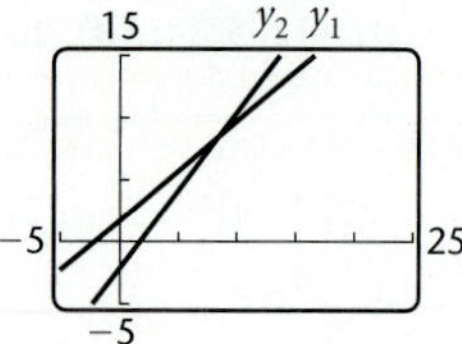

Domain and range of both f and f^{-1}: all real numbers

31.
$$y_1 = \tfrac{1}{2}x - 4,$$
$$y_2 = 2x + 8$$

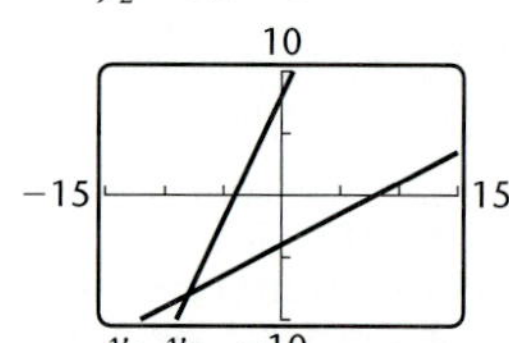

Domain and range of both f and f^{-1}: all real numbers

33.
$$y_1 = \sqrt{x - 3},$$
$$y_2 = x^2 + 3, x \geq 0$$

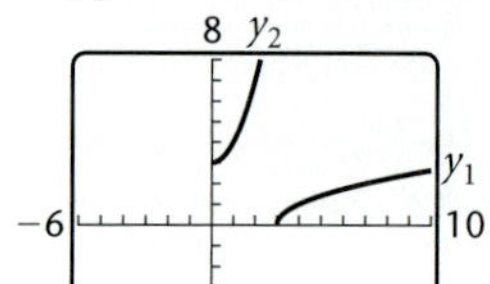

Domain of f: $[3, \infty]$; range of f: $[0, \infty)$; domain of f^{-1}: $[0, \infty)$; range of f^{-1}: $[3, \infty)$

35. $y_1 = x^2 - 4, x \geq 0;\ y_2 = \sqrt{4 + x}$

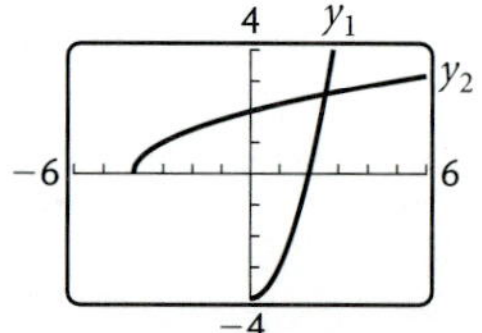

Domain of f: $[0, \infty)$; range of f: $[-4, \infty)$; domain of f^{-1}: $[-4, \infty)$; range of f^{-1}: $[0, \infty)$

37.
$$y_1 = (3x - 9)^3,\ y_2 = \frac{\sqrt[3]{x} + 9}{3}$$

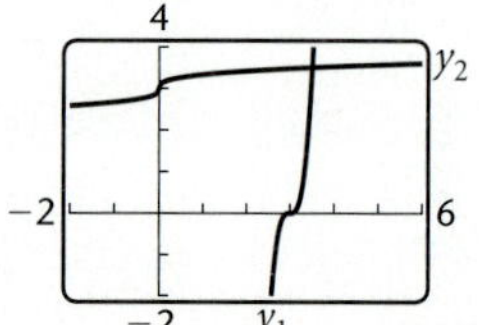

Domain and range of both f and f^{-1}: all real numbers

39. (a) One-to-one; **(b)** $f^{-1}(x) = x - 4$

41. (a) One-to-one; **(b)** $f^{-1}(x) = \dfrac{1}{2}x$

43. (a) One-to-one; **(b)** $f^{-1}(x) = \dfrac{4}{x} - 7$

45. (a) One-to-one; **(b)** $f^{-1}(x) = \dfrac{3x + 4}{x - 1}$

47. (a) One-to-one; **(b)** $f^{-1}(x) = \sqrt[3]{x + 1}$

49. (a) Not one-to-one

51. (a) One-to-one; **(b)** $f^{-1}(x) = \sqrt{\dfrac{x + 2}{5}}, \; x \geq 18$

53. (a) One-to-one; **(b)** $f^{-1}(x) = x^2 - 1, \; x \geq 0$

55. $\frac{1}{3}x$　　**57.** $-x$　　**59.** $x^3 + 5$

61. $(f^{-1} \circ f)(x) = f^{-1}(f(x))$

$$= f^{-1}\left(\tfrac{7}{8}x\right)$$
$$= \tfrac{8}{7}\left(\tfrac{7}{8}x\right) = x;$$
$$(f \circ f^{-1})(x) = f(f^{-1}(x))$$
$$= f\left(\tfrac{8}{7}x\right)$$
$$= \tfrac{7}{8}\left(\tfrac{8}{7}x\right) = x$$

63. $(f^{-1} \circ f)(x) = f^{-1}\left(\dfrac{1 - x}{x}\right)$

$$= \dfrac{1}{\dfrac{1 - x}{x} + 1}$$
$$= \dfrac{1}{\dfrac{1}{x}} = x;$$
$$(f \circ f^{-1})(x) = f\left(\dfrac{1}{x + 1}\right)$$
$$= \dfrac{1 - \dfrac{1}{x + 1}}{\dfrac{1}{x + 1}}$$
$$= \dfrac{\dfrac{x + 1 - 1}{x + 1}}{\dfrac{1}{x + 1}} = x$$

65. 5; a

67. (a) 36, 40, 44, 52, 60; **(b)** $g^{-1}(x) = \dfrac{x}{2} - 12$;

(c) 8, 10, 14, 18

69. (a) 0.5, 11.5, 22.5, 55.5, 72;

(b), (d) $\quad y_1 = \dfrac{11x + 5}{10}, \quad y_2 = \dfrac{10x - 5}{11}$

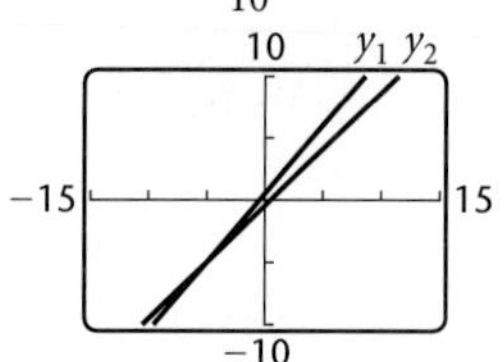

(c) $D^{-1}(r) = \dfrac{10r - 5}{11}$; the speed, in miles per hour, that the car is traveling when the reaction distance is r feet

71.

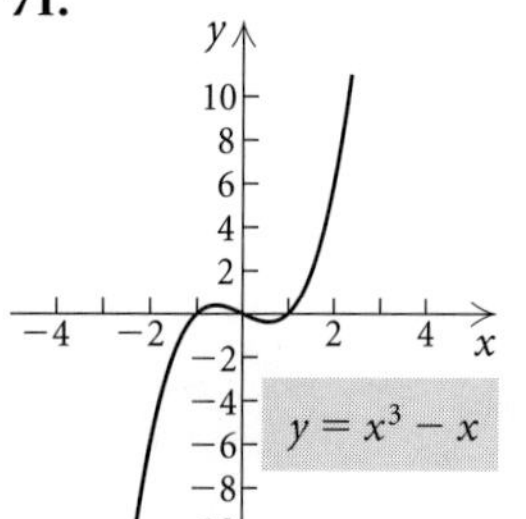

73.

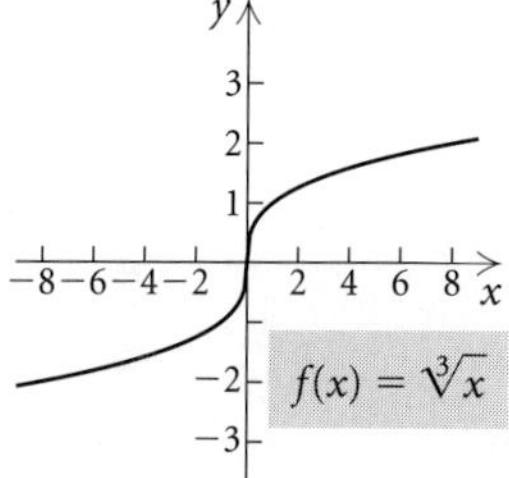

75. ◈　　**77.** Yes　　**79.** No

81. Answers may vary. $f(x) = 3/x, \; f(x) = 1 - x, \; f(x) = x$

EXERCISE SET 3.2

1. 54.5982　　**3.** 0.0856

5.

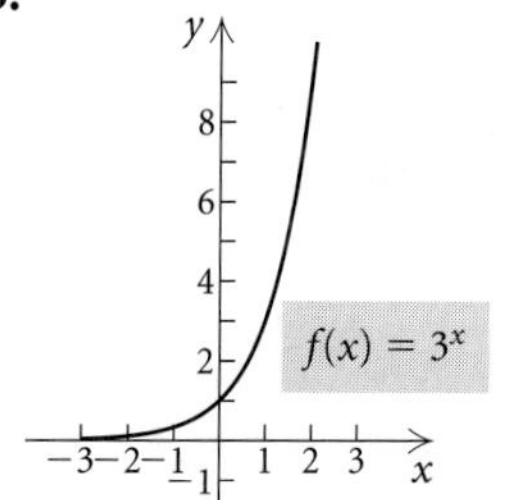

7.

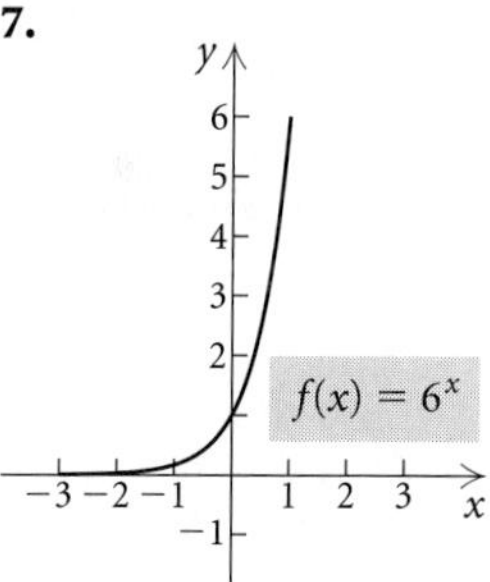

9.

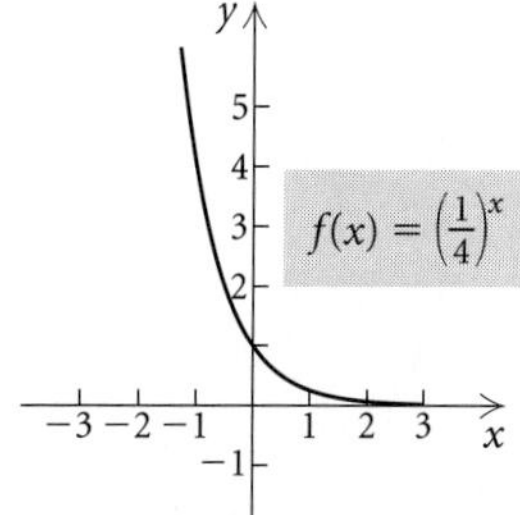

11.

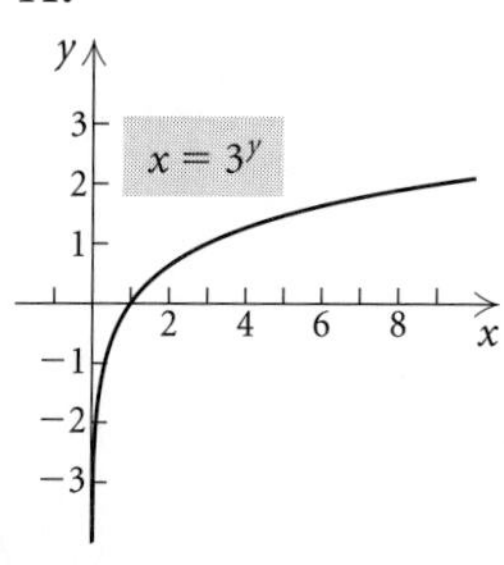

13.

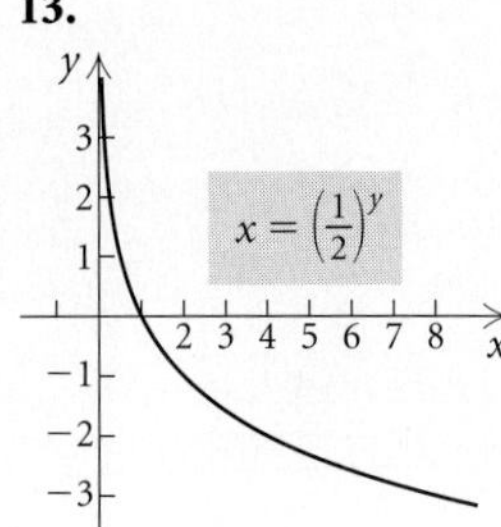

15.

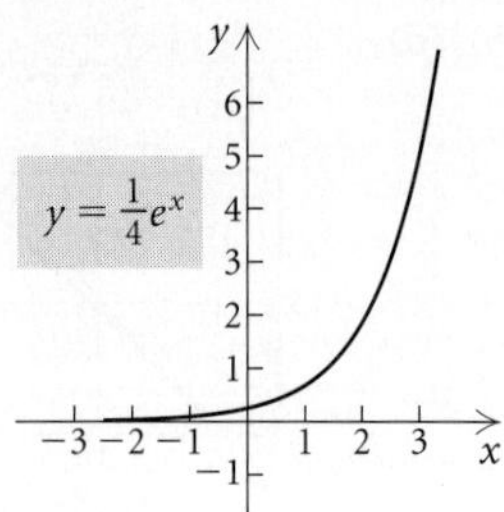

17.

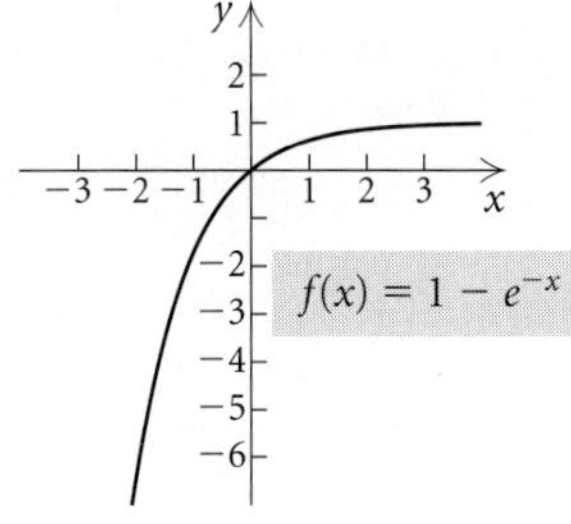

19. Shift the graph of $y = 2^x$ left 1 unit.

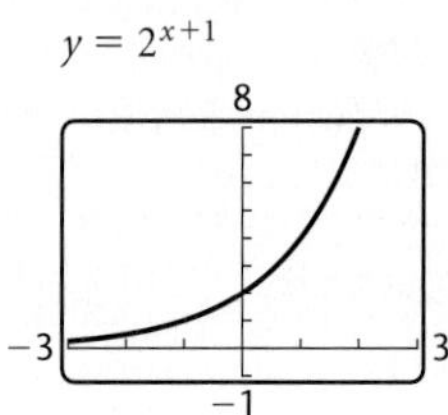

21. Shift the graph of $y = 2^x$ down 3 units.

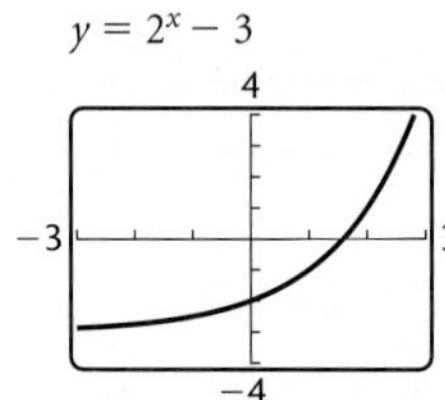

23. Reflect the graph of $y = 3^x$ across the y-axis, then across the x-axis, and then shift it up 4 units.

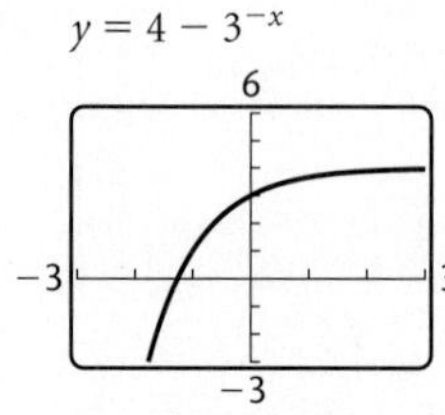

25. Shift the graph of $y = \left(\frac{3}{2}\right)^x$ right 1 unit.

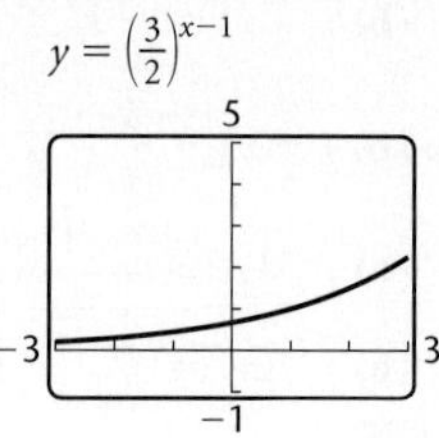

27. Shift the graph of $y = 2^x$ left 3 units, and then down 5 units.

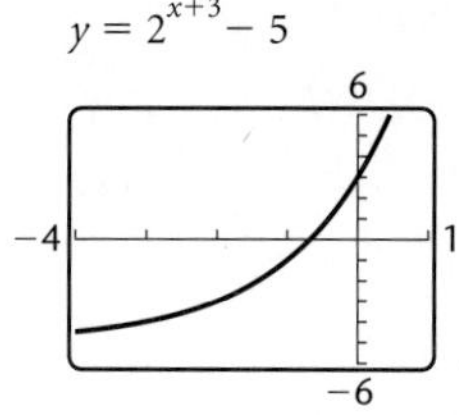

29. Shrink the graph of $y = e^x$ horizontally.

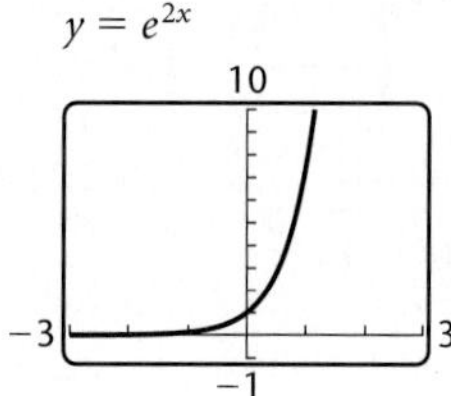

31. Shift the graph of $y = e^x$ left 1 unit, then reflect it across the y-axis.

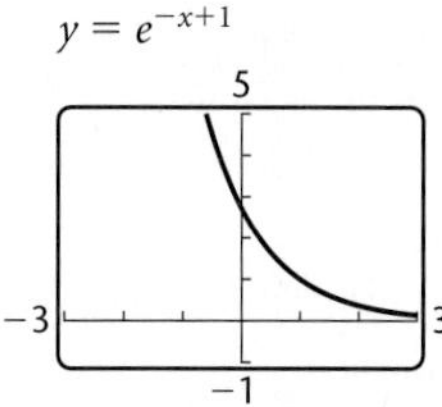

33. Reflect the graph of $y = e^x$ across the y-axis, then across the x-axis, then shift it up 1 unit, and then stretch it vertically.

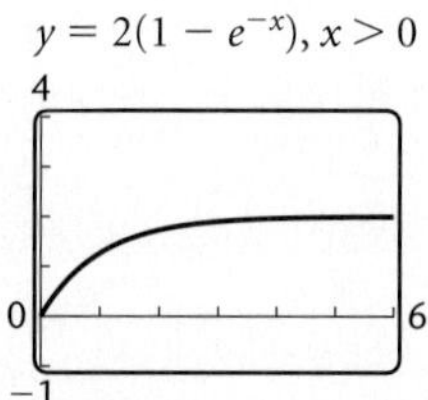

35. (a) 1,475,789; **(b)** 5,669,391;
(c) $y = 100,000(1.4)^x$ **(d)** 6.8 yr

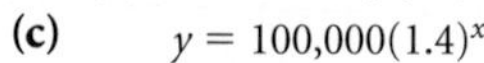
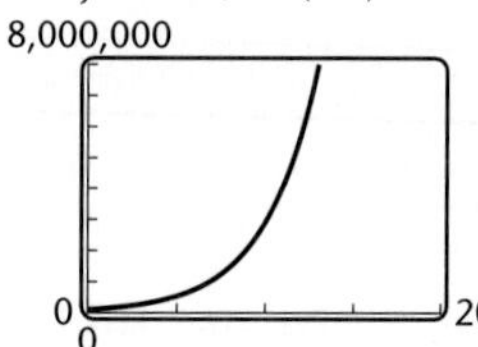

37. (a) 350,000; 233,333; 69,136; 6070;
(b) $y = 350,000\left(\frac{2}{3}\right)^x$ **(c)** after about 13 yr

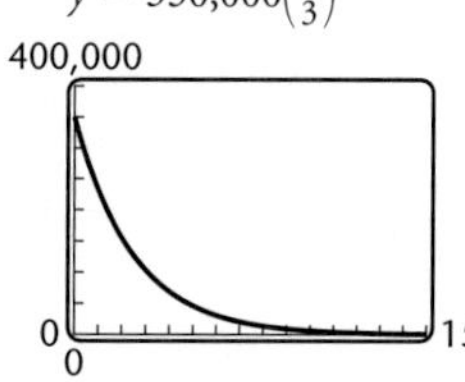

39. (a) \$5800, \$4640, \$3712, \$1900.54, \$622.77;
(b) $y = 5800(0.8)^x$ **(c)** after 11 yr

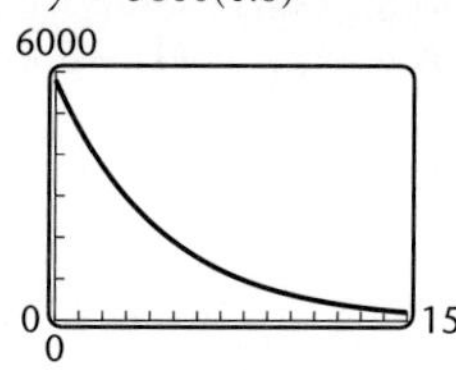

41. (a) About 50.0 billion ft^3, 62.0 billion ft^3, 78.2 billion ft^3; **(b)** $y = 46.6(1.018)^x$ **(c)** After about 39 yr

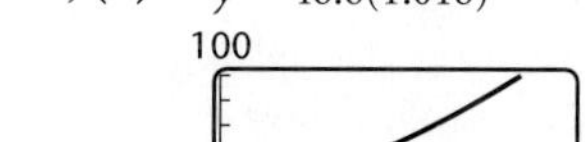
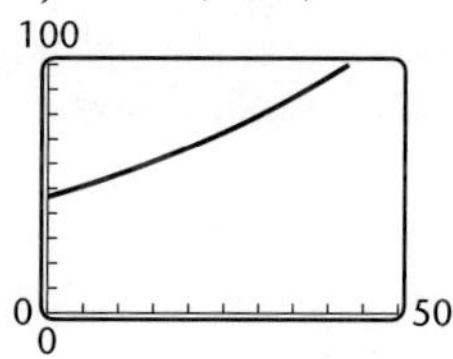

43. (a) About 63%; **(b)** $y = 100(1 - e^{-0.04x})$

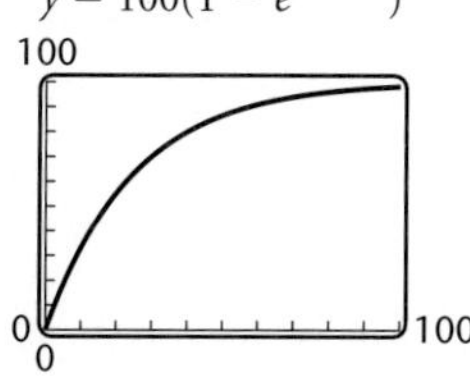

(c) after 58 days

45. (c)   **47.** (a)   **49.** (l)   **51.** (g)   **53.** (i)   **55.** (k)
57. (m)   **59.** (1.481, 4.090)
61. $(-0.402, -1.662)$, $(1.051, 2.722)$ **63.** $\{4.448\}$
65. $(0, \infty)$ **67.** $\{2.294, 3.228\}$ **69.** 1   **71.** 1000   **73.** 2
75. ◈   **77.** ◈   **79.** $y = 1$

81. (a) $y = e^{-x^2}$

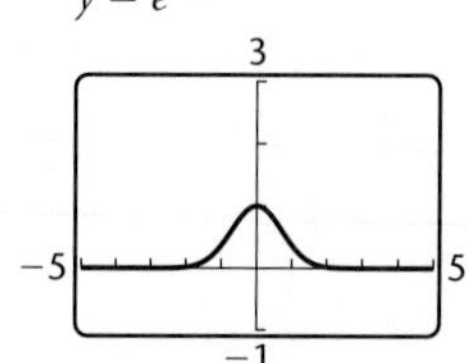

(b) none; **(c)** relative maximum: 1 at $x = 0$

EXERCISE SET 3.3

1.   **3.**

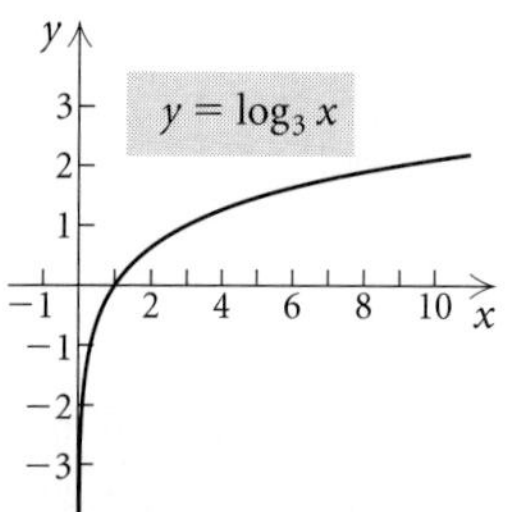

5. 4   **7.** 3   **9.** -3 **11.** -2 **13.** 0   **15.** 1
17. $\log_{10} 1000 = 3$ **19.** $\log_8 2 = \frac{1}{3}$ **21.** $\log_e t = 3$
23. $\log_e 7.3891 = 2$ **25.** $\log_p 3 = k$ **27.** $5^1 = 5$
29. $10^{-2} = 0.01$ **31.** $e^{3.4012} = 30$ **33.** $a^{-x} = M$
35. $a^x = T^3$ **37.** 0.4771   **39.** 2.7259   **41.** -0.2441
43. Does not exist   **45.** 0.6931   **47.** 6.6962
49. Does not exist
51. 3.3219   **53.** -0.2614 **55.** 0.7384
57. Translate the graph of $y = \log_2 x$ left 3 units. Domain: $(-3, \infty)$; vertical asymptote: $x = -3$;

$$y = \frac{\log (x + 3)}{\log 2}$$

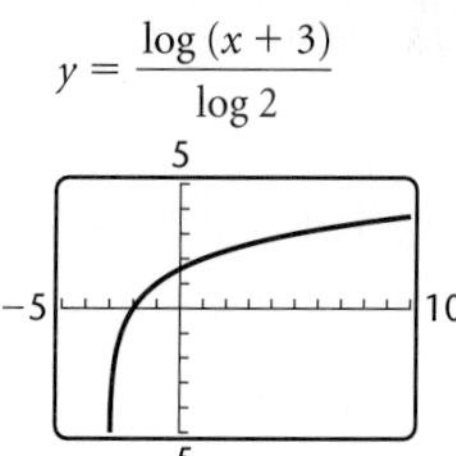

59. Translate the graph of $y = \log_3 x$ down 1 unit. Domain: $(0, \infty)$; vertical asymptote: $x = 0$;

$$y = \frac{\log x}{\log 3} - 1$$

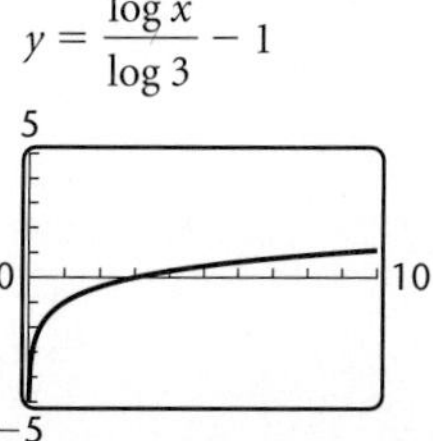

61. Stretch the graph of $y = \ln x$ vertically. Domain: $(0, \infty)$; vertical asymptote: $x = 0$;

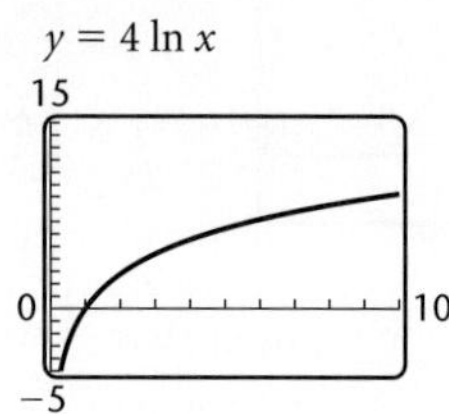

63. Reflect the graph of $y = \ln x$ across the x-axis and translate it up 2 units. Domain: $(0, \infty)$; vertical asymptote: $x = 0$;

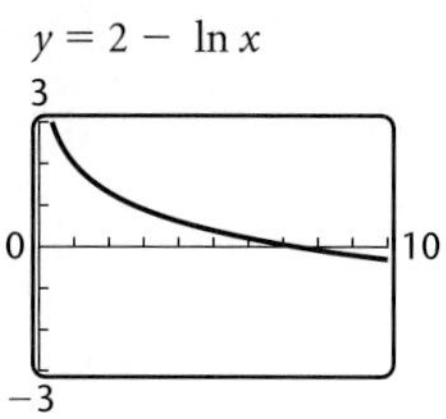

65. $y_1 = 3^x$, $y_2 = \dfrac{\log x}{\log 3}$

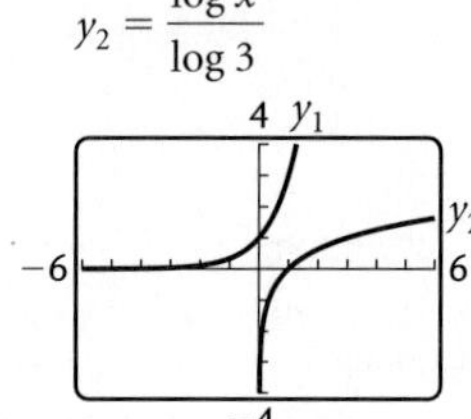

67. $y_1 = \log x$, $y_2 = 10^x$

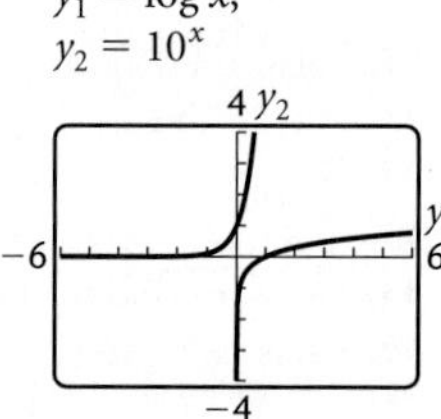

69. (a) 3.1 ft/sec; **(b)** 3.0 ft/sec; **(c)** 2.3 ft/sec; **(d)** 1.5 ft/sec
71. (a) 78%; **(b)** 67.5%, 57%;
(c) $y = 78 - 15 \log (x + 1), x \geq 0$ **(d)** after 73 months

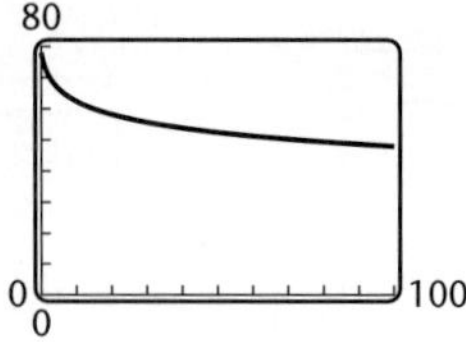

73. (a) 10^{-7}; **(b)** 4.0×10^{-6}; **(c)** 6.3×10^{-4}; **(d)** 1.6×10^{-5}
75. (a) 34 decibels; **(b)** 64 decibels; **(c)** 60 decibels;
(d) 90 decibels
77. y^{24} **79.** 10^{4t} **81.** ◈   **83.** 3   **85.** $(0, \infty)$
87. $(-\infty, 0) \cup (0, \infty)$ **89.** $\left(-\frac{5}{2}, -2\right)$ **91.** (d)   **93.** (b)

95. (a)

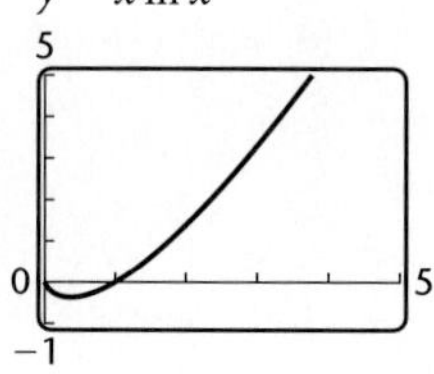

(b) 1; **(c)** relative minimum: -0.368 at $x = 0.368$
97. (a) $y = \dfrac{\ln x}{x^2}$

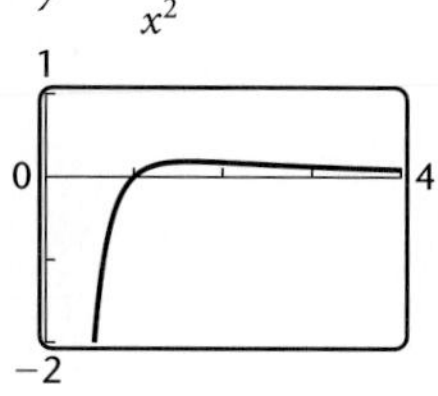

(b) 1; **(c)** relative maximum: 0.184 at $x = 1.649$

EXERCISE SET 3.4

1. $\log_3 81 + \log_3 27$ **3.** $\log_5 5 + \log_5 125$
5. $\log_t 8 + \log_t Y$ **7.** $3 \log_b t$ **9.** $8 \log y$ **11.** $-6 \log_c K$
13. $\log_t M - \log_t 8$ **15.** $\log_a x - \log_a y$
17. $\log_a 6 + \log_a x + 5 \log_a y + 4 \log_a z$
19. $2 \log_b p + 5 \log_b q - 4 \log_b m - 9$
21. $3 \log_a x - \frac{5}{2} \log_a p - 4 \log_a q$
23. $2 \log_a m + 3 \log_a n - \frac{3}{4} - \frac{5}{4} \log_a b$ **25.** $\log_a 150$
27. $\log 100 = 2$ **29.** $\log_a x^{-5/2}y^4$, or $\log_a \dfrac{y^4}{x^{5/2}}$ **31.** $\ln x$
33. $\ln (x - 2)$ **35.** $\ln \dfrac{x}{(x^2 - 25)^3}$ **37.** $\ln \dfrac{2^{11/5}x^9}{y^8}$
39. -0.5108 **41.** -1.6094 **43.** $\frac{1}{2}$ **45.** 2.6094
47. 4.3174 **49.** 3   **51.** $|x - 4|$ **53.** $4x$ **55.** w **57.** $8t$
59. $\{4\}$ **61.** $\{5\}$ **63.** ◈   **65.** $\{4\}$ **67.** $\log_a (x^3 - y^3)$
69. $\frac{1}{2} \log_a (x - y) - \frac{1}{2} \log_a (x + y)$ **71.** 7   **73.** True
75. True   **77.** True   **79.** -2 **81.** 3
83. $\log_a \left(\dfrac{x + \sqrt{x^2 - 5}}{5} \cdot \dfrac{x - \sqrt{x^2 - 5}}{x - \sqrt{x^2 - 5}}\right)$

$$= \log_a \dfrac{5}{5(x - \sqrt{x^2 - 5})}$$
$$= -\log_a (x - \sqrt{x^2 - 5})$$

EXERCISE SET 3.5

1. $\{4\}$ **3.** $\left\{\frac{3}{2}\right\}$ **5.** $\{5.044\}$ **7.** $\left\{\frac{5}{2}\right\}$ **9.** $\left\{-3, \frac{1}{2}\right\}$
11. $\{0.959\}$ **13.** $\{6.908\}$ **15.** $\{84.191\}$ **17.** $\{-1.710\}$
19. $\{2.844\}$ **21.** $\{-1.567, 1.567\}$ **23.** $\{0.347\}$ **25.** $\{625\}$
27. $\{0.0001\}$ **29.** $\{e\}$ **31.** $\left\{\frac{22}{3}\right\}$ **33.** $\{10\}$ **35.** $\left\{\frac{1}{63}\right\}$
37. $\{5\}$ **39.** $\left\{\frac{21}{8}\right\}$ **41.** $\{0.367\}$ **43.** $\{0.621\}$

45. $\{-1.532\}$ **47.** $\{7.062\}$ **49.** $\{2.444\}$
51. $(4.093, 0.786)$ **53.** $(7.586, 6.684)$
55. $\{-3i, 3i, -2, 2\}$
57. (a) $y = 400.27x + 10,655.15$, \$14,257.6 million;
(b) $y = -58.325x^2 + 691.895x + 10,363.525$,
\$11,866.3 million
59. ◈   **61.** $\{10\}$ **63.** $\{1, e^4\}$, or $\{1, 54.598\}$ **65.** $\left\{\frac{1}{3}, 27\right\}$
67. $\{1, e^2\}$, or $\{1, 7.389\}$ **69.** $\{0, 0.431\}$ **71.** $\{-9, 9\}$
73. $\{e^{-2}, e^2\}$, or $\{0.135, 7.389\}$ **75.** $\left\{\frac{7}{4}\right\}$ **77.** $(56.598, \infty)$
79. $\{5\}$ **81.** $a = \frac{2}{3}b$ **83.** 88

EXERCISE SET 3.6

1. (a) $P(t) = 6.6e^{0.028t}$; (b) 7.2 billion, 9.5 billion;
(c) in 6.9 yr; (d) 24.8 yr
3. (a) 19.8 yr; (b) 1% per year; (c) 6.93 yr; (d) 17.3 yr;
(e) 2.8% per year; (f) 3.5% per year
5. After about 69 yr
7. (a) $P(t) = 10,000e^{0.054t}$; (b) \$10,555; \$11,140; \$13,100;
\$17,160; (c) 12.8 yr
9. About 5135 yr
11. (a) 23.1% per minute; (b) 3.15% per year; (c) 7.2 days;
(d) 11 yr; (e) 2.8% per year; (f) 0.015% per year; (g) 0.003%
per year
13. (a) $k \approx 0.016$; $P(t) = 80e^{-0.016t}$; (b) about 63 lb per
person; (c) 86.6 yr
15. (a) 96; (b) 286, 1005, 1719, 1971, 1997;
(c)
$$y = \frac{2000}{1 + 19.9e^{-0.6x}}$$
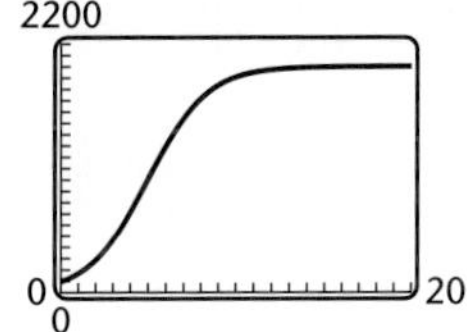

17. (a) 400, 520, 1214, 2059, 2396, 2478;
(b)
$$y = \frac{2500}{1 + 5.25e^{-0.32x}}$$
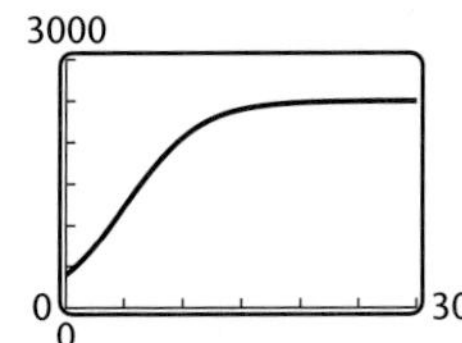

19. (f)   **21.** (b)   **23.** (c)
25. (a) Yes; (b) $y = 0.785(1.490)^x$, or $y = 0.785e^{0.399x}$;
(c) \$63.1 billion; (d) 1997

27. (a) Exponential function: $y = (107.9)(1.034)^x$, or
$y = (107.9)e^{0.033x}$; power function: $y = 110.8x^{0.077}$;
(b) $y_1 = (107.9)(1.034)^x$, $y_2 = 110.8x^{0.077}$
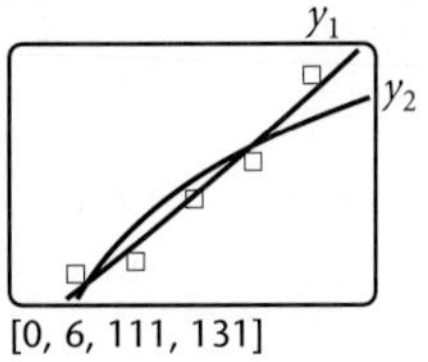
$[0, 6, 111, 131]$

(c) exponential function: \$178; power function: \$136;
(d) The exponential function has a higher correlation
coefficient and seems to be following data better.
29. $\sqrt{3}$ **31.** $\sqrt{3365}$ **33.** ◈   **35.** \$14,182.70
37. \$166.16   **39.** 46.7°F   **41.** $t = -\dfrac{L}{R}\left[\ln\left(1 - \dfrac{iR}{V}\right)\right]$
43. Linear

REVIEW EXERCISES, CHAPTER 3

1. [3.1] $\{(-2.7, 1.3), (-3, 8), (3, -5), (-3, 6), (-5, 7)\}$
2. [3.1] (a) $x = 3y^2 + 2y - 1$; (b) $0.8y^3 - 5.4x^2 = 3y$
3. [3.1] (a) Yes; (b) $f^{-1}(x) = x^2 + 6$, $x \geq 0$
4. [3.1] (a) Yes; (b) $f^{-1}(x) = \sqrt[3]{x} + 8$ **5.** [3.1] (a) No
6. [3.1] (a) Yes; (b) $f^{-1}(x) = \ln x$ **7.** [3.1] 657
8. [3.2] (c)   **9.** [3.3] (a)   **10.** [3.3] (b)   **11.** [3.2] (f)
12. [3.2] (e)   **13.** [3.3] (d)   **14.** [3.3] $4^2 = x$
15. [3.3] $\log_e 80 = x$ **16.** [3.5] $\{16\}$ **17.** [3.5] $\left\{\frac{1}{5}\right\}$
18. [3.5] $\{4.382\}$ **19.** [3.5] $\{2\}$ **20.** [3.5] $\left\{\frac{1}{2}\right\}$
21. [3.5] $\{5\}$ **22.** [3.5] $\{4\}$ **23.** [3.5] $\{9\}$ **24.** [3.5] $\{1\}$
25. [3.5] $\{3.912\}$ **26.** [3.4] $\log_b \dfrac{x^3\sqrt{z}}{y^4}$
27. [3.4] $\ln(x^2 - 4)$ **28.** [3.4] $\frac{1}{4}\ln w + \frac{1}{2}\ln r$
29. [3.4] $\frac{2}{3}\log M - \frac{1}{3}\log N$ **30.** [3.4] 0.477
31. [3.4] 1.699   **32.** [3.4] -0.699 **33.** [3.4] 0.233
34. [3.4] $-5k$ **35.** [3.4] $-6t$ **36.** [3.6] 12.8 yr
37. [3.6] 2.3%   **38.** [3.6] About 2623 yr   **39.** [3.3] 5.6
40. [3.3] 30 decibels
41. [3.6] (a) $P(t) = 52e^{0.028t}$; (b) 164 million, 287 million;
(c) in 62.6 yr; (d) 24.8 yr
42. [3.6] (a) $k = 0.304$; (b) $P(t) = 7e^{0.304t}$;
(c) $y = 7e^{0.304x}$
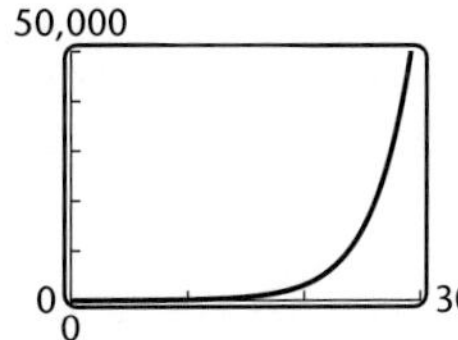

(d) 86.7 billion, 727.9 billion; (e) 1993

43. [3.3] **(a)** 2.3 ft/sec; **(b)** 8,553,143
44. [3.6] **(a)** Yes; **(b)** $y = (2.329)1.753^x$, or $y = 2.329e^{0.561x}$;
(c) 207.7 million; **(d)** in 6.7 yr
45. [3.4]   By the product rule, $\log_2 x + \log_2 5 = \log_2 5x$,
not $\log_2 (x + 5)$. Also, substituting various numbers for x
shows that both sides of the inequality are indeed unequal.
You could also graph each side and show that the graphs do
not coincide.
46. [3.1]   The inverse of a function $f(x)$ is written $f^{-1}(x)$,
whereas $[f(x)]^{-1}$ means $\dfrac{1}{f(x)}$.
47. [3.5] $\left\{\frac{1}{64}, 64\right\}$ **48.** [3.5] $\{1\}$ **49.** [3.5] $\{16\}$
50. [3.2], [3.3] **(a)** $y = 5e^{-x}\ln x$

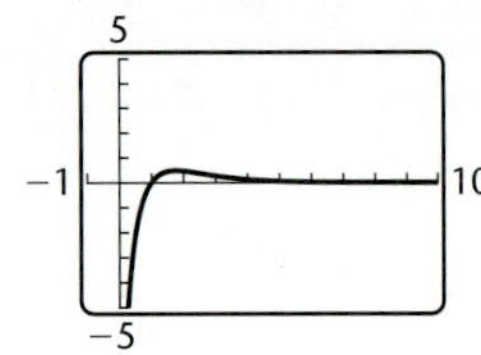

(b) Relative maximum: 0.486 at $x = 1.763$, (1.763, 0.486)
51. [3.1] No   **52.** [3.3] $(1, \infty)$
53. [3.2] $(-0.393, 1.143)$, $(0.827, 1.495)$

Chapter 4

EXERCISE SET 4.1

1. (c)   **3.** (f)   **5.** (b)   **7.** $(-1, 3)$ **9.** $(-0.5, 1.75)$
11. No solution   **13.** $(5, 4)$ **15.** $(1, -3)$ **17.** $(2, -2)$
19. $\left(\frac{39}{11}, -\frac{1}{11}\right)$ **21.** $(1, -1)$
23. $(1, 3)$; consistent, independent
25. $(-4, -2)$; consistent, independent
27. $(-3, 0)$; consistent, independent
29. $(4y + 2, y)$ or $\left(x, \frac{1}{4}x - \frac{1}{2}\right)$; consistent, dependent
31. $(10, 8)$; consistent, independent
33. $(1, 1)$; consistent, independent
35. 10 free rentals, 38 boxes of popcorn
37. Adult: $6, child: $3   **39.** (15, $100)   **41.** 140
43. 6000
45. $1\frac{1}{2}$ servings of spaghetti, 2 servings of lettuce
47. Boat: 20 km/h, stream: 3 km/h
49. $6000 at 7%, $9000 at 9%
51. 6 lb of French roast, 4 lb of Kenyan   **53.** $21,428.57
55. -19
57. $y = 2x^2 - 3x + 4$

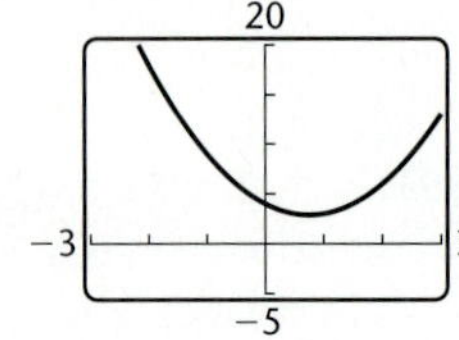

59. ◆   **61.** 4 km
63. First train: 36 km/h, second train: 54 km/h   **65.** $96
67. $m = -\frac{4}{3}$, $b = \frac{1}{3}$ **69.** City: 261 mi, highway: 204 mi

EXERCISE SET 4.2

1. $(3, -2, 1)$ **3.** $(-3, 2, 1)$ **5.** $\left(2, \frac{1}{2}, -2\right)$
7. No solution   **9.** $\left(\dfrac{11y + 19}{5}, y, \dfrac{9y + 11}{5}\right)$
11. $\left(\frac{1}{2}, \frac{2}{3}, -\frac{5}{6}\right)$ **13.** $(1, -2, 4, -1)$
15. Under 10 lb: 60; 10 lb up to 15 lb: 70; 15 lb or more: 20
17. $1\frac{1}{4}$ servings of beef, 1 baked potato, $\frac{3}{4}$ serving
strawberries
19. 4%: $1300; 6%: $900; 7%: $2800
21. Orange juice: $1; bagel: $1.25; coffee: $0.75
23. Airways: 351 billion; bus: 32 billion; railroad: 22 billion
25. Par-3: 4; par-4: 10; par-5: 4
27. **(a)** $f(x) = \frac{12}{247}x^2 - \frac{384}{247}x + 43$; **(b)** 58.5%
29. **(a)** $f(x) = -\frac{135}{238}x^2 + \frac{2659}{238}x + 539$; **(b)** 364 lb
31. $4x - 6y$ **33.** $x + 2y$ **35.**   **37.** $\left(-1, \frac{1}{5}, -\frac{1}{2}\right)$
39. 180°   **41.** $3x + 4y + 2z = 12$
43. $y = -4x^3 + 5x^2 - 3x + 1$
45. Adults: 5; students: 1; children: 94

EXERCISE SET 4.3

1. 3×2 **3.** 1×4 **5.** 3×3 **7.** $\begin{bmatrix} 2 & -1 & 7 \\ 1 & 4 & -5 \end{bmatrix}$
9. $\begin{bmatrix} 1 & -2 & 3 & 12 \\ 2 & 0 & -4 & 8 \\ 0 & 3 & 1 & 7 \end{bmatrix}$
11. $3x - 5y = 1,$
     $x + 4y = -2$
13. $2x + y - 4z = 12,$
     $3x \quad\;\; + 5z = -1,$
     $x - y + z = 2$
15. $\left(\frac{3}{2}, \frac{5}{2}\right)$ **17.** $\left(-\frac{63}{29}, -\frac{114}{29}\right)$ **19.** $\left(-1, \frac{5}{2}\right)$ **21.** $(0, 3)$
23. No solution   **25.** $(3y - 2, y)$ **27.** $(-1, 2, -2)$
29. $\left(\frac{3}{2}, -4, 3\right)$ **31.** $(-1, 6, 3)$
33. $\left(\frac{1}{2}z + \frac{1}{2}, -\frac{1}{2}z - \frac{1}{2}, z\right)$ **35.** $(r - 2, -2r + 3, r)$
37. No solution   **39.** $(1, -3, -2, -1)$ **41.** 1:00 A.M.
43. $8000 at 8%; $12,000 at 10%; $10,000 at 12%   **45.** 8
47. -9 **49.**   **51.** $y = 3x^2 + \frac{5}{2}x - \frac{15}{2}$
53. $\begin{bmatrix} 1 & 5 \\ 0 & 1 \end{bmatrix}, \begin{bmatrix} 1 & 0 \\ 0 & 1 \end{bmatrix}$ **55.** $\left(-\frac{4}{3}, -\frac{1}{3}, 1\right)$
57. $\left(-\frac{14}{13}z - 1, \frac{3}{13}z - 2, z\right)$ **59.** $(-3, 3)$

EXERCISE SET 4.4

1. $x = -3, y = 5$ **3.** $x = -1, y = 1$ **5.** $\begin{bmatrix} -2 & 7 \\ 6 & 2 \end{bmatrix}$

7. $\begin{bmatrix} 1 & 3 \\ 2 & 6 \end{bmatrix}$ **9.** $\begin{bmatrix} 9 & 9 \\ -3 & -3 \end{bmatrix}$ **11.** $\begin{bmatrix} 11 & 13 \\ 5 & 3 \end{bmatrix}$

13. $\begin{bmatrix} -4 & 3 \\ -2 & -4 \end{bmatrix}$ **15.** $\begin{bmatrix} 17 & 9 \\ -2 & 1 \end{bmatrix}$ **17.** $\begin{bmatrix} 0 & 0 \\ 0 & 0 \end{bmatrix}$

19. $\begin{bmatrix} 1 & 2 \\ 4 & 3 \end{bmatrix}$

21. (a) $[150 \quad 80 \quad 40]$; **(b)** $[157.5 \quad 84 \quad 42]$;
(c) $[307.5 \quad 164 \quad 82]$, the total budget for each area in June and July
23. (a) $C = [140 \quad 27 \quad 3 \quad 13 \quad 64]$,
$P = [180 \quad 4 \quad 11 \quad 24 \quad 662]$, $B = [50 \quad 5 \quad 1 \quad 82 \quad 20]$;
(b) $C + 2P + 3B = [650 \quad 50 \quad 28 \quad 307 \quad 1448]$, the total nutritional values of a meal of 1 serving of chicken, 1 cup of potato salad, and 3 broccoli spears

25. $\begin{bmatrix} 1 \\ 40 \end{bmatrix}$ **27.** $\begin{bmatrix} -10 & 28 \\ 14 & -26 \\ 0 & -6 \end{bmatrix}$ **29.** Not defined

31. $\begin{bmatrix} 3 & 16 & 3 \\ 0 & -32 & 0 \\ -6 & 4 & 5 \end{bmatrix}$

33. (a) $\begin{bmatrix} 45.29 & 6.63 & 10.94 & 7.42 & 8.01 \\ 53.78 & 4.95 & 9.83 & 6.16 & 12.56 \\ 47.13 & 8.47 & 12.66 & 8.29 & 9.43 \\ 51.64 & 7.12 & 11.57 & 9.35 & 10.72 \end{bmatrix}$;
(b) $[65 \quad 48 \quad 93 \quad 57]$;
(c) $[12{,}851.86 \quad 1862.1 \quad 3019.81 \quad 2081.9 \quad 2611.56]$;
(d) the total cost, in cents, for each item for the day's meals

35. (a) $\begin{bmatrix} 8 & 15 \\ 6 & 10 \\ 4 & 3 \end{bmatrix}$; **(b)** $[3 \quad 1.50 \quad 2]$; **(c)** $[41 \quad 66]$;
(d) the total cost, in dollars, of ingredients for each coffee shop
37. (a) $[6 \quad 4.50 \quad 5.20]$; **(b)** $\mathbf{PS} = [95.80 \quad 150.60]$

39. $\begin{bmatrix} 2 & -3 \\ 1 & 5 \end{bmatrix}\begin{bmatrix} x \\ y \end{bmatrix} = \begin{bmatrix} 7 \\ -6 \end{bmatrix}$ **41.** $\begin{bmatrix} 1 & 1 & -2 \\ 3 & -1 & 1 \\ 2 & 5 & -3 \end{bmatrix}\begin{bmatrix} x \\ y \\ z \end{bmatrix} = \begin{bmatrix} 6 \\ 7 \\ 8 \end{bmatrix}$

43. $\begin{bmatrix} 3 & -2 & 4 \\ 2 & 1 & -5 \end{bmatrix}\begin{bmatrix} x \\ y \\ z \end{bmatrix} = \begin{bmatrix} 17 \\ 13 \end{bmatrix}$

45. $\begin{bmatrix} -4 & 1 & -1 & 2 \\ 1 & 2 & -1 & -1 \\ -1 & 1 & 4 & -3 \\ 2 & 3 & 5 & -7 \end{bmatrix}\begin{bmatrix} w \\ x \\ y \\ z \end{bmatrix} = \begin{bmatrix} 12 \\ 0 \\ 1 \\ 9 \end{bmatrix}$ **47.** $\{9\}$ **49.** $\{-7\}$

51. ◈

53. $(\mathbf{A} + \mathbf{B})(\mathbf{A} - \mathbf{B}) = \begin{bmatrix} -2 & 1 \\ 2 & -1 \end{bmatrix}$; $\mathbf{A}^2 - \mathbf{B}^2 = \begin{bmatrix} 0 & 3 \\ 0 & -3 \end{bmatrix}$

55. $(\mathbf{A} + \mathbf{B})(\mathbf{A} - \mathbf{B}) = \begin{bmatrix} -2 & 1 \\ 2 & -1 \end{bmatrix}$
$$= \mathbf{A}^2 + \mathbf{BA} - \mathbf{AB} - \mathbf{B}^2$$

57. $\mathbf{A} + \mathbf{B} =$
$$\begin{bmatrix} a_{11} + b_{11} & a_{12} + b_{12} & a_{13} + b_{13} & \cdots & a_{1n} + b_{1n} \\ a_{21} + b_{21} & a_{22} + b_{22} & a_{23} + b_{23} & \cdots & a_{2n} + b_{2n} \\ a_{31} + b_{31} & a_{32} + b_{32} & a_{33} + b_{33} & \cdots & a_{3n} + b_{3n} \\ \vdots & \vdots & \vdots & & \vdots \\ a_{m1} + b_{m1} & a_{m2} + b_{m2} & a_{m3} + b_{m3} & \cdots & a_{mn} + b_{mn} \end{bmatrix}$$
$$= \begin{bmatrix} b_{11} + a_{11} & b_{12} + a_{12} & b_{13} + a_{13} & \cdots & b_{1n} + a_{1n} \\ b_{21} + a_{21} & b_{22} + a_{22} & b_{23} + a_{23} & \cdots & b_{2n} + a_{2n} \\ b_{31} + a_{31} & b_{32} + a_{32} & b_{33} + a_{33} & \cdots & b_{3n} + a_{3n} \\ \vdots & \vdots & \vdots & & \vdots \\ b_{m1} + a_{m1} & b_{m2} + a_{m2} & b_{m3} + a_{m3} & \cdots & b_{mn} + a_{mn} \end{bmatrix}$$
$$= \mathbf{B} + \mathbf{A}$$

59. $(kl)\mathbf{A} =$
$$\begin{bmatrix} (kl)a_{11} & (kl)a_{12} & (kl)a_{13} & \cdots & (kl)a_{1n} \\ (kl)a_{21} & (kl)a_{22} & (kl)a_{23} & \cdots & (kl)a_{2n} \\ (kl)a_{31} & (kl)a_{32} & (kl)a_{33} & \cdots & (kl)a_{3n} \\ \vdots & \vdots & \vdots & & \vdots \\ (kl)a_{m1} & (kl)a_{m2} & (kl)a_{m3} & \cdots & (kl)a_{mn} \end{bmatrix}$$
$$= \begin{bmatrix} k(la_{11}) & k(la_{12}) & k(la_{13}) & \cdots & k(la_{1n}) \\ k(la_{21}) & k(la_{22}) & k(la_{23}) & \cdots & k(la_{2n}) \\ k(la_{31}) & k(la_{32}) & k(la_{33}) & \cdots & k(la_{3n}) \\ \vdots & \vdots & \vdots & & \vdots \\ k(la_{m1}) & k(la_{m2}) & k(la_{m3}) & \cdots & k(la_{mn}) \end{bmatrix}$$
$$= k\begin{bmatrix} la_{11} & la_{12} & la_{13} & \cdots & la_{1n} \\ la_{21} & la_{22} & la_{23} & \cdots & la_{2n} \\ la_{31} & la_{32} & la_{33} & \cdots & la_{3n} \\ \vdots & \vdots & \vdots & & \vdots \\ la_{m1} & la_{m2} & la_{m3} & \cdots & la_{mn} \end{bmatrix}$$
$$= k(l\mathbf{A})$$

61. $(k + l)\mathbf{A} =$
$$\begin{bmatrix} (k+l)a_{11} & (k+l)a_{12} & (k+l)a_{13} & \cdots & (k+l)a_{1n} \\ (k+l)a_{21} & (k+l)a_{22} & (k+l)a_{23} & \cdots & (k+l)a_{2n} \\ (k+l)a_{31} & (k+l)a_{32} & (k+l)a_{33} & \cdots & (k+l)a_{3n} \\ \vdots & \vdots & \vdots & & \vdots \\ (k+l)a_{m1} & (k+l)a_{m2} & (k+l)a_{m3} & \cdots & (k+l)a_{mn} \end{bmatrix}$$
$$=$$
$$\begin{bmatrix} ka_{11}+la_{11} & ka_{12}+la_{12} & ka_{13}+la_{13} & \cdots & ka_{1n}+la_{1n} \\ ka_{21}+la_{21} & ka_{22}+la_{22} & ka_{23}+la_{23} & \cdots & ka_{2n}+la_{2n} \\ ka_{31}+la_{31} & ka_{32}+la_{32} & ka_{33}+la_{33} & \cdots & ka_{3n}+la_{3n} \\ \vdots & \vdots & \vdots & & \vdots \\ ka_{m1}+la_{m1} & ka_{m2}+la_{m2} & ka_{m3}+la_{m3} & \cdots & ka_{mn}+la_{mn} \end{bmatrix}$$
$$= k\mathbf{A} + l\mathbf{A}$$

EXERCISE SET 4.5

1. Yes **3.** No **5.** $\begin{bmatrix} -3 & 2 \\ 5 & -3 \end{bmatrix}$ **7.** $\begin{bmatrix} 2 & -3 \\ -7 & 11 \end{bmatrix}$

9. Does not exist **11.** $\begin{bmatrix} \frac{3}{8} & -\frac{1}{4} & \frac{1}{8} \\ -\frac{1}{8} & \frac{3}{4} & -\frac{3}{8} \\ -\frac{1}{4} & \frac{1}{2} & \frac{1}{4} \end{bmatrix}$

13. $\begin{bmatrix} \frac{1}{3} & 0 & \frac{1}{3} \\ -\frac{2}{5} & \frac{2}{5} & \frac{1}{5} \\ \frac{2}{15} & \frac{1}{5} & -\frac{1}{15} \end{bmatrix}$ **15.** Does not exist **17.** $\begin{bmatrix} 0.4 & -0.6 \\ 0.2 & -0.8 \end{bmatrix}$

19. $\begin{bmatrix} -1 & -1 & -6 \\ 1 & 0 & 2 \\ 0 & 1 & 3 \end{bmatrix}$ **21.** $\begin{bmatrix} 1 & 1 & 2 \\ 1 & 1 & 1 \\ 2 & 3 & 4 \end{bmatrix}$

23. $\begin{bmatrix} 0.5 & 0.5 & 0.5 \\ 0.5 & -0.5 & -0.5 \\ 0.5 & -0.5 & 0.5 \end{bmatrix}$ **25.** Does not exist

27. $\begin{bmatrix} 1 & -2 & 3 & 8 \\ 0 & 1 & -3 & 1 \\ 0 & 0 & 1 & -2 \\ 0 & 0 & 0 & -1 \end{bmatrix}$ **29.** $\begin{bmatrix} 0.25 & 0.25 & 1.25 & -0.25 \\ 0.5 & 1.25 & 1.75 & -1 \\ -0.25 & -0.25 & -0.75 & 0.75 \\ 0.25 & 0.5 & 0.75 & -0.5 \end{bmatrix}$

31. $(2, -2)$ **33.** $(0, 2)$ **35.** $(3, 4)$ **37.** $(3, -3, -2)$
39. $(-1, 0, 1)$ **41.** $(5, 7, -6)$ **43.** $(1, -1, 0, 1)$
45. $(2, 0, -3, 1)$ **47.** 50 sausages, 95 hot dogs
49. Topsoil: \$239; mulch: \$179; pea gravel: \$222

51.

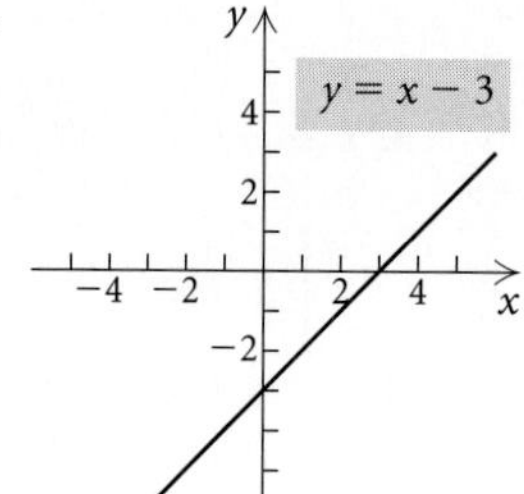

53.

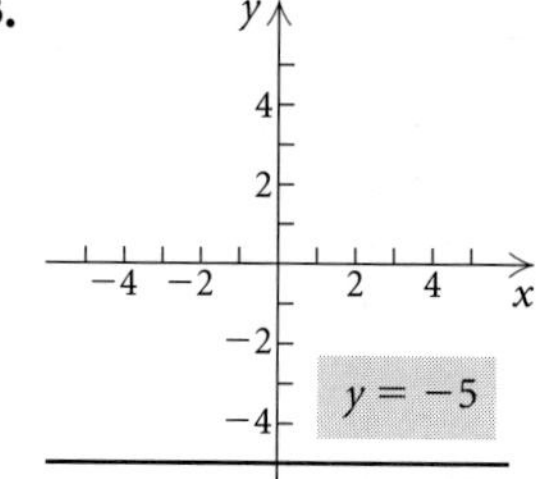

55. $\Diamond$ **57.** $\mathbf{A}^{-1}$ exists if and only if $x \neq 0$. $\mathbf{A}^{-1} = \begin{bmatrix} \frac{1}{x} \end{bmatrix}$

59. $\mathbf{A}^{-1}$ exists if and only if $xyz \neq 0$. $\mathbf{A}^{-1} = \begin{bmatrix} 0 & 0 & \frac{1}{z} \\ 0 & \frac{1}{y} & 0 \\ \frac{1}{x} & 0 & 0 \end{bmatrix}$

EXERCISE SET 4.6

1. (f) **3.** (h) **5.** (g) **7.** (b)

9.

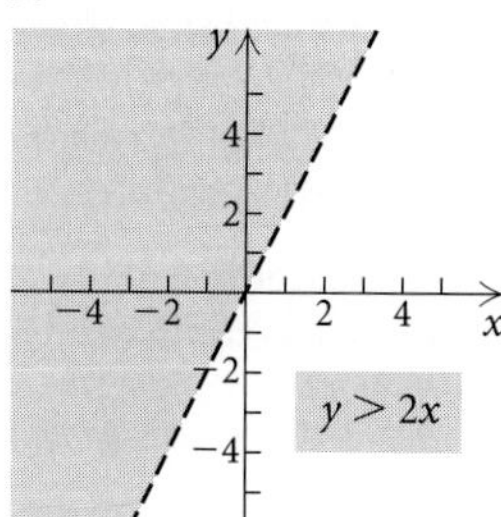

11.

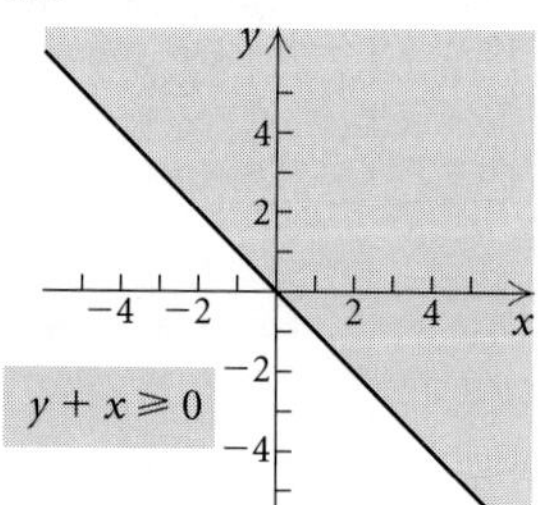

13.

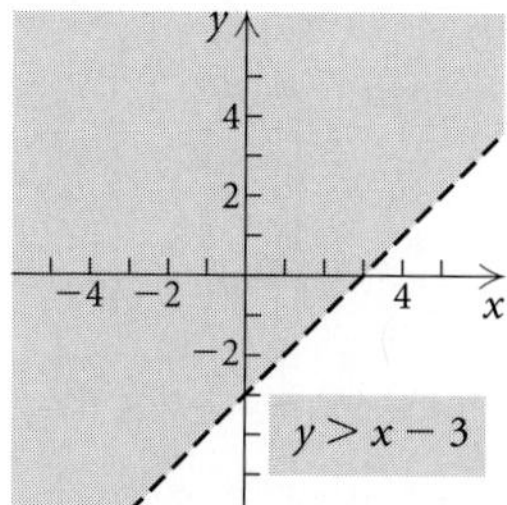

15.

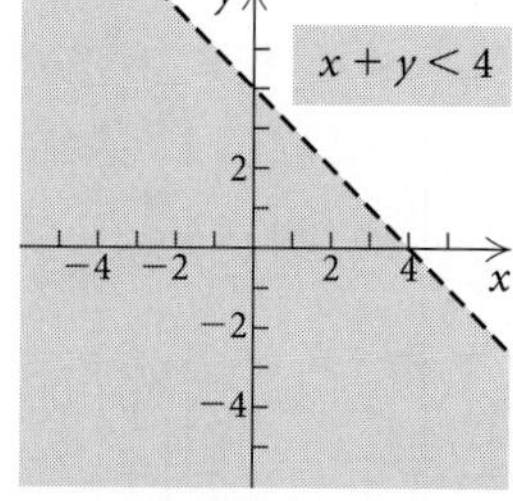

17. 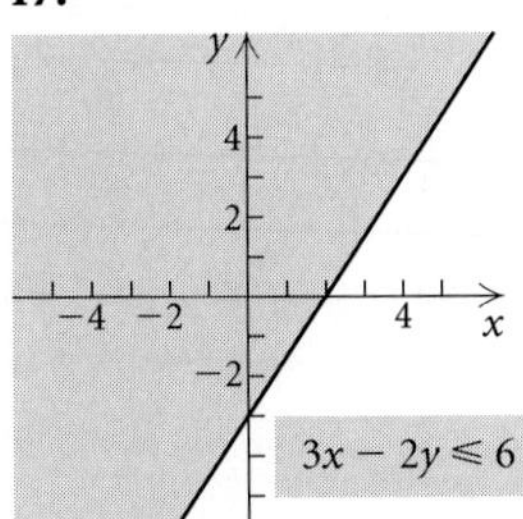
$$3x - 2y \le 6$$

19. 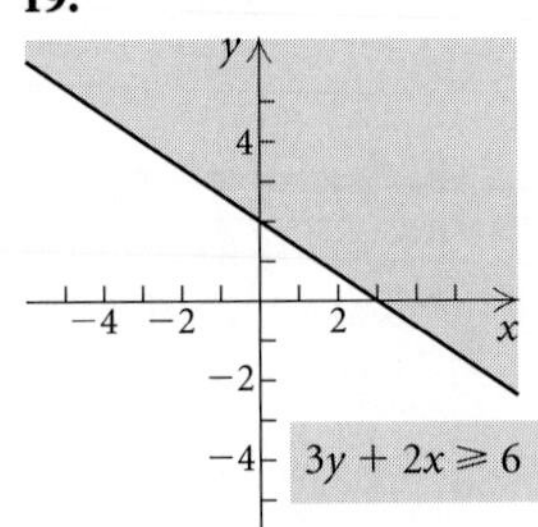
$$3y + 2x \ge 6$$

37. 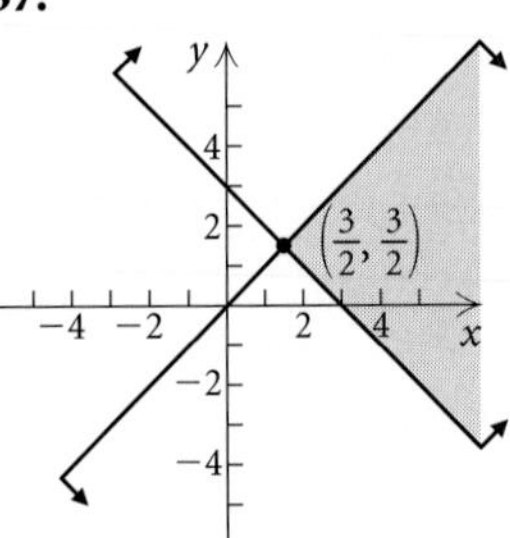

39. 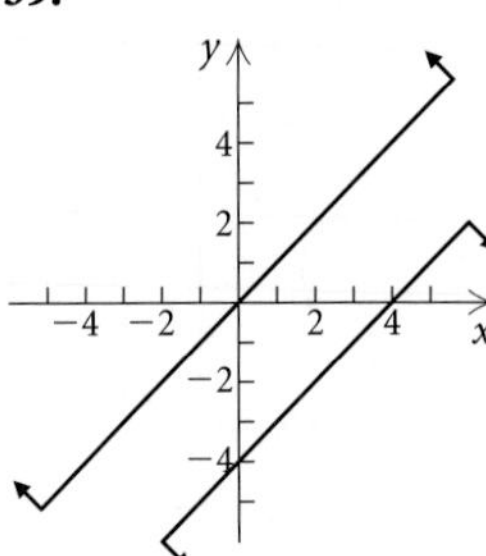

21. 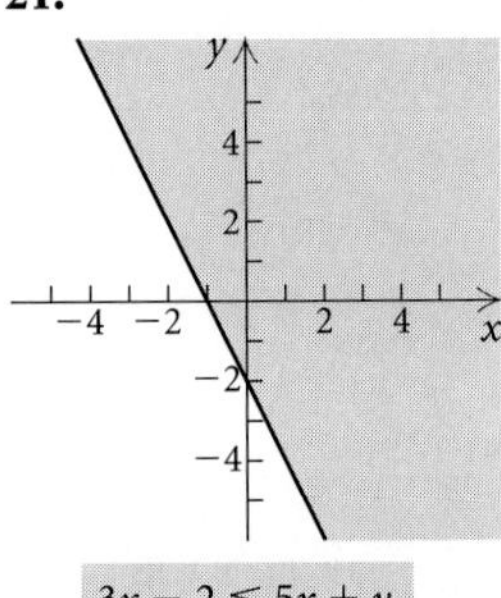
$$3x - 2 \le 5x + y$$

23. 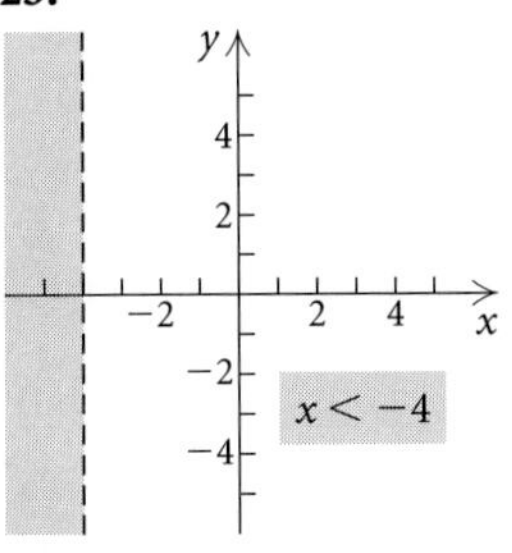
$$x < -4$$

41. 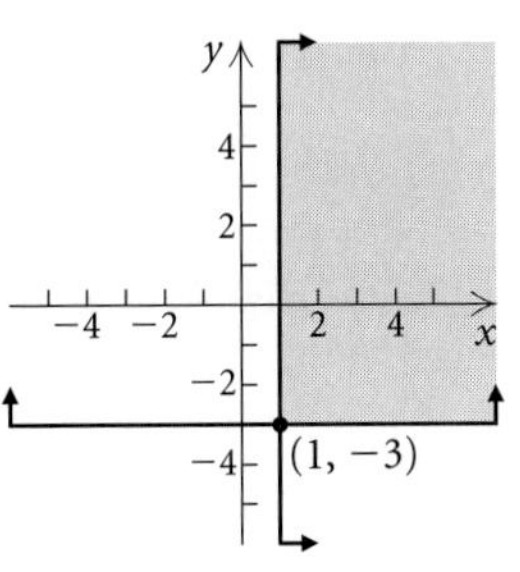

43. 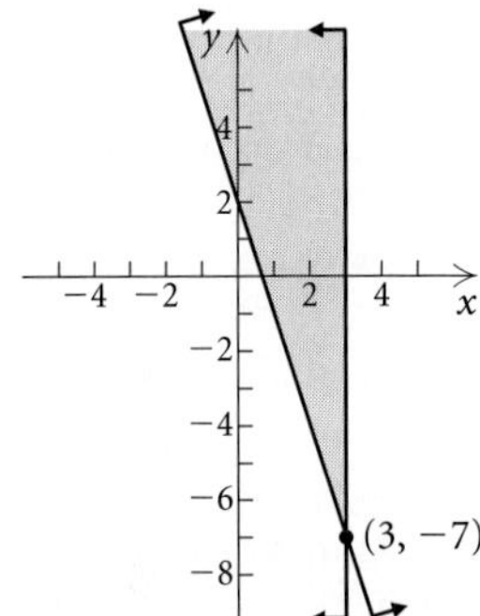

25. 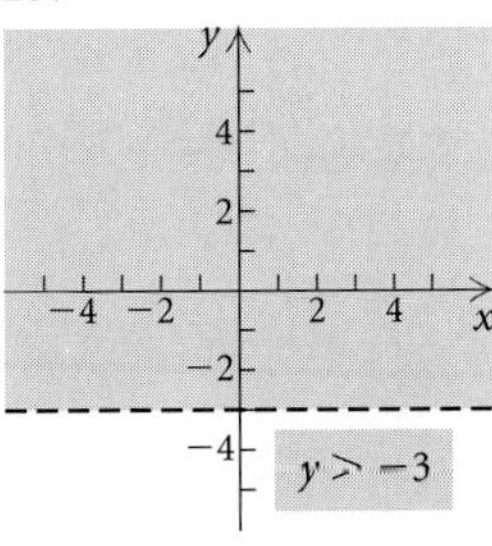
$$y > -3$$

27. 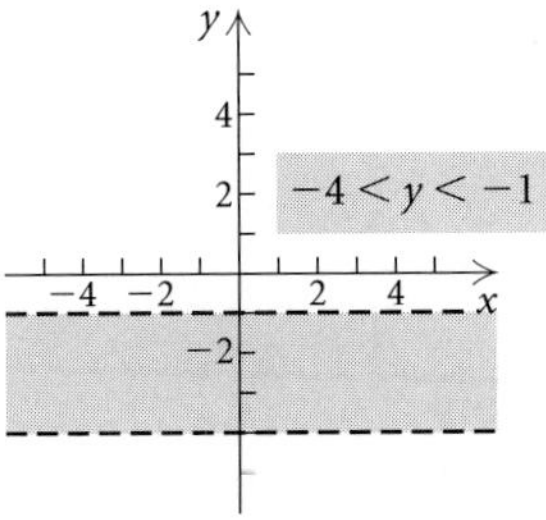
$$-4 < y < -1$$

45. 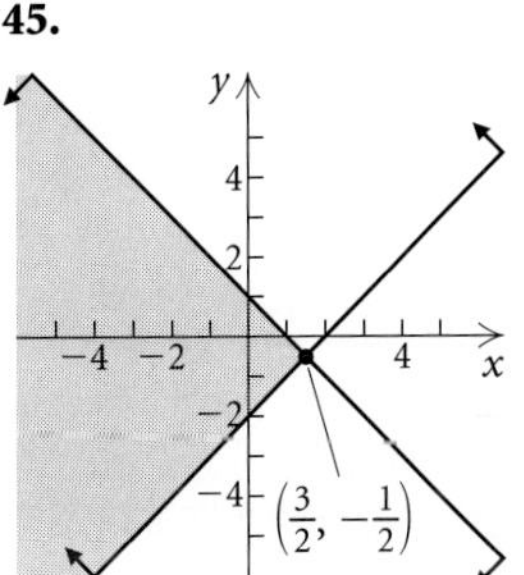

47. 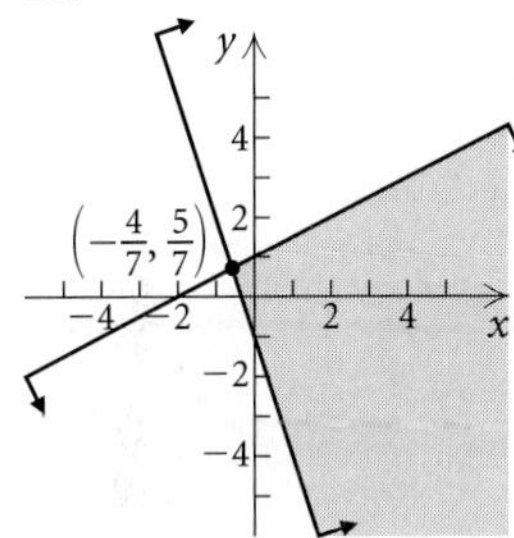

29. 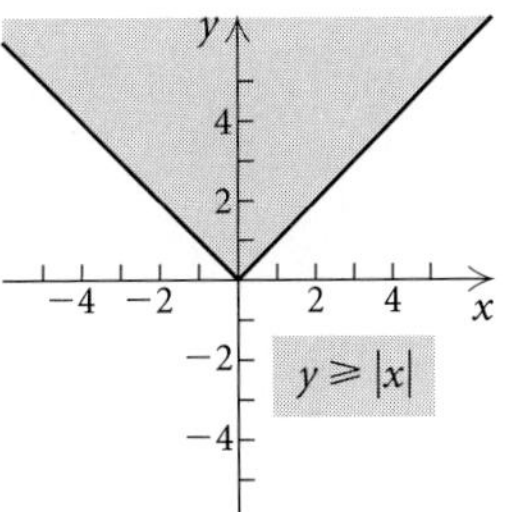
$$y \ge |x|$$

49. 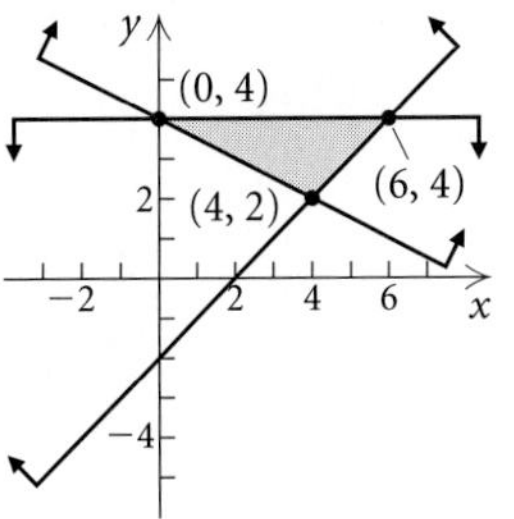

51. 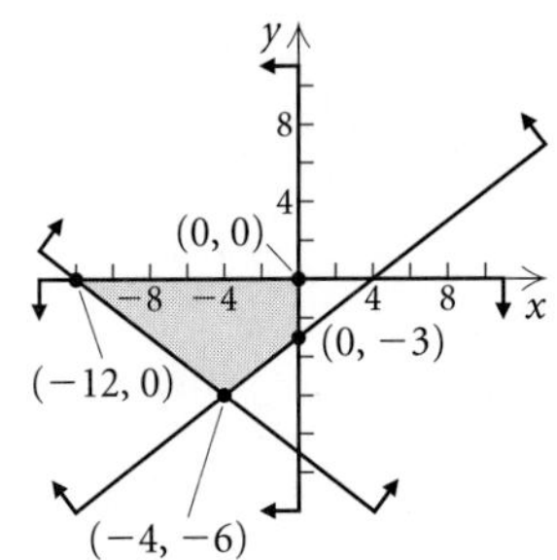

31. (f)   **33.** (a)   **35.** (b)

53.

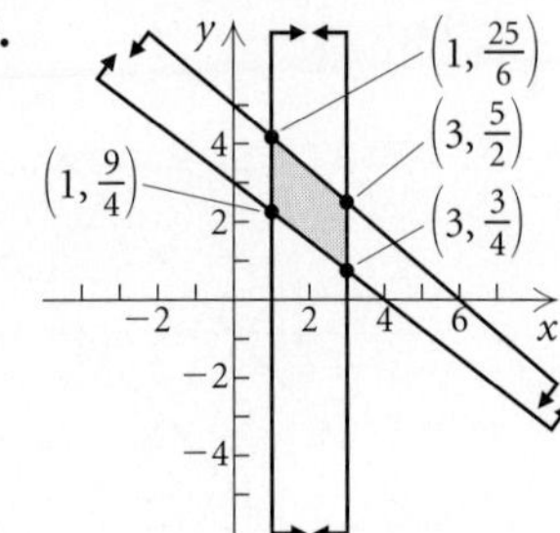

55. Maximum: 179 when $x = 7$ and $y = 0$; minimum: 48 when $x = 0$ and $y = 4$
57. Maximum: 216 when $x = 0$ and $y = 6$; minimum: 0 when $x = 0$ and $y = 0$
59. Maximum income of \$18 when 100 of each type of biscuit is made
61. Maximum profit of \$11,000 is achieved by producing 100 units of lumber and 300 units of plywood.
63. Minimum cost of $\$36\frac{12}{13}$ is achieved by using $1\frac{11}{13}$ sacks of soybean meal and $1\frac{11}{13}$ sacks of oats.
65. Maximum income of \$3110 is achieved when \$22,000 is invested in corporate bonds and \$18,000 is invested in municipal bonds.
67. Minimum cost of \$460 thousand is achieved using 30 P_1's and 10 P_2's.
69. Maximum profit per day of \$192 is achieved when 2 knit suits and 4 worsted suits are made.
71. Minimum weekly cost of \$19.05 when 1.5 lb of meat and 3 lb of cheese are used
73. Maximum total number is 800, or 550 of A and 250 of B.
75. $(x - 5)(x + 3)$ **77.** $\dfrac{5x - 10}{(x + 1)(x - 4)}$ **79.** ◈

81.

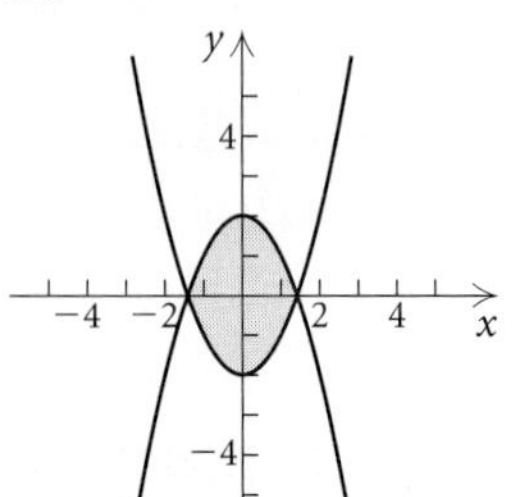

83.

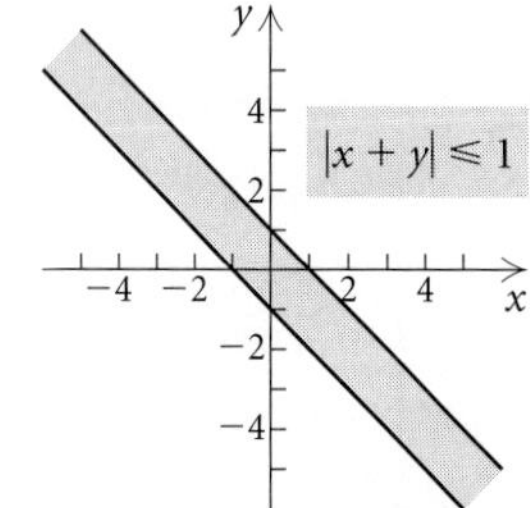

85.

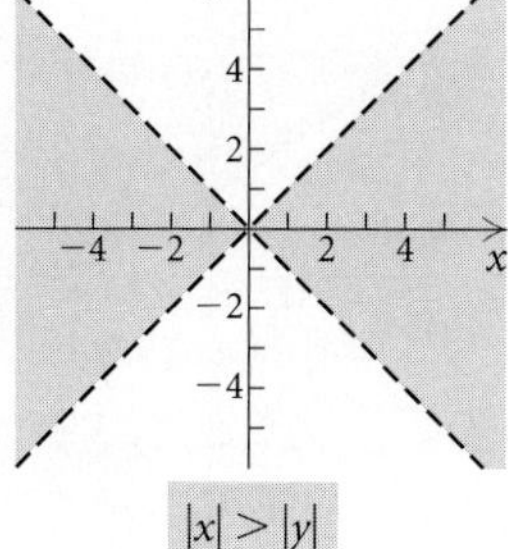

87. Maximum income of \$28,500 is achieved by making 30 less expensive assemblies and 30 more expensive assemblies.

EXERCISE SET 4.7

1. $\dfrac{2}{x - 3} - \dfrac{1}{x + 2}$ **3.** $-\dfrac{4}{3x - 1} + \dfrac{5}{2x - 1}$

5. $-\dfrac{3}{x - 2} + \dfrac{2}{x + 2} + \dfrac{4}{x + 1}$

7. $-\dfrac{3}{(x + 2)^2} - \dfrac{1}{x + 2} + \dfrac{1}{x - 1}$ **9.** $\dfrac{3}{x - 1} - \dfrac{4}{2x - 1}$

11. $x - 2 - \dfrac{\frac{11}{4}}{(x + 1)^2} + \dfrac{\frac{17}{16}}{x + 1} - \dfrac{\frac{17}{16}}{x - 3}$

13. $\dfrac{3x + 5}{x^2 + 2} - \dfrac{4}{x - 1}$ **15.** $-\dfrac{2}{x + 2} + \dfrac{10}{(x + 2)^2} + \dfrac{3}{2x - 1}$

17. $3x + 1 + \dfrac{2}{2x - 1} + \dfrac{3}{x + 1}$ **19.** $-\dfrac{1}{x - 3} + \dfrac{3x}{x^2 + 2x - 5}$

21. $\dfrac{5}{3x + 5} - \dfrac{3}{x + 1} + \dfrac{4}{(x + 1)^2}$ **23.** $\dfrac{8}{4x - 5} + \dfrac{3}{3x + 2}$

25. $\dfrac{2x - 5}{3x^2 + 1} - \dfrac{2}{x - 2}$ **27.** $(2x + 5y)(2x - 5y)$

29. $(x + 4)^2$ **31.** ◈   **33.** $-\dfrac{\frac{1}{2a^2}x}{x^2 + a^2} + \dfrac{\frac{1}{4a^2}}{x - a} + \dfrac{\frac{1}{4a^2}}{x + a}$

35. $-\dfrac{3}{25(\ln x + 2)} + \dfrac{3}{25(\ln x - 3)} + \dfrac{7}{5(\ln x - 3)^2}$
37. ◈

REVIEW EXERCISES, CHAPTER 4

1. [4.1] (a)   **2.** [4.1] (e)   **3.** [4.1] (h)   **4.** [4.1] (d)
5. [4.6] (b)   **6.** [4.6] (g)   **7.** [4.6] (c)   **8.** [4.6] (f)
9. [4.1] $(-2, -2)$ **10.** [4.1] $(-5, 4)$ **11.** [4.1] No solution
12. [4.1] $(y - 2, y)$, or $(x, x + 2)$ **13.** [4.2] No solution
14. [4.2] $(0, 0, 0)$ **15.** [4.2] $(-5, 13, 8, 2)$
16. [4.1], [4.2] Consistent: 9, 10, 12, 14, 15; the others are inconsistent.
17. [4.1], [4.2] Dependent: 12; the others are independent.
18. [4.3] $(1, 2)$ **19.** [4.3] $(-3, 4, -2)$
20. [4.3] $\left(\dfrac{z}{2}, -\dfrac{z}{2}, z\right)$ **21.** [4.3] $(-4, 1, -2, 3)$

22. [4.1] 31 nickels, 44 dimes
23. [4.1] \$1600 at 10%, \$3400 at 10.5%
24. [4.2] 1 bagel, $\frac{1}{2}$ serving cream cheese, 2 bananas
25. [4.2] 74.5, 68.5, 82
26. [4.2] **(a)** $f(x) = -\frac{23}{350}x^2 + \frac{171}{350}x + \frac{69}{10}$; **(b)** 2.1 lb

27. [4.4] $\begin{bmatrix} 0 & -1 & 6 \\ 3 & 1 & -2 \\ -2 & 1 & -2 \end{bmatrix}$ **28.** [4.4] $\begin{bmatrix} -3 & 3 & 0 \\ -6 & -9 & 6 \\ 6 & 0 & -3 \end{bmatrix}$

29. [4.4] $\begin{bmatrix} -1 & 1 & 0 \\ -2 & -3 & 2 \\ 2 & 0 & -1 \end{bmatrix}$ **30.** [4.4] $\begin{bmatrix} -2 & 2 & 6 \\ 1 & -8 & 18 \\ 2 & 1 & -15 \end{bmatrix}$

31. [4.4] Not possible   **32.** [4.4] $\begin{bmatrix} 2 & -1 & -6 \\ 1 & 5 & -2 \\ -2 & -1 & 4 \end{bmatrix}$

33. [4.4] $\begin{bmatrix} 3 & -2 & -6 \\ 3 & 8 & -4 \\ -4 & -1 & 5 \end{bmatrix}$ **34.** [4.4] $\begin{bmatrix} -2 & -1 & 18 \\ 5 & -3 & -2 \\ -2 & 3 & -8 \end{bmatrix}$

35. [4.4] **(a)** $\begin{bmatrix} 46.1 & 5.9 & 10.1 & 8.5 & 11.4 \\ 54.6 & 4.6 & 9.6 & 7.6 & 10.6 \\ 48.9 & 5.5 & 12.7 & 9.4 & 9.3 \\ 51.3 & 4.8 & 11.3 & 6.9 & 12.7 \end{bmatrix}$

(b) [32 19 43 38];
(c) [6564.7 695.1 1481.1 1082.8 1448.7];
(d) the total cost, in cents, for each item for the day's meal

36. [4.5] $\begin{bmatrix} -\frac{1}{2} & 0 \\ \frac{1}{6} & \frac{1}{3} \end{bmatrix}$ **37.** [4.5] $\begin{bmatrix} 0 & 0 & \frac{1}{4} \\ 0 & -\frac{1}{2} & 0 \\ \frac{1}{3} & 0 & 0 \end{bmatrix}$

38. [4.5] $\begin{bmatrix} 1 & 0 & 0 & 0 \\ 0 & \frac{1}{9} & \frac{5}{18} & 0 \\ 0 & -\frac{1}{9} & \frac{2}{9} & 0 \\ 0 & 0 & 0 & 1 \end{bmatrix}$

39. [4.4] $\begin{bmatrix} 3 & -2 & 4 \\ 1 & 5 & -3 \\ 2 & -3 & 7 \end{bmatrix}\begin{bmatrix} x \\ y \\ z \end{bmatrix} = \begin{bmatrix} 13 \\ 7 \\ -8 \end{bmatrix}$ **40.** [4.5] $(-8, 7)$

41. [4.5] $(1, -2, 5)$ **42.** [4.5] $(2, -1, 1, -3)$
43. [4.6]

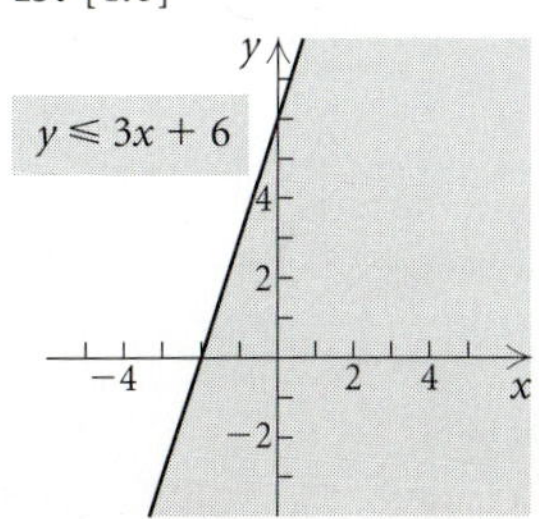

44. [4.6]

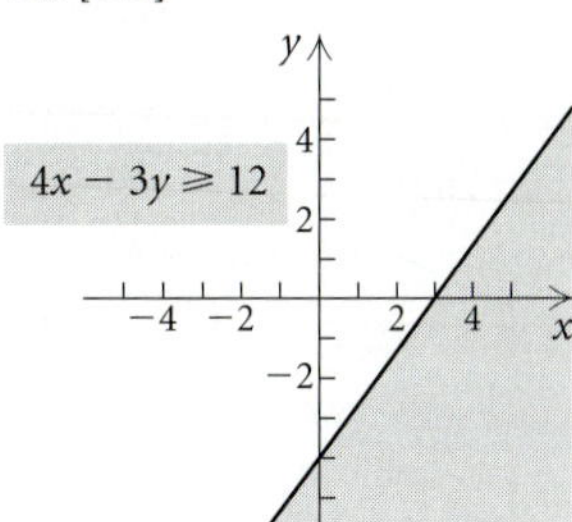

45. [4.6]

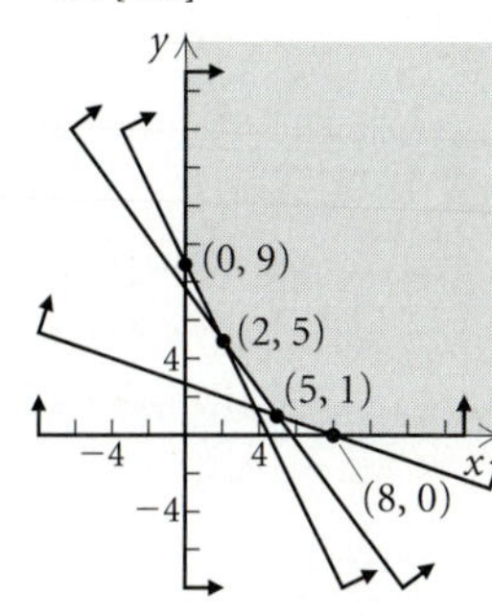

46. [4.6] Minimum $= 52$ at $(2, 4)$; maximum $= 92$ at $(2, 8)$
47. [4.6] Maximum score of 96 when 0 group A questions and 8 group B questions are answered
48. [4.7] $\dfrac{5}{x + 1} - \dfrac{5}{x + 2} - \dfrac{5}{(x + 2)^2}$
49. [4.7] $\dfrac{2}{2x - 3} - \dfrac{5}{x + 4}$
50. [4.1] ◈ During a holiday season, a caterer sold a total of 55 food trays. She sold 15 more seafood trays than cheese trays. How many of each were sold?
51. [4.4] ◈ In general, $(\mathbf{AB})^2 \ne \mathbf{A}^2\mathbf{B}^2$. $(\mathbf{AB})^2 = \mathbf{ABAB}$ and $\mathbf{A}^2\mathbf{B}^2 = \mathbf{AABB}$. Since matrix multiplication is not commutative, $\mathbf{BA} \ne \mathbf{AB}$, so $(\mathbf{AB})^2 \ne \mathbf{A}^2\mathbf{B}^2$.
52. [4.2] 12%: \$10,000; 13%: \$12,000; $14\frac{1}{2}$%: \$18,000
53. [4.1] $\left(\frac{5}{18}, \frac{1}{7}\right)$ **54.** [4.2] $\left(1, \frac{1}{2}, \frac{1}{3}\right)$
55. [4.6]   **56.** [4.6]

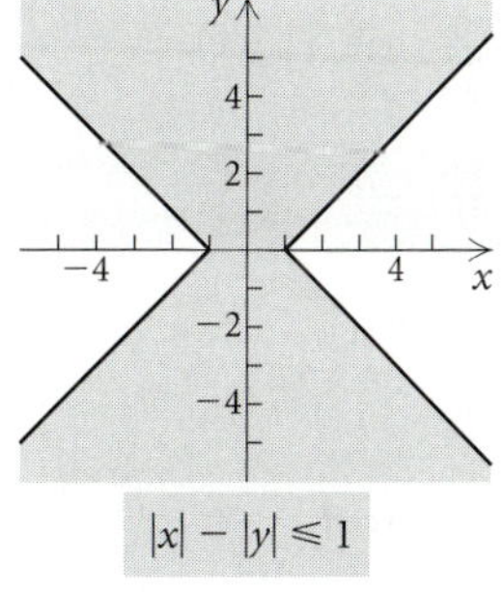

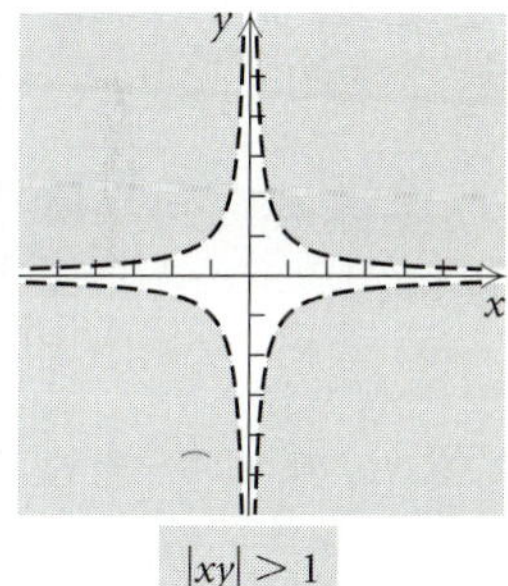

Chapter 5

EXERCISE SET 5.1

1. (f)   **3.** (b)   **5.** (d)

7. V: $(0, 0)$; F: $(0, 5)$; D: $y = -5$

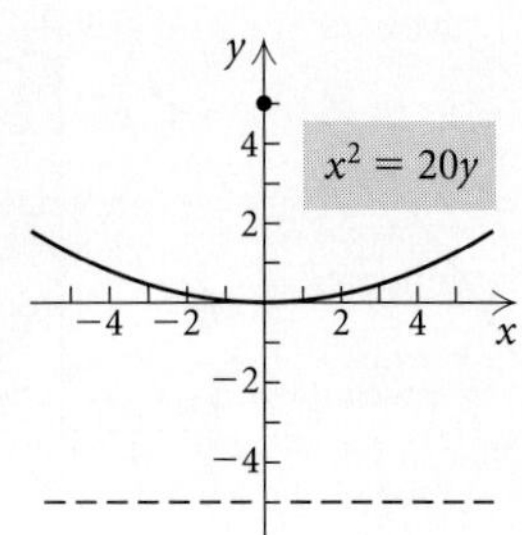

9. V: $(0, 0)$; F: $\left(-\frac{3}{2}, 0\right)$; D: $x = \frac{3}{2}$

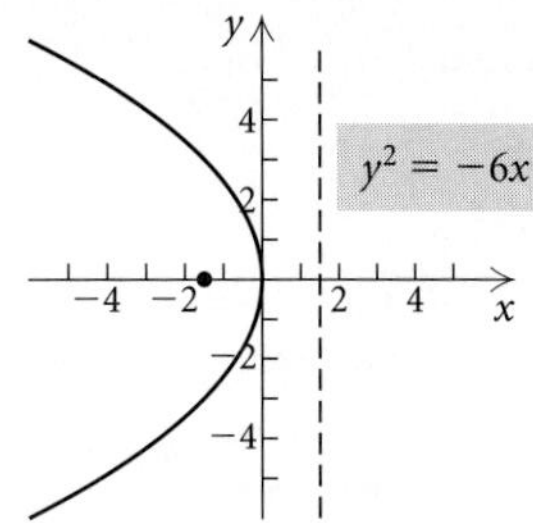

11. V: $(0, 0)$; F: $(0, 1)$; D: $y = -1$

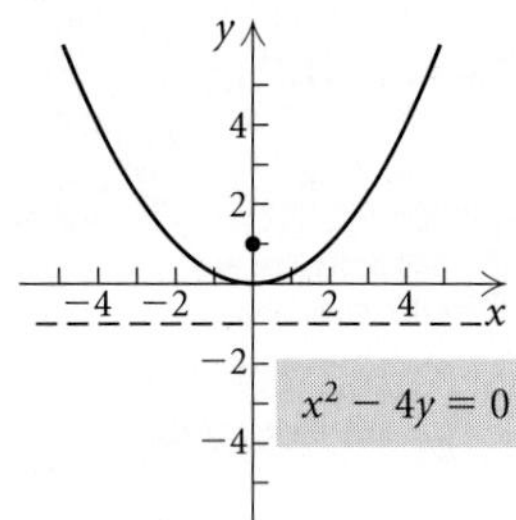

13. V: $(0, 0)$; F: $\left(\frac{1}{8}, 0\right)$; D: $x = -\frac{1}{8}$

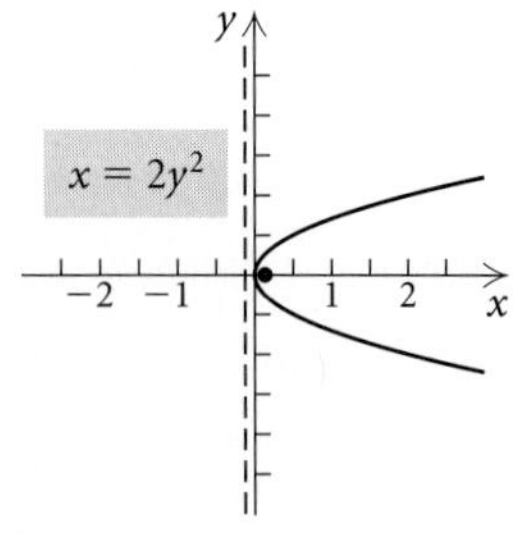

15. $y^2 = 16x$ **17.** $x^2 = -4\pi y$

19. $(y - 2)^2 = 14\left(x + \frac{1}{2}\right)$

21. V: $(-2, 1)$; F: $\left(-2, -\frac{1}{2}\right)$; D: $y = \frac{5}{2}$

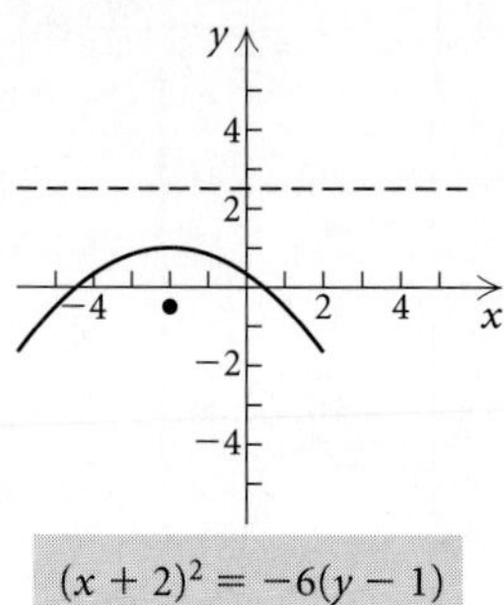

23. V: $(-1, -3)$; F: $\left(-1, -\frac{7}{2}\right)$; D: $y = -\frac{5}{2}$

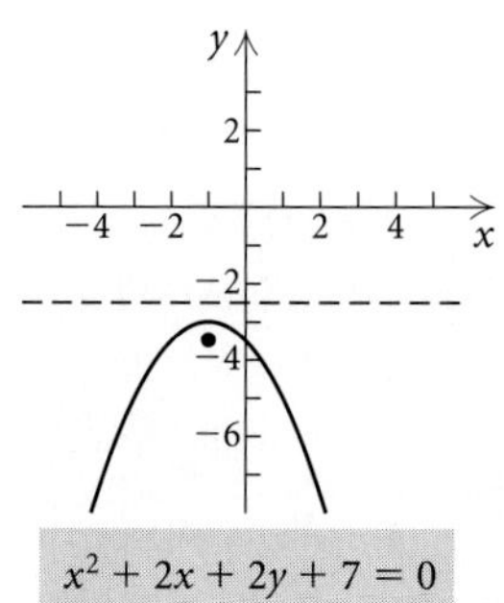

25. V: $(0, -2)$; F: $\left(0, -1\frac{3}{4}\right)$; D: $y = -2\frac{1}{4}$

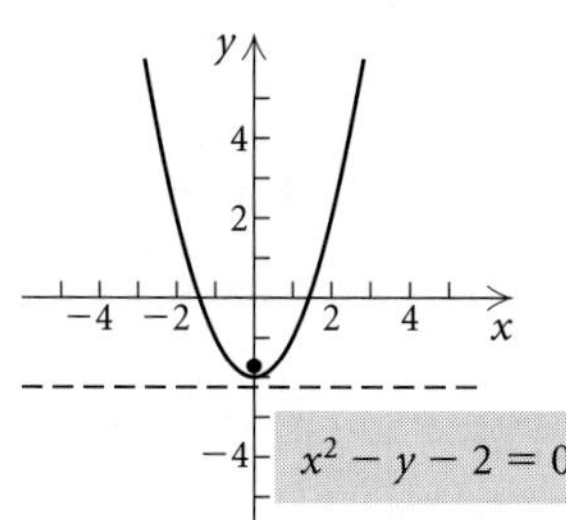

27. V: $(-2, -1)$; F: $\left(-2, -\frac{3}{4}\right)$; D: $y = -1\frac{1}{4}$

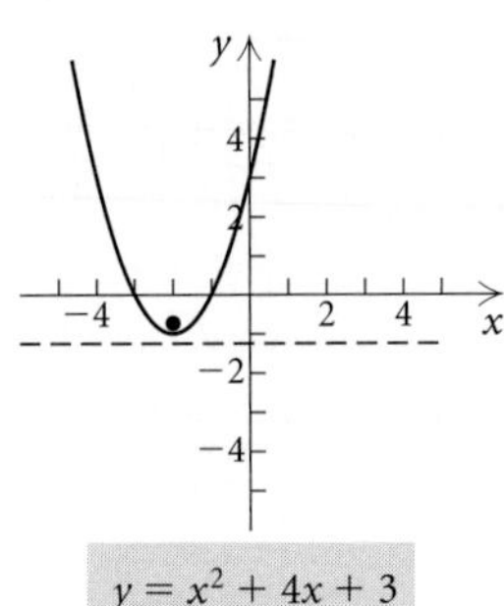

$$y = x^2 + 4x + 3$$

29. V: $\left(5\frac{3}{4}, \frac{1}{2}\right)$; F: $\left(6, \frac{1}{2}\right)$; D: $x = 5\frac{1}{2}$

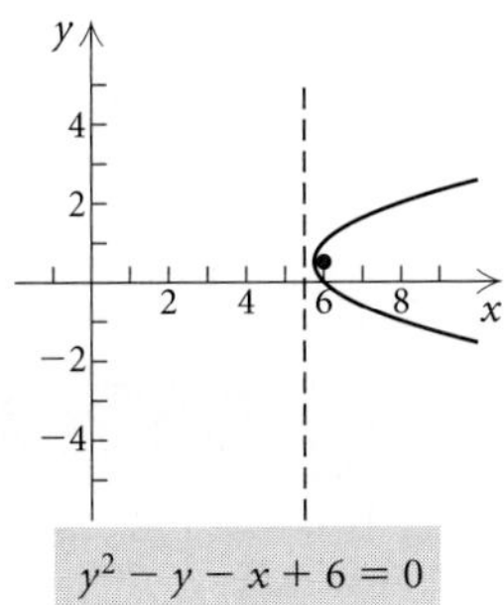

$$y^2 - y - x + 6 = 0$$

31. (a) $y^2 = 16x$; (b) $3\frac{33}{64}$ ft   **33.** $\frac{2}{3}$ ft, or 8 in.
35. $(x + 5)^2$ **37.** $(1, -2)$; 3   **39.** ◈
41. $(x + 1)^2 = -4(y - 2)$
43. V: $(0.867, 0.348)$; F: $(0.867, -0.191)$; D: $y = 0.887$
45. 10 ft, 11.6 ft, 16.4 ft, 24.4 ft, 35.6 ft, 50 ft

EXERCISE SET 5.2

1. (b)   **3.** (d)   **5.** (a)
7. $(7, -2)$; 8

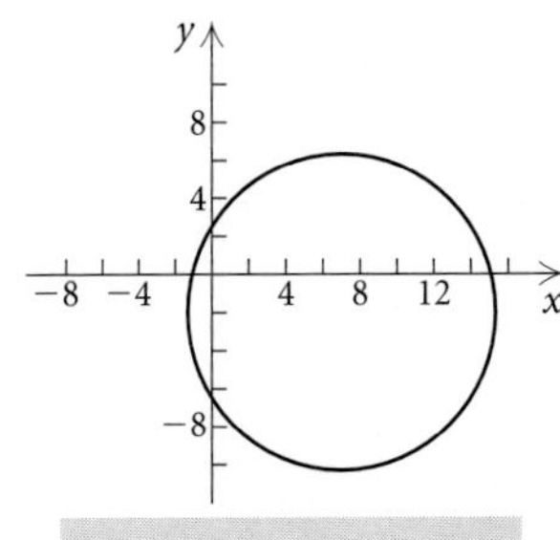

$$x^2 + y^2 - 14x + 4y = 11$$

9. $(-2, 3)$; 5

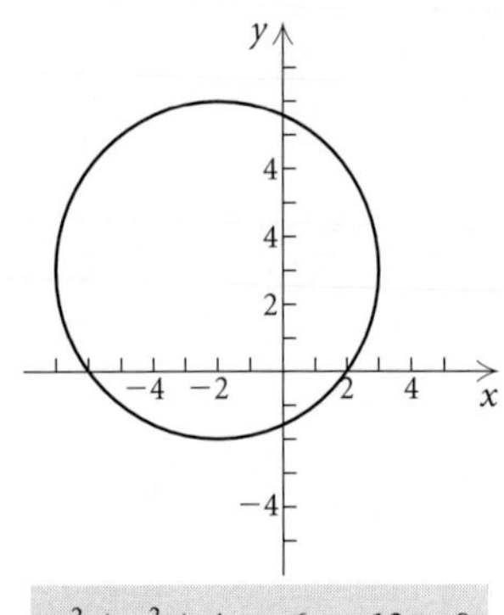

$$x^2 + y^2 + 4x - 6y - 12 = 0$$

11. $(-3, 5)$; $\sqrt{34}$ **13.** $\left(\frac{9}{2}, -2\right)$; $\dfrac{5\sqrt{5}}{2}$

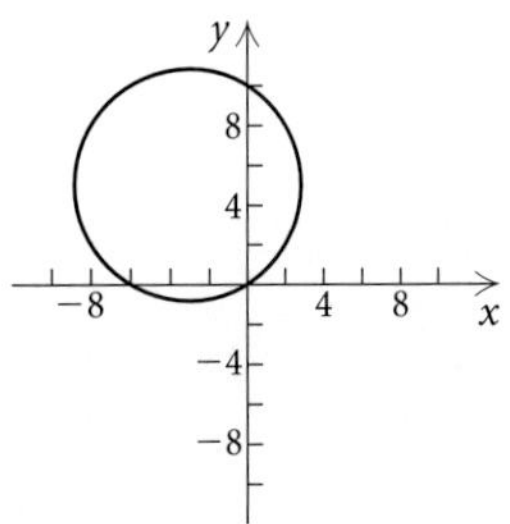

$$x^2 + y^2 + 6x - 10y = 0$$

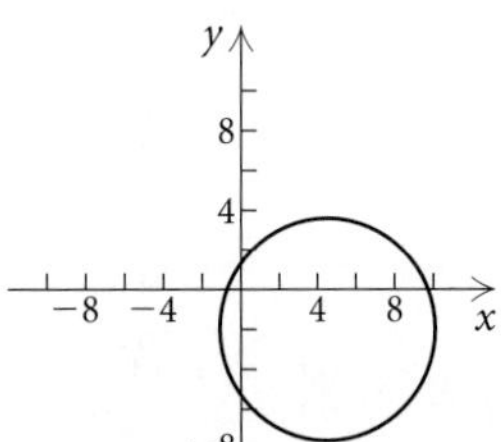

$$x^2 + y^2 - 9x = 7 - 4y$$

15. (c)   **17.** (d)
19. V: $(2, 0), (-2, 0)$;
     F: $(\sqrt{3}, 0), (-\sqrt{3}, 0)$

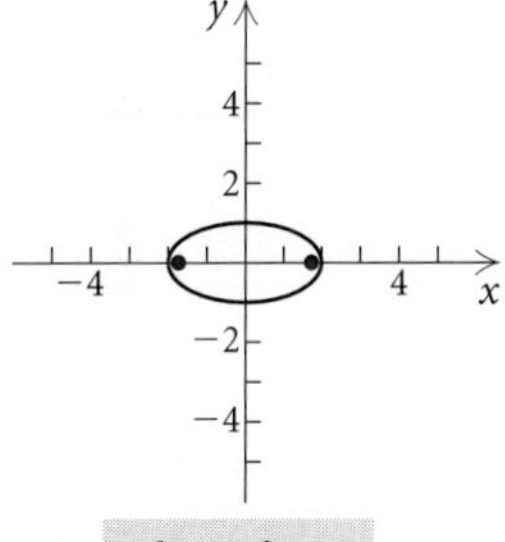

$$\frac{x^2}{4} + \frac{y^2}{1} = 1$$

21. V: $(0, 4), (0, -4)$;
     F: $(0, \sqrt{7}), (0, -\sqrt{7})$

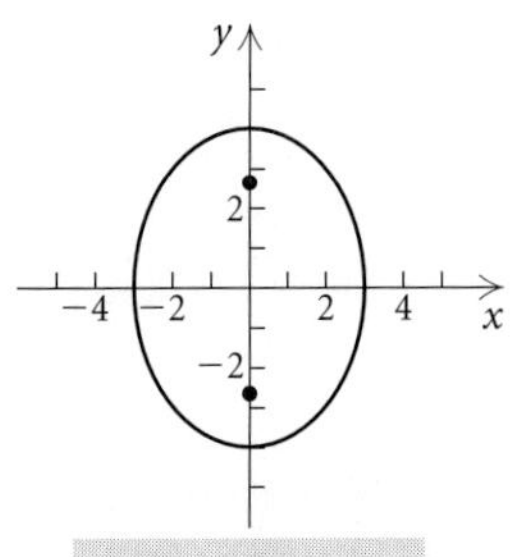

$$16x^2 + 9y^2 = 144$$

23. V: $(-\sqrt{3}, 0)$, $(\sqrt{3}, 0)$;
F: $(-1, 0)$, $(1, 0)$

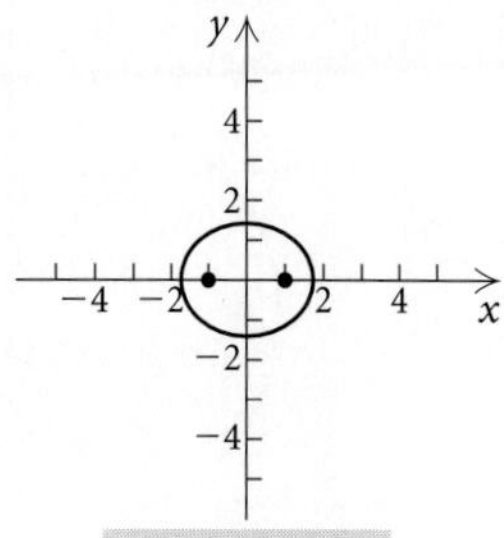

$$2x^2 + 3y^2 = 6$$

25. V: $\left(-\dfrac{1}{2}, 0\right)$, $\left(\dfrac{1}{2}, 0\right)$;
F: $\left(-\dfrac{\sqrt{5}}{6}, 0\right)$, $\left(\dfrac{\sqrt{5}}{6}, 0\right)$

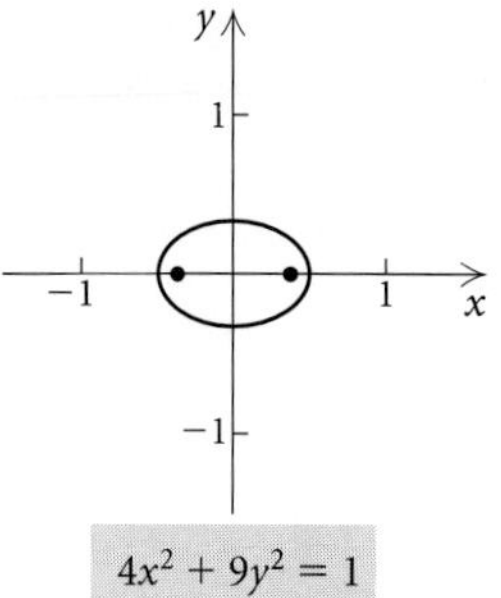

$$4x^2 + 9y^2 = 1$$

27. $\dfrac{x^2}{49} + \dfrac{y^2}{40} = 1$ **29.** $\dfrac{x^2}{25} + \dfrac{y^2}{64} = 1$ **31.** $\dfrac{x^2}{9} + \dfrac{y^2}{5} = 1$

33. C: $(1, 2)$; V: $(4, 2)$, $(-2, 2)$; F: $(1 + \sqrt{5}, 2)$, $(1 - \sqrt{5}, 2)$

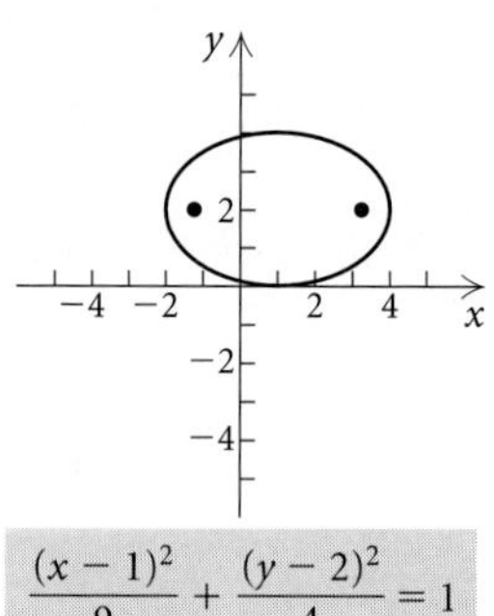

$$\dfrac{(x - 1)^2}{9} + \dfrac{(y - 2)^2}{4} = 1$$

35. C: $(-3, 5)$; V: $(-3, 11)$, $(-3, -1)$; F: $(-3, 5 + \sqrt{11})$, $(-3, 5 - \sqrt{11})$

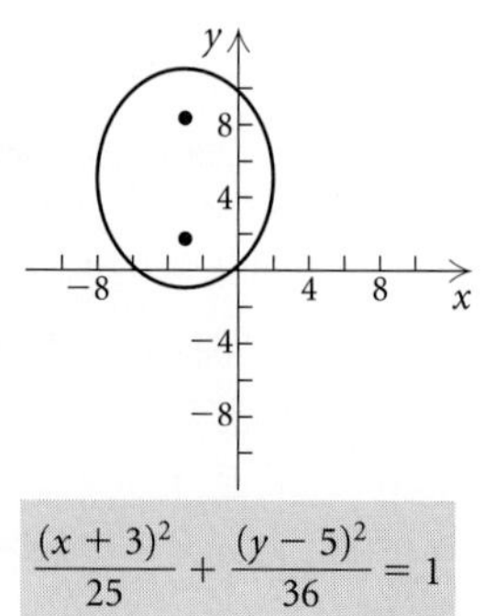

$$\dfrac{(x + 3)^2}{25} + \dfrac{(y - 5)^2}{36} = 1$$

37. C: $(-2, 1)$;
V: $(-10, 1)$, $(6, 1)$;
F: $(-6, 1)$, $(2, 1)$

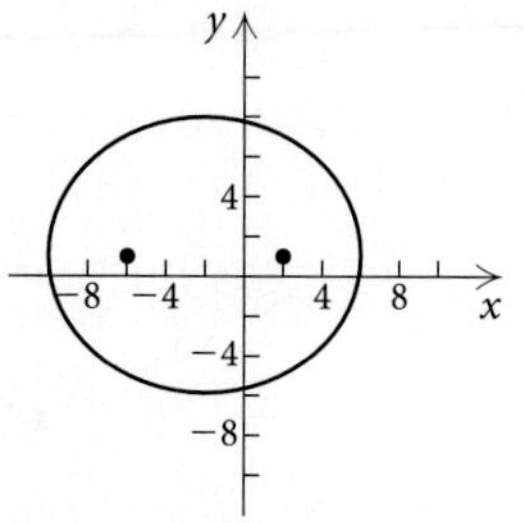

$$3(x + 2)^2 + 4(y - 1)^2 = 192$$

39. C: $(2, -1)$;
V: $(-1, -1)$, $(5, -1)$;
F: $(2 + \sqrt{5}, -1)$,
$(2 - \sqrt{5}, -1)$

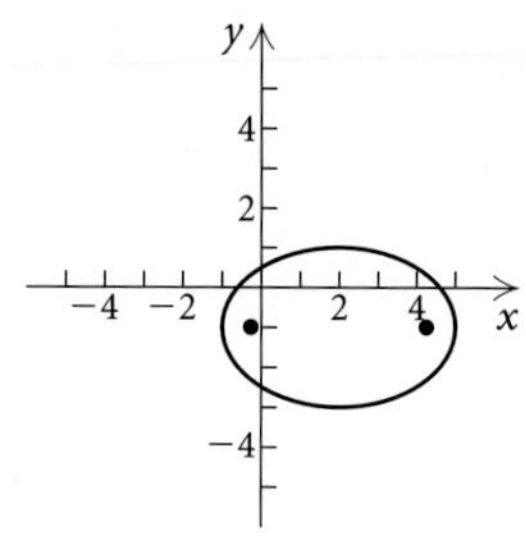

$$4x^2 + 9y^2 - 16x + 18y - 11 = 0$$

41. C: $(1, 1)$; V: $(1, 3)$, $(1, -1)$;
F: $(1, 1 + \sqrt{3})$, $(1, 1 - \sqrt{3})$

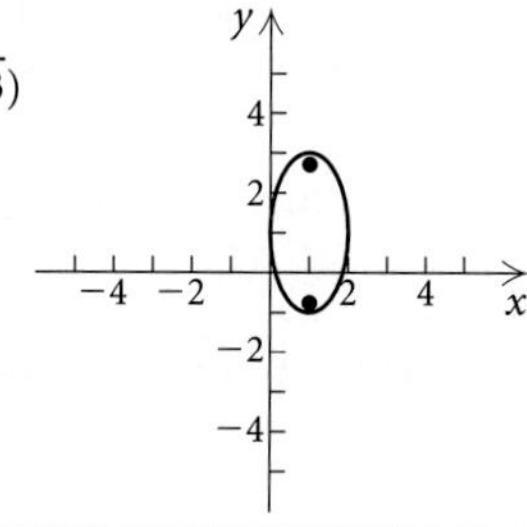

$$4x^2 + y^2 - 8x - 2y + 1 = 0$$

43. Example 2; $\dfrac{3}{5} < \dfrac{\sqrt{12}}{4}$ **45.** $\dfrac{x^2}{15} + \dfrac{y^2}{16} = 1$

47. $\dfrac{x^2}{2500} + \dfrac{y^2}{144} = 1$ **49.** 2×10^6 mi

51.

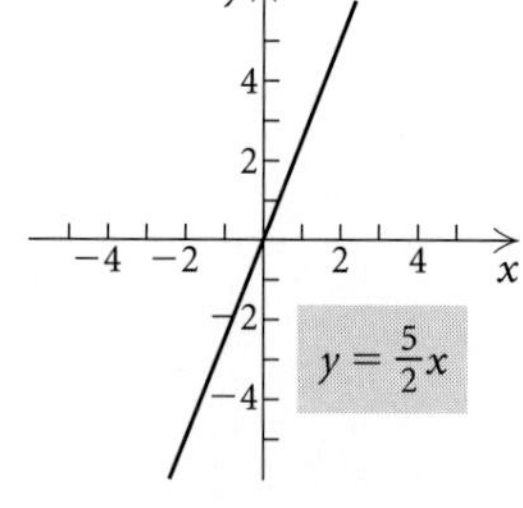

$$y = \dfrac{5}{2}x$$

53.

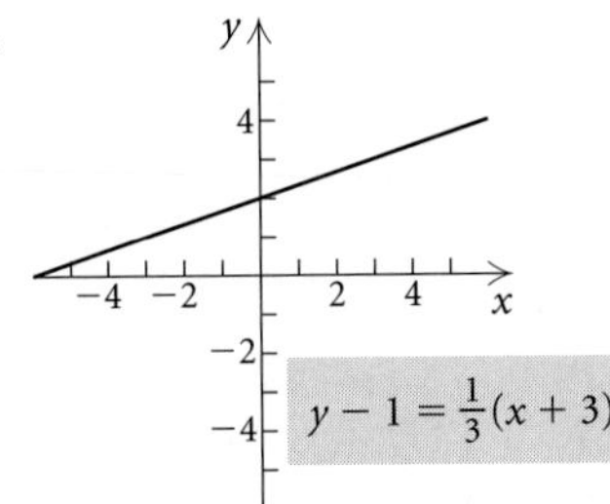

55. ◈   **57.** $\dfrac{(x-3)^2}{4} + \dfrac{(y-1)^2}{25} = 1$

59. $\dfrac{x^2}{9} + \dfrac{y^2}{484/5} = 1$

61. C: $(2.003, -1.005)$; V: $(-1.017, -1.005)$, $(5.023, -1.005)$

63. About 9.1 ft

EXERCISE SET 5.3

1. (b)   **3.** (c)   **5.** (a)   **7.** $\dfrac{y^2}{9} - \dfrac{x^2}{16} = 1$ **9.** $\dfrac{x^2}{4} - \dfrac{y^2}{9} = 1$

11. C: $(0, 0)$; V: $(2, 0)$, $(-2, 0)$; F: $(2\sqrt{2}, 0)$, $(-2\sqrt{2}, 0)$; A: $y = x, y = -x$

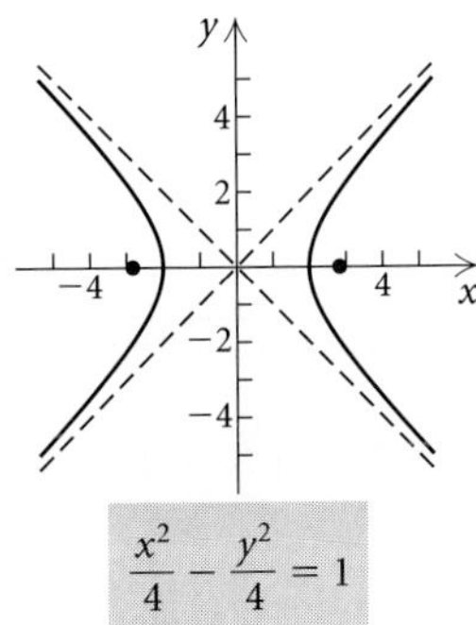

13. C: $(2, -5)$; V: $(-1, -5)$, $(5, -5)$; F: $(2 - \sqrt{10}, -5)$, $(2 + \sqrt{10}, -5)$; A: $y = -\dfrac{x}{3} - \dfrac{13}{3}, y = \dfrac{x}{3} - \dfrac{17}{3}$

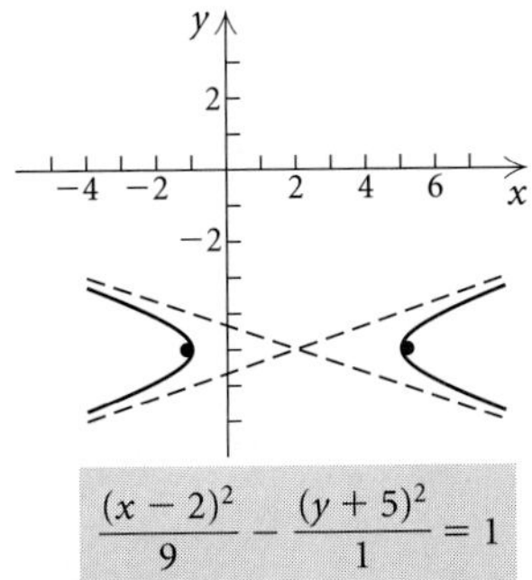

15. C: $(-1, -3)$; V: $(-1, -1)$, $(-1, -5)$; F: $(-1, -3 + 2\sqrt{5})$, $(-1, -3 - 2\sqrt{5})$; A: $y = \frac{1}{2}x - \frac{5}{2}$, $y = -\frac{1}{2}x - \frac{7}{2}$

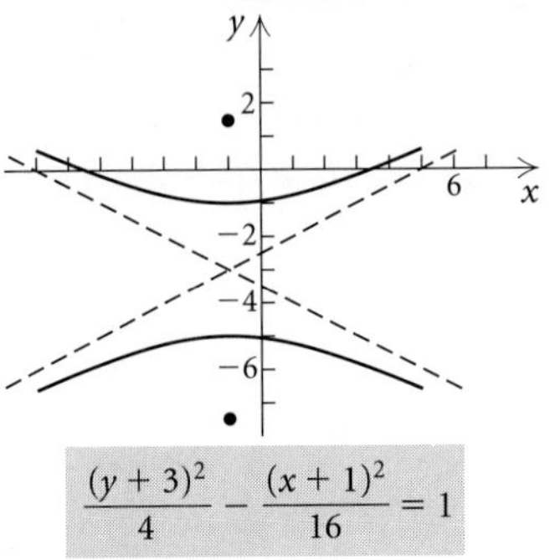

17. C: $(0, 0)$; V: $(-2, 0)$, $(2, 0)$; F: $(-\sqrt{5}, 0)$, $(\sqrt{5}, 0)$; A: $y = -\frac{1}{2}x, y = \frac{1}{2}x$

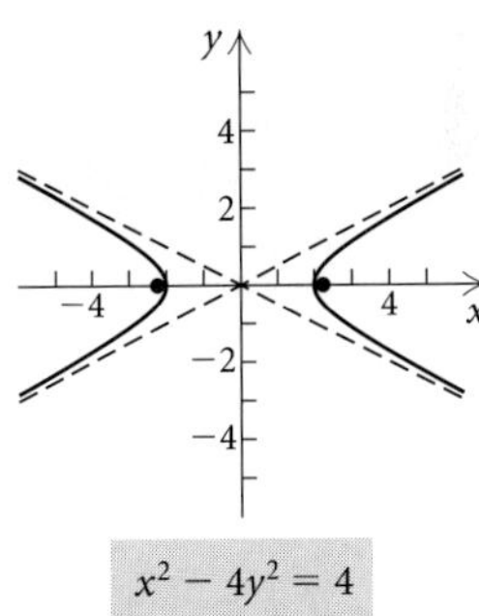

19. C: $(0, 0)$; V: $(0, -3)$, $(0, 3)$; F: $(0, -3\sqrt{10})$, $(0, 3\sqrt{10})$; A: $y = \frac{1}{3}x, y = -\frac{1}{3}x$

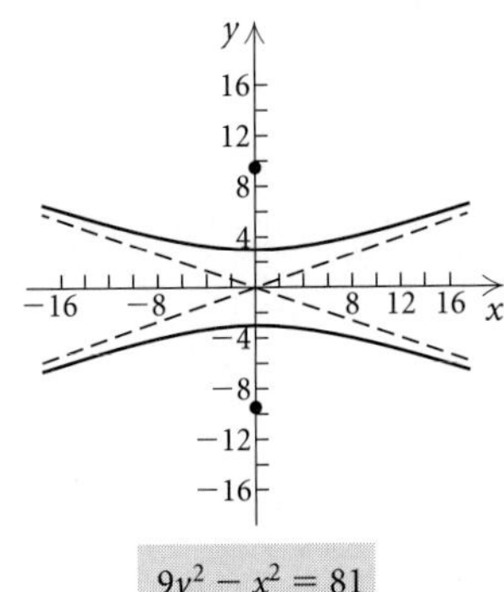

21. C: $(0, 0)$; V: $(-\sqrt{2}, 0)$, $(\sqrt{2}, 0)$; F: $(-2, 0)$, $(2, 0)$;
A: $y = x$, $y = -x$

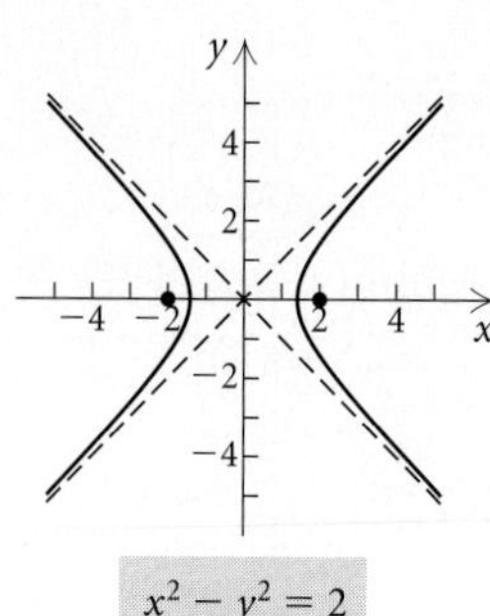

$$x^2 - y^2 = 2$$

23. C: $(0, 0)$; V: $\left(0, -\dfrac{1}{2}\right)$, $\left(0, \dfrac{1}{2}\right)$; F: $\left(0, -\dfrac{\sqrt{2}}{2}\right)$, $\left(0, \dfrac{\sqrt{2}}{2}\right)$; A: $y = x$, $y = -x$

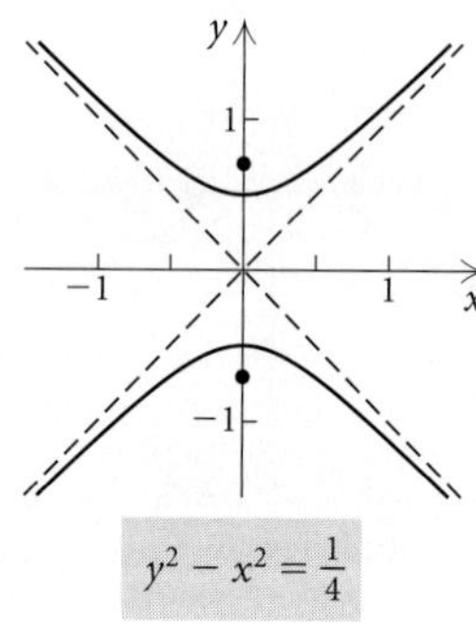

$$y^2 - x^2 = \frac{1}{4}$$

25. C: $(1, -2)$; V: $(0, -2)$, $(2, -2)$; F: $(1 - \sqrt{2}, -2)$, $(1 + \sqrt{2}, -2)$; A: $y = -x - 1$, $y = x - 3$

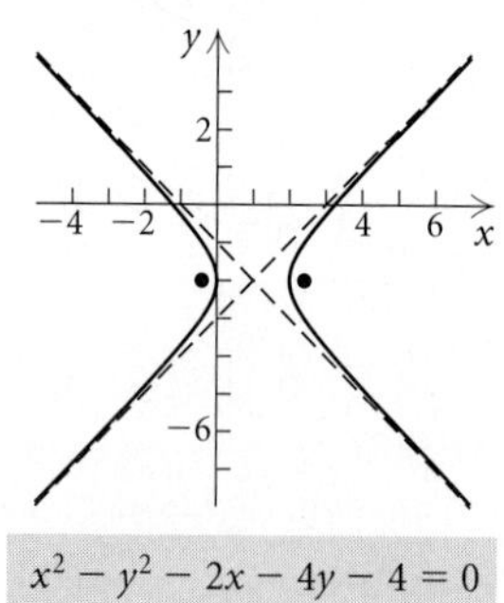

$$x^2 - y^2 - 2x - 4y - 4 = 0$$

27. C: $\left(\dfrac{1}{3}, 3\right)$; V: $\left(-\dfrac{2}{3}, 3\right)$, $\left(\dfrac{4}{3}, 3\right)$; F: $\left(\dfrac{1}{3} - \sqrt{37}, 3\right)$, $\left(\dfrac{1}{3} + \sqrt{37}, 3\right)$; A: $y = 6x + 1$, $y = -6x + 5$

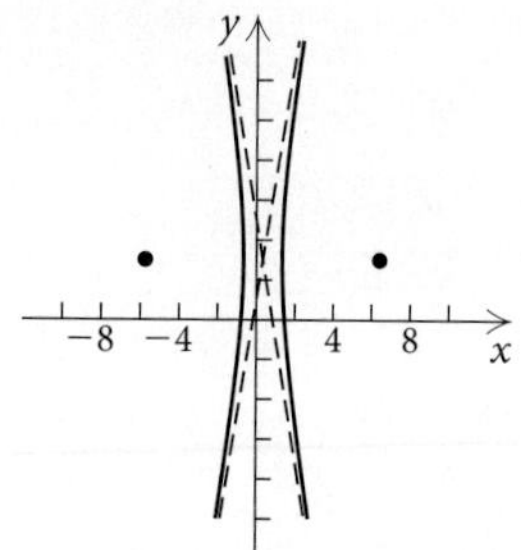

$$36x^2 - y^2 - 24x + 6y - 41 = 0$$

29. C: $(3, 1)$; V: $(3, 3)$, $(3, -1)$; F: $(3, 1 + \sqrt{13})$, $(3, 1 - \sqrt{13})$; A: $y = \dfrac{2}{3}x - 1$, $y = -\dfrac{2}{3}x + 3$

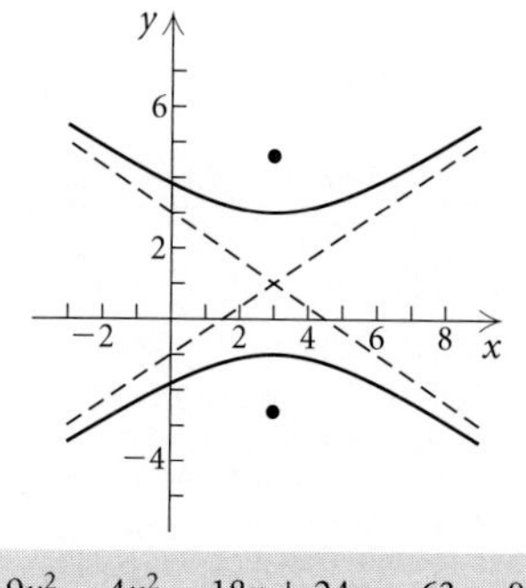

$$9y^2 - 4x^2 - 18y + 24x - 63 = 0$$

31. C: $(1, -2)$; V: $(2, -2)$, $(0, -2)$; F: $(1 + \sqrt{2}, -2)$, $(1 - \sqrt{2}, -2)$; A: $y = x - 3$, $y = -x - 1$

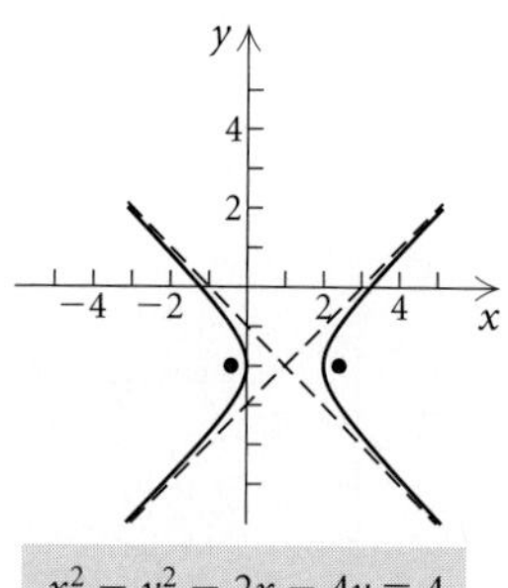

$$x^2 - y^2 - 2x - 4y = 4$$

33. C: $(-3, 4)$; V: $(-3, 10)$, $(-3, -2)$; F: $(-3, 4 + 6\sqrt{2})$, $(-3, 4 - 6\sqrt{2})$; A: $y = x + 7$, $y = -x + 1$

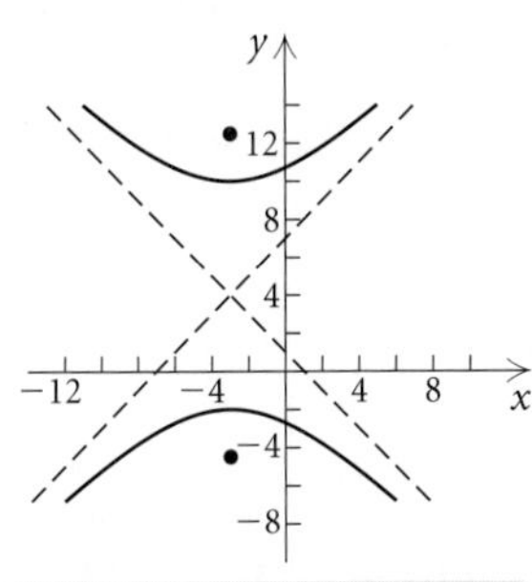

$$y^2 - x^2 - 6x - 8y - 29 = 0$$

35. Example 3; $\dfrac{\sqrt{5}}{1} > \dfrac{5}{4}$ **37.** $\dfrac{x^2}{9} - \dfrac{(y - 7)^2}{16} = 1$

39. $\dfrac{y^2}{25} - \dfrac{x^2}{11} = 1$ **41.** $(6, -1)$ **43.** $(2, -1)$ **45.** ◈

47. $\dfrac{(y + 5)^2}{9} - (x - 3)^2 = 1$

49. C: $(-1.460, -0.957)$; V: $(-2.360, -0.957)$, $(-0.560, -0.957)$; A: $y = -1.429x - 3.043$, $y = 1.429x + 1.129$

51. $\dfrac{x^2}{345.96} - \dfrac{y^2}{22{,}154.04} = 1$

EXERCISE SET 5.4

1. (e)   **3.** (c)   **5.** (b)   **7.** $(-4, -3)$, $(3, 4)$
9. $(0, 2)$, $(3, 0)$ **11.** $(-5, 0)$, $(4, 3)$, $(4, -3)$
13. $(3, 0)$, $(-3, 0)$ **15.** $(0, -3)$, $(4, 5)$ **17.** $(-2, 1)$
19. $(3, 4)$, $(-3, -4)$, $(4, 3)$, $(-4, -3)$
21. $\left(\dfrac{6\sqrt{21}}{7}, \dfrac{4i\sqrt{35}}{7}\right)$, $\left(\dfrac{6\sqrt{21}}{7}, -\dfrac{4i\sqrt{35}}{7}\right)$, $\left(-\dfrac{6\sqrt{21}}{7}, \dfrac{4i\sqrt{35}}{7}\right)$, $\left(-\dfrac{6\sqrt{21}}{7}, -\dfrac{4i\sqrt{35}}{7}\right)$
23. $(3, 2)$, $\left(4, \dfrac{3}{2}\right)$
25. $\left(\dfrac{5 + \sqrt{70}}{3}, \dfrac{-1 + \sqrt{70}}{3}\right)$, $\left(\dfrac{5 - \sqrt{70}}{3}, \dfrac{-1 - \sqrt{70}}{3}\right)$
27. $(\sqrt{2}, \sqrt{14})$, $(-\sqrt{2}, \sqrt{14})$, $(\sqrt{2}, -\sqrt{14})$, $(-\sqrt{2}, -\sqrt{14})$
29. $(1, 2)$, $(-1, -2)$, $(2, 1)$, $(-2, -1)$
31. $\left(\dfrac{15 + \sqrt{561}}{8}, \dfrac{11 - 3\sqrt{561}}{8}\right)$, $\left(\dfrac{15 - \sqrt{561}}{8}, \dfrac{11 + 3\sqrt{561}}{8}\right)$
33. $\left(\dfrac{7 - \sqrt{33}}{2}, \dfrac{7 + \sqrt{33}}{2}\right)$, $\left(\dfrac{7 + \sqrt{33}}{2}, \dfrac{7 - \sqrt{33}}{2}\right)$

35. $(3, 2)$, $(-3, -2)$, $(2, 3)$, $(-2, -3)$
37. $\left(\dfrac{5 - 9\sqrt{15}}{20}, \dfrac{-45 + 3\sqrt{15}}{20}\right)$, $\left(\dfrac{5 + 9\sqrt{15}}{20}, \dfrac{-45 - 3\sqrt{15}}{20}\right)$ **39.** $(3, -5)$, $(-1, 3)$
41. $(8, 5)$, $(-5, -8)$ **43.** $(3, 2)$, $(-3, -2)$
45. $(2, 1)$, $(-2, -1)$, $(1, 2)$, $(-1, -2)$
47. $\left(4 + \dfrac{3i\sqrt{6}}{2}, -4 + \dfrac{3i\sqrt{6}}{2}\right)$, $\left(4 - \dfrac{3i\sqrt{6}}{2}, -4 - \dfrac{3i\sqrt{6}}{2}\right)$
49. $(3, \sqrt{5})$, $(-3, -\sqrt{5})$, $(\sqrt{5}, 3)$, $(-\sqrt{5}, -3)$
51. $\left(\dfrac{8\sqrt{5}}{5}i, \dfrac{3\sqrt{105}}{5}\right)$, $\left(\dfrac{8\sqrt{5}}{5}i, -\dfrac{3\sqrt{105}}{5}\right)$, $\left(-\dfrac{8\sqrt{5}}{5}i, \dfrac{3\sqrt{105}}{5}\right)$, $\left(-\dfrac{8\sqrt{5}}{5}i, -\dfrac{3\sqrt{105}}{5}\right)$
53. $(2, 1)$, $(-2, -1)$, $\left(-i\sqrt{5}, \dfrac{2i\sqrt{5}}{5}\right)$, $\left(i\sqrt{5}, -\dfrac{2i\sqrt{5}}{5}\right)$
55. 6 cm by 8 cm   **57.** 4 in. by 5 in.   **59.** 1 m by $\sqrt{3}$ m
61. 30 yd by 75 yd   **63.** 16 ft, 24 ft   **65.** $\dfrac{24}{5}$ **67.** $\dfrac{23}{25}$
69. ◈   **71.** $(x - 2)^2 + (y - 3)^2 = 1$ **73.** $\dfrac{x^2}{4} + y^2 = 1$
75. $\left(x + \dfrac{5}{13}\right)^2 + \left(y - \dfrac{32}{13}\right)^2 = \dfrac{5365}{169}$

77. There is no number x such that $\dfrac{x^2}{a^2} - \dfrac{\left(\dfrac{b}{a}x\right)^2}{b^2} = 1$, because the left side simplifies to $\dfrac{x^2}{a^2} - \dfrac{x^2}{a^2}$, which is 0.
79. $\left(\dfrac{1}{2}, \dfrac{1}{4}\right)$, $\left(\dfrac{1}{2}, -\dfrac{1}{4}\right)$, $\left(-\dfrac{1}{2}, \dfrac{1}{4}\right)$, $\left(-\dfrac{1}{2}, -\dfrac{1}{4}\right)$
81. Factor: $x^3 + y^3 = (x + y)(x^2 - xy + y^2)$. We know that $x + y = 1$, so $(x + y)^2 = x^2 + 2xy + y^2 = 1$, or $x^2 + y^2 = 1 - 2xy$. We also know that $xy = 1$, so $x^2 + y^2 = 1 - 2 \cdot 1 = -1$. Then $x^3 + y^3 = 1 \cdot (-1 - 1) = -2$.
83. $(2, 4)$, $(4, 2)$ **85.** $(3, -2)$, $(-3, 2)$, $(2, -3)$, $(-2, 3)$
87. $\left(\dfrac{2 \log 3 + 3 \log 5}{3(\log 3 \cdot \log 5)}, \dfrac{4 \log 3 - 3 \log 5}{3(\log 3 \cdot \log 5)}\right)$
89. $(1.564, 2.448)$, $(0.138, 0.019)$ **91.** $(0.871, 1.388)$
93. $(1.146, 3.146)$, $(-1.841, 0.159)$
95. $(0.965, 4402.33)$, $(-0.965, -4402.33)$
97. $(2.112, -0.109)$, $(-13.041, -13.337)$
99. $(400, 1.431)$, $(-400, 1.431)$, $(400, -1.431)$, $(-400, -1.431)$

REVIEW EXERCISES, CHAPTER 5

1. [5.1] (d)   **2.** [5.2] (a)   **3.** [5.2] (e)   **4.** [5.3] (g)
5. [5.2] (b)   **6.** [5.2] (f)   **7.** [5.1] (h)   **8.** [5.3] (c)
9. [5.1] $x^2 = -6y$ **10.** [5.1] F: $(-3, 0)$; V: $(0, 0)$; D: $x = 3$
11. [5.1] V: $(-5, 8)$; F: $\left(-5, \dfrac{15}{2}\right)$; D: $y = \dfrac{17}{2}$

12. [5.2] C: $(2, -1)$; V: $(-3, -1)$, $(7, -1)$; F: $(-1, -1)$, $(5, -1)$;

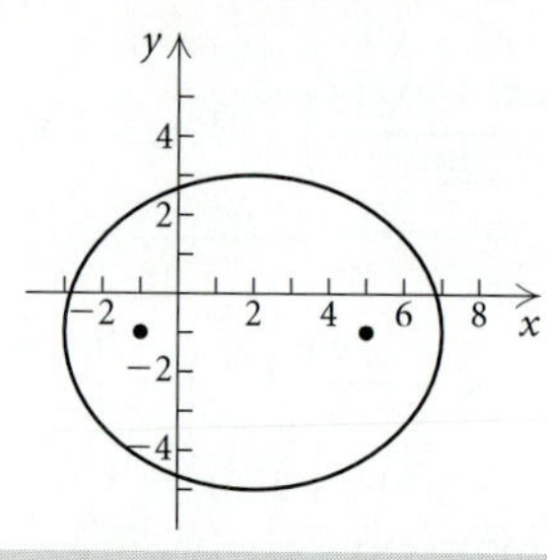

$$16x^2 + 25y^2 - 64x + 50y - 311 = 0$$

13. [5.2] $\dfrac{x^2}{9} + \dfrac{y^2}{16} = 1$

14. [5.3] C: $\left(-2, \dfrac{1}{4}\right)$; V: $\left(0, \dfrac{1}{4}\right)$, $\left(-4, \dfrac{1}{4}\right)$;

F: $\left(-2 + \sqrt{6}, \dfrac{1}{4}\right)$, $\left(-2 - \sqrt{6}, \dfrac{1}{4}\right)$;

A: $y - \dfrac{1}{4} = \dfrac{\sqrt{2}}{2}(x + 2)$, $y - \dfrac{1}{4} = -\dfrac{\sqrt{2}}{2}(x + 2)$

15. [5.1] 0.167 ft   **16.** [5.4] $(-8\sqrt{2}, 8)$, $(8\sqrt{2}, 8)$

17. [5.4] $\left(3, \dfrac{\sqrt{29}}{2}\right)$, $\left(-3, \dfrac{\sqrt{29}}{2}\right)$, $\left(3, -\dfrac{\sqrt{29}}{2}\right)$, $\left(-3, -\dfrac{\sqrt{29}}{2}\right)$ **18.** [5.4] $(7, 4)$ **19.** [5.4] $(2, 2)$, $\left(\dfrac{32}{9}, -\dfrac{10}{9}\right)$

20. [5.4] $(0, -3)$, $(2, 1)$

21. [5.4] $(4, 3)$, $(4, -3)$, $(-4, 3)$, $(-4, -3)$

22. [5.4] $(-\sqrt{3}, 0)$, $(\sqrt{3}, 0)$, $(-2, 1)$, $(2, 1)$

23. [5.4] $\left(-\dfrac{3}{5}, \dfrac{21}{5}\right)$, $(3, -3)$

24. [5.4] $(6, 8)$, $(6, -8)$, $(-6, 8)$, $(-6, -8)$

25. [5.4] $(2, 2)$, $(-2, -2)$, $(2\sqrt{2}, \sqrt{2})$, $(-2\sqrt{2}, -\sqrt{2})$

26. [5.4] 7, 4   **27.** [5.4] 7 m by 12 m   **28.** [5.4] 4, 8

29. [5.4] 32 cm, 20 cm   **30.** [5.4] 11 ft, 3 ft

31. [5.4] ◆ An algebraic solution of a nonlinear system of equations may be preferable to a graphical solution if there are complex-number solutions or if any of the equations is difficult to enter on a grapher. However, there are nonlinear systems that are difficult or impossible to solve using algebraic methods. Approximations of the real-number solutions of many of those systems may be found using graphical methods.

32. [5.2] ◆ The equation of a circle can be written as

$$\frac{(x - h)^2}{a^2} + \frac{(y - k)^2}{b^2} = 1,$$

where $a = b = r$, the radius of the circle. In an ellipse, $a > b$, so a circle is not a special type of ellipse.

33. [5.2] $x^2 + \dfrac{y^2}{9} = 1$ **34.** [5.4] $\dfrac{8}{7}, \dfrac{7}{2}$

35. [5.2], [5.4] $(x - 2)^2 + (y - 1)^2 = 100$

36. [5.3] $\dfrac{x^2}{778.41} - \dfrac{y^2}{39,221.59} = 1$

Chapter 6

EXERCISE SET 6.1

1. 3, 7, 11, 15; 39; 59   **3.** $2, \dfrac{3}{2}, \dfrac{4}{3}, \dfrac{5}{4}; \dfrac{10}{9}; \dfrac{15}{14}$

5. $0, \dfrac{3}{5}, \dfrac{4}{5}, \dfrac{15}{17}; \dfrac{99}{101}; \dfrac{112}{113}$ **7.** $-1, 4, -9, 16; 100; -225$

9. 7, 3, 7, 3; 3; 7   **11.** 34   **13.** 225   **15.** $-33,880$

17. 67   **19.** $2n$ **21.** $(-1)^n \cdot 2 \cdot 3^{n-1}$ **23.** $\dfrac{n + 1}{n + 2}$

25. $n(n + 1)$ **27.** $\log 10^{n-1}$, or $n - 1$ **29.** 6; 28

31. 20; 30   **33.** $\dfrac{1}{2} + \dfrac{1}{4} + \dfrac{1}{6} + \dfrac{1}{8} + \dfrac{1}{10} = \dfrac{137}{120}$

35. $1 + 2 + 4 + 8 + 16 + 32 + 64 = 127$

37. $\ln 7 + \ln 8 + \ln 9 + \ln 10 = \ln (7 \cdot 8 \cdot 9 \cdot 10) = \ln 5040 \approx 8.5252$

39. $\dfrac{1}{2} + \dfrac{2}{3} + \dfrac{3}{4} + \dfrac{4}{5} + \dfrac{5}{6} + \dfrac{6}{7} + \dfrac{7}{8} + \dfrac{8}{9} = \dfrac{15,551}{2520}$

41. $-1 + 1 - 1 + 1 - 1 = -1$

43. $3 - 6 + 9 - 12 + 15 - 18 + 21 - 24 = -12$

45. $2 + 1 + \dfrac{2}{5} + \dfrac{1}{5} + \dfrac{2}{17} + \dfrac{1}{13} + \dfrac{2}{37} = \dfrac{157,351}{40,885}$

47. $3 + 2 + 3 + 6 + 11 + 18 = 43$

49. $\dfrac{1}{2} + \dfrac{2}{3} + \dfrac{4}{5} + \dfrac{8}{9} + \dfrac{16}{17} + \dfrac{32}{33} + \dfrac{64}{65} + \dfrac{128}{129} + \dfrac{256}{257} + \dfrac{512}{513} + \dfrac{1024}{1025} \approx 9.736$

51. $\displaystyle\sum_{k=1}^{\infty} 5k$ **53.** $\displaystyle\sum_{k=1}^{6} (-1)^{k+1} 2^k$ **55.** $\displaystyle\sum_{k=1}^{6} (-1)^k \dfrac{k}{k + 1}$

57. $\displaystyle\sum_{k=2}^{n} (-1)^k k^2$ **59.** $\displaystyle\sum_{k=1}^{\infty} \dfrac{1}{k(k + 1)}$ **61.** $4, 1\dfrac{1}{4}, 1\dfrac{4}{5}, 1\dfrac{5}{9}$

63. $6561, -81, 9i, -3\sqrt{i}$ **65.** 2, 3, 5, 8

67.

n	U_n
1	2
2	2.25
3	2.3704
4	2.4414
5	2.4883
6	2.5216
7	2.5465
8	2.5658
9	2.5812
10	2.5937

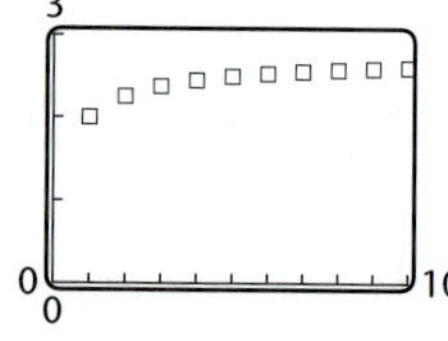

69.

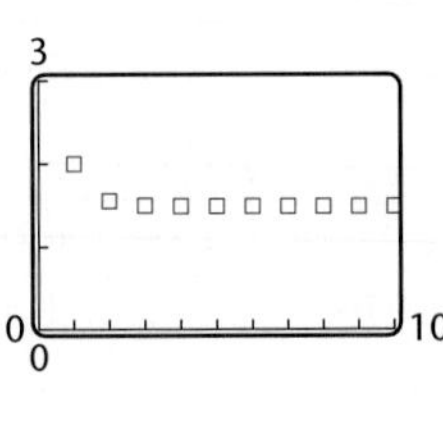

n	U_n
1	2
2	1.5538
3	1.4988
4	1.4914
5	1.4904
6	1.4902
7	1.4902
8	1.4902
9	1.4902
10	1.4902

71. (a) 2014.11, 2450.38, 2886.65, 3322.92, 3759.19, 4195.46, 4631.73, 5068, 5504.27, 5940.54; **(b)** \$39,773.25 million
73. (a) 1062, 1127.84, 1197.77, 1272.03, 1350.90, 1434.65, 1523.60, 1618.07, 1718.39, 1824.93; **(b)** \$3330.35
75. 1, 2, 4, 8, 16, 32, 64, 128, 256, 512, 1024, 2048, 4096, 8192, 16,384, 32,768, 65,536
77. (a) $a_n = -17.748n^2 + 1784.95n - 23,768.85$;
(b) \$246.86, \$4830.95, \$19,232.35, \$19,435.35, \$10,269.90
79. $2a_1 + a_2 + 3a_n$ **81.** 0   **83.** ◈   **85.** $i, -1, -i, 1, i; i$
87. $\ln(1 \cdot 2 \cdot 3 \cdot \cdots \cdot n)$

EXERCISE SET 6.2

1. $a_1 = 3, d = 5$ **3.** $a_1 = 9, d = -4$ **5.** $a_1 = \frac{3}{2}, d = \frac{3}{4}$
7. $a_1 = \$316, d = -\$3$ **9.** $a_{12} = 46$ **11.** $a_{14} = -\frac{17}{3}$
13. $a_{10} = \$7941.62$ **15.** 33rd   **17.** 46th   **19.** $a_1 = 5$
21. $n = 39$ **23.** $a_1 = \frac{1}{3}; d = \frac{1}{2}; \frac{1}{3}, \frac{5}{6}, \frac{4}{3}, \frac{11}{6}, \frac{7}{3}$ **25.** 670
27. 160,400   **29.** 735   **31.** 990   **33.** 1760   **35.** $\frac{65}{2}$
37. $-\frac{6026}{13}$ **39.** 1260   **41.** 3; 171   **43.** 4960¢, or \$49.60
45. 1320   **47.** $-1, 1, 3$ **49.** Yes; 3   **51.** $-\frac{1}{8}$
53. 10,737,418.23   **55.** ◈   **57.** n^2
59. $a_1 = -5p - 5q + 60, d = 5p + 2q - 20$
61. $-10, -4, 2, 8$
63. Sides are $a, a + d, a + 2d$; and $a^2 + (a + d)^2 = (a + 2d)^2$. Solving, we get $d = \frac{a}{3}$. Thus the sides are $a, \frac{4a}{3}$, and $\frac{5a}{3}$ in the ratio 3:4:5.
65. $a_n = a_1 + (n - 1)d; a_n = f(n) = d \cdot n + (a_1 - d)$, where $f(n)$ is a linear function of n with slope d and y-intercept $(a_1 - d)$.

EXERCISE SET 6.3

1. 2   **3.** -1 **5.** -2 **7.** 0.1   **9.** $\frac{a}{2}$ **11.** 128   **13.** 162
15. $7(25)^{20}$ **17.** 3^{n-1} **19.** $(-1)^{n-1}$ **21.** $\frac{1}{x^n}$ **23.** 762
25. $\frac{4921}{18}$ **27.** 8   **29.** 125   **31.** Does not exist   **33.** $\frac{2}{3}$
35. $29\frac{38,569}{59,049}$ **37.** 2   **39.** Does not exist   **41.** $\$4545.\overline{45}$

43. $\frac{160}{9}$ **45.** $\frac{13}{99}$ **47.** 9   **49.** $\frac{34,091}{9990}$
51. (a) $\frac{1}{256}$ ft; **(b)** $5\frac{1}{3}$ ft   **53. (a)** About 297 ft; **(b)** 300 ft
55. \$31,497.57   **57.** \$523,619.17   **59.** \$86,666,666,667
61. $k^2 + 2k + 1$ **63.** $2k^2 + 6k + 4$ **65.** $\frac{k + 1}{k + 2}$ **67.** ◈
69. $(4 - \sqrt{6})/(\sqrt{3} - \sqrt{2}) = 2\sqrt{3} + \sqrt{2}$, $(6\sqrt{3} - 2\sqrt{2})/(4 - \sqrt{6}) = 2\sqrt{3} + \sqrt{2}$; there exists a common ratio, $2\sqrt{3} + \sqrt{2}$; thus the sequence is geometric.
71. (a) $\frac{13}{3}; \frac{22}{3}, \frac{34}{3}, \frac{46}{3}, \frac{58}{3}$;
(b) $-\frac{11}{3}; -\frac{2}{3}, \frac{10}{3}, -\frac{50}{3}, \frac{250}{3}$ or 5; 8, 12, 18, 27
73. $S_n = \dfrac{x^2(1 - (-x)^n)}{x + 1}$
75. $\dfrac{a_{n+1}}{a_n} = r$, so $\ln \dfrac{a_{n+1}}{a_n} = \ln r$. But $\ln \dfrac{a_{n+1}}{a_n} = \ln a_{n+1} - \ln a_n = \ln r$. Thus, $\ln a_1, \ln a_2, \ldots$, is an arithmetic sequence with common difference $\ln r$.
77. 512 cm^2

EXERCISE SET 6.4

1. $1^2 < 1^3$, false; $2^2 < 2^3$, true; $3^2 < 3^3$, true; $4^2 < 4^3$, true; $5^2 < 5^3$, true   **3.** A polygon of 3 sides has $\dfrac{3(3 - 3)}{2}$ diagonals. True; A polygon of 4 sides has $\dfrac{4(4 - 3)}{2}$ diagonals. True; A polygon of 5 sides has $\dfrac{5(5 - 3)}{2}$ diagonals. True; A polygon of 6 sides has $\dfrac{6(6 - 3)}{2}$ diagonals. True; A polygon of 7 sides has $\dfrac{7(7 - 3)}{2}$ diagonals. True.
5.
S_n: $2 + 4 + 6 + \cdots + 2n = n(n + 1)$
S_1: $2 = 1(1 + 1)$
S_k: $2 + 4 + 6 + \cdots + 2k = k(k + 1)$
S_{k+1}: $2 + 4 + 6 + \cdots + 2k + 2(k + 1) = (k + 1)(k + 2)$
1. *Basis step:* S_1 true by substitution.
2. *Induction step:* Assume S_k. Deduce S_{k+1}.
    Starting with the left side of S_{k+1}, we have
$$2 + 4 + 6 + \cdots + 2k + 2(k + 1)$$
$$= k(k + 1) + 2(k + 1) \qquad \textbf{By } S_k$$
$$= (k + 1)(k + 2). \qquad \textbf{Distributive law}$$
7. S_n: $1 + 5 + 9 + \cdots + (4n - 3) = n(2n - 1)$
S_1: $1 = 1(2 \cdot 1 - 1)$
S_k: $1 + 5 + 9 + \cdots + (4k - 3) = k(2k - 1)$
S_{k+1}: $1 + 5 + 9 + \cdots + (4k - 3) + [4(k + 1) - 3]$
$$= (k + 1)[2(k + 1) - 1]$$
$$= (k + 1)(2k + 1)$$
1. *Basis step*: S_1 true by substitution.
2. *Induction step*: Assume S_k. Deduce S_{k+1}.

Starting with the left side of S_{k+1}, we have

$$\underbrace{1 + 5 + 9 + \cdots + (4k - 3)} + [4(k + 1) - 3]$$
$$= k(2k - 1) + [4(k + 1) - 3] \qquad \textbf{By } S_k$$
$$= 2k^2 - k + 4k + 4 - 3$$
$$= (k + 1)(2k + 1).$$

9. S_n: $2 + 4 + 8 + \cdots + 2^n = 2(2^n - 1)$

S_1: $2 = 2(2 - 1)$

S_k: $2 + 4 + 8 + \cdots + 2^k = 2(2^k - 1)$

S_{k+1}: $2 + 4 + 8 + \cdots + 2^k + 2^{k+1} = 2(2^{k+1} - 1)$

1. *Basis step*: S_1 is true by substitution.
2. *Induction step*: Assume S_k. Deduce S_{k+1}.

Starting with the left side of S_{k+1}, we have

$$\underbrace{2 + 4 + 8 + \cdots + 2^k} + 2^{k+1}$$
$$= \underbrace{2(2^k - 1)} + 2^{k+1} \qquad \textbf{By } S_k$$
$$= 2^{k+1} - 2 + 2^{k+1}$$
$$= 2 \cdot 2^{k+1} - 2$$
$$= 2(2^{k+1} - 1).$$

11. 1. *Basis step*: Since $1 < 1 + 1$, S_1 is true.
2. *Induction step*: Assume S_k. Deduce S_{k+1}. Now

$$k < k + 1 \qquad \textbf{By } S_k$$
$$k + 1 < k + 1 + 1; \qquad \textbf{Adding 1}$$
$$\therefore k + 1 < k + 2.$$

13. 1. *Basis step*: Since $2 = 2$, S_1 is true.
2. *Induction step*: Let k be any natural number. Assume S_k. Deduce S_{k+1}.

$$2k \le 2^k \qquad \textbf{By } S_k$$
$$2 \cdot 2k \le 2 \cdot 2^k \qquad \textbf{Multiplying by 2}$$
$$4k \le 2^{k+1}$$

Since $1 \le k$, $k + 1 \le k + k$, or $k + 1 \le 2k$.
Then $2(k + 1) \le 4k$.
Thus, $2(k + 1) \le 4k \le 2^{k+1}$, so $2(k + 1) \le 2^{k+1}$.

15.

S_n: $$\frac{1}{1 \cdot 2 \cdot 3} + \frac{1}{2 \cdot 3 \cdot 4} + \frac{1}{3 \cdot 4 \cdot 5} + \cdots$$
$$+ \frac{1}{n(n + 1)(n + 2)}$$
$$= \frac{n(n + 3)}{4(n + 1)(n + 2)}$$

S_1: $$\frac{1}{1 \cdot 2 \cdot 3} = \frac{1(1 + 3)}{4 \cdot 2 \cdot 3}$$

S_k: $$\frac{1}{1 \cdot 2 \cdot 3} + \frac{1}{2 \cdot 3 \cdot 4} + \cdots + \frac{1}{k(k + 1)(k + 2)}$$
$$= \frac{k(k + 3)}{4(k + 1)(k + 2)}$$

S_{k+1}: $$\frac{1}{1 \cdot 2 \cdot 3} + \frac{1}{2 \cdot 3 \cdot 4} + \cdots + \frac{1}{k(k + 1)(k + 2)}$$
$$+ \frac{1}{(k + 1)(k + 2)(k + 3)}$$
$$= \frac{(k + 1)(k + 1 + 3)}{4(k + 1 + 1)(k + 1 + 2)} = \frac{(k + 1)(k + 4)}{4(k + 2)(k + 3)}$$

1. *Basis step*: Since $\dfrac{1}{1 \cdot 2 \cdot 3} = \dfrac{1}{6}$ and $\dfrac{1(1 + 3)}{4 \cdot 2 \cdot 3} = \dfrac{1 \cdot 4}{4 \cdot 2 \cdot 3} = \dfrac{1}{6}$, S_1 is true.

2. *Induction step*: Assume S_k. Deduce S_{k+1}.

Add $\dfrac{1}{(k + 1)(k + 2)(k + 3)}$ on both sides of S_k and simplify the right side.
Only the right side is shown here.

$$\frac{k(k + 3)}{4(k + 1)(k + 2)} + \frac{1}{(k + 1)(k + 2)(k + 3)}$$
$$= \frac{k(k + 3)(k + 3) + 4}{4(k + 1)(k + 2)(k + 3)}$$
$$= \frac{k^3 + 6k^2 + 9k + 4}{4(k + 1)(k + 2)(k + 3)}$$
$$= \frac{(k + 1)^2(k + 4)}{4(k + 1)(k + 2)(k + 3)}$$
$$= \frac{(k + 1)(k + 4)}{4(k + 2)(k + 3)}$$

17.

S_n: $$1 + 2 + 3 + \cdots + n = \frac{n(n + 1)}{2}$$

S_1: $$1 = \frac{1(1 + 1)}{2}$$

S_k: $$1 + 2 + 3 + \cdots + k = \frac{k(k + 1)}{2}$$

S_{k+1}: $$1 + 2 + 3 + \cdots + k + (k + 1) = \frac{(k + 1)(k + 2)}{2}$$

1. *Basis step*: S_1 true by substitution.
2. *Induction step*: Assume S_k. Deduce S_{k+1}.

Starting with the left side of S_{k+1}, we have

$$\underbrace{1 + 2 + 3 + \cdots + k} + (k + 1)$$
$$= \frac{k(k + 1)}{2} + (k + 1) \qquad \textbf{By } S_k$$
$$= \frac{k(k + 1) + 2(k + 1)}{2} \qquad \textbf{Adding}$$
$$= \frac{(k + 1)(k + 2)}{2}. \qquad \textbf{Distributive law}$$

19. 1. *Basis step*. S_1: $1^3 = \dfrac{1^2(1 + 1)^2}{4} = 1$. True.

2. *Induction step*: Assume S_k. Deduce S_{k+1}.

$$S_k: \quad 1^3 + 2^3 + \cdots + k^3 = \frac{k^2(k+1)^2}{4}$$

$$1^3 + 2^3 + \cdots + (k+1)^3 = \frac{k^2(k+1)^2}{4} + (k+1)^3 \quad \textbf{By } S_k$$

$$= \frac{(k+1)^2}{4}[k^2 + 4(k+1)]$$

$$= \frac{(k+1)^2(k+2)^2}{4}$$

21.

1. *Basis step.* S_1: $1^5 = \dfrac{1^2(1+1)^2(2 \cdot 1^2 + 2 \cdot 1 - 1)}{12}$. True.

2. *Induction step.* Assume S_k: $1^5 + 2^5 + \cdots + k^5$
$$= \frac{k^2(k+1)^2(2k^2 + 2k - 1)}{12}.$$

Then $1^5 + 2^5 + \cdots + k^5 + (k+1)^5$

$$= \frac{k^2(k+1)^2(2k^2 + 2k - 1)}{12} + (k+1)^5$$

$$= \frac{k^2(k+1)^2(2k^2 + 2k - 1) + 12(k+1)^5}{12}$$

$$= \frac{(k+1)^2(2k^4 + 14k^3 + 35k^2 + 36k + 12)}{12}$$

$$= \frac{(k+1)^2(k+2)^2(2k^2 + 6k + 3)}{12}$$

$$= \frac{(k+1)^2(k+1+1)^2(2(k+1)^2 + 2(k+1) - 1)}{12}.$$

23.

1. *Basis step.* S_1: $1(1+1) = \dfrac{1(1+1)(1+2)}{3}$. True.

2. *Induction step.* Assume
S_k: $1(1+1) + 2(2+1) + \cdots + k(k+1)$
$$= \frac{k(k+1)(k+2)}{3}.$$

Then $1(1+1) + 2(2+1) + \cdots + k(k+1)$
$+ (k+1)(k+1+1)$

$$= \frac{k(k+1)(k+2)}{3} + (k+1)(k+2)$$

$$= \frac{(k+1)(k+2)(k+3)}{3}$$

$$= \frac{(k+1)(k+1+1)(k+1+2)}{3}.$$

25. 1. *Basis step:* Since $\frac{1}{2}[2a_1 + (1-1)d] = \frac{1}{2} \cdot 2a_1 = a_1$, S_1 is true.

2. *Induction step:* Assume S_k. Deduce S_{k+1}. Starting with the left side of S_{k+1}, we have

$$\underbrace{a_1 + (a_1 + d) + \cdots + [a_1 + (k-1)d]} + [a_1 + kd]$$

$$= \qquad \frac{k}{2}[2a_1 + (k-1)d] \qquad + [a_1 + kd]$$
$$\textbf{By } S_k$$

$$= \frac{k[2a_1 + (k-1)d]}{2} + \frac{2[a_1 + kd]}{2}$$

$$= \frac{2ka_1 + k(k-1)d + 2a_1 + 2kd}{2}$$

$$= \frac{2a_1(k+1) + k(k-1)d + 2kd}{2}$$

$$= \frac{2a_1(k+1) + (k-1+2)kd}{2}$$

$$= \frac{2a_1(k+1) + (k+1)kd}{2}$$

$$= \frac{k+1}{2}[2a_1 + kd].$$

27. 336   **29.** 90

31.

1. *Basis step.* S_1: $x + y$ is a factor of $x^2 - y^2$. True.
   S_2: $x + y$ is a factor of $x^4 - y^4$. True.

2. *Induction step.* Assume S_{k-1}: $x + y$ is a factor of $x^{2(k-1)} - y^{2(k-1)}$.
   Then $x^{2(k-1)} - y^{2(k-1)} = (x+y)Q(x)$ for some polynomial Q.
   Assume S_k: $x + y$ is a factor of $x^{2k} - y^{2k}$.
   Then $x^{2k} - y^{2k} = (x+y)P(x)$ for some polynomial P.

$$x^{2(k+1)} - y^{2(k+1)}$$
$$= (x^{2k} - y^{2k})(x^2 + y^2) - (x^{2(k-1)} - y^{2(k-1)})(x^2 y^2)$$
$$= (x+y)P(x)(x^2 + y^2) - (x+y)Q(x)(x^2 y^2)$$
$$= (x+y)[P(x)(x^2 + y^2) - Q(x)(x^2 y^2)]$$

so $x + y$ is a factor of $x^{2(k+1)} - y^{2(k+1)}$.

33.

1. *Basis step.* S_2: $\dfrac{x^2 - y^2}{x - y} = x + y$. True.

   S_3: $\dfrac{x^3 - y^3}{x - y} = x^2 + yx + y^2$. True.

2. *Induction step.* Assume S_{k-1}: $\dfrac{x^{k-1} - y^{k-1}}{x - y}$
$$= x^{k-2} + yx^{k-3} + \cdots + y^{k-3}x + y^{k-2}$$

   Assume S_k: $\dfrac{x^k - y^k}{x - y} = x^{k-1} + yx^{k-2}$
$$+ \cdots + y^{k-2}x + y^{k-1}.$$

$$\frac{x^{k+1} - y^{k+1}}{x - y} = \frac{(x^k - y^k)(x + y)}{x - y} - \frac{xy(x^{k-1} - y^{k-1})}{x - y}$$

$$= (x^{k-1} + yx^{k-2} + \cdots + y^{k-2}x + y^{k-1})(x + y)$$
$$- xy(x^{k-2} + yx^{k-3} + \cdots + y^{k-3}x + y^{k-2})$$
$$= x^k + yx^{k-1} + \cdots + y^{k-2}x^2 + y^{k-1}x$$
$$+ x^{k-1}y + y^2 x^{k-2} + \cdots + y^{k-1}x + y^k$$
$$- (x^{k-1}y + y^2 x^{k-2} + \cdots + y^{k-2}x^2 + y^{k-1}x)$$
$$= x^k + yx^{k-1} + \cdots + y^{k-1}x + y^k$$

35. ◈

37.

S_2: $\log_a (b_1 b_2) = \log_a b_1 + \log_a b_2$
S_k: $\log_a (b_1 b_2 \ldots b_k) = \log_a b_1 + \log_a b_2 + \cdots + \log_a b_k$
S_{k+1}: $\log_a (b_1 b_2 \ldots b_{k+1}) = \log_a b_1 + \log_a b_2 + \cdots$
      $+ \log_a b_{k+1}$

1. *Basis step:* S_2 is true by the properties of logarithms.
2. *Induction step:* Let k be a natural number $k \geq 2$. Assume
   S_k. Deduce S_{k+1}.

$\log_a (b_1 b_2 \ldots b_{k+1})$ **Left side of S_{k+1}**
      $= \log_a (b_1 b_2 \ldots b_k) + \log_a b_{k+1}$ **By S_2**
      $= \log_a b_1 + \log_a b_2 + \cdots + \log_a b_k + \log_a b_{k+1}$

39. 1. *Basis step:* $\dfrac{1}{\sqrt{1}} + \dfrac{1}{\sqrt{2}}$

$$= 1 + \frac{1}{\sqrt{2}}$$

$$= \frac{\sqrt{2} + 1}{\sqrt{2}} > \frac{2}{\sqrt{2}} = \sqrt{2} \quad \text{Since } \sqrt{2} > 1$$

2. *Induction step:* Assume S_k. Deduce S_{k+1}.

$$\frac{1}{\sqrt{1}} + \frac{1}{\sqrt{2}} + \frac{1}{\sqrt{3}} + \cdots + \frac{1}{\sqrt{k}} > \sqrt{k} \quad \textbf{By } S_k$$

Therefore, adding $\dfrac{1}{\sqrt{k+1}}$, we get

$$\frac{1}{\sqrt{1}} + \frac{1}{\sqrt{2}} + \frac{1}{\sqrt{3}} + \cdots + \frac{1}{\sqrt{k}} + \frac{1}{\sqrt{k+1}} >$$

$$\sqrt{k} + \frac{1}{\sqrt{k+1}}.$$

Then

$$\sqrt{k} + \frac{1}{\sqrt{k+1}} = \frac{\sqrt{k} \cdot \sqrt{k+1} + 1}{\sqrt{k+1}} > \frac{k+1}{\sqrt{k+1}} = \sqrt{k+1}.$$

41. S_2: $\overline{z_1 + z_2} = \bar{z}_1 + \bar{z}_2$:

$$\overline{(a + bi) + (c + di)} = \overline{(a + c) + (b + d)i}$$
$$= (a + c) - (b + d)i$$
$$\overline{(a + bi)} + \overline{(c + di)} = a - bi + c - di$$
$$= (a + c) - (b + d)i.$$

S_k: $\overline{z_1 + z_2 + \cdots + z_k} = \bar{z}_1 + \bar{z}_2 + \cdots + \bar{z}_k.$

$$\overline{(z_1 + z_2 + \cdots + z_k) + z_{k+1}}$$
$$= \overline{(z_1 + z_2 + \cdots + z_k)} + \overline{z_{k+1}} \quad \textbf{By } S_2$$
$$= \bar{z}_1 + \bar{z}_2 + \cdots + \bar{z}_k + \bar{z}_{k+1} \quad \textbf{By } S_k$$

43. S_1: i is either i or -1 or $-i$ or 1.
  S_k: i^k is either i or -1 or $-i$ or 1.
  $i^{k+1} = i^k \cdot i$ is then $i \cdot i = -1$ or $-1 \cdot i = -i$ or
  $-i \cdot i = 1$ or $1 \cdot i = i.$

45. S_1:   3 is a factor of $1^3 + 2 \cdot 1.$
  S_k: 3 is a factor of $k^3 + 2k$, i.e., $k^3 + 2k = 3 \cdot m.$
  S_{k+1}: 3 is a factor of $(k + 1)^3 + 2(k + 1).$
  Consider

$$(k + 1)^3 + 2(k + 1) = k^3 + 3k^2 + 5k + 3$$
$$= (k^3 + 2k) + 3k^2 + 3k + 3$$
$$= 3m + 3(k^2 + k + 1). \quad \textbf{A multiple}$$
$$\textbf{of 3}$$

47. S_1: 3 is a factor of $1(1 + 1)(1 + 2).$
  S_k: 3 is a factor of $k(k + 1)(k + 2)$, or $k(k^2 + 3k + 2).$

$$(k + 1)(k + 1 + 1)(k + 1 + 2)$$
$$= (k + 1)(k + 2)(k + 3)$$
$$= (k^2 + 3k + 2)(k + 3)$$
$$= k(k^2 + 3k + 2) + 3(k^2 + 3k + 2)$$

By S_k, 3 is a factor of $k(k^2 + 3k + 2)$; hence 3 is a
factor of the righthand side, so 3 is a factor of
$(k + 1)(k + 2)(k + 3).$

49. **(a)** *Basis step:* $1 = 1 + 1.$ Cannot be proved.
  (b) *Induction step:* Assume S_k: $k = k + 1.$ Deduce S_{k+1}.

$$k = k + 1 \quad \textbf{By } S_k$$
$$k + 1 = k + 2 \quad \textbf{Adding 1}$$
$$= (k + 1) + 1$$

(c) No

EXERCISE SET 6.5

1. 720   **3.** 604,800   **5.** 120   **7.** 1   **9.** 3024
11. 120   **13.** 120   **15.** 1   **17.** 6,497,400
19. $n(n - 1)(n - 2)$ **21.** n **23.** $6! = 720$
25. $9! = 362,880$ **27.** $_9P_4 = 3024$
29. $_5P_5 = 120$; $5^5 = 3125$
31. $\dfrac{8!}{3!} = 6720$; $\dfrac{7!}{2!} = 2520$; $\dfrac{11!}{2! \, 2! \, 2!} = 4,989,600$

33. $8 \cdot 10^6 = 8,000,000$; 8 million   **35.** $\dfrac{9!}{2! \, 3! \, 4!} = 1260$

37. (a) $_6P_5 = 720$; **(b)** $6^5 = 7776$; **(c)** $1 \cdot {_5P_4} = 120$;
(d) $1 \cdot 1 \cdot {_4P_3} = 24$
39. (a) 10^5, or 100,000; **(b)** 100,000
41. (a) $10^9 = 1,000,000,000$; **(b)** yes   **43.** 126
45. 7,059,052   **47.** ◈   **49.** 8   **51.** 11   **53.** $n - 1$

EXERCISE SET 6.6

1. 78   **3.** 78   **5.** 7   **7.** 10   **9.** 1   **11.** 15   **13.** 128
15. 270,725   **17.** 13,037,895   **19.** n **21.** 1
23. $_{23}C_4 = 8855$ **25.** $_{13}C_{10} = 286$

27. $\dbinom{8}{2} = 28$; $\dbinom{8}{3} = 56$ **29.** $\dbinom{52}{5} = 2,598,960$

31. (a) $_{31}P_2 = 930$; **(b)** $31^2 = 961$; **(c)** $_{31}C_2 = 465$
33. $a + b$ **35.** $a^2 + 3a^2b + 3ab^2 + b^3$
37. $a^4 + 4a^3b + 6a^2b^2 + 4ab^3 + b^4$ **39.** ◈

41. $\dbinom{13}{5} = 1287$ **43.** $\dbinom{n}{2}$; $2\dbinom{n}{2}$ **45.** 4   **47.** 7

49. $\dbinom{n}{k-1} + \dbinom{n}{k}$

$$= \frac{n!}{(k-1)!(n-k+1)!} \cdot \frac{k}{k} + \frac{n!}{k!(n-k)!} \cdot \frac{(n-k+1)}{(n-k+1)}$$
$$= \frac{n!(k + (n-k+1))}{k!(n-k+1)!}$$
$$= \frac{(n+1)!}{k!(n-k+1)!} = \dbinom{n+1}{k}$$

EXERCISE SET 6.7

1. $x^4 + 20x^3 + 150x^2 + 500x + 625$
3. $x^5 - 15x^4 + 90x^3 - 270x^2 + 405x - 243$
5. $x^5 - 5x^4y + 10x^3y^2 - 10x^2y^3 + 5xy^4 - y^5$
7. $15{,}625x^6 + 75{,}000x^5y + 150{,}000x^4y^2 +$
$160{,}000x^3y^3 + 96{,}000x^2y^4 + 30{,}720xy^5 + 4096y^6$
9. $128t^7 + 448t^5 + 672t^3 + 560t + 280t^{-1} + 84t^{-3}$
$+ 14t^{-5} + t^{-7}$
11. $x^{10} - 5x^8 + 10x^6 - 10x^4 + 5x^2 - 1$
13. $125 + 150\sqrt{5}t + 375t^2 + 100\sqrt{5}t^3 + 75t^4 +$
$6\sqrt{5}t^5 + t^6$
15. $a^9 - 18a^7 + 144a^5 - 672a^3 + 2016a - 4032a^{-1} +$
$5376a^{-3} - 4608a^{-5} + 2304a^{-7} - 512a^{-9}$
17. $140\sqrt{2}$ **19.** $x^{-8} + 4x^{-4} + 6 + 4x^4 + x^8$ **21.** $21a^5b^2$
23. $-252x^5y^5$ **25.** $-745{,}472a^3$ **27.** $1120x^{12}y^2$
29. $-1{,}959{,}552u^5v^{10}$ **31.** 128 **33.** 2^{24}, or $16{,}777{,}216$
35. 20 **37.** $-12 + 316i$ **39.** $-7 - 4\sqrt{2}\,i$
41. $\displaystyle\sum_{k=0}^{n}\binom{n}{k}(-1)^k a^{n-k}b^k$ **43.** $\displaystyle\sum_{k=1}^{n}\binom{n}{k}x^{n-k}h^{k-1}$ **45.** $\frac{1}{4}$
47. ◈   **49.** $-5 \pm 2\sqrt{2}$ **51.** $9, 3$
53. (a) 0.08386; **(b)** 0.89479 **55.** $10x^{3/2}y, -10xy^{3/2}$
57. $-4320x^6y^{9/2}$ **59.** $-\dfrac{35}{x^{1/6}}$ **61.** 2^{100} **63.** $[\log_a (xt)]^{23}$
65. 1. *Basis step:* Since $a + b = (a + b)^1$, S_1 is true.
  2. *Induction step:* Let S_k be the statement of the binomial
    theorem with n replaced by k. Multiply both sides of
    S_k by $(a + b)$ to obtain

$$(a + b)^{k+1}$$
$$= \left[a^k + \cdots + \binom{k}{r-1} a^{k-(r-1)}b^{r-1} \right.$$
$$\left. + \binom{k}{r} a^{k-r}b^r + \cdots + b^k \right](a + b)$$
$$= a^{k+1} + \cdots + \left[\binom{k}{r-1} + \binom{k}{r} \right] a^{(k+1)-r}b^r$$
$$+ \cdots + b^{k+1}$$
$$= a^{k+1} + \cdots + \binom{k+1}{r} a^{(k+1)-r}b^r + \cdots + b^{k+1}.$$

  This proves S_{k+1}, assuming S_k. Hence S_n is true for
  $n = 1, 2, 3, \ldots$.

EXERCISE SET 6.8

1. (a) $0.18, 0.24, 0.23, 0.23, 0.12$; **(b)** Opinions may vary, but
it seems that people tend not to pick the first or last numbers.
3. $11{,}700$ **5. (a)** T, S, R, N, L; **(b)** E; **(c)** yes
7. (a) $\frac{2}{7}$; **(b)** $\frac{5}{7}$; **(c)** 0; **(d)** 1 **9.** $\dfrac{350}{31{,}977}$ **11.** $\dfrac{1}{108{,}290}$
13. $\dfrac{33}{66{,}640}$ **15.** Answers will vary.   **17.** $\frac{9}{19}$ **19.** $\frac{18}{19}$
21. $\frac{1}{38}$ **23.** $\frac{9}{19}$

25. $P(\text{red}) = \frac{1}{3}$, $P(\text{green}) = \frac{2}{9}$, $P(\text{blue}) = \frac{1}{6}$, $P(\text{yellow}) = \frac{5}{18}$
27. ◈   **29. (a)** 36; **(b)** 1.39×10^{-5}
31. (a) $(13 \cdot {}_4C_3) \cdot (12 \cdot {}_4C_2) = 3744$; **(b)** 0.00144
33. (a) $4 \cdot \dbinom{13}{5} - 4 - 36 = 5108$; **(b)** 0.00197
35. (a) $\dbinom{10}{1}\dbinom{4}{1}\dbinom{4}{1}\dbinom{4}{1}\dbinom{4}{1}\dbinom{4}{1} - 4 - 36 = 10{,}200$;
(b) 0.00392

REVIEW EXERCISES, CHAPTER 6

1. [6.1] $-\frac{1}{2}, \frac{4}{17}, -\frac{9}{82}, \frac{16}{257}; -\frac{121}{14{,}642}; -\frac{529}{279{,}842}$
2. [6.1] $(-1)^{n+1}(n^2 + 1)$ **3.** [6.1] $\frac{3}{2} - \frac{9}{8} + \frac{27}{26} - \frac{81}{80} = \frac{417}{1040}$
4. [6.1]

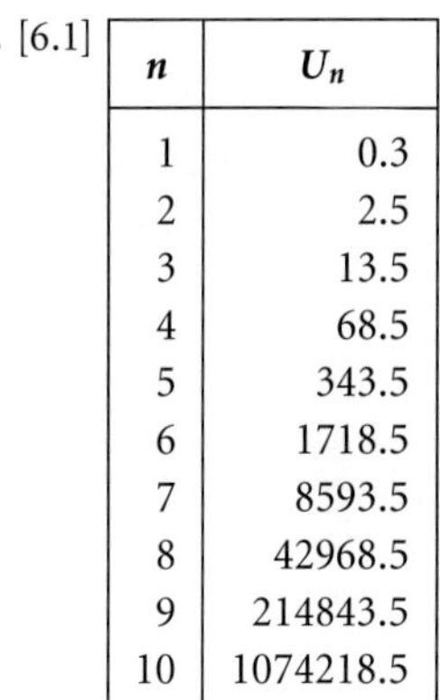

n	U_n
1	0.3
2	2.5
3	13.5
4	68.5
5	343.5
6	1718.5
7	8593.5
8	42968.5
9	214843.5
10	1074218.5

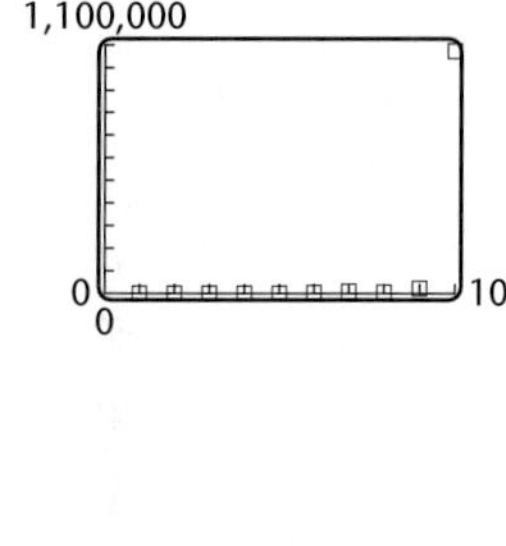

5. [6.1] $\displaystyle\sum_{k=1}^{7} k^2 - 1$ **6.** [6.2] $3\frac{3}{4}$ **7.** [6.2] $a + 4b$
8. [6.2] 531 **9.** [6.2] $20{,}100$ **10.** [6.2] 11 **11.** [6.2] -4
12. [6.3] $n = 6, S_n = -126$ **13.** [6.3] $a_1 = 8, a_5 = \frac{1}{2}$
14. [6.3] Does not exist   **15.** [6.3] $\frac{3}{11}$ **16.** [6.3] $\frac{3}{8}$
17. [6.3] $\frac{241}{99}$ **18.** [6.2] $5\frac{4}{5}, 6\frac{3}{5}, 7\frac{2}{5}, 8\frac{1}{5}$ **19.** [6.3] 167.3 ft
20. [6.3] $\$60{,}653.05$ **21.** [6.2] **(a)** $\$7.38$; **(b)** $\$1365.10$
22. [6.3] $\$88{,}888{,}888{,}889$

23. [6.4] S_n: $1 + 4 + 7 + \cdots + (3n - 2) = \dfrac{n(3n - 1)}{2}$
  S_1: $1 = \dfrac{1(3 - 1)}{2}$
  S_k: $1 + 4 + 7 + \cdots + (3k - 2) = \dfrac{k(3k - 1)}{2}$
  S_{k+1}: $1 + 4 + 7 + \cdots + [3(k + 1) - 2]$
    $= 1 + 4 + 7 + \cdots + (3k - 2) + (3k + 1)$
    $= \dfrac{(k + 1)(3k + 2)}{2}$

  1. *Basis step:* $1 = \dfrac{2}{2} = \dfrac{1(3 - 1)}{2}$ is true.

2. *Induction step:* Assume S_k. Add $(3k + 1)$ to both sides.

$$1 + 4 + 7 + \cdots + (3k - 2) + (3k + 1)$$
$$= \frac{k(3k - 1)}{2} + (3k + 1)$$
$$= \frac{k(3k - 1)}{2} + \frac{2(3k + 1)}{2}$$
$$= \frac{3k^2 - k + 6k + 2}{2}$$
$$= \frac{3k^2 + 5k + 2}{2}$$
$$= \frac{(k + 1)(3k + 2)}{2}$$

24. [6.4] S_1: $1 = \dfrac{3^1 - 1}{2}$; S_2: $1 + 3 = \dfrac{3^2 - 1}{2}$

S_k: $1 + 3 + 3^2 + \cdots + 3^{k-1} = \dfrac{3^k - 1}{2}$

2. *Induction step:* Assume S_k. Add 3^k on both sides.

$$1 + 3 + \cdots + 3^{k-1} + 3^k$$
$$= \frac{3^k - 1}{2} + 3^k = \frac{3^k - 1}{2} + 3^k \cdot \frac{2}{2}$$
$$= \frac{3 \cdot 3^k - 1}{2} = \frac{3^{k+1} - 1}{2}$$

25. [6.4]

S_n: $\left(1 - \dfrac{1}{2}\right)\left(1 - \dfrac{1}{3}\right) \cdots \left(1 - \dfrac{1}{n}\right) = \dfrac{1}{n}$

S_2: $\left(1 - \dfrac{1}{2}\right) = \dfrac{1}{2}$

S_k: $\left(1 - \dfrac{1}{2}\right)\left(1 - \dfrac{1}{3}\right) \cdots \left(1 - \dfrac{1}{k}\right) = \dfrac{1}{k}$

S_{k+1}: $\left(1 - \dfrac{1}{2}\right)\left(1 - \dfrac{1}{3}\right) \cdots \left(1 - \dfrac{1}{k}\right)\left(1 - \dfrac{1}{k + 1}\right)$
$$= \frac{1}{k + 1}$$

1. *Basis step:* S_2 is true by substitution.
2. *Induction step:* Assume S_k. Deduce S_{k+1}.
   Starting with the left side of S_{k+1}, we have

$$\underbrace{\left(1 - \frac{1}{2}\right)\left(1 - \frac{1}{3}\right) \cdots \left(1 - \frac{1}{k}\right)}\left(1 - \frac{1}{k + 1}\right)$$
$$= \frac{1}{k} \cdot \left(1 - \frac{1}{k + 1}\right). \qquad \textbf{By } S_k$$
$$= \frac{1}{k} \cdot \left(\frac{k + 1 - 1}{k + 1}\right)$$
$$= \frac{1}{k} \cdot \frac{k}{k + 1}$$
$$= \frac{1}{k + 1}. \qquad \textbf{Simplifying}$$

26. [6.5] $6! = 720$ **27.** [6.5] $9 \cdot 8 \cdot 7 \cdot 6 = 3024$

28. [6.6] $\dbinom{15}{8} = 6435$ **29.** [6.5] $24 \cdot 23 \cdot 22 = 12{,}144$

30. [6.5] $\dfrac{9!}{1!\,4!\,2!\,2!} = 3780$ **31.** [6.5] 36

32. [6.5] **(a)** $_6P_5 = 720$; **(b)** $6^5 = 7776$; **(c)** $_5P_4 = 120$;
(d) $_3P_2 = 6$

33. [6.7] 256

34. [6.7] $m^7 + 7m^6n + 21m^5n^2 + 35m^4n^3 + 35m^3n^4 + 21m^2n^5 + 7mn^6 + n^7$

35. [6.7] $x^5 - 5\sqrt{2}\,x^4 + 20x^3 - 20\sqrt{2}\,x^2 + 20x - 4\sqrt{2}$

36. [6.7] $x^8 - 12x^6y + 54x^4y^2 - 108x^2y^3 + 81y^4$

37. [6.7] $a^8 + 8a^6 + 28a^4 + 56a^2 + 70 + 56a^{-2} + 28a^{-4} + 8a^{-6} + a^{-8}$

38. [6.7] $-6624 + 16{,}280i$

39. [6.7] $220a^3x^9$ **40.** [6.7] $-\dbinom{18}{11}128a^7b^{11}$

41. [6.8] $\dfrac{86}{206} \approx 0.42$, $\dfrac{97}{206} \approx 0.47$, $\dfrac{23}{206} \approx 0.11$ **42.** [6.8] $\dfrac{1}{12}$, 0

43. [6.8] $\dfrac{1}{4}$ **44.** [6.8] $\dfrac{6}{5525}$

45. [6.1] **(a)** $a_n = 0.359n^2 - 6.965n + 151.913$;
(b) \$137 billion, \$164 billion

46. ◈ [6.3] Someone who has managed several sequences of hiring has a considerable income from the sales of the people in the lower levels. However, with a finite population, it will not be long before the salespersons in the lowest level have no one to hire and no one to sell to.

47. ◈ [6.6] A list of 9 candidates for an office is to be narrowed down to 4 candidates. In how many ways can this be done?

48. [6.4] S_1 fails for both (a) and (b).

49. [6.3] $\dfrac{a_{k+1}}{a_k} = r_1$, $\dfrac{b_{k+1}}{b_k} = r_2$, so $\dfrac{a_{k+1}b_{k+1}}{a_kb_k} = r_1r_2$, a constant

50. [6.2] **(a)** No (unless a_n is all positive or all negative);
(b) yes; **(c)** yes; **(d)** no (unless a_n is constant); **(e)** no (unless a_n is constant); **(f)** no (unless a_n is constant)

51. [6.2] $-2, 0, 2, 4$ **52.** [6.3] $\dfrac{1}{2}, -\dfrac{1}{6}, \dfrac{1}{18}$

53. [6.6] $\left(\log \dfrac{x}{y}\right)^{10}$ **54.** [6.6] 18   **55.** [6.6] 36

56. [6.7] -9

Appendixes

EXERCISE SET A

1. -11 **3.** $x^3 - 4x$ **5.** -109 **7.** $-x^4 + x^2 - 5x$
9. The answer checks.   **11.** The answer checks.
13. $M_{11} = 6$, $M_{32} = -9$, $M_{22} = -29$
15. $A_{11} = 6$, $A_{32} = 9$, $A_{22} = -29$ **17.** -10 **19.** -10
21. $M_{41} = -14$, $M_{33} = 20$ **23.** $A_{24} = 15$, $A_{43} = 30$
25. 110   **27.** 110   **29.** $\left(-\dfrac{25}{2}, -\dfrac{11}{2}\right)$ **31.** $(3, 1)$
33. $\left(\dfrac{1}{2}, -\dfrac{1}{3}\right)$ **35.** $(1, 1)$ **37.** $\left(\dfrac{3}{2}, \dfrac{13}{14}, \dfrac{33}{14}\right)$ **39.** $(3, -2, 1)$

41. $(1, 3, -2)$ **43.** $\left(\frac{1}{2}, \frac{2}{3}, -\frac{5}{6}\right)$ **45.** ◈   **47.** ± 2

49. $(-\infty, -\sqrt{3}] \cup [\sqrt{3}, \infty)$ **51.** -34 **53.** 4

55. Answers may vary.   **57.** Answers may vary.

$$\begin{vmatrix} L & -W \\ 2 & 2 \end{vmatrix} \qquad\qquad \begin{vmatrix} a & b \\ -b & a \end{vmatrix}$$

59. Answers may vary.

$$\begin{vmatrix} 2\pi r & 2\pi r \\ -h & r \end{vmatrix}$$

61. Evaluate the determinant and compare with the two-point equation of a line.

APPENDIX B

1. $y = 12x - 7, -\frac{1}{2} \le x \le 3$

3. $y = \sqrt[3]{x} - 4, -1 \le x \le 1000$ **5.** $x = y^4, 0 \le x \le 16$

7. $y = \frac{1}{x}, 1 \le x \le 5$ **9.** $y = \frac{1}{4}(x + 1)^2, -7 \le x \le 5$

11. $y = \frac{1}{x}, x > 0$

13. Answers may vary. $x = t, y = 4t - 3; x = \frac{t}{4} + 3,$ $y = t + 9$

15. Answers may vary. $x = t, y = (t - 2)^2 - 6t; x = t + 2,$ $y = t^2 - 6t - 12$

17. Answers may vary. $x = \frac{1}{4}t, y = -t - 4$

Index of Applications

Web Connection

Students and instructors can obtain more information about books written by Professor Bittinger and his co-authors by contacting *Marv's Math Corner* on the Internet at the following address:

> http://www.math.iupui.edu/~mbitting/

Included on this Web site is information about books and their supplements, as well as study tips and sample practice final examinations. Students and instructors are welcome to e-mail questions, comments, and constructive criticism.

Algebra (continued)

The Distance Formula

The distance from (x_1, y_1) to (x_2, y_2) is given by

$$d = \sqrt{(x_2 - x_1)^2 + (y_2 - y_1)^2}.$$

Midpoint Formula

The midpoint of the line segment from (x_1, y_1) to (x_2, y_2) is given by

$$\left(\frac{x_1 + x_2}{2}, \frac{y_1 + y_2}{2} \right).$$

Formulas Involving Lines

The slope of the line containing points (x_1, y_1) to (x_2, y_2) is given by

$$m = \frac{y_2 - y_1}{x_2 - x_1}.$$

Slope–intercept equation: $\quad y = f(x) = mx + b$

Horizontal line: $\quad y = b \quad$ or $\quad f(x) = b$

Vertical line: $\quad x = a$

Point–slope equation: $\quad y - y_1 = m(x - x_1)$

Two-point equation: $\quad y - y_1 = \dfrac{y_2 - y_1}{x_2 - x_1}(x - x_1)$

The Quadratic Formula

The solutions of $ax^2 + bx + c = 0$, $a \neq 0$, are given by

$$x = \frac{-b \pm \sqrt{b^2 - 4ac}}{2a}.$$

Compound Interest Formulas

Compounded n times per year: $\quad A = P\left(1 + \dfrac{i}{n}\right)^{nt}$

Compounded continuously: $\quad P(t) = P_0 e^{kt}$

Properties of Exponential and Logarithmic Functions

$$\log_a x = y \leftrightarrow x = a^y \qquad a^x = a^y \leftrightarrow x = y$$

$$\log_a MN = \log_a M + \log_a N \qquad \log_a M^p = p \log_a M$$

$$\log_a \frac{M}{N} = \log_a M - \log_a N$$

$$\log_b M = \frac{\log_a M}{\log_a b}$$

$$\log_a a = 1 \qquad\qquad \log_a 1 = 0$$

$$\log_a a^x = x \qquad\qquad a^{\log_a x} = x$$

Arithmetic Sequences and Series

$$a_1, a_1 + d, a_1 + 2d, a_1 + 3d, \ldots$$

$$a_{n+1} = a_n + d \qquad a_n = a_1 + (n - 1)d$$

$$S_n = \frac{n}{2}(a_1 + a_n)$$

Geometric Sequences and Series

$$a_1, a_1 r, a_1 r^2, a_1 r^3, \ldots$$

$$a_{n+1} = a_n r \qquad a_n = a_1 r^{n-1}$$

$$S_n = \frac{a_1(1 - r^n)}{1 - r} \qquad S_\infty = \frac{a_1}{1 - r}, \; |r| < 1$$

Conic Sections

Circle: $\quad (x - h)^2 + (y - k)^2 = r^2$

Ellipse:
$$\frac{(x - h)^2}{a^2} + \frac{(y - k)^2}{b^2} = 1,$$
$$\frac{(x - h)^2}{b^2} + \frac{(y - k)^2}{a^2} = 1$$

Parabola:
$$(x - h)^2 = 4p(y - k),$$
$$(y - k)^2 = 4p(x - h)$$

Hyperbola:
$$\frac{(x - h)^2}{a^2} - \frac{(y - k)^2}{b^2} = 1,$$
$$\frac{(y - k)^2}{a^2} - \frac{(x - h)^2}{b^2} = 1$$